W9-AOI-978

A MODERN COURSE IN AEROELASTICITY

SOLID MECHANICS AND ITS APPLICATIONS
Volume 32

Series Editor: G.M.L. GLADWELL
Solid Mechanics Division, Faculty of Engineering
University of Waterloo
Waterloo, Ontario, Canada N2L 3G1

Aims and Scope of the Series

The fundamental questions arising in mechanics are: *Why?*, *How?*, and *How much?* The aim of this series is to provide lucid accounts written by authoritative researchers giving vision and insight in answering these questions on the subject of mechanics as it relates to solids.

The scope of the series covers the entire spectrum of solid mechanics. Thus it includes the foundation of mechanics; variational formulations; computational mechanics; statics, kinematics and dynamics of rigid and elastic bodies; vibrations of solids and structures; dynamical systems and chaos; the theories of elasticity, plasticity and viscoelasticity; composite materials; rods, beams, shells and membranes; structural control and stability; soils, rocks and geomechanics; fracture; tribology; experimental mechanics; biomechanics and machine design.

The median level of presentation is the first year graduate student. Some texts are monographs defining the current state of the field; others are accessible to final year undergraduates; but essentially the emphasis is on readability and clarity.

For a list of related mechanics titles, see final pages.

A Modern Course in Aeroelasticity

Third revised and enlarged edition

EARL H. DOWELL, editor
Professor of Mechanical Engineering and Materials Science,
Duke University, Durham, North Carolina, U.S.A.

EDWARD F. CRAWLEY
Professor of Aeronautics and Astronautics,
Massachusetts Institute of Technology, Cambridge, Massachusetts, U.S.A.

HOWARD C. CURTISS, Jr.
Professor of Mechanical and Aerospace Engineering,
Princeton University, Princeton, New Jersey, U.S.A.

DAVID A. PETERS
Professor of Mechanical Engineering,
Washington University, St. Louis, Missouri, U.S.A.

ROBERT H. SCANLAN
Professor of Civil Engineering,
Johns Hopkins University, Baltimore, Maryland, U.S.A.

and

FERNANDO SISTO
Professor of Mechanical Engineering,
Stevens Institute of Technology, Hoboken, New Jersey, U.S.A.

KLUWER ACADEMIC PUBLISHERS
DORDRECHT / BOSTON / LONDON

Library of Congress Cataloging-in-Publication Data

A Modern course in aeroelasticity / Earl H. Dowell, editor ... [et al.]. -- 3rd rev. and enl. ed.
p. cm. -- (Solid mechanics and its applications ; v. 32)
Includes bibliographical references and index.
ISBN 0-7923-2788-8 (alk. paper)
1. Aeroelasticity. I. Dowell, E. H. II. Series.
TL574.A37M62 1994
629.132'362--dc20 94-9743

ISBN 0-7923-2788-8 (HB)
ISBN 0-7923-2789-6 (PB)

Published by Kluwer Academic Publishers,
P.O. Box 17, 3300 AA Dordrecht, The Netherlands.

Kluwer Academic Publishers incorporates
the publishing programmes of
D. Reidel, Martinus Nijhoff, Dr W. Junk and MTP Press.

Sold and distributed in the U.S.A. and Canada
by Kluwer Academic Publishers,
101 Philip Drive, Norwell, MA 02061, U.S.A.

In all other countries, sold and distributed
by Kluwer Academic Publishers Group,
P.O. Box 322, 3300 AH Dordrecht, The Netherlands.

This is the revised and enlarged edition of a book published in 1978 by Sijthoff & Noordhoff International Publishers, in the series 'Mechanics: Dynamical Systems' (Volume 3)

Printed on acid-free paper

All Rights Reserved
© 1995 Kluwer Academic Publishers
No part of the material protected by this copyright notice may be reproduced or utilized in any form or by any means, electronic or mechanical, including photocopying, recording or by any information storage and retrieval system, without written permission from the copyright owner.

Printed in the Netherlands

Contents

Preface xv

Preface to the second edition xvii

Preface to the third edition xix

Short bibliography xxi

Introduction 1

2. Static aeroelasticity 3

2.1 *Typical section model of an airfoil* 3
Typical section with control surface 8
Typical section—nonlinear effects 14

2.2 *One dimensional aeroelastic model of airfoils* 17
Beam-rod representation of large aspect ratio wing . . 17
Eigenvalue and eigenfunction approach 21
Galerkin's method 23

2.3 *Rolling of a straight wing* 25
Integral equation of equilibrium 25
Derivation of equation of equilibrium 27
Calculation of $C^{\alpha\alpha}$ 28
Sketch of function $S(y_1, \eta)$ 28
Aerodynamic forces (including spanwise induction) . . . 30
Aeroelastic equations of equilibrium and lumped element solution method 32
Divergence 34
Reversal and rolling effectiveness 34

Integral equation eigenvalue problem and the experimental determination of influence functions 38

2.4 *Two dimensional aeroelastic model of lifting surfaces* . . . 41
Two dimensional structures—integral representation . . . 41
Two dimensional aerodynamic surfaces—integral representation 43
Solution by matrix-lumped element approach 43

2.5 *Nonairfoil physical problems* 45
Fluid flow through a flexible pipe 45
(Low speed) fluid flow over a flexible wall 48

2.6 *Sweptwing divergence* 49

References for Chapter 2 52

3. Dynamic aeroelasticity 54

3.1 *Hamilton's principle* 55
Single particle 55
Many particles 57
Continuous body 57
Potential energy 57
Nonpotential forces 60

3.2 *Lagrange's equations* 61
Example—typical section equations of motion 62

3.3 *Dynamics of the typical section model of an airfoil* . . . 65
Sinusoidal motion 66
Periodic motion 68
Arbitrary motion 69
Random motion 75
Flutter—an introduction to dynamic aeroelastic instability 83
Quasi-steady, aerodynamic theory 87

3.4 *Aerodynamic forces for airfoils—an introduction and summary* 89
Aerodynamic theories available 94
General approximations 97
'Strip theory' approximation 98

'Quasi-steady' approximation 98
Slender body or slender (low aspect ratio) wing approximation 99

3.5 *Solutions to the aeroelastic equations of motion* 100
Time domain solutions 101
Frequency domain solutions. 102

3.6 *Representative results and computational considerations* . . 106
Time domain 106
Frequency domain 107
Flutter and gust response classification including parameter trends 110
Flutter 110
Gust response 123

3.7 *Generalized equations of motion for complex structures* . . 129
Lagrange's equations and modal methods 129
Kinetic energy 131
Strain (potential, elastic) energy 131
Examples 134
(a) Torsional vibrations of a rod. 134
(b) Bending-torsional motion of a beam-rod 135
Natural frequencies and modes-eigenvalues and eigenvectors 136
Evaluation of generalized aerodynamic forces 138
Equations of motion and solution methods 138
Integral equations of equilibrium 140
Natural frequencies and modes 14
Proof of orthogonality 14
Forced motion including aerodynamic forces 145
Examples 148
(a) Rigid wing undergoing translation responding to a gust 148
(b) Wing undergoing translation and spanwise bending . 154
(c) Random gusts—solution in the frequency domain . . 156

3.8 *Nonairfoil physical problems* 157
Fluid flow through a flexible pipe 157
(High speed) fluid flow over a flexible wall—a simple prototype for plate flutter 160

References for Chapter 3 165

4. Nonsteady aerodynamics of lifting and non-lifting surfaces . . 167

4.1 *Basic fluid dynamic equations* 167
Conservation of mass 168
Conservation of momentum 169
Irrotational flow, Kelvin's theorem and Bernoulli's equation 170
Derivation of single equation for velocity potential . . . 173
Small perturbation theory 175
Reduction to acoustics 176
Boundary conditions 177
Symmetry and anti-symmetry 179

4.2 *Supersonic flow* 182
Two-dimensional flow 182
Simple harmonic motion of the airfoil 183
Discussion of inversion 185
Discussion of physical significance of results 188
Gusts 189
Transient motion 190
Lift, due to airfoil motion 191
Lift, due to atmospheric gusts 192
Three-dimensional flow 195

4.3 *Subsonic flow* 201
Derivation of the integral equation by transform methods and solution by collocation 202
An alternative determination of the Kernel Function using Green's theorem 205
Incompressible, three-dimensional flow 207
Compressible, three-dimensional flow 212
Incompressible, two-dimensional flow 217
Simple harmonic motion of an airfoil 220
Transient motion 227
Evaluation of integrals 231

4.4 *Representative numerical results* 235

4.5 *Transonic flow* 242

References for Chapter 4 272

5. Stall flutter 275

5.1 *Background* 275
5.2 *Analytical formulation* 276
5.3 *Stability and aerodynamic work* 278
5.4 *Bending stall flutter* 279
5.5 *Nonlinear mechanics description* 281
5.6 *Torsional stall flutter* 282
5.7 *General comments* 285
5.8 *Reduced order models* 288
5.9 *Computational stalled flow* 289

References for Chapter 5 296

6. Aeroelastic problems of civil engineering structures 297

6.1 *Vortex shedding* 299
Introduction 299
Aspects of response to vortex shedding 302
Empirical models of vortex-induced oscillation 305
Uncoupled single-degree-of-freedom models 306
Coupled two-degree-of-freedom ('wake oscillator') models . . 309
Commentary on vortex excitation models 322

6.2 *Galloping* 327
Across-wind galloping 329
Wake galloping 334

6.3 *Divergence* 337

6.4 *Flutter and buffeting* 341
Basic concepts 341
Three-dimensional flutter and buffeting 350
Single-mode flutter 355
Indical function formulations 356

References for Chapter 6 363

7. Aeroelastic response of rotorcraft 370

7.1 *Blade dynamics* 371
Articulated, rigid blade motion 373
Elastic motion of hingeless blades 382
7.2 *Stall flutter* 395
7.3 *Rotor-body coupling* 400

7.4 *Unsteady aerodynamics* 422
Dynamic inflow 422
Frequency domain 428
Finite-state wake modeling 430
Summary 433
References for Chapter 7 434

8. Aeroelasticity in turbomachines 438

8.1 *Aeroelastic environment in turbomachines* 439
8.2 *The compressor performance map* 441
8.3 *Blade mode shapes and materials of construction* . . . 444
8.4 *Nonsteady potential flow in cascades* 446
8.5 *Compressible flow* 451
8.6 *Periodically stalled flow in turbomachines* 455
8.7 *Stall flutter in turbomachines* 459
8.8 *Choking flutter* 461
8.9 *Aeroelastic eigenvalues* 462
8.10 *Recent trends* 466

References for Chapter 8 469

9. Unsteady transonic aerodynamics and aeroelasticity . . 472

Summary 472

Nomenclature 473

9.1 *Introduction* 474

9.2 *Linear/nonlinear behavior in unsteady transonic aerodynamics* 475
Motivation and general background 475
NACA 64A006 airfoil 477
Mach number trends 482
Conclusions 486

9.3 *Viable alternative solution procedures to finite difference methods* 488
Hounjet 488
Cockey 489
A possible synthesis 490

9.4 *Nonuniqueness* 491
Early work 491
Recent work 491
Studies of Williams and Salas 598
Aileron buzz 599

9.5 *Effective, efficient computational approaches for determining aeroelastic response* 500
Various approaches and their merits 500
Time domain 501
Frequency domain 502
Summary comparison 503
Nonlinear flutter analysis 503

9.6 *Nonlinear flutter analysis in the frequency domain* . . 504
Motivation and background 504
Typical airfoil section 506
Aerodynamic describing function 507
Aeroelastic system equations 510
Extension of the describing function 513
Results and discussion 513
Conclusions 523

9.7 *Concluding remarks* 524
Some present answers 524
The physical phenomena associated with unsteady transonic flow 525
Future work 527

References for Chapter 9 528

10. Experimental aeroelasticity 533
10.1 *Review of structural dynamics experiments* 533
10.2 *Wind tunnel experiments* 534
Sub-critical testing 535
Approaching the flutter boundary 535
Safety devices 535
Research tests vs. clearance tests 536
Scaling laws 536

10.3 *Flight experiments* 536
Approaching the flutter boundary 536
When is flight flutter testing required? 540

Excitation 540
Examples of recent flight flutter test programs 540

10.4 *The role of experimentation and theory in design* 540
References for Chapter 10 541

11. Nonlinear aeroelasticity 542

Abstract 452
11.1 *Introduction* 452
11.2 *The physical domain of nonlinear aeroelasticity* 455
11.3 *The mathematical consequences of nonlinearity* 549
11.4 *Representative results* 550
Flutter of airfoils in transonic flow. 550
Flutter of airfoils at high angles of attack 554
Flutter of an airfoil with free-play structured nonlinearities . . . 563
Nonlinear fluid oscillator models in bluff body aeroelasticity . . 566
Flutter of plates and shells 567

11.5 *The future* 569

References for Chapter 11 570

12. Aeroelastic control 573

12.1 *Objectives and elements of aeroelastic control* 574
12.2 *Modeling for aeroelastic control* 578
Structural modeling 581
Aerodynamic modeling 583
Other modeling elements 588
Model summary 593
Model reduction 595

12.3 *Control modeling of the typical section* 602
Typical section governiong equations 602
Open loop poles and zeros 612
Reference typical section 614
Plant variation with airspeed 615
Plant variation with structural parameters 619
Plant variation with aerodynamic model 622

12.4 *Control design for the typical section* 624
Single-input static feedback 625
Full state static feedback 629
Full state single input static feedback 631
Full state multiple input feedback 634
Dynamic output feedback 635
Compensation for an aeroelastic servo 640
Compensation for an unstable system 643
Lessons learned from the typical section 646
Other issues in compensator design 648

References for Chapter 12 650

Appendix I A primer for structural response to random pressure fluctuations 653

References for Appendix I 659

Appendix II Some example problems 661

Preface

A reader who achieves a substantial command of the material contained in this book should be able to read with understanding most of the literature in the field. Possible exceptions may be certain special aspects of the subject such as the aeroelasticity of plates and shells or the use of electronic feedback control to modify aeroelastic behavior. The first author has considered the former topic in a separate volume. The latter topic is also deserving of a separate volume.

In the first portion of the book the basic physical phenomena of divergence, control surface effectiveness, flutter and gust response of aeronautical vehicles are treated. As an indication of the expanding scope of the field, representative examples are also drawn from the non-aeronautical literature. To aid the student who is encountering these phenomena for the first time, each is introduced in the context of a simple physical model and then reconsidered systematically in more complicated models using more sophisticated mathematics.

Beyond the introductory portion of the book, there are several special features of the text. One is the treatment of unsteady aerodynamics. This crucial part of aeroelasticity is usually the most difficult for the experienced practitioner as well as the student. The discussion is developed from the basic fluid mechanics and includes a comprehensive review of the fundamental theory underlying numerical lifting surface analysis. Not only the well known results for subsonic and supersonic flow are covered; but also some of the recent developments for transonic flow, which hold promise of bringing effective solution techniques to this important regime.

Professor Sisto's chapter on Stall Flutter is an authoritative account of this important topic. A difficult and still incompletely understood phenomenon, stall flutter is discussed in terms of its fundamental aspects as well as its significance in applications. The reader will find this chapter particularly helpful as an introduction to this complex subject.

Another special feature is a series of chapters on three areas of advanced application of the fundamentals of aeroelasticity. The first of these is a discussion of Aeroelastic Problems of Civil Engineering Structures by Professor Scanlan. The next is a discussion of Aeroelasticity of Helicopters and V/STOL aircraft by Professor Curtiss. The final chapter in this series treats Aeroelasticity in Turbomachines and is by Professor Sisto. This series of chapters is unique in the aeroelasticity literature and the first author feels particularly fortunate to have the contributions of these eminent experts.

The emphasis in this book is on fundamentals because no single volume can hope to be comprehensive in terms of applications. However, the above three chapters should give the reader an appreciation for the relationship between theory and practice. One of the continual fascinations of aeroelasticity is this close interplay between fundamentals and applications. If one is to deal successfully with applications, a solid grounding in the fundamentals is essential.

For the beginning student, a first course in aeroelasticity could cover Chapters 1–3 and selected portions of 4. For a second course and the advanced student or research worker, the remaining Chapters would be appropriate. In the latter portions of the book, more comprehensive literature citations are given to permit ready access to the current literature.

The reader familiar with the standard texts by Scanlan and Rosenbaum, Fung, Bisplinghoff, Ashley and Halfman and Bisplinghoff and Ashley will appreciate readily the debt the authors owe to them. Recent books by Petre* and Forsching† should also be mentioned though these are less accessible to an english speaking audience. It is hoped the reader will find this volume a worthy successor.

*Petre, A., *Theory of Aeroelasticity. Vol. I Statics, Vol. II Dynamics.* In Romanian. Publishing House of the Academy of the Socialist Republic of Romania, Bucharest, 1966.
†Forsching, H. W., *Fundamentals of Aeroelasticity*. In German. Springer-Verlag, Berlin, 1974.

Preface to the second edition

The authors would like to thank all those readers who have written with comments and errata for the First Edition. Many of these have been incorporated into the Second Edition. They would like to thank especially Professor Holt Ashley of Stanford University who has been most helpful in identifying and correcting various errata.

Also the opportunity has been taken in the Second Edition to bring up-to-date several of the chapters as well as add a chapter on unsteady transonic aerodynamics and aeroelasticity. Chapters 2, 5, 6, and 8 have been substantially revised. These cover the topics of Static Aeroelasticity, Stall Flutter, Aeroelastic Problems of Civil Engineering Structures and Aeroelasticity in Turbomachines, respectively. Chapter 9, Unsteady Transonic Aerodynamics and Aeroelasticity, is new and covers this rapidly developing subject in more breadth and depth than the First Edition. Again the emphasis is on fundamental concepts rather than, for example, computer code development per se. Unfortunately due to the press of other commitments, it has not been possible to revise Chapter 7, Aeroelastic Problems of Rotorcraft. However, the Short Bibliography has been expanded for this subject as well as for others. It is hoped that the readers of the First Edition and also new readers will find the Second Edition worthy of their study.

Preface to the Third Edition

The authors would like to thank all those readers of the first and second editions who have written with comments and suggestions. In the third edition the opportunity has been taken to revise and update Chapters 1 through 9. Also three new chapters have been added, i.e., Chapter 10, Experimental Aeroelasticity, Chapter 11, Nonlinear Aeroelasticity; and Chapter 12, Aeroelastic Control. Chapter 10 is a brief introduction to a vast subject: Chapter 11 is an overview of a frontier of research; and Chapter 12 is the first connected, authoritative account of the feedback control of aeroelastic systems. Chapter 12 meets a significant need in the literature. The authors of the first and second editions welcome the two new authors, David Peters who has provided a valuable revision of Chapter 7 on rotorcraft, and Edward Crawley who has provided Chapter 12 on aeroelastic control. It is a privilege and a pleasure to have them as members of the team. The author of Chapter 10 would also like to acknowledge the great help he has received over the years from his distinguished colleague, Wilmer H. "Bill" Reed, III, in the study of experimental aeroelasticity. Mr. Reed kindly provided the figures for Chapter 10. The author of Chapter 12 would like to acknowledge the significant scholarly contribution of Charrissa Lin and Ken Lazarus in preparing the chapter on aeroelastic control. Finally the readers of the first and second editions will note that author and subject indices have been omitted from this edition. If any reader finds this an inconvenience, please contact the editor and we will reconsider the matter for the next edition.

Short bibliography

Books

Bolotin, V. V., *Nonconservative Problems of the Elastic Theory of Stability*, Pergamon Press, 1963.

(BAH) Bisplinghoff, R. L., Ashley, H. and Halfman, R. L., *Aeroelasticity*, Addison-Wesley Publishing Company, Cambridge, Mass., 1955.

(BA) Bisplinghoff, R. L. and Ashley, H., *Principles of Aeroelasticity*, John Wiley and Sons, Inc., New York, N.Y., 1962. Also available in Dover Edition.

Fung, Y. C., *An Introduction to the Theory of Aeroelasticity*, John Wiley and Sons, Inc., New York, N.Y., 1955. Also available in Dover Edition.

Scanlan, R. H. and Rosenbaum, R., *Introduction to the Study of Aircraft Vibration and Flutter*, The Macmillan Company, New York, N.Y., 1951. Also available in Dover Edition.

(AGARD) AGARD Manual on Aeroelasticity, Vols. I–VII, Beginning 1959 with continual updating.

Ashley, H., Dugundji, J. and Rainey, A. G., *Notebook for Aeroelasticity*, AIAA Professional Seminar Series, 1969.

Dowell, E. H., *Aeroelasticity of Plates and Shells*, Noordhoff International Publishing, Leyden, 1975.

Simiu, E. and Scanlan, R. H., *Wind Effects on Structures—An Introduction to Wind Engineering*, John Wiley and Sons, 1978.

Johnson, W., *Helicopter Theory*, Princeton University Press, 1980.

Dowell, E. H. and Ilgamov, M., *Studies in Nonlinear Aeroelasticity*, Springer-Verlag, 1988.

In parentheses, abbreviations for the above books are indicated which are used in the text.

Survey articles

Garrick, I. E., 'Aeroelasticity—Frontiers and Beyond', 13th Von Karman Lecture, *J. of Aircraft*, Vol. 13, No. 9, 1976, pp. 641–657.

Several Authors, 'Unsteady Aerodynamics. Contribution of the Structures and Materials Panel to the Fluid Dynamics Panel Round Table Discussion on Unsteady Aerodynamics', Goettingen, May 1975, *AGARD Report* R-645, March 1976.

Rodden, W. P., *A Comparison of Methods Used in Interfering Lifting Surface Theory*, AGARD Report R-643, March 1976.

Short bibliography

Ashley, H., 'Aeroelasticity', *Applied Mechanics Reviews,* February 1970.
Abramson, H. N., 'Hydroelasticity: A Review of Hydrofoil Flutter', *Applied Mechanics Reviews,* February 1969.
Many Authors, 'Aeroelastic Effects From a Flight Mechanics Standpoint', AGARD, Conference Proceedings No. 46, 1969.
Landahl, M. T. and Stark, V. J. E., 'Numerical Lifting Surface Theory-Problems and Progress', *AIAA Journal,* No. 6, No. 11, November 1968, pp. 2049–2060.
Many Authors, 'Symposium on Fluid-Solid Interaction', *ASME Annual Winter Meeting,* November 1967.
Kaza, K. R. V., 'Development of Aeroelastic Analysis Methods for Turborotors and Propfans—Including Mistuning', in *Lewis Structures Technology*, Vol. 1, Proceedings, NASA Lewis Research Center, 1988.
Ericsson, L. E. and Reding, J. P., 'Fluid Mechanics of Dynamic Stall, Part I, Unsteady Flow Concepts, and Part II, Prediction of Full Scale Characteristics', *J. Fluids and Structures*, Vol. 2, No. 1 and 2, 1988, pp. 1–33 and 113–143, respectively.
Mabey, D. G., 'Some Aspects of Aircraft Dynamic Loads Due to Flow Separation', AGARD-R-750, February, 1988.
Yates, E. C., Jr. and Whitlow, W., Jr., 'Development of Computational Methods for Unsteady Aerodynamics at the NASA Langley Research Center', in AGARD-R-749, Future Research on Transonic Unsteady Aerodynamics and its Aeroelastic Applications, August 1987.
Gad-el-Hak, M., 'Unsteady Separation on Lifting Surfaces', *Applied Mechanics Reviews*, Vol. 40, No. 4, 1987, pp. 441–453.
Hajela, P. (Ed.), 'Recent Trends in Aeroelasticity, Structures and Structural Dynamics', University of Florida Press, Gainesville, 1987.
Jameson, A., 'The Evolution of Computational Methods in Aerodynamics', *J. Applied Mechanics*, Vol. 50, No. 4, 1983, pp. 1052–1070.
Seebass, R., 'Advances in the Understanding and Computation of Unsteady Transonic Flows', in *Recent Advances in Aerodynamics*, edited by A. Krothapalli and C. Smith, Springer-Verlag, 1984.
McCroskey, W. J., 'Unsteady Airfoils', in *Annual Review of Fluid Mechanics*, 1982, Vol. 14, pp. 285–311.
Tijdeman, H. and Seebass, R., 'Transonic Flow Past Oscillating Airfoils', in *Annual Review of Fluid Mechanics*, 1980, Vol. 12, pp. 181–222.
Ormiston, R.,Warmbrodt, W., Hodges, D. and Peters, D., 'Survey of Army/NASA Rotocraft Aeroelastic Stability Research', NASA TM 101026 and USAASCOM TR 88-A-005, 1988.

Journals

AHS Journal
AIAA Journal
ASCE Transactions, Engineering Mechanics Division
ASME Transactions, Journal of Applied Mechanics
International Journal of Solids and Structures
Journal of Aircraft
Journal of Fluids and Structures
Journal of Sound and Vibration

Other journals will have aeroelasticity articles, of course, but these are among those with the most consistent coverage.

The impact of aeroelasticity on design is not discussed in any detail in this book. For insight into this important area the reader may consult the following volumes prepared by the National Aeronautics and Space Administration in its series on SPACE VEHICLE DESIGN CRITERIA. Although these documents focus on space vehicle applications, much of the material is relevant to aircraft as well. The depth and breadth of coverage varies considerably from one volume to the next, but each contains at least a brief State-of-the-Art review of its topic as well as a discussion of Recommended Design Practices. Further some important topics are included which have not been treated at all in the present book. These include, as already mentioned in the Preface,

Aeroelasticity of plates and shells (panel flutter) (NASA SP-8004)
Aeroelastic effects on control system dynamics (NASA SP-8016, NASA SP-8036 NASA SP-8079)

as well as

Structural response to time-dependent separated fluid flows (buffeting) (NASA SP-8001)
Fluid motions inside elastic containers (fuel sloshing) (NASA SP-8009, NASA SP-8031)
Coupled structural—propulsion instability (POGO) (NASA SP-8055)

It is intended to revise these volumes periodically to keep them up-to-date.

NASA SP-8001 1970
Buffeting During Atmospheric Ascent
NASA SP-8002 1964
Flight Loads Measurements During Launch and Exit
NASA SP-8003 1964
Flutter, Buzz, and Divergence
NASA SP-8004 1972
Panel Flutter
NASA SP-8006 1965
Local Steady Aerodynamic Loads During Launch and Exit
NASA SP-8008 1965
Prelaunch Ground Wind Loads
NASA SP-8012 1968
Natural Vibration Modal Analysis
NASA SP-8016 1969
Effects of Structural Flexibility on Spacecraft Control Systems
NASA SP-8009 1968
Propellant Slosh Loads
NASA SP-8031 1969
Slosh Suppression
NASA SP-8035 1970
Wind Loads During Ascent
NASA SP-8036 1970
Effects of Structural Flexibility on Launch Vehicle Control Systems
NASA SP-8050 1970
Structural Vibration Prediction
NASA SP-8055 1970
Prevention of Coupled Structure-Propulsion Instability (POGO)
NASA SP-8079 1971
Structural Interaction with Control Systems.

1

Introduction

Several years ago, Collar suggested that aeroelasticity could be usefully visualized as forming a triangle of disciplines.

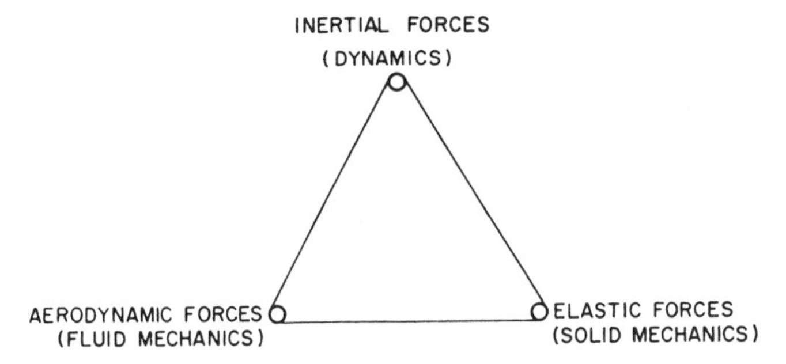

Aeroelasticity is concerned with those physical phenomena which involve significant mutual interaction among inertial, elastic and aerodynamic forces. Other important technical fields can be identified by pairing the several points of the triangle. For example,

Stability and control (flight mechanics) = dynamics + aerodynamics
Structural vibrations = dynamics + solid mechanics
Static aeroelasticity = fluid mechanics + solid mechanics

Conceptually, each of these technical fields may be thought of as a special aspect of aeroelasticity. For historical reasons only the last topic, static aeroelasticity, is normally so considered. However, the impact of aeroelasticity on stability and control (flight mechanics) has increased substantially in recent years, for example.

In modern aerospace vehicles, life can be even more complicated. For example, stresses induced by high temperature environments can be important in aeroelastic problems, hence the term

'aerothermoelasticity'

In other applications, the dynamics of the guidance and control system may significantly affect aeroelastic problems or vice versa, hence the term

'aeroservoelasticity'

For a historical discussion of aeroelasticity including its impact on aerospace vehicle design, consult Chapter I of Bisplinghoff and Ashley (BA) and AGARD C.P. No. 46, 'Aeroelastic Effects from a Flight Mechanics Standpoint'.

We shall first concentrate on the dynamics and solid mechanics aspects of aeroelasticity with the aerodynamic forces taken as given. Subsequently, the aerodynamic aspects of aeroelasticity shall be treated from first principles. Theoretical methods will be emphasized, although these will be related to experimental methods and results where this will add to our understanding of the theory and its limitations. For simplicity, we shall begin with the special case of static aeroelasticity.

Although the technological cutting edge of the field of aeroelasticity has centered in the past on aeronautical applications, applications are found at an increasing rate in civil engineering, e.g., flows about bridges and tall buildings; mechanical engineering, e.g., flows around turbomachinery blades and fluid flows in flexible pipes; and nuclear engineering; e.g., flows about fuel elements and heat exchanger vanes. It may well be that such applications will increase in both absolute and relative number as the technology in these areas demands lighter weight structures under more severe flow conditions. Much of the fundamental theoretical and experimental developments can be applied to these areas as well and indeed it is hoped that a common language can be used in these several areas of technology. To further this hope we shall discuss subsequently in some detail several nonairfoil examples, even though our principal focus shall be on aeronautical problems. Separate chapters on civil engineering, turbomachinery and helicopter (rotor systems) applications will introduce the reader to the fascinating phenomena which arise in these fields.

Since most aeroelastic phenomena are of an undesirable character, leading to loss of design effectiveness or even sometimes spectacular structural failure as in the case of aircraft wing flutter or the Tacoma Narrows Bridge disaster, the spreading importance of aeroelastic effects will not be warmly welcomed by most design engineers. However, the mastery of the material to be discussed here will permit these effects to be better understood and dealt with if not completely overcome.

2

Static aeroelasticity

2.1 Typical section model of an airfoil

We shall find a simple, somewhat contrived, physical system useful for introducing several aeroelastic problems. This is the so-called 'typical section' which is a popular pedagogical device.* This simplified aeroelastic system consists of a rigid, flat plate airfoil mounted on a torsional spring attached to a wind tunnel wall. See Figure 2.1; the airflow over the airfoil is from left to right.

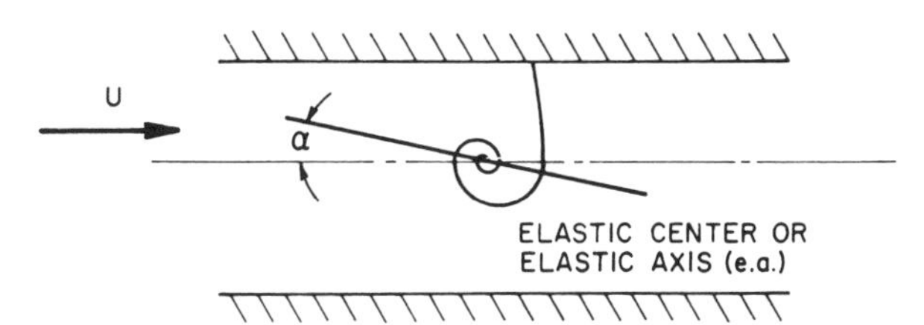

Figure 2.1 Geometry of typical section airfoil.

The principal interest in this model for the aeroelastician is the rotation of the plate (and consequent twisting of the spring), α, as a function of airspeed. If the spring were very stiff or airspeed were very slow, the rotation would be rather small; however, for flexible springs or high flow velocities the rotation may twist the spring beyond its ultimate strength and lead to structural failure. A typical plot of elastic twist, α_e, vs airspeed, U, is given in Figure 2.2. The airspeed at which the elastic twist increases rapidly to the point of failure is called the 'divergence airspeed', U_D. A major aim of any theoretical model is to accurately predict U_D. It should be emphasized that the above curve is representative not only of our typical section model but also of real aircraft wings. Indeed the

* See Chapter 6, BA, especially pp. 189–200.

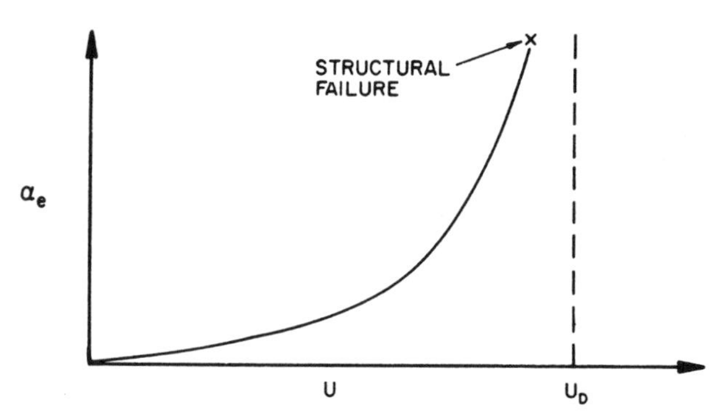

Figure 2.2 Elastic twist vs airspeed.

primary difference is not in the basic physical phenomenon of divergence but rather in the elaborateness of the theoretical analysis required to predict accurately U_D for an aircraft wing versus that required for our simple typical section model.

To determine U_D theoretically we proceed as follows. The equation of static equilibrium simply states that the sum of aerodynamic plus elastic moments about any point on the airfoil is zero. By convention, we take the point about which moments are summed as the point of spring attachment, the so-called 'elastic center' or 'elastic axis' of the airfoil.

The total aerodynamic angle of attack, α, is taken as the sum of some initial angle of attack, α_0 (with the spring untwisted), plus an additional increment due to elastic twist of the spring, α_e.

$$\alpha = \alpha_0 + \alpha_e \tag{2.1.1}$$

In addition, we define a point on the airfoil known as the 'aerodynamic center'.* This is the point on the airfoil about which the aerodynamic moment is independent of angle of attack, α. Thus, we may write the moment about the elastic axis as

$$M_y = M_{AC} + Le \tag{2.1.2}$$

where M_y moment about elastic axis or center
M_{AC} moment about aerodynamic center, both moments are positive nose up
L lift, net vertical force positive up
e distance from aerodynamic center to elastic axis, positive aft.

* For two dimensional, incompressible flow this is at the airfoil quarter-chord; for supersonic flow it moves back to the half-chord. See Ashley and Landahl [1]. References are given at the end of each chapter.

From aerodynamic theory [1] (or experiment plus dimensional analysis) one has

$$L = C_L qS$$
$$M_{AC} = C_{MAC} qSc \qquad (2.1.3a)$$

where

$$C_L = C_{L_0} + \frac{\partial C_L}{\partial \alpha}\alpha, \text{ lift coefficient} \qquad (2.1.3b)$$

$C_{MAC} = C_{MAC_0}$, a constant, aerodynamic center moment coefficient in which

$$q = \frac{\rho U^2}{2}, \text{ dynamic pressure and}$$

ρ air density
U air velocity
c airfoil chord
l airfoil span
S airfoil area, $c \times 1$

(2.1.3a) defines C_L and C_{MAC}. (2.1.3b) is a Taylor Series expansion of C_L for small α. C_{L_0} is the lift coefficient at $\alpha \equiv 0$. From (2.1.2), (2.1.3a) and (2.1.3b), we see the moment is also expanded in a Taylor series. The above forms are traditional in the aerodynamic literature. They are not necessarily those a nonaerodynamicist would choose.

Note that C_{L_0}, $\partial C_L/\partial\alpha$, C_{MAC_0} are nondimensional functions of airfoil shape, planform and Mach number. For a flat plate in two-dimensional incompressible flow [1]

$$\frac{\partial C_L}{\partial \alpha} = 2\pi, \qquad C_{MAC_0} = 0 = C_{L_0}$$

In what follows, we shall take $C_{L_0} \equiv 0$ for convenience and without any essential loss of information.

From (2.1.2), (2.1.3a) and (2.1.3b)

$$M_y = eqS\left[\frac{\partial C_L}{\partial \alpha}(\alpha_0 + \alpha_e)\right] + qScC_{MAC_0} \qquad (2.1.4)$$

Now consider the elastic moment. If the spring has linear moment-twist characteristics then the elastic moment (positive nose up) is $-K_\alpha \alpha_e$ where K_α is the elastic spring constant and has units of moment (torque) per angle of twist. Hence, summing moments we have

$$eqS\left[\frac{\partial C_L}{\partial \alpha}(\alpha_0+\alpha_e)\right]+qScC_{MAC_0}-K_\alpha\alpha_e=0 \tag{2.1.5}$$

which is the equation of static equilibrium for our 'typical section' airfoil.

Solving for the elastic twist (assuming $C_{MAC_0}=0$ for simplicity) one obtains

$$\alpha_e=\frac{qS}{K_\alpha}\frac{e\dfrac{\partial C_L}{\partial \alpha}\alpha_0}{1-q\dfrac{Se}{K_\alpha}\dfrac{\partial C_L}{\partial \alpha}} \tag{2.1.6}$$

This solution has several interesting properties. Perhaps most interesting is the fact that at a particular dynamic pressure the elastic twist becomes infinitely large. That is, when the denominator of the right-hand side of (2.1.6) vanishes

$$1-q\frac{Se}{K_\alpha}\frac{\partial C_L}{\partial \alpha}=0 \tag{2.1.7}$$

at which point $\alpha_e \to \infty$.

Equation (2.1.7) represents what is termed the 'divergence condition' and the corresponding dynamic pressure which may be obtained by solving (2.1.7) is termed the 'divergence dynamic pressure',

$$q_D \equiv \frac{K_\alpha}{Se(\partial C_L/\partial \alpha)} \tag{2.1.8}$$

Since only positive dynamic pressures are physically meaningful, note that only for $e>0$ will divergence occur, i.e., when the aerodynamic center is ahead of the elastic axis. Using (2.1.8), (2.1.6) may be rewritten in a more concise form as

$$\alpha_e=\frac{(q/q_D)\alpha_0}{1-q/q_D} \tag{2.1.9}$$

Of course, the elastic twist does not become infinitely large for any real airfoil; because this would require an infinitely large aerodynamic moment. Moreover, the linear relation between the elastic twist and the aerodynamic moment would be violated long before that. However, the elastic twist can become so large as to cause structural failure. For this reason, all aircraft are designed to fly below the divergence limits of all airfoil or lifting surfaces, e.g., wings, fins, control surfaces.

Now let us examine equations (2.1.5) and (2.1.9) for additional insight into our problem, again assuming $C_{MAC_0}=0$ for simplicity. Two special cases will be informative. First, consider $\alpha_0 \equiv 0$. Then (2.1.5) may be written

$$\alpha_e\left[qS\frac{\partial C_L}{\partial \alpha}e - K_\alpha\right] = 0 \tag{2.1.5a}$$

Excluding the trivial case $\alpha_e = 0$ we conclude from (2.1.5a) that

$$qS\frac{\partial C_L}{\partial \alpha}e - K_\alpha = 0 \tag{2.1.7a}$$

which is the 'divergence condition'. This will be recognized as an eigenvalue problem, the vanishing of the coefficient of α_e in (2.1.5a) being the condition for nontrivial solutions of the unknown, α_e.* Hence, 'divergence' requires only a consideration of elastic deformations.

Secondly, let us consider another special case of a somewhat different type, $\alpha_0 \neq 0$, but $\alpha_e \ll \alpha_0$. Then (2.1.5a) may be written approximately as

$$eqS\frac{\partial C_L}{\partial \alpha}\alpha_0 - K_\alpha \alpha_e = 0 \tag{2.1.10}$$

Solving

$$\alpha_e = \frac{qSe(\partial C_L/\partial \alpha)\alpha_0}{K_\alpha} \tag{2.1.11}$$

Note this solution agrees with (2.1.6) if the denominator of (2.1.6) can be approximated by

$$1 - q\frac{Se}{K_\alpha}\frac{\partial C_L}{\partial \alpha} = 1 - \frac{q}{q_D} \approx 1$$

Hence, this approximation is equivalent to assuming that the dynamic pressure is much smaller than its divergence value. Note that the term neglected in (2.1.5) is the aerodynamic moment due to the elastic twist. Without this term, solution (2.1.11) is valid only for $q/q_D \ll 1$; it cannot predict divergence, however. This term can be usefully thought of as the 'aeroelastic feedback'.† A feedback diagram of equation (2.1.5) is given

---

* Here in static aeroelasticity q plays the role of the eigenvalue; in dynamic aeroelasticity q will be a parameter and the (complex) frequency will be the eigenvalue. This is a source of confusion for some students when they first study the subject.

† For the reader with some knowledge of feedback theory as in, for example, Savant [2].

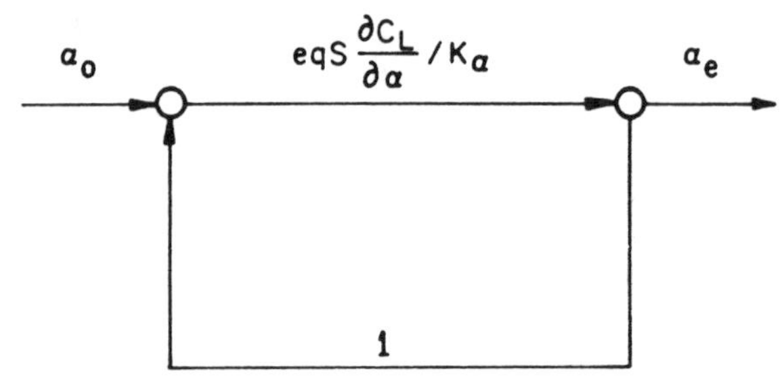

Figure 2.3 Feedback representation of aeroelastic divergence.

in Figure 2.3. Thus, when the forward loop gain exceeds unity, $qeS(\partial C_L/\partial\alpha)K_\alpha > 1$, the system is statically unstable, see equation (2.1.8). Hence, aeroelasticity can also be thought of as the study of aerodynamic + elastic feedback systems. One might also note the similarity of this divergence problem to conventional 'buckling' of structures.* Having exhausted the interpretations of this problem, we will quickly pass on to some slightly more complicated problems, but whose physical content is similar.

Typical section with control surface

We shall add a control surface to our typical section of Figure 2.1, as indicated in Figure 2.4. For simplicity, we take $\alpha_0 = C_{MAC_0} = 0$; hence, $\alpha = \alpha_e$. The aerodynamic lift is given by

$$L = qSC_L = qS\left(\frac{\partial C_L}{\partial\alpha}\alpha + \frac{\partial C_L}{\partial\delta}\delta\right) \text{ positive up} \tag{2.1.12}$$

the moment by

$$M_{AC} = qScC_{MAC} = qSc\frac{\partial C_{MAC}}{\partial\delta}\delta \text{ positive nose up} \tag{2.1.13}$$

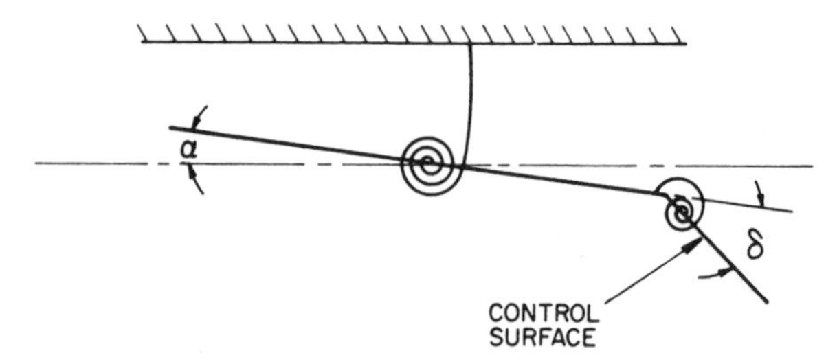

Figure 2.4 Typical section with control surface.

* Timoshenko and Gere [3].

the moment about the hinge line of the control surface by

$$H = qS_H c_H C_H = qS_H c_H \left(\frac{\partial C_H}{\partial \alpha} \alpha + \frac{\partial C_H}{\partial \delta} \delta \right) \text{ positive tail down} \tag{2.1.14}$$

where S_H is the area of control surface, c_H the chord of the control surface and C_H the (nondimensional) aerodynamic hinge moment coefficient. As before, $\frac{\partial C_L}{\partial \alpha}$, $\frac{\partial C_L}{\partial \delta}$, $\frac{\partial C_{MAC}}{\partial \delta}$, $\frac{\partial C_H}{\partial \alpha}$, $\frac{\partial C_H}{\partial \delta}$ are aerodynamic constants which vary with Mach and airfoil geometry. Note $\frac{\partial C_H}{\partial \delta}$ is typically negative.

The basic purpose of a control surface is to change the lift (or moment) on the main lifting surface. It is interesting to examine aeroelastic effects on this purpose.

To write the equations of equilibrium, we need the elastic moments about the elastic axis of the main lifting surface and about the hinge line of the control surface. These are $-K_\alpha \alpha$ (positive nose up), $-K_\delta(\delta - \delta_0)$ (positive tail down), and $\delta_e \equiv \delta - \delta_0$, where δ_e is the elastic twist of control surface in which δ_0 is the difference between the angle of zero aerodynamic control deflection and zero twist of the control surface spring.

The two equations of static moment equilibrium are

$$eqS\left(\frac{\partial C_L}{\partial \alpha} \alpha + \frac{\partial C_L}{\partial \delta} \delta \right) + qSc \frac{\partial C_{MAC}}{\partial \delta} \delta - K_\alpha \alpha = 0 \tag{2.1.15}$$

$$qS_H c_H \left(\frac{\partial C_H}{\partial \alpha} \alpha + \frac{\partial C_H}{\partial \delta} \delta \right) - K_\delta (\delta - \delta_0) = 0 \tag{2.1.16}$$

The above are two algebraic equations in two unknowns, α and δ, which can be solved by standard methods. For example, Cramer's rule gives

$$\alpha = \frac{\begin{vmatrix} 0 & eqS \frac{\partial C_L}{\partial \delta} + qSc \frac{\partial C_{MAC}}{\partial \delta} \\ -K_\delta \delta_0 & qS_H c_H \frac{\partial C_H}{\partial \delta} - K_\delta \end{vmatrix}}{\begin{vmatrix} eqS \frac{\partial C_L}{\partial \alpha} - K_\alpha & eqS \frac{\partial C_L}{\partial \delta} + qSc \frac{\partial C_{MAC}}{\partial \delta} \\ qS_H c_H \frac{\partial C_H}{\partial \alpha} & qS_H c_H \frac{\partial C_H}{\partial \delta} - K_\delta \end{vmatrix}} \tag{2.1.17}$$

and a similar equation for δ. To consider divergence we again set the denominator to zero. This gives a quadratic equation in the dynamic pressure q. Hence, there are two values of divergence dynamic pressure. Only the lower positive value of the two is physically significant.

In addition to the somewhat more complicated form of the divergence condition, there is a *new physical phenomenon* associated with the control surface called 'control surface reversal'. If the two springs were rigid, i.e., $K_\alpha \to \infty$ and $K_\delta \to \infty$, then $\alpha = 0$, $\delta = \delta_0$, and

$$L_r = qS\frac{\partial C_L}{\partial \delta}\delta_0 \tag{2.1.18}$$

With flexible springs

$$L = qS\left(\frac{\partial C_L}{\partial \alpha}\alpha + \frac{\partial C_L}{\partial \delta}\delta\right) \tag{2.1.19}$$

where α, δ are determined by solving the equilibrium equations (2.1.15), and (2.1.16). In general, the latter value of the lift will be smaller than the rigid value of lift. Indeed, the lift may actually become zero or even negative due to aeroelastic effects. Such an occurrence is called 'control surface reversal'. To simplify matters and show the essential character of control surface reversal, we will assume $K_\delta \to \infty$ and hence, $\delta \to \delta_0$ from the equilibrium condition (2.1.16). Solving the equilibrium equation (2.1.15), we obtain

$$\alpha = \delta_0 \frac{\dfrac{\partial C_L}{\partial \delta} + \dfrac{c}{e}\dfrac{\partial C_{MAC}}{\partial \delta}}{\dfrac{K_\alpha}{qSe} - \dfrac{\partial C_L}{\partial \alpha}} \tag{2.1.20}$$

But

$$\begin{aligned} L &= qS\left(\frac{\partial C_L}{\partial \delta}\delta_0 + \frac{\partial C_L}{\partial \alpha}\alpha\right) \\ &= qS\left(\frac{\partial C_L}{\partial \delta} + \frac{\partial C_L}{\partial \alpha}\frac{\alpha}{\delta_0}\right)\delta_0 \end{aligned} \tag{2.1.21}$$

so that, introducing (2.1.20) into (2.1.21) and normalizing by L_r, we obtain

$$\frac{L}{L_r} = \frac{1 + q\dfrac{Sc}{K_\alpha}\dfrac{\partial C_{MAC}}{\partial \delta}\left(\dfrac{\partial C_L}{\partial \alpha}\Big/\dfrac{\partial C_L}{\partial \delta}\right)}{1 - q\dfrac{Se}{K_\alpha}\dfrac{\partial C_L}{\partial \alpha}} \tag{2.1.22}$$

Control surface reversal occurs when $L/L_r = 0$

$$1 + q_R \frac{Sc}{K_\alpha} \frac{\partial C_{MAC}}{\partial \delta} \left(\frac{\partial C_L}{\partial \alpha} \bigg/ \frac{\partial C_L}{\partial \delta} \right) = 0 \tag{2.1.23}$$

where q_R is the dynamic pressure at reversal, or

$$q_R \equiv \frac{-\dfrac{K_\alpha}{Sc} \left(\dfrac{\partial C_L}{\partial \delta} \bigg/ \dfrac{\partial C_L}{\partial \alpha} \right)}{\dfrac{\partial C_{MAC}}{\partial \delta}} \tag{2.1.24}$$

Typically, $\partial C_{MAC}/\partial \delta$ is negative, i.e., the aerodynamic moment for positive control surface rotation is nose down. Finally, (2.1.22) may be written

$$\frac{L}{L_r} = \frac{1 - q/q_r}{1 - q/q_D} \tag{2.1.25}$$

where q_R is given by (2.1.22) and q_D by (2.1.8). It is very interesting to note that when K_δ is finite, the reversal dynamic pressure is still given by (2.1.24). However, q_D is now the lowest root of the denominator of (2.1.17). Can you reason physically why this is so?*

A graphical depiction of (2.1.25) is given in the Figure 2.5 where the two cases, $q_D > q_R$ and $q_D < q_R$, are distinguished. In the former case L/L_r decreases with increasing q and in the latter the opposite is true. Although the graphs are shown for $q > q_D$, our analysis is no longer valid when the divergence condition is exceeded without taking into account nonlinear effects.

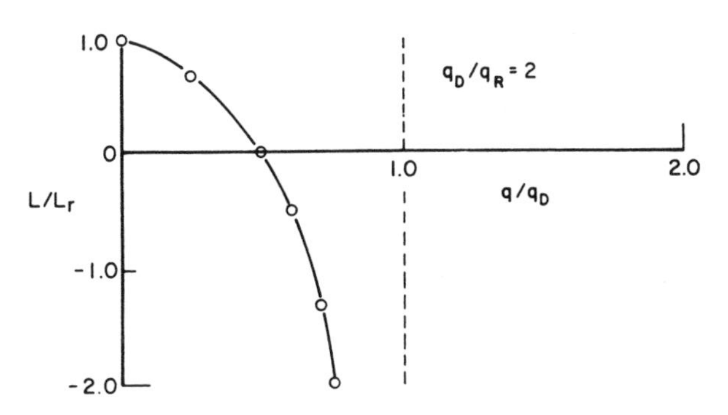

Figure 2.5 Lift vs dynamic pressure.

* See, BA, pp. 197–200.

It is interesting to note that the q_R given by (2.1.24) is still the correct answer even for finite K_δ. Consider (2.1.15). For reversal or zero lift, L = 0, (2.1.15) simplifies to

$$q_R S s \frac{\partial C_{\mathrm{MAC}}}{\partial \delta}\delta - K_\alpha \alpha = 0 \tag{2.1.15$_\mathrm{R}$}$$

and (2.1.12) becomes

$$\frac{\partial C_L}{\partial \alpha}\alpha + \frac{\partial C_L}{\partial \alpha}\delta = 0 \tag{2.1.12$_\mathrm{R}$}$$

Eliminating α, δ from these two equations (or setting the determinant to zero for nontrivial solutions) gives

$$K_\alpha \frac{\partial C_L}{\partial \delta} + \frac{\partial C_L}{\partial \alpha} q_R S c \frac{\partial C_{\mathrm{MAC}}}{\partial \delta} = 0 \tag{2.1.26}$$

Solving (2.1.26) for q_R gives (2.1.24). Note that by this approach an eigenvalue problem has been created. Also note the moment equilibrium about the control surface hinge line does not enter into this calculation. See Appendix II, Chapter 2 for a more conceptually straightforward, but algebraically more tedious approach. At the generalized reversal condition, when $\alpha_0 \neq 0$, $C_{\mathrm{MAC}}0 \neq 0$, the lift due to a *change* in δ is zero, by definition. In mathematical language,

$$\frac{dL}{d\delta} = 0 \text{ at } q \doteq q_R \tag{2.1.27}$$

To see how this generalized definition relates to our earlier definition of the reversal condition, consider again the equation for lift and also the equation for overall moment equilibrium of the main wing plus control surface, viz.

$$L = qS\left[\frac{\partial C_L}{\partial \alpha} + \frac{\partial C_L}{\partial \delta}\delta\right] \tag{2.1.19}$$

and

$$qScC_{MAC_0} + qSc\frac{\partial C_{MAC}}{\partial \delta}\delta + eqS\left[\frac{\partial C_L}{\partial \alpha}\alpha + \frac{\partial C_L}{\partial \delta}\delta\right] - K_\alpha(\alpha - \alpha_0) = 0 \tag{2.1.15}$$

From (2.1.19)

$$\frac{\mathrm{d}L}{\mathrm{d}\delta} = qs\left[\frac{\partial C_L}{\partial \alpha}\frac{\mathrm{d}\alpha}{\mathrm{d}\delta} + \frac{\partial C_L}{\partial \delta}\right] \tag{2.1.28}$$

where $\frac{d\alpha}{d\delta}$ may be calculated from (2.1.15) as

$$\frac{d\alpha}{d\delta} = \frac{-\left[qSc\frac{\partial C_{MAC}}{\partial \delta} + qSe\frac{\partial C_L}{\partial \delta}\right]}{eqS\frac{\partial C_L}{\partial \delta} - K_\alpha} \tag{2.1.29}$$

Note that neither C_{MAC_0} nor α_0 appear in (2.1.29). Moreover when (2.1.29) is substituted into (2.1.28) and $dL/d\delta$ is set to zero, the same expression for q_R is obtained as before. (2.1.24), when reversal was defined as $L = 0$ (for $\alpha_0 = C_{MAC_0} = 0$).

This result may be given a further physical interpretation. Consider a Taylor series expansion for L in terms of δ about the reference condition, $\delta \equiv 0$. Note that $\delta \equiv 0$ corresponds to a wing without any control surface deflection relative to the main wing. Hence the condition, $\delta \equiv 0$, may be thought of as a wing *without* any control surface.

The lift at any δ may then be expressed as

$$L(\delta) = L(\delta = 0) + \frac{\partial L}{\partial \delta}|_{\delta=0}\delta + ... \tag{2.1.30}$$

Because a linear model is used, it is clear that higher order terms in thies expansion vanish. Moreover, it is clear that $dL/d\delta$ is the same for any δ, cf. (2.1.28) and (2.1.29).

Now consider $L(\delta = 0)$. From (2.1.19)

$$L(\delta = 0) = qS\frac{\partial C_L}{\partial \alpha}\alpha(\delta = 0) \tag{2.1.31}$$

But from (2.1.15)

$$\alpha(\delta = 0) = \frac{K_\alpha \alpha_0 + qSC_{MAC_0}}{K_\alpha - eqS\frac{\partial C_L}{\partial \alpha}} \tag{2.1.32}$$

Note that $\alpha(\delta = 0) = 0$ for $\alpha_0 = C_{MAC_0} = 0$. Thus, in this special case, $L(\delta = 0) = 0$, and

$$L(\delta) = \frac{dL}{d\delta}|_{\delta=0)}\delta = \frac{dL}{d\delta}|_{\text{any}\delta}\delta \tag{2.1.33}$$

and hence

$$L(\delta) = 0 \text{ or } \frac{dL}{d\delta}|_{\text{any}}\delta = 0 \tag{2.1.34}$$

are equivalent statements when $\alpha_0 = C_{MAC_0} = 0$.

For $\alpha_0 \neq 0$ and/or $C_{MAC_0} \neq 0$, however, the reversal condition is more meaningfully defined as the condition when the lift due to $\delta \neq 0$ is zero, i.e.

$$\frac{dL}{d\delta} = 0 \text{ at } q = q_R \tag{2.1.27}$$

In this case, at the reversal condition from (2.1.31) and (2.1.32),

$$L(\delta)|_{\text{at reversal}} = L(\delta = 0)|_{\text{at reversal}} = qS\frac{\partial C_L}{\partial \alpha}\left[\frac{\alpha_0 + \frac{qSc}{K_\alpha}C_{MAC_0}}{1 - \frac{eqS\frac{\partial C_L}{\partial \alpha}}{K_\alpha}}\right] \tag{2.1.35}$$

and hence the lift at reversal per se is indeed not zero in general unless $\alpha_0 = C_{MAC_0} = 0$.

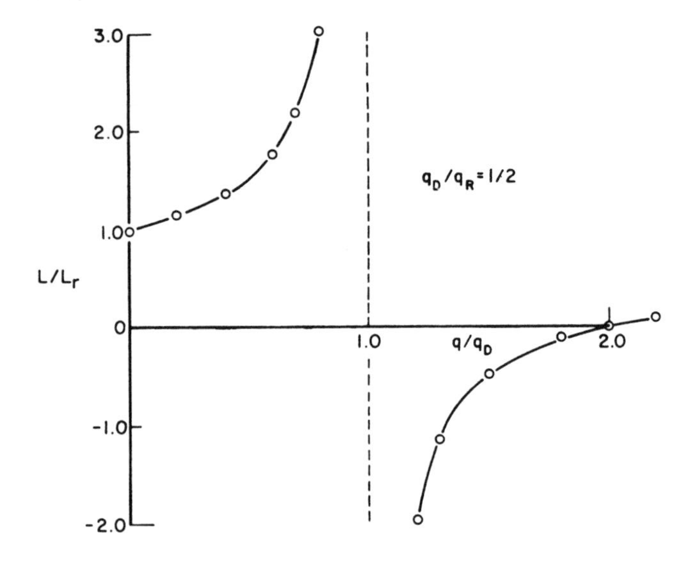

Typical section—nonlinear effects

For sufficiently large twist angles, the assumption of elastic and/or aerodynamic moments proportional to twist angle becomes invalid. Typically the elastic spring becomes stiffer at larger twist angles; for example the elastic moment-twist relation might be

$$M_E = -K_\alpha \alpha_e - K_{\alpha_3} \alpha_e^3$$

where $K_\alpha > 0$, $K_{\alpha_3} > 0$. The lift-angle of attack relation might be

$$L = qS[(\partial C_L/\partial\alpha)\alpha - (\partial C_L/\partial\alpha)_3 \alpha^3]$$

where $\partial C_L/\partial\alpha$ and $(\partial C_L/\partial\alpha)_3$ are positive quantities. Note the lift decreases for large α due to flow separation from the airfoil. Combining the above in a moment equation of equilibrium and assuming for simplicity that $\alpha_0 = C_{MAC} = 0$, we obtain (recall (2.1.5))

$$eqS[(\partial C_L/\partial\alpha)\alpha_e - (\partial C_L/\partial\alpha)_3 \alpha_e^3] - [K_\alpha \alpha_e + K_{\alpha_3} \alpha_e^3] = 0$$

Rearranging,

$$\alpha_e[eq(S\, \partial C_L/\partial\alpha) - K_\alpha] - \alpha_e^3[eqS(\partial C_L/\partial\alpha)_3 + K_{\alpha_3}] = 0$$

Solving, we obtain the trivial solution $\alpha_e \equiv 0$, as well as

$$\alpha_e^2 = \frac{\left[eqS\dfrac{\partial C_L}{\partial\alpha} - K_\alpha\right]}{\left[eqS\left(\dfrac{\partial C_L}{\partial\alpha}\right)_3 + K_{\alpha_3}\right]}$$

To be physically meaningful α_e must be a real number; hence the right hand side of the above equation must be a positive number for the nontrivial solution $\alpha_e \neq 0$ to be possible.

For simplicity let us first assume that $e > 0$. Then we see that only for $q > q_D$ (i.e., for $eqS(\partial C_L/\partial\alpha) > K_\alpha$) are nontrivial solutions possible. See Figure 2.6. For $q < q_D$, $\alpha_e \equiv 0$ as a consequence of setting $\alpha_0 \equiv C_{MAC} \equiv 0$. Clearly for $e > 0$, $\alpha_e \equiv 0$ when $q < q_D$ where

$$q_D \equiv \frac{K_\alpha}{eS\, \partial C_L/\partial\alpha}$$

Note that two (symmetrical) equilibrium solutions are possible for $q > q_D$. The actual choice of equilibrium position would depend upon how the airfoil is disturbed (by gusts for example) or possibly upon imperfections in the spring or airfoil geometry. α_0 may be thought of as an initial imperfection and its sign would determine which of the two equilibria positions occurs. Note that for the nonlinear model α_e remains finite for any finite q.

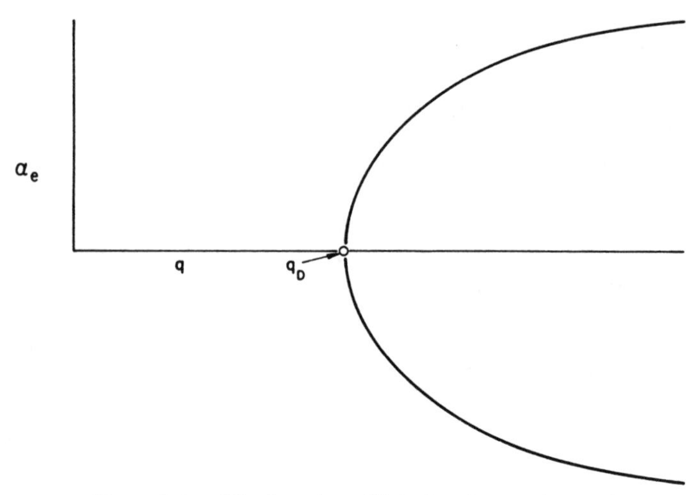

Figure 2.6a (Nonlinear) equilibria for elastic twist: $e > 0$.

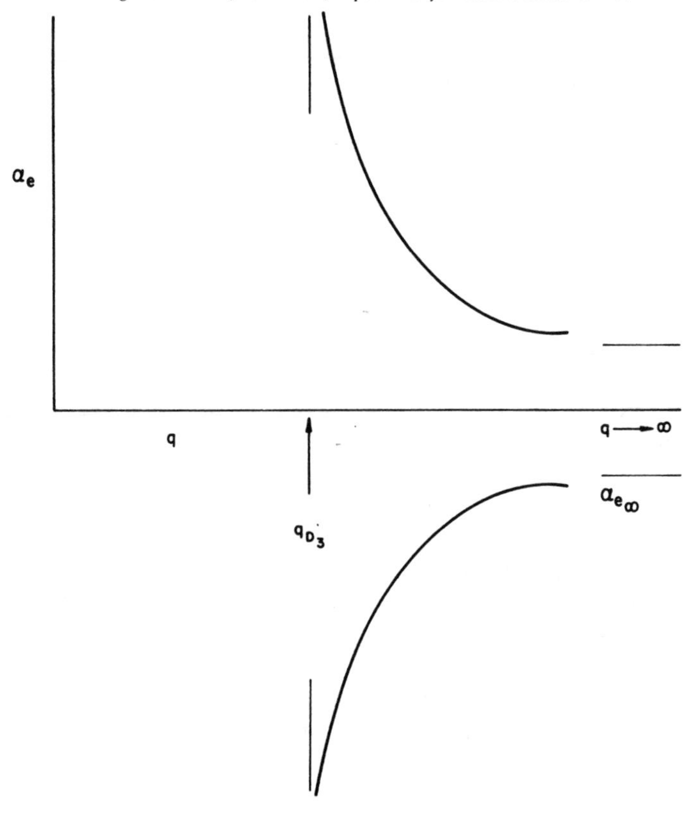

Figure 2.6b (Nonlinear) equilibria for elastic twist: $e < 0$.

For $e < 0$, the equilibrium configurations would be as shown in Figure 2.6. where

$$q_{D_3} = -K_{\alpha_3}/eS(\partial C_L/\partial\alpha)_3$$

and

$$\alpha_{e_\infty}^2 = \partial C_L/\partial\alpha/(\partial C_L/\partial\alpha)_3$$

As far as the author is aware, the behavior indicated in Figure 2.6 has never been observed experimentally. Presumably structural failure would occur for $q > q_{D_3}$ even though α_{e_∞} is finite. It would be most interesting to try to achieve the above equilibrium diagram experimentally.

The above discussion does not exhaust the possible types of nonlinear behavior for the typical section model. Perhaps one of the most important nonlinearities in practice is that associated with the control surface spring and the elastic restraint of the control surface connection to the main lifting surface.*

2.2 One dimensional aeroelastic model of airfoils

Beam-rod representation of large aspect ratio wing†

We shall now turn to a more sophisticated, but more realistic beam-rod model which contains the same basic physical ingredients as the typical section. A beam-rod is here defined as a flat plate with rigid chordwise sections whose span, l, is substantially larger than its chord, c. See Figure 2.7. The airflow is in the x direction. The equation of static moment equilibrium for a beam-rod is

$$\frac{d}{dy}\left(GJ\frac{d\alpha_e}{dy}\right) + M_y = 0 \tag{2.2.1}$$

$\alpha_e(y)$ nose up twist about the elastic axis, e.a., at station y
M_y nose up aerodynamic moment about e.a. per unit distance in the spanwise, y, direction
G shear modulus
J polar moment of inertia ($= ch^3/3$ for a rectangular cross-section of thickness h, $h \ll c$)
GJ torsional stiffness

---

* Woodcock [4].
† See Chapter 7, BA, pp. 280–295, especially pp. 288–295.
‡ Housner, and Vreeland [5].

Equation (2.2.1) can be derived by considering a differential element dy (see Figure 2.8) The internal elastic moment is $GJ(d\alpha_e/dy)$ from the theory of elasticity.‡ Note for $d\alpha_e/dy > 0$, $GJ(d\alpha_e/dy)$ is positive nose

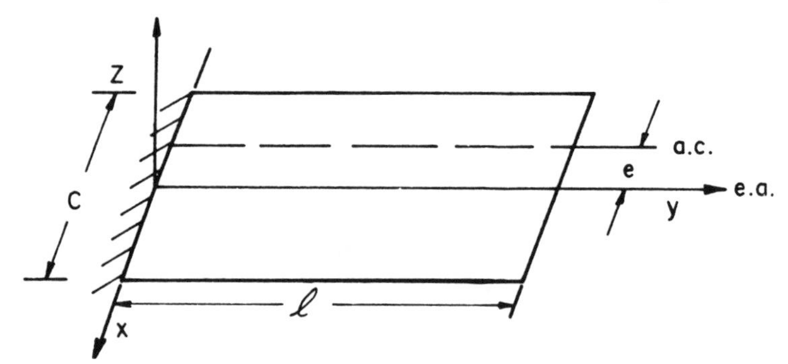

Figure 2.7 Beam-rod representation of wing.

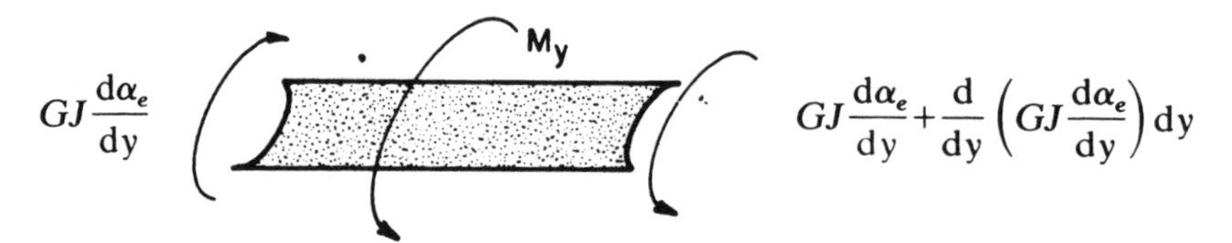

Figure 2.8 Differential element of beam-rod.

down. Summing moments on the differential element, we have

$$-GJ\frac{d\alpha_e}{dy}+GJ\frac{d\alpha_e}{dy}+\frac{d}{dy}\left(GJ\frac{d\alpha_e}{dy}\right)dy+\text{H.O.T.}+M_y\,dy=0$$

In the limit, as $dy \to 0$,

$$\frac{d}{dy}\left(GJ\frac{d\alpha_e}{dy}\right)+M_y=0 \tag{2.2.1}$$

Equation (2.2.1) is a second order differential equation in y. Associated with it are two boundary conditions. The airfoil is fixed at its root and free at its tip, so that the boundary conditions are

$$\alpha_e=0 \quad \text{at} \quad y=0 \qquad GJ\frac{d\alpha_e}{dy}=0 \quad \text{at} \quad y=l \tag{2.2.2}$$

Turning now to the aerodynamic theory, we shall use the 'strip theory' approximation. That is, *we shall assume that the aerodynamic lift and moment at station* y *depends only on the angle of attack at station* y (and is independent of the angle of attack at other spanwise locations).

Thus moments and lift per unit span are, as before,

$$M_y = M_{AC} + Le \tag{2.2.3a}$$

$$L \equiv qcC_L \tag{2.2.3b}$$

where now the lift and moment coefficients are given by

$$C_L(y) = \frac{\partial C_L}{\partial \alpha}[\alpha_0(y) + \alpha_e(y)]^* \tag{2.2.3c}$$

$$M_{AC} = qc^2 C_{MAC} \tag{2.2.3d}$$

(2.2.3b) and (2.2.3d) define C_L and C_{MAC} respectively.

Using (2.2.3) in (2.2.1) and nondimensionalizing (assuming for simplicity, constant wing properties)

$$\bar{y} \equiv \frac{y}{l}$$

$$\lambda^2 \equiv \frac{ql^2}{GJ} c \frac{\partial C_L}{\partial \alpha} e$$

$$K \equiv -\frac{qcl^2}{GJ}\left(e\frac{\partial C_L}{\partial \alpha}\alpha_0 + C_{MAC_0}c\right)$$

(2.2.1) becomes

$$\frac{d^2\alpha_e}{d\bar{y}^2} + \lambda^2\alpha_e = K \tag{2.2.4}$$

which is subject to boundary conditions (2.2.2). These boundary conditions have the nondimensional form

$$\alpha_e = 0 \quad \text{at} \quad \bar{y} = 0$$
$$\frac{d\alpha_e}{d\bar{y}} = 0 \quad \text{at} \quad \bar{y} = 1 \tag{2.2.5}$$

The general solution to (2.2.4) is

$$\alpha_e = A \sin \lambda\bar{y} + B \cos \lambda\bar{y} + \frac{K}{\lambda^2} \tag{2.2.6}$$

---

* A more complete aerodynamic model would allow for the effect of angle of attack at one spanwise location, say η, on (nondimensional) lift at another, say y. This relation would then be replaced by $C_L(y) = \int A(y-\eta)[\alpha_0(\eta) + \alpha_e(\eta)]\,d\eta$ where A is an aerodynamic influence function which must be measured or calculated from an appropriate theory. More will be said of this later.

Applying boundary conditions (2.2.5), we obtain

$$B+\frac{K}{\lambda^2}=0, \qquad \lambda[A\cos\lambda - B\sin\lambda]=0 \tag{2.2.7}$$

Solving equation (2.2.7), $A=-(K/\lambda^2)\tan\lambda$, $B=-K/\lambda^2$, so that

$$\alpha_e=\frac{K}{\lambda^2}[1-\tan\lambda\sin\lambda\bar{y}-\cos\lambda\bar{y}] \tag{2.2.8}$$

Divergence occurs when $\alpha_e\to\infty$, i.e., $\tan\lambda\to\infty$, or $\cos\lambda\to 0$.* Thus, for $\lambda=\lambda_m=(2m-1)\frac{\pi}{2}(m=1,2,3,\dots)$, $\alpha_e\to\infty$. The lowest of these, $\lambda_1=\frac{\pi}{2}$, is physically significant. Using the definition of λ preceding equation (2.2.4), the divergence dynamic pressure is

$$q=(\pi/2)^2\frac{GJ}{l}\Big/ lce(\partial C_L/\partial\alpha) \tag{2.2.9}$$

Recognizing that $S=lc$, we see that (2.2.9) is equivalent to the typical section value, (2.1.8), with

$$K_\alpha=\left(\frac{\pi}{2}\right)^2\frac{GJ}{l} \tag{2.2.10}$$

Consider again (2.2.8). A further physical interpretation of this result may be helpful. For simplicity, consider the case when $C_{MAC_0}=0$ and thus $K=-\lambda^2\alpha_0]$. Then the expression for α_e, (2.2.8), may be written as

$$\alpha_e=\alpha_0[-1+\tan\lambda\sin\lambda\bar{y}+\cos\lambda\bar{y}] \tag{2.2.8$_a$}$$

The tip twist of $\bar{y}=1$ may be used to characterize the variation of α_e with λ, i.e.

$$\alpha_e(\bar{y}=1)=\alpha_0\left[\frac{1}{\cos\lambda}-1\right] \tag{2.2.8$_b$}$$

and thus

$$\alpha=\alpha_0+\alpha_e=\alpha_0/\cos\lambda \tag{2.2.8$_c$}$$

---

* Note $\lambda\equiv 0$ is not a divergence condition! Expanding (2.2.8) for $\lambda\ll 1$, we obtain $\alpha_e=\frac{K}{\lambda^2}\left[1-\lambda^2\bar{y}-\left(1-\frac{\lambda^2\bar{y}^2}{2}\right)+\cdots\right]\to K\left[\frac{\bar{y}^2}{2}-\bar{y}\right]$ as $\lambda\to 0$.

From (2.2.8)$_c$, we see that for low flow speeds or dynamic pressure, $\lambda \to 0, \alpha = \alpha_0$. As $\lambda \to \pi/2, \alpha$ monotonically increases and $\alpha \to \infty$ as $\lambda \to \pi/2$. For a given wing design, a certain twist might be allowable. From (2.2.8)$_c$ or its counterpart for more complex physical and mathematical models, the corresponding allowable or design λ may be determined.

Another design allowable might be the allowable structural moment, $T \equiv GJd\alpha_e/dy$. Using (2.2.8) and the definition of T, for a given allowable T the corresponding allowable λ or q may be determined.

Eigenvalue and eigenfunction approach

One could have treated divergence from the point of view of an eigenvalue problem. Neglecting those terms which do *not* depend on the elastic twist, i.e., setting $\alpha_0 = C_{MAC_0} = 0$, we have $K = 0$ and hence

$$\frac{d^2\alpha}{d\bar{y}^2} + \lambda^2\alpha = 0 \qquad (2.2.11)$$

with

$$\begin{aligned} \alpha &= 0 \quad \text{at} \quad y = 0 \\ \frac{d\alpha}{d\bar{y}} &= 0 \quad \text{at} \quad y = 1 \end{aligned} \qquad (2.2.12)$$

The general solution is

$$\alpha = A \sin \lambda\bar{y} + B \cos \lambda\bar{y} \qquad (2.2.13)$$

Using (2.2.12), (2.2.13)

$$B = 0$$

$$\lambda[A \cos \lambda - B \sin \lambda] = 0$$

we conclude that

$$A = 0$$

or

$$\lambda \cos \lambda = 0 \quad \text{and} \quad A \neq 0 \qquad (2.2.14)$$

The latter condition, of course, is 'divergence'. Can you show that $\lambda = 0$, does not lead to divergence? What does (2.2.13) say? For each eigenvalue, $\lambda = \lambda_m = (2m-1)\frac{\pi}{2}$ there is an eigenfunction, $A \neq 0$, $B = 0$,

$$\alpha_m \sim \sin \lambda_m\bar{y} = \sin (2m-1)\frac{\pi}{2}\bar{y} \qquad (2.2.15)$$

These eigenfunctions are of interest for a number of reasons:

(1) They give us the twist distribution at the divergence dynamic pressure as seen above in (2.2.15).

(2) They may be used to obtain a series expansion of the solution for any dynamic pressure.

(3) They are useful for developing an approximate solution for variable property wings.

Let us consider further the second of these. Now we let $\alpha_0 \neq 0$, $C_{MAC_0} \neq 0$ and begin with (2.2.4).

$$\frac{d^2\alpha_e}{d\bar{y}} + \lambda^2 \alpha_e = K \tag{2.2.4}$$

Assume a series solution of the form

$$\alpha_e = \sum_n a_n \alpha_n(\bar{y}) \tag{2.2.16}$$

$$K = \sum_n A_n \alpha_n(\bar{y}) \tag{2.2.17}$$

where a_n, A_n are to be determined. Now it can be shown that

$$\begin{aligned} \int_0^1 \alpha_n(\bar{y})\alpha_m(\bar{y})\, d\bar{y} &= \tfrac{1}{2} \quad \text{for} \quad m = n \\ &= 0 \quad \text{for} \quad m \neq n \end{aligned} \tag{2.2.18}$$

This is the so-called 'orthogonality condition'. We shall make use of it in what follows. First, let us determine A_n. Multiply (2.2.17) by α_m and $\int_0^1 \cdots d\bar{y}$.

$$\begin{aligned} \int_0^1 K\alpha_m(\bar{y})\, d\bar{y} &= \sum_n A_n \int_0^1 \alpha_n(\bar{y})\alpha_m(\bar{y})\, d\bar{y} \\ &= A_m \tfrac{1}{2} \end{aligned}$$

using (2.2.18). Solving for A_m,

$$A_m = 2\int_0^1 K\alpha_m(\bar{y})\, d\bar{y} \tag{2.2.19}$$

Now let us determine a_n. Substitute (2.2.16) and (2.2.17) into (2.2.4) to obtain

$$\sum_n \left[a_n \frac{d^2\alpha_n}{d\bar{y}^2} + \lambda^2 a_n \alpha_n \right] = \sum_n A_n \alpha_n \tag{2.2.20}$$

Now each eigenfunction, α_n, satisfies (2.2.11).

$$\frac{d^2\alpha_n}{d\bar{y}^2} + \lambda_n^2 \alpha_n = 0 \tag{2.2.11}$$

Therefore, (2.2.20) may be written

$$\sum_n a_n[-\lambda_n^2 + \lambda^2]\alpha_n = \sum A_n \alpha_n \tag{2.2.21}$$

Multiplying (2.2.21) by α_m and $\int_0^1 \cdots d\bar{y}$,

$$[\lambda^2 - \lambda_m^2] a_m \tfrac{1}{2} = A_m \tfrac{1}{2}$$

Solving for a_m,

$$a_m = \frac{A_m}{[\lambda^2 - \lambda_m^2]} \tag{2.2.22}$$

Thus,

$$\alpha_e = \sum a_n \alpha_n = \sum_n \frac{A_n}{[\lambda^2 - \lambda_n^2]} \alpha_n(\bar{y}) \tag{2.2.23}$$

where A_n is given by (2.2.19).*

Similar calculations can be carried out for airfoils whose stiffness, chord, etc., are *not* constants but vary with spanwise location. One way to do this is to first determine the eigenfunction expansion *for the variable property wing* as done above for the constant property wing. The determination of such eigenfunctions may itself be fairly complicated, however. An alternative procedure can be employed which expands the solution for *the variable property* wing in terms of the eigenfunctions of the *constant property wing*. This is the last of the reasons previously cited for examining the eigenfunctions.

Galerkin's method

The equation of equilibrium for a *variable* property wing may be obtained by substituting (2.2.3) into (2.2.1). In dimensional terms

$$\frac{d}{dy}\left(GJ\frac{d}{dy}\alpha_e\right) + eqc\frac{\partial C_L}{\partial \alpha}\alpha_e = -eqc\frac{\partial C_L}{\partial \alpha}\alpha_0 - qc^2 C_{MAC_0} \tag{2.2.24}$$

* For a more detailed mathematical discussion of the above, see Hildebrand [6], pp. 224–234. This problem is one of a type known as 'Sturm–Liouville Problems'.

In nondimensional terms

$$\frac{d}{d\bar{y}}\left(\gamma \frac{d\alpha_e}{d\bar{y}}\right) + \lambda^2 \alpha_e \beta = K \tag{2.2.25}$$

where

$$\gamma \equiv \frac{GJ}{(GJ)_{ref}} \qquad K = -\frac{qcl^2}{(GJ)_{ref}}\left[e\frac{\partial C_L}{\partial \alpha}\alpha_0 + C_{MAC_0}c\right]$$

$$\lambda^2 \equiv \frac{ql^2 c_{ref}}{(GJ)_{ref}}\left(\frac{\partial C_L}{\partial \alpha}\right)_{ref} e_{ref} \qquad \beta = \frac{c}{c_{ref}}\frac{e}{e_{ref}}\frac{\left(\dfrac{\partial C_L}{\partial \alpha}\right)}{\left(\dfrac{\partial C_L}{\partial \alpha}\right)_{ref}}$$

Let

$$\alpha_e = \sum_n a_n \alpha_n(\bar{y})$$

$$K = \sum_n A_n \alpha_n(\bar{y})$$

as before. Substituting the series expansions into (2.2.25), multiplying by α_m and $\int_0^1 \cdots d\bar{y}$,

$$\sum a_n \left\{\int_0^1 \frac{d}{d\bar{y}}\left(\gamma \frac{d\alpha_n}{d\bar{y}}\right)\alpha_m \, d\bar{y} + \lambda^2 \int_0^1 \beta \alpha_n \alpha_m \, d\bar{y}\right\}$$

$$= \sum_n A_n \int_0^1 \alpha_n \alpha_m \, d\bar{y} = \frac{A_m}{2} \tag{2.2.26}$$

The first and second terms cannot be simplified further unless the eigenfunctions or 'modes' employed are eigenfunctions for the variable property wing. Hence, a_n is not as simply related to A_n as in the constant property wing example. (2.2.26) represents a system of equations for the a_n. In matrix notation

$$[C_{mn}]\{a_n\} = \{A_m\}\tfrac{1}{2} \tag{2.2.27}$$

where

$$C_{mn} \equiv \int_0^1 \frac{d}{d\bar{y}}\left(\gamma \frac{d\alpha_n}{d\bar{y}}\right)\alpha_m \, d\bar{y} + \lambda^2 \int_0^1 \beta \alpha_n \alpha_m \, d\bar{y}$$

By truncating the series to a finite number of terms, we may formally solve for the a_n,

$$\{a_n\} = \tfrac{1}{2}[C_{mn}]^{-1}\{A_m\} \tag{2.2.28}$$

The divergence condition is simply that the determinant of C_{mn} vanish (and hence $a_n \to \infty$)

$$|C_{mn}| = 0 \tag{2.2.29}$$

which is a polynomial in λ^2. It should be emphasized that for an 'exact' solution, (2.2.27), (2.2.28) etc., are infinite systems of equations (in an infinite number of unknowns). In practice, some large but finite number of equations is used to obtain an accurate approximation. By systematically increasing the terms in the series, the convergence of the method can be assessed. This procedure is usually referred to as Galerkin's method or as a 'modal' method.* The modes, α_n, used are called 'primitive modes' to distinguish them from eigenfunctions, i.e., they are 'primitive functions' for a variable property wing even though eigenfunctions for a constant property wing.

2.3 Rolling of a straight wing

We shall now consider a more complex physical and mathematical variation on our earlier static aeroelastic lifting surface (wing) studies. For variety, we treat a new physical situation, the rolling of a wing (rotation about the root axis). Nevertheless, we shall meet again our old friends, 'divergence' and 'control surface effectiveness' or 'reversal'.

The present analysis differs from the previous one as follows:

(a) integral equation formulation vs. differential equation formulation

(b) aerodynamic induction effects vs. 'strip' theory

(c) 'lumped element' method of solution vs. modal (or eigenfunction) solution.

The geometry of the problem is shown in Figure 2.9.

Integral equation of equilibrium

The integral equation of equilibrium is

$$\alpha(y) = \int_0^l C^{\alpha\alpha}(y, \eta) M_y(\eta)\, \mathrm{d}\eta \tag{2.3.1}$$†

* Duncan [7].

† For simplicity, $\alpha_0 \equiv 0$ in what follows.

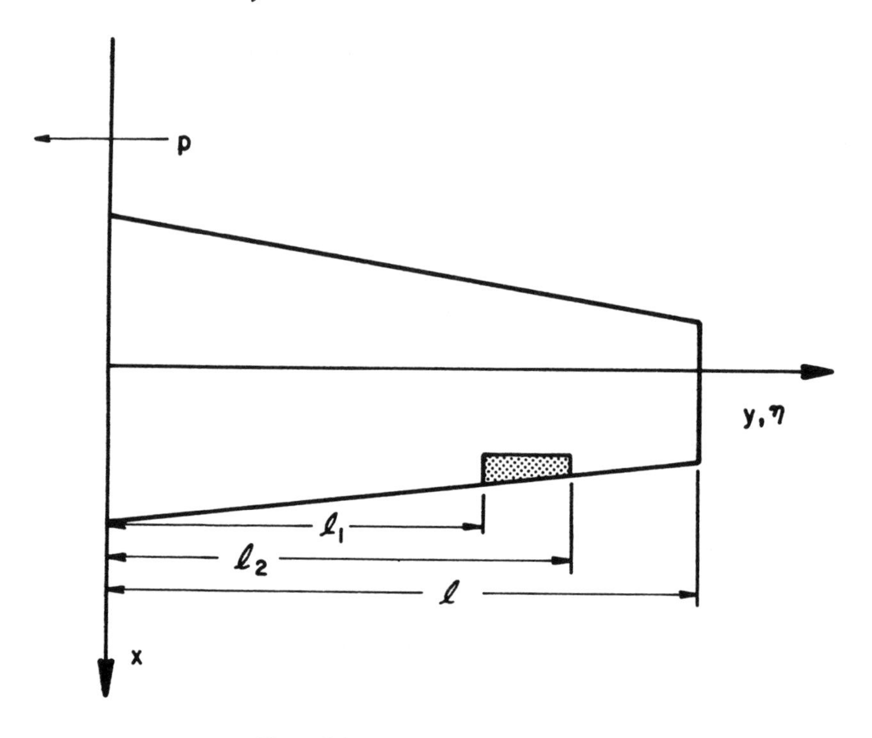

Figure 2.9 Rolling of a straight wing.

Before deriving the above equation, let us first consider the physical interpretation of $C^{\alpha\alpha}$:

Apply a unit point moment at some point, say $y=\gamma$, i.e.,

$$M_y(\eta)=\delta(\eta-\gamma).$$

Then (2.3.1) becomes

$$\alpha(y)=\int_0^l C^{\alpha\alpha}(y,\eta)\,\delta(\eta-\gamma)\,\mathrm{d}\eta$$

$$=C^{\alpha\alpha}(y,\gamma) \qquad (2.3.2)$$

Thus, $C^{\alpha\alpha}(y,\gamma)$ is the twist at y due to a unit moment at γ, or alternatively, $C^{\alpha\alpha}(y,\eta)$ is the twist at y due to a unit moment at η. $C^{\alpha\alpha}$ is called a *structural influence function.*

Also note that (2.3.1) states that to obtain the total twist, one multiplies the actual distributed torque, M_y, by $C^{\alpha\alpha}$ and sums (integrates) over the span. This is physically plausible.

$C^{\alpha\alpha}$ plays a central role in the integral equation formulation.* The physical interpretation of $C^{\alpha\alpha}$ suggests a convenient means of measuring $C^{\alpha\alpha}$ in a laboratory experiment. By successively placing unit couples at various locations along the wing and measuring the twists of all such stations for each loading position we can determine $C^{\alpha\alpha}$. This capability for measuring $C^{\alpha\alpha}$ gives the integral equation a preferred place in aeroelastic analysis where $C^{\alpha\alpha}$ and/or GJ are not always easily determinable from purely theoretical considerations.

Derivation of equation of equilibrium. Now consider a derivation of (2.3.1) taking as our starting point the differential equation of equilibrium. We have, you may recall,

$$\frac{d}{dy}\left(GJ\frac{d\alpha}{dy}\right) = -M_y \tag{2.3.3}$$

with

$$\alpha(0) = 0 \quad \text{and} \quad \frac{d\alpha}{dy}(l) = 0 \tag{2.3.4}$$

as boundary conditions.

As a special case of (2.3.3) and (2.3.4) we have for a unit torque applied at $y = \eta$,

$$\frac{d}{dy}GJ\frac{dC^{\alpha\alpha}}{dy} = -\delta(y-\eta) \tag{2.3.5}$$

with

$$C^{\alpha\alpha}(0, \eta) = 0 \quad \text{and} \quad \frac{dC^{\alpha\alpha}}{dy}(l, \eta) = 0 \tag{2.3.6}$$

Multiply (2.3.5) by $\alpha(y)$ and integrate over the span,

$$\int_0^l \alpha(y)\frac{d}{dy}\left(GJ\frac{dC^{\alpha\alpha}}{dy}\right)dy = -\int_0^l \delta(y-\eta)\alpha(y)\,dy$$

$$= -\alpha(\eta) \tag{2.3.7}$$

* For additional discussion, see the following selected references: Hildebrand [6] pp. 388–394 and BAH, pp. 39–44.

Integrate LHS of (2.3.7) by parts,

$$\alpha GJ\frac{dC^{\alpha\alpha}}{dy}\bigg|_0^l - GJ\frac{d\alpha}{dy}C^{\alpha\alpha}\bigg|_0^l + \int_0^l C^{\alpha\alpha}\frac{d}{dy}\left(GJ\frac{d\alpha}{dy}\right)dy = -\alpha(\eta) \quad (2.3.8)$$

Using boundary conditions, (2.3.4) and (2.3.6), the first two terms of LHS of (2.3.8) vanish. Using (2.3.3) the integral term may be simplified and we obtain,

$$\alpha(\eta) = \int_0^l C^{\alpha\alpha}(y, \eta)M_y(y)\, dy \quad (2.3.9)$$

Interchanging y and η,

$$\alpha(y) = \int_0^l C^{\alpha\alpha}(\eta, y)M_y(\eta)\, d\eta \quad (2.3.10)$$

(2.3.10) is identical to (1), *if*

$$C^{\alpha\alpha}(\eta, y) = C^{\alpha\alpha}(y, \eta) \quad (2.3.11)$$

We shall prove (2.3.11) subsequently.

Calculation of $C^{\alpha\alpha}$. We shall calculate $C^{\alpha\alpha}$ from (2.3.5) using (2.3.6). Integrating (2.3.5) with respect to y from 0 to y_1,

$$GJ(y_1)\frac{dC^{\alpha\alpha}}{dy}(y_1, \eta) - GJ(0)\frac{dC^{\alpha\alpha}}{dy}(0, \eta)$$
$$\left.\begin{aligned} &= -1 \quad \text{if} \quad y_1 > \eta \\ &= \ \ 0 \quad \text{if} \quad y_1 < \eta \end{aligned}\right\} \equiv S(y_1, \eta) \quad (2.3.12)$$

Sketch of function $S(y_1, \eta)$.

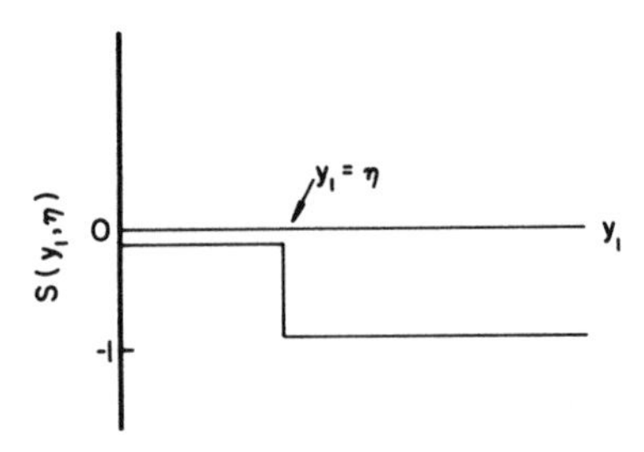

Dividing (2.3.12) by $GJ(y_1)$ and integrating with respect to y_1 from 0 to y_2,

$$C^{\alpha\alpha}(y_2,\eta)-C^{\alpha\alpha}(0,\eta)-GJ(0)\frac{dC^{\alpha\alpha}}{dy}(0,\eta)\int_0^{y_2}\frac{1}{GJ}dy_1$$

$$=\int_0^{y_2}\frac{S(y_1,\eta)}{GJ(y_1)}dy_1$$

$$=-\int_\eta^{y_2}\frac{1}{GJ(y_1)}dy_1 \quad \text{for} \quad y_2>\eta$$

$$=0 \quad \text{for} \quad y_2<\eta \tag{2.3.13}$$

From boundary conditions, (2.3.6),

(a) $C^{\alpha\alpha}(0,\eta)=0$

(b) $\dfrac{dC^{\alpha\alpha}}{dy}(l,\eta)=0$

These may be used to evaluate the unknown terms in (2.3.12) and (2.3.13). Evaluating (2.3.12) at $y_1=l$

(c) $GJ(l)\cancelto{0}{\dfrac{dC^{\alpha\alpha}}{dy}}(l,\eta)-GJ\dfrac{dC^{\alpha\alpha}}{dy}(0,\eta)=-1$

Using (a) and (c), (2.3.13) may be written,

$$C^{\alpha\alpha}(y_2,\eta)=\int_0^{y_2}\frac{1}{GJ}dy_1-\int_\eta^{y_2}\frac{1}{GJ}dy_1$$

$$=\int_0^{\eta}\frac{1}{GJ}dy_1 \quad \text{for} \quad y_2>\eta$$

$$=\int_0^{y_2}\frac{1}{GJ}dy_1 \quad \text{for} \quad y_2<\eta$$

One may drop the dummy subscript on y_2, of course. Thus

$$C^{\alpha\alpha}(y,\eta)=\int_0^{y}\frac{1}{GJ}dy_1 \quad \text{for} \quad y<\eta$$

$$=\int_0^{\eta}\frac{1}{GJ}dy_1 \quad \text{for} \quad y>\eta \tag{2.3.14}$$

Note from the above result we may conclude by interchanging y and η that

$C^{\alpha\alpha}(y,\eta)=C^{\alpha\alpha}(\eta,y)$

This is a particular example of a more general principle known as Maxwell's Reciprocity Theorem* which says that all structural influence

* Bisplinghoff, Mar, and Pian [8], p. 247.

functions for linear elastic bodies are symmetric in their arguments. In the case of $C^{\alpha\alpha}$ these are y and η, of course.

Aerodynamic forces (including spanwise induction)

First, let us identify the aerodynamic angle of attack; i.e., the angle between the airfoil chord and relative airflow. See Figure 2.10. Hence, the total angle of attack due to twisting and rolling is

$$\alpha_{\text{Total}} = \alpha(y) - \frac{py}{U}$$

The control surface will be assumed rigid and its rotation is given by

$$\begin{aligned} \delta(y) &= \delta_R \quad \text{for} \quad l_1 < y < l_2 \\ &= 0 \quad \text{otherwise} \end{aligned}$$

From aerodynamic theory or experiment

$$C_L \equiv \frac{L}{qc} = \int_0^l A^{L\alpha}(y,\eta)\alpha_T(\eta)\frac{\mathrm{d}\eta}{l} + \int_0^l A^{L\delta}(y,\eta)\,\delta(\eta)\frac{\mathrm{d}\eta}{l} \tag{2.3.15}$$

Here $A^{L\alpha}$, $A^{L\delta}$ are aerodynamic influence functions; as written, they are nondimensional. Thus, $A^{L\alpha}$ is nondimensional lift at y due to unit angle of attack at η. Substituting for α_T and δ, (2.3.15) becomes,

$$C_L = \int_0^l A^{L\alpha}\,\alpha\frac{\mathrm{d}\eta}{l} - \frac{pl}{U}\int_0^l A^{L\alpha}\frac{\eta}{l}\frac{\mathrm{d}\eta}{l} + \delta_R\int_{l_1}^{l_2} A^{L\delta}\frac{\mathrm{d}\eta}{l}$$

$$C_L = \int_0^l A^{L\alpha}\alpha\frac{\mathrm{d}\eta}{l} + \frac{pl}{U}\frac{\partial C_L}{\partial\left(\frac{pl}{U}\right)} + \delta_R\frac{\partial C_L}{\partial \delta_R} \tag{2.3.16}$$

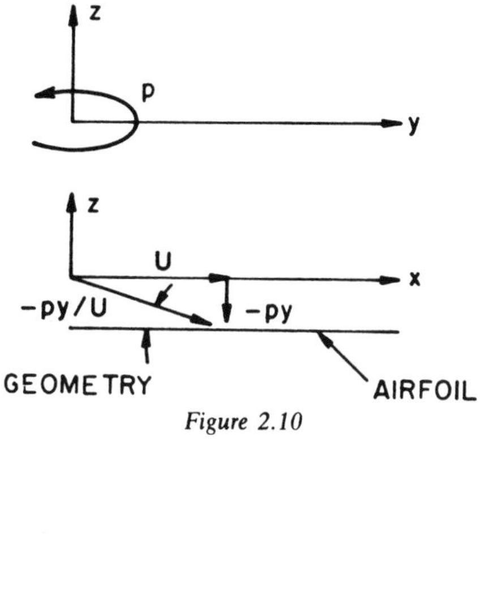

Figure 2.10

where

$$\frac{\partial C_L}{\partial\left(\frac{pl}{U}\right)}(y) \equiv -\int_0^l A^{L\alpha}\frac{\eta}{l}\frac{\mathrm{d}\eta}{l}$$

and

$$\frac{\partial C_L}{\partial \delta_R}(y) \equiv \int_{l_1}^{l_2} A^{L\delta}\frac{\mathrm{d}\eta}{l}$$

Physical Interpretation of $A^{L\alpha}$ and $A^{L\delta}$: $A^{L\alpha}$ is the lift coefficient at y due to unit angle of attack at η. $A^{L\delta}$ is the lift coefficient at y due to unit rotation of control surface at η.

Physical Interpretation of $\partial C_L/\partial(pl/U)$ and $\partial C_L/\partial\delta_R$: $\partial C_L/\partial(pl/U)$ is the lift coefficient at y due to unit rolling velocity, pl/U. $\partial C_L/\partial\delta_R$ is the lift coefficient at y due to unit control surface rotation, δ_R.

As usual

$$C_{MAC} \equiv \frac{M_{AC}}{qc^2} = \frac{\partial C_{MAC}}{\partial \delta_R}\delta_R \tag{2.3.17}$$

is the aerodynamic coefficient moment (about a.c.) at y due to control surface rotation. Note

$$\partial C_{MAC}/\partial\alpha_T \equiv 0$$

by definition of the aerodynamic center. Finally the total moment loading about the elastic axis is

$$\begin{aligned} M_y &= M_{AC} + Le \\ &= qc[C_{MAC}c + C_L e] \end{aligned} \tag{2.3.18}$$

Using (2.3.16) and (2.3.17), the above becomes

$$M_y = qc\left[c\frac{\partial C_{MAC}}{\partial \delta_R}\delta_R + e\left\{\int_0^l A^{L\alpha}\alpha\frac{\mathrm{d}\eta}{l} + \frac{\partial C_L}{\partial\left(\frac{pl}{U}\right)}\left(\frac{pl}{U}\right) + \frac{\partial C_L}{\partial \delta_R}\delta_R\right\}\right] \tag{2.3.19}$$

Note that $A^{L\alpha}$, $A^{L\delta}$ are more difficult to measure than their structural counterpart, $C^{\alpha\alpha}$. One requires an experimental model to which one can apply unit angles of attack at various discrete points along the span of the wing. This requires a rather sophisticated model and also introduces experimental difficulties in establishing and maintaining a smooth flow over the airfoil. Conversely

$$\frac{\partial C_L}{\partial \frac{pl}{U}}, \frac{\partial C_L}{\partial \delta_R} \quad \text{and} \quad \frac{\partial C_{MAC}}{\partial \delta_R}$$

are relatively easy to measure since they only require a rolling or control surface rotation of *a rigid wing with the same geometry* as the flexible airfoil of interest.

Aeroelastic equations of equilibrium and lumped element solution method

The key relations are (2.3.1) and (2.3.19). The former describes the twist due to an aerodynamic moment load, the latter the aerodynamic moment due to twist as well as rolling and control surface rotation.

By substituting (2.3.19) in (2.3.1), one could obtain a single equation for α. However, this equation is not easily solved analytically except for some simple cases, which are more readily handled by the differential equation approach. Hence, we seek an approximate solution technique. Perhaps the most obvious and convenient method is to approximate the integrals in (2.3.1) and (2.3.19) by sums, i.e., the wing is broken into various spanwise segments or 'lumped elements'. For example, (2.3.1) would be approximated as:

$$\alpha(y_i) \cong \sum_{j=1}^{N} C^{\alpha\alpha}(y_i, \eta_j) M_y(\eta_j)\, \Delta\eta \qquad i = 1, \ldots, N \tag{2.3.20}$$

where $\Delta\eta$ is the segment width and N the total number of segments. Similarly, (2.3.19) may be written:

$$M_y(y_i) \cong qc\left\{\left[c\frac{\partial C_{MAC}}{\partial \delta_R}\delta_R + e\frac{\partial C_L}{\partial \frac{pl}{U}}\frac{pl}{U} + e\frac{\partial C_L}{\partial \delta_R}\delta_R\right] + e\sum_{j=1}^{N} A^{L\alpha}(y_i, \eta_j)\alpha(\eta_j)\frac{\Delta\eta}{l}\right\} \qquad i = 1, \ldots, N \tag{2.3.21}$$

To further manipulate (2.3.20) and (2.3.21), it is convenient to use matrix notation. That is,

$$\{\alpha\} = \Delta\eta[C^{\alpha\alpha}]\{M_y\} \tag{2.3.20}$$

and

$$\{M_y\} = q\left[\diagdown c^2 \diagdown\right]\left\{\frac{\partial C_{MAC}}{\partial \delta_R}\right\}\delta_R$$

$$+ q\left[\diagdown ce \diagdown\right]\left\{\frac{\partial C_L}{\partial \frac{pl}{U}}\right\}\frac{pl}{U}$$

$$+ q\left[\diagdown ce \diagdown\right]\left\{\frac{\partial C_L}{\partial \delta_R}\right\}\delta_R$$

$$+ q\left[\diagdown ce \diagdown\right][A^{L\alpha}]\{\alpha\}\frac{\Delta\eta}{l} \tag{2.3.21}$$

All full matrices are of order $N \times N$ and row or column matrices of order N. Substituting (2.3.21) into (2.3.20), and rearranging terms gives,

$$\left[\left[\diagdown 1 \diagdown\right] - q\frac{(\Delta\eta)^2}{l}[E][A^{L\alpha}]\right]\{\dot{\alpha}\} = \{f\} \tag{2.3.22}$$

where the following definitions apply

$$\{f\} \equiv q[E]\left\{\left\{\frac{\partial C_L}{\partial \delta_R}\right\}\delta_R + \left\{\frac{\partial C_L}{\partial\left(\frac{pl}{U}\right)}\right\}\frac{pl}{U}\right\}\Delta\eta$$

$$+ q[F]\left\{\frac{\partial C_{MAC}}{\partial \delta_R}\right\}\delta_R\,\Delta\eta$$

$$[E] \equiv [C^{\alpha\alpha}]\left[\diagdown ce \diagdown\right]$$

$$[F] \equiv [C^{\alpha\alpha}]\left[\diagdown c^2 \diagdown\right]$$

Further defining

$$[D] \equiv \left[\diagdown 1 \diagdown\right] - q\frac{(\Delta\eta)^2}{l}[E][A^{L\alpha}]$$

we may formally solve (2.2.22) as

$$\{\alpha\} = [D]^{-1}\{f\} \tag{2.3.23}$$

Now let us interpret this solution.

Divergence

Recall that the inverse does not exist if

$$|D| = 0 \tag{2.3.24}$$

and hence,

$$\{\alpha\} \rightarrow \{\infty\}.$$

(2.3.24) gives rise to an eigenvalue problem for the divergence dynamic pressure, q_D. Note (2.3.24) is a polynomial in q.

The lowest positive root (eigenvalue) of (2.3.24) gives the q of physical interest, i.e., $q_{\text{Divergence}}$. Rather than seeking the roots of the polynomial we might more simply plot $|D|$ versus q to determine the values of dynamic pressure for which the determinant is zero. A schematic of such results for various choices of N is shown below in Figure 2.11. From the above results we may plot q_D (the lowest positive q for which $|D| = 0$) vs. N as shown below in Figure 2.12. The 'exact' value of q_D is obtained as $N \rightarrow \infty$. Usually reasonably accurate results can be obtained for small values of N, say 10 or so. The divergence speed calculated above does not depend upon the rolling of the wing, i.e., p is considered prescribed, e.g., $p = 0$.

Reversal and rolling effectiveness

In the above we have taken pl/U as known; however, in reality it is a function of δ_R and the problem parameters through the requirement that the wing be in static rolling equilibrium, i.e., it is an additional degree of

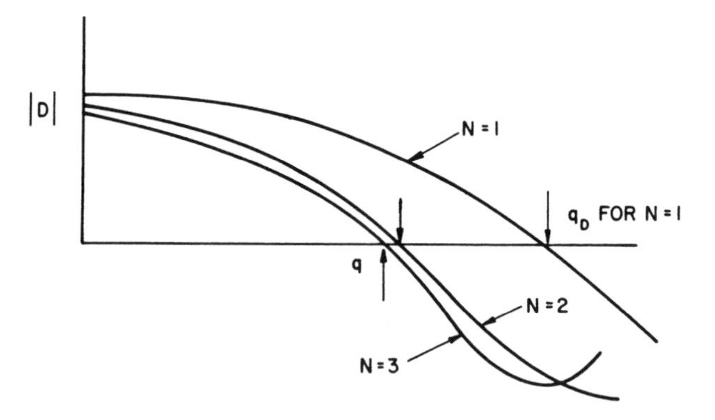

Figure 2.11 Characteristic determinant vs dynamic pressure.

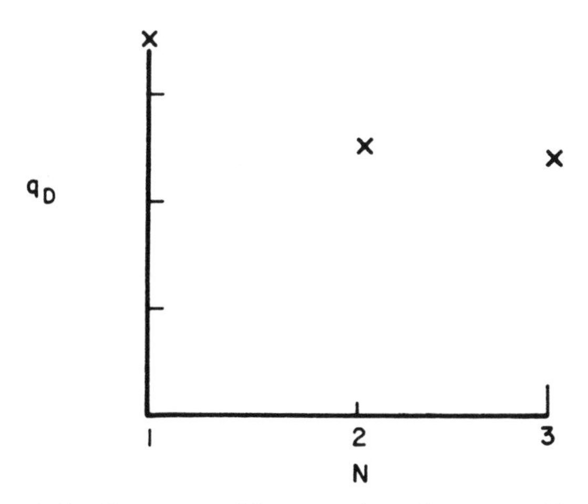

Figure 2.12 Convergence of divergence dynamic pressure with modal number.

freedom. For rolling equilibrium at a steady roll rate, p, the rolling moment about the x-axis is zero.

$$M_{\text{Rolling}} \equiv 2\int_0^l Ly\, dy = 0 \tag{2.3.25}$$

Approximating (2.3.25),

$$\sum_i L_i y_i\, \Delta y = 0 \tag{2.3.26}$$

or, in matrix notation,

$$2\lfloor y \rfloor \{L\}\, \Delta y = 0 \tag{2.3.27}$$

or

$$2q\lfloor cy \rfloor \{C_L\}\, \Delta y = 0$$

From (2.3.16), using the 'lumped element' approximation and matrix notation,

$$\{C_L\} = \frac{\Delta\eta}{l}[A^{L\alpha}]\{\alpha\} + \left\{\frac{\partial C_L}{\partial \delta_R}\right\}\delta_R + \left\{\frac{\partial C_L}{\partial\left(\frac{pl}{U}\right)}\right\}\frac{pl}{U} \tag{2.3.16}$$

Substitution of (2.3.16) into (2.3.27) gives

$$\lfloor cy \rfloor \left\{\frac{\Delta\eta}{l}[A^{L\alpha}]\{\alpha\} + \left\{\frac{\partial C_L}{\partial \delta_R}\right\}\delta_R + \left\{\frac{\partial C_L}{\partial\left(\frac{pl}{U}\right)}\right\}\frac{pl}{U}\right\} = 0 \tag{2.3.28}$$

Note that (2.3.28) is a single algebraic equation. (2.3.28) plus (2.3.20) and (2.3.21) are $2N+1$ linear algebraic equations in the $N(\alpha)$ plus $N(M_y)$ plus $1(p)$ unknowns. As before $\{M_y\}$ is normally eliminated using (2.3.21) in (2.3.20) to obtain N, equation (2.3.22), plus 1, equation (2.3.28), equations in $N(\alpha)$ plus $1(p)$ unknowns. In either case the divergence condition may be determined by setting the determinant of coefficients to zero and determining the smallest positive eigenvalue, $q = q_D$.

For $q < q_D$, pl/U (and $\{\alpha\}$) may be determined from (2.3.22) and (2.3.28). Since our mathematical model is linear

$$pl/U \sim \delta_R$$

and hence a convenient plot of the results is as shown in Figure 2.13. As

$$q \to q_D, \frac{pl}{U} (\text{and } \{\alpha\}) \to \infty.$$

Another qualitatively different type of result may sometimes occur. See Figure 2.14. If

$$\frac{pl}{U/\delta_R} \to 0 \quad \text{for} \quad q \to q_R < q_D$$

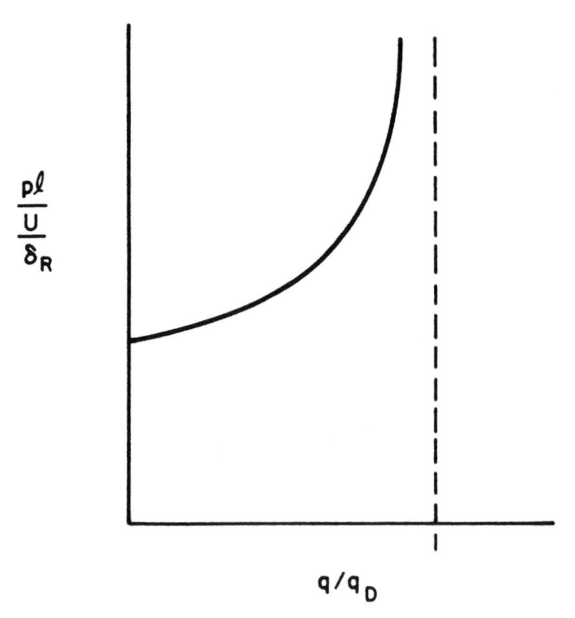

Figure 2.13 Roll rate vs dynamic pressure.

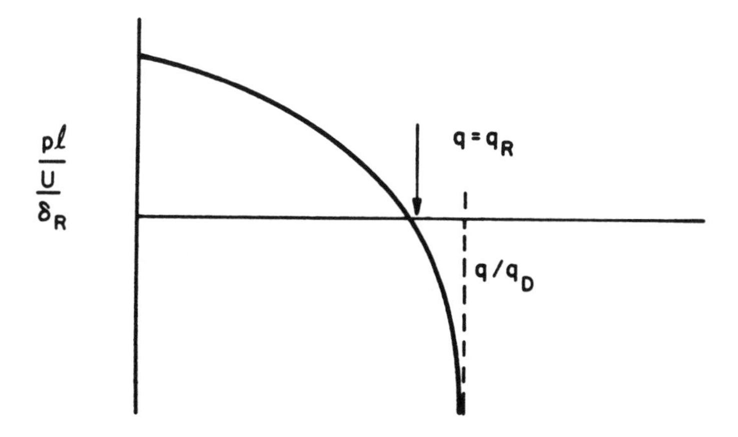

Figure 2.14 Roll rate vs dynamic pressure.

then 'rolling reversal' is said to have occurred and the corresponding $q = q_R$ is called the 'reversal dynamic pressure'. The basic phenomenon is the same as that encountered previously as 'control surface reversal'. Figures 2.13 and 2.14 should be compared to Figures 2.5a,b.

It is worth emphasizing that the divergence condition obtained above by permitting p to be determined by (static) rolling equilibrium will be different from that obtained previously by assuming $p = 0$. The latter physically corresponds to an aircraft constrained not to roll, as might be the case for some wind tunnel models. The former corresponds to no restraint with respect to roll, either structural or pilot induced.*

The above analysis has introduced the simple yet powerful idea of structural and aerodynamic influence functions. While the utility of the concept has been illustrated for a one-dimensional aeroelastic model, not the least advantage of such an approach is the conceptual ease with which the basic notion can be extended to two-dimensional models, e.g., plate-like structures, or even three-dimensional ones (though the latter is rarely needed for aeroelastic problems).

In a subsequent section we briefly outline the generalization to two-dimensional models. Later this subject will be considered in more depth in the context of dynamic aeroelasticity.

* This distinction between the two ways in which the aircraft may be restrained has recently received new emphasis in the context of the oblique wing concept. Weisshaar and Ashley [9].

Integral equation eigenvalue problem and the experimental determination of influence functions

For the special case of a constant section wing with 'strip theory' aerodynamics one may formulate a standard integral equation eigenvalue problem for the determination of divergence. In itself this problem is of little interest. However, it does lead to some interesting results with respect to the determination of the structural and aerodynamic influence functions by experimental means.

For such a wing,

$$M_y = Le + M_{AC}$$
$$= eqc\frac{\partial C_L}{\partial \alpha}\alpha + \cdots$$

where the omitted terms are independent of twist and may therefore be ignored for the divergence (eigenvalue) problem. Also the coefficients of α may be taken as constants for a constant section wing. Substituting the above expression into the integral equation of structural equilibrium we have

$$\alpha(y) = eqc\frac{\partial C_L}{\partial \alpha}\int_0^l C^{\alpha\alpha}(y, \eta)\alpha(\eta)\,\mathrm{d}\eta$$

This is an eigenvalue problem in integral form where the eigenvalue is

$$\lambda \equiv eqc\frac{\partial C_L}{\partial \alpha}$$

One may solve this problem for the corresponding eigenvalues and eigenfunctions which satisfy the equation

$$\alpha_n(y) = \lambda_n\int_0^l C^{\alpha\alpha}(y, \eta)\alpha_n(\eta)\,\mathrm{d}\eta$$

Incidentally, the restriction to a constant section wing was unnecessary and with a moderate amount of effort one could even use a more sophisticated aerodynamic model. Such complications are not warranted here.

These eigenfunctions or similar functions may be usefully employed to determine by experimental means the structural, $C^{\alpha\alpha}$, and aerodynamic, $A^{L\alpha}$, influence functions. The former is not as attractive as the use of point unit structural loads as we shall see; however, the procedure outlined below for the determination of $A^{L\alpha}$ probably deserves more attention than it has previously received.

Assume the structural influence function can be expanded in terms of the eigenfunctions

$$C^{\alpha\alpha}(y, \eta) = \sum_n C_n(y)\alpha_n(\eta) \tag{2.3.29}$$

where the C_n are to be determined. Also recall that

$$\alpha_n(y) = \lambda_n \int_0^l C^{\alpha\alpha}(y, \eta)\alpha_n(\eta)\,\mathrm{d}\eta \tag{2.3.30}$$

and the α_n are the eigenfunctions and λ_n the eigenvalues of $C^{\alpha\alpha}$ satisfying (2.3.30) and an orthogonality condition

$$\int \alpha_n\alpha_m\,\mathrm{d}y = 0 \quad \text{for} \quad m \neq n$$

Then multiply (2.3.29) by $\alpha_m(\eta)$ and integrate over the span of the wing; the result is

$$C_m(y) = \frac{\int_0^l C^{\alpha\alpha}(y, \eta)\alpha_m(\eta)\,\mathrm{d}\eta}{\int_0^l \alpha_m^2(\eta)\,\mathrm{d}\eta}$$

from (2.3.30)

$$= \frac{\alpha_m(y)}{\lambda_m \int_0^l \alpha_m^2(\eta)\,\mathrm{d}\eta} \tag{2.3.31}$$

Hence (2.3.31) in (2.3.29) gives

$$C^{\alpha\alpha}(y, \eta) = \sum_n \frac{\alpha_n(y)\alpha_n(\eta)}{\lambda_n \int_0^l \alpha_n^2(\eta)\,\mathrm{d}\eta} \tag{2.3.32}$$

Thus if the eigenfunctions are known then the Green's function is readily determined from (2.3.32). Normally this holds no special advantage since the determination of the α_n, theoretically or experimentally, is at least as difficult as determining the Green's function, $C^{\alpha\alpha}$, directly. Indeed as discussed previously if we apply unit moments at various points along the span the resulting twist distribution is a direct measure of $C^{\alpha\alpha}$. A somewhat less direct way of measuring $C^{\alpha\alpha}$ is also possible which makes use of the expansion of the Green's (influence) function. Again using (2.3.29)

$$C^{\alpha\alpha}(y, \eta) = \sum_n C_n \alpha_n(\eta) \tag{2.3.29}$$

and assuming the α_n are orthogonal (although not necessarily eigenfunctions of the problem at hand) we have

$$C_n(y) = \frac{\int_0^l C^{\alpha\alpha}(y, \eta)\alpha_n(\eta)\,\mathrm{d}\eta}{\int_0^l \alpha_n^2(\eta)\,\mathrm{d}\eta} \tag{2.3.33}$$

Now we have the relation between twist and moment

$$\alpha(y) = \int_0^l C^{\alpha\alpha}(y, \eta) M_y(\eta)\,\mathrm{d}\eta \tag{2.3.34}$$

Clearly if we use a moment distribution

$$M_y(\eta) = \alpha_n(\eta)$$

the resulting twist distribution will be (from 2.3.33))

$$\alpha(y) = C_n(y) \int_0^l \alpha_n^2(\eta)\,\mathrm{d}\eta \tag{2.3.35}$$

Hence we may determine the expansion of the Green's function by successively applying moment distributions in the form of the expansion functions and measuring the resultant twist distribution. For the structural influence function this offers no advantage in practice since it is easier to apply point moments rather than moment distributions.

However, for the aerodynamic Green's functions the situation is different. In the latter case we are applying a certain twist to the wing and measuring the resulting aerodynamic moment distribution. It is generally desirable to maintain a smooth (if twisted) aerodynamic surface to avoid complications of flow separation and roughness and hence the application of a point twist distribution is less desirable than a distributed one. We quickly summarize the key relations for the aerodynamic influence function. Assume

$$A^{L\alpha}(y, \eta) = \sum_n A_n^{L\alpha}(y)\alpha_n(\eta) \tag{2.3.36}$$

We know that

$$C_L(y)=\int_0^l A^{L\alpha}(y,\eta)\alpha(\eta)\,\mathrm{d}\eta \tag{2.3.37}$$

For orthogonal functions, α_n, we determine from (2.3.36) that

$$A_n^{L\alpha}(y)=\frac{\int_0^l A^{L\alpha}(y,\eta)\alpha_n(\eta)\,\mathrm{d}\eta}{\int_0^l \alpha_n^2(\eta)\,\mathrm{d}\eta} \tag{2.3.38}$$

Applying the twist distribution $\alpha=\alpha_n(\eta)$ to the wing, we see from (2.3.37) and (2.3.38) that the resulting lift distribution is

$$C_L(y)=A_n^{L\alpha}(y)\int_0^l \alpha_n^2(\eta)\,\mathrm{d}\eta \tag{2.3.39}$$

Hence, by measuring the lift distributions on 'warped wings' with twist distributions $\alpha_n(\eta)$ we may completely determine the aerodynamic influence function in terms of its expansion (2.3.36). This technique or a similar one has been used occasionally,* but not as frequently as one might expect, possibly because of the cost and expense of testing the number of wings sufficient to establish the convergence of the series. In this regard, if one uses the α_n for a Galerkin or modal expansion solution for the complete aeroelastic problem one can show that the number of C_n, $A_n^{L\alpha}$ required is equal to the number of modes, α_n, employed in the twist expansion.

2.4 Two dimensional aeroelastic model of lifting surfaces

We consider in turn, structural modeling, aerodynamic modeling, the combining of the two into an aeroelastic model, and its solution.

Two dimensional structures—integral representation

The two dimensional or plate analog to the one-dimensional or beam-rod model is

$$w(x,y)=\iint C^{wp}(x,y;\xi,\eta)p(\xi,\eta)\,\mathrm{d}\xi\,\mathrm{d}\eta \tag{2.4.1}$$

* Covert [10].

where w vertical deflection at a point, x, y, on plate
p force/area (pressure) at point ξ, η on plate
C^{wp} deflection at x, y due to unit pressure at ξ, η

Note that w and p are taken as positive in the same direction. For the special case where

$$w(x, y) = h(y) + x\alpha(y) \tag{2.4.2}$$

and

$$C^{wp}(x, y; \xi, \eta) = C^{hF}(y, \eta) + xC^{\alpha F}(y, \eta) + \xi C^{hM}(y, \eta) + x\xi C^{\alpha M}(y, \eta) \tag{2.4.3}$$

with the definitions,

C^{hF} is the deflection of y axis at y due to unit force F
$C^{\alpha F}$ is the twist about y axis at y due to unit force F etc.,

we may retrieve our beam-rod result. Note that (2.4.2) and (2.4.3) may be thought of as polynomial (Taylor Series) expansions of deflections.

Substituting (2.4.2), (2.4.3) into (2.4.1), we have

$$\begin{aligned} h(y) + x\alpha(y) = &\left[\int C^{hF}\left(\int p(\xi, \eta)\, \mathrm{d}\xi\right) \mathrm{d}\eta\right. \\ &\left. + \int C^{hM}\left(\int \xi p(\xi, \eta)\, \mathrm{d}\xi\right) \mathrm{d}\eta\right] \\ &+ x\left[\int C^{\alpha F}\left(\int p(\xi, \eta)\, \mathrm{d}\xi\right) \mathrm{d}\eta\right. \\ &\left. + \int C^{\alpha M}\left(\int \xi p(\xi, \eta)\, \mathrm{d}\xi\right) \mathrm{d}\eta\right] \end{aligned} \tag{2.4.4}$$

If y, η lie along an elastic axis, then $C^{hM} = C^{\alpha F} = 0$. Equating coefficients of like powers of x, we obtain

$$h(y) = \int C^{hF}(y, \eta) F(\eta)\, \mathrm{d}\eta \tag{2.4.5a}$$

$$\alpha(y) = \int C^{\alpha M}(y, \eta) M(\eta)\, \mathrm{d}\eta \tag{2.4.5b}$$

where

$$F \equiv \int p\, \mathrm{d}\xi, \qquad M \equiv \int p\xi\, \mathrm{d}\xi$$

(2.4.5b) is our previous result. Since for static aeroelastic problems, M is only a function of α (and not of h), (2.4.5b) may be solved independently of (2.4.5a). Subsequently (2.4.5b) may be solved to determine h if desired. (2.4.5a) has no effect on divergence or control surface reversal, of course, and hence we were justified in neglecting it in our previous discussion.

Two dimensional aerodynamic surfaces—integral representation

In a similar manner (for simplicity we only include deformation dependent aerodynamic forces to illustrate the method),

$$\frac{p(x, y)}{q} = \iint A^{pw_x}(x, y; \xi, \eta)\frac{\partial w}{\partial \xi}(\xi, \eta)\frac{\mathrm{d}\xi}{c_r}\frac{\mathrm{d}\eta}{l} \tag{2.4.6}$$

where A^{pw_x} nondimensional aerodynamic pressure at x, y due to unit $\partial w/\partial \xi$ at point ξ, η

c_r reference chord, l reference span

For the special case

$$w = h + x\alpha$$

and, hence,

$$\frac{\partial w}{\partial x} = \alpha$$

we may retrieve our beam-rod aerodynamic result.

For example, we may compute the lift as

$$L \equiv \int p\,\mathrm{d}x = qc_r \int_0^l A^{L\alpha}(y, \eta)\alpha(\eta)\frac{\mathrm{d}\eta}{l} \tag{2.4.7}$$

where

$$A^{L\alpha} \equiv \iint A^{pw_x}(x, y; \xi, \eta)\frac{\mathrm{d}\xi}{c_r}\frac{\mathrm{d}x}{c_r}$$

Solution by matrix-lumped element approach

Approximating the integrals by sums and using matrix notation, (2.4.1) becomes

$$\{w\} = \Delta\xi\,\Delta\eta[C^{wp}]\{p\} \tag{2.4.8}$$

and (2.4.6) becomes

$$\{p\} = q\frac{\Delta\xi\,\Delta\eta}{c_r\;\; l}[A^{pw_x}]\left(\frac{\partial w}{\partial \xi}\right) \tag{2.4.9}$$

Now

$$\left(\frac{\partial w}{\partial \xi}\right)_i \cong \frac{w_{i+1} - w_{i-1}}{2\,\Delta\xi}$$

is a difference representation of the surface slope. Hence

$$\left(\frac{\partial w}{\partial \xi}\right) = \frac{1}{2\,\Delta\xi}[W]\{w\} = \frac{1}{2\,\Delta\xi}\begin{bmatrix} [W][0][0][0] \\ [W][0][0] \\ [W][0] \\ [W] \end{bmatrix}\{w\}^* \tag{2.4.10}$$

is the result shown for *four* spanwise locations, where

$$[W] = \underbrace{\begin{bmatrix} 0 & 1 & 0 & 0 & \cdot \\ -1 & 0 & 1 & 0 & \cdot \\ 0 & -1 & 0 & 1 & \cdot \\ & & \cdots & & \\ \cdots & 0 & 0 & -1 & 0 \end{bmatrix}}_{\text{number of chordwise locations}}$$

is a numerical weighting matrix. From (2.4.8), (2.4.9), (2.4.10), we obtain an equation for w,

$$[D]\{w\} \equiv \left[\left[\diagdown 1 \diagdown\right] - q\frac{(\Delta\xi)^2}{c_r}\frac{(\Delta\eta)^2}{l}\frac{1}{2\,\Delta\xi}[C^{wp}][A^{pw_x}][W]\right]\{w\} = \{0\} \tag{2.4.11}$$

For divergence

$$|D| = 0$$

which permits the determination of q_D.

---

* For definiteness consider a rectangular wing divided up into small (rectangular) finite difference boxes. The weighting matrix $[(W)]$ is for a given spanwise location and various chordwise boxes. The elements in the matrices, $\{\partial w/\partial\xi\}$ and $\{w\}$, are ordered according to fixed spanwise location and then over all chordwise locations. This numerical scheme is only illustrative and not necessarily that which one might choose to use in practice.

2.5 Nonairfoil physical problems

Fluid flow through a flexible pipe

Another static aeroelastic problem exhibiting divergence is encountered in long slender pipes with a flowing fluid.* See Figure 2.15. We shall assume the fluid is incompressible and has no significant variation across the cross-section of the pipe. Thus, the aerodynamic loading per unit length along the pipe is (invoking the concept of an equivalent fluid added mass moving with the pipe and including the effects of convection velocity),† U,

$$-L = \rho A\left[\frac{\partial}{\partial t} + U\frac{\partial}{\partial x}\right]^2 w = \rho A\left[\frac{\partial^2 w}{\partial t^2} + 2U\frac{\partial^2 w}{\partial x\,\partial t} + U^2\frac{\partial^2 w}{\partial x^2}\right] \tag{2.5.1}$$

where $A \equiv \pi R^2$, open area for circular pipe
ρ, U fluid density, axial velocity
w transverse deflection of the pipe
x axial coordinate
t time

The equation for the beam-like slender pipe is

$$EI\frac{\partial^4 w}{\partial x^4} + m_p\frac{\partial^2 w}{\partial t^2} = L \tag{2.5.2}$$

where $m_p \equiv \rho_p\, 2\pi R h$ for a thin hollow circular pipe of thickness h, mass per unit length
EI beam bending stiffness

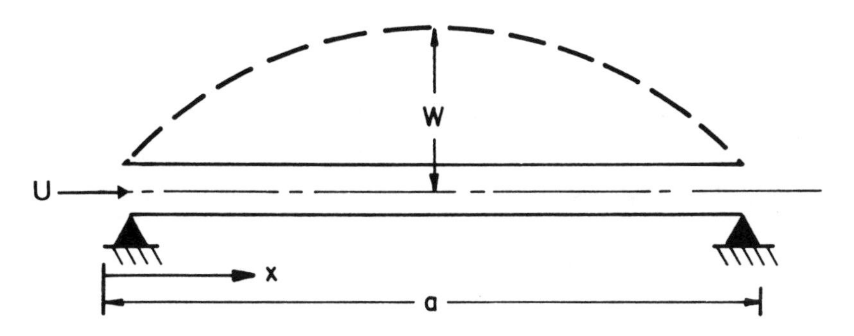

Figure 2.15 Fluid flow through a flexible pipe.

* Housner [11].
† See Section 3.4.

Both static and dynamic aeroelastic phenomena are possible for this physical model but for the moment we shall only consider the former. Further we shall consider for simplicity simply supported or pinned boundary conditions, i.e.,

$$w = 0$$

and

$$M \equiv EI\frac{\partial^2 w}{\partial x^2} = 0 \quad \text{at} \quad x = 0, a \tag{2.5.3}$$

where M is the elastic bending moment and, a, the pipe length.

Substituting (2.5.1) into (2.5.2) and dropping time derivatives consistent with limiting our concern to static phenomena, we have

$$EI\frac{\partial^4 w}{\partial x^4} + \rho A U^2 \frac{\partial^2 w}{\partial x^2} = 0 \tag{2.5.4}$$

subject to boundary conditions

$$w = \frac{\partial^2 w}{\partial x^2} = 0 \quad \text{at} \quad x = 0, a \tag{2.5.5}$$

The above equations can be recognized as the same as those governing the buckling of a beam under a compressive load of magnitude,* P. The equivalence is

$$P = \rho U^2 A$$

Formally we may compute the buckling or divergence dynamic pressure by assuming†

$$w = \sum_{i=1}^{4} A_i e^{p_i x}$$

where the p_i are the four roots of the characteristic equation associated with (2.5.4),

$$EIp^4 + \rho U^2 A p^2 = 0$$

Thus

$$p_{1,2} = 0$$

$$p_3, p_4 = \pm i\left(\frac{\rho U^2 A}{EI}\right)^{\frac{1}{2}}$$

* Timoshenko and Gere [3].

† Alternatively one could use Galerkin's method for (2.5.4) and (2.5.5) or convert them into an integral equation to be solved by the 'lumped element' method.

and

$$w = A_1 + A_2 x + A_3 \sin \frac{\lambda x}{a} + A_4 \cos \frac{\lambda x}{a} \tag{2.5.6}$$

where

$$\lambda^2 \equiv \left(\frac{\rho U^2 A}{EI}\right) a^2$$

Using the boundary conditions (2.5.5) with (2.5.6) we may determine that

$$A_1 = A_2 = A_4 = 0$$

and either $A_3 = 0$ or $\sin \lambda = 0$

For nontrivial solutions

$$A_3 \neq 0$$

and

$$\sin \lambda = 0$$

or

$$\lambda = \pi, 2\pi, 3\pi, \text{etc.} \tag{2.5.7}$$

Note that $\lambda = 0$ is a trivial solution, e.g., $w \equiv 0$.

Of the several eigenvalue solutions the smallest nontrivial one is of greatest physical interest, i.e.,

$$\lambda = \pi$$

The corresponding divergence or buckling dynamic pressure is

$$\rho U^2 = \frac{EI}{Aa^2} \pi^2 \tag{2.5.8}$$

Note that λ^2 is a nondimensional ratio of aerodynamic to elastic stiffness; we shall call it and similar numbers we shall encounter an 'aeroelastic stiffness number'. It is as basic to aeroelasticity as Mach number and Reynolds number are to fluid mechanics. Recall that in our typical section study we also encountered an 'aeroelastic stiffness number', namely,

$$\frac{qS \dfrac{\partial C_L}{\partial \alpha}}{K_\alpha} e$$

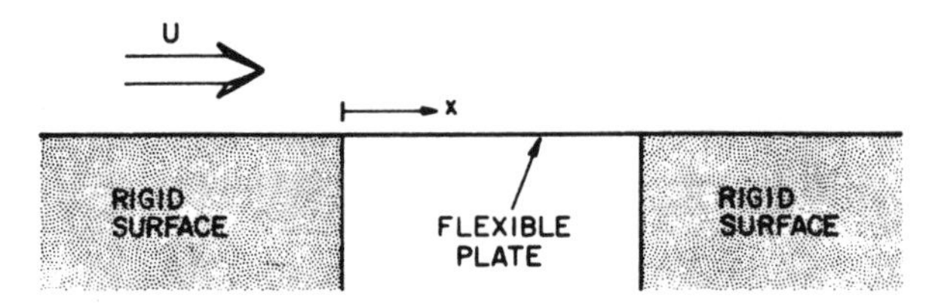

Figure 2.16 Fluid flow over a flexible wall.

as well as in the (uniform) beam-rod wing model,

$$\frac{q(lc)e\dfrac{\partial C_L}{\partial \alpha}}{\dfrac{GJ}{l}}$$

(Low speed) fluid flow over a flexible wall

A mathematically similar problem arises when a flexible plate is embedded in an otherwise rigid surface. See Figure 2.16. This is a simplified model of a physical situation which arises in nuclear reactor heat exchangers, for example. Aeronautical applications may be found in the local skin deformations on aircraft and missiles. Early airships may have encountered aeroelastic skin buckling.*

For a one dimensional (beam) structural representation of the wall, the equation of equilibrium is, as in our previous example,

$$EI\frac{\partial^4 w}{\partial x^4} = L$$

Also, as a rough approximation, it has been shown that the aerodynamic loading may be written†

$$L \sim \rho U^2 \frac{\partial^2 w}{\partial x^2}$$

Hence using this aerodynamic model, there is a formal mathematical analogy to the previous example and the aeroelastic calculation is the same. For more details and a more accurate aerodynamic model, the cited references should be consulted.

* Shute [12], p. 95.

† Dowell [13], p. 19, Kornecki [14], Kornecki, Dowell and O'Brien [15].

2.6 Sweptwing divergence

A swept wing, one whose elastic axis is at an oblique angle to an oncoming fluid stream, offers an interesting variation on the divergence phenomenon. Consider Figure 2.17. The angle of sweep is that between the axis perpendicular to the oncoming stream (y axis) and the elastic axis ($\bar{y}$ axis). It is assumed that the wing can be modeled by the bending-torsion deformation of a beam-rod. Thus the two structural equations of equilibrium are

Bending equilibrium of a beam-rod

$$\frac{d^2}{d\bar{y}^2}\left(EI\frac{d^2h}{d\bar{y}^2}\right) = -\bar{L} \tag{2.6.1}$$

Torsional equilibrium of a beam-rod

$$\frac{d^2}{d\bar{y}^2}\left(GJ\frac{d\alpha_e}{d\bar{y}}\right) + \bar{M}_y = 0 \tag{2.6.2}$$

Here h is the bending displacement of the elastic axis and is assumed positive downward. α_e, the elastic twist about the $\bar{y}$ axis, is positive nose up.

Now consider the aerodynamic model. Consider the velocity diagram, Figure 2.18. A strip theory aerodynamic model will be invoked with respect to chords perpendicular to the $\bar{y}$ axis. Thus the lift and aerodynamic moment per unit span are given by

$$\bar{L} = \bar{C}_L\bar{c}\bar{q} \tag{2.6.3}$$

and

$$\begin{aligned}\bar{M}y &= \bar{L}\bar{e} + \bar{M}_{AC}\\ &= \bar{C}_L\bar{c}\bar{q}\bar{e} + \bar{C}_{MAC}\bar{c}^2\bar{q}\end{aligned} \tag{2.6.4}$$

where $\bar{q} = \frac{1}{2}\rho(U\cos\Lambda)^2 = q\cos^2\Lambda$.

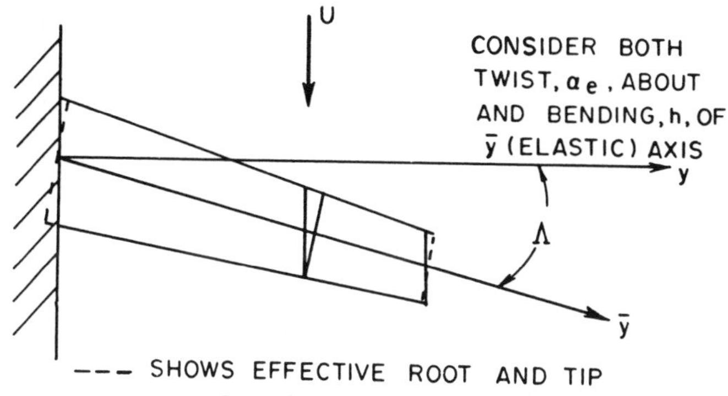

Figure 2.17 *Sweptwing geometry.*

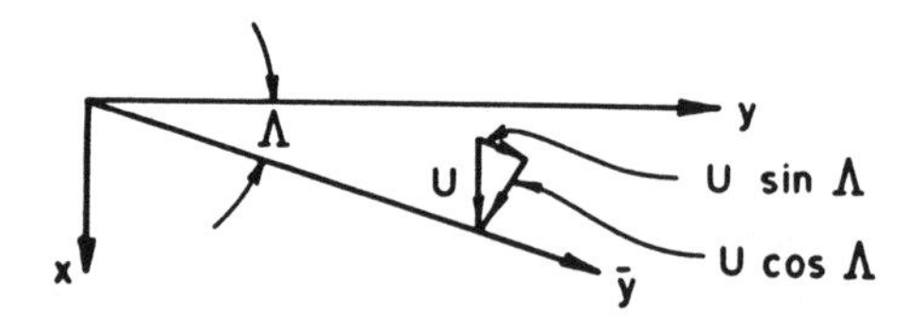

Figure 2.18 Velocity diagram in the x, y(x̄, ȳ) plane.

Also $\bar{C}_L$ is related to the (total) angle of attack, α_T, by

$$\bar{C}_L(\bar{y}) = \frac{\partial \bar{C}_L}{\partial \alpha} \alpha_T(\bar{y}) \tag{2.6.5}$$

where

$$\alpha_T = \alpha_e + \frac{\mathrm{d}h}{\mathrm{d}\bar{y}} \tan \Lambda \tag{2.6.6}$$

To understand the basis of the second term in (2.6.6), consider the velocity diagram of Figure 2.19. From this figure we see the fluid velocity normal to the wing is $U \sin \Lambda \, \mathrm{d}h/\mathrm{d}\bar{y}$ and thus the effective angle of attack due to bending of a swept wing is

$$U \sin \Lambda \frac{\mathrm{d}h}{\mathrm{d}\bar{y}} \Big/ U \cos \Lambda = U \frac{\mathrm{d}h}{\mathrm{d}\bar{y}} \tan \Lambda \tag{2.6.7}$$

From (2.6.1)–(2.6.6), the following form of the equations of equilibrium is obtained.

$$\frac{\mathrm{d}^2}{\mathrm{d}\bar{y}^2}\left(EI \frac{\mathrm{d}^2 h}{\mathrm{d}\bar{y}^2}\right) = -\frac{\partial \bar{C}_L}{\partial \alpha}\left[\alpha_e + \frac{\mathrm{d}h}{\mathrm{d}\bar{y}} \tan \Lambda\right] \bar{c} q \cos^2 \Lambda \tag{2.6.7}$$

$$\frac{\mathrm{d}}{\mathrm{d}\bar{y}}\left(GJ \frac{\mathrm{d}\alpha_e}{\mathrm{d}\bar{y}}\right) + \frac{\partial \bar{C}_L}{\partial \alpha}\left[\alpha_e + \frac{\mathrm{d}h}{\mathrm{d}\bar{y}} \tan \Lambda\right] \bar{c} q \cos^2 \Lambda \bar{e} + \bar{C}_{MAC} \bar{c}^2 q \cos^2 \Lambda = 0 \tag{2.6.8}$$

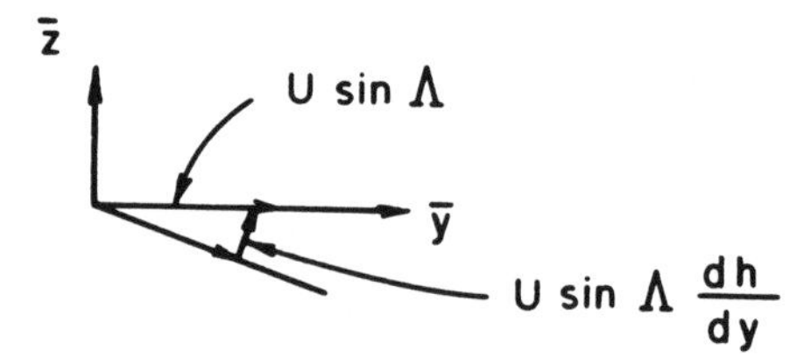

Figure 2.19 Velocity diagram in ȳ, z̄ plane.

SPECIAL CASES;

- If the beam is very stiff in bending, $EI \rightarrow \infty$, then from (2.6.7), $h \rightarrow 0$. (2.6.8) then is very similar to the torsional equation for an unswept wing with slightly modified coefficeints.
- If the beam-rod is very stiff in torsion, $GJ \rightarrow \infty$, then from (2.6.8), $\alpha \rightarrow 0$. (2.6.7) then reduces to

$$\frac{d^2}{d\bar{y}^2}\left(EI\frac{d^2h}{d\bar{y}^2}\right) + \frac{\partial C_L}{\partial \alpha}\sin\Lambda\cos\Lambda\bar{c}q\frac{dh}{d\bar{y}} = 0 \tag{2.6.9}$$

As we shall see, divergence in bending alone is possible even for a swept wing which is very stiff in torsion. This is not possible for an unswept wing.

To illustrate this, consider a further special case, namely a beam with constant spanwise properties. Introducing appropriate non-dimensionalization then (2.6.9) becomes

$$\frac{d^4h}{d\tilde{y}^4} + \lambda\frac{dh}{d\tilde{y}} = 0 \tag{2.6.10}$$

where $\tilde{y} \equiv \bar{y}/l$

$$\lambda \equiv \frac{\dfrac{\partial \bar{C}_L}{\partial \alpha} q\bar{c}\bar{l}^3}{EI}\sin\Lambda\cos\Lambda$$

The boundary conditions associated with this differential equation are zero deflection and slope at the root:

$$h = \frac{dh}{d\tilde{y}} = 0 \quad @ \quad \tilde{y} = 0 \tag{2.6.11}$$

and zero bending moment and shear force at the tip

$$EI\frac{d^2h}{d\tilde{y}^2} = EI\frac{d^3h}{d\tilde{y}^3} = 0 \quad @ \quad \tilde{y} = 1 \tag{2.6.12}$$

(2.6.10)–(2.6.12) constitute an eigenvalue problem. The eigenvalues of λ are all negative and the lowest of these provides the divergence condition.

$$\lambda_D = -6.33 = \frac{\partial \bar{C}_L}{\partial \alpha}\frac{\sin\Lambda\cos\Lambda\bar{c}\bar{l}^3q}{EI} \tag{2.6.13}$$

The only way the right hand side of (2.6.13) can be less than zero is if $\sin\Lambda < 0$ or $\Lambda < 0$.

Thus only swept forward wings can diverge in bending without torsional deformation. This suggests that swept forward wings are more susceptible to divergence than swept back wings. This proves to be the case when both bending and torsion are present as well.

For many years, the divergence tendency of swept forward wings precluded their use. In recent years composite materials provide a mechanism for favorable bending-torsion coupling which alleviates this divergence. For a modern treatment of these issues including the effects of composite structures two reports by Weisshaar [16, 17] are recommended reading.

A final word on how the eigenvalues are calculated. For (2.6.10)–(2.6.12), classical techniques for constant coefficient differential equations may be employed. See BAH, pp. 479–489. Even when both bending and torsion are included (2.6.7, 2.6.8), if the wing properties are independent of spanwise location, then classical techniques may be applied. Although the calculation does become more tedious. Finally, for variable spanwise properties Galerkin's method may be invoked, in a similar though more elaborate manner to that used for unswept wing divergence.

References for Chapter 2

[1] Ashley, H. and Landahl, M., *Aerodynamics of Wings and Bodies*, Addison-Wesley, 1965.

[2] Savant, Jr., C. J., *Basic Feedback Control System Design*, McGraw-Hill, 1958.

[3] Timoshenko, S. P., and Gere, J., *Theory of Elastic Stability*, McGraw-Hill, 1961.

[4] Woodcock, D. L., 'Structural Non-linearities', Vol. I, Chapter 6, *AGARD Manual on Aeroelasticity*.

[5] Housner, G. W. and Vreeland, T., Jr., *The Analysis of Stress and Deformation*, The MacMillan Co., 1966.

[6] Hildebrand, F. B., *Advance Calculus for Engineers*, Prentice-Hall, Inc. 1961.

[7] Duncan, W. J. 'Galerkin's Methods in Mechanics and Differential Equations', Br. A.R.C., R&M., 1798, 1937.

[8] Bisplinghoff, R. L., Mar, J. W. and Pian, T. H. H., *Statics of Deformable Solids*, Addison-Wesley, 1965.

[9] Weisshaar, T. A. and Ashley, H., 'Static Aeroelasticity and the Flying Wing, Revisited', *J. Aircraft*, Vol. 11 (Nov. 1974) pp. 718–720.

[10] Covert, E. E., 'The Aerodynamics of Distorted Surfaces', *Proceedings of Symposium on Aerothermoelasticity*. ASD TR 61–645, 1961, pp. 369–398.

[11] Housner, G. W., 'Bending Vibrations of a Pipe Line Containing Flowing Fluid', *Journal of Applied Mechanics*, Vol. 19 (June 1952) p. 205.

[12] Shute, N., *Slide Rule*, Wm. Morrow & Co., Inc., New York, N.Y. 10016.

[13] Dowell, E. H. *Aeroelasticity of Plates and Shells*, Noordhoff International Publishing, 1974.

[14] Kornecki, A., 'Static and Dynamic Instability of Panels and Cylindrical Shells in Subsonic Potential Flow', *J. Sound Vibration*, Vol. 32 (1974) pp. 251–263.

[15] Kornecki, A., Dowell, E. H., and O'Brien, J., 'On the Aeroelastic Instability of Two-Dimensional Panels in Uniform Incompressible Flow', *J. Sound Vibration*, Vol. 47 (1976) pp. 163–178.

[16] Weisshaar, T. A., 'Aeroelastic Stability and Performance Characteristics of Aircraft with Advanced Composite Sweptforward Wing Structures', AFFDL TR-78-116, Sept. 1978.

[17] Weisshaar, T. A., 'Forward Swept Wing Static Aeroelasticity', AFFDL TR-79-3087, June 1979.

3

Dynamic aeroelasticity

In static aeroelasticity we have considered various mathematical models of aeroelastic systems. In all of these, however, the fundamental physical content consisted of two distinct phenomena, 'divergence' or static instability, and loss of aerodynamic effectiveness as typified by 'control surface reversal'. Turning to dynamic aeroelasticity we shall again be concerned with only a few distinct fundamental physical phenomena. However, they will appear in various theoretical models of increasing sophistication. The principal phenomena of interest are (1) 'flutter' or dynamic instability and (2) response to various dynamic loadings as modified by aeroelastic effects. In the latter category primary attention will be devoted to (external) aerodynamic loadings such as atmospheric turbulence or 'gusts'. These loadings are essentially random in nature and must be treated accordingly. Other loadings of interest may be impulsive or discrete in nature such as the sudden loading due to maneuvering of a flight vehicle as a result of control surface rotation.

To discuss these phenomena we must first develop the dynamic theoretical models. This naturally leads us to a discussion of how one obtains the equations of motion for a given aeroelastic system including the requisite aerodynamic forces. Our initial discussion of aerodynamic forces will be conceptual rather than detailed. Later, in Chapter 4, these forces are developed from the fundamentals of fluid mechanics. We shall begin by using the 'typical section' as a pedagogical device for illustrating the physical content of dynamic aeroelasticity. Subsequently using the concepts of structural and aerodynamic influence and impulse functions, we shall discuss a rather general model of an aeroelastic system. The solution techniques for our aeroelastic models are for the most part standard for the modern treatment of the dynamics of linear systems and again we use the typical section to introduce these methods.

We now turn to a discussion of energy and work methods which have proven very useful for the development of structural equations of motion.

In principle, one may use Newton's Second Law (plus Hooke's Law) to obtain the equations of motion for any elastic body. However, normally an alternative procedure based on Hamilton's Principle or Lagrange's Equations is used.* For systems with many degrees of freedom, the latter are more economical and systematic.

We shall briefly review these methods here by first deriving them from Newton's Second Law for a single particle and then generalizing them for many particles and/or a continuous body. One of the major advantages over the Newtonian formulation is that we will deal with work and energy (scalars) as contrasted with accelerations and forces (vectors).

3.1 Hamilton's principle

Single particle

Newton's Law states

$$\vec{F} = m\frac{\mathrm{d}^2\vec{r}}{\mathrm{d}t^2} \tag{3.1.1}$$

where $\vec{F}$ is the force vector and $\vec{r}$ is the displacement vector, representing the actual path of particle.

Consider an adjacent path, $\vec{r}+\delta\vec{r}$, where $\delta\vec{r}$ is a 'virtual displacement' which is small in some appropriate sense. If the time interval of interest is $t = t_1 \rightarrow t_2$ then we shall require that

$$\delta\vec{r} = 0 \quad \text{at} \quad t = t_1, t_2$$

although this can be generalized. Thus, the actual and adjacent paths coincide at $t = t_1$ or t_2.

Now form the dot product of (3.1.1) with $\delta\vec{r}$ and $\int_{t_1}^{t_2} \cdots \mathrm{d}t$. The result is

$$\int_{t_1}^{t_2} \left(m\frac{\mathrm{d}^2\vec{r}}{\mathrm{d}t^2} \cdot \delta\vec{r} - \vec{F} \cdot \delta\vec{r} \right) \mathrm{d}t = 0 \tag{3.1.2}$$

The second term in brackets can be identified as work or more precisely the 'virtual work'. The 'virtual work' is defined as the work done by the actual forces being moved through the virtual displacement. We assume that the force remains fixed during the virtual displacement or, equivalently, the virtual displacement occurs instantaneously, i.e., $\delta t = 0$.

It follows that the first term must also have the dimensions of work

* See, for example, Meirovitch [1].

(or energy). To see this more explicitly, we manipulate the first term by an integration by parts as follows:

$$
\begin{aligned}
m\int_{t_1}^{t_2}\frac{\mathrm{d}^2\vec{r}}{\mathrm{d}t^2}\cdot\delta\vec{r}\,\mathrm{d}t &= m\frac{\mathrm{d}\vec{r}}{\mathrm{d}t}\cdot\delta\vec{r}\Big|_{t_1}^{t_2}\nearrow 0\\
&\quad - m\int_{t_1}^{t_2}\frac{\mathrm{d}\vec{r}}{\mathrm{d}t}\cdot\frac{\mathrm{d}}{\mathrm{d}t}(\delta r)\,\mathrm{d}t\\
&= -m\int_{t_1}^{t_2}\frac{\mathrm{d}\vec{r}}{\mathrm{d}t}\cdot\delta\frac{\mathrm{d}\vec{r}}{\mathrm{d}t}\,\mathrm{d}t\\
&= -\frac{m}{2}\int_{t_1}^{t_2}\delta\left(\frac{\mathrm{d}\vec{r}}{\mathrm{d}t}\cdot\frac{\mathrm{d}\vec{r}}{\mathrm{d}t}\right)\mathrm{d}t
\end{aligned}
\tag{3.1.3}
$$

Hence (3.1.2) becomes

$$
\int_{t_1}^{t_2}\left[\frac{1}{2}m\delta\left(\frac{\mathrm{d}\vec{r}}{\mathrm{d}t}\cdot\frac{\mathrm{d}\vec{r}}{\mathrm{d}t}\right)+F\cdot\delta\vec{r}\right]\mathrm{d}t=0
$$

or

$$
\int_{t_1}^{t_2}\delta[T+W]\,\mathrm{d}t=0 \tag{3.1.4}
$$

where

$$
\delta T\equiv\delta\frac{1}{2}m\frac{\mathrm{d}\vec{r}}{\mathrm{d}t}\cdot\frac{\mathrm{d}\vec{r}}{\mathrm{d}t} \tag{3.1.5}
$$

is defined as the 'virtual kinetic energy' and

$$
\delta W\equiv\vec{F}\cdot\delta\vec{r} \tag{3.1.6}
$$

is the 'virtual work'. Hence, the problem is cast in the form of scalar quantities, work and energy. (3.1.4) is Hamilton's Principle. It is equivalent to Newton's Law.

Before proceeding further it is desirable to pause to consider whether we can reverse our procedure, i.e., starting from (3.1.4), can we proceed to (3.1.1)? It is not immediately obvious that this is possible. After all, Hamilton's Principle represent an integrated statement over the time interval of interest while Newton's Second Law holds at every instant in time. By formally reversing our mathematical steps however, we may proceed from (3.1.4) to (3.1.2). To take the final step from (3.1.2) to (3.1.1) we must recognize that our choice of $\delta\vec{r}$ is arbitrary. Hence, if (3.1.2) is to hold for any possible choice of $\delta\vec{r}$, (3.1.2) must follow. To demonstrate this we note that, if $\delta\vec{r}$ is arbitrary and (3.1.1) were not true, then it would be possible select $\delta\vec{r}$ such that (3.1.2) would not be true. Hence (3.1.2) implies (3.1.1) if $\delta\vec{r}$ is arbitrary.

Many particles

The previous development is readily generalized to many particles. Indeed, the basic principle remains the same and only the work and energy expressions are changed as follows:

$$\delta T = \sum_i \frac{m_i}{2}\,\delta\left(\frac{\mathrm{d}\vec{r}_i}{\mathrm{d}t}\cdot\frac{\mathrm{d}\vec{r}_i}{\mathrm{d}t}\right) \tag{3.1.7}$$

$$\delta W = \sum_i \vec{F}_i \cdot \delta\vec{r}_i \tag{3.1.8}$$

where m_i is the mass of ith particle,
$\vec{r}_i$ is the displacement of ith particle, (3.1.9)
and $\vec{F}_i$ is the force acting on ith particle.

Continuous body

For a continuous body (3.1.7) and (3.1.8) are replaced by (3.1.10) and (3.1.11).

$$\delta T = \iiint_{\text{volume}} \frac{\rho}{2}\,\delta\frac{\mathrm{d}\vec{r}}{\mathrm{d}t}\cdot\frac{\mathrm{d}\vec{r}}{\mathrm{d}t}\,\mathrm{d}V \tag{3.1.10}$$

where ρ is the density (mass per unit volume), V is the volume, and δW is the virtual work done by external applied forces and internal elastic forces. For example, if $\vec{f}$ is the vector body force per unit volume and $\vec{p}$ the surface force per unit area then

$$\delta W = \iiint_{\text{volume}} \vec{f}\cdot\delta\vec{r}\,\mathrm{d}V + \iint_{\text{surface area}} \vec{p}\cdot\delta\vec{r}\,\mathrm{d}A \tag{3.1.11}$$

Potential energy

In a course on elasticity* it would be shown that the work done by internal elastic forces is the negative of the virtual elastic potential

* Bisplinghoff, Mar, and Pian [2], Timoshenko and Goodier [3].

energy. The simplest example is that of an elastic spring. See sketch below.

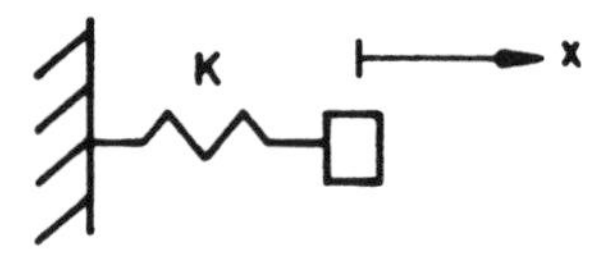

The force in the spring is

$$-Kx$$

where the minus sign arises from the fact that the force of the spring on the mass opposes the displacement, x. The virtual work is

$$\delta W = -Kx\,\delta x$$
$$= -\delta \frac{Kx^2}{2}$$

The virtual change in potential energy is

$$\delta U \equiv -\delta W$$
$$= \delta \frac{Kx^2}{2} = \delta\left(\frac{Fx}{2}\right) \tag{3.1.12}$$

Considering the other extreme, the most complete description of the potential energy of an elastic body which satisfies Hooke's Law is (see Bisplinghoff, Mar and Pian [2])

$$U = \tfrac{1}{2} \iiint_V [\sigma_{xx}\varepsilon_{xx} + \sigma_{xy}\varepsilon_{xy} + \sigma_{yx}\varepsilon_{yx} + \cdots]\,\mathrm{d}V \tag{3.1.13}$$

where σ_{xx} is the stress component (analogous to F) and ε_{xx} is the strain component (analogous to x).

From this general expression for potential (strain) energy of an elastic body we may derive some useful results for the bending and twisting of beams and plates. For the bending of a beam, the usual assumption of plane sections over the beam cross-section remaining plane leads to a strain-displacement relation of the form

$$\varepsilon_{yy} = -z\frac{\partial^2 w}{\partial y^2}$$

where z is the vertical coordinate through beam, w is the vertical displacement of beam, Hooke's Law reads,

$$\sigma_{yy} = E\varepsilon_{yy} = -Ez\frac{\partial^2 w}{\partial y^2}$$

and we assume all other stresses are negligible

$$\sigma_{yz} = \sigma_{xy} = \sigma_{xz} = \sigma_{xx} = \sigma_{zz} = 0$$

If we further assume $w(x, y, z) = h(y)$ where y is the lengthwise coordinate axis of the beam, then

$$U = \frac{1}{2}\int EI\left(\frac{\partial^2 h}{\partial y^2}\right)^2 \mathrm{d}y$$

where

$$I \equiv \int z^2\,\mathrm{d}z \int \mathrm{d}x$$

For the twisting of a thin beam, analogous reasoning leads to similar results.

$$\varepsilon_{xy} = -z\frac{\partial^2 w}{\partial x\,\partial y}$$

$$\sigma_{xy} = \frac{E}{(1+\nu)}\varepsilon_{xy} = -\frac{E}{(1+\nu)}z\frac{\partial^2 w}{\partial x\,\partial y}$$

Thus

$$U = \frac{1}{2}\int GJ\left(\frac{\partial \alpha}{\partial y}\right)^2 \mathrm{d}y$$

where

$$G \equiv \frac{E}{2(1+\nu)}, \qquad J \equiv 4\int z^2\,\mathrm{d}z \int \mathrm{d}x \quad \text{and} \quad w = x\alpha(y)$$

The above can be generalized to the bending of a plate in two dimensions.

$$\varepsilon_{yy} = -z\frac{\partial^2 w}{\partial y^2}$$

$$\varepsilon_{xx} = -z\frac{\partial^2 w}{\partial x^2}$$

$$\varepsilon_{xy} = -z\frac{\partial^2 x}{\partial x\,\partial y}$$

$$\sigma_{xx} = \frac{E}{(1-\nu^2)}[\varepsilon_{xx} + \nu\varepsilon_{yy}]$$

$$\sigma_{yy} = \frac{E}{(1-\nu^2)}[\varepsilon_{yy} + \nu\varepsilon_{xx}]$$

$$\sigma_{xy} = \frac{E}{(1+\nu)}\varepsilon_{xy}$$

and

$$U = \frac{1}{2}\iint D\left[\left(\frac{\partial^2 w}{\partial x^2}\right)^2 + \left(\frac{\partial^2 w}{\partial y^2}\right)^2 + 2\nu\frac{\partial^2 w}{\partial x^2}\frac{\partial^2 w}{\partial y^2} + 2(1-\nu)\left(\frac{\partial^2 w}{\partial x\,\partial y}\right)^2\right]\mathrm{d}x\,\mathrm{d}y$$

where

$$D \equiv \frac{E}{(1-\nu^2)}\int_{-h/2}^{+h/2} z^2\,\mathrm{d}y, \text{ plate bending stiffness}$$

and

$$w = w(x, y)$$

Nonpotential forces

Now, if one divides the virtual work into potential and nonpotential contributions, one has Hamilton's Principle in the form

$$\int [(\delta T - \delta U) + \underbrace{\vec{F}_{NC} \cdot \vec{\delta r}}_{\delta W_{NC}}]\,\mathrm{d}t = 0 \tag{3.1.14}$$

where F_{NC} includes only the nonpotential (or nonconservative) forces.

In our aeroelastic problems the nonconservative virtual work is a result of aerodynamic loading. For example, the virtual work due to the aerodynamic pressure (force per unit area) on a two-dimensional plate is clearly

$$\delta W_{NC} = \iint p\,\delta w\,\mathrm{d}x\,\mathrm{d}y$$

Note that if the deflection is taken to be a consequence of a chordwise rigid rotation about and bending of a spanwise elastic axis located at, say $x = 0$, then

$$w = -h(y) - x\alpha(y)$$

and hence

$$\delta W = \int \left[-\int p\,\mathrm{d}x \right] \delta h\,\mathrm{d}y + \int \left[-\int px\,\mathrm{d}x \right] \delta\alpha\,\mathrm{d}y$$

where $L = \int p\,\mathrm{d}x$ net vertical force/per unit span

$M_y \equiv -\int px\,\mathrm{d}x$ net moment about y axis per unit span

Thus, for this special case,

$$\delta W = \int -L\,\delta h\,\mathrm{d}y + \int M_y\,\delta\alpha\,\mathrm{d}y$$

3.2 Lagrange's equations

Lagrange's equations may be obtained by reversing the process by which we obtained Hamilton's Principle. However to obtain a more general result than simply a retrieval of Newton's Second Law we introduce the notion of 'generalized' coordinates. A 'generalized' coordinate is one which is arbitrary and independent (of other coordinates). A set of 'generalized' coordinates is sufficient* to describe the motion of a dynamical system. That is, the displacement of a particle or point in a continuous body may be written

$$\vec{r} = \vec{r}(q_1, q_2, q_3, \ldots, t) \tag{3.2.1}$$

where q_i is the ith generalized coordinate. From (3.2.1) it follows that

$$\begin{aligned} T &= T(\dot{q}_i, q_i, t) \\ U &= U(\dot{q}_i, q_i, t) \end{aligned} \tag{3.2.2}$$

Thus Hamilton's Principle may be written

$$\int_{t_1}^{t_2} [\delta(T-U) + \delta W_{NC}]\,\mathrm{d}t = 0 \tag{3.1.14}$$

Using (3.2.2) in (3.1.14)

$$\sum_i \int_{t_1}^{t_2} \left[\frac{\partial(T-U)}{\partial \dot{q}_i} \delta\dot{q}_i + \frac{\partial(T-U)}{\partial q_i} \delta q_i + Q_i\,\delta q_i \right] \mathrm{d}t = 0 \tag{3.2.3}$$

* and necessary, i.e., they are independent.

where the generalized forces, Q_i, are known from

$$\delta W_{NC} \equiv \sum_i Q_i \, \delta q_i \tag{3.2.4}$$

As we will see (3.2.4) *defines the* Q_i as the coefficients of δq_i in an expression for δW_{NC} which must be obtained independently of (3.2.4). Integrating the first term of (3.2.3) by parts (noting that $\delta q_i = 0$ $t = t_1, t_2$) we have

$$\sum_i \overbrace{\frac{\partial (T-U)}{\partial \dot{q}_i} \delta q_i \bigg|_{t_1}^{t_2}}^{0} + \int_{t_1}^{t_2} \left[-\frac{\mathrm{d}}{\mathrm{d}t} \frac{\partial (T-U)}{\partial \dot{q}_i} \delta q_i + \frac{\partial (T-U)}{\partial q_i} \delta q_i + Q_i \, \delta q_i \right] \mathrm{d}t = 0 \tag{3.2.5}$$

Collecting terms

$$\sum_i \int_{t_1}^{t_2} \left[-\frac{\mathrm{d}}{\mathrm{d}t} \frac{\partial (T-U)}{\partial \dot{q}_i} + \frac{\partial (T-U)}{\partial q_i} + Q_i \right] \delta q_i \, \mathrm{d}t = 0 \tag{3.2.6}$$

Since the δq_i are independent and arbitrary it follows that each bracketed quantity must be zero, i.e.,

$$-\frac{\mathrm{d}}{\mathrm{d}t} \frac{\partial (T-U)}{\partial \dot{q}_i} + \frac{\partial (T-U)}{\partial q_i} + Q_i = 0 \qquad i = 1, 2, \ldots \tag{3.2.7}$$

These are Lagrange's equations.

Example — Typical section equations of motion

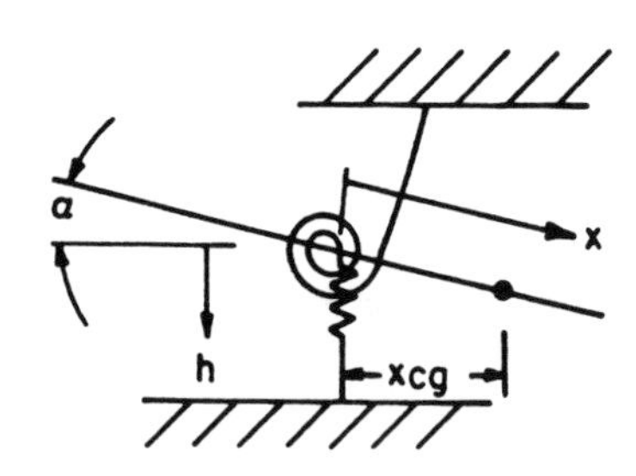

Figure 3.1 Geometry of typical section airfoil

x is measured along chord from e.a.; note that x is *not* a generalized coordinate, e.g., it cannot undergo a virtual change.

$$\begin{matrix}\text{generalized} \\ \text{coordinates}\end{matrix} \begin{cases} q_1 = h, \\ q_2 = \alpha, \end{cases}$$

The displacement of any point on the airfoil is

$$\vec{r} = u\vec{i} + w\vec{k} \tag{3.2.8}$$

where u is the horizontal displacement component, w is the vertical displacement component, and $\vec{i}$, $\vec{k}$ are the unit, cartesian vectors.

From geometry

$$\left.\begin{aligned} u &= x[\cos\alpha - 1] \simeq 0 \\ w &= -h - x\sin\alpha \simeq -h - x\alpha \end{aligned}\right\} \text{for } \alpha \ll 1 \tag{3.2.9}$$

Hence,

$$\begin{aligned} T &= \frac{1}{2}\int\left[\left(\frac{\mathrm{d}w}{\mathrm{d}t}\right)^2 + \left(\frac{\mathrm{d}u}{\mathrm{d}t}\right)^2\right]\rho\,\mathrm{d}x \\ &\simeq \frac{1}{2}\int\left(\frac{\mathrm{d}w}{\mathrm{d}t}\right)^2\rho\,\mathrm{d}x \\ &= \tfrac{1}{2}\int(-\dot{h} - \dot{\alpha}x)^2\rho\,\mathrm{d}x \\ &= \tfrac{1}{2}\dot{h}^2\int\rho\,\mathrm{d}x + \tfrac{1}{2}2\dot{h}\dot{\alpha}\int x\rho\,\mathrm{d}x + \tfrac{1}{2}\dot{\alpha}^2\int x^2\rho\,\mathrm{d}x \\ &= \tfrac{1}{2}\dot{h}^2 m + \tfrac{1}{2}2\dot{h}\dot{\alpha}S_\alpha + \tfrac{1}{2}\dot{\alpha}^2 I_\alpha \end{aligned} \tag{3.2.10}$$

$$m \equiv \int\rho\,\mathrm{d}x$$

$$S_\alpha \equiv \int\rho x\,\mathrm{d}x \equiv x_{\text{c.g.}}m$$

$$I_\alpha \equiv \int\rho x^2\,\mathrm{d}x$$

The potential energy is

$$U = \tfrac{1}{2}K_h h^2 + \tfrac{1}{2}K_\alpha\alpha^2 \tag{3.2.11}$$

For our system, Lagrange's equations are

$$\begin{aligned} -\frac{\mathrm{d}}{\mathrm{d}t}\left(\frac{\partial(T-U)}{\partial\dot{h}}\right) + \frac{\partial(T-U)}{\partial h} + Q_h &= 0 \\ -\frac{\mathrm{d}}{\mathrm{d}t}\left(\frac{\partial(T-U)}{\partial\dot{\alpha}}\right) + \frac{\partial(T-U)}{\partial\alpha} + Q_\alpha &= 0 \end{aligned} \tag{3.2.12}$$

where

$$\delta W_{NC} = Q_h\,\delta h + Q_\alpha\,\delta\alpha \tag{3.2.13}$$

Now let us evaluate the terms in (3.2.12) and (3.2.13). Except for Q_h, Q_α these are readily obtained by using (3.2.10) and (3.2.11) in (3.2.12). Hence, let us first consider the determination of Q_h, Q_α. To do this we calculate *independently* the work done by the aerodynamic forces.

$$\begin{aligned}
\delta W_{NC} &= \int p\,\delta w\,\mathrm{d}x \\
&= \int p(-\delta h - x\,\delta\alpha)\,\mathrm{d}x \\
&= \delta h\left(-\int p\,\mathrm{d}x\right) + \delta\alpha\left(-\int px\,\mathrm{d}x\right) \\
&= \delta h(-L) + \delta\alpha(M_y)
\end{aligned} \tag{3.2.14}$$

where we identify from (3.2.13) and (3.2.14)

$$L \equiv \int p\,\mathrm{d}x = -Q_h$$

$$M_y \equiv -\int px\,\mathrm{d}x = Q_\alpha$$

Note the sign convention is that p is positive up, L is positive up and M_y is positive nose up. Putting it all together, noting that

$$\frac{\partial(T-U)}{\partial h} = -K_h h \quad \text{etc.}$$

we have from Lagrange's equations

$$-\frac{\mathrm{d}}{\mathrm{d}t}(m\dot{h} + S_\alpha\dot{\alpha}) - K_h h - L = 0$$

$$-\frac{\mathrm{d}}{\mathrm{d}t}(S_\alpha\dot{h} + I_\alpha\dot{\alpha}) - K_\alpha\alpha + M_y = 0 \tag{3.2.15}$$

These are the equations of motion for the 'typical section' in terms of the particular coordinates h and α.

Other choices of generalized coordinates are possible; indeed, one of the principal advantages of Lagrange's equations is this freedom to make various choices of generalized coordinates. The choice used above simplifies the potential energy but not the kinetic energy. If the generalized coordinates were chosen to be the translation of and rotation about the center of mass the kinetic energy would be simplified, viz.

$$T = \frac{m}{2}\dot{h}_{cm}^2 + \frac{I_{cm}}{2}\dot{\alpha}_{cm}^2$$

but the potential energy would be more complicated. Also the relevant aerodynamic moment would be that about the center of mass axis rather than that about the elastic axis (spring attachment point).

Another choice might be the translation of and rotation about the aerodynamic center axis though this choice is much less often used than those discussed above.

Finally we note that there is a particular choice of coordinates which leads to a maximum simplification of the inertial and elastic terms (though not necessarily the aerodynamic terms). These may be determined by making some arbitrary initial choice of coordinates, e.g., h and α, and then determining the 'normal modes' of the system in terms of these.* These 'normal modes' provide us with a coordinate transformation from the initial coordinates, h and α, to the coordinates of maximum simplicity. We shall consider this matter further subsequently.

3.3 Dynamics of the typical section model of an airfoil

To study the dynamics of aeroelastic systems, we shall use the 'typical section'† as a device for exploring mathematical tools and the physical content associated with such systems. To simplify matters, we begin by assuming the aerodynamic forces *are given* where $p(x, t)$ is the aerodynamic pressure, L, the resultant (lift) force and M_y the resultant moment about the elastic axis. See Figure 3.2. The equations of motion are

$$m\ddot{h} + K_h h + S_\alpha \ddot{\alpha} = -L \tag{3.3.1}$$

$$S_\alpha \ddot{h} + I_\alpha \ddot{\alpha} + K_\alpha \alpha = M_y \tag{3.3.2}$$

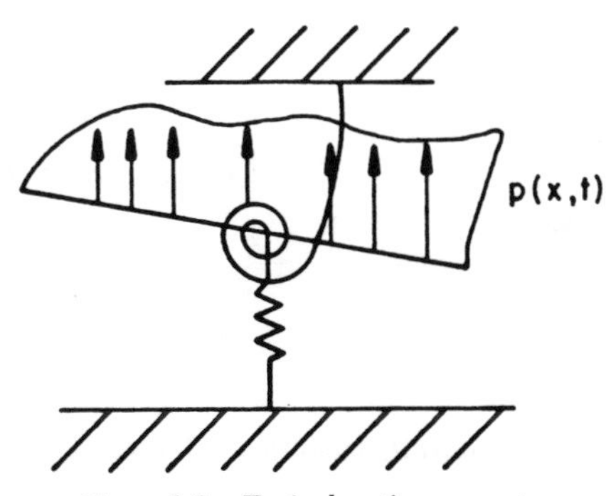

Figure 3.2 Typical section geometry

* Meirovitch [4].
† BA, pp. 201–246.

where

$$L \equiv \int p \, dx$$

$$M_y \equiv \int px \, dx$$

We will find it convenient also to define the 'uncoupled natural frequencies',

$$\omega_h^2 \equiv K_h/m, \qquad \omega_\alpha^2 \equiv K_\alpha/I_\alpha \tag{3.3.3}$$

These are 'natural frequencies' of the system for $S_\alpha \equiv 0$ as we shall see in a moment.

Sinusoidal motion

This is the simplest type of motion; however, as we shall see, we can exploit it systematically to study more complicated motions.

Let

$$\begin{aligned} L &= \bar{L}e^{i\omega t}, \qquad M_y = \bar{M}_y e^{i\omega t} \\ h &= \bar{h}e^{i\omega t}, \qquad \alpha = \bar{\alpha}e^{i\omega t} \end{aligned} \tag{3.3.4}$$

Substituting (3.3.4) and (3.3.3) into (3.3.2) we have in matrix notation

$$\begin{bmatrix} m(-\omega^2+\omega_h^2) & -S_\alpha\omega^2 \\ -S_\alpha\omega^2 & I_\alpha(-\omega^2+\omega_\alpha^2) \end{bmatrix} \begin{Bmatrix} \bar{h}e^{i\omega t} \\ \bar{\alpha}e^{i\omega t} \end{Bmatrix} = \begin{Bmatrix} -\bar{L}e^{i\omega t} \\ \bar{M}_y e^{i\omega t} \end{Bmatrix} \tag{3.3.5}$$

Solving for $\bar{h}$, $\bar{\alpha}$ we have

$$\frac{\bar{h}}{\bar{L}} = \frac{-[1-(\omega/\omega_\alpha)^2] + d/b\dfrac{x_\alpha}{r_\alpha^2}\left(\dfrac{\omega}{\omega_\alpha}\right)^2}{K_h\left\{[1-(\omega/\omega_\alpha)^2][1-(\omega/\omega_h)^2] - \dfrac{x_\alpha^2}{r_\alpha^2}\left(\dfrac{\omega}{\omega_\alpha}\right)^2\left(\dfrac{\omega}{\omega_h}\right)^2\right\}}$$

$$\equiv H_{hL}\left(\omega/\omega_\alpha; \frac{\omega_h}{\omega_\alpha}, d/b, x_\alpha, r_\alpha\right) \tag{3.3.6}$$

where

$$d \equiv \bar{M}_y/\bar{L}$$

where b is the reference length (usually selected as half-chord by tradition),

$$x_\alpha \equiv \frac{S_\alpha}{mb} = \frac{x_{c.g.}}{b}$$

and

$$r_\alpha^2 \equiv \frac{I_\alpha}{mb^2}$$

A plot of H_{hL} is shown below in Figure 3.3. $\frac{\omega_1}{\omega_\alpha}, \frac{\omega_2}{\omega_\alpha}$ are the roots of the denominator, the system 'natural frequencies'.

$$\frac{\omega_1^2}{\omega_h \omega_\alpha}, \frac{\omega_2^2}{\omega_h \omega_\alpha} = \frac{\left[\frac{\omega_h}{\omega_\alpha} + \frac{\omega_\alpha}{\omega_h}\right] \pm \left\{\left[\frac{\omega_h}{\omega_\alpha} + \frac{\omega_\alpha}{\omega_h}\right]^2 - 4\left[1 - \frac{x_\alpha^2}{r_\alpha^2}\right]\right\}^{\frac{1}{2}}}{2[1 - x_\alpha^2/r_\alpha^2]} \tag{3.3.7}$$

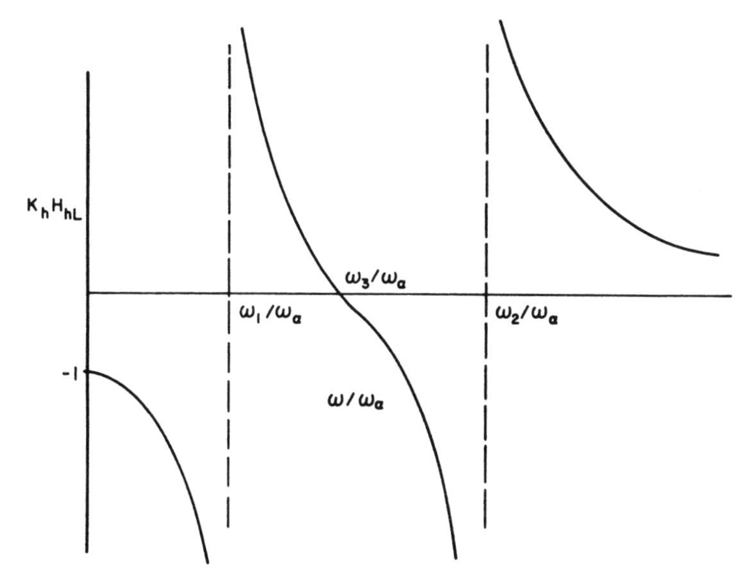

Figure 3.3 Transfer function.

A similar equation may be derived for

$$\frac{\bar{\alpha}}{\bar{L}} \equiv H_{\alpha L}\left(\omega/\omega_\alpha; \frac{\omega_h}{\omega_\alpha}, d/b, x_\alpha, r_\alpha\right) \tag{3.3.8}$$

ω_1 and ω_2 are again the natural frequencies. H_{hF}, $H_{\alpha F}$ are so-called 'transfer functions'; they are 'mechanical' or 'structural transfer functions' as they describe the motion of the structural system under specified loading. Later on we shall have occasion to consider 'aerodynamic transfer functions' and also 'aeroelastic transfer functions'. ω_3/ω_α is the root of the numerator of H_{hL} (but not in general of $H_{\alpha L}$ which will vanish at a different frequency),

$$\left(\frac{\omega_3}{\omega_\alpha}\right)^2 = \frac{1}{1+(d/b)x_\alpha/r_\alpha^2} \tag{3.3.9}$$

Note that infinite response occurs at the natural frequencies, ω_1 and ω_2, for both H_{hL} and $H_{\alpha L}$. This is not an instability; it is 'resonance' with the infinite response due to the absence of any damping in the system. Had structural or aerodynamic damping been included as will be done in later examples, then the transfer functions would become complex numbers which is a mathematical complication. However, the magnitude of the transfer functions would remain finite though large at $\omega = \omega_1$, ω_2 which is an improvement in the realism of the physical model. With L and M assumed given, which admittedly is somewhat artificial, the question of instability does not arise, We will elaborate on this point later when we discuss the notion of instability in a more precise way.

From sinusoidal motion we may proceed to periodic (but not necessarily sinusoidal) motion.

Periodic motion

The above analysis can be generalized to any periodic motion by expanding the motion into a Fourier (sinusoidal) series. Define:

$T_0 \equiv$ basic period

$\omega_0 \equiv 2\pi/T_0$, fundamental frequency

Then a periodic force, $L(t)$, may be written as

$$L(t) = \sum_{n=-\infty}^{\infty} L_n e^{+in\omega_0 t} \tag{3.3.10}$$

where

$$L_n = \frac{1}{T_0}\int_{-T_0/2}^{T_0/2} L(t)e^{-in\omega_0 t}\,dt \tag{3.3.11}$$

Using (3.3.10) and (3.3.6),

$$h(t) = \sum_n H_{hL}\left(\frac{\omega_0 n}{\omega_\alpha}\right) L_n e^{in\omega_0 t} \tag{3.3.12}$$

From periodic motion we may proceed to arbitrary time dependent motion.

Arbitrary motion

By taking the limit as the basic period becomes infinitely long, $T_0 \to \infty$, we obtain results for nonperiodic motion.

Define

$$\omega \equiv n\omega_0$$

$$\Delta\omega \equiv \Delta n \omega_0{}^{\dagger} = \omega_0 = 2\pi/T_0 \quad \text{frequency increment}$$

$$L^*(\omega) \equiv \frac{L_n}{\Delta\omega} = \frac{L_n T_0}{2\pi} \quad \text{force per frequency increment}$$

Then (3.3.10) becomes

$$L(t) = \int_{-\infty}^{\infty} L^*(\omega) e^{+i\omega t} \, d\omega \tag{3.3.10}$$

(3.3.11) becomes

$$L^*(\omega) = \frac{1}{2\pi} \int_{-\infty}^{\infty} L(t) e^{-i\omega t} \, dt \tag{3.3.11}$$

(3.3.12) becomes

$$h(t) = \int_{-\infty}^{\infty} H_{hL}(\omega/\omega_\alpha) L^*(\omega) e^{i\omega t} \, d\omega \tag{3.3.12}$$

An interesting alternate form of (3.3.12) can be obtained by substituting (3.3.11) into (3.3.12). Using a dummy time variable, τ, in (3.3.11) and interchanging order of integration in (3.3.12), gives

$$h(t) = \int_{-\infty}^{\infty} I_{hL}(t-\tau) L(\tau) \, d\tau \tag{3.3.13}$$

where

$$I_{hL}(t) \equiv \frac{1}{2\pi} \int_{-\infty}^{\infty} H_{hL}(\omega/\omega_\alpha) e^{i\omega t} \, d\omega \tag{3.3.14}$$

Comparing (3.3.12) and (3.3.14), note that I_{hL} is response to $L^*(\omega) = \dfrac{1}{2\pi}$ or from (3.3.10) and (3.3.11), $L(t) = \delta(t)$. Hence, I is the response to an impulse force and is thus called the impulse function.

(3.3.10)–(3.3.12) are Fourier transform relations and (3.3.13) is a so-called convolution integral.

† Note $\Delta n = 1$ since any n is an integer.

Note (3.3.13) is suitable for treating transient motion; however, a special case of the Fourier transform is often used for transient motion. This is the Laplace transform.

Laplace transform. Consider

$$L(\tau)=0 \quad \text{for} \quad \tau<0$$

also

$$I_{hL}(t-\tau)=0 \quad \text{for} \quad t-\tau<0$$

The latter will be true for any physically realizable system since the system cannot respond before the force is applied.

Define

$$p \equiv i\omega; \quad \text{thus} \quad \omega=-ip$$

and

$$L^{\dagger} \equiv 2\pi L^{*}(-ip)$$

then (3.3.10) becomes

$$L(t)=\frac{1}{2\pi i}\int_{-i\infty}^{i\infty} L^{\dagger}e^{pt}\,dp \tag{3.3.15}$$

(3.3.11) becomes

$$L^{\dagger}=\int_{0}^{\infty} L(t)e^{-pt}\,dt$$

(3.3.13) becomes

$$h(t)=\int_{0}^{t} I_{hL}(t-\tau)L(\tau)\,d\tau$$

where

$$I_{hL}(t)=\frac{1}{2\pi i}\int_{-i\infty}^{i\infty} H_{hL}\left(\frac{-ip}{\omega_\alpha}\right)e^{pt}\,dp$$

Utilization of transform integral approach for arbitrary motion. There are several complementary approaches in practice. In one the transfer function, H_{hL}, is first determined through consideration of simple sinusoidal motion. Then the impulse function is evaluated from

$$I_{hL}(t)=\frac{1}{2\pi}\int_{-\infty}^{\infty} H_{hL}(\omega)e^{i\omega t}\,d\omega \tag{3.3.14}$$

and the response is obtained from

$$h(t) = \int_0^t I_{hL}(t-\tau)L(\tau)\,d\tau \tag{3.3.13}$$

Alternatively, knowing the transfer function, $H_{hL}(\omega)$, the transform of the input force is determined from

$$L^*(\omega) = \frac{1}{2\pi}\int_{-\infty}^{\infty} L(t)e^{-i\omega t}\,d\omega \tag{3.3.11}$$

and the response from

$$h(t) = \int_{-\infty}^{\infty} H_{hL}(\omega)L^*(\omega)e^{i\omega t}\,d\omega \tag{3.3.12}$$

Both approaches give the same result, of course.

As a simple example we consider the translation of our typical section for $S_\alpha \equiv 0$, i.e., the center of mass coincides with the elastic axis or spring attachment point. This uncouples the rotation from translation and we need only consider

$$m\ddot{h} + K_h h = -L \tag{3.3.1}$$

We assume a force of the form

$$\begin{aligned} L &= e^{-at} \quad \text{for} \quad t>0 \\ &= 0 \quad \text{for} \quad t<0 \end{aligned} \tag{3.3.16}$$

From our equation of motion (or (3.3.6) for $S_\alpha = x_\alpha = 0$) we determine the transfer function as

$$H_{hL}(\omega) = \frac{-1}{m[\omega_h^2 - \omega^2]}, \qquad \omega_h^2 \equiv K_h/m \tag{3.3.6}$$

From (3.3.14), using the above and evaluating the integral, we have

$$\begin{aligned} I_{hL}(t) &= -\frac{1}{m\omega_h}\sin\omega_h t \quad \text{for} \quad t>0 \\ &= 0 \qquad\qquad\quad \text{for} \quad t<0 \end{aligned} \tag{3.3.17}$$

From (3.3.13), using above (3.3.17) for I_{hL} and given L, we obtain

$$h(t) = -\frac{1}{m\omega_h}\left\{\frac{\omega_h e^{-at} - \omega_h\cos\omega_h t + a\sin\omega_h t}{a^2+\omega_h^2}\right\} \tag{3.3.18}$$

We can obtain the same result using our alternative method. Calculating L^* from (3.3.11) for our given L, we have

$$L = \frac{1}{2\pi}\frac{1}{a + i\omega}$$

Using above and the previously obtained transfer function in (3.3.12) we obtain the response. The result is, of course, the same as that determined before. Note that in accordance with our assumption of a system initially at rest, $h = \dot{h} = 0$ at $t = 0$. Examining our solution, (3.3.18), for large time we see that

$$h \to -\frac{1}{m\omega_h}\left\{\frac{-\omega_h \cos \omega_h t + a \sin \omega_h t}{a^2 + \omega_h^2}\right\} \quad \text{as} \quad t \to \infty$$

This indicates that the system continues to respond even though the force L, approaches zero for large time! This result is quite unrealistic physically and is a consequence of our ignoring structural damping in our model. Had we included this effect in our equation of motion using a conventional analytical damping model†

$$m[\ddot{h} + 2\zeta_n \omega_h \dot{h}] + K_h h = -L \tag{3.3.1}$$

the response would have been

$$h = -\frac{1}{m\omega_h}\left\{\frac{\omega_h e^{-at} + [-\omega_h \cos \omega_h t + a \sin \omega_h t]e^{-\zeta_h \omega_h t}}{a^2 + \omega_h^2}\right\} \tag{3.3.19}$$

for small damping, $\zeta_h \ll 1$, which is the usual situation. *Now* $h \to 0$, for $t \to \infty$. Furthermore, if the force persists for a long time, i.e., $a \to 0$, then

$$h(t) \to -\frac{1}{m\omega_h}\left\{\frac{\omega_h}{\omega_h^2}\right\} = -\frac{1}{K_h}$$

which is the usual static or steady state response to a force of unit amplitude. The terms which approach zero for large time due to structural damping are usually termed the transient part of the solution. If

$$a \ll \zeta_h \omega_h$$

the transient solution dies out rapidly compared to the force and we usually are interested in the steady state response. If

$$a \gg \zeta_h \omega_h$$

the 'impulsive' force dies out rapidly and we are normally interested in

† Meirovitch [4].

the transient response. Frequently the maximum response is of greatest interest. A well known result is that the peak dynamic response is approximately twice the static response if the force persists for a long time and the damping is small. That is, if

$$\zeta_h \ll 1$$

$$a \ll \omega_h$$

then h_{max} occurs when (see (3.3.19))

$$\cos \omega_n t \cong -1 \quad \text{or} \quad t = \frac{\pi}{\omega_h}$$

$$\sin \omega_n t \cong 0$$

and

$$h_{max} = -\frac{1}{m\omega_h}\frac{\omega_h}{\omega_h^2}[1-(-1)]$$

$$= -\frac{2}{K_h}$$

The reader may wish to consider other special combinations of the relative sizes of

a	force time constant
ω_h	system natural time constant
$\zeta_h\omega_h$	damping time constant

A great deal of insight into the dynamics of linear systems can be gained thereby.

The question arises which of the two approaches is to be preferred. The answer depends upon a number of factors, including the computational efficiency and physical insight desired. Roughly speaking the second approach, which is essentially a frequency domain approach, is to be preferred when analytical solutions are to be attempted or physical insight based on the degree of frequency 'matching' or 'mis-matching' of H_{hL} and L^* is desired. In this respect larger response obviously will be obtained if the maxima of H and L^* occur near the same frequencies, i.e., they are 'matched', and lesser response otherwise, 'mismatched'. The first approach, which is essentially a time approach, is generally to be preferred when numerical methods are attempted and quantitative accuracy is of prime importance.

Other variations on these methods are possible. For example the transfer function, H_{hL}, and impulse function, I_{hL}, may be determined experimentally. Also the impulse function may be determined directly from the equation of motion, bypassing any consideration of the transfer function. To illustrate this latter remark, consider our simple example

$$m\ddot{h} + K_h h = -L \tag{3.3.1}$$

The impulse function is the response for h due to $L(t) = \delta(t)$. Hence, it must satisfy

$$m\ddot{I}_{hL} + K_h I_{hL} = -\delta(t) \tag{3.3.20}$$

Let us integrate the above from $t = 0$ to ε.

$$\int_0^\varepsilon [m\ddot{I}_{hL} + K_h I_{hL}]\,dt = -\int_0^\varepsilon \delta(t)\,dt$$

or

$$m\dot{I}_{hL}\Big|_0^\varepsilon + K_h \int_0^\varepsilon I_{hL}\,dt = -1$$

In the limit as $\varepsilon \to 0^+$, we obtain the 'initial condition',

$$\dot{I}_{hL}(0^+) = -\frac{1}{m} \tag{3.3.21}$$

and also

$$I_{hL}(0^+) = 0$$

Hence, solving (3.3.20) and using the initial velocity condition, (3.3.21), we obtain

$$I_{hL} = -\frac{1}{m\omega_h} \sin \omega_h t \quad \text{for} \quad t > 0 \tag{3.3.17}$$

which is the same result obtained previously.

Finally, all of these ideas can be generalized to many degrees of freedom. In particular using the concept of 'normal modes' any multi-degree-of-freedom system can be reduced to a system of uncoupled single-degree-of-freedom systems.† As will become clear, when aerodynamic forces are present the concept of normal modes which decouple the various degrees of freedom is not as easily applied and one must usually deal with all the degrees of freedom which are of interest simultaneously.

† Meirovitch [4].

Random motion

A random motion is by definition one whose detailed behavior is neither repeatable nor of great interest. Hence attention is focused on certain averages, usually the mean value and also the mean square value. The mean value may be treated as a static loading and response problem and hence we shall concentrate on the mean square relations which are the simplest characterization of *random, dynamic* response.

Relationship between mean values. To see the equivalence between mean value dynamic response and static response, consider

$$h(t) = \int_{-\infty}^{\infty} I_{hL}(t-\tau)L(\tau)\,\mathrm{d}\tau \tag{3.3.13}$$

and take the mean of both sides (here a bar above the quantity denotes its mean, which should not be confused with that symbol's previous use in our discussion of sinusoidal motion)

$$\begin{aligned}\bar{h} &\equiv \lim_{T\to\infty} \frac{1}{2T}\int_{-T}^{T} h(t)\,\mathrm{d}t \\ &= \lim_{T\to\infty} \frac{1}{2T}\int_{-T}^{T}\int_{-\infty}^{\infty} I_{hL}(t-\tau)L(\tau)\,\mathrm{d}\tau\,\mathrm{d}t\end{aligned}$$

Interchanging the order of integration and making a change of variables, the right hand side becomes

$$\begin{aligned}&= \int_{-\infty}^{\infty}\left\{\lim_{T\to\infty}\frac{1}{2T}\int_{-T}^{T} L(t-\tau)\,\mathrm{d}t\right\} I_{hL}(\tau)\,\mathrm{d}\tau \\ &= \bar{L}\int_{-\infty}^{\infty} I_{hL}(\tau)\,\mathrm{d}\tau \\ &= \bar{L}H_{hL}(\omega=0) \\ &= -\frac{\bar{L}}{K_h}\end{aligned}$$

which is just the usual static relation. Unfortunately, no such simple relation exists between the *mean square* values. Instead all frequency components of the transfer function, H_{hL}, contribute. Because of this it proves useful to generalize the definition of a mean square.

Relationship between mean square values. A more general and informative quantity than the mean square, the correlation function, ϕ, can be defined as

$$\phi_{LL}(\tau) \equiv \lim_{T\to\infty} \frac{1}{2T} \int_{-T}^{T} L(t)L(t+\tau)\,\mathrm{d}t \tag{3.3.22}$$

The mean square of L, $\overline{L^2}$, is given by

$$\overline{L^2} = \phi_{LL}(\tau = 0) \tag{3.3.23}$$

As $\tau \to \infty$, $\phi_{LL} \to 0$ if L is truly a random function since $L(t)$ and $L(t+\tau)$ will be 'uncorrelated'. Indeed, a useful check on the randomness of L is to examine ϕ for large τ. Analogous to (3.3.22), we may define

$$\begin{aligned} \phi_{hh}(\tau) &\equiv \lim_{T\to\infty} \frac{1}{2T} \int_{-T}^{T} h(t)h(t+\tau)\,\mathrm{d}t \\ \phi_{hL}(\tau) &\equiv \lim_{T\to\infty} \frac{1}{2T} \int_{-T}^{T} h(t)L(t+\tau)\,\mathrm{d}t \end{aligned} \tag{3.3.24}$$

ϕ_{hL} is the 'cross-correlation' between h and L. ϕ_{hh} and ϕ_{LL} are 'autocorrelations'. The Fourier Transform of the correlation function is also a quantity of considerable interest, the 'power spectra',

$$\Phi_{LL}(\omega) \equiv \frac{1}{\pi} \int_{-\infty}^{\infty} \phi_{LL}(\tau) \mathrm{e}^{-i\omega\tau}\,\mathrm{d}\tau \tag{3.3.25}$$

(Note that a factor of two difference exists in (3.3.25) from the usual Fourier transform definition. This is by tradition.) From (3.3.25), we have

$$\begin{aligned} \phi_{LL}(\tau) &= \tfrac{1}{2} \int_{-\infty}^{\infty} \Phi_{LL}(\omega) \mathrm{e}^{i\omega\tau}\,\mathrm{d}\omega \\ &= \int_{0}^{\infty} \Phi_{LL}(\omega) \cos \omega\tau \,\mathrm{d}\omega \end{aligned} \tag{3.3.26}$$

The latter follows since $\Phi_{LL}(\omega)$ is a real even function of ω. Note

$$\overline{L^2} = \phi_{LL}(0) = \int_{0}^{\infty} \Phi_{LL}(\omega)\,\mathrm{d}\omega \tag{3.3.27}$$

Hence a knowledge of Φ_{LL} is sufficient to determine the mean square. It turns out to be most convenient to relate the power spectra of L to that of h and use (3.3.27) or its counterpart for h to determine the mean square values.

To relate the power spectra, it is useful to start with a substitution of (3.3.13) into the first of (3.3.24).

$$\phi_{hh}(\tau) = \lim_{T\to\infty} \frac{1}{2T} \int_{-T}^{T} \left\{ \int_{-\infty}^{\infty} L(\tau_1) I_{hL}(t-\tau_1)\, \mathrm{d}\tau_1 \right\}$$
$$\times \left\{ \int_{-\infty}^{\infty} L(\tau_2) I_{hL}(t+\tau-\tau_2)\, \mathrm{d}\tau_2 \right\} \mathrm{d}t$$

Interchanging order of integrations and using a change of integration variables

$$t' \equiv t - \tau_1; \qquad \tau_1 = t - t'$$
$$t'' \equiv t + \tau - \tau_2; \qquad \tau_2 = t + \tau - t''$$

we have

$$\phi_{hh} = \int_{-\infty}^{\infty} \int_{-\infty}^{\infty} I_{hL}(t') I_{hL}(t'') \phi_{LL}(\tau + t' - t'')\, \mathrm{d}t'\, \mathrm{d}t'' \tag{3.3.28}$$

One could determine $\overline{h^2}$ from (3.3.28)

$$\overline{h^2} = \phi_{hh}(\tau = 0) = \int_{-\infty}^{\infty} \int_{-\infty}^{+\infty} I_{hL}(t') I_{hL}(t'') \phi_{LL}(t' - t'')\, \mathrm{d}t'\, \mathrm{d}t'' \tag{3.3.29}$$

However we shall proceed by taking the Fourier Transform of (3.3.28).

$$\Phi_{hh} \equiv \frac{1}{\pi} \int_{-\infty}^{\infty} \phi_{hh}(\tau) \mathrm{e}^{-i\omega\tau}\, \mathrm{d}\tau$$
$$= \frac{1}{\pi} \iiint_{-\infty}^{\infty} I_{hL}(t') I_{hL}(t'') \phi_{LL}(\tau + t' - t'') \mathrm{e}^{-i\omega\tau}\, \mathrm{d}t'\, \mathrm{d}t''\, \mathrm{d}\tau$$
$$= \frac{1}{\pi} \iiint_{-\infty}^{\infty} I_{hL}(t') \mathrm{e}^{+i\omega t'} I_{hL}(t'') \mathrm{e}^{-i\omega t''}$$
$$\times \phi_{LL}(\tau + t' - t'') \exp - i\omega(\tau + t' - t'')\, \mathrm{d}t'\, \mathrm{d}t''\, \mathrm{d}\tau$$

Defining a new variable

$$\tau' \equiv \tau + t' - t''$$
$$\mathrm{d}\tau' = \mathrm{d}\tau$$

we see that

$$\boxed{\Phi_{hh}(\omega) = H_{hL}(\omega) H_{hL}(-\omega) \Phi_{LL}(\omega)} \tag{3.3.30}$$

One can also determine that

$$\begin{aligned} \Phi_{hL}(\omega) &= H_{hL}(-\omega)\Phi_{LL}(\omega) \\ \Phi_{hh}(\omega) &= H_{hL}(-\omega)\Phi_{hL}(\omega) \end{aligned} \tag{3.3.31}$$

(3.3.30) is a powerful and well-known relation.* The basic procedure is to determine Φ_{LL} by analysis or measurement, compute Φ_{hh} from (3.3.30) and $\overline{h^2}$ from an equation analogous to (3.3.26).

$$\overline{h^2} = \int_0^\infty \Phi_{hh}(\omega)\, d\omega = \int_0^\infty |H_{hL}(\omega)|^2\, \Phi_{LL}(\omega)\, d\omega \tag{3.3.32}$$

Let us illustrate the utility of the foregoing discussion by an example.

Example: airfoil response to a gust. Again for simplicity consider translation only.

$$m\ddot{h} + K_h h = -L \tag{3.3.1}$$

Also for simplicity assume quasi-steady aerodynamics.†

$$L = qS\frac{\partial C_L}{\partial \alpha}\left[\frac{\dot{h}}{U} + \frac{w_G}{U}\right] \tag{3.3.33}$$

w_G taken as positive up, is a vertical fluid 'gust' velocity, which varies randomly with time but is assumed here to be uniformly distributed spatially over the airfoil chord. Various transfer functions may be defined and calculated. For example

$$\frac{\bar{h}}{\bar{L}} \equiv H_{hL} = \frac{-1}{m[-\omega^2 + \omega_h^2]}, \qquad \omega_h^2 \equiv K_h/m \tag{3.3.34}$$

structural transfer function‡ (motion due to lift) (cf. (3.3.6))

$$\frac{\bar{L}}{\bar{h}} \equiv H_{Lh} = qS\frac{\partial C_L}{\partial \alpha}\frac{i\omega}{U} \tag{3.3.35}$$

---

*Crandall and Mark [5].

† $\dfrac{\dot{h}}{U} + \dfrac{w_G}{U}$ is an effective angle of attack, α.

‡ Here we choose to use a dimensional rather than a dimensionless transfer function.

aerodynamic transfer function (lift due to motion)

$$\frac{\bar{L}}{\bar{w}_G} \equiv H_{Lw_G} = qS\frac{\partial C_L}{\partial \alpha}\frac{1}{U} \tag{3.3.36}$$

aerodynamic transfer function* (lift due to gust velocity field)

$$H_{hw_G} \equiv \frac{\bar{h}}{\bar{w}_G} = \frac{-H_{Lw_G}}{\left[-\dfrac{1}{H_{hL}} + H_{Lh}\right]} \tag{3.3.37}$$

aeroelastic transfer function (motion due to gust velocity field).

The most general of these is the aeroelastic transfer function which may be expressed in terms of the structural and aerodynamic transfer functions, (3.3.37). Using our random force-response relations, we have from (3.3.32)

$$\begin{aligned} \overline{h^2} &= \int_0^\infty |H_{hw_G}|^2 \, \Phi_{w_G w_G} \, d\omega \\ &= \int_0^\infty \frac{\left[qS\dfrac{\partial C_L}{\partial \alpha}\dfrac{1}{U}\right]^2}{[-m\omega^2 + K_h]^2 + \left[qS\dfrac{\partial C_L}{\partial \alpha}\dfrac{\omega}{U}\right]^2} \Phi_{w_G w_G} \, d\omega \end{aligned}$$

Define an effective damping constant as

$$\zeta \equiv \frac{qS\dfrac{\partial C_L}{\partial \alpha}\dfrac{1}{U}}{2\sqrt{mK_h}} \tag{3.3.38}$$

then

$$\bar{h}^2 = \frac{\left[qS\dfrac{\partial C_L}{\partial \alpha}\dfrac{1}{U}\right]^2}{m^2} \int_0^\infty \frac{\Phi_{w_G w_G} \, d\omega}{[-\omega^2 + \omega_h^2]^2 + 4\zeta^2\omega_h^2\omega^2}$$

which, for small ζ, may be evaluated as†

$$\bar{h}^2 \cong \frac{qS\dfrac{\partial C_L}{\partial \alpha}\pi}{K_h} \frac{\Phi_{w_G w_G}(\omega = \omega_h)}{U} \tag{3.3.39}$$

* We ignore a subtlety here in the interest of brevity. For a 'frozen gust', we must take $w_G = \bar{w}_G \exp i\omega(t - x/U_\infty)$ in determining this transfer function. See later discussion in Sections 3.6, 4.2 and 4.3.

† Crandall and Mark; the essence of the approximation is that for small ζ, $\Phi_{w_G w_G}(\omega) \cong \Phi_{w_G w_G}(\omega_h)$ and may be taken outside the integral. See subsequent discussion of graphical analysis.

Typically

$$\Phi_{w_G w_G}(\omega) = \bar{w}_G^2 \frac{L_G}{\pi U} \frac{1 + 3\left(\frac{\omega L_G}{U}\right)^{2*}}{\left[1 + \left(\frac{\omega L_G}{U}\right)^2\right]^2} \tag{3.3.40}$$

as determined from experiment or considerations of the statistical theory of atmospheric turbulence. Here, L_G is the 'scale length of turbulence'; which is not to be confused with the lift force. Nondimensionalizing and using (3.3.39) and (3.3.40), we obtain

$$\frac{\bar{h}^2/b^2}{\bar{w}_G^2/U^2} = qS \frac{\frac{\partial C_L}{\partial \alpha} \frac{\omega_h L_G}{U}}{K_h b \frac{\omega_h b}{U}} \left\{ \frac{1 + 3\left(\frac{\omega_h L_G}{U}\right)^2}{\left[1 + \left(\frac{\omega_h L_G}{U}\right)^2\right]^2} \right\} \tag{3.3.41}$$

Note as $\frac{\omega_h L_G}{U} \to 0$ or ∞, $\bar{h}^2/b^2 \to 0$. Recall L_G is the characteristic length associated with the random gust field. Hence, for very large or very small characteristic lengths the airfoil is unresponsive to the gust. For what $\frac{\omega_h L_G}{U}$ does the largest response occur?

As an alternative to the above discussion, a correlation function approach could be taken where one uses the time domain and the aeroelastic impulse function,

$$\frac{I_{hw_G}}{b} = -\frac{qS \frac{\partial C_L}{\partial \alpha} \frac{1}{U} e^{-\zeta \omega_h t}}{mb\omega_h^2 \sqrt{1-\zeta^2}} \sin \sqrt{1-\zeta^2}\, \omega_h t \tag{3.3.42}$$

but we shall not pursue this here. Instead the frequency domain analysis is pursued further.

It is useful to consider the preceding calculation in graphical form for a moment. The (square of the) transfer function is plotted in Figure 3.4, and the gust power spectral density in Figure 3.5.

We note that the power spectral density is slowly varying with ω relative to the square of the transfer function which peaks sharply near $\omega = \omega_h$. Hence one may, to a close approximation, take the power spectral density as a constant with its value determined at $\omega = \omega_h$ in

* Houbolt, Steiner and Pratt [6]. Also see later discussion in Section 3.6.

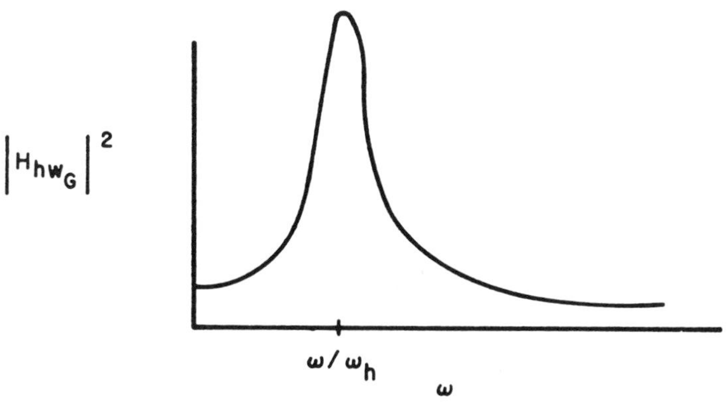

Figure 3.4 Aeroelastic transfer function.

computing the mean square response. This is a simple but powerful idea which carries over to many degrees-of-freedom, and hence many resonances, provided the resonant frequencies of the transfer function are known. For some aeroelastic systems, locating the resonances may prove difficult.

There are other difficulties with the approach which should be pointed out. First of all we note that including the (aerodynamic) damping due to motion is necessary to obtain a physically meanful result. Without it the computed response would be infinite! Hence, an accurate evaluation of the effective damping for an aeroelastic system is essential in random response studies. It is known that in general the available aerodynamic theories are less reliable for evaluating the (out-of-phase with displacement) damping forces than those forces in-phase with displacement.* Another difficulty may arise if instead of evaluating the mean square displacement response we instead seek to determine the mean

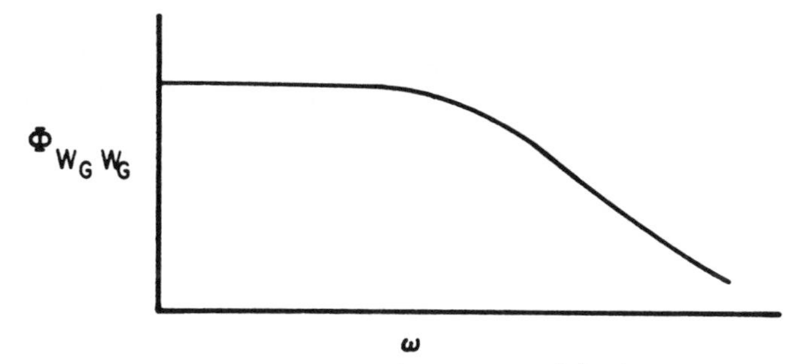

Figure 3.5 Gust (auto) power spectral density.

* Acum [7].

square of acceleration. The latter quantity is frequently of greater interest from the standpoint of design. The relevant transfer function is given by

$$H_{\ddot{h}w_G} = (i\omega)^2 H_{hw_G} \tag{3.3.43}$$

and the mean square is therefore

$$\bar{\ddot{h}}^2 = \frac{\displaystyle\int_0^\infty \omega^4 \left[qS \frac{\partial C_L}{\partial \alpha} \frac{1}{U} \right]^2 \Phi_{w_G w_G} \, \mathrm{d}\omega}{[-m\omega^2 + K_n]^2 + \left[qS \dfrac{\partial C_L}{\partial \alpha} \dfrac{\omega}{U} \right]^2} \tag{3.3.44}$$

If we make the same approximation as before that $\Phi_{w_G w_G}$ is a constant, we are in difficulty because $|H_{\ddot{h}w_G}|^2$ does not approach zero as $\omega \to \infty$ and hence the integral formally diverges. This means greater care must be exercised in evaluating the integral and in particular considering the high frequency behavior of the gust power spectral density. Also, one may need to use a more elaborate aerodynamic theory. In the present example we have used a quasi-steady aerodynamic theory which is reasonably accurate for low frequencies;* however, to evaluate the acceleration response it will frequently be necessary to use a full unsteady aerodynamic theory in order to obtain accurate results at high frequencies in (3.3.44).

Measurement of power spectra. We briefly digress to consider an important application of (3.3.27) to the experimental determination of power spectra. For definiteness consider the measurement of gust power spectra. Analogous to (3.3.27) we have

$$\bar{w}_G^2 = \int_0^\infty \Phi_{w_G w_G}(\omega) \, \mathrm{d}\omega \tag{3.3.45}$$

It is assumed that a device is available to measure w_G over a useful range of frequencies. The electronic signal from this device is then sent to an electronic 'filter'. The latter, in its most ideal form, has a transfer function given by

$$\begin{aligned} H_{Fw_G} &= 1 \quad \text{for} \quad \omega_c - \frac{\Delta\omega}{2} \omega < \omega_c + \frac{\Delta\omega}{2} \\ &= 0 \quad \text{otherwise} \end{aligned} \tag{3.3.46}$$

where $\omega_c \equiv$ center frequency of the filter

$\Delta\omega \equiv$ frequency bandwidth of the filter

*Acum [7].

Now if we assume that the power spectrum varies slowly with ω and we choose a filter with $\Delta\omega \ll \omega_c$, then (3.3.45) may be approximated by taking $\Phi_{w_G w_G}(\omega) \cong \Phi_{w_G w_G}(\omega_c)$ and moving it outside the integral. The result is

$$\overline{w_G^2} \cong \Phi_{w_G w_G}(\omega_c)\,\Delta\omega$$

Solving for the power spectrum,

$$\Phi_{w_G w_G}(\omega_c) = \frac{\overline{w_G^2}}{\Delta\omega} \tag{3.3.47}$$

By systematically changing the filter center frequency, the power spectrum may be determined over the desired range of frequency. The frequency bandwidth, $\Delta\omega$, and the time length over which $\overline{w_G^2}$ is calculated must be chosen with care. For a discussion of these matters, the reader may consult Crandall and Mark [5], and references cited therein.

Extension of discussion on random motion to two-dimensional plate-like structures with many degrees of freedom: This extension is considered briefly in Appendix I, 'A Primer for Structural Response to Random Pressure Fluctuations'.

Flutter — an introduction to dynamic aeroelastic instability

The most dramatic physical phenomenon in the field of aeroelasticity is flutter, a dynamic instability which often leads to catastrophic structural failure. One of the difficulties in studying this phenomenon is that it is not one but many. Here we shall introduce one type of flutter using the typical section structural model and a *steady flow* aerodynamic model. The latter is a highly simplifying assumption whose accuracy we shall discuss in more detail later. From (3.3.1) and with a steady aerodynamic model, $L = qS\dfrac{\partial C_L}{\partial \alpha}\alpha$, $M_y = eL$, the equations of motion are

$$m\ddot{h} + S_\alpha\ddot{\alpha} + K_h h + qS\frac{\partial C_L}{\partial \alpha}\alpha = 0 \tag{3.3.48}$$

$$I_\alpha\ddot{\alpha} + S_\alpha\ddot{h} + K_\alpha\alpha - qSe\frac{\partial C_L}{\partial \alpha}\alpha = 0$$

To investigate the stability of this system we assume solutions of the form

$$h = \bar{h}e^{pt}$$
$$\alpha = \bar{\alpha}e^{pt} \tag{3.3.49}$$

and determine the possible values of p, which are in general complex numbers. If the real part of any value of p is positive, then the motion diverges exponentially with time, cf. (3.3.49), and the typical section is unstable.

To determine p, substitute (3.3.49) into (3.3.48) and use matrix notation to obtain

$$\begin{bmatrix} [mp^2+K_h] & S_\alpha p^2 + qS\dfrac{\partial C_L}{\partial \alpha} \\ S_\alpha p^2 & I_\alpha p^2 + K_\alpha - qSe\dfrac{\partial C_L}{\partial \alpha} \end{bmatrix} \begin{Bmatrix} \bar{h}e^{pt} \\ \bar{\alpha}e^{pt} \end{Bmatrix} = \begin{Bmatrix} 0 \\ 0 \end{Bmatrix} \tag{3.3.50}$$

For nontrivial solutions the determinant of coefficients is set to zero which determines p, viz.

$$Ap^4 + Bp^2 + C = 0 \tag{3.3.51}$$

where

$$A \equiv mI_\alpha - S_\alpha^2$$

$$B \equiv m\left[K_\alpha - qSe\frac{\partial C_L}{\partial \alpha}\right] + K_h I_\alpha - S_\alpha qS\frac{\partial C_L}{\partial \alpha}$$

$$C \equiv K_h\left[K_\alpha - qSe\frac{\partial C_L}{\partial \alpha}\right]$$

Solving (3.3.51),

$$p^2 = \frac{-B \pm [B^2 - 4AC]^{\frac{1}{2}}}{2A} \tag{3.3.52}$$

and taking the square root of (3.3.52) determines p.

The signs of A, B and C determine the nature of the solution. A is always positive for any distribution of mass; C is positive as long as q is less than its divergence value, i.e.

$$\left[K_\alpha - qSe\frac{\partial C_L}{\partial \alpha}\right] > 0$$

which is the only case of interest as far as flutter is concerned. B may be either positive or negative; re-writing

$$B = mK_\alpha + K_h I_\alpha - [me + S_\alpha]qS\frac{\partial C_L}{\partial \alpha} \tag{3.3.53}$$

If $[me + S_\alpha] < 0$ then $B > 0$ for all q.

Otherwise $B<0$ when

$$K_\alpha + \frac{K_h I_\alpha}{m} - \left[1+\frac{S_\alpha}{me}\right] qSe\frac{\partial C_L}{\partial \alpha} < 0$$

Consider in turn the two possibilities, $B>0$ and $B<0$.

$B>0$: Then the values of p^2 from (3.3.52) are real and negative provided

$$B^2 - 4AC > 0$$

and hence the values of p are purely imaginary, representing neutrally stable oscillations. On the other hand if

$$B^2 - 4AC < 0$$

the values of p^2 are complex and hence at least one value of p will have a positive real part indicating an unstable motion. Thus

$$B^2 - 4AC = 0 \tag{3.3.54}$$

gives the boundary between neutrally stable and unstable motion. From (3.3.54) one may compute an explicit value of q at which the dynamic stability, 'flutter', occurs.

From (3.3.54) we have

$$Dq_F^2 + Eq_F + F = 0$$

$$q_F = \frac{-E \pm [E^2 - 4DF]^{\frac{1}{2}}}{2D} \tag{3.3.55}$$

where

$$D \equiv \left\{[me + S_\alpha]S\frac{\partial C_L}{\partial \alpha}\right\}^2$$

$$E \equiv \{-2[me + S_\alpha][mK_\alpha + K_h I_\alpha] + 4[mI_\alpha - S_\alpha^2]eK_h\}S\frac{\partial C_L}{\partial \alpha}$$

$$F = [mK_\alpha + K_h I_\alpha]^2 - 4[mI_\alpha - S_\alpha^2]K_h K_\alpha$$

In order for flutter to occur at least one of the q_F determined by (3.3.55) must be real and positive. If both are, the smaller of the two is the more critical; if neither are, flutter does not occur. Pines* has studied this example in some detail and derived a number of interesting results.

* Pines [8].

Perhaps the most important of these is that for

$$S_\alpha \leq 0$$

i.e., the center of gravity is ahead of the elastic axis, no flutter occurs. Conversely as S_α increases in a positive sense the dynamic pressure at which flutter occurs, q_F, is decreased. In practice, mass is often added to a flutter prone structure so as to decrease S_α and raise q_F. Such a structure is said to have been 'mass balanced'. Now consider the other possibility for B.

$B < 0$: B is positive for $q \equiv 0$ (cf. (3.3.51) et. seq.) and will only become negative for sufficiently large q. However,

$$B^2 - 4AC = 0$$

will occur before

$$B = 0$$

since $A > 0$, $C > 0$. Hence, to determine when flutter occurs, only $B > 0$ need be considered.

In concluding this discussion, let us study the effect of S_α in more detail following Pines.

Consider the first special case $S_\alpha = 0$. Then

$$D = \left[meS\frac{\partial C_L}{\partial \alpha}\right]^2$$

$$E = 2me\{I_\alpha K_h - mK_\alpha\}S\frac{\partial C_L}{\partial \alpha}$$

$$F = \{mK_\alpha - K_h I_\alpha\}^2$$

and one may show that

$$E^2 - 4DF = 0$$

Using this result and also (3.3.55) and (2.18), it is determined that

$$q_F/q_D = 1 - \omega_h^2/\omega_\alpha^2 \tag{3.3.56}$$

Thus if $q_d < 0$ and $\omega_h/\omega_\alpha < 1, q_F < 0$, i.e. no flutter will occur. Conversely if $q_D > 0$ and $\omega_h/\omega_\alpha > 1$, then $q_F < 0$ and again no flutter will occur.

Now consider the general case, $S_\alpha \neq 0$. Note that $D > 0$ and $F > 0$ for all parameter values. Thus from (3.3.55), $q_F < 0$ if $E > 0$ and no flutter will occur. After some rearrangement of the expression for E, it is found that (in non-dimensional form)

$$\bar{E} \equiv E/\left(2m^2 I_\alpha \omega_\alpha^2 S\frac{\partial C_L}{\partial \alpha}\right) =$$
$$e\left[-1 + (\omega_h/\omega_\alpha)^2 - 2\frac{x_{cg}^2}{r_{cg}^2}(\omega_h/\omega_\alpha)^2\right]$$
$$-x_{cg}\left[1 + (\omega_h/\omega_\alpha)^2\right] \tag{3.3.57}$$

From this equation, the condition for no flutter, $E > 0$ or $\bar{E} > 0$, gives the following results.

- If $x_{cg} = 0$, then no flutter occurs for $e > 0$ and $\omega_h/\omega_\alpha > 1$ or for $e < 0$ and $\omega_h/\omega_\alpha < 1$.
- If $e = 0$, then no flutter occurs for $x_{cg} < 0$ and any ω_h/ω_α.
- For small x_{cg}, i.e. if

  $$x_{cg}^2/r_{cg}^2 \ll 1$$

  then $\bar{E} > 0$ implies
  $$e\frac{\left[-1 + (\omega_h/\omega_\alpha)^2\right]}{\left[1 + (\omega_h/\omega_\alpha)^2\right]} - x_{cg} > 0$$

  as the condition for no flutter

  For ω_h/ω_α small (the usual case), this implies

  $$-e - x_{cg} > 0$$

  while for ω_h/ω_α large, this implies

  $$e - x_{cg} > 0$$

  as the conditions for no flutter.

Quasi-steady, aerodynamic theory

Often it is necessary to determine p by numerical methods as a function of q in order to evaluate flutter. For example, if one uses the slightly more complex 'quasi-steady' aerodynamic theory which includes the effective angle of attack contribution, $\dot{h}/U$, so that

$$qS\frac{\partial C_L}{\partial \alpha}\alpha$$

becomes

$$qS\frac{\partial C_L}{\partial \alpha}\left[\alpha + \frac{\dot{h}}{U}\right] = \rho\frac{US}{2}\frac{\partial C_L}{\partial \alpha}[U\alpha + \dot{h}]$$

then (3.3.51) will contain terms proportional to p and p^3 and the values of p must be determined numerically. An example of such a calculation is given in Figure 6.30 of B.A. which is reproduced below as Figure 3.6.

Denote

$p = p_R + i\omega$

$\omega_h^2 \equiv K_h/m, \ \omega_\alpha^2 \equiv K_\alpha/I_\alpha$

$x_\alpha \equiv S_\alpha/mb, \ r_\alpha^2 \equiv I_\alpha/mb^2$

b = a reference length

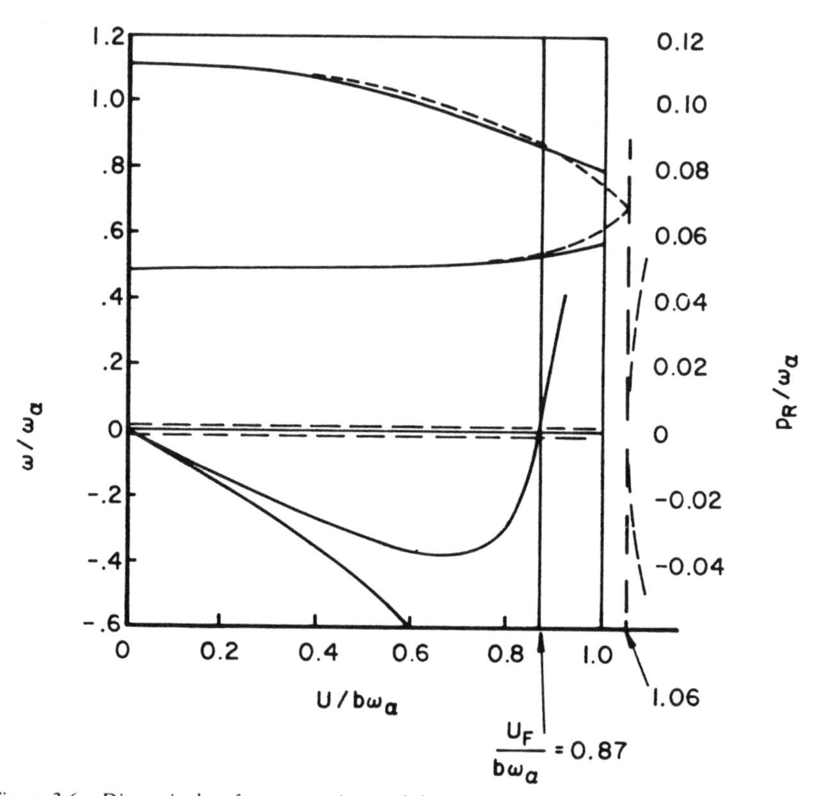

Figure 3.6 Dimensionless frequency ω/ω_α and damping p_R/ω_α of the aeroelastic modes of the typical section, estimated using steady-state aerodynamic operators and plotted vs. reduced airspeed $U/b\omega_\alpha$. System parameters are $x_\alpha = 0.05$, $r_\alpha = 0.5$, $\omega_h/\omega_\alpha = 0.5$, $(2m/\pi\rho_\infty bS) = 10$,

$e/b = 0.4$, $\frac{\partial C_L}{\partial \alpha} = 2\pi$. Solid curves — with aerodynamic damping. Dashed curves — without aerodynamic damping.

Since the values of p are complex conjugate only half of them are shown. The solid lines are for the $\dot{h}/U$ or aerodynamic damping effect included and the dash lines for it omitted. There are several interesting points to be made.

(1) With aerodynamic damping omitted the typical section model is neutrally stable until $U = U_F$. For $U = U_F$ the bending and torsion frequencies merge and for $U > U_F$ the system is unstable.

(2) With aerodynamic damping included, for small U all values of p are stable and flutter occurs at sufficiently large U where p_R changes sign from negative to positive. There is a tendency for the frequencies to merge but complete merging does not occur.

(3) The addition of aerodynamic damping reduces, *in this example for this approximate aerodynamic theory*, the flutter velocity U_F. This last result has been a source of consternation (and research papers). Whether it occurs in the real physical problem or whether it is a consequence of our simplified theoretical model is not known. No experiment has yet been performed where the aerodynamic (or structural) damping has been systematically varied to verify or refute this result.

Finally we mention one further general complication which commonly occurs in analysis. When even more elaborate, fully unsteady aerodynamic theories are employed, the aerodynamic forces are usually only conveniently known for neutrally stable motion, i.e.,

$$p = i\omega, \qquad p_R \equiv 0$$

Hence, indirect or iterative methods are usually required to effect a solution for $U = U_F$ and often no information is obtained for $U < U_F$ or $U > U_F$. We shall return to this issue later.

3.4 Aerodynamic forces for airfoils — an introduction and summary

Having developed the mathematical tools for treating the dynamics of our aeroelastic system, we now turn to a topic previously deferred, the determination of the aerodynamic forces. Usually, we wish to relate the aerodynamic lift and moment to the motion of the airfoil. In order not to break unduly the continuity of our discussion of aeroelastic phenomena, we give a brief summary of known results here and defer a discussion of the aerodynamic theory from first principles until Chapter 4.

From aerodynamic theory we know that the motion appears in the aerodynamic force relations through the 'downwash',

$$w_a \equiv \frac{\partial z_a}{\partial t} + U_\infty \frac{\partial z_a}{\partial x} \tag{3.4.1}$$

where z_a is vertical displacement of airfoil at point x, y at time t. We shall not give a formal derivation of (3.4.1) here but shall indicate the physical basis from which it follows. For an inviscid fluid the boundary condition at a fluid-solid interface, e.g., at the surface of an airfoil, requires that the fluid velocity component normal to the surface be equal to the normal velocity of the surface on the instantaneous position of the surface. (If we have a nearly planar solid surface undergoing small motions relative to its own dimensions we may apply the boundary condition on some average position of the body, say $z = 0$, rather than on the instantaneous position of the surface, $z = z_a$.) In a coordinate system fixed with respect to the fluid the boundary condition would read

$$w_a = \frac{\partial z_a}{\partial t}$$

where w_a is the normal fluid velocity component, the so-called 'downwash', and $\frac{\partial z_a}{\partial t}$ is the normal velocity of the body surface. In a coordinate system fixed with respect to the body there is an additional convection term as given in (3.4.1). This may be derived by a formal transformation from fixed fluid to fixed body axes.

Finally if in addition to the mean flow velocity, U_∞, we also have a vertical gust velocity, w_G, then the boundary condition is that the total normal fluid velocity at the body surface be equal to the normal body velocity, i.e.,

$$w_{\text{total}} \equiv w_a + w_G = \frac{\partial z_a}{\partial t} + U_\infty \frac{\partial z_a}{\partial x}$$

where w_a is the additional fluid downwash due to the presence of the airfoil beyond that given by the prescribed gust downwash w_G. The pressure loading on the airfoil is

$$p + p_G$$

where p is the pressure due to

$$w_a = -w_G(x, t) + \frac{\partial z_a}{\partial t} + U_\infty \frac{\partial z_a}{\partial x}$$

and p_G is the prescribed pressure corresponding to the given w_G. Note, however, that p_G is continuous through $z = 0$ and hence gives no net pressure loading on the airfoil. Thus, only the pressure p due to downwash w_a is of interest in most applications.

For the typical section

$$z_a = -h - \alpha x \tag{3.4.2}$$

and

$$w_a = -w_G - \dot{h} - \dot{\alpha}x - U_\infty \alpha$$

From the first and last terms we note that $\frac{w_G}{U_\infty}$ is in some sense equivalent to an angle of attack, although it is an angle of attack which varies with position along the airfoil, $w_G = w_G(x, t)$!

Using the concept of aerodynamic impulse functions, we may now relate lift and moment to h, α and w_G. For simplicity let us neglect w_G for the present.

The aerodynamic force and moment can be written

$$\begin{aligned} L(t) \sim & \int_{-\infty}^{\infty} I_{L\dot{h}}(t-\tau)[\dot{h}(\tau) + U_\infty \alpha(\tau)]\,\mathrm{d}\tau \\ & + \int_{-\infty}^{\infty} I_{L\dot{\alpha}}(t-\tau)\dot{\alpha}(\tau)\,\mathrm{d}\tau \end{aligned} \tag{3.4.3}$$

(3.4.3) is the aerodynamic analog to (3.3.13). Note that $\dot{h} + U_\infty \alpha$ always appear in the same combination in w_a from (3.4.2). It is conventional to express (3.4.3) in nondimensional form. Thus,

$$\begin{aligned} \frac{L}{qb} = & \int_{-\infty}^{\infty} I_{L\dot{h}}(s-\sigma)\left[\frac{\mathrm{d}\frac{h}{b}(\sigma)}{\mathrm{d}\sigma} + \alpha(\sigma)\right]\mathrm{d}\sigma \\ & + \int_{-\infty}^{\infty} I_{L\dot{\alpha}}(s-\sigma)\left[\frac{\mathrm{d}\alpha(\sigma)}{\mathrm{d}\sigma}\right]\mathrm{d}\sigma \end{aligned} \tag{3.4.4}$$

and

$$\frac{M_y}{qb^2} = \int_{-\infty}^{\infty} I_{M\dot{h}}(s-\sigma)\left[\frac{\mathrm{d}\frac{h}{b}(\sigma)}{\mathrm{d}\sigma} + \alpha(\sigma)\right]\mathrm{d}\sigma$$

$$+\int_{-\infty}^{\infty} I_{M\dot{\alpha}}(s-\sigma)\left[\frac{\mathrm{d}\alpha(\sigma)}{\mathrm{d}\sigma}\right]\mathrm{d}\sigma$$

where

$$s \equiv \frac{tU_\infty}{b}, \qquad \sigma \equiv \frac{\tau U_\infty}{b}$$

For the typical section, the 'aerodynamic impulse functions', $I_{L\dot{h}}$, etc., depend also upon Mach number. More generally, for a wing they vary with wing geometry as well.

(3.4.4) may be used to develop relations for sinusoidal motion by reversing the mathematical process which led to (3.3.13). Taking the Fourier Transform of (3.4.4),

$$\frac{\bar{L}(k)}{qb} \equiv \int_{-\infty}^{\infty} \frac{L(s)}{qb}\mathrm{e}^{-iks}\,\mathrm{d}s = \int_{-\infty}^{\infty}\int_{-\infty}^{\infty} I_{L\dot{h}}(s-\sigma)\left[\frac{\mathrm{d}\frac{h}{b}}{\mathrm{d}\sigma}+\alpha\right]\mathrm{e}^{-iks}\,\mathrm{d}\sigma\,\mathrm{d}s+\cdots \tag{3.4.5}$$

where the *reduced frequency* is given by

$$k \equiv \frac{\omega b}{U_\infty}$$

Defining

$$\gamma \equiv s-\sigma, \qquad \mathrm{d}\gamma = \mathrm{d}s$$

$$\frac{\bar{L}(k)}{qb} = \int_{-\infty}^{\infty}\int_{-\infty}^{\infty} I_{L\dot{h}}(\gamma)\left[\frac{\mathrm{d}\frac{h}{b}}{\mathrm{d}\sigma}+\alpha\right]\mathrm{e}^{-ik\gamma}\,\mathrm{e}^{-ik\sigma}\,\mathrm{d}\sigma\,\mathrm{d}\gamma+\cdots$$

$$= H_{L\dot{h}}(k)\left[ik\frac{\bar{h}}{b}+\bar{\alpha}\right]+\cdots \tag{3.4.6}$$

where

$$H_{L\dot{h}}(k) \equiv \int_{-\infty}^{\infty} I_{L\dot{h}}(\gamma)\mathrm{e}^{-ik\gamma}\,\mathrm{d}\gamma$$

$$\frac{\bar{h}}{b} \equiv \int_{-\infty}^{\infty} \frac{h(\sigma)}{b}\mathrm{e}^{-ik\sigma}\,\mathrm{d}\sigma$$

$$\bar{\alpha} \equiv \int_{-\infty}^{\infty} \alpha(\sigma)\mathrm{e}^{-ik\sigma}\,\mathrm{d}\sigma$$

$H_{L\dot{h}}$, etc., are 'aerodynamic transfer functions'. From (3.4.4), (3.4.6) we may write

$$\frac{\bar{L}}{qb} = H_{L\dot{h}}\left[ik\frac{\bar{h}}{b} + \bar{\alpha}\right] + H_{L\dot{\alpha}}ik\bar{\alpha}$$
$$\frac{\bar{M}_y}{qb^2} = H_{M\dot{h}}\left[ik\frac{\bar{h}}{b} + \bar{\alpha}\right] + H_{M\dot{\alpha}}ik\bar{\alpha} \tag{3.4.7}$$

Remember that 'transfer functions', aerodynamic or otherwise, may be determined from a consideration of sinusoidal motion only. Also note that (3.4.2), (3.4.3) and (3.4.7) are written for pitching about an axis $x = 0$. That is, the origin of the coordinate system is taken at the pitch axis. By convention, in aerodynamic analyses the axis is usually taken at mid-chord. Hence

$$\begin{aligned} z_a &= -h - \alpha(x - x_{\mathrm{e.a.}}) \\ w &= -\dot{h} - \dot{\alpha}(x - x_{\mathrm{e.a.}}) - U_\infty \alpha \\ &= (-\dot{h} - U_\infty \alpha) - \dot{\alpha}(x - x_{\mathrm{e.a.}}) \\ &= (-\dot{h} - U_\infty \alpha + \dot{\alpha} x_{\mathrm{e.a.}}) - \dot{\alpha} x \end{aligned} \tag{3.4.2}$$

where

$$x_{\mathrm{e.a.}} = \text{distance from mid-chord to e.a.}$$

(3.4.4) and (3.4.7) should be modified accordingly, i.e.,

$$\frac{d\frac{h}{b}}{d\sigma} + \alpha$$

is replaced by

$$\frac{d\frac{h}{b}}{d\sigma} + \alpha - \dot{\alpha} a \quad \text{where} \quad a \equiv \frac{x_{\mathrm{e.a.}}}{b}$$

In the following table we summarize general state-of-the-art as far as available aerodynamic theories in terms of Mach number range and geometry. All assume inviscid, linearized flow models. The transonic range, $M \approx 1$, is a currently active area of research.

Aerodynamic theories available

Mach number	Geometry	
	Two dimensional	Three dimensional
$M \ll 1$	Available	Rather elaborate numerical methods available for determing transfer functions.
$M \approx 1$	Available but of limited utility because of inherent three dimensionality of flow	Rather elaborate numerical methods available for determining (linear, inviscid) transfer functions; nonlinear and/or viscous effects may be important, however.
$M \gg 1$	Available and simple because of weak memory effect.	Available and simple because of weak three dimensional effects.

The results for high speed ($M \gg 1$) flow are particularly simple. In the limit of large Mach number the (perturbation) pressure loading on an airfoil is given by

$$p = \rho \frac{U_\infty^2}{M} \left[\frac{\dfrac{\partial z_a}{\partial t} + U_\infty \dfrac{\partial z_a}{\partial x}}{U_\infty} \right]$$

or

$$p = \rho a_\infty \left[\frac{\partial z_a}{\partial t} + U_\infty \frac{\partial z_a}{\partial x} \right]$$

This is a local, zero memory relation in that the pressure at position x, y at time t depends only on the motion at the same position and time and does *not* depend upon the motion at other positions (local effect) or at previous times (zero memory effect). This is sometimes referred to as aerodynamic 'piston theory'* since the pressure is that on a piston in a tube with velocity

$$w_a = \frac{\partial z_a}{\partial t} + U_\infty \frac{\partial z_a}{\partial x}$$

This pressure-velocity relation has been widely used in recent years in aeroelasticity and is also well known in one-dimensional plane wave acoustic theory. Impulse and transfer functions are readily derivable using aerodynamic 'piston theory'.

The 'aerodynamic impulse functions' and 'aerodynamic transfer func-

* Ashley, and Zartarian [9]. Also see Chapter 4.

tions' for two-dimensional, incompressible flow, although not as simple as those for $M \gg 1$, are well-known.* The forms normally employed are somewhat different from the notation of (3.4.4) and (3.4.7). For example, the lift due to transient motion is normally written

$$\frac{L}{qb} = 2\pi\left[\frac{d^2\frac{h}{b}}{ds^2} + \frac{d\alpha}{ds} - a\frac{d^2\alpha}{ds^2}\right]$$

$$+4\pi\left\{\phi(0)\left[\frac{d\frac{h}{b}}{ds} + \alpha + \left(\frac{1}{2} - a\right)\frac{d\alpha}{ds}\right]\right.$$

$$\left. + \int_0^s \left(\frac{d\frac{h}{b}}{d\sigma} + \alpha + \left(\frac{1}{2} - a\right)\frac{d\alpha}{d\sigma}\right)\dot{\phi}(s-\sigma)\, d\sigma\right\} \tag{3.4.8}$$

One can put (3.4.8) into the form of (3.4.4) where

$$\begin{aligned} I_{L\dot{h}} &= 2\pi D + 4\pi\dot{\phi} + 4\pi\phi(0)\,\delta \\ I_{L\dot{\alpha}} &= 4\pi\left(\frac{1}{2} - a\right)\dot{\phi} + 4\pi\left(\frac{1}{2} - a\right)\phi(0)\,\delta - 2\pi a D \end{aligned} \tag{3.4.9}$$

Here δ is the delta function and D the doublet function, the latter being the derivative of a delta function. In practice, one would use (3.4.8) rather than (3.4.4) since delta and doublet functions are not suitable for numerical integration, etc. However, (3.4.8) and (3.4.4) are formally equivalent using (3.4.9) Note that (3.4.8) is more amenable to physical interpretation also. The terms outside the integral involving $\ddot{h}$ and $\ddot{\alpha}$ may be identified as inertial terms, sometimes called 'virtual mass' terms. These are usually negligible compared to the inertial terms of the airfoil itself if the fluid is air.† The quantity

$$-\left[\frac{d\frac{h}{b}}{ds} + \alpha + \left(\frac{1}{2} - a\right)\frac{d\alpha}{ds}\right]$$

may be identified as the downwash at the $\frac{3}{4}$ chord, Hence, the $\frac{3}{4}$ chord has been given a special place for two-dimensional, incompressible flow.

---

* See Chapter 4.

† For light bodies or heavy fluids, e.g., lighter-than-airships or submarines, they may be important.

Finally, note that the 'aerodynamic impulse functions', I_{Lh}, I_{Li}, can be expressed entirely in terms of a single function ϕ, the so-called Wagner function.* This function is given below in Figure 3.7. A useful approximate formulae is

$$\phi(s) = 1 - 0.165e^{-0.0455s} - 0.335e^{-0.3s} \tag{3.4.10}$$

For Mach numbers greater than zero, the compressibility of the flow smooths out the delta and doublet functions of (3.4.9) and no such simple

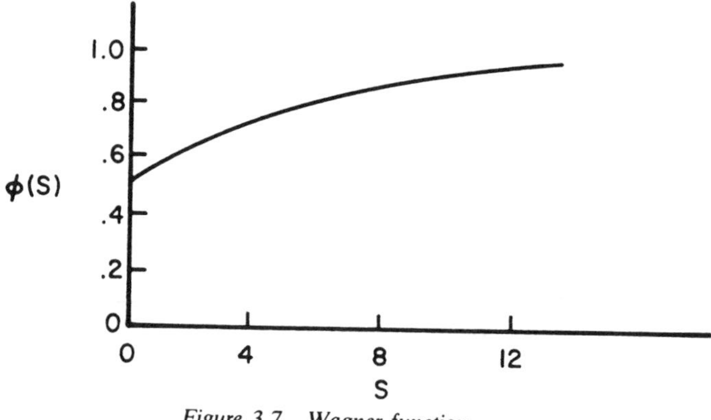

Figure 3.7 Wagner function.

form as (3.4.8) exists. Hence, only for incompressible flow is the form, (3.4.8), particularly useful.

Finally, we should mention that analogous impulse functions exist for gust loading due to gust vertical velocity, w_G.

$$\frac{L_G}{qb} = \oint_{-\infty}^{\infty} I_{LG}(s-\sigma)\frac{w_G(\sigma)}{U}\,d\sigma$$
$$\frac{M_{yG}}{qb^2} = \int_{-\infty}^{\infty} I_{MG}(s-\sigma)\frac{w_G(\sigma)}{U}\,d\sigma \tag{3.4.11}$$

For incompressible flow

$$I_{LG} = 4\pi\dot{\psi}$$
$$I_{MG} = I_{LG}(\tfrac{1}{2}+a)$$

where ψ, the Kussner function, is given by (see Figure 3.8)

$$\psi(s) = 1 - 0.5e^{-0.13s} - 0.5e^{-s} \tag{3.4.12}$$

* For a clear, concise discussion of *transient*, two-dimensional, incompressible aerodynamics, see Sears [10], and the discussion of Sears' work in BAH, pp. 288–293.

The Wagner and Kussner functions have been widely employed for transient aerodynamic loading of airfoils. Even for compressible, subsonic flow they are frequently used with empirical corrections for Mach number effects. Relatively simple, exact formulae exist for two-dimensional, supersonic flow also.* However, for subsonic and/or three-dimensional flow the aerodynamic impulse functions must be determined by fairly elaborate numerical means.

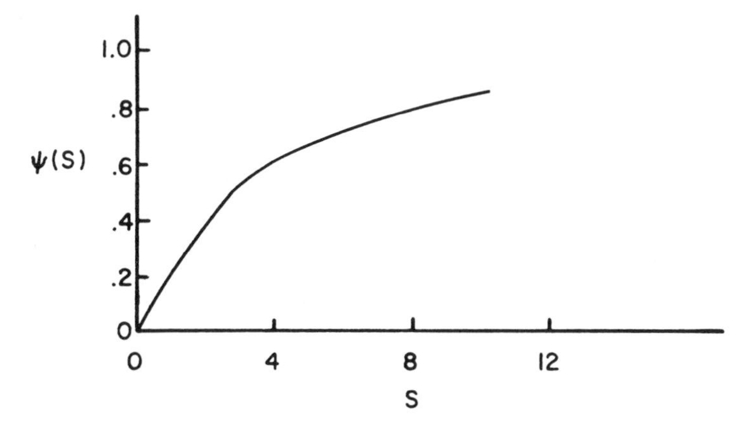

Figure 3.8 Kussner function.

Finally we note that (3.4.11) may be written in the frequency domain as

$$\frac{\bar{L}_G}{qb} = H_{LG}(\omega)\frac{\bar{w}_G}{U}$$
$$\frac{\bar{M}_y}{qb^2} = H_{MG}(\omega)\frac{\bar{w}_G}{U} \tag{3.4.13}$$

(3.4.7) and (3.4.13) will be useful when we treat the gust problem as a random process and make use of power spectral techniques. For further discussion of gust aerodynamics, see Sections 4.2 and 4.3.

General approximations

Frequently simplifying assumptions are made with respect to the spatial or temporal dependence of the aerodynamic forces. Here we discuss three widely used approximations.

---

* See BAH, pp. 367–375, for a traditional approach and Chapter 4 for an approach via Laplace and Fourier Transforms.

'Strip theory' approximation. In this approximation, one employs the known results for two-dimensional flow (infinite span airfoil) to calculate the aerodynamic forces on a lifting surface of finite span. The essence of the approximation is to consider each spanwise station as though it were a portion of an infinite span wing with uniform spanwise properties. Therefore the lift (or, more generally, chordwise pressure distribution) at any spanwise station is assumed to depend only on the downwash at that station as given by two-dimensional aerodynamic theory and to be independent of the downwash at any other spanwise station.

'Quasisteady' approximation:

The strip theory approximation discussed above is unambiguous and its meaning is generally accepted. Unfortunately, this is not true for the quasi-steady approximation. Its qualitative meaning is generally accepted, i.e., one ignores the temporal memory effect in the aerodynamic model and assumes the aerodynamic forces at any time depend only on the motion of the airfoil at that same time and are independent of the motion at earlier times. That is, the history of the motion is neglected as far as determining aerodynamic forces. For example, the piston theory aerodynamic approximation is inherently a quasi-steady approximation.

As an example of the ambiguity that can develop in constructing a quasi-steady approximation, consider the aerodynamic forces for two-dimensional, incompressible flow, e.g., see (3.4.8). One such approximation which is sometimes used is to approximate the Wagner function by

$$\phi = 1$$

and hence

$$\phi(0) = 1, \qquad \dot{\phi} = 0$$

This is clearly a quasi-steady model since the convolution integral in (3.4.8) may now be evaluated in terms of the airfoil motion at the present time, $s \equiv \frac{tU_\infty}{b}$, and thus the aerodynamic forces are independent of the history of the airfoil motion. An alternate quasi-steady approximation which is used on occasion is to first obtain the aerodynamic forces for steady motion, e.g., only those terms which involve α in (3.4.8) and then to define an equivalent unsteady angle of attack

$$\alpha + \frac{\mathrm{d}h}{\mathrm{d}t}\frac{1}{U_\infty}$$

to replace α everywhere in the steady aerodynamic theory. Clearly this second quasi-steady approximation is different from the first. (An interesting and relatively short exercise for the reader is to work out and compare these two approximations in detail using (3.4.8).) However, both are used in practice and the reader should be careful to determine exactly what a given author means by 'quasi-steady approximation'.

The ambiguity could be removed if there were general agreement that what is meant by the quasi-steady approximation is an expansion in reduced frequency for sinusoidal airfoil motion. However, even then, there would have to be agreement as to the number of terms to be retained in the expansion. (Recall that powers of frequency formally correspond to time derivatives.)

Slender body or slender (low aspect ratio) wing approximation. Another approximation based upon spatial considerations is possible when the lifting surface is of low aspect ratio or one is dealing with a slender body. In such cases the chordwise spatial rates of change (derivatives) may be neglected compared to spanwise rates of change and hence the chordwise coordinate effectively becomes a parameter rather than an independent coordinate. This approach is generally attributed to R. T. Jones.* It is useful as an asymptotic check on numerical methods for slender bodies and low aspect ratio wings. However it is useful for quantitative predictions for only a modest range of practical lifting surfaces.

A particularly interesting result is available for the external flow about a slender body when the body has rigid cross-sections and deforms only in the flow direction, i.e.,

$$z_a(x, y, t) = z_a(x, t)$$

The lift force per unit chordwise distance is given by†

$$L = -\rho_\infty \frac{\mathrm{d}S}{\mathrm{d}x} U\left[U\frac{\partial z_a}{\partial x} + \frac{\partial z_a}{\partial t}\right] - \rho_\infty S\left[U^2\frac{\partial^2 z_a}{\partial x^2} + 2U\frac{\partial^2 z_a}{\partial x\,\partial t} + \frac{\partial^2 z_a}{\partial t^2}\right] \tag{3.4.14}$$

For a cylinder of constant, circular cross-section

$$S = \pi R^2, \qquad \frac{\mathrm{d}S}{\mathrm{d}x} = 0$$

and (3.4.14) becomes

* Jones [11].

† BAH, p. 418.

$$L = -\rho_\infty S\left[U^2\frac{\partial^2 z_a}{\partial x^2} + 2U\frac{\partial^2 z_a}{\partial x\, \partial t} + \frac{\partial^2 z_a}{\partial t^2}\right] \tag{3.4.15}$$

It is interesting to note that (3.4.15) is the form of the lift force used by Paidoussis and others for *internal* flows. Recall Section 2.5, equation (2.5.2). Dowell and Widnall, among others, have shown under what circumstances (3.4.15) is a rational approximation for external and internal flows.*

3.5 Solutions to the aeroelastic equations of motion

With the development of the aerodynamic relations, we may now turn to the question of solving the aeroelastic equations of motion. Substituting (3.4.4) into (3.3.1) and (3.3.2), these equations become:

$$m\ddot{h} + S_\alpha\ddot{\alpha} + K_h h = -L = \left\{ -\int_0^s I_{L\ddot{h}}(s-\sigma)\left[\frac{\mathrm{d}\frac{h}{b}}{\mathrm{d}\sigma} + \alpha\right]\mathrm{d}\sigma - \int_0^s I_{L\dot{\alpha}}(s-\sigma)\frac{\mathrm{d}\alpha}{\mathrm{d}\sigma}\,\mathrm{d}\sigma - \int_0^s I_{LG}(s-\sigma)\frac{w_G}{U}\,\mathrm{d}\sigma\right\} qb$$

and

$$I_\alpha\ddot{\alpha} + S_\alpha\ddot{h} + K_\alpha\alpha = M_y = \left\{\int_0^s I_{M\ddot{h}}(s-\sigma)\left[\frac{\mathrm{d}\frac{h}{b}}{\mathrm{d}\sigma} + \alpha\right]\mathrm{d}\sigma + \int_0^s I_{M\dot{\alpha}}(s-\sigma)\frac{\mathrm{d}\alpha}{\mathrm{d}\sigma}\,\mathrm{d}\sigma + \int_0^s I_{MG}(s-\sigma)\frac{w_G}{U}\,\mathrm{d}\sigma\right\} qb^2 \tag{3.5.1}$$

$$s \equiv \frac{tU_\infty}{b}$$

* Dowell and Widnall [12], Widnall and Dowell [13], Dowell [14].

$I_{L\dot{h}}$, etc., nondimensional impulse functions. (3.5.1) are linear, differential-integral equations for h and α. They may be solved in several ways, all of which involve a moderate amount of numerical work. Basically, we may distinguish between those methods which treat the problem in the time domain and those which work in the frequency domain. The possibilities are numerous and we shall discuss representative examples of solution techniques rather than attempt to be exhaustive.

Time domain solutions

In this day and (computer) age, perhaps the most straightforward way of solving (3.5.1) (and similar equations which arise for more complicated aeroelastic systems) is by numerical integration using finite differences. Such integration is normally done on a digital computer. A simplified version of the procedure follows:

Basically, we seek a step by step solution for the time history of the motion. In particular, given the motion at some time, t, we wish to be able to obtain the motion at some later time, $t+\Delta t$. In general Δt must be sufficiently small; just how small we will discuss in a moment. In relating the solution at time, $t+\Delta t$, to that at time, t, we use the idea of a Taylor series, i.e.,

$$\begin{aligned} h(t+\Delta t) &= h(t) + \frac{\mathrm{d}h(t)}{\mathrm{d}t}\Delta t + \frac{1}{2}\frac{\mathrm{d}^2 h(t)}{\mathrm{d}t^2}(\Delta t)^2 + \cdots \\ \alpha(t+\Delta t) &= \alpha(t) + \frac{\mathrm{d}\alpha(t)}{\mathrm{d}t}\Delta t + \frac{1}{2}\frac{\mathrm{d}^2\alpha(t)}{\mathrm{d}t^2}(\Delta t) + \cdots \end{aligned} \tag{3.5.2}$$

If we think of starting the solution at the initial instant, $t=0$, we see that normally $h(0)$, $\mathrm{d}h(0)/\mathrm{d}t$, $\alpha(0)$, $\mathrm{d}\alpha(0)/\mathrm{d}t$, are given as initial conditions since we are dealing with (two) second order equations for h and α. However, in general, $\mathrm{d}^2h(0)/\mathrm{d}t^2$, $\mathrm{d}^2\alpha(0)/\mathrm{d}t^2$ and all higher order derivatives are *not* specified. They can be determined though from equations of motion themselves, (3.5.1). (3.5.1) are two *algebraic* equations for $\mathrm{d}^2h/\mathrm{d}t^2$, $\mathrm{d}^2\alpha/\mathrm{d}t^2$, in terms of lower order derivatives. Hence, they may readily be solved for $\mathrm{d}^2h/\mathrm{d}t^2$, $\mathrm{d}^2\alpha/\mathrm{d}t^2$. Moreover, by differentiating (3.5.1) successively the higher order derivatives may also be determined, e.g., $\mathrm{d}^3h/\mathrm{d}t^3$, etc. Hence, by using the equations of motion themselves the terms in the Taylor Series may be evaluated, (3.5.2), and h at $t=\Delta t$ determined. Repeating this procedure, the time history may be determined at $t=2\,\Delta t$, $3\,\Delta t$, $4\,\Delta t$, etc.

The above is the essence of the procedure. However, there are many variations on this basic theme and there are almost as many numerical integration schemes as there are people using them.* This is perhaps for two reasons: (1) an efficient scheme is desired (this involves essentially a trade-off between the size of Δt and the number of terms retained in the series, (3.5.2) or more generally a trade-off between Δt and the complexity of the algorithm relating $h(t+\Delta t)$ to $h(t)$); (2) some schemes including the one outlined above, are numerically unstable (i.e., numerical errors grow exponentially) if Δt is too large. This has led to a stability theory for difference schemes to determine the critical Δt and also the development of difference schemes which are inherently stable for all Δt. Generally speaking, a simple difference scheme such as the one described here will be stable if Δt is small compared to the shortest natural period of the system, say one-tenth or so. A popular method which is inherently stable for all Δt is due to Houbolt.†

An alternative but somewhat similar method to stepwise numerical integration is based upon the use of an analog computer. In such a device one again solves for the highest derivatives which are integrated by electrical devices and fed back to form the differential equation. The difficulty with such a device in the present context is having electrical components to perform the aerodynamic integrations of (3.5.1) using electronic function generators to obtain $I_{L\dot{h}}$, etc. There are also hybrid computers, combination analog-digital, which have their devotees for problems of this type. Generally, the digital computer is simplest to program and most flexible. However, if a large number of computations are contemplated then there may be economic advantages to using analog or hybrid computers.‡

Finally, analytical solutions or semi-analytical solutions may be obtained under certain special circumstances given sufficient simplification of the system dynamics and aerodynamics. These are usually obtained via a Laplace Transform. Since the Laplace Transform is a special case of the Fourier Transform, we defer a discussion of this topic to the following section on frequency domain solutions.

Frequency domain solutions

An alternative procedure to the time domain approach is to treat the problem in the frequency domain. This approach is more popular and widely used today than the time domain approach. Perhaps the most important reason for this is the fact that the aerodynamic theory is much

* Hamming [15].
† Houbolt [16].
‡ Hausner [17].

more completely developed for simple harmonic motion than for arbitrary time dependent motion. That is, the unsteady aerodynamicist normally provides $H_{L\dot{h}}$, for example, rather than $I_{L\dot{h}}$. Of course, these two quantities form a Fourier Transform pair,

$$\begin{aligned} H_{L\dot{h}}(k) &= \int_{-\infty}^{\infty} I_{L\dot{h}}(s) \mathrm{e}^{-iks} \, \mathrm{d}s \\ I_{L\dot{h}}(s) &= \frac{1}{2\pi} \int_{-\infty}^{\infty} H_{L\dot{h}}(k) \mathrm{e}^{iks} \, \mathrm{d}k \end{aligned} \tag{3.5.3}$$

where

$$k \equiv \frac{\omega b}{U}, \qquad s \equiv \frac{tU}{b}$$

and, in principle, given $H_{L\dot{h}}$ one can compute $I_{L\dot{h}}(s)$. However, for the more complex (and more accurate) aerodynamic theories $H_{L\dot{h}}$ is a highly oscillatory function which is frequently only known numerically at a relatively small number of frequencies, k. Hence, although there have been attempts to obtain $I_{L\dot{h}}$ by a numerical integration of $H_{L\dot{h}}$ over all frequency, they have not been conspicuously successful. Fortunately, for a determination of the stability characteristics of a system, e.g., flutter speed, one need only consider the frequency characteristics of the system dynamics, per se, and may avoid such integrations.

Another reason for the popularity of the frequency domain method is the powerful power spectral description of random loads such as gust loads, landing loads (over randomly rough surfaces), etc. These require a frequency domain description. Recall (3.3.25) and (3.3.40).

The principal disadvantage of the frequency domain approach is that one performs two separate calculations; one, to assess the system stability, 'flutter', and a second, to determine the response to external loads such as gusts, etc. This will become clearer as we discuss the details of the procedures.

Let us now turn to the equations of motion, (3.5.1), and convert them to the frequency domain by taking the Fourier Transform of these equations. The result is

$$\begin{aligned} & -\omega^2 m\bar{h} - \omega^2 S_\alpha \bar{\alpha} + K_n \bar{h} = -\bar{L} \\ & = \left\{ -H_{L\dot{h}}(k) \left[\frac{i\omega\bar{h}}{U} + \bar{\alpha} \right] - H_{L\dot{\alpha}}(k) \frac{i\omega b}{U} \bar{\alpha} \right. \\ & \quad \left. - H_{LG}(k) \frac{\bar{w}_G}{U} \right\} qb \end{aligned}$$

$$-\omega^2 I_\alpha \bar{\alpha} - \omega^2 S_\alpha \bar{h} + K_\alpha \bar{\alpha} = \bar{M}_y \tag{3.5.4}$$

$$= \left\{ H_{M\dot{h}}(k) \left[\frac{i\omega \bar{h}}{U} + \bar{\alpha} \right] + H_{M\dot{\alpha}}(k) \frac{i\omega b}{U} \bar{\alpha} + H_{MG} \frac{\bar{w}_G}{U} \right\} q b^2$$

where

$$\bar{h} \equiv \int_{-\infty}^{\infty} h(t) \mathrm{e}^{-i\omega t} \, \mathrm{d}t, \text{ etc.}$$

Collecting terms and using matrix notation,

$$\begin{bmatrix} \left[-\omega^2 m + K_h + \left(H_{L\dot{h}} \frac{i\omega}{U} \right) q b \right] & \left[-\omega^2 S_\alpha + \left(H_{L\dot{h}} + H_{L\dot{\alpha}} \frac{i\omega b}{U} \right) q b \right] \\ \left[-\omega^2 S_\alpha - \left(H_{M\dot{h}} \frac{i\omega}{U} \right) q b^2 \right] & \left[-\omega^2 I_\alpha + K_\alpha - \left(H_{M\dot{h}} + H_{M\dot{\alpha}} \frac{i\omega b}{U} \right) q b^2 \right] \end{bmatrix} \begin{Bmatrix} \bar{h} \\ \bar{\alpha} \end{Bmatrix}$$

$$= q b \frac{\bar{w}_G}{U} \begin{Bmatrix} -H_{LG} \\ H_{MG} b \end{Bmatrix} \tag{3.5.5}$$

Formally, we may solve for $\bar{h}$ and $\bar{\alpha}$ by matrix inversion. The result will be:

$$\frac{\frac{\bar{h}}{b}}{\frac{\bar{w}_G}{U}} \equiv H_{hG}$$

which is one of the aeroelastic transfer functions to a gust input and

$$\frac{\bar{\alpha}}{\frac{\bar{w}_G}{U}} \equiv H_{\alpha G} \tag{3.5.6}$$

It is left to the reader to evaluate these transfer functions explicitly from (3.5.5). Note these are aeroelastic transfer functions as opposed to the purely mechanical or structural transfer functions, H_{hF} and $H_{\alpha F}$, considered previously or the purely aerodynamic transfer functions, $H_{L\dot{h}}$, etc. That is, H_{hG} include not only the effects of structural inertia and stiffness, but also the aerodynamic forces due to structural motion.

With the aeroelastic transfer functions available one may now formally write the solutions in the frequency domain

$$\frac{h(t)}{b}=\frac{1}{2\pi}\int_{-\infty}^{\infty} H_{hG}(\omega)\mathscr{F}\left(\frac{w_G}{U}\right)\mathrm{e}^{-i\omega t}\,\mathrm{d}\omega \tag{3.5.7}$$

where the Fourier Transform of the gust velocity is written as

$$\mathscr{F}w_G\equiv\int_{-\infty}^{\infty} w_G(t)\mathrm{e}^{i\omega t}\,\mathrm{d}t \tag{3.5.8}$$

Compare (3.5.7) with (3.3.12).

Alternatively, one could write

$$\frac{h(t)}{b}=\int_{-\infty}^{\infty} I_{hG}(t-\tau)\frac{w_G(\tau)}{U}\,\mathrm{d}\tau \tag{3.5.9}$$

where

$$I_{hG}(t)\equiv\frac{1}{2\pi}\int_{-\infty}^{\infty} H_{hG}(\omega)\mathrm{e}^{i\omega t}\,\mathrm{d}\omega \tag{3.5.10}$$

Compare (3.5.9) and (3.5.10) with (3.3.13) and (3.3.14). As mentioned in our discussion of time domain solutions, the integrals over frequency may be difficult to evaluate because of the oscillatory nature of the aerodynamic forces.

Finally, for random gust velocities one may write

$$\Phi_{(h/b)(h/b)}=|H_{hG}(\omega)|^2\Phi_{(w_G/U)(w_G/U)} \tag{3.5.11}$$

where $\Phi_{(h/b)(h/b)}$, $\Phi_{(w_G/U)(w_G/U)}$, are the (auto) power spectra of $\dfrac{h}{b}$ and $\dfrac{w_G}{U}$, respectively. Thus

$$\left(\frac{\bar{h}}{b}\right)^2=\int_0^{\infty} |H_{hG}|^2\Phi_{(w_G/U)(w_G/U)}\,\mathrm{d}\omega \tag{3.5.12}$$

Compare (3.5.12) with (3.3.25). Since the transfer function is squared, the integral (3.5.12) may be somewhat easier to evaluate than (3.5.7) or (3.5.10). The gust velocity power spectra is generally a smoothly varying function. (3.5.12) is commonly used in applications.

To evaluate stability, 'flutter', of the system one need not evaluate any integrals over frequency. It suffices to consider the eigenvalues (or poles) of the transfer function. A pole of the transfer function, ω_p, will give rise to an aeroelastic impulse function of the form

$$I_{hG} \sim e^{i\omega_p t} = e^{i(\omega_p)_R t} e^{-(\omega_p)_I t}$$

see (3.5.10). Hence, the system will be stable if the imaginary part, $(\omega_p)_I$, of all poles is positive. If any one (or more) pole has a negative imaginary part, the system is unstable, i.e., it flutters. The frequency of oscillation is $(\omega_p)_R$, the real part of the pole. Note that the poles are also the eigenvalues of the determinant of coefficients of $\bar{h}$ and $\bar{\alpha}$ in (3.5.5).

Having developed the mathematical techniques for treating dynamic aeroelastic problems we will now turn to a discussion of results and some of the practical aspects of such calculations.

3.6 Representative results and computational considerations

We will confine ourselves to two important types of motion, 'flutter' and 'gust response'.

Time domain

If we give the typical section (or any aeroelastic system) an initial disturbance due to an impulsive force, the resultant motion may take one of two possible forms as shown in Figures 3.9 and 3.10. 'Flutter' is the more interesting of these two motions, since, if it is present, it will normally lead to catastrophic structural failure which will result in the loss of the flight vehicle. All flight vehicles are carefully analyzed for flutter and frequently the structure is stiffened to prevent flutter inside the flight envelope of the vehicle.

Even if flutter does not occur, however, other motions in response to continuous external forces may be of concern with respect to possible structural failure. An important example is the gust response of the flight vehicle. Consider a vertical gust velocity time history as shown in Figure 3.11. The resulting flight vehicle motion will have the form shown in Figure 3.12. Note that the time history of the response has a certain well defined average period or frequency with modulated, randomly varying amplitude. The more random input has been 'filtered' by the aeroelastic transfer function and only that portion of the gust velocity signal which has frequencies near the natural frequencies of the flight vehicle will be

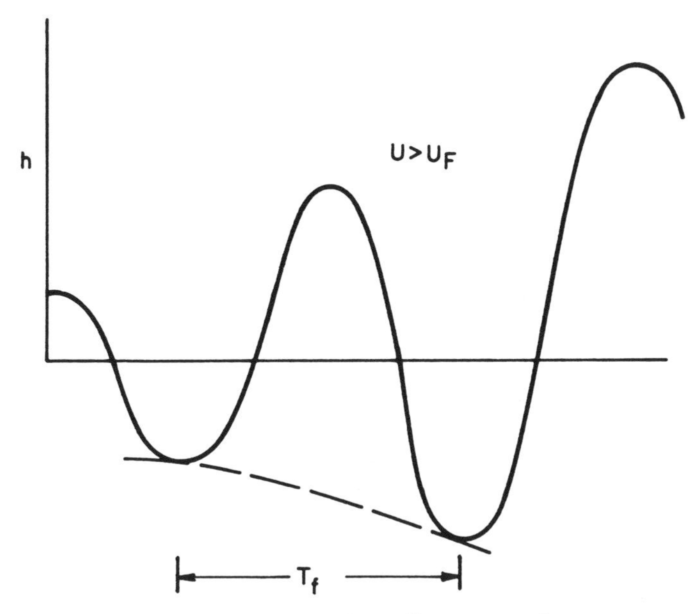

Figure 3.9 Time history of unstable motion or "flutter".

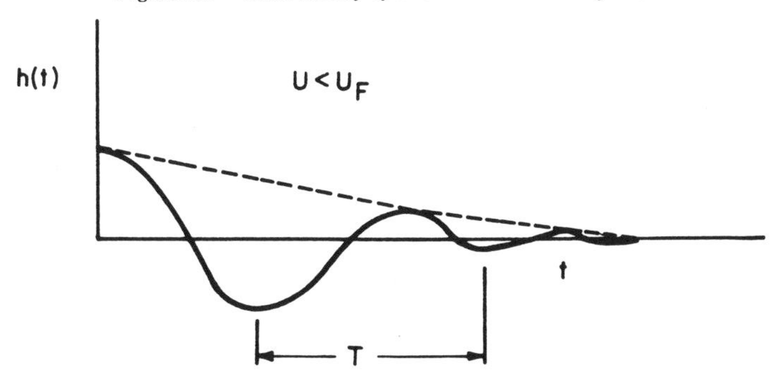

Figure 3.10 Time history of stable motion.

identifiable in the response. This characteristic is perhaps more readily seen in the frequency domain than in the time domain.

Frequency domain

To examine flutter, we need only examine the poles of the transfer function. This is similar to a 'root locus' plot.* Typically, the real, ω_R, and imaginary, ω_I, parts of the complex frequency are plotted versus flight speed. For the typical section there will be two principal poles correspond-

* Savant [18].

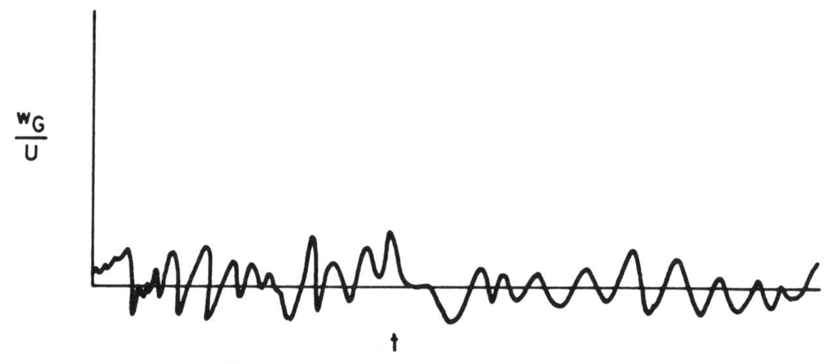

Figure 3.11 Time history of gust velocity.

ing to two degrees of freedom and, at small flight speed or fluid velocity, these will approach the natural frequencies of the mechanical or structural system. See Figure 3.13. Flutter is identified by the lowest airspeed for which one of the ω_I becomes negative. Note the coming together or 'merging' of the ω_R of the two poles which is typical of some types of flutter. There are many variations on the above plot in practice but we shall defer a more complete discussion for the moment.

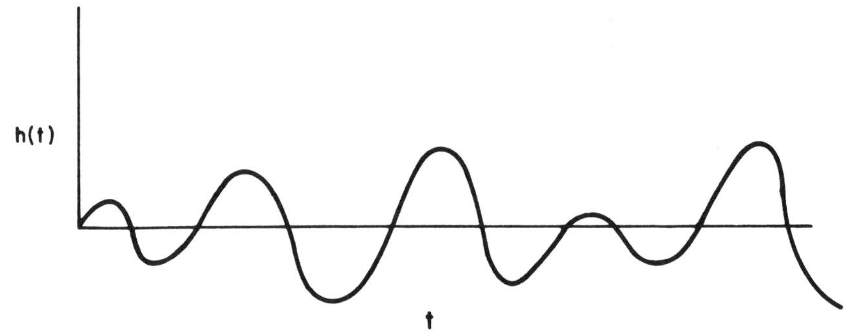

Figure 3.12 Time history of motion due to gust velocity.

Next, let us turn to the gust problem. A typical gust spectrum would be as in Figure 3.14. The transfer function (at some flight speed) would be as shown in Figure 3.15. Thus, the resultant response spectrum would appear as in Figure 3.16. As U approaches U_F, the resonant peaks of $|H_{hG}|^2$ and Φ_{hh} would approach each other for the system whose poles were sketched previously. For $U = U_F$ the two peaks would essentially collapse into one and the amplitude become infinite. For $U > U_F$ the

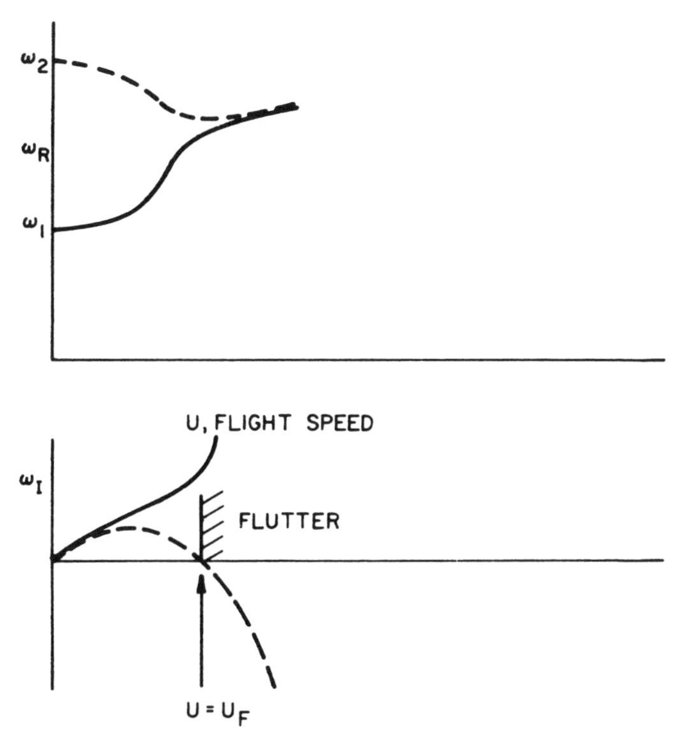

Figure 3.13 Real and imaginary components of frequency vs air speed.

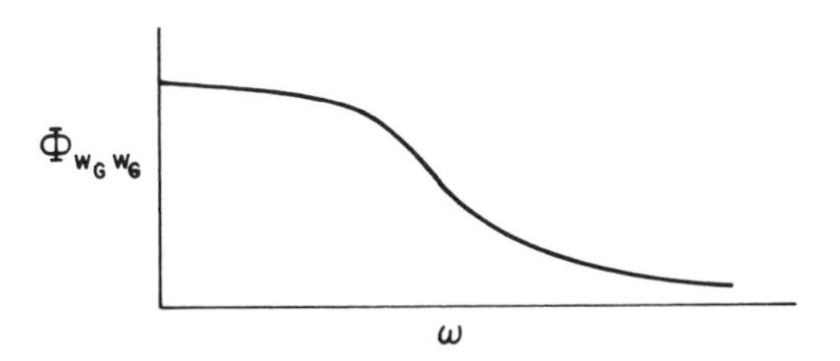

Figure 3.14 Gust power spectra.

amplitude predicted by the analytical model would become finite again for the power spectral approach and this physically unrealistic result is a possible disadvantage of the method.

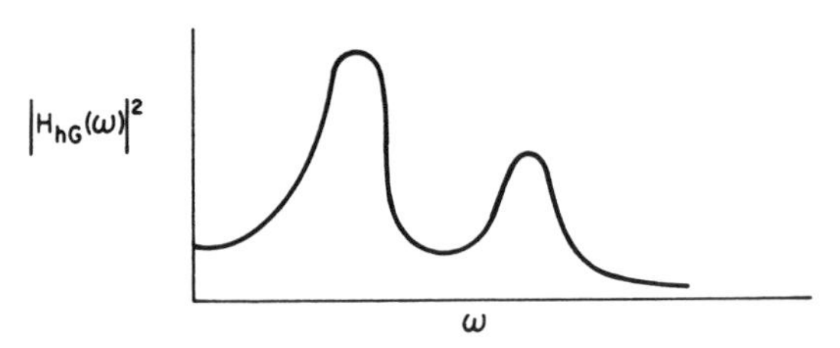

Figure 3.15 Transfer function.

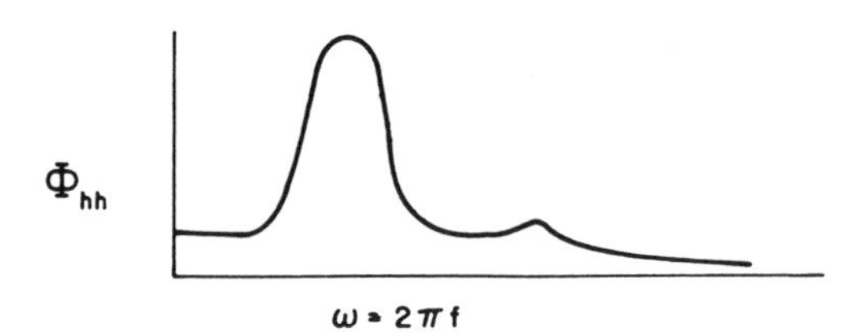

Figure 3.16 Power spectra of motion.

Flutter and gust response classification including parameter trends

Here we shall study some of the important parameters which affect flutter and gust response of the typical section as well as more complex flight vehicles.

Flutter. If one nondimensionalizes the typical section equations of motion, one finds that the motion can be expressed formally as

$$\frac{h}{b} = F_1\left(\omega_\alpha t; \frac{S_\alpha}{mb}, \frac{I_\alpha}{mb^2}, \frac{m}{\rho(2b)^2}, \frac{e}{b}, \frac{\omega_h}{\omega_\alpha}, M, \frac{U}{b\omega_\alpha}\right)$$
$$\alpha = F_2(\omega_\alpha t; \ldots) \tag{3.6.1}$$

where the functions F_1, F_2, symbolize the results of a calculated solution using one of the several methods discussed earlier.

The choice of nondimensional parameters is not unique but a matter of convenience. Some authors prefer a nondimensional dynamic pressure, or 'aeroelastic stiffness number'

$$\lambda \equiv \frac{1}{\mu k_\alpha^2} = \frac{4\rho U^2}{m\omega_\alpha^2},$$

to the use of a nondimensional velocity, $U/b\omega_\alpha$.

The following short-hand notation will be employed:

$\omega_\alpha t$ nondimensional time

$x_\alpha \equiv \dfrac{S_\alpha}{mb}$ static unbalance

$r_\alpha^2 \equiv \dfrac{I_\alpha}{mb^2}$ radius of gyration (squared)

$\mu \equiv \dfrac{m}{\rho(2b)^2}$ mass ratio

$a \equiv \dfrac{e}{b}$ location of elastic axis measured from aerodynamic center or mid-chord.

$\dfrac{\omega_h}{\omega_\alpha}$ frequency ratio

M Mach number

$k_\alpha = \dfrac{\omega_\alpha b}{U}$

Time is an independent variable which we do not control; however, in some sense we can control the parameters, x_α, r_α, etc., by the design of our airfoil and choice of where and how we fly it. For some combination of parameters the airfoil will be dynamically unstable, i.e., it will 'flutter'.

An alternative parametric representation would be to assume sinusoidal motion

$$h = \bar{h}e^{i\omega t}$$
$$\alpha = \bar{\alpha}e^{i\omega t}$$

and determine the eigenvalues, ω. Formally, recalling $\omega = \omega_R + i\omega_I$,

$$\frac{\omega_R}{\omega_\alpha} = G_R\left(x_\alpha, r_\alpha, \mu, a, \frac{\omega_h}{\omega_\alpha}, M, \frac{U}{b\omega_\alpha}\right)$$
$$\frac{\omega_I}{\omega_\alpha} = G_I\left(x_\alpha, r_\alpha, \mu, a, \frac{\omega_h}{\omega_\alpha}, M, \frac{U}{b\omega_\alpha}\right) \tag{3.6.2}$$

If for some combination of parameters, $\omega_I < 0$, the system flutters.

Several types of flutter are possible. Perhaps these are most easily distinguished on the basis of the eigenvalues, ω_R/ω_α, ω_I/ω_α, and their variation with airspeed, $U/b\omega_\alpha$. Figures are shown below of the several possibilities with brief discussions of each.

In this type of flutter (also called coupled mode or bending-torsion flutter) the distinguishing feature is the coming together of two (or more) frequencies, ω_R, near the flutter condition, $\omega_I \to 0$ and $U \to U_F$. For

Types of flutter

'Coalescense' or 'merging frequency' flutter

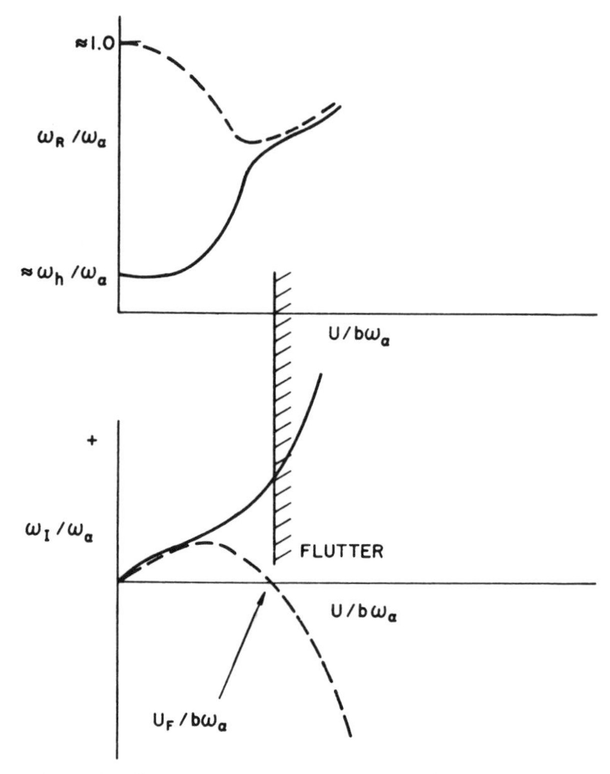

Figure 3.17 Real and imaginary components of frequency vs air speed.

$U > U_F$ one of ω_I becomes large and positive (stable pole) and the other which gives rise to flutter becomes large and negative (unstable pole) while the corresponding ω_R remain nearly the same. Although one usually speaks of the torsion mode as being unstable and the bending mode stable, the airfoil normally is undergoing a flutter oscillation composed of important contributions of both h and α. For this type of flutter the out-of-phase or damping forces of the structure or fluid are not qualitatively important. Often one may neglect structural damping entirely in the model and use a quasi-steady or even a quasi-static aerodynamic assumption. This simplifies the analysis and, perhaps more importantly, leads to generally accurate and reliable results based on theoretical calculations.

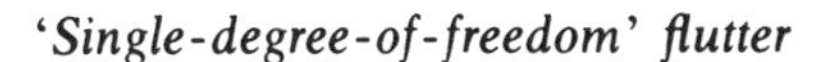

'Single-degree-of-freedom' flutter

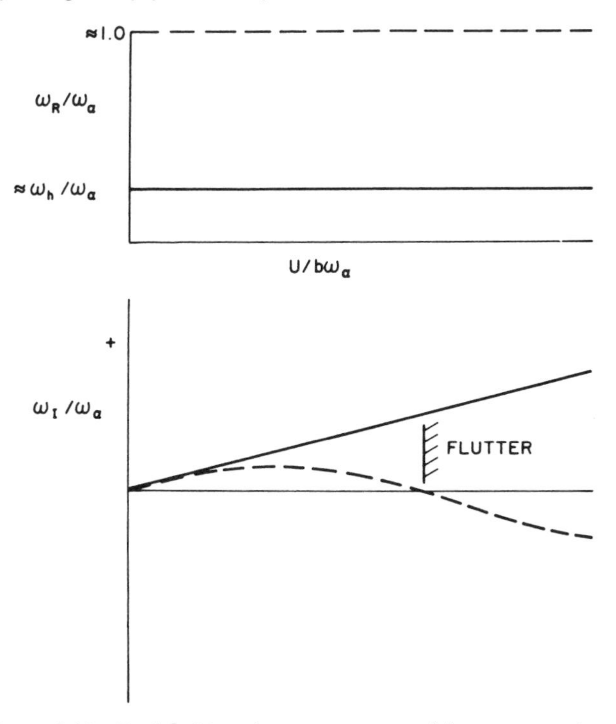

Figure 3.18 Real and imaginary components of frequency vs air speed.

In this type of flutter, the distinguishing feature is the virtual independence of the frequencies, ω_R, with respect to variations in airspeed, $U/b\omega_\alpha$. Moreover the change in the true damping, ω_I, with airspeed is also moderate. However, above some airspeed one of the modes (usually torsion) which has been slightly positively damped becomes slightly negatively damped leading to flutter. This type of flutter is very sensitive to structural and aerodynamic out-of-phase or damping forces. Since these forces are less well described by theory than the in-phase forces, the corresponding flutter analysis generally gives less reliable results. One simplification for this type of flutter is the fact that the flutter mode is virtually the same as one of the system natural modes at zero airspeed and thus the flutter mode and frequency (though not flutter speed!) are predicted rather accurately by theory. Airfoil blades in turbomachinery and bridges in a wind usually encounter this type of flutter.

This is also a one-degree-of-freedom type of flutter, but of a very special type. The flutter frequency is zero and hence this represents the

'Divergence' or 'zero frequency' flutter

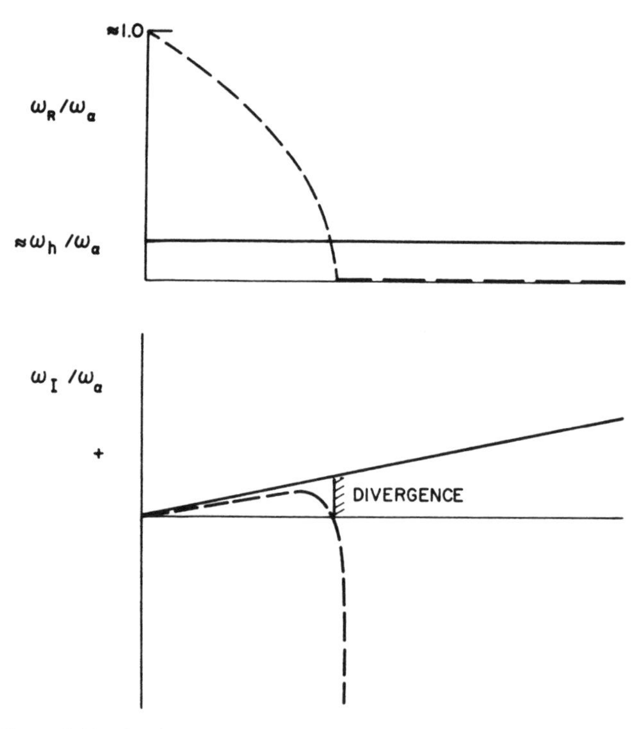

Figure 3.19 Real and imaginary components of frequency vs air speed.

static instability which we have previously analyzed in our discussion of static aeroelasticity under the name of 'divergence'. Because it is a static type of instability, out-of-phase forces are again unimportant and the theory is generally reliable.

We note that in all of the above we have considered only positive ω_R even though there are negative ω_R as well and these *are* physically meaningful. There are at least two reasons why this practice is usually followed. For those models where the aerodynamic transfer functions can be (approximately) expressed as a polynominal in $p \equiv i\omega$, the negative ω_R plane is (nearly) the mirror image of the positive ω_R plane and the ω_I are identical, i.e., all poles are complex conjugates in p. Secondly, some of the structural damping models employed in flutter analysis are only valid for $\omega_R > 0$; hence, the $\omega_R < 0$ in such cases cannot be interpreted in a physically valid way. However, there are some types of traveling wave flutter in plates and shells for which a consideration of negative ω_R is essential. In such cases a change in sign of ω_R represents a change in direction of the traveling wave.

Flutter calculations in practice

At this point it should be emphasized that, in practice, one or another of several *indirect* methods is often used to compute the flutter velocity, e.g., the so called 'V–g method'. In this approach structural damping is introduced by multiplying the structural frequencies squared,

$$\omega_h^2, \omega_\alpha^2$$

by $1+ig$ where g is a structural coefficient and pure sinusoidal motion is assumed, i.e., $\omega = \omega_R$ with $\omega_I \equiv 0$. For a given U the g required to sustain pure sinusoidal motion for each aeroelastic mode is determined. The computational advantage of this approach is that the aerodynamic forces only need be determined for real frequencies. The disadvantage is the loss of physical insight. For example, if a system (with no structural damping) is stable at a given airspeed, U, all the values of g so determined will be negative, but these values of g cannot be interpreted directly in terms of ω_I. Moreover, for a given system with some prescribed damping, only at one airspeed $U = U_F$ (where $\omega = \omega_R$ and $\omega_I \equiv 0$) will the mathematical solution be physically meaningful. The limitations of the 'V–g method' are fully appreciated by experienced practitioners and it is a measure of the difficulty of determining the aerodynamic forces for other than pure sinusoidal motion, that this method remains very popular. Here we digress from our main discussion to consider this and related methods in some detail.

For sinusoidal motion

$$\begin{aligned} h &= \bar{h}e^{i\omega t} \\ \alpha &= \bar{\alpha}e^{i\omega t} \\ L &= \bar{L}e^{i\omega t} \\ M_y &= \bar{M}_y e^{i\omega t} \end{aligned}$$

The aerodynamic forces (due to motion *only*) can be expressed as

$$\begin{aligned} \bar{L} &= 2\rho_\infty b^2\omega^2(2b)\left\{[L_1 + iL_2]\frac{\bar{h}}{b} + [L_3 + iL_4]\bar{\alpha}\right\} \\ \bar{M}_y &= -2\rho_\infty b^3\omega^2(2b)\left\{[M_1 + iM_2]\frac{\bar{h}}{b} + [M_3 + iM_4]\bar{\alpha}\right\} \end{aligned} \tag{3.6.3}$$

This form of the aerodynamic forces is somewhat different from that previously used in this text and is only one of several alternative forms employed in the literature. Here L_1, L_2, L_3, L_4 are (nondimensional) real aerodynamic coefficients which are functions of reduced frequency and Mach number. L_1, L_2, L_3, L_4 are the forms in which the coefficients are generally tabulated for supersonic flow.*

Using the above aerodynamic forms for $\bar{L}$ and $\bar{M}_y$ in (74) and setting the determinant of coefficients of $\bar{h}$ and $\bar{\alpha}$ to zero to determine nontrivial solutions, one obtains

$$\Delta(\omega) \equiv \left\{\frac{m}{2\rho_\infty b(2b)}\left[1+\left(\frac{\omega_\alpha}{i\omega}\right)^2\left(\frac{\omega_h}{\omega_\alpha}\right)^2\right]-[L_1+iL_2]\right\}$$
$$\times\left\{\frac{m}{2\rho_\infty b(2b)}r_\alpha^2\left[1+\left(\frac{\omega_\alpha}{i\omega}\right)^2\right]-[M_3+iM_4]\right\}$$
$$-\left\{\frac{mx_\alpha}{2\rho_\infty b(2b)}-[L_3+iL_4]\right\}\left\{\frac{mx_\alpha}{2\rho_\infty b(2b)}-[M_1+iM_2]\right\}=0 \quad (3.6.4)$$

Because L_1, L_2, etc. are complicated, transcendental functions of k (and M) which are usually only known for real values of k (and hence real values of ω), often one does not attempt to determine from (3.6.4) the complex eigenvalue, $\omega = \omega_R + i\omega_I$. Instead one seeks to determine the conditions of neutral stability when ω is purely real. Several alternative procedures are possible; two are described below.

In the first the following parameters are chosen.

$$\frac{\omega_h}{\omega_\alpha},\ r_\alpha,\ x_\alpha,\ M \text{ and (a real value of) } k$$

(3.6.4) is then a complex equation whose real and imaginary parts may be used independently to determine the two (real) unknowns

$$\left(\frac{\omega}{\omega_\alpha}\right)^2 \quad\text{and}\quad \frac{m}{2\rho_\infty bS}$$

From the imaginary part of (3.6.4), which is a linear equation in these two unknowns, one may solve for $(\omega/\omega_\alpha)^2$ in terms of $m/2\rho_\infty bS$. Substituting this result into the real part of (3.6.4) one obtains a quadratic equation in $m/2\rho_\infty bS$ which may be solved in the usual manner. Of course, only real positive values of $m/2\rho_\infty bS$ are meaningful and if negative or complex values are obtained these are rejected. By choosing various values of the parameters one may determine under what physically meaningful conditions flutter (neutrally stable oscillations) may occur. This procedure is

* Garrick [19].

not easily extendable to more than two degrees of freedom and it is more readily applied for determining parameter trends than the flutter boundary of a specific structure. Hence, a different method which is described below is frequently used.

This method has the advantage of computational efficiency, though from a physical point of view it is somewhat artificial. Structural damping is introduced as an additional parameter by multiplying ω_α^2 and ω_h^2 by $1+ig$ where g is the structural damping coefficient. The following parameters are selected ω_h/ω_α, r_α, x_α, M, (a real value) of k, and $m/2\rho_\infty bS$. (3.6.4) is then identified as a complex polynomial in the complex unknown

$$\left(\frac{\omega_\alpha}{\omega}\right)^2(1+ig)$$

Efficient numerical algorithms have been devised for determining the roots of such polynomials. A complex root determines

$$\frac{\omega_\alpha}{\omega} \quad \text{and} \quad g$$

From ω_α/ω and the previously selected value of $k \equiv \omega b/U_\infty$ one may compute

$$\frac{\omega_\alpha b}{U_\infty} = \frac{\omega_\alpha}{\omega} k$$

One may then plot g vs $U_\infty/b\omega_\alpha$*. A typical result is shown in Figure 3.20

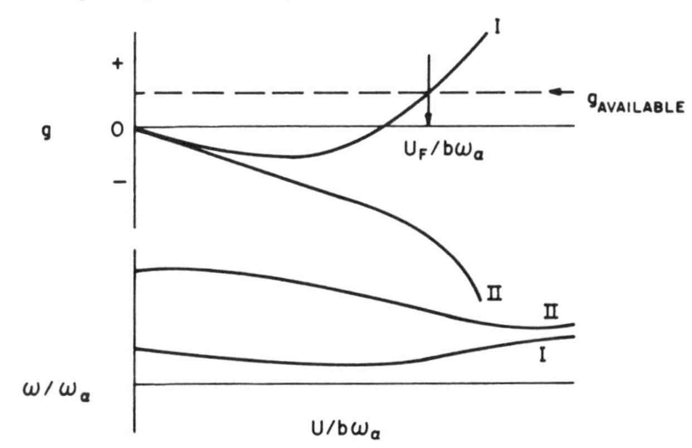

Figure 3.20. Structural damping and frequency required for neutrally stable motion vs air speed.

* (For each complex root of the polynomial.)

for two roots (two degrees of freedom). g is the value of structural damping ***required*** for neutral stability. If the ***actual*** structural damping is $g_{\text{AVAILABLE}}$ then flutter occurs when (see Figure 3.20)

$$g = g_{\text{AVAILABLE}}$$

It is normally assumed in this method that for $g < g_{\text{AVAILABLE}}$ and $U < U_F$, no flutter will occur. Sometimes more complicated velocity-damping or V–g curves are obtained, however. See Figure 3.21. Given the uncertainty as to what $g_{\text{AVAILABLE}}$ may be for a real physical system, it may then be prudent to define the flutter speed as the minimum value of $U_\infty/b\omega_\alpha$ for any $g > 0$. Here the physical interpretation of the result becomes more difficult, particularly when one recalls that the factor $1 + ig$ is only an approximate representation of damping in a structure. Despite this qualification, the V–g method remains a very popular approach to flutter analysis and is usually only abandoned or improved upon when the physical interpretation of the result becomes questionable.

One alternative to the V–g method is the so-called p–k method.* In this approach time dependence of the form h, $\alpha \sim e^{pt}$ is assumed where $p = \sigma + i\omega$. In the aerodynamic terms *only a* $k \equiv \omega b/U$ *is assumed.* The eigenvalues p are computed and the new ω used to compute a new k and the aerodynamic terms re-evaluated. The iteration continues until the process converges. For small σ, i.e., $|\sigma| \ll |\omega|$, the σ so computed may be interpreted as true damping of the system.

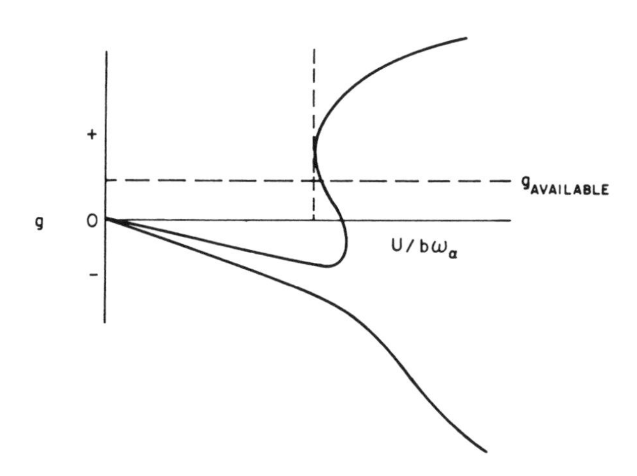

Figure 3.21 Structural damping required for flutter vs air speed.

* Hassig [20].

Nonlinear flutter behavior

There are two other types of flutter which are of importance, 'transonic buzz' and 'stall flutter'. Both of these involve significant aerodynamic nonlinearities and are, therefore, not describable by our previous models. Indeed, both are poorly understood theoretically and recourse to experiment and/or empirical rules-of-thumb is normally the only possibility. Recent advances in numerical solution of the nonlinear equations of fluid mechanics (computational fluid dynamics) have provided an improved methodology for modeling these types of flutter. See Chapter 9.

'Transonic buzz'

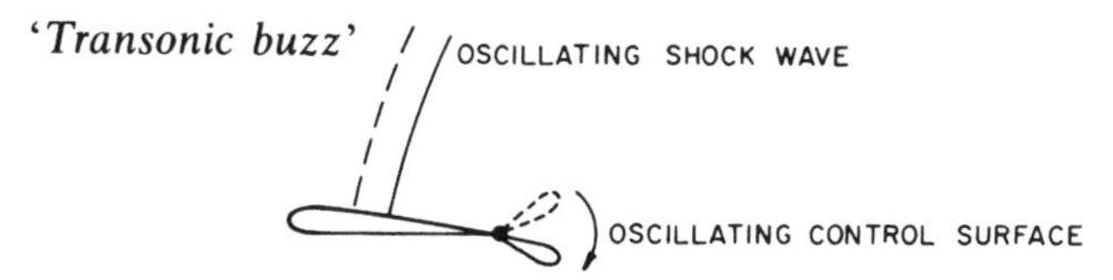

Figure 3.22 Schematic of transonic buzz geometry.

Typically an oscillating control surface gives rise to an oscillating shock which produces an oscillating pressure field which gives rise to an oscillating control surface which gives rise to an oscillating shock and so on and so forth.

The airfoil profile shape is known to be an important parameter and this fact plus the demonstrated importance of the shock means that any aerodynamic theory which hopes to successfully predict this type of flutter must accurately account for the nonuniform mean steady flow over the airfoil and its effect on the small dynamic motions which are superimposed due to control surface and shock oscillation. Perhaps the best theoretical model to date is that of Eckhaus; also see the discussion by Landahl. Lambourne has given a valuable summary of the experimental and theoretical evidence.*

'Stall' flutter

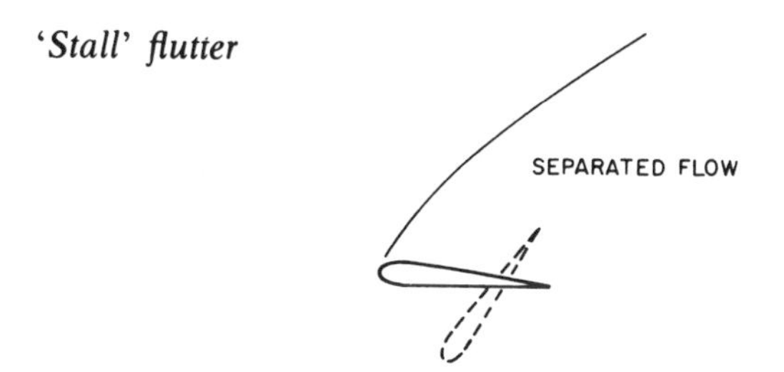

Figure 3.23 Schematic of separated flow geometry.

* Eckhaus [21], Landahl [22], Lambourne [23].

An airfoil oscillating through large angles of attack will create a time lag in the aerodynamic moment which may give rise to ***negative*** aerodynamic damping in pitch and, hence, flutter, even though for small angles of attack the aerodynamic damping would be positive. Compressor, turbine and helicopter blades are particularly prone to this type of flutter, since they routinely operate through large ranges of angle of attack. A later chapter discusses this type of flutter in some detail.

Parameter trends

The coalescence flutter is perhaps most common for airfoils under conventional flow conditions (no shock oscillation and no stall). It is certainly the best understood. Hence, for this type of flutter, let us consider the variation of (nondimensional) flutter velocity with other important parameters.

Static unbalance. x_α:

If $x_\alpha < 0$ (i.e., c.g. is ahead of e.a.) frequently no flutter occurs. If $x_\alpha < 0$ the surface is said to be 'mass balanced'.

Frequency ratio. $\frac{\omega_h}{\omega_\alpha}$:

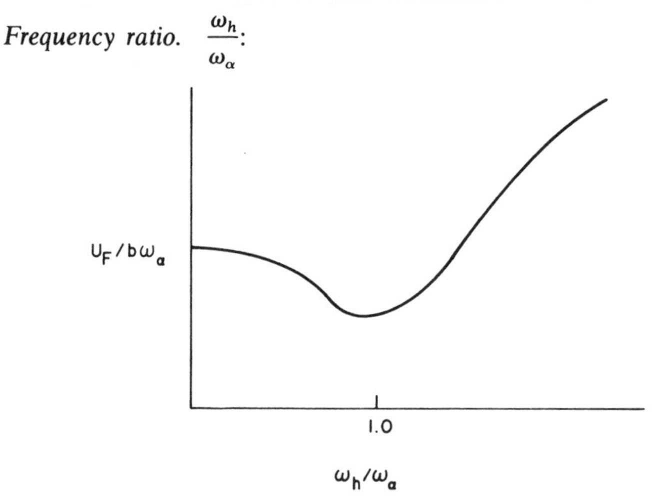

Figure 3.24 Flutter airspeed vs frequency ratio.

Not unexpectedly, for coalescence flutter $U_F/b\omega_\alpha$ is a minimum when $\omega_h/\omega_\alpha \simeq 1$.

Mach number. M:

The aerodynamic pressure on an airfoil is normally greatest near Mach number equal to one* and hence, the flutter speed tends to be a

* See Chapter 4.

minimum there. For $M \gg 1$ the aerodynamic piston theory predicts that the aerodynamic pressure is proportional to

$$p \sim \rho \frac{U^2}{M}$$

Hence, $U_F \sim M^{\frac{1}{2}}$ for $M \geqslant 1$ and constant μ. Also

$$\lambda_F \sim (\rho U^2)_F \sim M$$

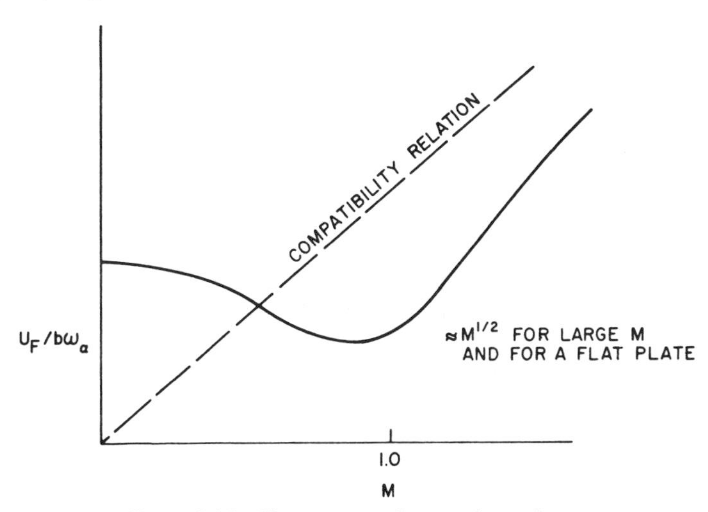

Figure 3.25 Flutter air speed vs mach number.

Note that for flight at constant altitude of a specific aircraft ρ (hence, μ) and a_∞ (speed of sound) are fixed. Since

$$U = Ma_\infty$$

$U/b\omega_\alpha$ and M are not independent, but are related by

$$\left(\frac{U}{b\omega_\alpha}\right) = M\left(\frac{a_\infty}{b\omega_\alpha}\right)$$

Thus, a compatibility relation must also be satisfied as indicated by dashed line in Figure 3.25. By repeating the flutter calculation for various altitudes (various ρ, a_∞ and hence various μ and $a_\infty/b\omega_\alpha$), one may obtain a plot of flutter Mach number versus altitude as given in Figure 3.26.

Mass ratio. μ:

For large μ the results are essentially those of a constant flutter dynamic pressure; for small μ they are often those of constant flutter

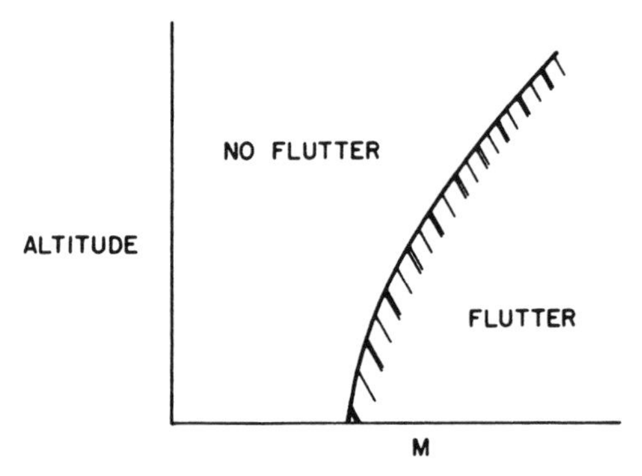

Figure 3.26 Altitude vs mach number.

velocity as indicated by dashed line. However, for $M \equiv 0$ and two-dimensional airfoils theory predicts $U_F \to \infty$ for some small but finite μ (solid line). This is contradicted by the experimental evidence and remains a source of some controversy in the literature.* Crisp† has recently suggested that the rigid airfoil chord assumption is untenable for small μ and that by including elastic chordwise bending the discrepancy between theory and experiment may be resolved. See Figure 3.27.

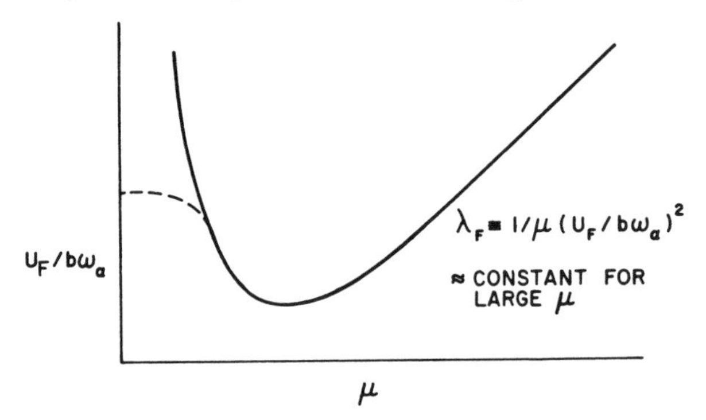

Figure 3.27 Flutter air speed vs mass ratio.

Flutter prevention

After one has ascertained that there is a flutter problem then there is more than a casual curiosity as to how to fix it, i.e., increase U_F, without adding any weight, of course. There is no universal solution, but frequently one or more of the following are tried.

---

* Abramson [24]. Viscous fluid effects are cited as the source of the difficulty.
† Crisp [25].

(1) add mass or redistribute mass so that $x_\alpha < 0$, 'mass balance'
(2) increase torsional stiffness, i.e., increase ω_α
(3) increase (or decrease) $\dfrac{\omega_h}{\omega_\alpha}$ if it is near one (for fixed ω_α)
(4) add damping to the structure, particularly for single-degree-of-freedom flutter or stall flutter
(5) require the aircraft to be flown below its critical Mach number (normally used as a temporary expedient while one of the above items is studied).

The above discussion was in the context of the typical section. For more complex aerospace vehicles, additional degrees of freedom, equations of motion and parameters will appear. Basically, these will have the form of additional frequency ratios (stiffness distribution) and inertial constants (mass distribution). Hence, *for example*, we might have

$$\frac{\omega_h}{\omega_\alpha} \quad \text{replaced by} \quad \frac{\omega_1}{\omega_\alpha}, \frac{\omega_2}{\omega_\alpha}, \frac{\omega_3}{\omega_\alpha}, \quad \text{etc.}$$

and x_α, r_α replaced by

$$\int \rho x \, \mathrm{d}x, \int \rho x^2 \, \mathrm{d}x, \int \rho x^3 \, \mathrm{d}x, \text{ etc.}$$

$$\int \rho xy \, \mathrm{d}x \, \mathrm{d}y, \int \rho y \, \mathrm{d}y, \int \rho y^2 \, \mathrm{d}y, \text{ etc.}$$

*Gust response.** To the parameters for flutter we add

$$\frac{w_G}{U}$$

for gust response. Since w_G is a time history (deterministic or random) we actually add a function parameter rather than a constant. Hence, various gust responses will be obtained depending on the nature of the assumed gust time history.

The several approaches to gust response analysis can be categorized by the type of atmospheric turbulence model adopted. The simplest of these is the sharp edged gust; a somewhat more elaborate model is the 1-COSINE gust. Both of these are deterministic; in recent years a third gust model has been increasingly used where the gust velocity field is treated as a random process.

* Houbolt, Steiner and Pratt [6].

Discrete deterministic gust

An example of a gust time history is a *sharp edged gust*,

$$w_G = 50 \text{ ft/sec.} \quad \text{for} \quad \left.\begin{array}{l} x < Ut \\ \text{or} \quad t > \dfrac{x}{U} \end{array}\right\}, \; x' < 0$$

$$= 0 \qquad \text{for} \quad x > Ut \,, \; x' > 0$$

x', t' fixed in atmosphere

x, t fixed with aircraft

(Galilean transformation) $x' = x - Ut$ (if $x' = x = 0$ at $t = t' = 0$)

$t' = t$

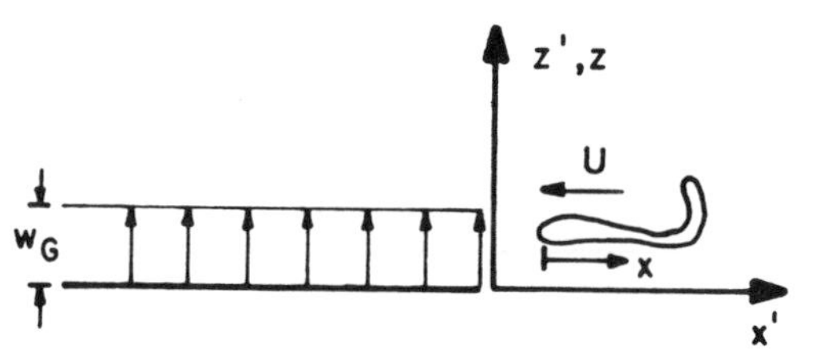

Figure 3.28 Sharp edged gust.

In this model w_G is constant with respect to space *and* time in the atmospheric fixed coordinate system for all $x' < 0$. We shall deal with the aerodynamic consequences of this property in the next chapter.

A somewhat more realistic gust model allows for the spatial scale of the gust field. In this model w_G is independent of time, t', but varies with distance, x', in the atmospheric fixed coordinate system, x', t'.

$$w_G = \frac{w_{G_{\max}}}{2}\left[1 - \cos\frac{2\pi x'}{x_G}\right]$$

$$\text{for} \quad t < \frac{x_G}{U}, \qquad x' < 0$$

$$= 0 \quad \text{for} \quad t > \frac{x_G}{U}, \qquad x' > 0$$

Recall

$$x' = x - U_\infty t$$

x_G is normally varied to obtain the most critical design condition and typically $w_{G_{\max}} \simeq 50$ ft/sec. See sketch below.

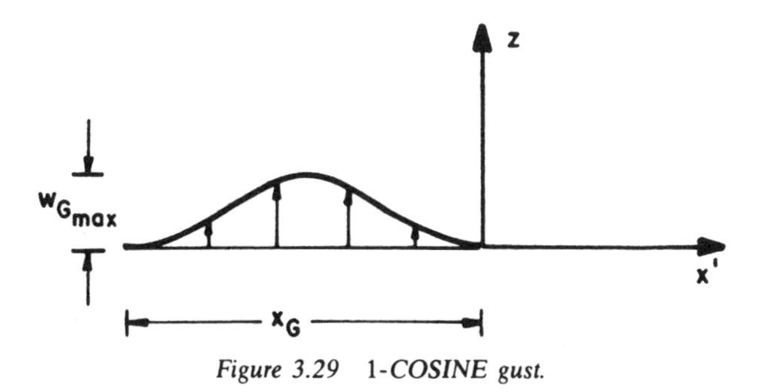

Figure 3.29 1-*COSINE gust.*

Schematic results for flight vehicle response to these deterministic gust models are shown below.

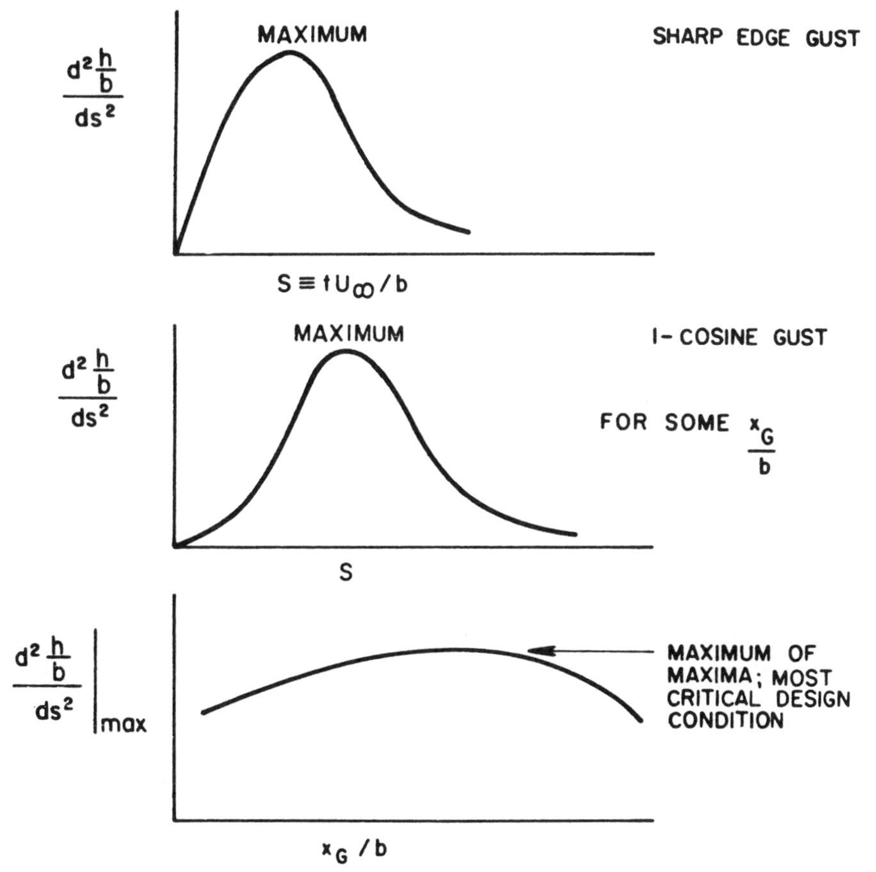

Figure 3.30 *Response to deterministic gusts.*

Random gust

In a random gust field, we still adopt the assumption that w_G, though now a random variable, varies only with x' and is independent of t'. In the theory of isotropic turbulence this is usually referred to as Taylor's hypothesis* or the 'frozen gust' assumption. Thus

$$w_G(x') = w_G(x - U_\infty t)$$

Since x and t only appear in the above combination, we may consider the alternative functional form

$$w_G = w_G\left(t - \frac{x}{U_\infty}\right)$$

The correlation function may then be defined as

$$\phi_{w_G w_G}(\tau) \equiv \lim_{T\to\infty} \frac{1}{2T}\int_{-\infty}^{\infty} w_G\left(t - \frac{x}{U_\infty}\right) w_G\left(t - \frac{x}{U_\infty} + \tau\right) \mathrm{d}t$$

and the power spectral density as

$$\Phi_{w_G w_G}(\omega) \equiv \frac{1}{\pi}\int_{-\infty}^{\infty} \phi_{w_G w_G}(\tau) \mathrm{e}^{-i\omega\tau}\, \mathrm{d}\tau$$

The power spectral density is given in Figure 3.31. A useful approximate formula which is in reasonable agreement with measurements is†

$$\Phi_{w_G w_G} = \overline{w_G^2}\, \pi U \frac{1 + 3\left(\dfrac{\omega L_G}{U}\right)^2}{\left[1 + \left(\dfrac{\omega L_G}{U}\right)^2\right]^2}$$

Typically,

$$\overline{w_G^2} \simeq 33 \text{ ft/sec.}$$

$$L_G \simeq 50\text{–}500 \text{ ft; gust scale length}$$

We conclude this discussion with representative vehicle responses to random gust fields drawn from a variety of sources.‡ The analytical

* Houbolt, Steiner and Pratt [6]. The basis for the frozen gust assumption is that in the time interval for any part of the gust field to pass over the flight vehicle (the length/U_∞) the gust field does not significantly change its (random) spatial distribution. Clearly this becomes inaccurate as U_∞ becomes small.

† Houbolt, Steiner and Pratt [6].

‡ These particular examples were collected and discussed in Ashley, Dugundji and Rainey, [24].

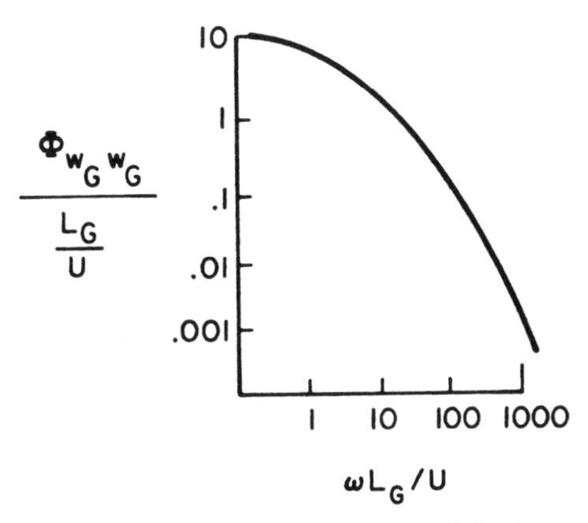

Figure 3.31 *Gust power spectral density.*

results are from mathematical models similar to those described above, but with more elaborate structural and aerodynamic ingredients as described in succeeding pages in this chapter and Chapter 4.

In the Figure 3.32, the measured and calculated power spectral densities for acceleration at the pilot station of the XB-70 aircraft are shown. The theoretical structural model allows for rigid body and elastic degrees of freedom using methods such as those described later in this chapter. The aerodynamic theory is similar to those described in Chapter

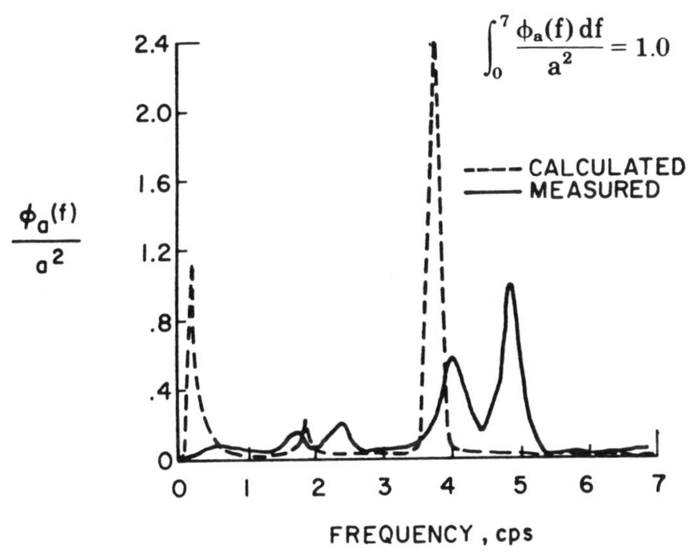

Figure 3.32 *Acceleration power spectral density. From Stenton* [26].

4. The dramatic conclusion drawn from this figure is that theory and experiment do not necessarily agree closely! If one assumes the peaks in the measured and calculated spectra are associated with resonances at natural frequencies of the (aeroelastic) system, then one concludes the theoretical model is not predicting these adequately. Since the resonances are determined primarily by mass and stiffness (springs), one concludes that for real vehicles even these characteristics may be difficult to model mathematically. This is quite aside from other complications such as structural damping and aerodynamic forces.

Usually when one is dealing with a real vehicle, physical small scale models are built and with these (as well as the actual vehicle when it is available) the resonant frequencies are measured (in the absence of any airflow). The results are then used to 'correct' the mathematical model, by one method or another, including a possible direct replacement of calculated resonant frequencies by their measured counterparts in the equations of motion. When this is done the peak frequencies in the measured and calculated spectra will then agree (necessarily so) and the question then becomes one of how well the peak levels agree.

A comparison for another aircraft, B-47, is shown in Figure 3.33. Here the measured and calculated resonant frequencies are in good agreement. Moreover the peak levels and indeed all levels are in good correspondence. The particular comparison shown is for the system transfer function which relates the acceleration at a point on the aircraft to the random gust input. The calculated transfer function has been obtained from an aeroelastic mathematical model. The measured transfer function (from flight test) is inferred from a measurement of gust power spectra and cross-spectra between the vehicle acceleration and gust velocity field using the relation (c.f. e.g. (3.3.31))

$$H_{\ddot{h}w_G} = \frac{\Phi_{\ddot{h}w_G}}{\Phi_{w_G w_G}}$$

Both the amplitude and phase of the transfer function are shown as a function of frequency for various positions along the wing span ($\bar{y} = 0$ is at the wing rcot and $\bar{y} = 1$ at the wing tip). Such good agreement between theory and experiment is certainly encouraging. However, clearly there is a major combined theoretical-experimental effort required to determine accurately the response of structures to gust loading. It should be noted that according to [6], Figure 3.33 is the bending strain transfer function. 'The dimensions of the ordinates . . . are those for acceleration because the responses of the strain gages were calibrated in terms of the strain per unit normal acceleration experienced during a shallow pull-up maneuver.'

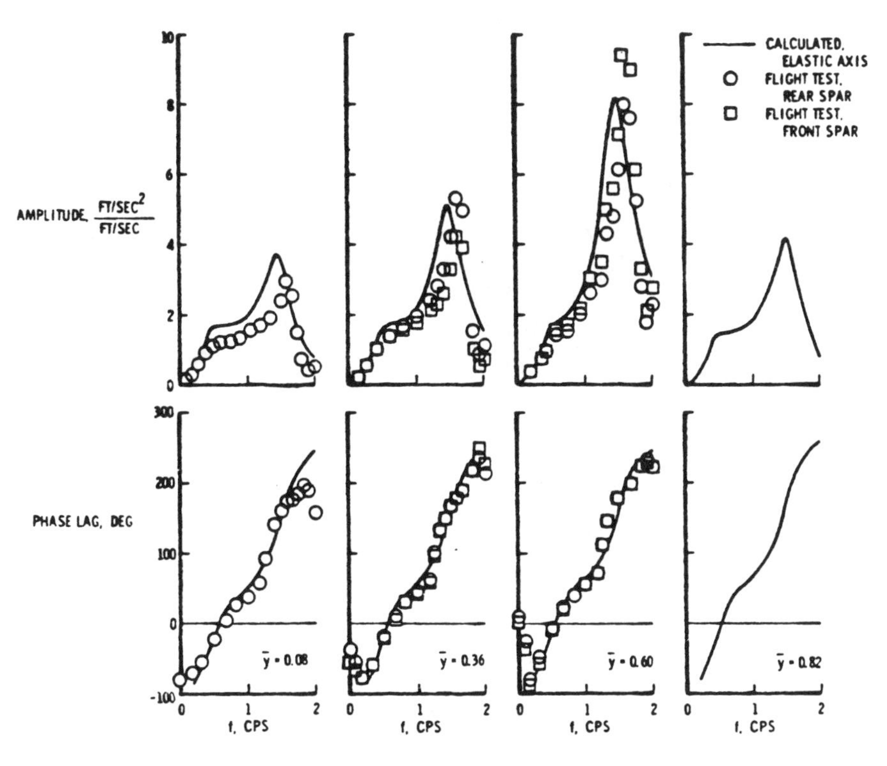

Figure 3.33 $\bar{y}$ = *nondimensional distance along span. From Houbolt* [6].

3.7 Generalized equations of motion for complex structures

Lagrange's equations and modal methods (*Rayleigh-Ritz*)

The most effective method for deriving equations of motion for many complex dynamical systems is to use Lagrange's Equations.*

$$\frac{\mathrm{d}}{\mathrm{d}t}\frac{\partial L}{\partial \dot{q}_i} - \frac{\partial L}{\partial q_i} = Q_i$$

where

$L \equiv T - U$, Lagrangian
$T \equiv$ kinetic energy
$U \equiv$ potential energy
$Q_i \equiv$ generalized forces
$q_i \equiv$ generalized coordinates

* Recall Section 3.2.

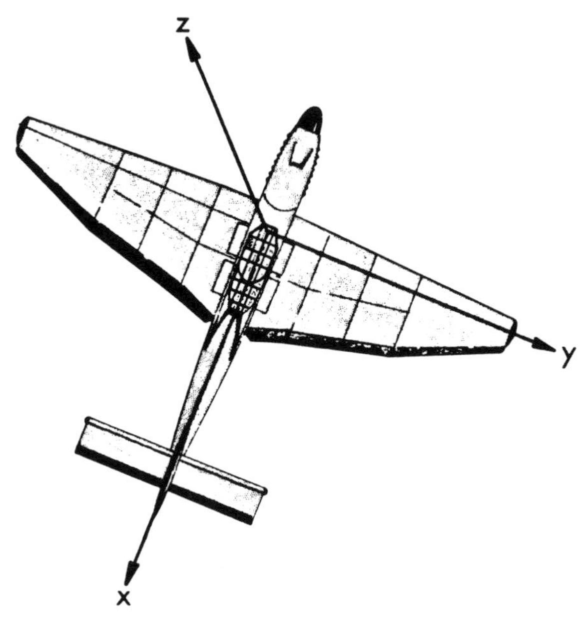

Figure 3.34 Two-dimensional (planar) representation of a flight vehicle.

The essential steps in the method are, first, a suitable choice of q_i and then an evaluation of T, U and Q_i in terms of q_i and $\dot{q}_i$.

Lagrange's equations have, as one of their principal advantages, the ability to obtain the equations of motion for complex systems with little or no more difficulty than that required for rather simple ones, such as the 'typical section'. Here we shall consider a two-dimensional (planar) representation of a flight vehicle. (See Figure 3.34).

We note that this formulation can include 'rigid' body as well as flexible body modes. For example, the following choices of modal functions, z_m, include rigid body vertical translation, pitching (rotation about y axis) and rolling (rotation about x axis), respectively

$z_1 = 1$ vertical translation
$z_2 = x$ pitching
$z_3 = y$ rolling

For such modes the potential elastic or strain energy is zero; however, in general, strain energy must be included for the flexible body modes.

The use of Lagrange's equations, while formally compact, does not reveal explicitly all of the complications which may arise in deriving equations of motion for an unrestrained vehicle or structure. These are seen more clearly in the discussion in a later section of integral equations of equilibrium.

Kinetic energy

The x–y plane is the plane of the (aircraft) structure. We consider deformations perpendicular to the x–y plane (in the z direction). The normal displacement with respect to a fixed inertial reference plane we call $z_a(x, y, t)$. We may then express the kinetic energy as

$$T = \tfrac{1}{2} \iint m \dot{z}_a^2 \, dx \, dy \tag{3.7.1}$$

where m – mass/area and $\dot{z} \equiv \dfrac{\partial}{\partial t}$. If we expand the displacement in a modal series, say

$$z_a = \sum_m q_m(t) z_m(x, y) \tag{3.7.2}$$

then the kinetic energy may be written as

$$T = \tfrac{1}{2} \sum_m \sum_n \dot{q}_m \dot{q}_n M_{mn} \tag{3.7.3}$$

where the *generalized mass* is given by

$$M_{mn} \equiv \iint m z_m z_n \, dx \, dy$$

For small motions the above integral over the body may be taken as over the undeformed structure.

If the chosen modes, z_m, satisfy an orthogonality relation

$$M_{mn} = M_m \delta_{mn} \qquad \begin{aligned} \delta_{mn} &= 1 \quad \text{for} \quad m = m \\ &= 0 \quad \text{for} \quad m \neq n \end{aligned}$$

then (3.7.3) simplifies to

$$T = \tfrac{1}{2} \sum_m \dot{q}_m^2 M_m \tag{3.7.4}$$

Strain (potential, elastic) energy

For the strain energy, we may write a similar relation to (3.7.3).

$$U = \tfrac{1}{2} \sum_m \sum q_m q_n K_{mn} \tag{3.7.5}$$

where K_{mn} is a generalized spring constant which is determined from an appropriate structural theory.* Indeed if the z_m are the 'natural' or 'normal' modes of the structure, one may show that

$$K_{mn} = \omega_m^2 M_m \delta_{mn} \tag{3.7.6}$$

where ω_m is the mth 'natural frequency'.†

Equations (3.7.3)–(3.7.6) are the keys to the Lagrangian approach. Before continuing, we pause to consider K_{mn} in more detail.

Alternative determination of K_{mn}. A stiffness influence function, $K(x, y; \xi, \eta)$, may be defined which is the (static) force/area required at point x, y to give a unit deflection at point ξ, η. Hence

$$p(x, y) = \iint K(x, y; \xi, \eta) z_a(\xi, \eta)\, d\xi\, d\eta \tag{3.7.7}$$

$$F = K\,d \quad \text{simple spring analog}$$

The potential energy stored in the elastic body is thus

$$U = \tfrac{1}{2} \iint p(x, y) z_a(x, y)\, dx\, dy$$

$$U = \tfrac{1}{2} F d \quad \text{simple spring analog}$$

Using (3.7.8) in (3.7.7),

$$U = \tfrac{1}{2} \iiiint K(x, y; \xi, \eta) z_a(\xi, \eta) z_a(x, y)\, d\xi\, d\eta\, dx\, dy \tag{3.7.9}$$

$$U = \tfrac{1}{2} K d^2 \quad \text{simple spring analog}$$

Using our modal expansion

$$z_a(x, y, t) = \sum_m q_m(t) z_m(x, y)$$

in (3.7.9) we obtain

$$U = \tfrac{1}{2} \sum_m \sum_n K_{mn} q_m q_n$$

where

$$K_{mn} \equiv \iiiint K(x, y; \xi, \eta) z_m(\xi, \eta) z_n(x, y)\, d\xi\, d\eta\, dx\, dy$$

$$U = \tfrac{1}{2} K d^2 \quad \text{simple spring analog} \tag{3.7.10}$$

---

* Recall Section 3.1.

† Meirovitch [4].

From Maxwell's Reciprocity Theorem

$$K(x, y; \xi, \eta) = K(\xi, \eta; x, y)$$

and hence

$$K_{mn} = K_{nm} \tag{3.7.11}$$

$K(x, y; \xi, \eta)$ can be determined by a suitable theoretical analysis or it can be inferred from experiment. For the additional insight to be gained, let us consider the latter alternative. It is a difficult experiment to measure K directly since we must determine a distribution of force/area which gives unit deflection at one point and zero deflection elsewhere. Instead it is much easier to measure the inverse of K, a flexibility influence, $C(x, y; \xi, \eta)$, which is the deflection at x, y due to a unit force/area at ξ, η. For $C(x, y; \xi, \eta)$ we have the following relation (recall Section 2.4)

$$z_a(x, y) = \iint C(x, y; \xi, \eta) p(\xi, \eta)\, \mathrm{d}\xi\, \mathrm{d}\eta \tag{3.7.12}$$

Using (3.7.6) and (3.7.1) it can be shown that

$$\iint C(x, y; \xi, \eta) K(\xi, \eta; r, s)\, \mathrm{d}\xi\, \mathrm{d}\eta = \delta(r - x, s - y) \tag{3.7.13}$$

where δ is a Dirac delta function. (3.7.13) is an integral equation for C or K given the other. However, it is rarely, if ever, used. Instead (3.7.6) and (3.7.1) are attacked directly by considering a finite number of loads and deflections over small (finite) areas of size $\Delta x\, \Delta y = \Delta\xi\, \Delta\eta$. Hence (3.7.7) and (3.7.12) are written

$$p(x_i, y_i) = \sum_j K(x_i, y_i; \xi_j, \eta_j) z_a(\xi_j, \eta_j)\, \Delta\xi\, \Delta\eta \tag{3.7.7}$$

$$z_a(x_j, y_j) = \sum_i C(x_j, y_j; \xi_i, \eta_i) p(\xi_i, \eta_i)\, \Delta\xi\, \Delta\eta \tag{3.7.12}$$

In matrix notation

$$\{p\} = [K]\{z_a\}\Delta\xi\, \Delta\eta \tag{3.7.7}$$

$$\{z_a\} = [C]\{p\}\Delta\xi\, \Delta\eta \tag{3.7.12}$$

Substitution of (3.7.12) into (3.7.7) and solving, gives

$$[K] = [C]^{-1}/(\Delta\xi)^2(\Delta\eta)^2 \tag{3.7.14}$$

(3.7.14) is essentially a finite difference solution to (3.7.13). Hence, *in practice*, if (3.7.10) is used to compute K_{mn}, one measures C, computes K from (3.7.14) and then evaluates K_{mn} by numerical integration of

(3.7.10). For a fuller discussion of influence functions, the reader may wish to consult Bisplinghoff, Mar and Pian [2].

There is one further subtlety which we have not discussed as yet. If rigid body motions of the structure are possible, then one may wish to use a C measured with respect to a fixed point. For example, it may be convenient to measure C with the center of mass fixed with respect to translation and rotation. This matter is discussed more fully later in this chapter when integral equations of equilibrium are reviewed.

We now continue the general discussion from which we digressed to consider K_{mn}. Two examples will be considered next.

Examples

(a) *Torsional vibrations of a rod.* To illustrate the key relations (3.7.3)–(3.7.6) in a more familiar situation, consider the torsional vibrations of a rod.

Here

$$z_a = -x\alpha(y, t) \qquad \text{cf. (3.7.2)}$$

and thus (3.7.1) becomes

$$T = \tfrac{1}{2}\int I_\alpha \dot{\alpha}^2 \, \mathrm{d}y \tag{3.7.15}$$

where

$$I_\alpha \equiv \int m x^2 \, \mathrm{d}x$$

$$\alpha \equiv \text{angle of twist}$$

From structural theory [2],

$$U = \tfrac{1}{2}\int GJ\left(\frac{\mathrm{d}\alpha}{\mathrm{d}y}\right)^2 \mathrm{d}y \tag{3.7.16}$$

Let

$$\alpha = \sum_{m=1}^{M} q_m^\alpha \alpha_m(y) \tag{3.7.17}$$

then

$$T = \tfrac{1}{2}\sum_m \sum_n \dot{q}_m^\alpha \dot{q}_n^\alpha M_{mn} \tag{3.7.18}$$

where

$$M_{mn} \equiv \int I_\alpha \alpha_m \alpha_n \, \mathrm{d}y \qquad \text{cf. (3.7.3)}$$

and

$$U = \tfrac{1}{2}\sum_m \sum_n q_m^\alpha q_n^\alpha K_{mn} \tag{3.7.19}$$

where

$$K_{mn} = \iint GJ \frac{\mathrm{d}\alpha_m}{\mathrm{d}y}\frac{\mathrm{d}\alpha_n}{\mathrm{d}y}\,\mathrm{d}y \qquad \text{cf. (3.7.5)}$$

The specific structural model chosen determines the accuracy with which the generalized masses and stiffnesses are determined, but they always exist.

(b) *Bending-torsional motion of a beam-rod.* The above is readily generalized to include bending as well as torsional vibrations of a beam-rod.

Let

$$z_a(x, y, t) = -x\alpha(y, t) - h(y, t) \qquad \text{cf. (3.7.2)}$$

$\alpha \equiv$ twist about elastic axis

$h \equiv$ bending deflection of elastic axis

and thus (3.7.1) becomes

$$T = \tfrac{1}{2}\left\{\int M\dot{h}^2\,\mathrm{d}y + 2\int S_\alpha \dot{h}\dot{\alpha}\,\mathrm{d}y + \int I_\alpha \dot{\alpha}^2\,\mathrm{d}y\right\} \tag{3.7.15}$$

where

$$M \equiv \int m\,\mathrm{d}x, \qquad S_\alpha \equiv \int mx\,\mathrm{d}x, \qquad I_\alpha \equiv \int mx^2\,\mathrm{d}x$$

Also from structural theory [2],

$$U = \frac{1}{2}\left\{\int GJ\left(\frac{\partial\alpha}{\partial y}\right)^2 \mathrm{d}y + \int EI\left(\frac{\partial^2 h}{\partial y^2}\right)^2\right\}\mathrm{d}y \tag{3.7.16}$$

Let

$$h = \sum_{r=1}^{R} q_r^h h_r(y)$$

$$\alpha = \sum_{m=1}^{M} q_m^\alpha \alpha_m(y) \tag{3.7.17}$$

Then

$$T = \tfrac{1}{2}\sum_m \sum_n \dot{q}_m^\alpha \dot{q}_n^\alpha M_{mn}^{\alpha\alpha} + 2\sum_m \sum_r \dot{q}_m^\alpha \dot{q}_r^h M_{mr}^{\alpha h} + \sum_r \sum_s \dot{q}_r^h \dot{q}_s^h M_{rs}^{hh} \qquad \text{cf. (3.7.3)}$$

where

$$M_{mn}^{\alpha\alpha} \equiv \int I_\alpha \alpha_m \alpha_n \, \mathrm{d}y, \qquad M_{mr}^{\alpha h} \equiv \int S_\alpha \alpha_m h_r \, \mathrm{d}y, \qquad M_{rs}^{hh} \equiv \int m h_r h_s \, \mathrm{d}y \tag{3.7.18}$$

and

$$U = \tfrac{1}{2}\left\{\sum_m \sum_n q_m^\alpha q_n^\alpha K_{mn}^{\alpha\alpha} + \sum_r \sum_s q_r^h q_s^h K_{rs}^{hh}\right\} \qquad \text{cf. (3.7.5)}$$

where

$$K_{mn}^{\alpha\alpha} \equiv \int GJ \frac{\mathrm{d}\alpha_m}{\mathrm{d}y}\frac{\mathrm{d}\alpha_n}{\mathrm{d}y}\,\mathrm{d}y, \qquad K_{rs}^{hh} \equiv \int EI \frac{\mathrm{d}^2 h_r}{\mathrm{d}y^2}\frac{\mathrm{d}^2 h_s}{\mathrm{d}y^2}\,\mathrm{d}y \tag{3.7.19}$$

Of all possible choices of modes, the 'free vibration, natural modes' are often the best choice. These are discussed in more detail in the next section.

Natural frequencies and modes-eigenvalues and eigenvectors

Continuing with our general discussion, consider Lagrange's equations with the generalized forces set to zero,

$$\frac{\mathrm{d}}{\mathrm{d}t}\left(\frac{\partial (T-U)}{\partial \dot{q}_i}\right) + \frac{\partial U}{\partial q_i} = 0 \qquad i = 1, 2, \ldots, M$$

and thus obtain, using (3.7.3) and (3.7.5) in the above,

$$\sum M_{mi}\ddot{q}_m + K_{mi} q_m = 0 \qquad i = 1, \ldots, M \tag{3.7.20}$$

Consider sinusoidal motion

$$q_m = \bar{q}_m \mathrm{e}^{i\omega t} \tag{3.7.21}$$

then, in matrix notation, (3.7.20) becomes

$$-\omega^2 [M]\{q\} + [K]\{q\} = \{0\} \tag{3.7.22}$$

This is an eigenvalue problem, for the eigenvalues, ω_j, $j = 1, \ldots, M$ and corresponding eigenvectors, $(q)_j$. If the function originally chosen, z_m or α_m and h_r, were 'natural modes' of the system then the M and K matrices will be diagonal and the eigenvalue problem simplifies.

$$-\omega^2 \left[\diagdown M \diagdown\right]\{q\} + \left[\diagdown M\omega_j^2 \diagdown\right]\{q\} = \{0\} \tag{3.7.23}$$

and

$$\omega_1^2, \begin{Bmatrix} q_1 \\ 0 \\ 0 \\ 0 \\ 0 \end{Bmatrix}_1$$

$$\omega_2^2, \begin{Bmatrix} 0 \\ q_2 \\ 0 \\ 0 \\ 0 \end{Bmatrix}_2$$

etc.

$$\omega_M^2, \begin{Bmatrix} 0 \\ 0 \\ 0 \\ q_M \end{Bmatrix}_M$$

If this is not so then the eigenvalues may be determined from (3.7.22) and a linear transformation may be made to diagonalize the M and K matrices. The reader may wish to determine the eigenvalues and eigenvectors of the typical section as an exercise.

For our purposes, the key point is that expressions like (3.7.3)–(3.7.6) *exist.* For a more extensive discussion of these matters, the reader may consult Meirovitch [4].

Evaluation of generalized aerodynamic forces

The generalized forces in Lagrange's equations are evaluated from their definition in terms of virtual work.

$$\delta W_{NC} = \sum_m Q_m \delta q_m \tag{3.7.24}$$

Now the virtual work may be evaluated independently from

$$\delta W_{NC} = \iint p \delta z_\alpha \, dx \, dy \tag{3.7.25}$$

where p is the net aerodynamic pressure on an element of the structure with (differential) area $dx\,dy$. Using (3.7.2) in (3.7.25)

$$\delta W_{NC} = \sum_m \delta q_m \iint p z_m \, dx \, dy \tag{3.7.26}$$

and we may identify from (3.7.24) and (3.7.26)

$$Q_m \equiv \iint p z_m \, \mathrm{d}x \, \mathrm{d}y \tag{3.7.27}$$

From aerodynamic theory,* one can establish a relation of the form

$$p(x, y, t) = \iiint_0^t A(x-\xi, y-\eta, t-\tau) \times \underbrace{\left[\frac{\partial z_a}{\partial \tau}(\xi, \eta, \tau) + U \frac{\partial z_a}{\partial \xi}(\xi, \eta, \tau)\right]}_{\text{'downwash'}} \mathrm{d}\xi \, \mathrm{d}\eta \, \mathrm{d}\tau \tag{3.7.28}$$

A may be physically interpreted as the pressure at point x, y at time t due to a unit impulse of downwash at point ξ, η at time τ. Using (3.7.2) and (3.7.28) in (3.7.27) we may evaluate Q_m in more detail,

$$Q_m = \sum_n \int_0^t [\dot{q}_n(\tau) I_{nm\dot{q}}(t-\tau) + q_n(\tau) I_{nmq}(t-\tau)] \, \mathrm{d}\tau \tag{3.7.29}$$

where

$$I_{nm\dot{q}}(t-\tau) \equiv \iiiint A(x-\xi, y-\eta, t-\tau) z_n(\xi, \eta) z_m(x, y) \, \mathrm{d}x \, \mathrm{d}y \, \mathrm{d}\xi \, \mathrm{d}\eta$$

$$I_{nmq}(t-\tau) \equiv \iiiint A(x-\xi, y-\eta, t-\tau) \times U \frac{\partial z_n}{\partial \xi}(\xi, \eta) z_m(x, y) \, \mathrm{d}x \, \mathrm{d}y \, \mathrm{d}\xi \, \mathrm{d}\eta$$

$I_{nm\dot{q}}$, I_{nmq} may be thought of as generalized aerodynamic impulse functions.

Equations of motion and solution methods

Finally applying Lagrange's equations, using 'normal mode' coordinates for simplicity,

$$M_m[\ddot{q}_m + \omega_m^2 q_m] = \sum_{n=1}^{M} \int_0^t [\dot{q}_n(\tau) I_{nm\dot{q}}(t-\tau) + q_n(\tau) I_{nmq}(t-\tau)] \, \mathrm{d}\tau \qquad m = 1, \ldots, M \tag{3.7.30}$$

* See Chapter 4, and earlier discussion in Section 3.4.

Note the form of (3.7.30). It is identical, mathematically speaking, to the earlier results for the typical section.* Hence similar mathematical solution techniques may be applied.

Time domain solutions. Taylor Series expansion

$$q_n(t+\Delta t) = q_n(t) + \dot{q}_n\Big|_t \Delta t + \frac{\ddot{q}_n}{2}\Big|_t (\Delta t)^2$$

One may solve for $\ddot{q}_n$ from (3.7.30) and hence $q_n(t+\Delta t)$ is determined. $q_n(t)$, $\dot{q}_n(t)$ are known from initial conditions and

$$\dot{q}_n(t+\Delta t) = \dot{q}_n(t) + \ddot{q}_n(t)\Delta t + \cdots \tag{3.7.31}$$

Frequency domain solutions. Taking a Fourier Transform of (3.7.30)

$$M_m[-\omega^2 + \omega_m^2]\bar{q}_m = \sum_n^M [i\omega H_{nm\dot{q}} + H_{nmq}]\bar{q}_n$$

where

$$\bar{q}_m \equiv \int_{-\infty}^{\infty} q_m \mathrm{e}^{-i\omega t}\, \mathrm{d}t$$

In matrix notation

$$\left[\left[\diagdown M_m(-\omega^2 + \omega_m^2) \diagdown\right] - [i\omega H_{nm\dot{q}} + H_{nmq}]\right] \{\bar{q}_n\} = \{0\} \tag{3.7.32}$$

By examining the condition for nontrivial solutions

$$\|[\cdots]\| = 0$$

we may find the 'poles' of the aeroelastic transfer functions and assess the stability of the system.

Response to gust excitation. If we wish to examine the gust response problem then we must return to (3.7.28) and add the aerodynamic pressure due to the gust loading

$$p_G(x, y, t) = \iiint A(x-\xi, y-\eta, t-\tau) w_G(\xi, \eta, \tau)\, \mathrm{d}\xi\, \mathrm{d}\eta\, \mathrm{d}\tau$$

* Provided $S_\alpha \equiv 0$ so that h, α are normal mode coordinates for the typical section.

The resulting generalized forces are

$$Q_{mG}(t) = \iiiint\!\!\int A(x-\xi, y-\eta, t-\tau) \times w_G(\xi, \eta, \tau) z_m(x, y)\, d\xi\, d\eta\, dx\, dy\, d\tau \quad (3.7.33)$$

Adding (3.7.33) to (3.7.30) does not change the mathematical technique for the time domain solution. In the frequency domain, the right hand column of (3.7.32) is now ($\bar{Q}_{mG}$)

$$\bar{Q}_{mG} = \int_{-\infty}^{\infty} Q_{mG} e^{-i\omega t}\, dt$$

Hence by solving (3.7.32) we may obtain generalized aeroelastic transfer functions

$$\frac{\bar{q}_n}{\bar{Q}_{mG}} \equiv H_{q_n Q_{mG}}(\omega; \cdots) \quad (3.7.34)$$

and employ the usual techniques of the frequency domain calculus including power spectral methods.

Integral equations of equilibrium

As an alternative approach to Lagrange's Equations, we consider an integral equation formulation using the concept of a structural influence (Green's) function. We shall treat a flat (two-dimensional) structure which deforms under (aerodynamic) loading in an arbitrary way. We shall assume a symmetrical vehicle and take the origin of our coordinate system at the vehicle center of mass with the two axes in the plane of the vehicle as principal axes, x, y. See Figure 3.34. Note the motion is assumed sufficiently small so that no distinction is made between the deformed and undeformed axes of the body. For example the inertia and elastic integral properties are evaluated using the (undeformed) axes x, y. The axes x, y are inertial axes, i.e., fixed in space. If we consider small deflections normal to the x, y plane, the x, y axes are approximately the principal axes of the deformed vehicle.

It will be useful to make several definitions.

z_a absolute vertical displacement of a point from x, y plane, positive up

m mass/area

p_E external applied force/area, e.g., aerodynamic forces due to gust, p_G

p_M force/area due to motion, e.g., aerodynamic forces (but not including inertial forces)

$$p_Z = p_E + p_M - m\frac{\partial^2 z_a}{\partial t^2}$$

total force/area, including inertial forces. Let us first consider equilibrium of rigid body motions.

$$\text{Translation:} \quad \iint p_Z \, \mathrm{d}x \, \mathrm{d}y = 0 \tag{3.7.35}$$

$$\text{Pitch:} \quad \iint x p_Z \, \mathrm{d}x \, \mathrm{d}y = 0 \tag{3.7.36}$$

$$\text{Roll:} \quad \iint y p_Z \, \mathrm{d}x \, \mathrm{d}y = 0 \tag{3.7.37}$$

Now consider equilibrium of deformable or elastic motion.

$$\begin{aligned} z_a^{\text{elastic}} &\equiv z_a(x, y, t) - z_a(0, 0, t) - x\frac{\partial z_a}{\partial x}(0, 0, t) - y\frac{\partial z_a}{\partial y}(0, 0, t) \\ &= \iint C(x, y; \xi, \eta) p_Z(\xi, \eta, t) \, \mathrm{d}\xi \, \mathrm{d}\eta \end{aligned} \tag{3.7.38}$$

where

$z_a^{\text{elastic}} \equiv$ deformation (elastic) of a point on vehicle

$C \equiv$ structural influence or Green's function; the (static) elastic deformation at x, y due to unit force/area at ξ, η for a vehicle fixed* at origin, $x = y = 0$.

Since the method of obtaining the subsequent equations of motion involves some rather extensive algebra, we outline the method here.

(1) Set $p_E = p_M = 0$.

(2) Obtain 'natural frequencies and modes'; prove orthogonality of modes.

(3) Expand deformation, z_a, for nonzero p_E and p_M in terms of normal modes or natural modes and obtain a set of equations for the (time dependent) coefficients of the expansion. The final result will again be (3.7.30).

---

* By fixed we mean 'clamped' in the sense of the structural engineer, i.e., zero displacement and slope. It is sufficient to use a static influence function, since invoking D'Alambert's Principle the inertial contributions are treated as equivalent forces.

Natural frequencies and modes

Set $p_E = p_M = 0$. Assume sinusoidal motion, i.e.,

$$z_a(x, y, t) = \bar{z}_a(x, y)e^{i\omega t} \tag{3.7.39}$$

then (3.7.38) becomes

$$\bar{z}_a(x, y) - \bar{z}_a(0, 0) - x\frac{\partial \bar{z}_a}{\partial x}(0, 0) - y\frac{\partial \bar{z}_a}{\partial y}(0, 0) = \omega^2 \iint C(x, y; \xi, \eta) m(\xi, \eta) \bar{z}_a(\xi, \eta)\, d\xi\, d\eta \tag{3.7.40}$$

The frequency, ω, has the character of an eigenvalue. (3.7.40) can be put into the form of a standard eigenvalue problem by solving for $\bar{z}_a(0, 0)$, $\frac{\partial \bar{z}_a}{\partial x}(0, 0)$, $\frac{\partial \bar{z}_a}{\partial y}(0, 0)$ and substituting into (3.7.40). For example, consider the determination of $\bar{z}_a(0, 0)$. Multiply (3.7.40) by m and integrate over the flight vehicle area. The result is:

$$\iint m\bar{z}_a\, dx\, dy - \bar{z}_a(0, 0)\iint m\, dx\, dy - \frac{\partial \bar{z}_a}{\partial x}(0, 0)\iint mx\, dx\, dy - \frac{\partial \bar{z}_a}{\partial y}(0, 0)\iint my\, dx\, dy = \omega^2 \iint m(x, y)\left[\iint C(x, y; \xi, \eta)\bar{z}_a(\xi, \eta) m(\xi, \eta)\, d\xi\, d\eta\right] \cdot dx\, dy \tag{3.7.41}$$

Examining the left-hand side of (3.7.41), the first integral is zero from (3.7.35), the third and fourth integrals are zero because of our use of the center-of-mass as our origin of coordinates. The second integral is identifiable as the total mass of the vehicle.

$$M \equiv \iint m\, dx\, dy$$

Hence,

$$\bar{z}_a(0, 0) = -\frac{\omega^2}{M}\iint m(x, y)\left[\iint Cm\bar{z}_a\, d\xi\, d\eta\right] dx\, dy = -\frac{\omega^2}{M}\iint m(\xi, \eta)\bar{z}_a(\xi, \eta) \times \left[\iint C(x, y; \xi, \eta) m(x, y)\, dx\, dy\right] d\xi\, d\eta \tag{3.7.42}$$

where the second line follows by change of order of integration. In a similar fashion $\frac{\partial \bar{z}_a}{\partial x}(0,0)$, $\frac{\partial \bar{z}_a}{\partial y}(0,0)$ may be determined by multiplying (3.7.40) by mx and my respectively with integration over the flight vehicle. The results are

$$\frac{\partial \bar{z}_a}{\partial x}(0,0) = -\frac{\omega^2}{I_y}\iint m(\xi,\eta)\bar{z}_a(\xi,\eta)\left[\iint C(x,y;\xi,\eta)xm(x,y)\,\mathrm{d}x\,\mathrm{d}y\right] \cdot \mathrm{d}\xi\,\mathrm{d}\eta \tag{3.7.43}$$

etc. where

$$I_y \equiv \iint x^2 m(x,y)\,\mathrm{d}x\,\mathrm{d}y$$

$$I_x \equiv \iint y^2 m(x,y)\,\mathrm{d}x\,\mathrm{d}y$$

In (3.7.42), and (3.7.43) note that x, y are now dummy integration variables, not to be confused with the x, y which appear in (3.7.40). Using (3.7.43) and (3.7.44) in (3.7.40) we have

$$\bar{z}_a(x,y) = \omega^2 \iint G(x,y;\xi,\eta)m(\xi,\eta)\bar{z}_a(\xi,\eta)\,\mathrm{d}\xi\,\mathrm{d}\eta \tag{3.7.44}$$

where

$$G(x,y;\xi,\eta) \equiv C(x,y;\xi,\eta) - \iint C(r,s;\xi,\eta)\left[\frac{1}{M}+\frac{xr}{I_y}+\frac{ys}{I_x}\right]m(r,s)\,\mathrm{d}r\,\mathrm{d}s$$

(3.7.44) has the form of a standard eigenvalue problem. In general, there are infinite number of nontrivial solutions (eigenfunctions), ϕ_m, with corresponding eigenvalues, ω_m, such that

$$\phi_m(x,y) = \omega_m^2 \iint G(x,y;\xi,\eta)m(\xi,\eta)\phi_m(\xi,\eta)\,\mathrm{d}\xi\,\mathrm{d}\eta \tag{3.7.45}$$

These eigenfunctions could be determined in a number of ways; perhaps the most efficient method being the replacement of (3.7.45) by system of linear algebraic equations through approximation of the integral in (3.7.45) by a sum.

$$\phi_m(x_i,y_i) = \omega_m^2 \sum_j G(x_i,y_i;\xi_j,\eta_j)m(\xi_j,\eta_j)\phi_m(\xi_j,\eta_j)\Delta\xi\,\Delta\eta \tag{3.7.46}$$

In matrix notation,

$$\{\phi\} = \omega^2 [G_{ij}\, \Delta\xi\, \Delta\eta] \left[\diagdown m \diagdown\right] \{\phi\}$$

or

$$\left[\left[\diagdown 1 \diagdown\right] - \omega^2 [G_{ij}\, \Delta\xi\, \Delta\eta] \left[\diagdown m \diagdown\right]\right] \{\phi\} = \{0\} \tag{3.7.47}$$

Setting the determinant of coefficients equal to zero, we obtain a polynomial in ω^2 which gives us (approximate) eigenvalues as roots. The related eigenvector of (3.7.47) is an approximate description of the eigenfunctions of (3.7.44) or (3.7.45).

An important and useful property of eigenfunctions is their orthogonality, i.e.,

$$\iint \phi_m(x, y)\phi_n(x, y) m(x, y)\, \mathrm{d}x\, \mathrm{d}y = 0 \quad \text{for} \quad m \neq n \tag{3.7.48}$$

We shall digress briefly to prove (3.7.48).

Proof of orthogonality. Consider two different eigenvalues and eigenfunctions.

$$\phi_m(x, y) = \omega_m^2 \iint G m \phi_m \, \mathrm{d}\xi\, \mathrm{d}\eta \tag{3.7.49a}$$

$$\phi_n(x, y) = \omega_n^2 \iint G m \phi_n \, \mathrm{d}\xi\, \mathrm{d}\eta \tag{3.7.49b}$$

Multiply (3.7.49a) and (3.7.49b) by $m\phi_n(x, y)$ and $m\phi_m(x, y)$ respectively and $\iint \cdots \mathrm{d}x\, \mathrm{d}y$.

$$\frac{1}{\omega_m^2} \iint \phi_n \phi_m m \, \mathrm{d}x\, \mathrm{d}y = \iint \phi_n m \left[\iint G \phi_m m \, \mathrm{d}\xi\, \mathrm{d}\eta\right] \cdot \mathrm{d}x\, \mathrm{d}y \tag{3.7.49c}$$

$$\frac{1}{\omega_n^2} \iint \phi_m \phi_n m \, \mathrm{d}x\, \mathrm{d}y = \iint \phi_m m \left[\iint G \phi_n m \, \mathrm{d}\xi\, \mathrm{d}\eta\right] \cdot \mathrm{d}x\, \mathrm{d}y \tag{3.7.49d}$$

Interchanging the order of integration in (3.7.49c) and transferring x, y to ξ, η, and vice versa on the right-hand side gives:

$$\frac{1}{\omega_m^2} \iint \phi_m \phi_n m \, \mathrm{d}x\, \mathrm{d}y = \iint \phi_m m \left[\iint G(\xi, \eta; x, y) \cdot \phi_n(\xi, \eta) m(\xi, \eta)\, \mathrm{d}\xi\, \mathrm{d}\eta\right] \mathrm{d}x\, \mathrm{d}y \tag{3.7.50}$$

If G were symmetric, i.e.,

$$G(\xi, \eta; x, y) = G(x, y; \xi, \eta) \tag{3.7.51}$$

then the right-hand side of (3.7.49d) and (3.7.50) would be equal and hence one could conclude that

$$\left[\frac{1}{\omega_m^2} - \frac{1}{\omega_n^2}\right] \iint \phi_m \phi_n m \, dx \, dy = 0$$

or

$$\iint \phi_m \phi_n m \, dx \, dy = 0 \quad \text{for} \quad m \neq n \tag{3.7.52}$$

Unfortunately, the situation is more complicated since G is *not* symmetric. However, from (3.7.44), et. seq., one can write

$$\begin{aligned} G(\xi, \eta; x, y) - G(x, y; \xi, \eta) \\ &= \iint C(r, s; \xi, \eta)\left[\frac{1}{M} + \frac{ys}{I_x} + \frac{xr}{I_y}\right] m(r, s) \, dr \, ds \\ &\quad - \iint C(r, s; x, y)\left[\frac{1}{M} + \frac{\eta s}{I_x} + \frac{\xi r}{I_y}\right] m(r, s) \, dr \, ds \end{aligned} \tag{3.7.53}$$

Using the above to substitute for $G(\xi, \eta; x, y)$ in (3.7.50) and using (3.7.35)–(3.7.37) to simplify the result, one sees that the terms on the right-hand side of (3.7.53) contribute nothing. Hence, the right-hand sides of (3.7.49d) and (3.7.50) are indeed equal.

The orthogonality result follows. Note that the rigid body modes

$$\begin{aligned} \omega_1 &= 0 \qquad \phi_1 = 1 \\ \omega_2 &= 0 \qquad \phi_2 = x \\ \omega_3 &= 0 \qquad \phi_3 = y \end{aligned} \tag{3.7.54}$$

are orthogonal as well. One can verify readily that the above satisfy the equations of motion, (3.7.35)–(3.7.38), and that the orthogonality conditions follow from (3.7.35)–(3.7.37).

Forced motion including aerodynamic forces

We will simplify the equations of motion to a system of ordinary integral-differential equations in time by expanding the deformation in terms of normal modes.

$$z_a(x, y, t) = \sum_{m=1}^{\infty} q_m(t) \phi_m(x, y) \tag{3.7.55}$$

Recall the natural modes, ϕ_m, must satisfy the equations of motion with $p_E = p_M = 0$ and

$$z_a \sim e^{i\omega_m t}$$

Substituting (3.7.55) in (3.7.35)–(3.7.37) and using orthogonality, (3.7.52), and (3.7.54),

$$\ddot{q}_1 \iint m \, dx \, dy = \iint [p_E + p_M] \, dx \, dy \tag{3.7.56}$$

$$\ddot{q}_2 \iint x^2 m \, dx \, dy = \iint x[p_E + p_M] \, dx \, dy \tag{3.7.57}$$

$$\ddot{q}_3 \iint y^2 m \, dx \, dy = \iint y[p_E + p_M] \, dx \, dy \tag{3.7.58}$$

The reader should be able to identify readily the physical significance of the several integrals in the above equations. Substituting (3.7.55) into (3.7.38) gives

$$\sum_{m=1}^{\infty} q_m \left[\phi_m(x, y) - \phi_m(0, 0) - x \frac{\partial \phi_m}{\partial x}(0, 0) - y \frac{\partial \phi_m}{\partial y}(0, 0) \right]$$
$$= \iint C(x, y; \xi, \eta) \left[p_E + p_M - m \sum_{m=1}^{\infty} \ddot{q}_m \phi_m(\xi, \eta) \right] d\xi \, d\eta \tag{3.7.59}$$

Now the normal modes, ϕ_m, satisfy

$$\phi_m(x, y) - \phi_m(0, 0) - x \frac{\partial \phi_m}{\partial x}(0, 0) - y \frac{\partial \phi_m}{\partial y}(0, 0)$$
$$= \omega_m^2 \iint C(x, y; \xi, \eta) m(\xi, \eta) \phi_m(\xi, \eta) \, d\xi \, d\eta \quad m = 1, \ldots, \infty \tag{3.7.60}$$

Also the left-hand side of (3.7.59) is identically zero for the rigid body modes, $m = 1, 2, 3$. Further using (3.7.60) in the right-hand side of (3.7.59) for $m = 4, 5, \ldots$, gives finally

$$\sum_{m=4}^{\infty} \left(q_m + \frac{\ddot{q}_m}{\omega_m^2} \right) \left[\phi_m(x, y) - \phi_m(0, 0) - x \frac{\partial \phi_m}{\partial x}(0, 0) - y \frac{\partial \phi_m}{\partial y}(0, 0) \right]$$
$$= \iint C(x, y; \xi, \eta)[p_E + p_M - m\ddot{q}_1 - m\xi\ddot{q}_2 - m\eta\ddot{q}_3] \, d\xi \, d\eta \tag{3.7.61}$$

Multiplying (3.7.61) by $m(x, y)\phi_n(x, y)$ and $\iint \cdots \mathrm{d}x\,\mathrm{d}y$, invoking orthogonality, gives

$$M_n\left(q_n + \frac{\ddot{q}_n}{\omega_n^2}\right) = \iint \phi_n m \left\{ \iint C[p_E + p_M - m\ddot{q}_1 - m\xi\ddot{q}_2 - m\eta\ddot{q}_3]\,\mathrm{d}\xi\,\mathrm{d}\eta \right\} \mathrm{d}x\,\mathrm{d}y \quad (3.7.62)$$

where the 'generalized mass', M_n, is defined as

$$M_n \equiv \iint \phi_n^2 m\,\mathrm{d}x\,\mathrm{d}y$$

Now the structural influence function, C, is symmetric, i.e.,

$$C(x, y; \xi, \eta) = C(\xi, \eta; x, y) \quad (3.7.63)$$

This follows from Maxwell's reciprocity theorem* which states that the deflection at x, y due to a unit load at ξ, η is equal to the deflection at ξ, η due to a unit load at x, y.

Using (3.7.63) and interchanging the order of integration in (3.7.62), one obtains

$$M_n\left(q_n + \frac{\ddot{q}_n}{\omega_n^2}\right) = \iint [p_E + p_M - m\ddot{q}_1 - m\xi\ddot{q}_2 - m\eta\ddot{q}_3] \cdot \left\{ \iint C(\xi, \eta; x, y)\phi_n(x, y)m(x, y)\,\mathrm{d}x\,\mathrm{d}y \right\} \cdot \mathrm{d}\xi\,\mathrm{d}\eta \quad (3.7.64)$$

Using (3.7.60) in (3.7.64),

$$M_n\left(q_n + \frac{\ddot{q}_n}{\omega_n^2}\right) = \frac{1}{\omega_n^2} \iint [p_E + p_M - m\ddot{q}_1 - m\xi\ddot{q}_2 - m\eta\ddot{q}_3] \cdot \left[\phi_n(\xi, \eta) - \phi_n(0, 0) - \xi \frac{\partial \phi_n}{\partial \xi}(0, 0) - \eta \frac{\partial \phi_n}{\partial \eta}(0, 0) \right] \cdot \mathrm{d}\xi\,\mathrm{d}\eta \quad (3.7.65)$$

By using orthogonality, (3.7.52), and the equations of rigid body equilibrium, (3.7.56)–(3.7.58), one may show that the right-hand side of (3.7.65) can be simplified as follows:

$$M_n\left(q_n + \frac{\ddot{q}_n}{\omega_n^2}\right) = \frac{1}{\omega_n^2} \iint [p_E + p_M]\phi_n\,\mathrm{d}\xi\,\mathrm{d}\eta \quad (3.7.66)$$

---

* Bisplinghoff, Mar and Pian [2].

Defining the generalized force,

$$Q_n \equiv \iint [p_E + p_M]\phi_n \, d\xi \, d\eta$$

one has

$$M_n[\ddot{q}_n + \omega_n^2 q_n] = Q_n \qquad n = 1, 2, 3, 4, \ldots \tag{3.7.67}$$

Note that there is no inertial or structural coupling in the equations (3.7.67). However, p_M generally depends upon $q_1, q_2, \ldots$ and hence the equations are aerodynamically coupled.* The lack of inertial and structural coupling is due to our use of natural or normal modes. Finally, note that the rigid body equation of motions, (3.7.56)–(3.7.58), also have the form of (3.7.67). Hence n may run over all integer values.

Examples

(a) *Rigid wing undergoing translation responding to a gust.* One mode only, $\phi_1 = 1$, q_1 ($\equiv -h$ was notation used previously in typical section model) and thus

$$M_1\ddot{q}_1 = Q_1^M + Q_1^E \tag{3.7.68}$$

$$Q_1^M = \iint p_M\phi_1 \, dx \, dy = \int L_M \, dy \tag{3.7.69}$$

$$Q_1^E = \iint p_E\phi_1 \, dx \, dy = \int L_G \, dy \tag{3.7.70}$$

where

$$L_M \equiv \int p_M \, dx \qquad \text{lift/span} \tag{3.7.71}$$

$$L_G \equiv \int p_E \, dx \qquad \text{lift/span} \tag{3.7.72}$$

Introducing nondimensional time, $s \equiv tU/b$, (3.7.68) may be written

$$\frac{U^2}{b^2} M_1 q'' = \int_0^l L_M \, dy + \int_0^l L_G \, dy \tag{3.7.73}$$

where $\quad ' \equiv \dfrac{d}{ds}$

* cf. (3.7.29).

Assuming strip-theory aerodynamics, two dimensional, incompressible flow, one has (recall Section 3.4 and see Chapter 4)

$$L_M(s) = -\pi\rho U_\infty^2\left[q''(s) + 2\int_0^s q''(\sigma)\phi(s-\sigma)\,\mathrm{d}\sigma\right] \tag{3.7.74}$$

Note we have assumed $q_1'(0) = 0$ in the above. Similarly

$$\begin{aligned} L_G &= 2\pi\rho U_\infty b\left[w_G(0)\psi(s) + \int_0^s \frac{\mathrm{d}w_G(\sigma)}{\mathrm{d}\sigma}\,\psi(s-\sigma)\,\mathrm{d}\sigma\right] \\ &= 2\pi\rho U^2 b\left[\int_0^s \frac{w_G(\sigma)}{U}\,\psi'(s-\sigma)\,\mathrm{d}\sigma\right] \end{aligned} \tag{3.7.75}$$

$$\psi'(s) \equiv \frac{\mathrm{d}\psi}{\mathrm{d}s}$$

Here we have assumed w_G is independent of y for simplicity. Substituting (3.7.74) and (3.7.75) into (3.7.73) we have

$$\begin{aligned} \frac{U_\infty^2}{b^2} M q_1''(s) = \pi\rho U_\infty^2(2bl)\Bigg[-\frac{q_1''}{2b} - \frac{1}{b}\int_0^s q_1''(\sigma)\phi(s-\sigma)\,\mathrm{d}\sigma \\ + \int_0^s \frac{w_G(\sigma)}{U_\infty}\,\psi'(s-\sigma)\,\mathrm{d}\sigma\Bigg] \end{aligned} \tag{3.7.76}$$

$$M \equiv \iint m\phi_1\,\mathrm{d}x\,\mathrm{d}y, \qquad \text{total mass of wing}$$

Note $\int L\,\mathrm{d}y = lL$ since we have assumed b a constant and $l \equiv$ half-span of wing. (3.7.76) may be solved in several ways which have previously been discussed in the context of the typical section. Here we shall pursue the method of Laplace Transforms. Transforming (3.7.76) (p is the Laplace Transform variable) gives

$$\frac{U^2}{b^2} M p^2 \bar{q}_1(p) = \pi\rho U^2(2bl)\left[\frac{\bar{w}_G}{U}\,p\bar{\psi} - \frac{p^2\bar{q}_1}{2b} - \frac{p^2\bar{q}_1}{b}\,\bar{\phi}\right] \tag{3.7.77}$$

We have taken $q(0) = q'(0) = 0$ while using the convolution theorem, i.e.,

$$\overline{\left\{\int_0^s w_G(\sigma)\psi'(s-\phi)\,\mathrm{d}\sigma\right\}} = \bar{w}_G p\bar{\psi}$$

$$\overline{\left\{\int_0^s q_1''(\sigma)\phi(s-\sigma)\,\mathrm{d}\sigma\right\}} = p^2\bar{q}_1\bar{\phi}$$

and a bar ($\bar{\ }$) denotes Laplace Transform. Solving (3.7.77) for $\bar{q}_1$ gives

$$\bar{q}_1(p) = \frac{\frac{b}{2}\frac{\bar{w}_G}{U}\bar{\psi}}{p\left(\frac{\mu}{2}+\frac{1}{4}+\frac{1}{2}\bar{\phi}\right)} \tag{3.7.78}$$

where

$$\mu \equiv \frac{M}{\pi(2bl)b\rho}, \quad \text{mass ratio.}$$

To complete the solution we must invert (3.7.78). To make this inversion tractable, ϕ and ψ are approximated by

$$\begin{aligned} \psi(s) &= 1 - 0.5\mathrm{e}^{-0.13s} - 0.5\mathrm{e}^{-s} \\ \phi(s) &= 1 - 0.165\mathrm{e}^{-0.0455s} - 0.335\mathrm{e}^{-0.3s} \end{aligned} \tag{3.7.79}$$

Thus

$$\begin{aligned} \bar{\psi} &= (0.565p + 0.13)/p(p+0.0455)(p+0.3) \\ \bar{\phi} &= \frac{0.5p^2 + 0.2805p + 0.01365}{p^3 + 0.3455p^2 + 0.01365p} \end{aligned} \tag{3.7.80}$$

and

$$\bar{q}_1 = \frac{b\frac{\bar{w}_G}{U}0.565(p^3 + 0.575p^2 + 0.093p + 0.003)}{(\mu+0.5)p(p+0.13)(p+1)(p^3 + a_1p^2 + a_2p + a_3)} \tag{3.7.81}$$

where

$$a_1 \equiv \frac{0.3455\mu + 0.67}{\mu + 0.5}$$

$$a_2 \equiv \frac{0.01365\mu + 0.28}{\mu + 0.5}$$

$$a_3 \equiv \frac{0.01365}{\mu + 0.5}$$

Often one is interested in the acceleration,

$$\ddot{q}_1 = \frac{U^2}{b^2}q_1''.$$

$$\ddot{q}_1 = \frac{U^2}{b^2}\mathscr{L}^{-1}\{p^2\bar{q}_1\}^*$$

---

* For $q_1(0) = \dot{q}_1(0) = 0$. $\mathscr{L}^{-1} \equiv$ inverse Laplace Transform.

$$= \frac{0.565}{\mu + 0.5} \int_0^s \frac{U_\infty}{b} w_G(\sigma)\{A_1 e^{-0.13(s-\sigma)}$$
$$+ A_2 e^{-(s-\sigma)} + B_1 e^{\gamma_1(s-\sigma)}$$
$$+ B_2 e^{\gamma_2(s-\sigma)} + B_3 e^{\gamma_3(s-\sigma)}\} \, d\sigma \tag{3.7.82}$$

where

$$A_1 = \frac{N(-0.13)}{D'(-0.13)}$$
$$A_2 = \frac{N(-1)}{D'(-1)}$$
$$B_{k=1\,2\,3} = \frac{N(\gamma_k)}{D'(\gamma_k)}$$

and

$$N(p) \equiv p(p^3 + 0.5756p^2 + 0.09315p + 0.003141)$$
$$D(p) \equiv (p + 0.13)(p + 1)(p^3 + a_1 p^2 + a_1 p + a_3)$$
$$\gamma_k \quad \text{roots of} \quad p^3 + a_1 p^2 + a_1 p + a_3 = 0$$

Note that bracketed term in (3.7.82) must be a real quantity though components thereof may be complex (conjugates). What does it mean physically if the real part of γ_1, γ_2, or γ_3 is positive?

An even simpler theory of gust response is availablė if one further approximates the aerodynamic forces. For example, using a quasi-static aerodynamic theory (recall Section 3.4), one has

$$\psi = 1 \quad \text{and thus} \quad L_G = 2\pi\rho U_\infty^2 b \frac{w_G}{U_\infty}$$

and

$$\phi = 0, \quad \text{and thus} \quad L_M = 0 \text{ (ignoring virtual inertia term)}$$

Hence

$$M_1 \ddot{q}_1 = \int L^G \, dy = 2\pi\rho U^2 bl \frac{w_G}{U} \tag{3.7.83}$$

$$\ddot{q}_{1_s} = \pi \frac{\rho U^2}{M} (2bl) \frac{w_G}{U_\infty} = \frac{U_\infty}{b} \frac{w_G}{\mu} \tag{3.7.83}$$

The subscripted quantity, $\ddot{q}_{1_s}$, is called the *static approximation to the gust response.* Figure 3.35 is a schematic of the result from the full theory, (3.7.82), referenced to the static result, (3.7.83). Here we have further assumed a sharp-edge gust, i.e., w_G = constant. After Figure 10.22 BAH.

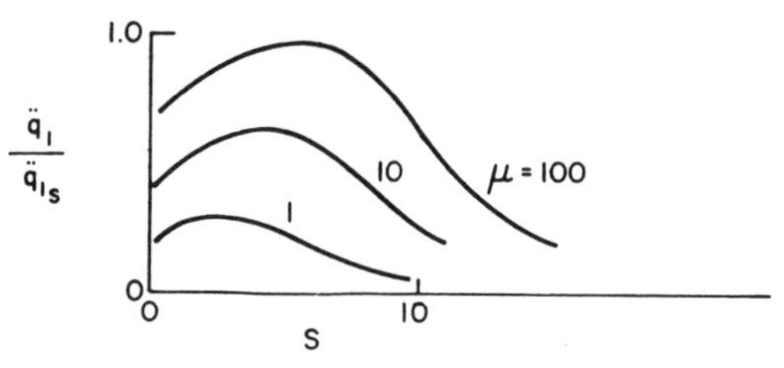

Figure 3.35 Acceleration time history.

The maxima of the above curves are presented in Figure 3.36. As can be seen the static approximation is a good approximation for large mass ratio, μ. For smaller μ the acceleration is less than the static result. Hence the quantity,

$$\frac{\ddot{q}_{1_{\max}}}{\ddot{q}_{1_s}}$$

is sometimes referred to as a 'gust alleviation' factor.

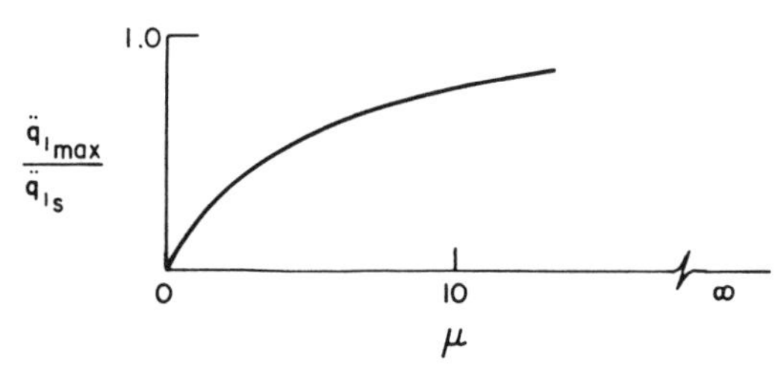

Figure 3.36 Maximum acceleration vs mass ratio.

A somewhat more sophisticated aerodynamic approximation is to let (again recall Section 3.4)

$$\begin{aligned} \psi &= 1 \quad \text{and thus} \quad L_G = 2\pi\rho U^2 b \frac{w_G}{U} \\ \phi &= 1 \quad \text{and thus} \quad L_M = -\pi\rho U^2[q''(s) + 2q'(s)] \end{aligned} \tag{3.7.84}$$

assuming $q'(0) = 0$. In the motion derived lift, the first term is a virtual inertial term which is generally negligible compared to the inertia of the flight vehicle. However, the second term is an aerodynamic damping term which provides the only damping in the system and hence may be important. It is this aerodynamic damping, even in the guise of the full (linear) aerodynamic theory, which gives results substantially different from the static approximation. (3.7.84) is termed a quasi-steady aerodynamic approximation.

Using the approximation (3.7.84), (3.7.68) becomes for a constant chord wing (span: l) and in nondimensional form

$$(\mu + 0.5)q_1''(s) + q_1'(s) = \frac{bw_G(s)}{U_\infty} \tag{3.7.85}$$

where

$$\mu \equiv \frac{M_1}{\pi\rho(2bl)\cdot b} \quad \text{mass ratio}$$

Taking the Laplace transform of (3.7.85) with initial conditions

$$q_1'(0) = q(0) = 0, \qquad w_G(0) = 0,$$

we have

$$(\mu + 0.5)p^2\bar{q}_1(p) + p\bar{q}_1(s) = \frac{b\bar{w}_G(p)}{U_\infty}$$

Solving

$$\bar{q}_1(p) = \frac{\dfrac{b}{U_\infty}\bar{w}_G(p)}{p\{(\mu + 0.5)p + 1\}}$$

and thus

$$q_1''(s) = \mathscr{L}^{-1}p^2\bar{q}_1(p)$$

$$= \frac{1}{(\mu + 0.5)}\mathscr{L}^{-1}\frac{b}{U_\infty}\bar{w}_G(p)\cdot\left[1 - \frac{\dfrac{1}{\mu + 0.5}}{p + \dfrac{1}{\mu + 0.5}}\right]$$

$$= \frac{1}{\mu + 0.5}\int_0^s \frac{b}{U_\infty}w_G(\sigma)\cdot\left\{\delta(s - \sigma) - \frac{1}{\mu + 0.5}\exp\left(-\frac{s - \sigma}{\mu + 0.5}\right)\right\}d\sigma \tag{3.7.86}$$

or

$$\ddot{q}_1 = \frac{U_\infty^2}{b^2}q_1''(s) = \frac{1}{\mu + 0.5}\int_0^s \frac{U_\infty}{b}w_G(\sigma)$$

$$\times\left\{\delta(s - \sigma) - \frac{1}{\mu + 0.5}\exp\left(-\frac{s - \sigma}{\mu + 0.5}\right)\right\}d\sigma$$

Since

$$\ddot{q}_{1s} = \frac{U_\infty}{b}\frac{w_G(s)}{\mu} \quad \text{(static result)},$$

$$\frac{\ddot{q}_1}{\ddot{q}_{1s}} = \frac{\mu}{\mu + 0.5}\frac{1}{w_G(s)}\int_0^s w_G(\sigma) \times \left\{\delta(s-\sigma) - \frac{1}{\mu+0.5}\exp\left(-\frac{s-\sigma}{\mu+0.5}\right)\right\} d\sigma \quad (3.7.87)$$

For a sharp edge gust,

$$w_G(s) = w_0 : \text{const} \qquad (s > 0),$$
$$= 0 \qquad (s < 0)$$

(3.7.87) becomes

$$\frac{\ddot{q}_1}{\ddot{q}_{1s}} = \frac{\mu}{\mu+0.5}\exp\left(-\frac{s}{\mu+0.5}\right) \quad (3.7.88)$$

(3.7.88) is presented graphically in the Figure 3.37. From (3.7.88) one may plot the maxima (which occur at $s = 0$ for the quasi-steady aerodynamic theory) vs. μ. These are shown in Figure 3.38 where the results are compared with those using the full unsteady aerodynamic theory and the static aerodynamic theory. What conclusion do you draw concerning the adequacy of the various aerodynamic theories?

(b) *Wing undergoing translation and spanwise bending*

$$M_n\ddot{q}_n + M_n\omega_n^2 q_n = Q_n^M + Q_n^G \qquad n = 1, 2, 3, \ldots \quad (3.7.89)$$

q_1 rigid body mode of translation

$q_2, q_3, \ldots$ beam bending modal amplitude of wing

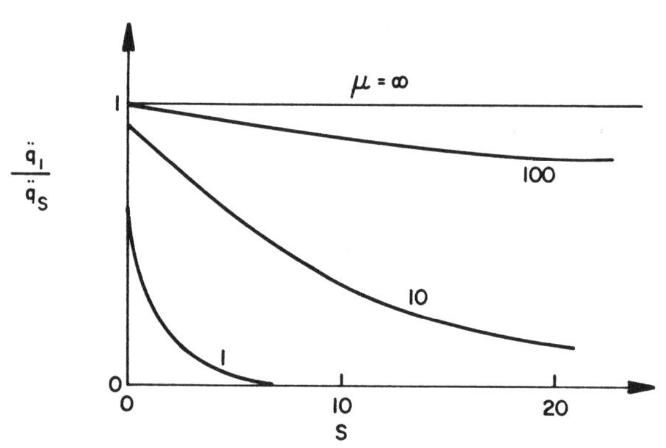

Figure 3.37 Acceleration time history: Quasi-steady aerodynamics.

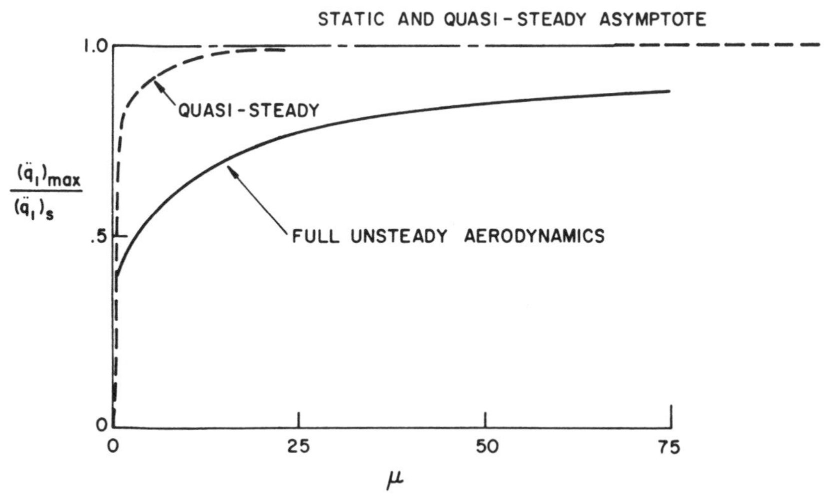

Figure 3.38 Maximum acceleration for wing in translation encountering a sharp edged (step function) gust as given by various aerodynamic models vs mass ratio.

The mode shapes are denoted by $\phi_n(y)$ and are normalized such that the generalized masses are given by

$$M_n \equiv \iint \phi_n^2 m \,\mathrm{d}x\,\mathrm{d}y = \int\left[\int m \,\mathrm{d}x\right]\phi_n^2 \,\mathrm{d}y = M \tag{3.7.90}$$

The generalized forces are given by

$$\begin{aligned} Q_n^M &= \iint P_M \phi_n \,\mathrm{d}x\,\mathrm{d}y = \int L_M \phi_n \,\mathrm{d}y \\ Q_n^G &= \iint P_G \phi_n \,\mathrm{d}x\,\mathrm{d}y = \int L_G \phi_n \,\mathrm{d}y \end{aligned} \tag{3.7.91}$$

Introduce $s \equiv \dfrac{Ut}{b_r}$ where b_r is reference half chord. Also let chord vary spanwise, i.e.,

$$b(y) = b_r g(y) \tag{3.7.92}$$

where g is given from wing geometry. (3.7.89) may be written

$$\frac{U^2}{b_r^2} M q_n'' + M\omega_n^2 q_n = Q_n^M + Q_n^G \tag{3.7.93}$$

Using two-dimensional aerodynamics in a 'strip theory' approximation and assuming the gust velocity is uniform spanwise, the aerodynamic lift forces are

$$L_M(y, s) = -\pi\rho(b_r g)^2 \frac{U^2}{b_r^2} \sum_m \phi_m q''_m$$
$$-2\pi\rho U\left(\frac{U}{b_r}\right)(b_r g)\int_0^s \left(\sum_m \phi_m q''_m(\sigma)\right)\phi(s-\sigma)\, d\sigma$$

and

$$L_G(y, s) = 2\pi\rho U(b_r g)\int_0^s w_G(\sigma)\psi'(s-\sigma)\, d\sigma \tag{3.7.94}$$

Substituting (3.7.94) into (3.7.91) and the result into (3.7.89) gives (when nondimensionalized)

$$\mu[q''_n + \Omega_n^2 q_n] + \sum_{m=1}^{\infty} A_{nm} q''_m + 2\sum_m B_{nm}\int_0^s q''_m(\sigma)\phi(s-\sigma)\, d\sigma$$
$$= 2b_r B_{1n}\int_0^s \frac{w_G(\sigma)}{U}\psi'(s-\sigma)\, d\sigma \qquad n = 1, 2, 3, \ldots \tag{3.7.95}$$

where

$$\mu \equiv \frac{M}{\pi\rho S b_r}, \qquad \Omega_n \equiv \frac{\omega_n b_r}{U}$$
$$A_{nm} \equiv \frac{b_r}{S}\int_{-l/2}^{l/2} g^2 \phi_n \phi_m\, dy$$
$$B_{nm} \equiv \frac{b_r}{S}\int_{-l/2}^{l/2} g \phi_n \phi_m\, dy \tag{3.7.96}$$
$$S \equiv \int_{-l/2}^{l/2} 2b\, dy = 2b_r \int_{-l/2}^{l/2} g\, dy, \quad \text{wing area}$$

(3.7.95) is a set of integral-differential equations in one variable, time. They are mathematically similar to the typical section equations. If we further restrict ourselves to consideration of translation plus the first wing bending mode, we have two equations in two unknowns. These may be solved as in Example (a) by Laplace Transformation. Alternatively, Examples (a) and (b) could be handled by numerical integration in the time domain. Yet another option is to work the problem in the frequency domain.

(c) *Random gusts—solution in the frequency domain.* Pursuing the latter option, we only need replace the Laplace transform variable, p, by $i\omega$ where ω is the Fourier frequency. For simplicity, consider again Example (a). (3.7.81) may be written

$$\frac{\bar{q}_1}{b} = H_{qG}(\omega)\frac{w_{G(\omega)}}{U}$$

where

$$H_{qG}(\omega) \equiv \frac{0.565[(i\omega)^3 + 0.5756(i\omega)^2 + 0.093i\omega + 0.003]}{(\mu + 0.5)(i\omega)[i\omega + 0.13][i\omega + 1][(i\omega)^3 + a_1(i\omega)^2 + a_2(i\omega)a_3]}$$

is a transfer function relating sinusoidal rigid body response to sinusoidal gust velocity. The poles of the transfer function can be examined for stability. The mean square response to a random gust velocity can be written as (cf. equation (3.7.40) in Section 3.3)

$$\overline{\left(\frac{q_1}{b}\right)^2} = \int_0^\infty |H_{qG}(\omega)|^2 \Phi_{(w_G/U)(w_G/U)}\, d\omega \tag{3.7.98}$$

Similar expressions can be obtained for two or more degrees of freedom.

3.8 Nonairfoil physical problems

Fluid flow through a flexible pipe

This problem has received a good deal of attention in the research literature. It has a number of interesting features, including some analogies to the flutter of plates. Possible technological applications include oil pipelines, hydraulic lines, rocket propellant fuel lines, and human lung airways.* The equation of motion is given by

$$EI\frac{\partial^4 w}{\partial x^4} + m\frac{\partial^2 w}{\partial t^2} + \rho A\left[\frac{\partial^2 w}{\partial t^2} + 2U\frac{\partial^2 w}{\partial x\,\partial t} + U^2\frac{\partial^2 w}{\partial x^2}\right] = 0 \tag{3.8.1}$$

EI	bending stiffness of pipe	A	open area of pipe
m	mass/length of pipe	w	transverse deflection of pipe
ρ	fluid density	a	pipe length
U	fluid velocity		

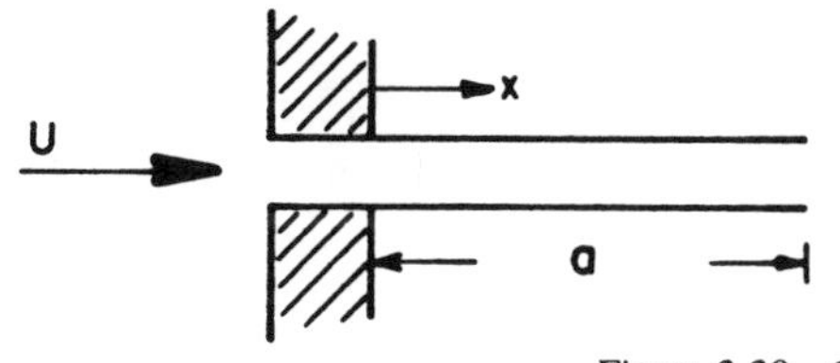

Figure 3.39 Geometry.

* Weaver and Paidoussis [27].

We consider a cantilevered pipe clamped at one end and free at the other. Previously we had considered a pipe pinned at both ends and discovered that a static instability occurred.* The present boundary conditions allow a dynamic instability, flutter. We shall consider a classical eigenvalue analysis of this differential equation. Let

$$w = \bar{w}(x)e^{i\omega t} \tag{3.8.2}$$

where the ω are to be determined by the requirement that nontrivial solutions, $\bar{w}(x) \not\equiv 0$, are sought. Substituting (3.8.2) into (3.8.1) we have

$$\left\{EI\frac{d^4\bar{w}}{dx^4} - m\omega^2\bar{w} + \rho A\left[-\omega^2\bar{w} + 2Ui\omega\frac{d\bar{w}}{dx} + U^2\frac{d^2\bar{w}}{dx^2}\right]\right\}e^{i\omega t} = 0 \tag{3.8.3}$$

This ordinary differential equation may be solved by standard methods. The solution has the form

$$\bar{w}(x) = \sum_{i=1}^{4} C_i e^{p_i x}$$

where $p_1, \ldots, p_4$ are the four roots of

$$EIp^4 - m\omega^2 + \rho A[-\omega^2 + 2Ui\omega p + U^2p^2] = 0 \tag{3.8.4}$$

The four boundary conditions give four equations for $C_1, \ldots, C_4$. They are

$$w(x=0) = 0 \Rightarrow C_1 + C_2 + C_3 + C_4 = 0$$

$$\frac{\partial w}{\partial x}(x=0) = 0 \Rightarrow C_1p_1 + C_2p_2 + C_3p_3 + C_4p_4 = 0$$

$$EI\frac{\partial^2 w}{\partial x^2}(x=a) = 0 \Rightarrow C_1p_1^2e^{p_1a} + C_2p_2^2e^{p_2a} + C_3p_3^2e^{p_3a} + C_4p_4^2e^{p_4a} = 0$$

$$EI\frac{\partial^3 w}{\partial x^3}(x=a) = 0 \Rightarrow C_1p_1^3e^{p_1a} + C_2p_2^3e^{p_2a} + C_3p_3^3e^{p_3a} + C_4p_4^3e^{p_4a} = 0 \tag{3.8.5}$$

Setting the determinant of coefficients of (3.8.5) equal to zero gives

$$D \equiv \begin{vmatrix} 1 & 1 & 1 & 1 \\ p_1 & p_2 & p_3 & p_4 \\ p_1^2e^{p_1a} & p_2^2e^{p_2a} & p_3^2e^{p_3a} & p_4^2e^{p_4a} \\ p_1^3e^{p_1a} & p_2^3e^{p_2a} & p_3^3e^{p_3a} & p_4^3e^{p_4a} \end{vmatrix} = 0 \tag{3.8.6}$$

* Section 2.5.

(3.8.6) is a transcendental equation for ω which has no known analytical solution. Numerical solutions are obtained as follows. For a given pipe at a given U one makes a guess for ω (in general a complex number with real and imaginary parts). The $p_1, \ldots, p_4$ are then evaluated from (3.8.4). D is evaluated from (3.8.6); in general it is not zero and one must improve upon the original guess for ω (iterate) until D is zero. A new U is selected and the process repeated. For $U = 0$, the ω will be purely real and correspond to the natural frequencies of the pipe including the virtual mass of the fluid. Hence, it is convenient to first set $U = 0$ and then systematically increase it. A sketch of ω vs U is shown below in nondimensional form. These results are taken from a paper by Paidoussis who has worked extensively on this problem. When the imaginary part of ω_I becomes negative, flutter occurs. The nondimensional variables used in presenting these results are (we have changed the notation from Paidoussis with respect to frequency)

$$\beta \equiv \rho A/(\rho A + m)$$

$$u \equiv \left(\rho \frac{AU^2}{EI}\right)^{\frac{1}{2}} a$$

$$\Omega \equiv [(m + \rho A)/EI]^{\frac{1}{2}} \omega a^2$$

Also shown are results obtained by a Galerkin procedure using the natural modes of a cantilevered beam.

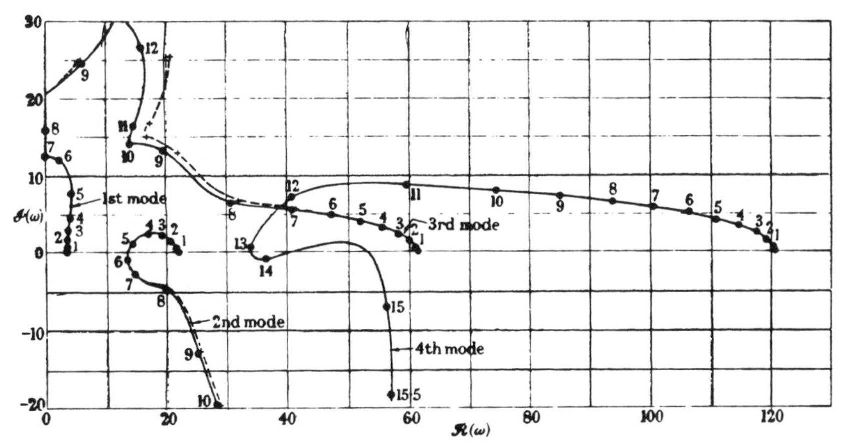

Figure 3.40a The dimensionless complex frequency of the four lowest modes of the system as a function of the dimensionless flow velocity for $\beta = 0.200$. ——, Exact analysis ----four-mode approximation (Galerkin). Numbers on graph are values of u.

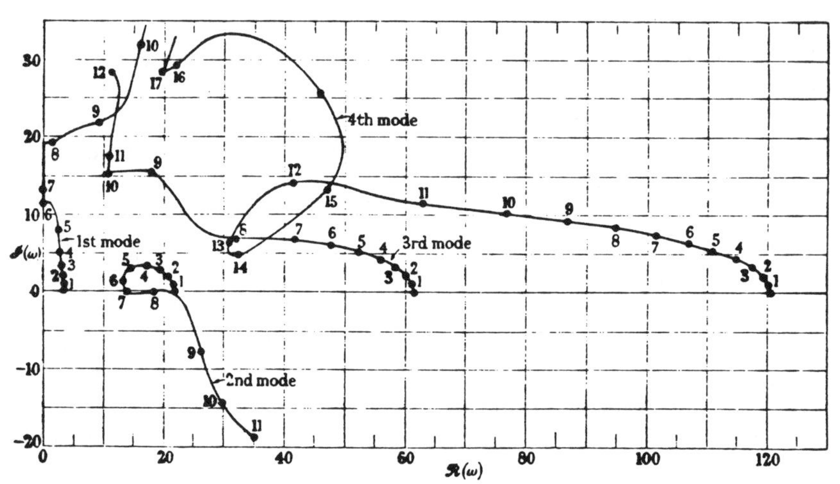

Figure 3.40b The dimensionless complex frequency of the four lowest modes of the system as a function of the dimensionless flow velocity for $\beta = 0.295$.

The stability boundary for this system may be presented in terms of u and β as given in Figure 3.41. Also shown is the frequency, Ω_F, of the flutter oscillation. These results have been verified experimentally by Gregory and Paidoussis.* For a very readable historical and technical review of this problem, see the paper by Paidoussis and Issid.† A similar physical problem arises in nuclear reactor fuel bundles where one has a pipe in an external flow. The work of Chen is particularly noteworthy.‡

(High speed) fluid flow over a flexible wall—a simple prototype for plate flutter

One type of flutter which became of considerable technological interest with the advent of supersonic flight is called 'panel flutter'. Here the concern is with a thin elastic panel supported at its edges. For simplicity consider two dimensional motion. The physical situation is sketched below. Over the top of the elastic plate, which is mounted flush in an otherwise rigid wall, there is an airflow. The elastic bending of the plate in the direction of the airflow (streamwise) is the essential difference between this type of flutter and classical flutter of an airfoil as exemplified by the typical section. It is not our purpose to probe deeply into this

* Gregory and Paidoussis [28].
† Paidoussis and Issid [29].
‡ Chen [30].

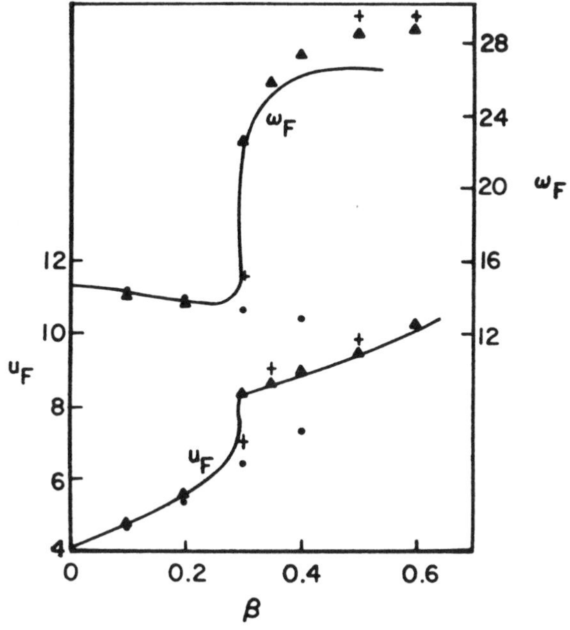

Figure 3.41 *Flutter boundary.*

problem here; for a thorough treatment the reader is referred to Dowell.* We shall instead be content to consider a highly simplified model (somewhat analogous to the typical section model for airfoil flutter) which will bring out some of the important features of this type of problem. Thus we consider the alternative physical model shown below.† Here our model consists of three rigid plates each hinged at both ends. The hinges between the first and second plates and also the second and third plates are supported by springs. The plates have mass per unit length, m, and are of length, l. At high supersonic Mach number, $M \gg 1$, the aerodynamic pressure change (perturbation) p, due to plate motion is

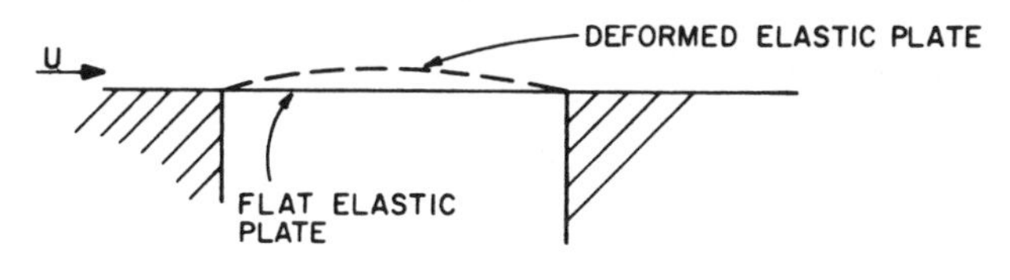

Figure 3.42

* Dowell [31]. Also see Bolotin [32].
† This was suggested by Dr. H. M. Voss.

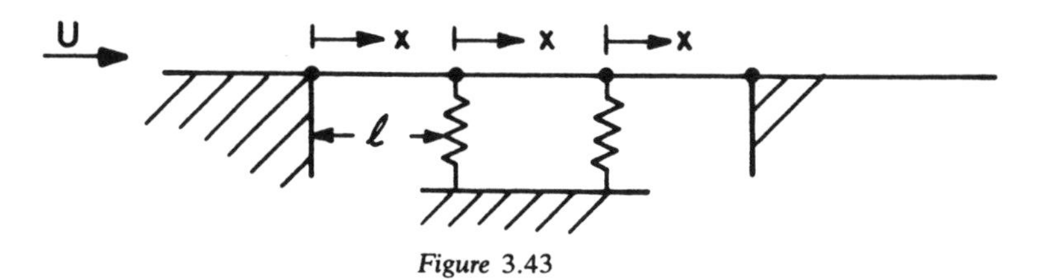

Figure 3.43

given by (see Chapter 4)

$$p = \frac{\rho_\infty U_\infty^2}{M_\infty} \frac{\partial w}{\partial x} \tag{3.8.7}$$

where $w(x, t)$, $\dfrac{\partial w}{\partial x}$ are deflection and slope of any one of the rigid plates.

To write the equations of motion for this physical model we must recognize that there are two degrees of freedom. It is convenient to choose as generalized coordinates, q_1, q_2, the vertical deflections of the springs.

The potential energy of the model is then

$$U = \tfrac{1}{2}kq_1^2 + \tfrac{1}{2}kq_2^2 \tag{3.8.8}$$

The kinetic energy requires expressions for w in terms of q_1 and q_2 since the mass is distributed. For each plate we have, in turn,

$$\text{Plate 1:} \quad w = q_1 \frac{x}{l}, \qquad \frac{\partial w}{\partial x} = q_1/l$$

$$\text{Plate 2:} \quad w = q_1\left[1 - \frac{x}{l}\right] + q_2 x/l, \qquad \frac{\partial w}{\partial x} = \frac{q_2 - q_1}{l}$$

$$\text{Plate 3:} \quad w = q_2\left[1 - \frac{x}{l}\right], \qquad \frac{\partial w}{\partial x} = \frac{-q_2}{l} \tag{3.8.9}$$

Because the plates are rigid, the slopes are constant within each plate. x is measured from the front (leading) edge of each plate. The kinetic energy is

$$T = \frac{1}{2}\int m\left(\frac{\partial w}{\partial t}\right)^2 \mathrm{d}x \tag{3.8.10}$$

Using (3.8.9) in (3.8.10), we obtain after integration

$$T = \tfrac{1}{2}ml[(\tfrac{2}{3})\dot{q}_1^2 + (\tfrac{2}{3})\dot{q}_2^2 + \tfrac{2}{6}\dot{q}_1\dot{q}_2] \tag{3.8.11}$$

The virtual work done by the aerodynamic pressure is given by

$$\delta W = \int (-p)\delta w \,\mathrm{d}x \tag{3.8.12}$$

and using (3.8.9) in (3.8.12) we obtain

$$\delta W = Q_1 \delta q_1 + Q_2 \delta q_2 \tag{3.8.13}$$

where

$$Q_1 \equiv -\frac{\rho_\infty U_\infty^2}{M_\infty} q_2/2$$

$$Q_2 \equiv \frac{\rho_\infty U_\infty^2}{M_\infty} q_1/2$$

Using Lagrange's Equations and (3.8.8), (3.8.11), (3.8.13) the equations of motion are

$$\begin{aligned} &\frac{2}{3} ml\ddot{q}_1 + \frac{ml}{6}\ddot{q}_2 + kq_1 + \frac{\rho_\infty U_\infty^2}{2M_\infty} q_2 = 0 \\ &\frac{ml}{6}\ddot{q}_1 + \frac{2}{3} ml\ddot{q}_2 + kq_2 - \frac{\rho_\infty U_\infty^2}{2M_\infty} q_1 = 0 \end{aligned} \tag{3.8.14}$$

In the usual way we seek an eigenvalue solution to assess the stability of the system, i.e., let

$$q_1 = \bar{q}_1 e^{i\omega t}$$

$$q_2 = \bar{q}_2 e^{i\omega t}$$

then (3.8.14) becomes (in matrix notation)

$$\left[-\omega^2 ml \begin{bmatrix} \frac{2}{3} & \frac{1}{6} \\ \frac{1}{6} & \frac{2}{3} \end{bmatrix} + \begin{bmatrix} k & 0 \\ 0 & k \end{bmatrix} + \frac{\rho_\infty U_\infty^2}{2M_\infty} \begin{bmatrix} 0 & 1 \\ -1 & 0 \end{bmatrix}\right] \begin{Bmatrix} \bar{q}_1 e^{i\omega t} \\ \bar{q}_2 e^{i\omega t} \end{Bmatrix} = \begin{Bmatrix} 0 \\ 0 \end{Bmatrix} \tag{3.8.15}$$

We seek nontrivial solutions by requiring the determinant of coefficients to vanish which gives the following (nondimensional) equation after some algebraic manipulation

$$\tfrac{15}{36}\Omega^4 - \tfrac{4}{3}\Omega^2 + 1 + \lambda^2 = 0 \tag{3.8.16}$$

where

$$\Omega^2 \equiv \frac{\omega^2 ml}{k}, \qquad \lambda \equiv \frac{\rho_\infty U_\infty^2}{2M_\infty k}$$

Solving (3.8.16) for Ω^2 we obtain

$$\Omega^2 = \tfrac{8}{5} \pm \tfrac{2}{5}[1 - 15\lambda^2]^{\frac{1}{2}} \tag{3.8.17}$$

When the argument of the square root becomes negative, Ω^2 becomes complex conjugate and hence one solution for Ω will have a negative imaginary part corresponding to unstable motion. Hence, flutter will occur for

$$\lambda^2 > \lambda_F^2 \equiv \tfrac{1}{15} \tag{3.8.18}$$

The frequency at this λ_F is given by (3.8.17).

$$\Omega_F = \left[\tfrac{8}{5}\right]^{\frac{1}{2}}$$

For reference the natural frequencies ($\lambda \equiv 0$) are from (3.8.17)

$$\Omega_1 = \left(\tfrac{6}{5}\right)^{\frac{1}{2}} \quad \text{and} \quad \Omega_2 = (2)^{\frac{1}{2}}$$

From (3.8.15) (say the first of the equations) the eigenvector ratio may be determined

1st Natural Mode: $\dfrac{\bar{q}_1}{\bar{q}_2} = +1 \quad \text{for} \quad \Omega = \Omega_1 \quad \text{at} \quad \lambda = 0$

2nd Natural Mode: $\dfrac{\bar{q}_1}{\bar{q}_2} = -1 \quad \text{for} \quad \Omega = \Omega_2 \quad \text{at} \quad \lambda = 0$

and at flutter

Flutter Mode: $\dfrac{\bar{q}_1}{\bar{q}_2} = -4 + 15^{\frac{1}{2}} \quad \text{for} \quad \Omega = \Omega_F, \quad \lambda = \lambda_F$

Sketches of the corresponding plate shapes are given below. The important features of this hinged rigid plate model which carry over to an elastic plate are:

(1) The flutter mechanism is a convergence of natural frequencies with increasing flow velocity. The flutter frequency is between the first and second natural frequencies. In this respect it is similar to classical bending-torsion flutter of an airfoil.

(2) The flutter mode shape shows a maximum nearer the rear edge of the plate (rather than the front edge).

There are some oversimplifications in the rigid plate model. For example,

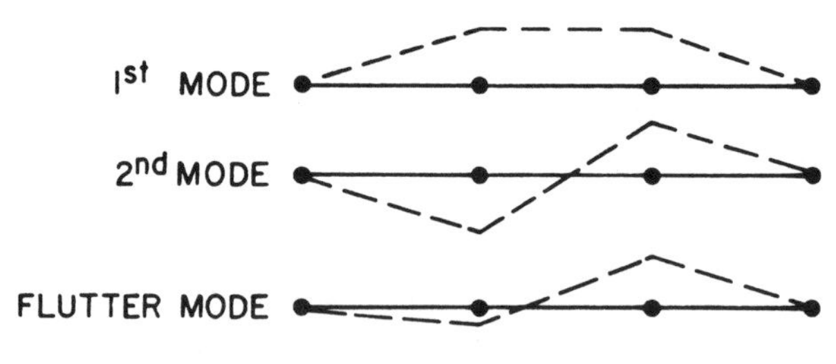

Figure 3.44

the plate length does not affect the flow velocity at which flutter occurs. For an elastic plate, it would. Also in subsonic flow the curvature of the plate has a strong influence on the aerodynamic pressure. In the rigid plate model, the curvature is identically zero, of course. Nevertheless the model serves a useful purpose in introducing this type of flutter problem.

References for Chapter 3

[1] Meirovitch, L., *Methods of Analytical Dynamics*, McGraw-Hill Book Co., New York, 1970.
[2] Bisplinghoff, R. L., Mar, J. W. and Pian, T. H. H., *Statics of Deformable Solids*, Addison-Wesley, 1965.
[3] Timoshenko, S. P. and Goodier, J. N., *Theory of Elasticity*, McGraw-Hill, 1951.
[4] Meirovitch, L., *Elements of Vibration Analysis*, McGraw-Hill, 1975.
[5] Crandall, S. and Mark, W. D., *Random Vibration in Mechanical Systems*, Academic Press, 1963.
[6] Houbolt, J. C., Steiner, R. and Pratt, K. G., 'Dynamic Response of Airplanes to Atmospheric Turbulence Including Flight Data on Input and Response', *NASA TR R*-199 (June 1964).
[7] Acum, W. E. A., *The Comparison of Theory and Experiment of Oscillating Wings*, Vol. II, Chapter 10, AGARD Manual on Aeroelasticity.
[8] Pines, S., 'An Elementary Explanation of the Flutter Mechanism', *Proceedings Nat. Specialists Meeting on Dynamics and Aeroelasticity*, Institute of the Aeronautical Sciences, Ft. Worth, Texas (November 1958) pp. 52–58.
[9] Ashley, H. and Zartarian, G., 'Piston Theory—A New Aerodynamic Tool for the Aeroelastician', *J. Aero. Sci.* Vol. 23, No. 12 (December 1956) pp. 1109–1118.
[10] Sears, W. R., 'Operational Methods in the Theory of Airfoils in Non-uniform Motion', *J. of the Franklin Institute*, Vol. 230, pp. 95–111, 1940.
[11] Jones, R. T., 'Properties of Low Aspect-Ratio Pointed Wings at Speeds Below and Above the Speed of Sound', *NACA Report* 835, 1946.
[12] Dowell, E. H. and Widnall, S. E., 'Generalized Aerodynamic Forces on an Oscillating Cylindrical Shell: Subsonic and Supersonic Flow', *AIAA Journal*, Vol. 4, No. 4 (April 1966) pp. 607–610.
[13] Widnall, S. E. and Dowell, E. H., 'Aerodynamic Forces on an Oscillating Cylindrical Duct with an Internal Flow', *J. Sound Vibration*, Vol. 1, No. 6 (1967) pp. 113–127.
[14] Dowell, E. H., 'Generalized Aerodynamic Forces on a Flexible Cylindrical Shell Undergoing Transient Motion', *Quarterly of Applied Mathematics*, Vol. 26, No. 3 (October 1968) pp. 343–353.
[15] Hamming, R. W., *Numerical Methods for Scientists and Engineers*, McGraw-Hill, 1973.
[16] Houbolt, J. C., 'A Recurrence Matrix Solution for the Dynamic Response of Elastic Aircraft', *J. Aero. Sc.*, Vol. 17, No. 9 (September 1950) pp. 540–550.
[17] Hausner, A., *Analog and Analog/Hybrid Computer Programming*, Prentice-Hall, Inc. 1971.
[18] Savant, C. J., *Basic Feedback Control System Design*, McGraw-Hill, 1958.
[19] Garrick, I. E. and Rubinow, S. L., 'Flutter and Oscillating Air Force Calculations for an Airfoil in a Two-Dimensional Supersonic Flow', *NACA TR* 846, 1946.

[20] Hassig, H. J., 'An Approximate True Damping Solution of the Flutter Equation by Iteration', *J. Aircraft,* Vol. 8, No. 11 (November 1971) pp. 885–889.
[21] Eckhaus, W., 'Theory of Transonic Aileron Buzz, Neglecting Viscous Effects', *J. Aerospace Sciences,* Vol. 29 (June 1962) pp. 712–718.
[22] Landahl, M., *Unsteady Transonic Flow,* Pergamon Press, 1961.
[23] Lambourne, N. C., *Flutter in One Degree of Freedom,* Vol. V., Chapter 5, AGARD Manual on Aeroelasticity.
[24] Abramson, H. N., 'Hydroelasticity: A Review of Hydrofoil Flutter', *Applied Mechanics Reviews,* Vol. 22, No. 2, p. 115, 1969.
[25] Crisp, J. D. C., 'On the Hydrodynamic Flutter Anomaly' *Noise, Shock, and Vibration Conference,* Monash University, Melbourne, Australia, 1974.
[26] Stenton, T. E., 'Turbulence Response Calculations for the XB-70 Airplane and Preliminary Comparison with Flight Data', *presented at the Meeting on Aircraft Response to Turbulence,* NASA Langley Research Center, Sept. 24–25, 1968.
[27] Weaver, D. S. and Paidoussis, M. P., 'On Collapse and Flutter Phenomena in Thin Tubes Conveying Fluid', *J. of Sound Vibration,* 50 (Jan. 8, 1977) pp. 117–132.
[28] Gregory, R. W. and Paidoussis, M. P., 'Unstable Oscillation of Tubular Cantilevers Conveying Fluid. I. Theory. II. Experiments'. *Proc. of the Royal Society A* 293, pp. 512–527, 528–542, 1966.
[29] Paidoussis, M. P. and Issid, N. T., 'Dynamic Instability of Pipes Conveying Fluid', *J. Sound and Vibration,* Vol. 33, No. 3, pp. 267–294, 1974.
[30] Chen, S. S., *Vibration of Nuclear Fuel Bundles,* Nuclear Engineering Design, Vol. 35, pp. 399–422, 1975.
[31] Dowell, E. H., *Aeroelasiticty of Plates and Shells,* Noordhoff International Publishing, Leyden, The Netherlands, 1974.
[32] Bolotin, V. V., *Non-conservative Problems of the Elastic Theory of Stability,* Pergamon Press, 1963.

4

Nonsteady aerodynamics of lifting and non-lifting surfaces

4.1 Basic fluid dynamic equations

Nonsteady aerodynamics is concerned with the *time dependent* fluid motion generated by (solid) bodies moving in a fluid. Normally (and as distinct from classical acoustics) the body motion is composed of a (large) steady motion plus a (small) time dependent motion. In classical acoustics no (large) steady motions are examined. On the other hand, it should be said, in most of classical aerodynamic theory small time dependent motions are ignored, i.e., only small *steady* perturbations from the original steady motion are usually examined. However, in a number of problems arising in aeroelasticity, such as flutter and gust analysis, and also in fluid generated noise, such as turbulent boundary layers and jet wakes, the more general problem must be attacked. It shall be our concern here.* The basic assumptions concerning the nature of fluid are that it be inviscid and its thermodynamic processes be isentropic. We shall first direct our attention to a derivation of the equations of motion, using the apparatus of vector calculus and, of course, allowing for a large mean flow velocity.

Let us recall some purely mathematical relations developed in the vector calculus. These are all variations of what is usually termed Gauss' theorem.†

---

* References: Chapter 7, Liepmann [1].
Chapter 4, BA pp. 70–81, Brief Review of Fundamentals; pp. 82–152, Catalog of available results with some historical perspective (1962).
Chapters 5, 6, 7, BAH, Detailed discussion of the state-of-the art (1955) now largely of interest to aficionados. Read pp. 188–200 and compare with Chapter 4, BA.
AGARD, Vol. II.

† Hildebrand [2].

$$\text{I} \quad \iint c\vec{n}\,\mathrm{d}A = \iiint \nabla c\,\mathrm{d}V$$

$$\text{II} \quad \iint \vec{b}\cdot\vec{n}\,\mathrm{d}A = \iiint \nabla\cdot b\,\mathrm{d}V$$

$$\text{III} \quad \iint \vec{a}(\vec{b}\cdot\vec{n})\,\mathrm{d}A = \iiint [\vec{a}(\nabla\cdot\vec{b})+(\vec{b}\cdot\nabla)\vec{a}]\,\mathrm{d}V$$

Also

$$\text{IV} \quad \nabla(\vec{a}\cdot\vec{a}) = 2(\vec{a}\cdot\nabla)\vec{a} + 2\vec{a}\times(\nabla\times\vec{a})$$

In the above, V is an arbitrary closed volume, A its surface area and $\vec{n}$, the unit normal to the surface, positive outward. $\vec{a}$ and $\vec{b}$ are arbitrary vectors and c an arbitrary scalar.

Conservation of mass

Consider an arbitrary but fixed volume of fluid, V, enclosed by a surface, A. $\vec{q}$ is the (vector) fluid velocity, $\mathrm{d}A$ is the surface elemental area, $\vec{n}$ is surface normal, $\vec{q}\cdot\vec{n}$ is the (scalar) velocity component normal to surface, $\iint \rho\vec{q}\cdot\vec{n}\,\mathrm{d}A$ is the rate of mass flow (mass flux) through surface, positive outward, $\partial/\partial t \iiint \rho\,\mathrm{d}V$ is the rate of mass increase inside volume and $= \iiint (\partial\rho/\partial t)\,\mathrm{d}V$ since V, though arbitrary, is fixed.

The physical principle of continuity of mass states that the fluid increase inside volume = rate of mass flow *into* volume through surface.

$$\iiint \frac{\partial\rho}{\partial t}\mathrm{d}V = -\iint \rho\vec{q}\cdot\vec{n}\,\mathrm{d}A \tag{4.1.1}$$

Using II, the area integral may be transformed to a volume integral. (4.1.1) then reads:

$$\iiint \frac{\partial\rho}{\partial t}\mathrm{d}V = -\iiint \nabla\cdot(\rho\vec{q})\,\mathrm{d}V$$

or

$$\iiint \left[\frac{\partial\rho}{\partial t} + \nabla\cdot(\rho\vec{q})\right]\mathrm{d}V = 0 \tag{4.1.2}$$

Since V is arbitrary, (4.1.2) can only be true for any and all V, if the integrand is zero.

$$\frac{\partial\rho}{\partial t} + \nabla\cdot(\rho\vec{q}) = 0 \tag{4.1.3}$$

This is the conservation of mass, differential equation in three dimensions. Alternative forms are:

$$\frac{\partial \rho}{\partial t} + \rho \nabla \cdot \vec{q} + (\vec{q} \cdot \nabla)\rho = 0$$

$$\frac{D\rho}{Dt} + \rho(\nabla \cdot \vec{q}) = 0 \tag{4.1.4}$$

where

$$\frac{D}{Dt} \equiv \frac{\partial}{\partial t} + (\vec{q} \cdot \nabla)$$

Conservation of momentum

The conservation or balance of momentum equation may be derived in a similar way.

$$\iiint \frac{\partial}{\partial t}(\rho \vec{q})\, \mathrm{d}V$$

is the rate of momentum increase inside the volume

$$\iint \rho \vec{q}(\vec{q} \cdot \vec{n})\, \mathrm{d}A$$

is the rate of momentum flow (momentum flux) through surface, positive outward

$$\iint -p\vec{n}\, \mathrm{d}A$$

is the force acting on volume (recall $\vec{n}$ is positive outward)

The physical principle is that the total rate of change of momentum = force acting on V.

$$\iiint \frac{\partial(\rho \vec{q})}{\partial t}\, \mathrm{d}V + \iint \rho \vec{q}(\vec{q} \cdot \vec{n})\, \mathrm{d}A = \iint -p\vec{n}\, \mathrm{d}A \tag{4.1.5}$$

Using I and III to transform the area integrals and rearranging terms,

$$\iiint \left\{ \frac{\partial}{\partial t}(\rho \vec{q}) + \rho \vec{q}(\nabla \cdot \vec{q}) + (\vec{q} \cdot \nabla)\rho \vec{q} + \nabla p \right\} \mathrm{d}V = 0 \tag{4.1.6}$$

Again because V is arbitrary,

$$\frac{\partial}{\partial t}(\rho\vec{q})+\rho\vec{q}(\nabla\cdot\vec{q})+(\vec{q}\cdot\nabla)\rho\vec{q}=-\nabla p \tag{4.1.7}$$

Alternative forms are

$$\frac{D}{Dt}(\rho\vec{q})+\rho\vec{q}(\nabla\cdot\vec{q})=-\nabla p$$

or

$$\rho\frac{D\vec{q}}{Dt}+\vec{q}\left[\rho\nabla\cdot\vec{q}+\frac{D\rho}{Dt}\right]=-\nabla p \tag{4.1.8}$$

where the bracketed term vanishes from (4.1.8).

Finally to complete our system of equations we have the isentropic relation,

$$p/\rho^{\gamma}=\text{constant} \tag{4.1.9}$$

(4.1.3), (4.1.8) and (4.1.9) are five scalar equations (or two scalar plus one vector equations) in five scalar unknowns: p, ρ, and three scalar components of (vector) velocity, $\vec{q}$.

Irrotational flow, Kelvin's theorem and Bernoulli's equation

To solve these nonlinear, partial differential equations we must integrate them. Generally, this is an impossible task except by numerical procedures. However, there is one integration which may be performed which is both interesting theoretically and useful for applications.

Consider the momentum equation which may be written

$$\frac{D\vec{q}}{Dt}=\frac{-\nabla p}{\rho} \tag{4.1.10}$$

On the right-hand side, using Leibnitz' Rule,* we may write

$$\frac{\nabla p}{\rho}=\nabla\int_{p_{\text{ref}}}^{p}\frac{dp_1}{\rho_1(p_1)} \tag{4.1.11}$$

* Hildebrand [2], pp. 348–353.

where ρ_1, p_1 are dummy integration variables, and p_{ref} some constant reference pressure On the left-hand side

$$\frac{D\vec{q}}{Dt} \equiv \frac{\partial \vec{q}}{\partial t} + (\vec{q} \cdot \nabla)\vec{q}$$

In the above the second term may be written

$$(\vec{q} \cdot \nabla)\vec{q} = \nabla \frac{(\vec{q} \cdot \vec{q})}{2} \quad \text{from IV}$$

and, if we assume the flow is irrotational,

$$\vec{q} = \nabla \phi \tag{4.1.12}$$

where ϕ is the scalar velocity potential. (4.1.12) implies and is implied by

$$\nabla \times \vec{q} = 0 \tag{4.1.13}$$

The vanishing of the curl of velocity is a consequence of Kelvin's Theorem which states that a flow which is initially irrotational, $\nabla \times \vec{q} = 0$, remains so at all subsequent time in the absence of dissipation, e.g., viscosity or shock waves. It can be proven using (4.1.3), (4.1.8) and (4.1.9). No additional assumptions are needed.

Let us pause to prove this result. We shall begin with the momentum equation.

$$\frac{D\vec{q}}{Dt} = -\frac{\nabla p}{\rho}$$

First form $\nabla \times$ and then dot the result into $\vec{n}_A\, dA$ and integrate over A. $\vec{n}_A$ is a unit normal to A and A itself is an arbitrary area of the fluid. The result is

$$\frac{D}{Dt} \iint (\nabla \times \vec{q}) \cdot \vec{n}_A \, dA = -\iint \left[\nabla \times \left(\frac{\nabla p}{\rho}\right)\right] \cdot \vec{n}_A \, dA$$

From Stokes Theorem,*

$$\begin{aligned} -\iint \left[\nabla \times \left(\frac{\nabla p}{\rho}\right)\right] \cdot \vec{n}_A \, dA &= -\iint \frac{\nabla p}{\rho} \cdot d\vec{r} \\ &= -\oint \left(\frac{\partial p}{\partial r} \Big/ \rho\right) dr \\ &= -\oint \frac{dp}{\rho} \end{aligned}$$

* Hildebrand [2], p. 318.

$d\vec{r} \equiv$ arc length along contour of bounding arc of A. Since the bounding contour is closed, and ρ is solely a function of p,

$$\oint \frac{dp}{\rho} = 0$$

Hence

$$\frac{D}{Dt} \iint (\nabla \times \vec{q}) \cdot \vec{n}_A \, dA = 0$$

Since A is arbitrary

$$\nabla \times \vec{q} = \text{constant}$$

and if $\nabla \times \vec{q} = 0$ initially, it remains so thereafter.

Now let us return to the integration of the momentum equation, (4.1.10). Collecting the several terms from (4.1.10)–(4.1.12), we have

$$\frac{\partial}{\partial t}(\nabla \phi) + \nabla \frac{(\nabla \phi \cdot \nabla \phi)}{2} + \nabla \int_{p_{ref}}^{p} \frac{dp_1}{\rho_1} = 0 \tag{4.1.14}$$

or

$$\nabla \left[\frac{\partial \phi}{\partial t} + \frac{\nabla \phi \cdot \nabla \phi}{2} + \int_{p_{ref}}^{p} \frac{dp_1}{\rho_1} \right] = 0$$

or

$$\frac{\partial \phi}{\partial t} + \frac{\nabla \phi \cdot \nabla \phi}{2} + \int_{p_{ref}}^{p} \frac{dp_1}{\rho_1} = F(t) \tag{4.1.15}$$

We may evaluate $F(t)$ by examining the fluid at some point where we know its state. For example, if we are considering an aircraft or missile flying at constant velocity through the atmosphere we know that far away from the body

$$\vec{q} = U_\infty \vec{i}$$

$$\phi = U_\infty x$$

$$p = p_\infty$$

If we choose as the lower limit, $p_{ref} = p_\infty$, then (4.1.15) becomes

$$0 + \frac{U_\infty^2}{2} + 0 = F(t)$$

and we find that F is a constant independent of space *and* time. Hence finally

$$\frac{\partial\phi}{\partial t}+\frac{\nabla\phi\cdot\nabla\phi}{2}+\int_{p_\infty}^{p}\frac{\mathrm{d}p_1}{\rho_1}=\frac{U_\infty^2}{2} \tag{4.1.16}$$

(4.1.16) is usually referred to as Bernoulli's equation although the derivation for nonsteady flow is due to Kelvin.

The practical value of Bernoulli's equation is that it allows one to relate p to ϕ. Using

$$\frac{p}{p_\infty}=\left(\frac{\rho}{\rho_\infty}\right)^{\gamma}$$

one may compute from (4.1.16) (the reader may do the computation)

$$C_p\equiv\frac{p-p_\infty}{\frac{\gamma}{2}p_\infty M^2}$$

$$=\frac{2}{\gamma M^2}\left\{\left[1+\frac{\gamma-1}{2}M^2\left(1-\frac{\left(\vec{q}\cdot\vec{q}+2\frac{\partial\phi}{\partial t}\right)}{U_\infty^2}\right)\right]^{\gamma/(\gamma-1)}-1\right\} \tag{4.1.17}$$

where the Mach number is

$$M^2\equiv\frac{U_\infty^2}{a_\infty^2}$$

and

$$a^2\equiv\frac{\mathrm{d}p}{\mathrm{d}\rho}=\frac{\gamma p}{\rho}$$

is the speed of sound.

Derivation of single equation for velocity potential

Most solutions are obtained by solving this equation.

We shall begin with the conservation of mass equation (4.1.4).

$$\frac{1}{\rho}\frac{\partial\rho}{\partial t}+\frac{\vec{q}\cdot\nabla\rho}{\rho}+\nabla\cdot\vec{q}=0 \tag{4.1.4}$$

Consider the first term. Using Leibnitz' rule we may write

$$\frac{\partial}{\partial t}\int_{p_\infty}^{p}\frac{\mathrm{d}p_1}{\rho_1}=\frac{\partial\rho}{\partial t}\frac{\mathrm{d}p}{\mathrm{d}\rho}\frac{\mathrm{d}}{\mathrm{d}p}\int_{p_\infty}^{p}\frac{\mathrm{d}p_1}{\rho_1}$$

$$=\frac{\partial\rho}{\partial t}a^2\frac{1}{\rho}$$

Thus

$$\frac{1}{\rho}\frac{\partial\rho}{\partial t}=\frac{1}{a^2}\frac{\partial}{\partial t}\int_{p_\infty}^{p}\frac{\mathrm{d}p_1}{\rho_1}$$

$$=-\frac{1}{a^2}\frac{\partial}{\partial t}\left[\frac{\partial\phi}{\partial t}+\frac{\nabla\phi\cdot\nabla\phi}{2}\right] \qquad (4.1.18)$$

from Bernouilli's equation (4.1.16).

In a similar fashion, the second term may be written

$$\vec{q}\cdot\frac{\nabla\rho}{\rho}=\frac{-\vec{q}\cdot\nabla}{a^2}\left[\frac{\partial\phi}{\partial t}+\frac{\nabla\phi\cdot\nabla\phi}{2}\right] \qquad (4.1.19)$$

Finally, the third term

$$\nabla\cdot\vec{q}=\nabla\cdot\nabla\phi=\nabla^2\phi \qquad (4.1.20)$$

Collecting terms, and rearranging

$$-\frac{1}{a^2}\left\{\frac{\partial^2\phi}{\partial t^2}+\frac{\partial}{\partial t}\left(\frac{\nabla\phi\cdot\nabla\phi}{2}\right)+\nabla\phi\cdot\frac{\partial}{\partial t}\nabla\phi+\nabla\phi\cdot\nabla\left(\frac{\nabla\phi\cdot\nabla\phi}{2}\right)\right\}+\nabla^2\phi=0$$

$$\nabla^2\phi-\frac{1}{a^2}\left[\frac{\partial}{\partial t}(\nabla\phi\cdot\nabla\phi)+\frac{\partial^2\phi}{\partial t^2}+\nabla\phi\cdot\nabla\left(\frac{\nabla\phi\cdot\nabla\phi}{2}\right)\right]=0 \qquad (4.1.21)$$

Note we have not yet accomplished what we set out to do, since (4.1.21) is a single equation with *two* unknowns, ϕ and a. A second independent relation between ϕ and a is needed.

The simplest method of obtaining this is to use

$$a^2\equiv\frac{\mathrm{d}p}{\mathrm{d}\rho}$$

and

$$\frac{p}{\rho^\gamma}=\text{constant}$$

in Bernoulli's equation. The reader may verify that

$$\frac{a^2-a_\infty^2}{\gamma-1}=\frac{U_\infty^2}{2}-\left(\frac{\partial\phi}{\partial t}+\frac{\nabla\phi\cdot\nabla\phi}{2}\right) \tag{4.1.22}$$

Small perturbation theory

(4.1.21) and (4.1.22) are often too difficult to solve. Hence a simpler approximate theory is sought.

As in acoustics we shall linearize about a uniform equilibrium state. Assume

$$\begin{aligned} a &= a_\infty + \hat{a} \\ p &= p + \hat{p} \\ \rho &= \rho_\infty + \hat{\rho} \\ \vec{q} &= U_\infty \vec{i} + \vec{q} \qquad \nabla\phi = U_\infty \vec{i} + \nabla\hat{\phi} \\ \phi &= U_\infty x + \hat{\phi} \end{aligned} \tag{4.1.23}$$

Note in the present case we linearize about a uniform flow with velocity, U_∞. Using (4.1.23) in (4.1.21) and retaining lowest order terms:

First term:

$$\nabla^2\phi \rightarrow \nabla^2\hat{\phi}$$

Second term:

$$\begin{aligned} \frac{\partial}{\partial t}(\nabla\phi\cdot\nabla\phi)+\frac{\partial^2\phi}{\partial t^2}+\nabla\phi\cdot\nabla\left(\frac{\nabla\phi\cdot\nabla\phi}{2}\right) &= 2[U_\infty\vec{i}+\nabla\hat{\phi}]\cdot\frac{\partial}{\partial t}[U_\infty\vec{i}+\nabla\hat{\phi}]+\frac{\partial^2\hat{\phi}}{\partial t^2} \\ &\quad +[U_\infty\vec{i}+\nabla\hat{\phi}]\cdot\nabla\left[\frac{U_\infty^2}{2}+U_\infty\vec{i}\cdot\nabla\hat{\phi}+\frac{1}{2}\nabla\hat{\phi}\cdot\nabla\hat{\phi}\right] \\ &= 2U_\infty\frac{\partial^2\hat{\phi}}{\partial x\,\partial t}+\frac{\partial^2\hat{\phi}}{\partial t^2}+U_\infty^2\frac{\partial^2\hat{\phi}}{\partial x^2}+0(\hat{\phi}^2) \end{aligned}$$

Thus the linear or small perturbation equation becomes

$$\nabla^2\hat{\phi}-\frac{1}{a_\infty^2}\left[\frac{\partial^2\hat{\phi}}{\partial t^2}+2U_\infty\frac{\partial^2\hat{\phi}}{\partial x\,\partial t}+U_\infty^2\frac{\partial^2\hat{\phi}}{\partial x^2}\right]=0 \tag{4.1.24}$$

Note that we have replaced a by a_∞ which is correct to lowest order. By examining (4.1.22) one may show that

$$\hat{a} = -\frac{\gamma - 1}{2} \frac{\left[\dfrac{\partial \hat{\phi}}{\partial t} + U_\infty \dfrac{\partial \hat{\phi}}{\partial x}\right]}{a_\infty} \tag{4.1.25}$$

Hence it is indeed consistent to replace a by a_∞ as long as M is not too large where $M \equiv U_\infty / a_\infty$.

In a similar fashion the relationship between pressure and velocity potential, (4.1.17), may be linearized

$$C_p \simeq \frac{\hat{p}}{\dfrac{\gamma}{2} p_\infty M^2} = -\frac{2}{U_\infty} \frac{\partial \hat{\phi}}{\partial x} - \frac{2}{U_\infty^2} \frac{\partial \hat{\phi}}{\partial t}$$

or

$$\hat{p} = -\rho_\infty \left[\frac{\partial \hat{\phi}}{\partial t} + U_\infty \frac{\partial \hat{\phi}}{\partial x}\right] \tag{4.1.26}$$

Reduction to acoustics. By making a transformation of coordinates to a system at rest with respect to the fluid, we may formally reduce the problem to that of classical acoustics.

Define

$$x' \equiv x - U_\infty t$$

$$y' \equiv y$$

$$z' \equiv z$$

$$t' \equiv t$$

then

$$\frac{\partial}{\partial x} = \frac{\partial}{\partial x'}$$

$$\frac{\partial}{\partial t} = \frac{\partial x'}{\partial t} \frac{\partial}{\partial x'} + \frac{\partial t'}{\partial t} \frac{\partial}{\partial t'}$$

$$= -U_\infty \frac{\partial}{\partial x'} + \frac{\partial}{\partial t'}$$

and (4.1.24) becomes the classical wave equation

$$\nabla'^2 \hat{\phi} - \frac{1}{a_\infty^2} \frac{\partial^2 \hat{\phi}}{\partial t'^2} = 0 \tag{4.1.27}$$

as well as (4.1.26) becomes

$$\hat{p} = -\rho_\infty \frac{\partial \hat{\phi}}{\partial t'}$$

The general solution to (4.1.27) is

$$\hat{\phi} = f(\alpha x' + \beta y' + \varepsilon z' + a_\infty t') \\ + g(\alpha x' + \beta y' + \varepsilon z' - a_\infty t')$$

where

$$\alpha^2 + \beta^2 + \varepsilon^2 = 1$$

Unfortunately the above solution is not very useful, nor is the primed coordinate system, as it is difficult to satisfy the boundary conditions on the moving body in a coordinate system at rest with respect to the air (and hence moving with respect to the body). That is, obtaining solutions of (4.1.24) or (4.1.27) is not especially difficult per se. It is obtaining solutions subject to the boundary conditions of interest which is challenging.

Boundary conditions

We shall need to consider boundary conditions of various types and also certain continuity conditions as well. In general we shall see that, at least in the small perturbation theory, it is the boundary conditions, rather than the equations of motion per se, which offer the principal difficulty.

The BODY BOUNDARY CONDITION states the normal velocity of the fluid at the body surface equals the normal velocity of the body.

Consider a body whose surface is described by $F(x, y, z, t) = 0$ at some time, t, and at some later time, $t + \Delta t$, by $F(x + \Delta x, y + \Delta y, z + \Delta z, t + \Delta t) = 0$.

Now

$$\Delta F \equiv F(\vec{r} + \Delta \vec{r}, t + \Delta t) - F(\vec{r}, t) = 0$$

also

$$\Delta F = \frac{\partial F}{\partial x}\Delta x + \frac{\partial F}{\partial y}\Delta y + \frac{\partial F}{\partial z}\Delta z + \frac{\partial F}{\partial t}\Delta t$$

$$= \nabla F \cdot \Delta \vec{r} + \frac{\partial F}{\partial t}\Delta t$$

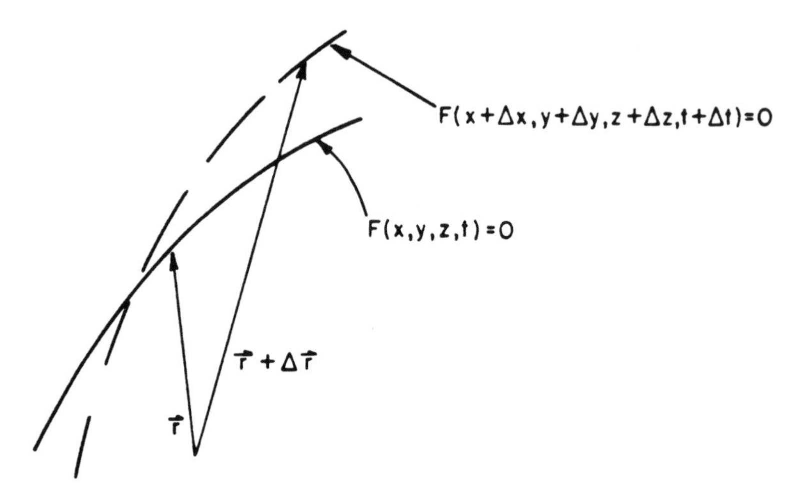

Figure 4.1 Body geometry

Thus

$$\nabla F \cdot \Delta \vec{r} + \frac{\partial F}{\partial t} \Delta t = 0 \tag{4.1.28}$$

Now

$$\vec{n} = \frac{\nabla F}{|\nabla F|} \quad \text{unit normal} \tag{4.1.29}$$

also

$$\vec{V} \equiv \lim_{\Delta t \to 0} \frac{\Delta \vec{r}}{\Delta t} \equiv \text{body velocity}$$

Thus the body *normal* velocity is

$$\begin{aligned} \vec{V} \cdot \vec{n} &= \frac{\Delta \vec{r}}{\Delta t} \cdot \frac{\nabla F}{|\nabla F|} \\ &= -\frac{\partial F}{\partial t} \frac{1}{|\nabla F|} \quad \text{from (4.1.28)} \end{aligned} \tag{4.1.30}$$

The boundary condition on the body is, as stated before, the normal fluid velocity equals the normal body velocity on the body. Thus, using (4.1.28) and (4.1.29) one has

$$\vec{q} \cdot \vec{n} = \vec{q} \cdot \frac{\nabla F}{|\nabla F|} = -\frac{\partial F}{\partial t} \frac{1}{|\nabla F|} \tag{4.1.31}$$

or

$$\frac{\partial F}{\partial t} + \vec{q} \cdot \nabla F = 0 \tag{4.1.32}$$

on the body surface

$$F = 0$$

Example. Planar (airfoil) surface

$$F(x, y, z, t) \equiv z - f(x, y, t)$$

where f is height above the plane, $z = 0$, of the airfoil surface. (4.1.32) may be written:

$$-\frac{\partial f}{\partial t} + [(U_\infty + u)\vec{i} + v\vec{j} + w\vec{k}] \cdot \left[-\frac{\partial f}{\partial x}\vec{i} - \frac{\partial f}{\partial y}\vec{j} + \vec{k} \right] = 0$$

or

$$\frac{\partial f}{\partial t} + (U_\infty + u)\frac{\partial f}{\partial x} + v\frac{\partial f}{\partial y} = w \tag{4.1.33}$$

on

$$z = f(x, y, t) \tag{4.1.34}$$

One may approximate (4.1.33) and (4.1.34) using the concept of a Taylor series about $z = 0$ and noting that $u \ll U_\infty$.

$$\frac{\partial f}{\partial t} + U_\infty \frac{\partial f}{\partial x} = w \quad \text{on} \quad z = 0 \tag{4.1.35}$$

Note

$$w_{\text{on} z = f} = w_{z=0} + \left.\frac{\partial w}{\partial z}\right|_{z=0} f + \text{H.O.T.}$$

$$\simeq w_{z=0}$$

to a consistent approximation within the context of small perturbation theory.

Symmetry and anti-symmetry

One of the several advantages of linearization is the ability to divide the aerodynamic problem into two distinct cases, symmetrical (thickness) and

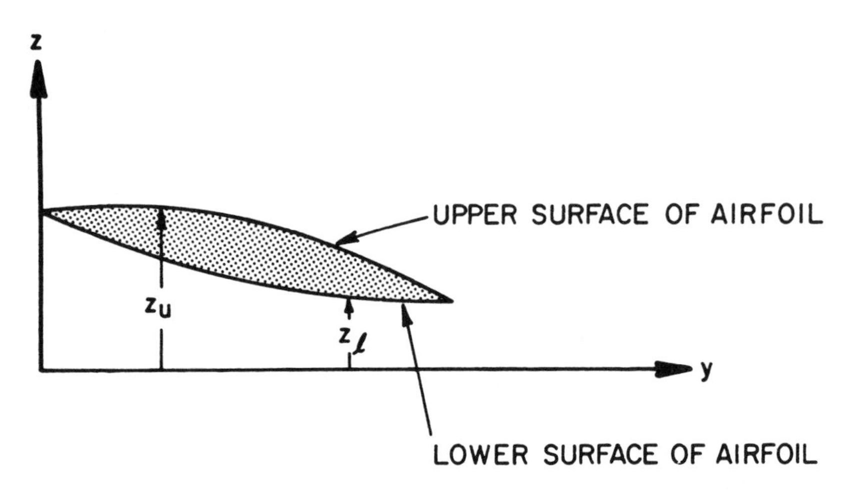

Figure 4.2 Airfoil geometry.

anti-symmetrical (lifting) flow. If one denotes the upper surface by

$$f_{\text{upper}} = z_u(x, y, t)$$

and the lower surface by

$$f_{\text{lower}} = z_l(x, y, t)$$

then it is useful to write

$$\begin{aligned} z_u &\equiv z_t + z_L \\ z_l &\equiv -z_t + z_L \end{aligned} \tag{4.1.36}$$

where (4.1.36) defines z_t, thickness, and z_L, lifting, contributions to z_u and z_l.

One may treat the thickness and lifting cases separately (due to linearity) and superimpose their results for any z_u and z_l. The thickness case is much simpler than the lifting case as we shall see.

Recall (4.1.35), (we henceforward drop the ˆ on ϕ, p)

$$\frac{\partial f}{\partial t} + U_\infty \frac{\partial f}{\partial x} = \frac{\partial \phi}{\partial z}\bigg|_{z=0^+ \text{ or } 0^-} \tag{4.1.35}$$

where + denotes upper surfaces and − denotes lower. From (4.1.35) and (4.1.36), one sees that

Thickness case

$\frac{\partial \phi}{\partial z}$ is anti-symmetric with respect to z (discontinuous across airfoil)

hence ϕ is symmetric (and also p).

Lifting case

$\frac{\partial \phi}{\partial z}$ is symmetric with respect to z (continuous across airfoil)

hence ϕ is anti-symmetric (and also p).

Consider now the pressure difference across the airfoil.

$$\Delta p \equiv p_l - p_u = -\rho\left[\frac{\partial\, \Delta\phi}{\partial t} + U_\infty \frac{\partial\, \Delta\phi}{\partial x}\right]$$

Thus $\Delta p = 0$ for the thickness case, i.e., there is no lift on the airfoil.

The OFF-BODY BOUNDARY CONDITIONS (these are really continuity conditions), state that p and $\vec{q}\cdot\vec{n}$ are continuous across any fluid surface. In particular, for $z = 0$,

$$p_u = p_l \quad \text{and} \quad \left.\frac{\partial \phi}{\partial z}\right|_u = \left.\frac{\partial \phi}{\partial z}\right|_l \tag{4.1.37}$$

(4.1.37) may be used to prove some interesting results.

Thickness case

$$\frac{\partial \phi}{\partial z} = 0 \quad \text{off wing}$$

This follows from the fact that since $\partial\phi/\partial z$ is anti-symmetric, one has

$$\left.\frac{\partial \phi}{\partial z}\right|_{0^+} = -\left.\frac{\partial \phi}{\partial z}\right|_{0^-}$$

But from the second of (4.1.37), this can only be true if

$$\left.\frac{\partial \phi}{\partial z}\right|_{0^+} = \left.\frac{\partial \phi}{\partial z}\right|_{0^-} = 0$$

Lifting case

$p = 0$ off wing

This follows in a similar way using the anti-symmetry of p and the first of (4.1.37).

The BOUNDARY CONDITIONS AT INFINITY are conditions of finiteness or outwardly propagating waves (Sommerfeld radiation condition) which will be imposed at infinity, $z \to \pm\infty$.

4.2 Supersonic flow

It is convenient to distinguish between various flow regimes on the basis of geometry (two or three dimensions) and Mach number (subsonic or supersonic). We shall not give a historical development, but shall instead begin with the simplest regime and proceed to the more difficult problems. Our main focus will be the determination of pressure distributions on airfoils.

*Two-dimensional flow**

This flow regime is the simplest as the fluid ahead of the body remains undisturbed and that behind the body does not influence the pressure distribution on the body. These results follow from the mathematics, but they also can be seen from reasonably simple physical considerations. Take a body moving with velocity, U_∞, through a fluid whose undisturbed speed of sound is a_∞, where $M \equiv U_\infty / a_\infty > 1$. At any point in the fluid disturbed by the passage of the body, disturbances will propagate to the right with velocity, $+a_\infty$, and to the left, $-a_\infty$, *with respect to the fluid.* That is, as viewed in the prime coordinate system. The corresponding propagation velocities as seen with respect to the body or airfoil will be:

$U_\infty - a_\infty$ and $U_\infty + a_\infty$

Note these are both positive, hence the fluid ahead of the airfoil is never disturbed; also disturbances behind the airfoil never reach the body. For subsonic flow, $M < 1$, life is more complicated. Even for *three-dimensional, supersonic* flow one must consider possible effects of disturbances off side edges in the third dimension. Hence the two-dimensional, supersonic problem offers considerable simplification.

* See van der Vooren [3].

One of the consequences of the simplicity, as we will see, is that no distinction between thickness and lifting cases need be made as far as the mathematics is concerned. Hence the body boundary condition is (considering $z > 0$)

$$\left.\frac{\partial \phi}{\partial z}\right|_{z=0} = \frac{\partial z_a}{\partial t} + U_\infty \frac{\partial z_a}{\partial x} \equiv w_a \tag{4.2.1}$$

where one may use $z_a \equiv f$ interchangably and the equation of motion is

$$\nabla^2_{x,z} \phi - \frac{1}{a_\infty^2}\left[\frac{\partial}{\partial t} + U_\infty \frac{\partial}{\partial x}\right]^2 \phi = 0 \tag{4.2.2}$$

Simple harmonic motion of the airfoil. Almost all of the available literature is for simple harmonic motion, that is:

$$\begin{aligned} z_a &= \bar{z}_a(x)e^{i\omega t} \\ w_a &= \bar{w}_a(x)e^{i\omega t} \\ \phi &= \bar{\phi}(x, z)e^{i\omega t} \\ p &= \bar{p}(x, z)e^{i\omega t} \end{aligned} \tag{4.2.3}$$

Hence we shall consider this case first. Thus (4.2.1) becomes:

$$\frac{\partial \bar{\phi}}{\partial z} e^{i\omega t} = \bar{w}_a e^{i\omega t} \tag{4.2.4}$$

and (4.2.2)

$$\bar{\phi}_{xx} + \bar{\phi}_{zz} - \frac{1}{a_\infty^2}\left[-\omega^2 \bar{\phi} + 2i\omega U_\infty \frac{\partial \bar{\phi}}{\partial x} + U_\infty^2 \frac{\partial^2 \bar{\phi}}{\partial x^2}\right] = 0 \tag{4.2.5}$$

Since $\bar{\phi}$, $\partial\bar{\phi}/\partial x$, etc., are zero for $x < 0$, this suggests the possibility of using a Laplace Transform with respect to x, i.e.,

$$\Phi(p, z) \equiv \mathscr{L}\{\bar{\phi}\} = \int_0^\infty \bar{\phi} e^{-px}\, dx \tag{4.2.6}$$

$$W(p) \equiv \mathscr{L}\{\bar{w}_a\} = \int_0^\infty \bar{w}_a e^{-px}\, dx \tag{4.2.7}$$

Taking a transform of (4.2.4) and (4.2.5) gives:

$$\left.\frac{d\Phi}{dz}\right|_{z=0} = W \tag{4.2.8}$$

$$\frac{d^2\Phi}{dz^2} = \mu^2 \Phi \tag{4.2.9}$$

where

$$\mu^2 \equiv (M^2-1)p^2 + 2Mi\frac{\omega p}{a_\infty} - \frac{\omega^2}{a_\infty^2}$$

$$= (M^2-1)\left\{\left[p + \frac{iM\omega}{a_\infty(M^2-1)}\right]^2 + \frac{\omega^2}{a_\infty^2(M^2-1)^2}\right\}$$

Note $M \equiv U_\infty/a_\infty$. (4.2.8) and (4.2.9) are now equations we can solve.

$$\Phi = Ae^{\mu z} + Be^{-\mu z} \tag{4.2.10}$$

Select $A \equiv 0$ to keep Φ finite as $z \to +\infty$. Hence

$$\Phi = Be^{-\mu z}$$

where B can be determined using (4.2.8). From the above,

$$\left.\frac{d\Phi}{dz}\right|_{z=0} = -\mu B$$

Using this result and (4.2.8), one has

$$-\mu B = W$$

or

$$B = -W/\mu$$

and hence

$$\Phi = -(W/\mu)e^{-\mu z} \tag{4.2.11}$$

Inverting (4.2.11), using the convolution theorem,

$$\bar{\phi} = -\int_0^x \bar{w}_a(\xi)\mathscr{L}^{-1}\left\{\frac{e^{-\mu z}}{\mu}\right\} d\xi \tag{4.2.12}$$

and, in particular,

$$\bar{\phi}(x, z=0) = -\int_0^x \bar{w}_a(\xi)\mathscr{L}^{-1}\left\{\frac{1}{\mu}\right\} d\xi$$

From H. Bateman, 'Table of Integral Transforms', McGraw-Hill, 1954,

$$\mathscr{L}^{-1}\left\{\frac{1}{\sqrt{p^2+\alpha^2}}\right\} = J_0(\alpha x)$$

$$\mathscr{L}^{-1}\{F(p+a)\} = e^{-ax}f(x)$$

where $\mathscr{L}^{-1}\{F(p)\} \equiv f(x)$. Thus

$$\mathscr{L}^{-1}\left\{\frac{1}{\mu}\right\} = \frac{\exp\left[-\dfrac{iM\omega}{a_\infty(M^2-1)}(x-\xi)\right]}{(M^2-1)^{\frac{1}{2}}} J_0\left[\frac{\omega}{a_\infty(M^2-1)}(x-\xi)\right] \quad (4.2.13)$$

$\mathscr{L}^{-1}\{e^{-\mu z}/\mu\}$ may be computed by similar methods. In nondimensional terms,

$$\bar{\phi}(x^*, 0) = -\frac{2b}{(M^2-1)^{\frac{1}{2}}} \int_0^{x^*} \bar{w}(\xi^*)\exp[-i\bar{\omega}(x^*-\xi^*)] J_0\left[\frac{\bar{\omega}}{M}(x^*-\xi^*)\right] d\xi^* \quad (4.2.14)$$

where

$$\bar{\omega} \equiv \frac{kM^2}{M^2-1}, \qquad k \equiv \frac{2b\omega}{U_\infty} \text{ is a reduced frequency and}$$

$$x^* \equiv x/2b, \qquad \xi^* \equiv \xi/2b$$

One can now use Bernoulli's equation to compute p.

$$p = -\rho_\infty\left[\frac{\partial\phi}{\partial t} + U_\infty \frac{\partial\phi}{\partial x}\right]$$

or

$$\bar{p} = -\rho_\infty\left[i\omega\bar{\phi} + U_\infty \frac{\partial\bar{\phi}}{\partial x}\right]$$

$$= -\frac{\rho_\infty U_\infty}{2b}\left[\frac{\partial\bar{\phi}}{\partial x^*} + ik\bar{\phi}\right]$$

Using Leibnitz' rule,

$$\bar{p} = -\frac{\rho_\infty U_\infty^2}{(M^2-1)^{\frac{1}{2}}}\left\{\int_0^{x^*}\left[ik\frac{\bar{w}_a}{U_\infty} + \frac{1}{U_\infty}\frac{d\bar{w}}{d\xi^*}\right] e^{-\cdots} J_0[\cdots] d\xi^* + \frac{\bar{w}(0)}{U_\infty} e^{-i\omega x^*} J_0\left[\frac{\bar{\omega}}{M}x^*\right]\right\} \quad (4.2.15)$$

Discussion of inversion. The above inversion was something less than rigorous and, what is more important, in at least one substantial aspect it was misleading. Let us reconsider it, therefore, now that the general outline of the analysis is clear.

Formally the inversion formula reads:

$$\bar{\phi}(x, z) = \frac{1}{2\pi i}\int_{-i\infty}^{i\infty} \Phi(p, z) e^{px} \, dp \quad (4.2.16)$$

Define $\alpha \equiv ip$, (α can be thought of as a Fourier Transform variable), then

$$\bar{\phi}(x, z) = \frac{1}{2\pi}\int_{-\infty}^{\infty} \Phi(-i\alpha, z)\mathrm{e}^{-i\alpha x}\,\mathrm{d}\alpha \tag{4.2.17}$$

and

$$\mu = \sqrt{M^2-1}\sqrt{-\left[-\alpha + \frac{M\omega}{a_\infty(M^2-1)}\right]^2 + \frac{\omega^2}{a_\infty^2(M^2-1)^2}}$$

where

$$\Phi = \pm\frac{W}{\mu}\mathrm{e}^{\pm\mu z} \tag{4.2.18}$$

Consider μ as $\alpha = -\infty \rightarrow +\infty$.

The quantity under the radical changes sign at

$$\alpha = \alpha_1, \ \alpha_2 = \frac{\omega}{a_\infty}\frac{1}{M \pm 1}$$

where $\mu = 0$. Thus

$$\begin{aligned} \mu &= \pm i\,|\mu| \quad \text{for} \quad \alpha < \alpha_1 \quad \text{or} \quad \alpha > \alpha_2 \\ &= \pm|\mu| \quad \text{for} \quad \alpha_1 < \alpha < \alpha_2 \end{aligned}$$

where

$$|\mu| = (M^2-1)\left|-\left[-\alpha + \frac{M\omega}{a_\infty(M^2-1)}\right]^2 + \frac{\omega^2}{a_\infty^2(M^2-1)^2}\right|^{\frac{1}{2}}$$

In the interval, $\alpha_1 < \alpha < \alpha_2$, we have seen we must select the minus sign so that Φ is finite at infinity. What about elsewhere? In particular, when $\alpha < \alpha_1$ and/or $\alpha > \alpha_2$?

The solution for $\phi = \bar{\phi}\mathrm{e}^{i\omega t}$ has the form

$$\phi = -\frac{1}{2\pi}\int_{-\infty}^{\infty} \pm\frac{W}{\mu}\exp\left(\pm\mu z - i\alpha x + i\omega t\right)\mathrm{d}\alpha \tag{4.2.19}$$

In the intervals $\alpha < \alpha_1$ and/or $\alpha > \alpha_2$, (4.2.19) reads:

$$\phi = -\frac{1}{2\pi}\int_{-\infty}^{\infty} \pm i\frac{W}{|\mu|}\exp\left(\pm i\,|\mu|\,z - i\alpha x + i\omega t\right)\mathrm{d}\alpha \tag{4.2.20}$$

To determine the proper sign, we require that solution represent an outgoing wave in the fluid fixed coordinate system, i.e., in the prime system. In the prime system $x' = x - U_\infty t$, $z' = z$, $t' = t$ and thus

$$\phi = -\frac{1}{2\pi}\int_{-\infty}^{\infty} \pm i\frac{W}{|\mu|}\exp[\pm i|\mu|z' - i\alpha x' + i(\omega - U_\infty\alpha)t']\mathrm{d}\alpha \qquad (4.2.21)$$

Consider a z', t' wave for fixed x'. For a wave to be outgoing, if $\omega - U_\infty\alpha > 0$ then one must choose $-$ sign while if $\omega - U_\infty\alpha < 0$ then choose $+$ sign. Note that

$$\omega - U_\infty\alpha = 0$$

when

$$\alpha = \alpha_3 \equiv \frac{\omega}{U_\infty} = \frac{\omega}{a_\infty M}$$

also note that

$$\frac{\omega}{a_\infty(M+1)} \equiv \alpha_1 < \alpha_3 < \alpha_2 \equiv \frac{\omega}{a_\infty(M-1)}$$

Thus the signs are chosen as sketched below.

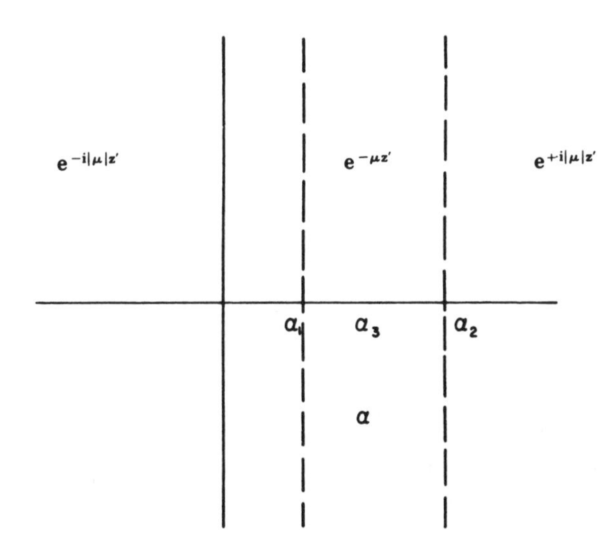

Here again

$$\alpha_1 \equiv \frac{\omega}{a_\infty}\frac{1}{M+1} \qquad \alpha_2 \equiv \frac{\omega}{a_\infty}\frac{1}{M-1} \qquad \alpha_3 \equiv \frac{\omega}{a_\infty}\frac{1}{M}$$

The reader may find it of interest to consider the subsonic case, $M<1$, using similar reasoning.

Knowing the appropriate choice for μ in the several intervals, (4.2.19)–(4.2.21) may be integrated numerically, or by contour integration. The inversion formulae used previously were obtained by contour integration.

Discussion of physical significance of results. Because of the complicated mathematical form of our solution, it is difficult to understand its physical significance. Perhaps it is most helpful for this purpose to consider the limits of low and high frequency.

One may show that (from (4.2.11) et. seq. or (4.2.15))*
$\omega \to 0$: *steady flow*

$$p(x) \to \frac{\rho_\infty U_\infty^2}{\sqrt{M^2-1}} \frac{w}{U}(x), \; p(x, z) = \frac{\rho_\infty U_\infty^2}{\beta} \frac{w(x-\beta z)}{U_\infty}$$

$$\frac{w}{U_\infty} \to \frac{\partial f}{\partial x} \qquad \beta \equiv \sqrt{M^2-1}$$

$\omega \to \infty$: *highly unsteady flow*

$$p(x, t) \to \frac{\rho_\infty U_\infty^2}{M} \frac{w(x, t)}{U_\infty}, \; p(x, z, t) = \frac{\rho_\infty U_\infty^2}{M} \frac{w}{U_\infty}(x - Mz, t)$$

$$\frac{w}{U_\infty} \to \frac{1}{U_\infty} \frac{\partial f}{\partial t} + \frac{\partial f}{\partial x}$$

The latter result may be written as

$$p = \rho_\infty a_\infty w$$

which is the pressure on a piston in a long, narrow (one-dimensional) tube with w the velocity of the piston. It is, therefore, termed 'piston theory' for obvious reasons. Note that in the limits of low and high frequency the pressure at point x depends only upon the downwash at that same point. For arbitrary ω, the pressure at one point depends in general upon the downwash at all other points. See (4.2.15). Hence the flow has a simpler behavior in the limits of small and large ω than for intermediate ω. Also recall that low and high frequency may be interpreted in the time domain for transient motion as long and short time respectively. This follows from the initial and final value Laplace Transform theorems.† For example, if

* See the appropriate example problem in Appendix II for details.
† Hildebrand [2].

we consider a motion which corresponds to a step change in angle of attack, α, we have

$$
\begin{aligned}
f &= -x\alpha \quad \text{for} \quad t>0 \\
&= 0 \quad \text{for} \quad t<0 \\
w &= -\alpha \quad \text{for} \quad t>0 \\
w/U_\infty &= 0 \quad \text{for} \quad t<0
\end{aligned}
$$

Hence for short time, (large ω)

$$p = \frac{-\rho_\infty U_\infty^2}{M}\alpha$$

and long time, (small ω)

$$p = \frac{-\rho_\infty U_\infty^2 \alpha}{\sqrt{M^2-1}}$$

The result for short time may also be deduced by applying a Laplace Transform with respect to time and taking the limit $t \to 0$ of the formal inversion.

General comments. A few general comments should be made about the solution. First of all, the solution has been obtained for simple harmonic motion. In principle, the solution for arbitrary time dependent motion may be obtained via Fourier superposition of the simple harmonic motion result. Actually it is more efficient to use a Laplace Transform with respect to time and invert the time variable *prior* to inverting the spatial variable, x. Secondly, with regard to the distinction between the lifting and thickness cases, one can easily show by direct calculation and using the method applied previously that

thickness	$z=0^+$	$w=w_a$	$p=p^+$
	$z=0^-$	$w=-w_a$	$p=p^+$
lifting	$z=0^+$	$w=w_a$	$p=p^+$
	$z=0^-$	$w=w_a$	$p=-p^+$

where p^+ is the solution previously obtained. Of course these results also follow from our earlier general discussion of boundary conditions.

Gusts. Finally it is of interest to consider how aerodynamic pressures develop on a body moving through a nonuniform flow, i.e., a 'gust'. If the body is motionless, the body boundary condition is that the total fluid velocity be zero on the body.

$$\left.\frac{\partial \phi}{\partial z}\right|_{z=0} + w_G = 0$$

where w_G is the specified vertical 'gust' velocity and $\partial\phi/\partial z$ is the perturbation fluid velocity resulting from the body passing through the gust field. Hence in our previous development we may replace w by $-w_G$ and the same analysis then applies. Frequently one assumes that the gust field is 'frozen', i.e., fixed with respect to the fluid fixed coordinates, x', y', z', t'. Hence

$$\begin{aligned} w_G &= w_G(x', y') \\ &= w_G(x - U_\infty t, y) \end{aligned}$$

Further a special case is a 'sharp edge' gust for which one simply has

$$\begin{aligned} w_G &= w_0 \quad \text{for} \quad x' < 0 \\ &= 0 \quad \text{for} \quad x' > 0 \end{aligned}$$

or

$$\begin{aligned} w_G &= w_0 \quad \text{for} \quad t > x/U_\infty \\ &= 0 \quad \text{for} \quad t < x/U_\infty \end{aligned}$$

These special assumptions are frequently used in applications.

Solutions for the sharp edge gust can be obtained through superposition of (simple harmonic motion) sinusoidal gusts. However, it is more efficient to use methods developed for transient motion. Hence before turning to three-dimensional supersonic flow, we consider transient motion. Transient solutions can be obtained directly (in contrast to Fourier superposition of simple harmonic motion results) for a two-dimensional, supersonic flow.

Transient motion. Taking a Laplace transform with respect to *time* and a Fourier transform with respect to the *streamwise coordinate*, x, the analog of (4.2.11) is

$$\mathcal{L}F\{\phi\}_{\text{at } z=0} = -\frac{\mathcal{L}Fw}{\mu} \tag{4.2.22}$$

$i\omega \equiv s$ is the Laplace Transform variable (where ω was the frequency in the simple harmonic motion result), α is the Fourier Transform variable (where $i\alpha \equiv p$ was the Laplace transform variable used in the previous simple harmonic motion result), $\mathscr{L} \equiv$ Laplace transform, $F \equiv$ Fourier transform, and

$$\mu^2 \equiv -(M^2-1)\alpha^2 + 2\frac{Msi}{a_\infty}\alpha + \frac{s^2}{a_\infty^2}$$

Inverting the Laplace Transform, and using * to denote a Fourier transform

$$\phi^*|_{\text{at } z=0} = -\int_0^t w^*(\tau)\mathscr{L}^{-1}\left\{\frac{1}{\mu}\right\}\bigg|_{t-\tau} \mathrm{d}\tau$$

$$= -a_\infty \int_0^t w^*(\tau) \exp[-i\alpha M a_\infty(t-\tau)] J_0[a_\infty \alpha(t-\tau)] \,\mathrm{d}\tau \quad (4.2.23)$$

Now from (4.1.26),

$$p^* = -\rho_\infty\left[\frac{\partial\phi^*}{\partial t} + U_\infty i\alpha\phi^*\right]$$

Thus using (4.2.23) and the above,

$$p^* = \rho_\infty\left\{a_\infty w^*(t) - a_\infty^2 \int_0^t w^*(\tau)\alpha \exp[-i\alpha M a_\infty(t-\tau)] J_1[\alpha a_\infty(t-\tau)] \,\mathrm{d}\tau\right\}$$

$$\equiv p_0^* + p_1^* \quad (4.2.24)$$

Finally, a formal solution is obtained using

$$p = \frac{1}{2\pi}\int_{-\infty}^{\infty} p^* \mathrm{e}^{i\alpha x} \,\mathrm{d}\alpha \quad (4.2.25)$$

The lift is obtained by using (4.2.24) and (4.2.25) in its definition below.

$$L \equiv -2\int_0^{2b} p \,\mathrm{d}x = -2\rho_\infty a_\infty \int_0^{2b} w \,\mathrm{d}x - \frac{1}{\pi}\int_{-\infty}^{\infty} p_1^*\left[\frac{\mathrm{e}^{i\alpha 2b} - 1}{i\alpha}\right] \mathrm{d}\alpha \quad (4.2.26)$$

In the second term the integration over x has been carried out explicitly.

Lift, due to airfoil motion. Considering a translating airfoil, $w = -\mathrm{d}h/\mathrm{d}t$, for example, we have

$$w^* = -\frac{\mathrm{d}h}{\mathrm{d}t}\frac{[\mathrm{e}^{-i\alpha 2b} - 1]}{-i\alpha}$$

and

$$L = 2\rho_\infty a_\infty \frac{dh}{dt}(2b) + \rho_\infty a_\infty^2 \int_0^t \frac{dh}{dt}(\tau) K(t-\tau)\, d\tau \tag{4.2.27}$$

where

$$K(t-\tau) \equiv -\frac{1}{\pi}\int_{-\infty}^{\infty} \frac{\exp[-i\alpha M a_\infty (t-\tau)]}{\alpha} J_1[e^{i\alpha 2b} - 1][e^{-i\alpha 2b} - 1]\, d\alpha$$

K may be simplified to

$$K(t-\tau) = -\frac{4}{\pi}\int_0^{\infty} \frac{J_1[a_\infty \alpha (t-\tau)] \cos[\alpha M a_\infty (t-\tau)]}{\alpha} \cdot [1 - \cos \alpha 2b]\, d\alpha$$

One can similarly work out aerodynamic lift (and moment) for pitching and other motions.

Lift, due to atmospheric gusts. For a 'frozen gust',

$$w_G(x - U_\infty t) = w_G(x')$$

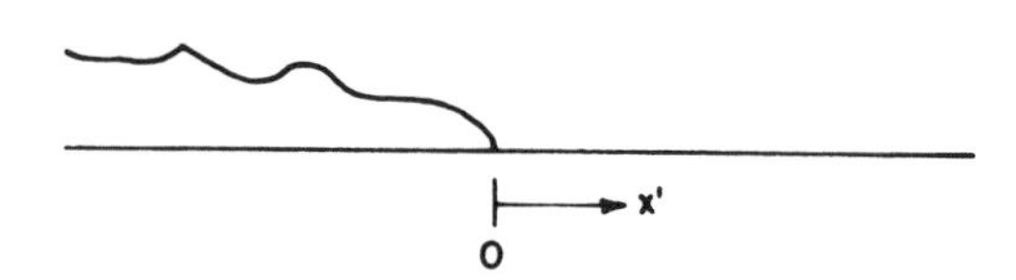

Figure 4.3 Frozen gust geometry in fluid fixed coordinate system.

x, t are coordinates fixed with respect to airfoil and x', t' are coordinates fixed with respect to atmosphere. At $t = t' = 0$ the airfoil enters the gust; the boundary condition is $w_a + w_G = 0$; $w_a = -w_G$ on airfoil. Short and long time correspond to high and low frequency; hence it is of interest to use our previously developed approximate theories for these limits. Subsequently we treat the full transient case.

(i) Piston Theory (short t) on the upper and lower airfoil surfaces

$$p_u = -\rho a_\infty w_G$$

and

$$p_l = +\rho a_\infty w_G$$

Thus

$$L(t) = \int (p_l - p_u)\,dx$$

$$= 2\rho_\infty a_\infty \int_0^{2b} w_G(x - U_\infty t)\,dx$$

For simplicity, we first consider a sharp edge gust.

Let

$$w_G = w_0 \quad \text{for} \quad x' < 0 \quad \text{or} \quad x < Ut,\ t > x/U_\infty$$

$$= 0 \quad \text{for} \quad x' > 0 \quad \text{or} \quad x > Ut,\ t < x/U_\infty$$

Thus

$$L(t) = 2\rho_\infty a_\infty w_0 \int_0^{U_\infty t} dx$$

$$= 2\rho_\infty a_\infty w_0 U_\infty t \quad \text{for} \quad t < \frac{2b}{U_\infty}$$

$$= 2\rho_\infty a_\infty w_0 2b \quad \text{for} \quad t > \frac{2b}{U_\infty} \tag{4.2.28}$$

(ii) Static theory (large t)

$$L(t) = \frac{2\rho_\infty U_\infty^2}{\sqrt{M^2-1}} \frac{w_0}{U_\infty} \int_0^{2b} dx = 4b \frac{\rho_\infty a_\infty w_0 M}{\sqrt{M^2-1}} \tag{4.2.29}$$

(iii) Full Transient Theory from (4.2.24),

$$p = \rho_\infty a_\infty \left[w_a(x, t) - a_\infty \int_0^t \alpha w_a^*(\alpha, \tau) e^{-(\cdots)} J_1(\cdots)\,d\tau \right] \tag{4.2.24}$$

Special case. Sharp Edge Gust

$$w_a = -w_G(x - U_\infty t) = -w_0 \quad \text{for} \quad x < U_\infty t$$

$$= 0 \quad \text{for} \quad x > U_\infty t$$

Thus

$$w_a^*(\alpha, \tau) = -\int_{-\infty}^{\infty} e^{-i\alpha x} w_G(x - U_\infty \tau)\,dx$$

$$= -w_0 \int_{-\infty}^{U_\infty \tau} e^{-i\alpha x}\,dx$$

$$= \frac{-w_0}{-i\alpha} e^{-i\alpha x} \Big|_{-\infty}^{U_\infty \tau}$$

$$= \frac{w_0}{i\alpha} e^{i\alpha U_\infty \tau}$$

Using the above and (4.2.24),

$$p = \rho_\infty a_\infty \left[-w_G(x - U_\infty t) - \frac{a_\infty w_0}{2\pi} \times \int_{-\infty}^{\infty} \int_0^t \alpha \frac{e^{-i\alpha U_\infty \tau}}{i\alpha} e^{-(\cdot)} J_1(\,)\, d\tau e^{i\alpha x}\, d\alpha \right] \quad (4.2.30)$$

Again one may proceed further by computing the lift.

$$L = 2\rho_\infty a_\infty w_0 \begin{matrix} U_\infty t, & \text{for} & U_\infty t < 2b \\ 2b, & \text{for} & U_\infty t > 2b \end{matrix}$$

$$+ 2\rho_\infty \frac{a_\infty^2 w_0}{2\pi} \int_0^{2b} \int_{-\infty}^{\infty} \int_0^t \cdots d\tau\, d\alpha\, dx$$

Integrating over x first, and introducing nondimensional notation

$$s \equiv \frac{tU_\infty}{2b} \qquad \alpha^* \equiv \alpha 2b$$

$$\sigma \equiv \frac{\tau U_\infty}{2b}$$

one obtains

$$\frac{L}{2\rho_\infty U_\infty^2 2b} = \left[\frac{w_0}{U_\infty} \frac{s}{M} - \frac{1}{M^2} \int_0^s F(s, \sigma)\, d\sigma \right] \quad (4.2.31)$$

where

$$F(s, \sigma) \equiv \frac{1}{\pi} \int_0^\infty \frac{[-\cos \alpha^* s + \cos \alpha^*(1 - s)] J_1 \left[\alpha^* \frac{(s - \sigma)}{M} \right]}{\alpha^*} d\alpha^*$$

General case. Arbitrary Frozen Gust

$$w_a^*(\alpha, \tau) = -\int_{-\infty}^{\infty} e^{-i\alpha x} w_G(x - U_\infty \tau)\, dx$$

$$= -\int_{-\infty}^{U\tau} e^{-i\alpha x} w_G(x - U_\infty \tau)\, dx$$

Let $x' = x - U_\infty t$, $dx' = dx$, then

$$w_a^* = -e^{-i\alpha U_\infty \tau} \int_{-\infty}^{0} e^{-i\alpha x'} w_G(x')\, dx$$

$$= -e^{-i\alpha U_\infty \tau} w_G^*(\alpha)$$

Using above in (4.2.24), the pressure is

$$p = \rho_\infty a_\infty \bigg[-w_G(x - U_\infty t) + \frac{a_\infty}{2\pi} \int_{-\infty}^{\infty} \int_0^t w_G^* \alpha e^{-i\alpha U_\infty \tau} e^{-(\,)} J_1(\,)\, d\tau e^{i\alpha x}\, d\alpha \bigg]$$

and the lift,

$$L = 2\rho_\infty a_\infty \int_0^{2b} w_G(x - U_\infty t)\, dx - \frac{2\rho_\infty a_\infty^2}{2\pi} \int_0^{2b} \int_{-\infty}^{\infty} \int_0^t \cdots d\tau\, d\alpha\, dx$$

Integrating over x first,

$$\frac{L}{2\rho_\infty U_\infty^2 2b} = \int_0^{2b} \frac{w_G/U_\infty}{M} \frac{dx}{2b} - \frac{1}{M^2} \int_0^s F(s, \sigma)\, d\sigma \tag{4.2.32}$$

where now

$$F(s, \sigma) \equiv \frac{1}{\pi} \int_0^\infty \bigg\{ W_I^* \{\cos[\alpha^*(1 - s)] - \cos \alpha^* s\} J_1\bigg[\frac{\alpha^*(s - \sigma)}{M}\bigg] + W_R^* \{\sin[\alpha^*(1 - s)] + \sin \alpha^* s\} J_1\bigg[\frac{\alpha^*(s - \sigma)}{M}\bigg]\bigg\}\, d\alpha^*$$

and

$$W^* \equiv \frac{w^*}{U_\infty 2b}$$

For an alternative approach to transient motion which makes use of an analogy between two-dimensional time dependent motion and three-dimensional steady motion, the reader may consult Lomax [4].

This completes our development for two-dimensional, supersonic flow. We now have the capability for determining the aerodynamic pressures necessary for flutter, gust and even, in principle, acoustic analyses for this type of flow. For the latter the pressure in the 'far field' (large z) is usually of interest. Now let us consider similar analyses for three-dimensional, supersonic flow.

Three dimensional flow †

We shall now add the third dimension to our analysis. As we shall see there is no essential complication with respect to solving the governing

† References: BA, pp. 134–139; Landahl and Stark [5], Watkins [6].

differential equation; the principal difficulty arises with respect to satisfying all of the relevant boundary conditions.

The convected wave equation reads in three spatial dimensions and time

$$\nabla^2\phi - \frac{1}{a_\infty^2}\left[\frac{\partial^2\phi}{\partial t^2} + 2U_\infty\frac{\partial^2\phi}{\partial x\,\partial t} + U_\infty^2\frac{\partial^2\phi}{\partial x^2}\right] = 0 \tag{4.2.33}$$

As before we assume simple harmonic time dependence.

$$\phi = \bar{\phi}(x, y, z)e^{i\omega t}$$

Further taking a Laplace transform with respect to x, gives

$$\frac{\partial^2\Phi}{\partial z^2} + \frac{\partial^2\Phi}{\partial y^2} = \mu^2\Phi \tag{4.2.34}$$

where

$$\Phi \equiv \mathscr{L}\bar{\phi} = \int_0^\infty \bar{\phi}e^{-px}\,dx$$

$$\mu = \sqrt{M^2-1}\left[\left(p + \frac{i\omega M}{a_\infty(M^2-1)}\right)^2 + \frac{\omega^2}{a^2(M^2-1)^2}\right]^{\frac{1}{2}}$$

To reduce (4.2.33) to an ordinary differential equation in z, we take a Fourier transform with respect to y. Why would a Laplace transform be inappropriate? The result is:

$$\frac{d^2\Phi^*}{dz^2} = (\mu^2 + \gamma^2)\Phi^* \tag{4.2.35}$$

where

$$\Phi^* \equiv F\Phi = \int_{-\infty}^{\infty} \Phi e^{-i\gamma y}\,dy$$

The solution to (4.2.34) is

$$\Phi^* = A\exp[+(\mu^2+\gamma^2)^{\frac{1}{2}}z] + B\exp[-(\mu^2+\gamma^2)^{\frac{1}{2}}z]$$

Selecting the appropriate solution for finiteness and/or radiation as $z \to +\infty$, we have

$$\Phi^* = B\exp[-(\mu^2+\gamma^2)^{\frac{1}{2}}z] \tag{4.2.36}$$

Applying the body boundary condition (as transformed)

$$\left.\frac{d\Phi^*}{dz}\right|_{z=0} = W^* \tag{4.2.37}$$

we have from (4.2.36) and (4.2.37)

$$B = -\frac{W^*}{(\mu^2+\gamma^2)^{\frac{1}{2}}}$$

and hence

$$\Phi^*_{z=0} = -\frac{W^*}{(\mu^2+\gamma^2)^{\frac{1}{2}}}$$

Using the convolution theorem

$$\bar{\phi}(x, y, z=0) = -\int_0^x \int_{-\infty}^{\infty} \bar{w}(\xi, \eta)\mathscr{L}^{-1}F^{-1}\frac{1}{(\mu^2+\gamma^2)^{\frac{1}{2}}}\,\mathrm{d}\xi\,\mathrm{d}\eta \qquad (4.2.38)$$

Now let us consider the transform inversions. The Laplace inversion is essentially the same as for the two-dimensional case.

$$\mathscr{L}^{-1}\frac{1}{(\mu^2+\gamma^2)^{\frac{1}{2}}} = \frac{\exp\left[-\dfrac{iM\omega x}{a_\infty(M^2-1)}\right]}{\sqrt{M^2-1}} J_0\left(\left[\frac{\omega^2}{a_\infty^2(M^2-1)^2}+\frac{\gamma^2}{(M^2-1)}\right]^{\frac{1}{2}} x\right)$$

To perform the Fourier inversion, we write

$$F^{-1}\left\{\mathscr{L}^{-1}\left\{\frac{1}{(\mu^2+\gamma^2)^{1/2}}\right\}\right\}$$

$$= \frac{\exp\left[-\dfrac{iM\omega x}{a_\infty(M^2-1)}\right]}{2\pi\sqrt{M^2-1}} \int_{-\infty}^{\infty} J_0\left(\left[\frac{\omega^2}{a_\infty^2(M^2-1)^2}+\frac{\gamma^2}{(M^2-1)}\right]^{\frac{1}{2}} x\right) e^{i\gamma y}\,\mathrm{d}\gamma$$

$$= \frac{\exp\left[-\dfrac{iM\omega x}{a_\infty(M^2-1)}\right]}{\pi\sqrt{M^2-1}} \int_0^{\infty} J_0(\cdots)\cos\gamma y\,\mathrm{d}\gamma$$

where the last line follows from the evenness of the integrand with respect to γ. The integral has been evaluated in Bateman, [7], p. 55.

$$\int_0^{\infty} J_0(\cdots)\cos\gamma y\,\mathrm{d}\gamma = \left[\frac{x^2}{M^2-1}-y^2\right]^{-\frac{1}{2}} \cos\left[\frac{\omega}{a_\infty(M^2-1)^{\frac{1}{2}}}\left(\frac{x^2}{M^2-1}-y^2\right)^{\frac{1}{2}}\right]$$

$$\text{for} \quad |y| < \frac{x}{\sqrt{M^2-1}}$$

$$= 0 \quad \text{for} \quad |y| > \frac{x}{\sqrt{M^2-1}}$$

Thus finally

$$F^{-1}\mathscr{L}^{-1}\left\{\frac{1}{(\mu^2+\gamma^2)^{\frac{1}{2}}}\right\}$$

$$= \frac{1}{\pi}\frac{\exp\left[-\frac{iM\omega}{a_\infty(M^2-1)}\right]}{\sqrt{M^2-1}} x \cos\frac{\left[\frac{\omega}{a_\infty(M^2-1)^{\frac{1}{2}}}\left(\frac{x^2}{M^2-1}-y^2\right)^{\frac{1}{2}}\right]}{\left[\frac{x^2}{M^2-1}-y^2\right]^{\frac{1}{2}}}$$

$$\text{for} \quad |y| < \frac{x}{\sqrt{M^2-1}}$$

$$= 0 \quad \text{for} \quad |y| > \frac{x}{\sqrt{M^2-1}}$$

Using the above in (4.2.37) and nondimensionalizing by $s \equiv$ wing semi-span and $b \equiv$ reference semi-chord,

$$\bar{\phi}(x^*, y^*, z=0)$$

$$= \frac{-s}{\pi}\int_0^{x^*}\int_{y^*-(2b/s)(x^*-\xi^*)/\beta}^{y^*+(2b/s)(x^*-\xi^*)/\beta} \bar{w}(\xi^*, \eta^*)\exp[-i\bar{\omega}(x^*-\xi^*)]\frac{\cos\frac{\bar{\omega}r^*}{M}}{r^*}\,\mathrm{d}\xi^*\,\mathrm{d}\eta^* \tag{4.2.39}$$

where

$$r^* \equiv [(x^*-\xi^*)^2 - \beta^2\left(\frac{s}{2b}\right)^2(y^*-\eta^*)^2]^{\frac{1}{2}}$$

$$\beta \equiv \sqrt{M^2-1}$$

$$x^*, \xi^* \equiv x/2b, \xi/2b \qquad y^*, \eta^* \equiv y/s, \eta/s$$

$$k \equiv \frac{\omega 2b}{U_\infty}, \qquad \bar{\omega} \equiv \frac{kM^2}{(M^2-1)}$$

If $\bar{w}$ is known everywhere in the region of integration then (4.2.39) is a solution to our problem. Unfortunately, in many case of interest, $\bar{w}$ is unknown over some portion of the region of interest. Recall that $\bar{w}$ is really $\left.\frac{\partial\bar{\phi}}{\partial z}\right|_{z=0}$. In general this vertical fluid velocity is unknown off the wing. There are three principal exceptions to this:

(1) If we are dealing with a thickness problem then $\left.\frac{\partial\phi}{\partial z}\right|_{z=0} = 0$ everywhere off the wing and no further analysis is required.

(2) Certain wing geometries above a certain Mach number will have undisturbed flow off the wing even in the lifting case. For these so-called 'supersonic planforms', $\left.\frac{\partial \phi}{\partial z}\right|_{z=0} = 0$ off wing as well.

(3) Even in the most general case, there will be no disturbance to the flow ahead of the rearward facing Mach lines, $\eta = \pm \xi/\beta$, which originate at the leading most point of the lifting surface.

To make case (2) more explicit and in order to discuss what must be done for those cases where the flow off the wing is disturbed, let us consider the following figure; Figure 4.4. Referring first to case (2), we see that if the tangents of the forward facing Mach lines (integration limits of (4.2.39))

$$\eta = y \pm \frac{(x-\xi)}{\beta}$$

are sufficiently small, i.e., $\left|\frac{1}{\beta}\right| \to 0$ then the regions where $\bar{w}_a$ unknown, $\bar{\phi} = 0$

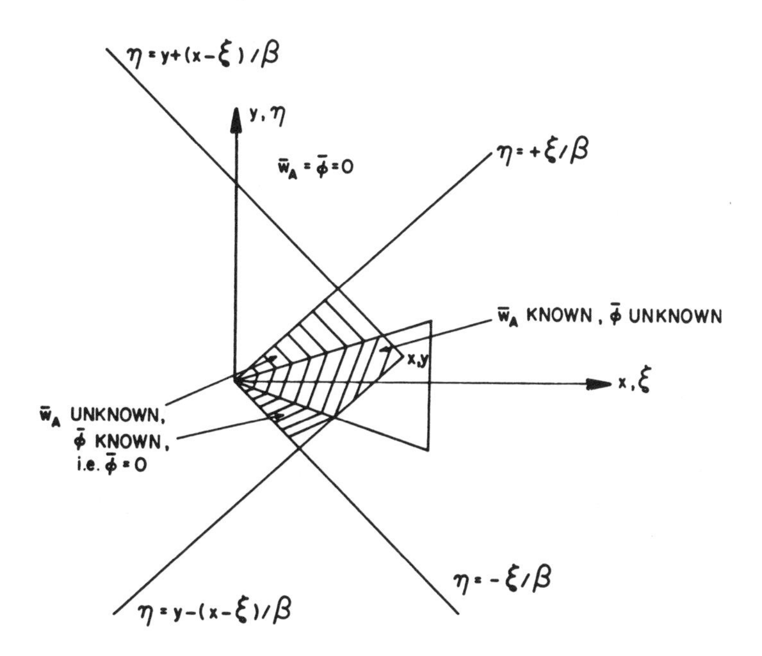

Figure 4.4 Lifting surface geometry. A representative delta wing is shown.

will vanish. This what we mean by a 'supersonic planform'. The mathematical problem for these planforms is essentially the same as for a 'thickness problem' whether or not lift is being produced.

Finally let us consider the most difficult case where we have a mixed boundary condition problem. In general analytical solutions are not possible and we resort to numerical methods. The technique which has received the largest amount of attention and development to date is the 'box' method. In this approach, the integral equation (4.2.39) is approximated by differences and sums, i.e.,

$$\frac{\bar{\phi}(x_i^*, y_j^*)}{U_\infty s} = \sum_{k=1}^{K} \sum_{l=1}^{L} A_{(ij)(kl)} \frac{\bar{w}_a(\xi_k^*, \eta_l^*)}{U_\infty} \tag{4.2.40}$$

where

$$A_{(ij)(kl)} \equiv -\frac{1}{\pi} \exp[-i\bar{\omega}(x_i^* - \xi_k^*)] \frac{\cos \dfrac{\bar{\omega}}{M} r^*_{(ij)(kl)}}{r^*_{(ij)(kl)}} \Delta\xi^* \Delta\eta^*$$

and

$$r^*_{(ij)(kl)} \equiv \left[(x_i^* - \xi_k^*)^2 - \beta^2 \left(\frac{s}{2b}\right)^2 (y_j^* - \eta_l^*)^2\right]^{\frac{1}{2}}$$

$\Delta\xi^*, \Delta\eta^* \equiv$ dimensions of aerodynamic box

$A_{(ij)(kl)}$ aerodynamic influence coefficients; the velocity potential at point, *ij*, due to a unit 'downwash', $\bar{w}_a$, at point *kl*

Equation (4.2.40) can be written in matrix notation as:

$$\left\{\bar{\phi}\right\} = \left[A\right]\left\{\bar{w}_a\right\} \tag{4.2.41}$$

The system of linear equation may be separated as follows:

$$\begin{matrix} N1 \\ — \\ N2 \end{matrix} \left\{\begin{matrix} \bar{\phi} \\ \text{unknown} \\ \hline \bar{\phi} \\ \text{known} \end{matrix}\right\} = \left[\begin{array}{c|c} \begin{matrix} A_1 \\ N1 \times N1 \end{matrix} & \begin{matrix} N2 \times N1 \\ A_2 \end{matrix} \\ \hline \begin{matrix} N1 \times N2 \\ A_3 \end{matrix} & \begin{matrix} N2 \times N2 \\ A_4 \end{matrix} \end{array}\right] \left\{\begin{matrix} \bar{w}_a \\ \text{known} \\ \hline \bar{w}_a \\ \text{unknown} \end{matrix}\right\} \begin{matrix} N1 \\ \\ N2 \end{matrix}$$

where $N1$ number of boxes where $\bar{w}_a$ known, $\bar{\phi}$ unknown (on wing)
$N2$ number of boxes where $\bar{w}_a$ unknown, $\bar{\phi}$ known (off wing)

Using last $N2$ equations of (4.2.42)

$$\bar{\phi}_{\text{known}} = [A_3]\{\bar{w}_{a\text{known}}\} + [A_4]\{\bar{w}_{a\text{unknown}}\} \tag{4.2.43}$$

Solving for $\bar{w}_{a\text{unknown}}$,

$$\begin{aligned}\{\bar{w}_{a\text{unknown}}\} &= [A_4]^{-1}\{\{\bar{\phi}_{\text{known}}\} - [A_3]\{\bar{w}_{a\text{known}}\}\} \\ &= -[A_4]^{-1}[A_3]\{\bar{w}_{a\text{known}}\}\end{aligned} \tag{4.2.44}$$

where we have noted that $\bar{\phi}_{\text{known}} = 0$. Using (4.2.44) in the first $N1$ equations of (4.2.42),

$$\begin{aligned}\{\bar{\phi}_{\text{unknown}}\} &= [A_1]\{\bar{w}_{a\text{known}}\} + [A_2]\{\bar{w}_{a\text{unknown}}\} \\ &= [[A_1] - [A_2][A_4]^{-1}[A_3]]\{\bar{w}_{a\text{known}}\}\end{aligned} \tag{4.2.45}$$

Computer programs have been written to perform the various computations.* Also it should be pointed out that in the evaluation of the 'aerodynamic' influence coefficients it is essential to account for the singular nature of the integrand along the Mach lines. This requires an analytical integration of (4.2.39) over each box with $\bar{w}$ assumed constant and taken outside the integral.

Recent advances in this technique have been made to include more complicated geometries, e.g., nonplanar and multiple surfaces,* and also preliminary efforts have been made to include other physical effects.†

4.3 Subsonic flow‡

Subsonic flow is generally a more difficult problem area since all parts of the flow are disturbed due to the motion of the airfoil. To counter this difficulty an inverse method of solution has been evolved, the so-called 'Kernel Function' approach. To provide continuity with our previous development we shall formulate and solve the problem in a formal way through the use of Fourier Transforms. Historically, however, other methods were used. These will be discussed after we have obtained our formal solution. To avoid repetition, we shall treat the three-dimensional problem straight away.

---

* Many Authors, Oslo AGARD Symposium [8].
† Landahl and Ashley [9].
‡ BA, pp. 125–133; Landahl and Stark, [5], Williams [10].

Bernoulli's equation reads:

$$p = -\rho_\infty\left[\frac{\partial \phi}{\partial t} + U_\infty \frac{\partial \phi}{\partial x}\right] \tag{4.3.1}$$

It will prove convenient to use this relation to formulate our solution in terms of pressure directly rather than velocity potential.

Derivation of the integral equation by transform methods and solution by collocation

As before we will use the transform calculus. Since there is no limited range of influence in subsonic flow we employ Fourier transforms with respect to x *and* y. We shall also assume, as before, simple harmonic time dependent motion. Thus

$$\phi = \bar{\phi}(x, y, z)e^{i\omega t} \tag{4.3.2}$$

and transformed

$$\Phi^* = \int_{-\infty}^{\infty}\int \bar{\phi}(x, y, z) \exp(-i\alpha x - i\gamma y)\, dx\, dy \tag{4.3.3}$$

Hence (4.3.1) may be transformed

$$P^* = -\rho[i\omega + U_\infty i\alpha]\Phi^* \tag{4.3.4}$$

where

$$p = \bar{p}(x, y, z)e^{i\omega t}$$

$$P^* \equiv \int_{-\infty}^{\infty}\int \bar{p} \exp(-i\alpha x - i\gamma y)\, dx\, dy \tag{4.3.5}$$

As in supersonic flow we may relate the (transformed) velocity potential to the (transformed) 'upwash' (see (4.2.36) et. seq.)

$$\Phi^*\Big|_{z=0} = \frac{-W^*}{(\mu^2 + \gamma^2)^{\frac{1}{2}}} \tag{4.3.6}$$

Substituting (4.3.6) into (4.3.4),

$$P^* = \rho_\infty \frac{[i\omega + U_\infty i\alpha]}{(\mu^2 + \gamma^2)^{\frac{1}{2}}} W^*$$

or

$$\frac{\bar{W}^*}{U_\infty} = \frac{P^*}{\rho_\infty U_\infty^2} \frac{(\mu^2+\gamma^2)^{\frac{1}{2}}}{\left[\dfrac{i\omega}{U_\infty} + i\alpha\right]} \tag{4.3.7}$$

Inverting

$$\frac{\bar{w}}{U_\infty}(x, y) = \int_{-\infty}^{\infty} \int K(x-\xi, y-\eta) \frac{\bar{p}}{\rho_\infty U_\infty^2}(\xi, \eta)\, \mathrm{d}\xi\, \mathrm{d}\eta \tag{4.3.8}$$

where

$$K(x, y) \equiv \frac{1}{(2\pi)^2} \int_{-\infty}^{\infty} \int \frac{(\mu^2+\gamma^2)^{\frac{1}{2}}}{\left[\dfrac{i\omega}{U_\infty} + i\alpha\right]} \exp\,(i\alpha x + i\gamma y)\, \mathrm{d}\alpha\, \mathrm{d}\gamma$$

K is physically interpreted as the (non-dimensional) 'upwash', $\bar{w}/U_\infty$ at x, y due to a unit (delta-function) of pressure, $\bar{p}/\rho_\infty U_\infty^2$, at ξ, η. For lifting flow (subsonic or supersonic), $\bar{p} = 0$ off the wing; hence in (4.3.8) the (double) integral can be confined to the wing area. This is the advantage of the present formulation.

Now we are faced with the problem of extracting the pressure from beneath the integral in (4.3.9). By analogy to the supersonic 'box' approach we might consider approximating the integral equation by a double sum

$$\frac{\bar{w}_{ij}}{U_\infty} \simeq \Delta\xi\, \Delta\eta \sum_k \sum_l K_{(ij)(kl)} \frac{\bar{p}_{kl}}{\rho_\infty U_\infty^2} \tag{4.3.9}$$

In matrix notation

$$\left\{\frac{\bar{w}}{U_\infty}\right\} = [\bar{K}\, \Delta\xi\, \Delta\eta] \left\{\frac{\bar{p}}{\rho_\infty U_\infty^2}\right\}$$

and formally inverting

$$\left\{\frac{\bar{p}}{\rho_\infty U_\infty^2}\right\} = \{K\, \Delta\xi\, \Delta\eta\}^{-1} \left\{\frac{\bar{w}}{U_\infty}\right\} \tag{4.3.10}$$

This solution is mathematically incorrect; worse, it is useless. The reason is that it is not unique unless an additional restriction is made, the so-called 'Kutta Condition'.* This restriction states that the pressure on the trailing edge of a thin airfoil must remain finite. For a lifting airfoil

* See Landahl and Stark or Williams, ibid.

this is tantamount to saying it must be zero. This constraint is empirical in nature being suggested by experiment. Other constraints such as zero pressure at the leading edge would also make the mathematical solution unique; however, this would not agree with available experimental data. Indeed these data suggest a pressure maxima at the leading edge; the theory with trailing edge Kutta condition gives a square root singularity at the leading edge.

Although, in principle, one could insure zero pressure at the trailing edge by using a constraint equation to supplement (4.3.9) and/or (4.3.10), another approach has gained favor in practice. In this approach the pressure is expanded in a series of (given) modes

$$\bar{p} = \sum_k \sum_l p_{kl} F_k(\xi) G_l(\eta) \tag{4.3.11}$$

where the functions $F_k(\xi)$ are chosen to satisfy the Kutta condition. (If the wing planform is other than rectangular, a coordinate transformation may need to be made in order to choose such functions readily.) The p_{kl} are, as yet, unknown.

Substituting (4.3.11) into (4.3.8) and integrating over the wing area

$$\frac{\bar{w}}{U_\infty}(x, y) = \sum_k \sum_l \frac{p_{kl}}{\rho_\infty U_\infty^2} \tilde{K}_{kl}(x, y) \tag{4.3.12}$$

where

$$\tilde{K}_{kl}(x, y) \equiv \iint K(x - \xi, y - \eta) F_k(\xi) G_l(\eta) \, d\xi \, d\eta$$

K is singular at $x = \xi$, $y = \eta$ (as we shall see later) and the above integral must be evaluated with some care.

The question remains how to evaluate the unknown coefficient, p_{kl}, in terms of $\bar{w}/U_\infty(x, y)$? The most common procedure is collocation. (4.3.12) is evaluated at a number of points, x_i, y_j, equal to the number of coefficients, p_{kl}. Thus (4.3.12) becomes

$$\frac{\bar{w}(x_i, y_j)}{U_\infty} = \sum_k \sum_l \frac{p_{kl}}{\rho_\infty U_\infty^2} \tilde{K}_{kl}(x_i, y_j) \tag{4.3.13}$$

Defining $\tilde{K}_{ijkl} \equiv \tilde{K}_{kl}(x_i, y_j)$, (4.3.13) becomes

$$\left\{\frac{\bar{w}_{ij}}{U_\infty}\right\} = [\tilde{K}_{(ij)(kl)}]\left\{\frac{p_{kl}}{\rho_\infty U_\infty^2}\right\}$$

Inverting

$$\left\{\frac{\bar{p}}{\rho_\infty U_\infty^2}\right\} = [\tilde{K}]^{-1}\left\{\frac{\bar{w}}{U_\infty}\right\} \tag{4.3.14}$$

This completes our formal solution. Relative to the supersonic 'box' method, the above procedure, the so-called 'Kernel Function' method, has proven to be somewhat delicate. In particular, questions have arisen as to:

(1) the 'optimum' selection of pressure modes
(2) the 'best' method for computing $\tilde{K}$
(3) convergence of the method as the number of pressure modes becomes large

It appears, however, that as experience is acquired these questions are being satisfactorily answered at least on an 'ad hoc' basis.

More recently an alternative approach for solving (4.3.8) has gained popularity which is known as the 'doublet lattice' method. In this method the lifting surface is divided into boxes and collocation is used.*

An alternative determination of the Kernel Function using Green's Theorem

The transform methods are most efficient at least for formal derivations, however historically other approaches were first used. Many of these are now only of interest to history, however we should mention one other approach which is a powerful tool for nonsteady aerodynamic problems. This is the use of Green's Theorem.

First let us review the nature of Green's Theorem.† Our starting point is the Divergence Theorem or Gauss' Theorem.

$$\iiint \nabla \cdot \vec{b}\, \mathrm{d}V = \iint \vec{b} \cdot \vec{n}\, \mathrm{d}S \tag{4.3.15}$$

S surface area enclosing volume V
$\vec{n}$ outward normal
$\vec{b}$ arbitrary vector

* Albano and Rodden [11]. The downwash is placed at the box three-quarters chord and pressure concentrated at the one-quarter chord. For two-dimensional steady flow this provides an exact solution which satisfies the Kutta condition. Lifanov, T. K. and Polanski, T. E., 'Proof of the Numerical Method of "Discrete Vortices" for Solving Singular Integral Equations', PMM (1975), pp. 742–746.
† References: Hildebrand [2] p. 312, Stratton [12], pp. 165–169.

Let $\vec{b} = \phi_1 \nabla \phi_2$ where ϕ_1, ϕ_2 are arbitrary scalars. Then (4.3.15) may be written as:

$$\iiint \nabla \cdot \phi_1 \nabla \phi_2 \, \mathrm{d}V = \iint \vec{n} \cdot \phi_1 \nabla \phi_2 \, \mathrm{d}S$$

Now use the vector calculus identity

$$\nabla \cdot c\vec{a} = c\nabla \cdot \vec{a} + \vec{a} \cdot \nabla c$$

c arbitrary scalar
$\vec{a}$ arbitrary vector

then $\nabla \cdot \phi_1 \nabla \phi_2 = \phi_1 \nabla^2 \phi_2 + \nabla \phi_2 \cdot \nabla \phi_1$ and (4.3.15) becomes

$$\iiint [\phi_1 \nabla^2 \phi_2 + \nabla \phi_2 \cdot \nabla \phi_1] \, \mathrm{d}V = \iint \vec{n} \cdot \phi_1 \nabla \phi_2 \, \mathrm{d}S \tag{4.3.16}$$

This is the first form of Green's Theorem. Interchanging ϕ_1 and ϕ_2 in (4.3.16) and subtracting the result from (4.3.16) gives

$$\iiint [\phi_1 \nabla^2 \phi_2 - \phi_2 \nabla^2 \phi_1] \, \mathrm{d}V = \iint \vec{n} \cdot (\phi_1 \nabla \phi_2 - \phi_2 \nabla \phi_1) \, \mathrm{d}S$$

$$= \iint \left(\phi_1 \frac{\partial \phi_2}{\partial n} - \phi_2 \frac{\partial \phi_1}{\partial n} \right) \mathrm{d}S \tag{4.3.17}$$

This is the second (and generally more useful) form of Green's Theorem. $\partial/\partial n$ denotes a derivative in the direction of the normal. Let us consider several special but informative cases.

(a) $\phi_1 = \phi_2 = \phi$ in (4.3.16)

$$\iiint [\phi \nabla^2 \phi + \nabla \phi \cdot \nabla \phi] \, \mathrm{d}V = \iint \phi \frac{\partial \phi}{\partial n} \, \mathrm{d}S \tag{4.3.18}$$

(b) $\phi_1 = \phi$, $\phi_2 = 1$ in (4.3.17)

$$\iiint \nabla^2 \phi \, \mathrm{d}V = \iint \frac{\partial \phi}{\partial n} \, \mathrm{d}S \tag{4.3.19}$$

(c) $\nabla^2 \phi_1 = 0$, $\phi_2 = 1/r$, $r \equiv \sqrt{(x - x_1)^2 + (y - y_1)^2 + (z - z_1)^2}$ in (4.3.17)

$$\iiint \phi_1 \nabla^2 (1/r) \, \mathrm{d}V = \iint \left[\phi_1 \frac{\partial}{\partial n} - \frac{\partial \phi_1}{\partial n} \right] \frac{1}{r} \, \mathrm{d}S \tag{4.3.20}$$

Now $\nabla^2(1/r)=0$ everywhere except at $r=0$. Thus

$$\iiint \phi_1 \nabla^2(1/r)\,\mathrm{d}V = \phi_1(r=0)\iiint \nabla^2(1/r)\,\mathrm{d}V$$

$$= \phi_1(r=0)\iiint \nabla\cdot\nabla\frac{1}{r}\,\mathrm{d}V$$

from divergence theorem (4.3.15)

$$= \phi_1(r=0)\iint \nabla(1/r)\cdot\frac{\nabla r}{|\nabla r|}\,\mathrm{d}S$$

$$= \phi_1(r=0)\int_0^{2\pi}\int_0^{\pi} -\frac{1}{r^2}\,r^2\sin\theta\,\mathrm{d}\theta\,\mathrm{d}\phi$$

$$= -4\pi\phi_1(r=0)$$

where we consider a small sphere of radius ε, say, in evaluating surface integral. Now

$$\phi_1(r=0)=\phi_1(x_1=x,\ y_1=y,\ z_1=z)=\phi_1(x,y,z)$$

Thus (4.3.20) becomes

$$\phi_1(x,y,z) = -\frac{1}{4\pi}\iint\left[\phi_1\frac{\partial}{\partial n}-\frac{\partial\phi_1}{\partial n}\right]\frac{1}{r}\,\mathrm{d}S \tag{4.3.21}$$

The choice of $\phi_2=1/r$ may seem rather arbitrary. This can be motivated by noting that

$$\frac{\nabla^2\phi_2}{4\pi} = -\delta(x-x_1)\,\delta(y-y_1)\,\delta(z-z_1)$$

Hence we seek a ϕ_2 which is the response to a delta function. This is what leads to the simplification of the volume integral.

Incompressible, three-dimensional flow

To simplify matters we will first confine ourselves to $M=0$. However, similar, but more complex calculations subsequently will be carried out for $M\neq 0$.* For incompressible flow, the equation of motion is

$$\nabla^2\phi = 0$$

or

$$\nabla^2 p = 0$$

* Watkins, Woolston and Cunningham [13], Williams [14].

where ϕ and p are (perturbation) velocity potential and pressure respectively. Hence we may identify ϕ_1 in (4.3.21) with either variable as may be convenient. To conform to convention in the aerodynamic theory literature, we will take the normal positive *into* the fluid and introduce a minus sign into (4.3.21) which now reads:

$$\phi_1(x, y, z) = \frac{1}{4\pi} \iint \left[\phi_1 \frac{\partial}{\partial n} - \frac{\partial \phi_1}{\partial n} \right] \frac{1}{r} \, \mathrm{d}S \tag{4.3.22}$$

For example for a planar airfoil surface

n on S at $z_1 = 0^+$ is $+z_1$

n on S at $z_1 = 0^-$ is $-z_1$

$\mathrm{d}S = \mathrm{d}x_1 \, \mathrm{d}y_1$

Note x, y, z is any given point, while x_1, y_1, z_1 are (dummy) integration variables. See following sketch.

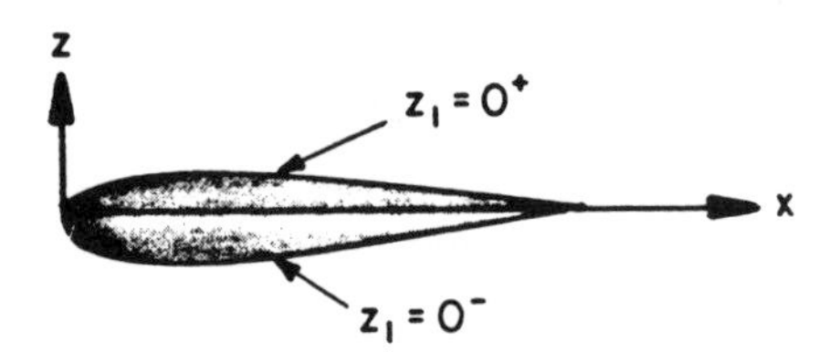

Let us identify the area S of the integral as enclosing the entire fluid, hence the area S is composed of two parts, the area of the airfoil plus wake, call it S_1, and the area of a sphere at infinity, call it S_2.

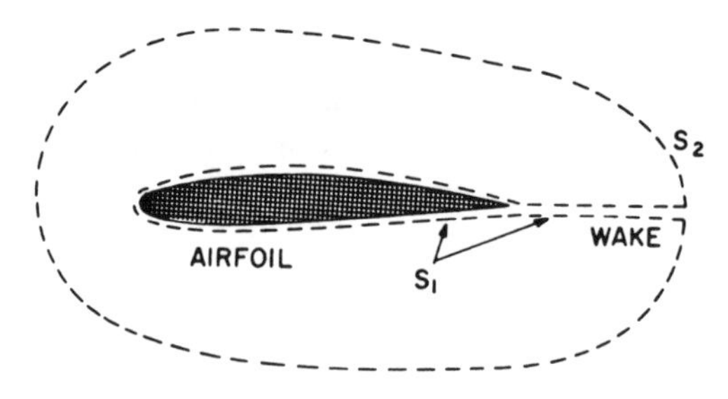

Figure 4.5 Airfoil and flow field geometry.

(i) *Thickness problem (nonlifting).* Let $\phi_1 = \phi$, velocity potential. Because ϕ is bounded at $r \to \infty$, there is no contribution from S_2. Hence

$$\phi(x, y, z) = \frac{1}{4\pi} \iint_{S_1 \text{ at } z_1 = 0^+} \left[\phi \frac{\partial}{\partial z_1} - \frac{\partial \phi}{\partial z_1} \right] \frac{1}{r} \, dS$$

$$+ \frac{1}{4\pi} \iint_{S_1 \text{ at } z_1 = 0^-} \left[\phi \left(-\frac{\partial}{\partial z_1} \right) - \left(-\frac{\partial \phi}{\partial z_1} \right) \right] \frac{dS}{r}$$

Now $\phi_{z_1=0^+} = \phi_{z_1=0^-}$ for thickness problem and

$$\left. \frac{\partial \phi}{\partial z_1} \right|_{z_1=0^+} = - \left. \frac{\partial \phi}{\partial z_1} \right|_{z_1=0^-}$$

Thus

$$\phi(x, y, z) = -\frac{1}{2\pi} \iint \left. \frac{\partial \phi}{\partial z_1} \right|_{z_1=0^+} \frac{dS}{r}$$

using body boundary condition

$$= -\frac{1}{2\pi} \iint w \frac{dS}{r} \tag{4.3.23}$$

where

$$w \equiv \frac{\partial z}{\partial t} + U_\infty \frac{\partial z}{\partial x}$$

Note this solution is valid for arbitrary time-dependent motion. Time only appears as a parameter in the solution $\phi(x, y, z) = \phi(x, y, z; t)$. This is a special consequence of $M \equiv 0$.

(ii) *Lifting problem.* For the lifting problem it again will prove convenient to use pressure rather than velocity potential. (4.3.22) becomes

$$p(x, y, z) = \frac{1}{4\pi} \iint \left[(p_{z=0^+} - p_{z=0^-}) \frac{\partial}{\partial z_1} \left(\frac{1}{r} \right) - \left(\left. \frac{\partial p}{\partial z_1} \right|_{z_1=0^+} - \left. \frac{\partial p}{\partial z_1} \right|_{z_1=0^-} \right) \frac{1}{r} \right] dS$$

Now

$$p_{z_1=0^+} = -p_{z_1=0^-}$$

for lifting problem and

$$\left.\frac{\partial p}{\partial z_1}\right|_{z_1=0^+} - \left.\frac{\partial p}{\partial z_1}\right|_{z_1=0^-} = 0$$

Thus

$$p(x, y, z) = \frac{1}{4\pi} \iint \Delta p \frac{\partial}{\partial z_1}\left(\frac{1}{r}\right) \mathrm{d}S \tag{4.3.24}$$

where

$$\Delta p \equiv p_{z=0^+} - p_{z=0^-}$$

(4.3.24) as it stands is not particularly helpful. We do not know either p or Δp. However we can relate p to something we do know, w. To simplify matters we shall specify simple harmonic motion,

$$p = \bar{p} \mathrm{e}^{i\omega t}$$
$$\phi = \bar{\phi} \mathrm{e}^{i\omega t}$$

hence from Bernoulli's equation

$$\bar{p} = -\rho_\infty \left[i\omega\bar{\phi} + U_\infty \frac{\partial \bar{\phi}}{\partial x} \right] \tag{4.3.25}$$

Solving (4.3.25), by variation of parameters,

$$\bar{\phi}(x, y, z) = -\int_{-\infty}^{x} \frac{\bar{p}}{\rho_\infty U_\infty}(\lambda, y, z) \exp\left[i\frac{\omega}{U_\infty}(\lambda - x) \right] \mathrm{d}\lambda \tag{4.3.26}$$

and using (4.3.24), one has

$$\bar{\phi}(x, y, z) = -\int_{-\infty}^{x} \exp\left[i\frac{\omega}{U_\infty}(\lambda - x) \right]$$
$$\cdot \left\{ \frac{1}{4\pi} \iint \frac{\Delta \bar{p}}{\rho_\infty U_\infty}(x_1, y_1, z_1 = 0) \frac{\partial}{\partial z_1}\left(\frac{1}{r(\lambda)}\right) \mathrm{d}S \right\} \mathrm{d}\lambda$$

where

$$r(\lambda) \equiv \sqrt{(\lambda - x_1)^2 + (y - y_1)^2 + (z - z_1)^2}$$
$$\mathrm{d}S \equiv \mathrm{d}x_1\, \mathrm{d}y_1$$

Define

$$\xi \equiv \lambda - x_1,\ \mathrm{d}\lambda = \mathrm{d}\xi,\ \lambda = \xi + x_1$$

and interchange order of integration with respect to ξ and S, then

$$\bar{\phi}(x, y, z) = -\frac{1}{4\pi} \iint \frac{\Delta \bar{p}}{\rho_\infty U_\infty}(x_1, y_1, z_1 = 0)$$

$$\cdot \left\{ \int_{-\infty}^{x-x_1} \frac{\partial}{\partial z_1}\left(\frac{1}{r(\xi)}\right) \exp\left\{ i \frac{\omega}{U_\infty}[\xi - (x - x_1)] \right\} \mathrm{d}\xi \right\} \mathrm{d}S$$

Compute $\partial\bar{\phi}/\partial z$ and set equal to $\bar{w}$ from body boundary condition, on $z = 0$.

$$\bar{w} = -\frac{1}{4\pi} \lim_{z \to 0} \iint \frac{\Delta \bar{p}}{\rho_\infty U_\infty}$$

$$\times \left\{ \frac{\partial}{\partial z} \int_{-\infty}^{x-x_1} \frac{\partial}{\partial z_1}\left(\frac{1}{r}\right) \exp\left\{ i \frac{\omega}{U_\infty}[\xi - (x - x_1)] \right\} \mathrm{d}\xi \right\} \mathrm{d}S$$

Now

$$\frac{\partial}{\partial z}\left(\frac{1}{r}\right) = -\frac{\partial}{\partial z_1}\left(\frac{1}{r}\right)$$

therefore

$$\frac{\bar{w}}{U_\infty} = \iint \frac{\Delta \bar{p}}{\rho_\infty U_\infty^2}(x_1, y_1, z_1 = 0) K(x - x_1, y - y_1, 0)\, \mathrm{d}x_1\, \mathrm{d}y_1 \qquad (4.3.27)$$

where

$$K(x - x_1, y - y_1, 0) \equiv \frac{1}{4\pi} \lim_{z \to 0} \frac{\partial^2}{\partial z^2} \int_{-\infty}^{x-x_1} \frac{\exp\left\{ \frac{i\omega}{U_\infty}[\xi - (x - x_1)] \right\} \mathrm{d}\xi}{r}$$

and where

$$r \equiv \sqrt{\xi^2 + z^2 + (y - y_1)^2}$$

(4.3.27), of course, has the same form as we had previously derived by transform methods.

The expression for the Kernel function may be simplified.

$$K(x - x_1, y - y_1, 0) = \frac{\exp\left[-\frac{i\omega}{U_\infty}(x - x_1) \right]}{4\pi} \int_{-\infty}^{x-x_1} \mathrm{e}^{+\frac{i\omega\xi}{U_\infty}} \lim_{z \to 0} \frac{\partial^2}{\partial z^2} \frac{1}{r}\, \mathrm{d}\xi$$

Now

$$\frac{\partial^2}{\partial z^2}\frac{1}{r} = -\frac{1}{2} r^{-3} 2 + (-1/2)(-3/2) r^{-5} (2z)^2$$

thus

$$\lim_{z\to 0}\frac{\partial^2}{\partial z^2}\frac{1}{r} = -[\xi^2+(y-y_1)^2]^{-3/2}$$

and finally

$$K = -\frac{\exp\left[-\dfrac{i\omega}{U_\infty}(x-x_1)\right]}{4\pi}\int_{-\infty}^{x-x_1}\frac{\exp\left[+\dfrac{i\omega\xi}{U_\infty}\right]}{[\xi^2+(y-y_1)^2]^{+3/2}}\,d\xi \qquad (4.3.28)$$

The integral in (4.3.28) must be evaluated numerically.

Compressible, three-dimensional flow

For the more general case of $M\neq 0$, we have an additional complication since

$$\nabla^2\phi\neq 0$$

For simple harmonic motion, the equation of motion reads

$$\nabla^2\bar{\phi}+\mathscr{L}\bar{\phi}=0 \qquad (4.3.29)$$

where

$$\mathscr{L}\equiv -\frac{1}{a^2}\left[(i\omega)+U\frac{\partial}{\partial x}\right]^2$$

By making a transformation we may reduce the compressible equation to a simpler form. Defining

$$x^1\equiv x,\ y^1\equiv\beta y,\qquad z^1\equiv\beta z$$

$$\beta\equiv\sqrt{1-M^2}$$

$$\bar{\phi}\equiv\exp\left[i\frac{M^2}{(1-M^2)}\frac{\omega}{U_\infty}x\right]\phi^*\ \dagger$$

The equation for ϕ^* is

$$\exp\left[i\frac{M^2}{(1-M^2)}\frac{\omega}{U_\infty}x^1\right][\nabla^{1^2}\phi^*+k^2\phi^*]=0 \qquad (4.3.30)$$

where

$$k\equiv\left[\frac{M}{(1-M^2)}\right]\frac{\omega}{U_\infty}$$

† By assuming a transformation of the form $e^{\Omega x}\phi^*=\bar{\phi}$, one can always determine Ω such that (4.3.29) reduces to (4.3.31).

Note this equation is essentially the reduced wave equation in its form. We shall use Green's Theorem on ϕ^* and then transform back to $\bar{\phi}$. Let

$$\nabla^2\phi_1^* + k^2\phi_1^* = 0 \tag{4.3.31}$$

$$\nabla^2\phi_2^* + k^2\phi_2^* = \delta(x^1 - x_1^1)\,\delta(y^1 - y_1^1)\,\delta(z^1 - z_1^1)$$

Solving for ϕ_2^*,

$$\phi_2^* = -\frac{e^{-ikr}}{4\pi r}$$

where

$$r = \sqrt{(x^1 - x_1^1)^2 + (y^1 - y_1^1)^2 + (z^1 - z_1^1)^2}$$

From (4.3.17)6,

$$\iiint [\phi_1^*(\delta - k^2\phi_2^*) - \phi_2^*(-k^2\phi_1^*)]\,dV = \iint \left[\phi_1^*\frac{\partial\phi_2}{\partial n} - \phi_2^*\frac{\partial\phi_1^*}{\partial n}\right] dS \tag{4.3.17a}$$

or

$$\phi_1^*(x, y, z) = -\frac{1}{4\pi}\iint \left[\phi_1^*\frac{\partial}{\partial n} - \frac{\partial\phi_1^*}{\partial n}\right]\frac{e^{-ikr}}{r}\,dS \tag{4.3.21a}$$

or

$$\phi_1^*(x, y, z) = +\frac{1}{4\pi}\iint \left[\phi_1^*\frac{\partial}{\partial n} - \frac{\partial\phi_1^*}{\partial n}\right]\frac{e^{-ikr}}{r}\,dS \tag{4.3.22a}$$

(if we redefine positive normal). Using symmetry and anti-symmetry properties of $\dfrac{\partial\phi_1^*}{\partial n}$ and ϕ_1^*

$$\phi_1^*(x, y, z) = \frac{1}{4\pi}\iint \Delta\phi_1^*\frac{\partial}{\partial z_1}\left\{\frac{e^{-ikr}}{r}\right\} dS \tag{4.3.24a}$$

where

$$\Delta\phi_1^* = \phi^*_{1_{z_1=0^+}} - \phi^*_{1_{z_1=0}}$$

and

$$-\left.\frac{\partial\phi_1^*}{\partial z_1}\right|_{z_1=0^+} + \left.\frac{\partial\phi_1^*}{\partial z_1}\right|_{z_1=0^-} = 0$$

Note $dS \equiv dx_1\, dy_1$ and

$$\left(\frac{\partial}{\partial z_1}\right) dx_1\, dy_1 = \left(\frac{\partial}{\partial z_1^1}\right) dx_1^1\, dy_1^1; \qquad x_1 = x_1^1$$

From (4.3.24a) and the definition of ϕ^*

$$\begin{aligned}\bar{\phi}_1 &= \exp\left[i\frac{M^2}{(1-M^2)}\frac{\omega}{U_\infty}x\right]\phi_1^*(x, y, z)\\ &= \frac{\exp\left[i\dfrac{M^2}{(1-M^2)}\dfrac{\omega}{U_\infty}x\right]}{4\pi}\\ &\quad\times\iint \Delta\bar{\phi}_1 \exp\left[-\frac{M^2}{(1-M^2)}\frac{\omega}{U_\infty}x_1\right]\frac{\partial}{\partial z_1}\left\{\frac{e^{-ikr}}{r}\right\}dS\end{aligned} \tag{4.3.32}$$

Identifying $\bar{\phi}_1$ with $\bar{p}$ and using (4.3.32) in (4.3.26),

$$\begin{aligned}\bar{\phi}(x, y, z, \omega) = -\frac{1}{4\pi}\int_{-\infty}^{x} &\exp\left[i\frac{M^2}{(1-M^2)}\frac{\omega}{U_\infty}\lambda\right]\exp\left[i\frac{\omega}{U_\infty}(\lambda - x)\right]\\ &\cdot\left\{\iint \frac{\Delta\bar{p}}{\rho_\infty U_\infty}\exp\left[-i\frac{M^2}{(1-M^2)}\frac{\omega}{U_\infty}x_1\right]\frac{\partial}{\partial z_1}\left\{\frac{e^{-ikr}}{r}\right\}dS\right\}d\lambda\end{aligned}$$

Define $\xi \equiv \lambda - x_1$, $d\lambda = d\xi$, $\lambda = \xi + x_1$ and interchange order of integration with respect to ξ and S,

$$\begin{aligned}\bar{\phi}(x, y, z; \omega) = &-\frac{1}{4\pi}\iint_{z_1=0}\frac{\Delta\bar{p}}{\rho_\infty U_\infty}(x_1, y_1, z_1)\\ &\cdot\left\{\int_{-\infty}^{x-x_1}\frac{\partial}{\partial z_1}\left\{\frac{e^{-ikr}}{r}\right\}\exp\left[i\frac{M^2}{(1-M^2)}\frac{\omega}{U_\infty}\xi\right]\right.\\ &\left.\cdot\exp\left(i\frac{\omega}{U_\infty}\xi\right)\exp\left[-i\frac{\omega}{U_\infty}(x - x_1)\right]d\xi\right\}dS\\ = &-\frac{1}{4\pi}\iint_{z_1=0}\frac{\Delta\bar{p}}{\rho_\infty U_\infty}(x_1, y_1, z_1)\exp\left[-\frac{i\omega}{U_\infty}(x - x_1)\right]\\ &\cdot\left\{\int_{-\infty}^{x-x_1}\exp\left[i\frac{1}{(1-M^2)}\frac{\omega}{U_\infty}\xi\right]\frac{\partial}{\partial z_1}\left\{\frac{e^{-ikr}}{r}\right)d\xi\right\}dS\end{aligned}$$

Compute $\partial\bar{\phi}/\partial z$ and set equal to $\bar{w}$ from body boundary condition on $z=0$, noting that

$$\frac{\partial}{\partial z}\left\{\frac{e^{-ikr}}{r}\right\} = -\frac{\partial}{\partial z_1}\left\{\frac{e^{-ikr}}{r}\right\}$$

The final result is

$$\frac{\bar{w}}{U_\infty} = \iint \frac{\Delta\bar{p}}{\rho_\infty U_\infty^2}(x_1, y_1, z_1 = 0) K(x - x_1, y - y_1, 0)\, dx_1\, dy_1 \qquad (4.3.33)$$

where

$$K(x, y) = \lim_{z\to 0} \frac{\exp\left(-i\frac{\omega}{U_\infty}x\right)}{4\pi} \int_{-\infty}^{x} \exp\left[\frac{i}{(1-M^2)}\frac{\omega}{U_\infty}\xi\right]\frac{\partial^2}{\partial z^2}\left\{\frac{e^{-ikr}}{r}\right\} d\xi$$

$$r \equiv [\xi^2 + (1-M^2)(y^2+z^2)]^{\frac{1}{2}}$$

The expression for K may be simplified as follows: Define a new variable, τ, to replace ξ by

$$(1-M^2)\tau \equiv \xi - Mr(\xi, y, z)$$

where one will recall

$$r(\xi, y, z) \equiv [\xi^2 + \beta^2(y^2+z^2)]^{\frac{1}{2}}$$

and

$$\beta^2 \equiv 1 - M^2$$

After some manipulation one may show that

$$\frac{d\tau}{[\tau^2 + y^2 + z^2]^{\frac{1}{2}}} = \frac{d\xi}{r}$$

and

$$\exp\left(+\frac{i\omega}{U_\infty}\frac{\xi}{(1-M^2)}\right)e^{-ikr} = \exp\left[i\frac{\omega}{U_\infty}\tau\right]$$

Thus

$$K = \lim_{z\to 0} \frac{\exp\left(-i\frac{\omega x}{U_\infty}\right)}{4\pi}\frac{\partial^2}{\partial z^2}\int_{-\infty}^{[x-Mr(x,y,z)]/(1-M^2)} \frac{\exp\left(\frac{i\omega\tau}{U_\infty}\right)}{[\tau^2+y^2+z^2]^{\frac{1}{2}}}\, d\tau \qquad (4.3.34)$$

Taking the second derivative and limit as indicated in (4.3.34) and using the identity

$$\left[\frac{Mx+r}{(x^2+y^2)}\right]^2 \equiv \frac{1}{\left[\frac{x-Mr}{(1-M^2)}\right]^2 + y^2}$$

one finally obtains

$$K = -\frac{1}{4\pi}\left\{\frac{M(Mx+r)}{r(x^2+y^2)}\exp\left[i\frac{\omega}{U_\infty}\frac{M}{(1-M^2)}(Mx-r)\right]\right.$$
$$\left. + \exp\left(-i\frac{\omega x}{U_\infty}\right)\int_{-\infty}^{(x-Mr)/(1-M^2)}\frac{\exp\left(i\frac{\omega\tau}{U_\infty}\right)}{[\tau^2+y^2]^{\frac{3}{2}}}\,\mathrm{d}\tau\right\} \qquad (4.3.35)$$

This is one form often quoted in the literature. By expressing K in nondimensional form we see the strong singularity in K as $y \to 0$.

$$y^2K(x, y) = -\frac{1}{4\pi}\left\{\frac{M(Mx/y + r/y)}{r/y[(x/y)^2+1]}\exp\left[i\frac{\omega y}{U_\infty}\frac{M}{(1-M^2)}\left(M\frac{x}{y}-\frac{r}{y}\right)\right]\right.$$
$$\left. + \exp\left(-i\frac{\omega x}{U}\right)\int_{-\infty}^{[x/y-M(r/y)]/(1-M^2)}\frac{\exp\left(\frac{i\omega y}{U_\infty}z\right)}{[z^2+1]^{\frac{3}{2}}}\,\mathrm{d}z\right\}$$

$$z \equiv \tau/y$$

Note that the compressible Kernel, K, has the same strength singularity as for incompressible flow and is of no more fundamental complexity.

There is a vast literature on unsteady aerodynamics within the framework of linearized, potential flow models. Among recent references one may mention the work of A. Cunningham* on combined subsonic-supersonic Kernel Function methods including an empirical correction for transonic effects and also the work of Morino† using Green's Theorem in a more general form for both subsonic and supersonic flow. For a brief overview with authoritative suggestions for future work, the paper by Rodden‡ is recommended. The reader who has mastered the material presented so far should be able to pursue this literature with confidence.

* Cunningham [15].

† Morino, Chen and Suciu [16].

‡ Rodden [17], Ashley and Rodden [18].

Before turning to representative numerical results the historically important theory of incompressible, two-dimensional flow will be presented.

Incompressible, two-dimensional flow

A classical solution is due to Theodorsen* and others. Traditionally, the coordinate system origin is selected at mid-chord with $b \equiv$ half-chord. The governing differential equation for the velocity potential, ϕ, is

$$\nabla^2 \phi = 0 \tag{4.3.36}$$

with boundary conditions for a lifting, airfoil of

$$\left.\frac{\partial \phi}{\partial z}\right|_{z=0^{+,-}} = w_a \equiv \frac{\partial z_a}{\partial t} + U_\infty \frac{\partial z_a}{\partial x} \tag{4.3.37}$$

on airfoil, $-b < x < b$, on $z = 0$ and

$$p = -\rho_\infty \left[\frac{\partial \phi}{\partial t} + U_\infty \frac{\partial \phi}{\partial x}\right] = 0 \tag{4.3.38}$$

off airfoil, $x > b$ or $x < -b$, on $z = 0$ and

$$p, \phi \to 0 \quad \text{as} \quad z \to \infty \tag{4.3.39}$$

From (4.3.36), (4.3.37) and (4.3.39) one may construct an integral equation,

$$w_a = \left.\frac{\partial \phi}{\partial z}\right|_{z=0} = -\frac{1}{2\pi} \int_{-b}^{\infty} \frac{\gamma(\xi, t)}{x - \xi} \, d\xi \tag{4.3.40}$$

where

$$\gamma(x, t) \equiv \left.\frac{\partial \phi}{\partial x}\right|_U - \left.\frac{\partial \phi}{\partial x}\right|_L \tag{4.3.41}$$

and

$$U \Rightarrow z = 0^+, \qquad L \Rightarrow z = 0^-$$

* Theodorsen [19]. Although this work is of great historical importance, the details are of less compelling interest today and some readers may wish to omit this section on a first reading. The particular approach followed here is a variation on Theodorsen's original theme by Marten Landahl.

Further definitions include

$$\text{'Circulation'} \equiv \Gamma(x) \equiv \int_{-b}^{x} \gamma(\xi)\,\mathrm{d}\xi \Rightarrow \frac{\partial \Gamma}{\partial x} = \gamma(x)$$

$$\Delta\phi \equiv \phi_L - \phi_U$$

$$C_p \equiv \frac{p}{\frac{1}{2}\rho_\infty U_\infty^2}$$

$$\Delta C_p \equiv C_{p_L} - C_{p_U}$$

From above, (4.3.41),

$$\Gamma(x, t) = \int_{-b}^{x} \gamma(\xi)\,\mathrm{d}\xi = \int_{-b}^{x} \left[\frac{\partial \phi_U}{\partial \xi} - \frac{\partial \phi_L}{\partial \xi}\right] \mathrm{d}\xi = -\Delta\phi(x), \tag{4.3.42}$$

Note: $\Delta\phi(x = -b) = 0$. Also from (4.3.38) and (4.3.41),

$$\Delta C_p = -\frac{2}{U_\infty^2}\left[\frac{\partial \Delta\phi}{\partial t} + U_\infty \frac{\partial \Delta\phi}{\partial x}\right]$$

and using (4.3.42),

$$\Delta C_p = \frac{2}{U_\infty^2}\left[\frac{\partial \Gamma}{\partial t} + U_\infty \frac{\partial \Gamma}{\partial x}\right] \tag{4.3.43}$$

Thus once γ *(and hence* Γ*) is known,* ΔC_p *is readily computed.* We therefore seek to solve (4.3.40) for γ. The advantage of (4.3.40) over (4.3.36)–(4.3.39) is that we have reduced the problem by one variable, having eliminated z. A brief derivation of (4.3.40) is given below.

Derivation of integral equation (4.3.40). A Fourier transform of (4.3.36) gives

$$\frac{\mathrm{d}^2\phi^*}{\mathrm{d}z^2} - \alpha^2\phi^* = 0 \tag{4.3.36a}$$

where

$$\phi^*(\alpha, z, t) \equiv \int_{-\infty}^{\infty} \phi(x, z, t)\mathrm{e}^{-i\alpha x}\,\mathrm{d}x$$

(4.3.37) becomes

$$\left.\frac{\mathrm{d}\phi^*}{\mathrm{d}z}\right|_{z=0} = w_a^* \tag{4.3.37a}$$

The general solution to (4.3.36a) is

$$\phi^* = Ae^{+|\alpha|z} + Be^{-|\alpha|z} \tag{4.3.38a}$$

From the finiteness condition, (4.3.39), we see that one must require that $A=0$ for $z>0$ (and $B=0$ for $z<0$). Considering $z>0$ for definiteness, we compute from (4.3.38a)

$$\left.\frac{d\phi^*}{dz}\right|_{z=0} = -|\alpha|\,B \tag{4.3.39a}$$

From (4.3.39a) and (4.3.37a),

$$B = -\frac{w_a^*}{|\alpha|} \tag{4.3.40a}$$

and from (4.3.38a) and (4.3.40a)

$$\left.\phi^*\right|_{z=0^+} = \frac{-w_a^*}{|\alpha|} \tag{4.3.41a}$$

From (4.3.41)

$$\gamma^* = \left.\left(\frac{\partial\phi}{\partial x}\right)^*\right|_{z=0^+} - \left.\left(\frac{\partial\phi}{\partial x}\right)^*\right|_{z=0^-}$$

and using (4.3.41a)

$$\gamma^* = -2i\alpha\frac{w_a^*}{|\alpha|} \tag{4.3.42a}$$

Re-arranging (4.3.42a),

$$w_a^* = -\frac{|\alpha|}{2i\alpha}\gamma^*$$

and inverting back to physical domain (using the convolution theorem) we obtain the desired result.

$$w_a = -\frac{1}{2\pi}\int_{-b}^{\infty}\frac{\gamma(\xi,t)}{x-\xi}\,d\xi \tag{4.3.40}$$

where

$$\frac{1}{2\pi}\int_{-\infty}^{\infty} -\frac{|\alpha|}{2i\alpha}e^{+i\alpha x}\,d\alpha = -\frac{1}{2\pi x}$$

The lower limit $x=-b$ in (4.3.40) follows from the fact that $p=0$ for $x<-b$ (on $z=0$) implies that $\phi=\phi_x=0$ for $x<-b$. This will be made more explicit when we consider $x > b$ where $p = 0$ does *not* imply $\phi = \phi_x = 0$! See discussion below.

Also one can calculate γ for $x > b$ in terms of γ for $b < x < b$ by using the condition that $\Delta C_p = 0$ (continuous pressure) for $x > b$. This is helpful in solving (4.3.40) for γ in terms of w_a. From (4.3.43)

$$\Delta C_p = 0 \Rightarrow \frac{\partial \Gamma}{\partial t} + U_\infty \frac{\partial \Gamma}{\partial x} = 0$$

$$\therefore \quad \Gamma = \Gamma\left(t - \frac{x}{U_\infty}\right) \tag{4.3.44}$$

Simple harmonic motion of an airfoil. For the special case of simple harmonic motion, one has

$$\begin{aligned} w_a(x, t) &= \bar{w}_a(x) e^{i\omega t} \\ \gamma(x, t) &= \bar{\gamma}(x) e^{i\omega t} \\ \Gamma &= \bar{\Gamma} e^{i\omega t} \quad \text{etc.} \end{aligned} \tag{4.3.45}$$

(4.3.44) and (4.3.45) imply

$$\Gamma(x, t) = A \exp(i\omega[t - x/U_\infty])$$

The (integration) constant A may be evaluated by considering the solution at $x = b$.

$$\Gamma(x = b, t) = A \exp(i\omega[t - b/U_\infty])$$

$$\therefore \quad \Gamma(x, t) = \bar{\Gamma}(b) \exp\{i\omega[t - (x - b)/U_\infty]\}$$

and

$$\bar{\gamma} = \frac{\partial \bar{\Gamma}}{\partial x} = \frac{-i\omega}{U_\infty} \bar{\Gamma}(b) \exp[-i\omega(x - b)/U_\infty] \tag{4.3.46}$$

Introducing traditional nondimensionalization

$$x^* \equiv \frac{x}{b}, \qquad \xi^* \equiv \xi/b, \qquad k \equiv \frac{\omega b}{U_\infty}$$

a summary of the key relations is given below

$$\bar{w}_a(x^*) = -\frac{1}{2\pi} \int_{-1}^{\infty} \frac{\bar{\gamma}(\xi^*)}{x^* - \xi^*} \, d\xi^* \quad \text{from (4.3.40)}$$

where

$$\frac{\bar{\gamma}(x^*)}{U_\infty} = -ik\frac{\bar{\Gamma}(b)}{U_\infty b}\exp[-ik(x^*-1)]$$

for $x^* > 1$ from (4.3.46)

$$\frac{\bar{\Gamma}(x^*)}{U_\infty b} = \int_{-1}^{x^*}\frac{\bar{\gamma}(\xi^*)}{U_\infty}\,\mathrm{d}\xi^* \quad \text{definition}$$

$$\Delta\bar{C}_p = 2\left[\frac{\bar{\gamma}(x^*)}{U_\infty} + ik\frac{\bar{\Gamma}(x^*)}{U_\infty b}\right] \quad \text{from (4.3.43)} \qquad (4.3.47)$$

Special Case: steady motion. For simplicity let us first consider steady flow, $\omega \equiv 0$. From (4.3.46) or (4.3.47)

$$\gamma = 0 \quad \text{for} \quad x^* > 1$$

and hence we have

$$w_a(x^*) = -\frac{1}{2\pi}\int_{-1}^{1}\frac{\gamma(\xi^*)}{x^*-\xi^*}\,\mathrm{d}\xi^* \qquad (4.3.48)$$

To solve (4.3.48) for γ, we replace x^* by u, multiply both sides of (4.3.48) by the 'solving kernel'

$$\sqrt{\frac{1+u}{1-u}}\frac{1}{u-x^*}$$

and integrate $\int_{-1}^{1}\cdots\,\mathrm{d}u$. The result is

$$\int_{-1}^{1}\sqrt{\frac{1+u}{1-u}}\frac{w_a(u)}{u-x^*}\,\mathrm{d}u = -\frac{1}{2\pi}\int_{-1}^{1}\sqrt{\frac{1+u}{1-u}}\frac{1}{u-x^*}\int_{-1}^{1}\frac{\gamma(\xi^*)}{u-\xi^*}\,\mathrm{d}\xi^*\,\mathrm{d}u$$

Now write $\gamma(\xi^*) = \gamma(x^*) + [\gamma(\xi^*) - \gamma(x^*)]$, then above may be written as

$$\int_{-1}^{1}\sqrt{\frac{1+u}{1-u}}\frac{w_a(u)}{u-x^*}\,\mathrm{d}u = -\frac{\gamma(x^*)}{2\pi}\left\{\int_{-1}^{1}\sqrt{\frac{1+u}{1-u}}\frac{1}{u-x^*}\int_{-1}^{1}\frac{\mathrm{d}\xi^*}{u-\xi^*}\right\}\mathrm{d}u$$

$$-\frac{1}{2\pi}\left\{\int_{-1}^{1}\sqrt{\frac{1+u}{1-u}}\frac{1}{u-x^*}\int_{-1}^{1}\frac{(\xi^*-x^*)}{u-\xi^*}F(\xi^*, x^*)\,\mathrm{d}\xi^*\,\mathrm{d}u\right\} \qquad (4.3.49)$$

where

$$F(\xi^*, x^*) \equiv \frac{\gamma(\xi^*)-\gamma(x^*)}{\xi^*-x^*}$$

To simplify (4.3.49) we will need to know several integrals. To avoid a diversion, these are simply listed here and are evaluated in detail at the end of this discussion of incompressible, two-dimensional flow.

$$I_0 \equiv \int_{-1}^{1} \frac{d\xi^*}{x^* - \xi^*} = \ln\left(\frac{1+x^*}{1-x^*}\right) \quad \text{for} \quad x^* < 1$$

$$= \ln\left(\frac{x^*+1}{x^*-1}\right) \quad \text{for} \quad x^* > 1$$

$$I_1 \equiv \int_{-1}^{1} \sqrt{\frac{1+u}{1-u}} \frac{du}{u - x^*} = \pi \quad \text{for} \quad x^* < 1$$

$$= \pi\left[1 - \sqrt{\frac{x^*+1}{x^*-1}}\right] \quad \text{for} \quad x^* > 1$$

$$I_2 \equiv \int_{-1}^{1} \sqrt{\frac{1+u}{1-u}} \ln\left|\frac{1-u}{1+u}\right| \frac{du}{u - x^*} = -\pi^2 \sqrt{\frac{1+x^*}{1-x^*}} \quad \text{for} \quad -1 < x^* < 1 \tag{4.3.50}$$

Now we can proceed to consider the several terms on the RHS of (4.3.49)

1*st term.* Now

$$\int_{-1}^{1} \frac{d\xi^*}{u - \xi^*} = \ln\left|\frac{1+u}{1-u}\right| \quad \text{from} \quad I_0$$

$$\therefore \quad I_3 \equiv \oint_{-1}^{1} \sqrt{\frac{1+u}{1-u}} \frac{1}{u - x^*} \int_{-1}^{1} \frac{d\xi^*}{u - \xi^*} du$$

$$= \int_{-1}^{1} \sqrt{\frac{1+u}{1-u}} \frac{1}{u - x^*} \ln\left|\frac{1+u}{1-u}\right| du = +\pi^2 \sqrt{\frac{1+x^*}{1-x^*}} \quad \text{from } I_2$$

$$\therefore \quad \text{1st term} = -\frac{\gamma I_3}{2\pi} = \frac{-\gamma(x^*)}{2} \pi \sqrt{\frac{1+x^*}{1-x^*}}$$

2*nd term.* Interchange order of integration;

$$I_4 \equiv \oint_{-1}^{1} [\xi^* - x^*] F(\xi^*, x^*) \int_{-1}^{1} \sqrt{\frac{1+u}{1-u}} \frac{du}{(u - x^*)(u - \xi^*)} d\xi^*$$

Now

$$\frac{1}{(u-x^*)(u-\xi^*)} = \frac{1}{x^*-\xi^*}\left[\frac{1}{u-x^*} - \frac{1}{u-\xi^*}\right]$$

partial fractions expansion

$$\therefore \quad I_4 = -\int_{-1}^{1} F(\xi^*, x^*)\left\{\int_{-1}^{1}\sqrt{\frac{1+u}{1-u}}\left[\frac{1}{u-x^*} - \frac{1}{u-\xi^*}\right]\mathrm{d}u\right\}\mathrm{d}\xi^*$$

$$= -\oint_{-1}^{1} F(\xi^*, x^*)\overset{0}{[\pi - \pi]}\,\mathrm{d}\xi^* \quad \text{from} \quad I_1$$

Finally then, from above and (4.3.49),

$$\int_{1}^{1}\sqrt{\frac{1+u}{1-u}}\frac{w_a(u)}{u-x^*}\mathrm{d}u = -\frac{\pi}{2}\gamma(x^*)\sqrt{\frac{1+x^*}{1-x^*}}$$

$$\therefore \quad \gamma(x^*) = -\frac{2}{\pi}\sqrt{\frac{1-x^*}{1+x^*}}\int_{-1}^{1}\sqrt{\frac{1+u}{1-u}}\frac{w_a(u)}{u-x^*}\mathrm{d}u \tag{4.3.51}$$

Note: Other 'solving kernels' exist, but they do not satisfy the Kutta condition, $\gamma(x^*)$ finite at $x^*=1$, i.e., finite pressure at trailing edge.

One might reasonably inquire, how do we know what the solving kernel should be? Perhaps the most straightforward way to motivate the choice is to recognize that the solution for steady flow can be obtained by other methods. Probably the simplest of these alternative solution methods is to use the transformations $x^* = \cos\theta$, $\xi^* = \cos\phi$ and expand γ and w_a in Fourier series in ϕ and θ. See BAH, p. 216. Once the answer is known, i.e. (4.3.51), the choice of the solving kernel is fairly obvious. The advantage of the solving kernel approach over other methods is that it is capable of extension to unsteady airfoil motion where an analytical solution may be obtained as will be described below. On the other hand a method which is based essentially on the Fourier series approach is often employed to obtain numerical solutions for three-dimensional, compressible flow. This is the so-called Kernel Function approach discussed earlier.

In the above we have obtained the following integral relation: Given

$$f(x^*) = -\frac{1}{2\pi}\int_{-1}^{1}\frac{g(\xi^*)}{x^*-\xi^*}\mathrm{d}\xi^*$$

with $g(1)$ finite or zero, then

$$g(x^*) = -\frac{2}{\pi}\sqrt{\frac{1-x^*}{1+x^*}}\int_{-1}^{1}\sqrt{\frac{1+\xi^*}{1-\xi^*}}\frac{f(\xi^*)}{\xi^*-x^*}\,d\xi^* \tag{4.3.52}$$

General case: Oscillating motion. We may employ the solving kernel approach to attack the unsteady problem also. Recall from (4.3.40), (4.3.43), (4.3.46) one has

$$\bar{w}_a(x^*) = -\frac{1}{2\pi}\int_{-1}^{1}\frac{\gamma(\xi^*)}{x^*-\xi^*}\,d\xi^* - \frac{1}{2\pi}\int_{1}^{\infty}\frac{\gamma(\xi^*)}{x^*-\xi^*}\,d\xi^* \tag{4.3.53}$$

$$\overline{\Delta C_p} = \frac{2\bar{\gamma}(x^*)}{U_\infty} + 2ik\frac{\bar{\Gamma}(x^*)}{U_\infty b} = 2\frac{\bar{\gamma}(x^*)}{U_\infty} + 2ik\int_{-1}^{x^*}\frac{\bar{\gamma}(\xi^*)}{U_\infty}\,d\xi^* \tag{4.3.54}$$

$$\frac{\bar{\gamma}(x^*)}{U_\infty} = -ik\frac{\bar{\Gamma}(1)}{U_\infty b}\exp[-ik(x^*-1)] \quad \text{for} \quad x^*>1 \tag{4.3.55}$$

Substitute (4.3.55) into (4.3.53).

$$\bar{w}_a(x^*) = -\frac{1}{2\pi}\int_{-1}^{1}\frac{\bar{\gamma}(\xi^*)}{x^*-\xi^*}\,d\xi^* + \bar{G}(x^*) \tag{4.3.56}$$

where

$$\bar{G}(x^*) \equiv \frac{ik\bar{\Gamma}(1)}{2\pi b}\int_{+1}^{\infty}\frac{\exp[-ik(\xi^*-1)]}{x^*-\xi^*}\,d\xi^*$$

Invert (4.3.56) to determine $\gamma(x^*)$; recall the steady flow solution, (4.3.52).

$$\bar{\gamma}(x^*) = -\frac{2}{\pi}\sqrt{\frac{1-x^*}{1+x^*}}\int_{-1}^{1}\sqrt{\frac{1+\xi^*}{1-\xi^*}}\left\{\frac{w_a(\xi^*)-\bar{G}(\xi^*)}{\xi^*-x^*}\right\}d\xi^* = -\frac{2}{\pi}\sqrt{\frac{1-x^*}{1+x^*}}$$

$$\times\int_{-1}^{1}\sqrt{\frac{1+\xi^*}{1-\xi^*}}\left\{\frac{\bar{w}_a(\xi^*) - \frac{ik\bar{\Gamma}(1)}{2\pi b}\int_{1}^{\infty}\exp[-ik(u-1)]/(\xi^*-u)\,du}{\xi^*-x^*}\right\}d\xi^* \tag{4.3.57}$$

Interchanging the order of integration of the term involving $\bar{\Gamma}(1)$ on the RHS side of (4.3.57) we may evaluate the integral over ξ^* and obtain

$$\bar{\gamma}(x^*) = +\frac{2}{\pi}\sqrt{\frac{1-x^*}{1+x^*}}\left\{\int_{-1}^{1}\sqrt{\frac{1+\xi^*}{1-\xi^*}}\frac{\bar{w}_a(\xi^*)}{(x^*-\xi^*)}\,d\xi^* + ik\frac{\bar{\Gamma}(1)}{b}e^{ik}\int_{1}^{\infty}\frac{e^{-iku}}{x^*-u}\,du\right\} \tag{4.3.58}$$

(4.3.58) is not a complete solution until we determine $\bar{\Gamma}(1)$ which we do as follows. Integrating (4.3.58) with respect to x^* we obtain

$$\frac{\bar{\Gamma}(1)}{b} \equiv \int_{-1}^{1} \bar{\gamma}(x^*)\,\mathrm{d}x^* = -2\int_{-1}^{1}\sqrt{\frac{1+\xi^*}{1-\xi^*}}\,\bar{w}_a(\xi^*)\,\mathrm{d}\xi^*$$

$$-ik\frac{\bar{\Gamma}(1)}{b}\mathrm{e}^{ik}\int_{1}^{\infty}\left[\sqrt{\frac{u+1}{u-1}}-1\right]\mathrm{e}^{-iku}\,\mathrm{d}u \quad (4.3.59)$$

where the integrals on the right hand side with respect to x^* have been evaluated explicitly. We may now solve (4.3.59) for $\bar{\Gamma}(1)$. Recognizing that

$$\int_{1}^{\infty}\left[\sqrt{\frac{u+1}{u-1}}-1\right]\mathrm{e}^{-iku}\,\mathrm{d}u = \frac{-\pi}{2}[H_1^{(2)}(k)+iH_0^{(2)}(k)] - \frac{\mathrm{e}^{-ik}}{ik} \quad (4.3.60)$$

we determine from (4.3.59) and (4.3.60) that

$$\frac{\bar{\Gamma}(1)}{b} = 4\,\frac{\mathrm{e}^{-ik}\displaystyle\int_{-1}^{1}\sqrt{\frac{1+\xi^*}{1-\xi^*}}\,\bar{w}_a(\xi^*)\,\mathrm{d}\xi^*}{\pi ik[H_1^{(2)}(k)+iH_0^{2}(k)]} \quad (4.3.61)$$

$H_1^{(2)}$, $H_0^{(2)}$ are standard Hankel functions.† (4.3.58) and (4.3.61) constitute the solution for $\bar{\gamma}$ in terms of $\bar{w}_a$. From $\bar{\gamma}$, we may determine $\overline{\Delta C_p}$ by using

$$\overline{\Delta C_p} = 2\frac{\bar{\gamma}(x^*)}{U_\infty} + 2ik\int_{-1}^{x^*}\frac{\bar{\gamma}(\xi^*)}{U_\infty}\,\mathrm{d}\xi^*$$

After considerable, but elementary, algebra

$$\overline{\Delta C_p} = \frac{4}{\pi}\sqrt{\frac{1-x^*}{1+x^*}}\int_{-1}^{1}\sqrt{\frac{1+\xi^*}{1-\xi^*}}\left\{\frac{\bar{w}_a(\xi^*)/U_\infty}{x^*-\xi^*}\right\}\mathrm{d}\xi^*$$

$$+\frac{4}{\pi}ik\sqrt{1-x^{*2}}\oint_{-1}^{1}\frac{W(\xi^*)\,\mathrm{d}\xi^*}{U_\infty\sqrt{1-\xi^{*2}}\,(x^*-\xi^*)}$$

$$+\frac{4}{\pi}[1-C(k)]\sqrt{\frac{1-x^*}{1+x^*}}\int_{-1}^{1}\sqrt{\frac{1+\xi^*}{1-\xi^*}}\frac{\bar{w}_a(\xi^*)}{U_\infty}\,\mathrm{d}\xi^* \quad (4.3.62)$$

† Abramowitz and Stegun [20].

where

$$W(\xi^*) \equiv \int_{-1}^{\xi^*} \bar{w}_a(u)\,\mathrm{d}u$$

and

$$C(k) \equiv \frac{H_1^{(2)}}{[H_1^{(2)} + iH_0^{(2)}]}$$

is Theodorsen's well known Function.

The lift may be computed as the integral of the pressure.

$$\begin{aligned} \bar{L} &\equiv \frac{\rho U_\infty^2}{2} b \int_{-1}^{1} \overline{\Delta C_p}\,\mathrm{d}x^* \\ &= \frac{\rho U_\infty^2}{2} b \bigg\{ -C(k) \int_{-1}^{1} \sqrt{\frac{1+\xi^*}{1-\xi^*}}\, \frac{\bar{w}_a(\xi^*)}{U_\infty}\,\mathrm{d}\xi^* \\ &\quad - ik \int_{-1}^{1} \sqrt{1-\xi^{*2}}\, \frac{\bar{w}_a(\xi^*)}{U_\infty}\,\mathrm{d}\xi^* \bigg\} \end{aligned} \tag{4.3.63}$$

Similarly for the moment about the point $x = ba$,

$$\bar{M}_y \equiv \frac{\rho U_\infty^2}{2} b^2 \int_{-1}^{1} \overline{\Delta C_p}[x^* - a]\,\mathrm{d}x^* \tag{4.3.64}$$

In particular, for

$$z_a = -h - \alpha(x - ba)$$

$$\bar{z}_a = -\bar{h} - \bar{\alpha}(x - ba)$$

one has

$$\bar{w}_a = -i\omega\bar{h} - i\omega\bar{\alpha}(x - ba) - U_\infty\bar{\alpha} \tag{4.3.65}$$

Thus (4.3.65) in (4.3.63) and (4.3.64) give

$$\begin{aligned} \bar{L} &= \pi\rho b^2[-\omega^2\bar{h} + i\omega U_\infty\bar{\alpha} + ba\omega^2\bar{\alpha}] \\ &\quad + 2\pi\rho U_\infty b C(k)[i\omega\bar{h} + U_\infty\bar{\alpha} + b(\tfrac{1}{2} - a)\,i\omega\bar{\alpha}] \\ \bar{M}_y &= \pi\rho b^2[-ba\omega^2\bar{h} - U_\infty b(\tfrac{1}{2} - a)i\omega\bar{\alpha} + b^2(\tfrac{1}{8} + a^2)\omega^2\bar{\alpha}] \\ &\quad + 2\pi\rho U_\infty b^2(\tfrac{1}{2} + a)C(k)[i\omega\bar{h} + U_\infty\bar{\alpha} + b(\tfrac{1}{2} - a)\,i\omega\bar{\alpha}] \end{aligned} \tag{4.3.66}$$

Theodorsen's Function, $C(k) = F + iG$, is given below in Fig 4.6.

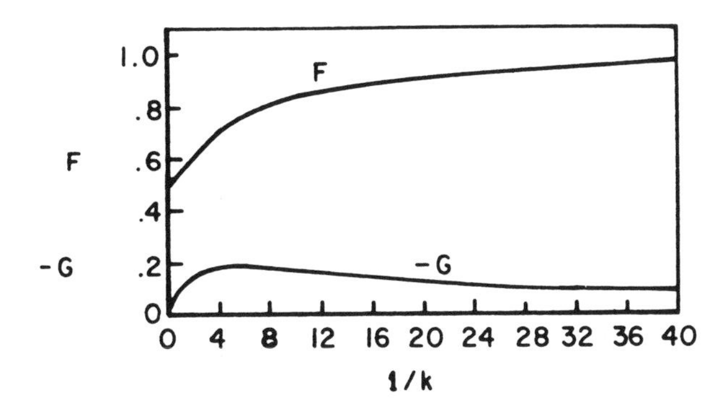

Figure 4.6 The functions F and G against $\frac{1}{k}$. *After Theodorsen* [19].

Transient motion. Using Fourier synthesis one may now obtain results for arbitrary time dependent motion from the simple harmonic motion results; using Fourier summation (integration) and (4.3.66),

$$L(t) = \frac{1}{2\pi}\int_{-\infty}^{\infty} \bar{L}(\omega)_{\text{due to } h}\bar{h}(\omega)e^{i\omega t}\,d\omega$$

$$+\frac{1}{2\pi}\int_{-\infty}^{\infty} \bar{L}(\omega)_{\text{due to } \alpha}\bar{\alpha}(\omega)e^{i\omega t}\,d\omega$$

$$= \frac{1}{2\pi}\int_{-\infty}^{\infty} \{\pi\rho b^2(-\omega^2) + 2\pi\rho U_\infty bC(k)(i\omega)\}\cdot \bar{h}(\omega)e^{i\omega t}\,d\omega$$

$$+\frac{1}{2\pi}\int_{-\infty}^{\infty} \{\pi\rho b^2(i\omega U_\infty + ba\omega^2) + 2\pi\rho U_\infty bC(k)(U_\infty + b(\tfrac{1}{2} - a)i\omega)\}$$

$$\bar{\alpha}(\omega)e^{i\omega t}\,d\omega \tag{4.3.67}$$

where

$$\bar{h}(\omega) = \int_{-\infty}^{\infty} h(t)e^{-i\omega t}\,d\omega$$

and

$$\bar{\alpha}(\omega) = \int_{-\infty}^{\infty} \alpha(t)e^{-i\omega t}\,d\omega \tag{4.3.68}$$

Now

$$\int_{-\infty}^{\infty} (i\omega)^n \bar{\alpha} e^{i\omega t} \, d\omega = \frac{d^n \alpha}{dt^n} \qquad n = 1, 2, \ldots \tag{4.3.69}$$

Thus

$$L = \pi \rho b^2 \left[\frac{d^2 h}{dt^2} + U_\infty \frac{d\alpha}{dt} - ba \frac{d^2 \alpha}{dt^2} \right] + \rho U_\infty b \int_{-\infty}^{\infty} C(k) f(\omega) e^{i\omega t} \, d\omega$$

where

$$f(\omega) \equiv i\omega \bar{h}(\omega) + U_\infty \bar{\alpha}(\omega) + b(\tfrac{1}{2} - a) i\omega \bar{\alpha}(\omega) = \int_{-\infty}^{\infty} \left[\frac{dh}{dt} + U_\infty \alpha + b(\tfrac{1}{2} - a) \frac{d\alpha}{dt} \right] e^{-i\omega t} \, dt \tag{4.3.70}$$

Physically,

$$\frac{dh}{dt} + U_\infty \alpha + b(\tfrac{1}{2} - a) \frac{d\alpha}{dt} = -w_a \quad \text{at} \quad x = b/2;$$

$x = b/2$ is $\frac{3}{4}$ chord of airfoil.

Similarly,

$$M_y = \pi \rho b^2 \left[ba \frac{d^2 h}{dt^2} - U_\infty b(\tfrac{1}{2} - a) \frac{d\alpha}{dt} - b^2(\tfrac{1}{8} + a^2) \frac{d^2 \alpha}{dt^2} \right] + \rho U_\infty b^2 (\tfrac{1}{2} + a) \int_{-\infty}^{\infty} C(k) f(\omega) e^{i\omega t} \, d\omega \tag{4.3.71}$$

Example I. Step change in angle of attack.

$$h \equiv 0$$

$$\alpha = 0 \quad \text{for} \quad t < 0$$

$$= \alpha_0 \equiv \text{constant} \quad \text{for} \quad t > 0$$

$$\therefore \frac{d\alpha}{dt} = \frac{d^2\alpha}{dt^2} = \frac{dh}{dt} = \frac{d^2 h}{dt^2} = 0 \quad \text{for} \quad t > 0$$

$$\therefore f(\omega) = U_\infty \alpha_0 \int_0^\infty e^{-i\omega t} \, dt = \frac{U_\infty \alpha_0}{-i\omega} e^{-i\omega t} \Big|_0^\infty = \frac{U_\infty \alpha_0}{i\omega}$$

$$\therefore\ L = \rho U_\infty^2 b\alpha_0 \int_{-\infty}^{\infty} \frac{C(k)}{i\omega} \mathrm{e}^{i\omega t}\, \mathrm{d}\omega$$

$$= \rho U_\infty^2 b\alpha_0 \int_{-\infty}^{\infty} \frac{C(k)}{ik} \mathrm{e}^{iks}\, \mathrm{d}k$$

where $s \equiv \dfrac{Ut}{b}$. Finally,

$$L = 2\pi\rho U^2 b\alpha_0 \left\{ \frac{1}{2\pi i} \int_{-\infty}^{\infty} \frac{C(k)}{k} \mathrm{e}^{iks}\, \mathrm{d}k \right\} \tag{4.3.72}$$

$\{\cdots\} \equiv \phi(s)$ is called the Wagner Function, see Figure 4.7. Note that if α is precisely a step function, then L has a singularity at $t = 0$ from (4.3.70). Also shown is the Kussner function, $\psi(s)$, to be discussed subsequently. Note also that ϕ is the lift of the airfoil due to step change in angle of attack or more generally due to step change in $-w_a/U_\infty$ at $\frac{3}{4}$ chord.

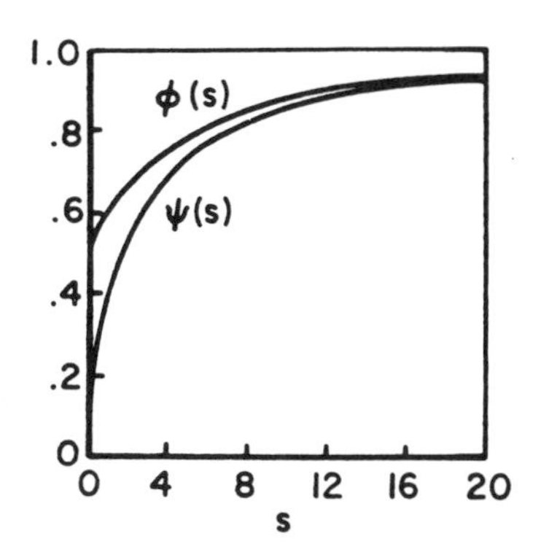

Figure 4.7 Wagner's function $\phi(s)$ for indicial lift and Küssner's function $\psi(s)$ for lift due to a sharp-edged gust, plotted as functions of distance traveled in semichordlengths. After BAH, Fig. 5.21.

Thus using Duhamel superposition formula

$$L(t) = \pi\rho b^2 \left[\frac{\mathrm{d}^2 h}{\mathrm{d}t^2} + U_\infty \frac{\mathrm{d}\alpha}{\mathrm{d}t} - ba \frac{\mathrm{d}^2\alpha}{\mathrm{d}t^2} \right]$$

$$-2\pi\rho U_\infty b \left[w_{a\frac{3}{4}}(0)\phi(s) + \int_0^s \frac{\mathrm{d}w_{a\frac{3}{4}}}{\mathrm{d}\sigma}(\sigma)\phi(s-\sigma)\, \mathrm{d}\sigma \right] \tag{4.3.73}$$

Example II. Entrance into a sharp edged gust. In the primed coordinate system, i.e., fixed with respect to the atmosphere, one has

$$\begin{aligned} w_G &= 0 \quad \text{for} \quad x' > 0 \\ &= w_0 \quad \text{for} \quad x' < 0 \end{aligned}$$

Note: The general transformation between fluid fixed and body fixed coordinate systems is

$$\begin{aligned} x' &= x + b - U_\infty t, \qquad & x + b &= x' + U_\infty t' \\ t' &= t & t &= t' \end{aligned}$$

The leading edge enters the gust at $t = t' = 0$ at

$$\begin{aligned} t &= 0, \qquad x' = x + b \\ t' &= 0. \end{aligned}$$

Thus in the coordinate system fixed with respect to the airfoil, one has

$$\begin{aligned} w_G &= 0 \quad \text{for} \quad x + b > U_\infty t \quad \text{or} \quad \frac{x+b}{U_\infty} > t \\ &= w_0 \quad \text{for} \quad x + b < U_\infty t \quad \text{or} \quad \frac{x+b}{U_\infty} < t \end{aligned} \tag{4.3.74}$$

$$\begin{aligned} \therefore \; w_G(\omega) &\equiv \int_{-\infty}^{\infty} w_G \mathrm{e}^{-i\omega t} \, \mathrm{d}t \\ &= w_0 \int_{(x+b)/U_\infty}^{\infty} \mathrm{e}^{-i\omega t} \, \mathrm{d}t \\ &= \frac{w_0}{-i\omega} \mathrm{e}^{-i\omega t} \bigg|_{(x+b)/U_\infty}^{\infty} \\ &= \frac{w_0}{i\omega} \exp\left[-i\omega \frac{(x+b)}{U_\infty}\right] = \frac{w_0}{i\omega} \mathrm{e}^{-ik} \mathrm{e}^{-ikx^*} \end{aligned}$$

where

$$x^* \equiv x/b \tag{4.3.75}$$

For

$$\bar{w}_a = -w_G \left(= -\frac{w_0}{i\omega} \mathrm{e}^{-ik} \mathrm{e}^{-ikx^*} \right)$$

one finds from the oscillating airfoil motion theory that

$$\bar{L} = 2\pi\rho U_\infty b\{C(k)[J_0(k) - iJ_1(k)] + iJ_1(k)\}\frac{w_0}{i\omega}\,e^{-ik}$$

and

$$\bar{M}_y = b(\tfrac{1}{2} + a)\bar{L}$$

$$\therefore \quad L(t) = \frac{1}{2\pi}\int_{\infty}^{\infty}\bar{L}(\omega)e^{i\omega t}\,d\omega$$

$$= \rho U_\infty b w_0 \int_{-\infty}^{\infty}\frac{\{\cdots\}}{ik}\,e^{-ik}e^{iks}\,dk \tag{4.3.76}$$

$$= 2\pi\rho U_\infty b w_0 \psi(s)$$

where

$$\psi(s) \equiv \frac{1}{2\pi i}\int_{-\infty}^{\infty}\frac{\{\cdots\}}{k}\exp[ik(s-1)]\,dk \tag{4.3.77}$$

is called the Kussner Function and was shown previously in Figure 4.7. Finally then, using Duhamel's integral,

$$L = \pi\rho U b\left\{w_G(0)\psi(s) + \int_0^s \frac{dw_G}{d\sigma}(\sigma)(s-\sigma)\,d\sigma\right\} \tag{4.3.78}$$

A famous controversy concerning the interpretation of Theodorsen's function for other than real frequencies (neutrally stable motion) took place in the 1950's. The issue has arisen again because of possible applications to feedback control of aeroelastic systems. For a modern view and discussion, the reader should consult Edwards, Ashley, and Breakwell [21]. Also see Sears, [10] in chapter 3.

Evaluation of integrals. I_0:

For $x^* < 1$.

$$I_0 \equiv \oint_{-1}^{1}\frac{d\xi^*}{x^* - \xi^*} = \lim_{\varepsilon\to 0}\left[\int_{-1}^{x^*-\varepsilon}\frac{d\xi^*}{x^*-\xi^*} + \int_{x^*+\varepsilon}^{1}\frac{d\xi^*}{x^*-\xi^*}\right]$$

$$= \lim_{\varepsilon\to 0}\left[-\int_{-1}^{x^*-\varepsilon}\frac{d(x^*-\xi^*)}{(x^*-\xi^*)} - \int_{x^*+\varepsilon}^{1}\frac{d(\xi^*-x^*)}{(\xi^*-x^*)}\right]$$

$$= -\ln(x^*-\xi^*)\Big|_{\xi^*=-1}^{x^*-\varepsilon} - \ln(\xi^*-x^*)\Big|_{x^*+\varepsilon}^{1}$$

$$= -[\ln\varepsilon - \ln(x^*+1)] - [\ln(1-x^*) - \ln\varepsilon] = \ln\left(\frac{1+x^*}{1-x^*}\right)$$

For $x^* > 1$, there is no need for a Cauchy Principal Value and

$$I_0 = \ln\left(\frac{x^*+1}{x^*-1}\right)$$

$$I_1\text{:}\quad I_1 \equiv \oint_{-1}^{1} \sqrt{\frac{1+u}{1-u}} \frac{du}{u-x^*}$$

Use contour integration. Define $w \equiv u + iv$ (a complex variable whose real part is u) and

$$F(w) \equiv \left(\frac{w+1}{w-1}\right)^{\frac{1}{2}} \frac{1}{w-x^*}$$

Choose a contour as follows

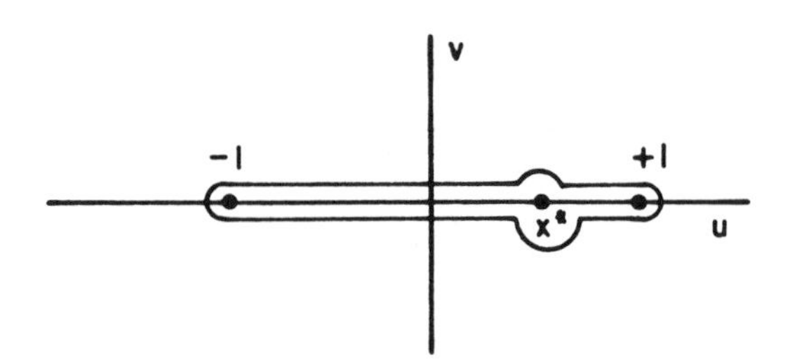

Figure 4.8 *Integral contour.*

Now

$$\frac{w+1}{w-1} = [R^2 + I^2]^{\frac{1}{2}} e^{-i\theta}$$

where

$$R \equiv \frac{(u+1)(u-1)+v^2}{(u-1)^2+v^2}, \qquad I \equiv \frac{-2v}{(u-1)^2+v^2}$$

$$\theta = \tan^{-1} I/R$$

(i) on top, $v = 0^+$, $\quad u-1<0$

$\therefore \quad R<0, \qquad I = 0^- \Rightarrow \theta = -\pi$

(ii) on bottom, $v = 0^-$, $u-1<0$

$\therefore \quad R<0, \qquad I = 0^+ \Rightarrow \theta = +\pi$

$$\therefore \quad \left(\frac{w+1}{w-1}\right)^{\frac{1}{2}} = \sqrt{\frac{1+u}{1-u}}\,e^{-i\pi/2} \quad \text{on top}$$

$$= \sqrt{\frac{1+u}{1-u}}\,e^{+i\pi/2} \quad \text{on bottom}$$

Now $dw = du$ on top or bottom and $w - x^* = u - x^*$ except on arcs near $u = x^*$. On arcs $w - x^* = \varepsilon e^{i\theta}$, $dw = \varepsilon e^{i\theta} i \, d\theta$ where ε is radius of arc. Also

$$\left(\frac{w+1}{w-1}\right)^{\frac{1}{2}} = \sqrt{\frac{1+u}{1-u}}(-i) \quad \text{on top}$$

and $= \cdots (+i)$ on bottom. Thus

$$\zeta_1 \equiv \int_C F(w)\,dw = \overbrace{\int_{-1}^{x^*-\varepsilon} + \int_{x^*+\varepsilon}^{1} i\sqrt{\frac{1+u}{1-u}}\frac{du}{u-x^*}}^{\text{bottom}}$$

$$+ \overbrace{\int_{1}^{x^*+\varepsilon} + \int_{x^*-\varepsilon}^{-1} -i\sqrt{\frac{1+u}{1-u}}\frac{du}{u-x^*}}^{\text{top}} +$$

$$+ \text{contributions from arcs which cancel each other}$$

$$\therefore \quad \lim_{\varepsilon \to 0} \zeta_1 = 2i \int_{-1}^{1} \sqrt{\frac{1+u}{1-u}}\frac{du}{u-x^*} = 2iI_1$$

ζ_1 can be simply evaluated by Cauchy's Theorem. As $w \to \infty$, $F(w) \to 1/w$.

$$\therefore \quad \zeta_1 = \int_{\substack{\text{around}\\ \text{arc at } \infty}} \frac{dw}{w} = 2\pi i$$

$$\therefore \quad I_1 = \frac{\zeta_1}{2i} = \frac{2\pi i}{2i} = \pi$$

For $x^* > 1$, I_1 is still equal to $\zeta_1/2\pi i$; however, now $\zeta_1 = \int_{\text{arc at infinity}} F(w)\,dw - \text{Residue of } F \text{ at } x^*$

$$= 2\pi i - 2\pi i \sqrt{\frac{x^*+1}{x^*-1}}$$

$$\therefore \quad I_1 = \frac{\zeta_1}{2\pi i} = \pi\left[1 - \sqrt{\frac{x^*+1}{x^*-1}}\right]$$

A similar calculation gives I_2.

Evaluation of I_2.

$$-I_2 \equiv \int_{-1}^{1} \sqrt{\frac{1+u}{1-u}} \ln \left|\frac{1+u}{1-u}\right| \frac{du}{u-x^*}$$

Define

$$w \equiv u + iv$$

and

$$F(w) \equiv \ln \left|\frac{w+1}{w-1}\right| \sqrt{\frac{w+1}{w-1}} \frac{1}{w-x^*}$$

The contour is the same as for I_1.

As before,

$$\left(\frac{w+1}{w-1}\right)^{\frac{1}{2}} = \sqrt{\frac{1+u}{1-u}} e^{-i\pi/2} \quad \text{on top}$$

$$= \sqrt{\frac{1+u}{1-u}} e^{+i\pi/2} \quad \text{on bottom}$$

Also

$$\ln\left(\frac{w+1}{w-1}\right) = \ln \sqrt{R^2 + I^2} + i\theta$$

$$= \ln \left|\frac{u+1}{u-1}\right| - i\pi \quad \text{on top}$$

$$= \ln \left|\frac{u+1}{u-1}\right| + i\pi \quad \text{on bottom}$$

Now $dw = du$ on top or bottom and $w - x^* = u - x^*$ *except* on arcs near $u = x^*$. On arcs $w - x^* = \varepsilon e^{i\theta}$, $dw = \varepsilon e^{i\theta} i \, d\theta$ where ε is radius of arc. Thus

$$\zeta_2 \equiv \int_C F_2(w) \, dw = \int_{-1}^{x^*-\varepsilon} + \int_{x^*+\varepsilon}^{1} i\left\{\sqrt{\frac{1+u}{1-u}}\left[\ln\left|\frac{1+u}{1-u}\right| - i\pi\right]\right\} \times \frac{du}{u-x^*} \quad \text{bottom}$$

$$+ \int_{1}^{x^*+\varepsilon} + \int_{x^*-\varepsilon}^{-1} -i\left\{\sqrt{\frac{1+u}{1-u}}\left[\ln\left|\frac{1+u}{1-u}\right| - i\pi\right]\right\} \frac{du}{u-x^*} \quad \text{top}$$

$$- \int_{0}^{\pi} i\pi \sqrt{\frac{1+x^*}{1-x^*}} \, d\theta - \int_{-\pi}^{0} i\pi \sqrt{\frac{1+x^*}{1-x^*}} \, d\theta \quad \text{arcs}$$

Note: ln terms cancel and thus are omitted in arc contributions. Cancelling π terms from bottom and top and adding arc terms, gives

$$\zeta_2 = \int_{-1}^{x^*-\varepsilon} + \int_{x^*+\varepsilon}^{1} i\sqrt{\frac{1+u}{1-u}} \ln\left|\frac{1+u}{1-u}\right| \frac{du}{u-x^*}$$

$$+\int_{1}^{x^*+\varepsilon} + \int_{x^*-\varepsilon}^{-1} -i\sqrt{\frac{1+u}{1-u}} \ln\left|\frac{1+u}{1-u}\right| \frac{du}{u-x^*}$$

$$-2i\pi^2\sqrt{\frac{1+x^*}{1-x^*}}$$

Adding bottom and top terms,

$$\lim_{\varepsilon\to 0} \zeta_2 = 2i\int_{-1}^{1}\sqrt{\frac{1+u}{1-u}} \ln\left|\frac{1+u}{1-u}\right| \frac{du}{u-x^*} - 2i\pi^2\sqrt{\frac{1+x^*}{1-x^*}}$$

$$= -2iI_2 - 2i\pi^2\sqrt{\frac{1+x^*}{1-x^*}}$$

ζ_2 can be simplv evaluated by Cauchy's Theorem. As $w \to \infty$, $F_2(w) \to 0$.

$$\therefore \quad \zeta_2 = 0 \Rightarrow I_2 = -\pi^2\sqrt{\frac{1+x^*}{1-x^*}}$$

4.4 Representative numerical results

Consider a flat plate airfoil, initially at zero angle of attack, which is given a step change in α, i.e.,

$$\begin{aligned} w &= -U_\infty \alpha \quad \text{for} \quad t > 0 \\ &= 0 \quad \text{for} \quad t < 0 \end{aligned}$$

Although most calculations in practice are carried out for sinusoidal time dependent motion, for our purposes examining aerodynamic pressures due to this step change leads to more insight into the nature of the physical system. Of course, *in principle*, the results for sinusoidal motion (or a step change) may be superposed to obtain results for arbitrary time dependent motion.

It is traditional to express the pressure in nondimensional form

$$\frac{p}{\frac{\rho_\infty U_\infty^2 \alpha}{2}} \equiv \frac{p}{q\alpha}$$

as a function of nondimensional time,

$$s \equiv \frac{tU_\infty}{c/2}$$

and M_∞. The results shown below are from an article by Lomax;* both subsonic and supersonic, two- and three dimensional results are displayed.

In Figure 4.9 the chord-wise pressure distribution for two-dimensional flow is shown at several times, s, for a representative subsonic Mach number. For $s = 0$, the result is given by piston theory (as in supersonic flow)†

$$p = \rho_\infty a_\infty w$$

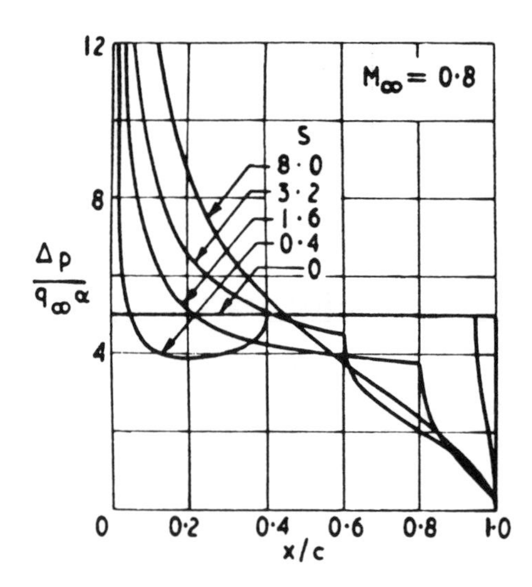

Figure 4.9 Chordwise lifting pressure distributions.

* Lomax [22].

† This can be shown by considering the transient analysis of Section 4.2 and noting it still applies for $t = 0^+$.

For a step change in α, piston theory gives

$$\frac{\Delta p}{\frac{\rho_\infty U_\infty^2 \alpha}{2}} = \frac{p_L - p_U}{\frac{\rho_\infty U_\infty^2 \alpha}{2}} = \frac{4}{M}$$

For $s \to \infty$, the result is also well known, with a square root singularity at the leading edge. Of course, the Kutta condition, $\Delta p = 0$, is enforced at the trailing edge for all s. As $s \to \infty$

$$\frac{\Delta p}{\frac{\rho_\infty U_\infty^2 \alpha}{2}} = \frac{4}{(1 - M^2)^{\frac{1}{2}}} \sqrt{\frac{c - x}{x}}$$

This result is implicit in the analysis of Section 4.3.

In Figure 4.10 the chord-wise pressure distribution is shown at several times, s, for a representative supersonic Mach number. For $s = 0$ the result is again that given by piston theory

$$\frac{\Delta p}{\frac{\rho_\infty U_\infty^2 \alpha}{2}} = \frac{4}{M}$$

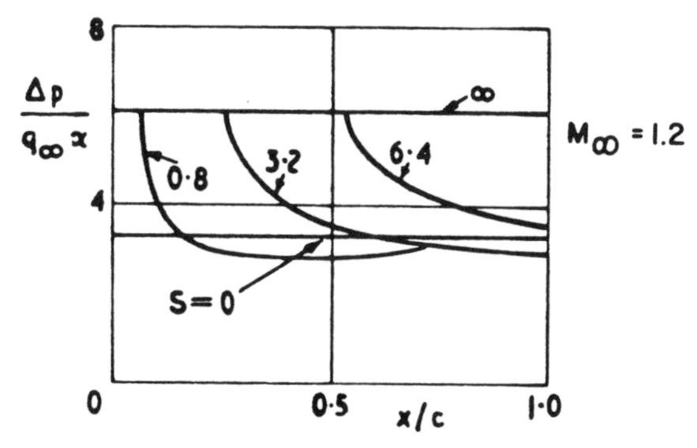

Figure 4.10 Chordwise lifting pressure distribution.

For $s \to \infty$, the result is (as previously cited in our earlier discussion, Section 4.2)

$$\frac{\Delta p}{\frac{\rho_\infty U_\infty^2 \alpha}{2}} = \frac{4}{(M^2 - 1)^{\frac{1}{2}}}$$

Indeed the pressure reaches this final steady state value at a finite s which can be determined as follows. All disturbances propagate *in the fluid* with the speed of sound, a_∞, but the airfoil moves faster with velocity $U_\infty > a_\infty$. Hence, the elapsed time for all disturbances (created by the step change of α for the airfoil) to move off the airfoil is the time required for a (forward propagating *in the fluid*) disturbance at the leading edge to move to the trailing edge, namely

$$t = c/(U_\infty - a_\infty)$$

or, in nondimensional form,

$$s \equiv \frac{tU_\infty}{c/2} = \frac{2M_\infty}{M_\infty - 1}$$

For

$$s > \frac{2M_\infty}{M_\infty - 1}$$

steady state conditions are obtained all along the airfoil. As can be seen from Figure 4.10 for $s = 0^+$ the leading edge pressure instantly reaches its final steady state value. As s increases the steady state is reached by increasing portions of the airfoil along the chord. Note that the initial results, $s = 0$, and steady state results,

$$s \geq \frac{2M_\infty}{M_\infty - 1}$$

have a constant pressure distribution; however, for intermediate s, the pressure varies along the chord.

The pressure distributions may be integrated along the chord to obtain the total force (lift) on the airfoil

$$L \equiv \int_0^c \Delta p \, \mathrm{d}x,$$

$$C_{L_\alpha} \equiv \frac{L}{\dfrac{\rho_\infty U_\infty^2 c\alpha}{2}}, \quad \text{lift curve slope}$$

Again the $s = 0$ result is that given by piston theory

$$C_{L_\alpha} = \frac{4}{M}$$

and the steady-state result is

$$C_{L_\alpha} = \frac{4}{(M^2 - 1)^{\frac{1}{2}}} \quad \text{for} \quad M_\infty > 1$$

and it is also known that

$$C_{L_\alpha} = \frac{2\pi}{(1 - M_\infty^2)^{\frac{1}{2}}} \quad \text{for} \quad M_\infty < 1$$

see Section 4.3. Results for C_{L_α} are shown in Figure 4.11 for various Mach number.

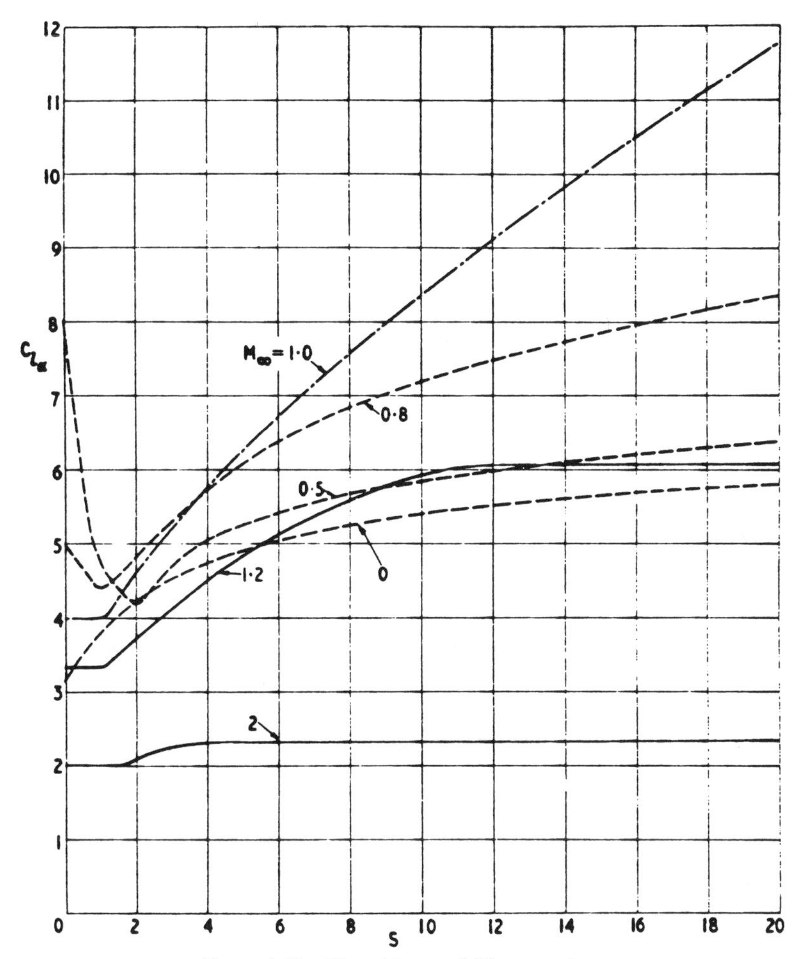

Figure 4.11 Time history of lift curve slope.

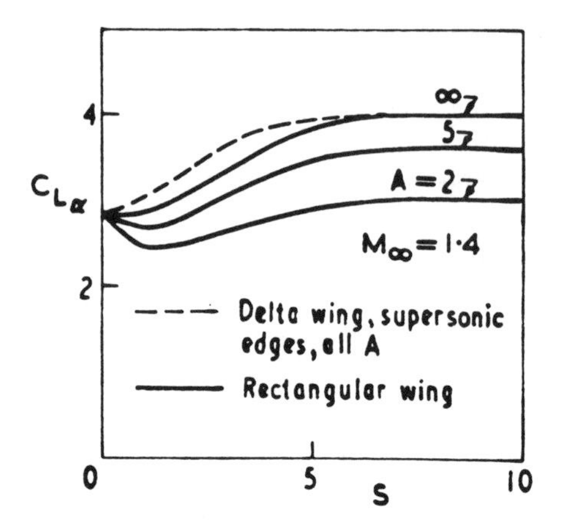

Figure 4.12 Time history of lift curve slope.

Finally some representative results for *three-dimensional*, supersonic flow are shown in Figure 4.12. The effect of three-dimensionality is to reduce the lift. For small aspect ratio, A, where

$$A \equiv \text{maximum span squared/wing area}$$

it is known from slender body theory* (an asymptotic theory for $A \to 0$) that

$$C_{L_\alpha} = \frac{\pi}{2} A$$

for $s \to \infty$. Note however, that the $s = 0^+$ result is independent of A and is that given by piston theory.

Hence, piston theory gives the correct result for $s = 0^+$ for two- and three-dimensional flows, subsonic as well as supersonic. However, only for relatively high supersonic and nearly two-dimensional flows does it give a reasonable approximation for *all s*.

For subsonic flows, the numerical methods are in an advanced state of development and results have been obtained for rather complex geometries including multiple aerodynamic surfaces. In Figures 4.13 to 4.17 representative data are shown. These are drawn from a paper by Rodden,† *et al.*, which contains an extensive discussion of such data and the numerical techniques used to obtain them. Simple harmonic motion is

* See Lomax, for example [22].
† Rodden, Giesing and Kálmán [23].

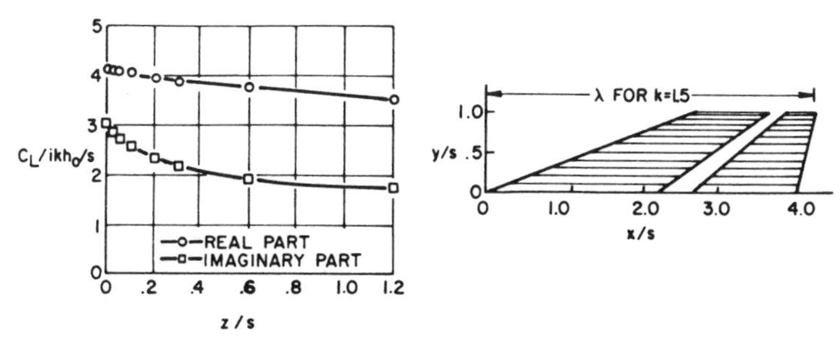

Figure 4.13 Lift coefficient of plunging wing-tail combination for various vertical separation distances; simple harmonic motion.

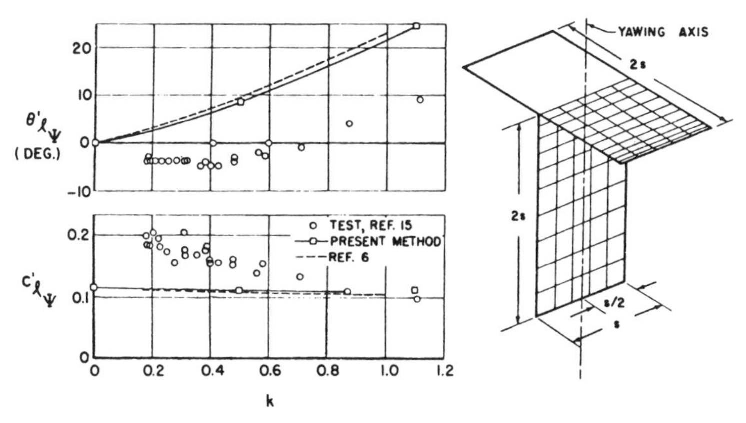

Figure 4.14 Rolling moment coefficient of horizontal stabilizer for simplified T-tail oscillating in yaw about fin mid-chord; simple harmonic motion.

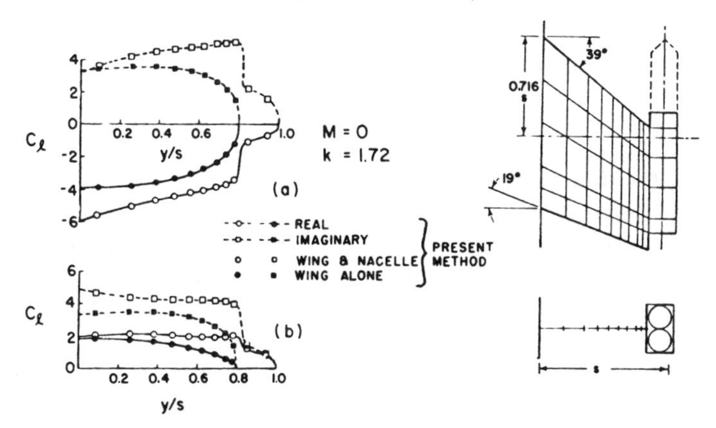

Figure 4.15 Distribution of span load for wing with and without engine nacelle. (a) *plunging* (b) *pitching; simple harmonic motion.*

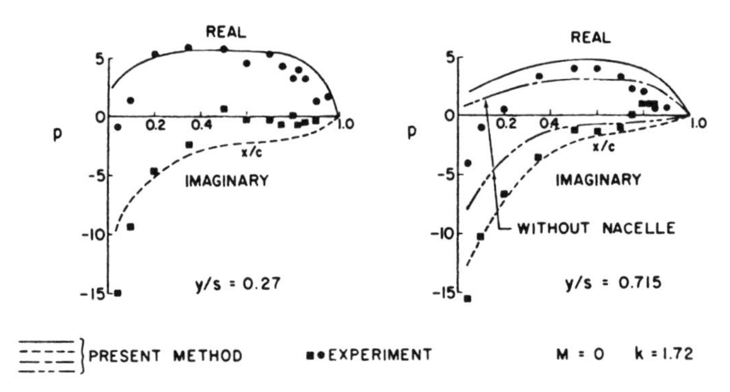

Figure 4.16 Comparison of experimental and calculated lifting pressure coefficient on a wing-nacelle combination in plunge; simple harmonic motion.

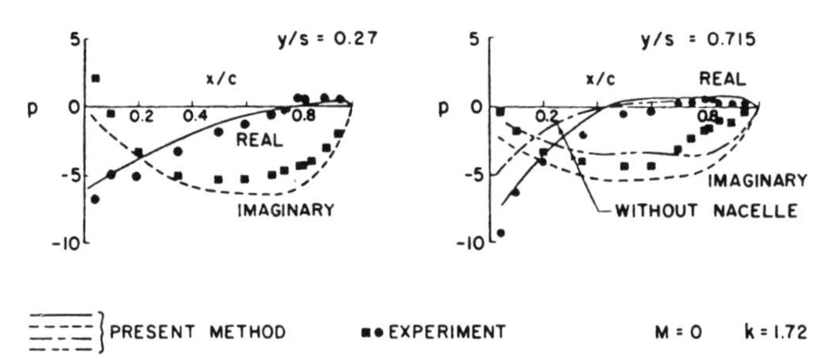

Figure 4.17 Comparison of experimental and calculated lifting pressure coefficient on a wing-nacelle combination in pitch; simple harmonic motion.

considered where k is a non-dimensional frequency of oscillation. Comparisons with experimental data are also shown.

4.5 Transonic flow

Major progress has been made in recent years on this important topic. Here we concentrate on the fundamental ideas and explore one simple approach to obtaining solutions using the same mathematical methods previously employed for subsonic and supersonic flow.

The failure of the classical linear, perturbation theory in transonic flow is well known and several attempts have been made to develop a theoretical model which will give consistent, accurate results. Among the more successful is the 'local linearization' concept of Spreiter which

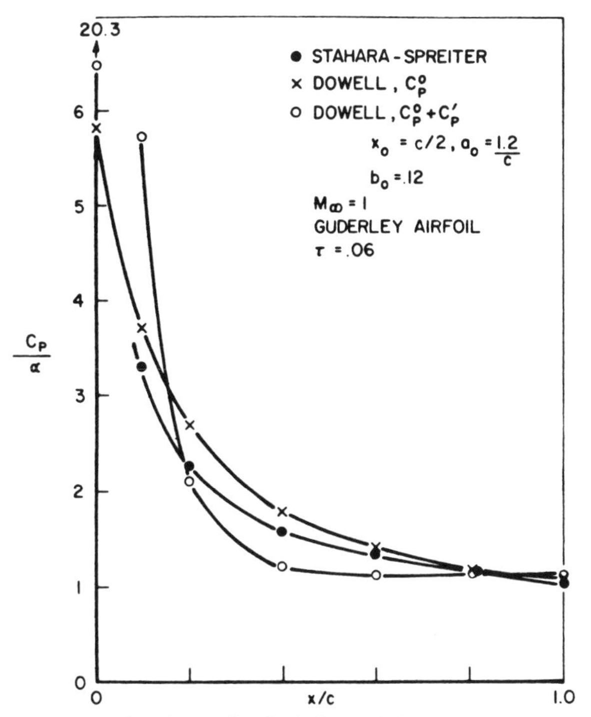

Figure 4.18 Pressure distribution for Guderley airfoil at constant angle of attack.

recently has been generalized to treat oscillating airfoils in transonic flow [24]. Another valuable method is that of parametric differentiation as developed by Rubbert and Landahl [25]. 'Local linearization' is an ad hoc approximation while parametric differentiation is a perturbation procedure from which the results of local linearization may be derived by making further approximations. Several authors [26–29] have attacked the problem in a numerical fashion using finite differences and results have been obtained for two-dimensional, high subsonic flow. Cunningham [30] has suggested a relatively simple, empirical modification of the classical theory.

In the present section a rational approximate method* is discussed which is broadly related to the local linearization concept. It has the advantages of (1) being simpler than the latter and (2) capable ot being systematically improved to obtain an essentially exact solution to the governing transonic equation. Although the method has been developed

* This section is a revised version of Dowell [31]. A list of nomenclature is given at the end of this section.

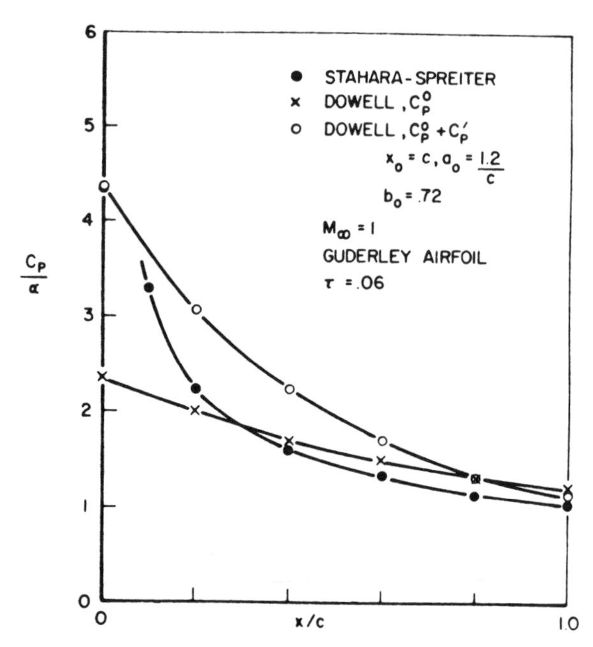

Figure 4.19 Pressure distribution for Guderley airfoil at constant angle of attack.

for treating infinitesimal dynamic motions of airfoils of finite thickness, it may also be employed (using the concept of parametric differentiation) to obtain solutions for nonlinear, steady nonlifting flows. This is the problem for which 'local linearization' was originally developed.

First, the basic idea will be explained for an infinitesimal steady motion of an airfoil of finite thickness in two-dimensional flow. Results will also be given for dynamic motion. The aerodynamic Green's functions for three-dimensional flow have also been derived. These are needed in the popular Mach Box and Kernel Function methods [32]. Using the Green's functions derived by the present method, three-dimensional calculations are effectively no more difficult than for the classical theory.

Analysis

From (4.1.21), Section 4.1, the full nonlinear equation for ϕ is

$$a^2\nabla^2\phi - \left[\frac{\partial}{\partial t}(\nabla\phi \cdot \nabla\phi) + \frac{\partial^2\phi}{\partial t^2} + \nabla\phi \cdot \nabla\left(\frac{\nabla\phi \cdot \nabla\phi}{2}\right)\right] = 0$$

In cartesian, scalar notation and re-arranging terms

$$\begin{aligned}&\phi_{xx}(a^2-\phi_x^2)+\phi_{yy}(a^2-\phi_y^2)+\phi_{zz}(a^2-\phi_z^2)\\&-2\phi_{yz}\phi_y\phi_z-2\phi_{xz}\phi_x\phi_z-2\phi_{xy}\phi_x\phi_y\\&-\frac{\partial}{\partial t}(\phi_x^2+\phi_y^2+\phi_z^2)-\frac{\partial^2\phi}{\partial t^2}=0\end{aligned} \tag{4.5.1}$$

Also we previously determined that ((4.1.22), Section 4.1)

$$\frac{a^2-a_\infty^2}{\gamma-1}=\frac{U_\infty^2}{2}-\left(\frac{\partial\phi}{\partial t}+\frac{\nabla\phi\cdot\nabla\phi}{2}\right) \tag{4.5.2}$$

Now let $\phi=U_\infty x+\hat{\phi}$, then (4.5.2) becomes

$$\frac{a^2-a_\infty^2}{\gamma-1}=-\left[\frac{\frac{\partial\hat{\phi}}{\partial t}+2U_\infty\frac{\partial\hat{\phi}}{\partial x}+\left(\frac{\partial\hat{\phi}}{\partial x}\right)^2+\left(\frac{\partial\hat{\phi}}{\partial y}\right)^2+\left(\frac{\partial\hat{\phi}}{\partial z}\right)^2}{2}\right]$$

$$\cong-\left[\frac{\partial\hat{\phi}}{\partial t}+U_\infty\frac{\partial\hat{\phi}}{\partial x}\right]$$

or

$$a^2\cong a_\infty^2-(\gamma-1)\left[\frac{\partial\hat{\phi}}{\partial t}+U_\infty\frac{\partial\hat{\phi}}{\partial x}\right] \tag{4.5.3}$$

(4.5.1) becomes

$$\begin{aligned}&\hat{\phi}_{xx}\left(a_\infty^2-(\gamma-1)\left[\frac{\partial\hat{\phi}}{\partial t}+U_\infty\frac{\partial\hat{\phi}}{\partial x}\right]-U_\infty^2-2U_\infty\frac{\partial\hat{\phi}}{\partial x}\right)\\&+\hat{\phi}_{yy}a_\infty^2+\hat{\phi}_{zz}a_\infty^2-\frac{\partial}{\partial t}\left(2U_\infty\frac{\partial\hat{\phi}}{\partial x}\right)-\frac{\partial^2\hat{\phi}}{\partial t^2}\cong0\end{aligned} \tag{4.5.4}$$

where obvious higher order terms have been neglected on the basis of $\hat{\phi}_x$, $\hat{\phi}_y$, $\hat{\phi}_z \ll U_\infty$ and a_∞.

The crucial distinction in transonic perturbation theory is in the coefficient of $\hat{\phi}_{xx}$. In the usual subsonic or supersonic small perturbation theory one approximates it as simply

$$a_\infty^2-U_\infty^2$$

However if $U_\infty=a_\infty$ or nearly so then the terms retained above become important. The time derivative term in the coefficient of $\hat{\phi}_{xx}$ may still be neglected compared to the next to last term in (4.5.4), but no further simplification is possible, in general. Hence, (4.5.4) becomes (dividing by a_∞^2)

$$\hat{\phi}_{xx}[1 - M_L^2] + \hat{\phi}_{yy} + \hat{\phi}_{zz} - \frac{1}{a_\infty^2}\left[2U_\infty \frac{\partial^2 \hat{\phi}}{\partial x\, \partial t} + \frac{\partial^2 \hat{\phi}}{\partial t^2}\right] = 0 \tag{4.5.5}$$

where

$$M_L^2 \equiv M_\infty^2\left[1 + \frac{(\gamma+1)\hat{\phi}_x}{U_\infty}\right], \; M_\infty \equiv U_\infty/a_\infty$$

It may be shown that M_L is the consistent transonic, small perturbation approximation to the local (rather than free stream) Mach number. Hence, the essence of *transonic* small perturbation theory is the allowance for variable, local Mach number rather than simply approximating the local Mach number by M_∞ as in the usual subsonic and supersonic theories.

We digress briefly to show that in (4.5.4) the term

$$\hat{\phi}_{xx}\left[-(\gamma - 1)\frac{\partial \hat{\phi}}{\partial t}\right] \tag{4.5.6}$$

may be neglected compared to

$$-2U_\infty \frac{\partial^2 \hat{\phi}}{\partial t\, \partial x} \tag{4.5.7}$$

This is done both for its interest in the present context as well as a prototype for estimation of terms in analyses of this general type.

We assume that a length scale, L, and a time scale, T, may be chosen so that

$x^* \equiv x/L$ 'is of order one'

$t^* \equiv t/T$ 'is of order one'

Hence, derivatives with respect to x^* or t^* do *not*, by assumption, change the order or size of a term. Thus (4.5.6) and (4.5.7) may be written (ignoring constants of order one like $\gamma - 1$ and 2) as

$$\frac{\hat{\phi}_{x^*x^*}}{L^2} \frac{\hat{\phi}_{t^*}}{T} \tag{4.5.6}$$

and

$$U_\infty \frac{\hat{\phi}_{t^*x^*}}{TL} \tag{4.5.7}$$

Hence

$$\frac{(A)}{(B)} \sim 0\left[\frac{\hat{\phi}}{U_\infty L}\right]$$

This ratio however, is much less than one by our original assumption of a small perturbation, viz.

$$\phi = U_\infty L x^* + \hat{\phi}$$

In the beginning we have assumed

$$\frac{\hat{\phi}}{U_\infty L x^*} \ll 1$$

Hence (4.5.6) may be neglected compared to (4.5.7).

(4.5.5) is a nonlinear equation even though we have invoked small perturbation ideas. One may develop a linear theory by considering a steady flow due to airfoil shape, $\hat{\phi}_s$, and an infinitesimal time dependent motion of the airfoil superimposed, $\hat{\phi}_d$. For definiteness, one may consider ϕ_s as due to an airfoil of symmetric thickness at zero angle of attack. Thus let

$$\hat{\phi}(x, y, z, t) = \hat{\phi}_s(x, y, z) + \hat{\phi}_d(x, y, z, t) \tag{4.5.6}$$

and substitute into (4.5.5). The equation for ϕ_s is (by definition)

$$\hat{\phi}_{s_{xx}}[1 - M_{L_s}^2] + \hat{\phi}_{s_{yy}} + \hat{\phi}_{s_{zz}} = 0 \tag{4.5.7}$$

where

$$M_{L_s}^2 \equiv M_\infty^2 \left[1 + (\gamma + 1)\frac{\hat{\phi}_{s_x}}{U_\infty}\right]$$

The equation for $\hat{\phi}_d$ (neglecting products of $\hat{\phi}_d$ and its derivatives which is acceptable for sufficiently small time dependent motions) is

$$\hat{\phi}_{d_{zz}} + \hat{\phi}_{d_{yy}} - \frac{1}{a_\infty^2}\hat{\phi}_{d_{tt}} - 2\frac{U_\infty}{a_\infty^2}\hat{\phi}_{d_{xt}} - b\hat{\phi}_{d_{xx}} - a\hat{\phi}_{d_x} = 0 \tag{4.5.8}$$

where

$$b \equiv \left[M_\infty^2 - 1 + (\gamma + 1)\frac{\hat{\phi}_{s_x}}{U_\infty}\right]$$

$$a \equiv (\gamma + 1) M_\infty^2 \frac{\hat{\phi}_{s_{xx}}}{U_\infty}$$

From Bernoulli's equation

$$C_{p_{ms}} \equiv \frac{\hat{p}_s}{\frac{\rho_\infty U_\infty^2}{2}} = -\frac{2\hat{\phi}_{s_x}}{U_\infty}$$

Hence, a and b may be written as

$$b \equiv \left[M_\infty^2 - 1 - \frac{(\gamma+1)M_\infty^2 C_{p_{ms}}(x)}{2} \right]$$

$$a \equiv -(\gamma+1)\frac{M_\infty^2}{2}\frac{dC_{p_{ms}}(x)}{dx}$$

ϕ is velocity potential due to the infinitesimal motion (henceforth ^ and d are dropped for simplicity). $C_{p_{ms}}$ is the mean steady pressure coefficient due to airfoil finite thickness and is taken as known. In general, it is a function of x, y, z and the method to be described will, in principle, allow for such dependence. However, all results have been obtained ignoring the dependence on y and z. See Refs. [24], [25] and [33] for discussion of this point.

The (perturbation) pressure, p, is related to ϕ by the Bernoulli relation

$$p = -\rho_\infty \left[\frac{\partial \phi}{\partial t} + U_\infty \frac{\partial \phi}{\partial x} \right]$$

and the boundary conditions are

$$\left.\frac{\partial \phi}{\partial z}\right|_{z=0} = w \equiv \frac{\partial f}{\partial t} + U_\infty \frac{\partial f}{\partial x}$$

on airfoil where

$f(x, y, t) \equiv$ vertical displacement of point x, y on airfoil
$w \equiv$ upwash velocity

and

$p|_{z=0} = 0$ off airfoil

plus appropriate finiteness or radiation conditions as $z \to \infty$.

Note that equation (4.5.7) is nonlinear in $\hat{\phi}_s$. If one linearizes, as for example in the classical *supersonic* theory, one would set $M_L = M_\infty$ and obtain as a solution to (4.5.7)

$$\hat{p}_s = \frac{\rho_\infty U_\infty^2}{(M_\infty^2 - 1)^{\frac{1}{2}}} \frac{\partial f}{\partial x}$$

where $\partial f/\partial x$ is the slope of airfoil shape. As $M_\infty \to 1$, $\hat{p} \to \infty$ which is a unrealistic physical result of the linear theory. On the other hand if one uses

$$M_L = M_\infty \left[1 + (\gamma - 1) \frac{\hat{\phi}_{s_x}}{U_\infty} \right]^{\frac{1}{2}}$$

a finite result is obtained for $\hat{p}_s$ as $M_\infty \to 1$ which is in reasonable agreement with the experimental data.*

Equation (4.5.7) with the full expression for M_L is a nonlinear partial differential equation which is much more difficult to solve than its linear counterpart. However two types of methods have proven valuable, the numerical finite difference methods† and various techniques associated with the name 'local linearization' as pioneered by Oswatitsch and Spreiter [34].

Once $\hat{\phi}_s$ is known (either from theory or experiment) (4.5.8) may be used to determine $\hat{\phi}_d$. (4.5.8) is a linear differential equation with variable coefficients which depend upon $\hat{\phi}_s$. Hence, the solution for the lifting problem, $\hat{\phi}_d$, depends upon the thickness solution, $\hat{\phi}_s$, unlike the classical linear theory where the two may be calculated separately and the results superimposed. Again either finite difference methods or 'local linearization' may be employed to solve (4.5.8). Here we pursue an improved analytical technique to determine $\hat{\phi}_d$, which has been recently developed in the spirit of 'local linearization' ideas [31].

To explain the method most concisely, let $\phi_y = \phi_t = 0$ in equation (4.5.8), i.e., consider two-dimensional, steady flow.

Assume

$$a = \sum_{m=0}^{\infty} a_m (x - x_0)^m$$

$$b = \sum_{n=0}^{\infty} b_n (x - x_0)^n \ddagger$$

and $\phi = \phi^0 + \phi'$ where, by *definition*,

$$\phi^0_{zz} - b_0 \phi^0_{xx} - a_0 \phi^0_x = 0 \qquad (4.5.8a)$$

---

* Spreiter [34].

† Ballhaus, Magnus and Yoshihara [35].

‡ We expand in a *power* series about $x = x_0$; however, other series might be equally or more useful for some applications. Results suggest the details of a and b are unimportant.

and ϕ^0 satisfies any nonhomogeneous boundary conditions on ϕ. The equation for ϕ' is thus from (4.5.8) and using the above

$$\phi'_{zz} - b_0\phi'_{xx} - a_0\phi'_x = \sum_{n=1}^{\infty} b_n(x-x_0)^n[\phi^0_{xx} + \phi'_{xx}]$$

$$+ \sum_{m=1}^{\infty} a_m(x-x_0)^m[\phi^0_x + \phi'_x] \quad (4.5.8b)$$

with homogeneous boundary conditions on ϕ'.

If $\phi' \ll \phi^0$, i.e., ϕ^0 is a good approximation to the solution, then ϕ' may be computed from (4.5.8b) by neglecting ϕ' in the right hand side. The retention of a_0 (but not b_1!) in (4.5.8a) is the key to the method, even though this may seem inconsistent at first.

We begin our discussion with steady airfoil motion in a two-dimensional flow. This is the simplest case from the point of view of computation, of course; however, it is also the most critical in the sense that, as Landahl [33] and others have pointed out, unsteadiness and/or three-dimensionality alleviate the nonlinear transonic effects. Indeed, if the flow is sufficiently unsteady and/or three-dimensional, the classical linear theory gives accurate results transonically.

Steady airfoil motion in two-dimensional, 'supersonic' ($b_0 > 0$) flow

Solution for ϕ^0. For $b_0 > 0$, x is a time-like variable and the flow is undisturbed ahead of the airfoil (as far as ϕ^0 is concerned). Hence, solutions may be obtained using a Laplace transform with respect to x. Defining

$$\phi^{0*} \equiv \int_0^{\infty} \phi^0(x, z)\mathrm{e}^{-px}\,\mathrm{d}x$$

(4.5.8a) becomes

$$\phi^{0*}_{zz} - \mu^2\phi^{0*} = 0 \quad (4.5.9)$$

with

$$\mu^2 \equiv [b_0p^2 + a_0p]$$

Solving (4.5.9)

$$\phi^{0*} = A^0_1\mathrm{e}^{-\mu z} + A^0_2\mathrm{e}^{+\mu z} \quad (4.5.10)$$

In order to satisfy finiteness/radiation condition at infinity, one selects $A^0_2 \equiv 0$. A^0_1 is determined from the (transformed) boundary condition,

$$\phi^{0*}_z|_{z=0} = w^* \quad (4.5.11)$$

From (4.5.10) and (4.5.11),

$$\phi^{0*}|_{z=0} = \frac{-w^*}{\mu} \tag{4.5.12}$$

Inverting (4.5.12),

$$\phi^0|_{z=0} = -\int_0^x b_0^{-\frac{1}{2}} \exp\left(\frac{-a_0\xi}{2b_0}\right) I_0\left[\frac{a_0\xi}{2b_0}\right] w(x-\xi)\,\mathrm{d}\xi \tag{4.5.13}$$

It is of interest to note two limiting cases. As $a_0\xi/2b_0 \to 0$,

$$\phi^0|_{z=0} = -\int_0^x b_0^{-\frac{1}{2}} w(x-\xi)\,\mathrm{d}\xi \tag{4.5.14}$$

the classical result. But, more importantly, as $a_0\xi/2b_0 \to \infty$,

$$\phi^0|_{z=0} = -\int_0^x (\pi a_0 \xi)^{-\frac{1}{2}} w(x-\xi)\,\mathrm{d}\xi \tag{4.5.15}$$

Hence, even when the effective Mach number at $x = x_0$ is transonic, i.e., $b_0 \equiv 0$, the present model gives a finite result. Before computing the correction, ϕ', to the velocity potential we shall exploit ϕ^0 to obtain several interesting results. For this purpose we further restrict ourselves to an airfoil at angle of attack, $w = -U_\infty \alpha$. From (4.5.15),

$$\frac{\phi^0_{z=0}}{U_\infty \alpha} = \frac{2b_0^{\frac{1}{2}}}{a_0} \bar{x} \mathrm{e}^{-\bar{x}}[I_0(\bar{x}) + I_1(\bar{x})]; \qquad \bar{x} \equiv \frac{a_0 x}{2b_0} \tag{4.5.16}$$

and the pressure on the *lower* aerodynamic surface is

$$\frac{C_p}{\alpha} \equiv \frac{p^0}{\frac{\rho_\infty U_\infty^2 \alpha}{2}} = \frac{2\phi^0_x}{U_\infty \alpha}\bigg|_{z=0} = 2b_0^{-\frac{1}{2}} \mathrm{e}^{-\bar{x}} I_0(\bar{x}) \tag{4.5.17}$$

The lift, moment and center of pressure may be computed.

$$\begin{aligned} L^0 &\equiv \int_0^C 2p^0\,\mathrm{d}x = \rho_\infty U_\infty^2 \alpha c 4(\pi a_0 c)^{-\frac{1}{2}} \tilde{L}^0 \\ \tilde{L}^0 &\equiv (\pi/2)^{\frac{1}{2}} \bar{c}^{\frac{1}{2}} \mathrm{e}^{-\bar{c}}[I_0(\bar{c}) + I_1(\bar{c})]; \qquad \bar{c} \equiv \frac{a_0 c}{2b_0} \end{aligned} \tag{4.5.18}$$

$$\begin{aligned} M^0 &= \int_0^c 2p^0 x\,\mathrm{d}x = L^0 c - \rho_\infty U_\infty^2 c^2 \tfrac{8}{3} (\pi a_0 c)^{-\frac{1}{2}} \tilde{M}^0 \\ \tilde{M}_0 &\equiv \tfrac{3}{4}(2\pi)^{\frac{1}{2}} \{\mathrm{e}^{-\bar{c}} I_1(\bar{c})[\bar{c}^{-\frac{1}{2}} + \tfrac{2}{3}\bar{c}^{\frac{1}{2}}] + \tfrac{2}{3}\mathrm{e}^{-\bar{c}} I_2(\bar{c})\bar{c}^{\frac{1}{2}}\} \end{aligned} \tag{4.5.19}$$

The center of pressure may be obtained from L^0 and M^0 in the usual way. We shall use and discuss these results for a particular airfoil later. But first let us consider the computation of ϕ'.

Solution for ϕ'. For *simplicity*, we shall consider only a *linear* variation in mean pressure, $C_{p_{ms}}$, along the airfoil chord. Hence, a_0, b_0 and b_1 are not zero and $b_1 = a_0$. All other a_m and b_n are zero. Assuming $\phi' \ll \phi^0$, the equation for ϕ' is

$$\phi'_{zz} - a_0\phi'_x - b_0\phi'_{xx} = b_1(x - x_0)\phi^0_{xx} \tag{4.5.20}$$

Taking a Laplace transform of (4.5.20),

$$\phi'^*_{zz} - \mu^2\phi'^* = -b_1\left[2p\phi^{0*} + p^2\frac{d\phi^{0*}}{dp} + x_0p^2\phi^{0*}\right] \tag{4.5.21}$$

A particular solution of (4.5.21) is

$$\phi'^*_p = (C_0z + C_1z^2)e^{-\mu z} \tag{4.5.22}$$

where

$$C_0 \equiv b_1\left[\frac{A}{2\mu} + \frac{B}{4\mu^2}\right]; \qquad C_1 \equiv \frac{b_1}{4\mu}B$$

$$A \equiv \frac{-2pw^*}{\mu} + \frac{p^2w^*}{\mu^3}\frac{[2b_0p + a_0]}{2} - x_0\frac{p^2w^*}{\mu} - \frac{p^2}{\mu}\frac{dw^*}{dp}$$

$$B \equiv \frac{p^2w^*}{\mu^2}\frac{[2b_0p + a_0]}{2}$$

The homogeneous solution for ϕ' is of the same form as for ϕ^0. After some calculation, applying homogeneous boundary conditions to ϕ', we determine

$$\phi'^*|_{z=0} = \frac{C_0}{\mu} \tag{4.5.23}$$

Inverting (4.5.23) using the definitions of C_0, A, B above, and assuming $w = -U_\infty\alpha$ for simplicity, we have

$$\left.\frac{\phi'}{U_\infty\alpha}\right|_{z=0} = \frac{b_0^{\frac{1}{2}}}{a_0}\left\{2e^{-\bar{x}}\bar{x}I_1(\bar{x}) - \left[\frac{d^2}{d\bar{x}^2} + \frac{d}{d\bar{x}}\right][e^{-\bar{x}}\bar{x}^2I_2(\bar{x})]\right.$$
$$\left. + \frac{\bar{c}2x_0}{c}e^{-\bar{x}}\bar{x}[I_0(\bar{x}) + I_1(\bar{x})]\right\}; \qquad \bar{c} \equiv \frac{a_0c}{2b_0} \tag{4.5.24}$$

The pressure coefficient corresponding to ϕ' is given by

$$C'_p = C'_{p_1} + C'_{p_2}$$

where

$$\frac{b_0^{\frac{1}{2}} C'_{p_1}}{\alpha} \equiv \mathrm{e}^{-\bar{x}}\{(2I_1 - I_0)(\bar{x} - \bar{x}^2) + I_2 \bar{x}^2\} \tag{4.5.25}$$

$$\frac{b_0^{\frac{1}{2}} C'_{p_2}}{\alpha} \equiv \bar{c}\frac{2x_0}{c}\mathrm{e}^{-\bar{x}}\{2\bar{x}(I_1 - I_0) + I_0\}$$

As may be seen C'_{p_1} is always a small correction to C_p^0; however, C'_{p_2} may be large or small (particularly near the leading edge as $\bar{x}\to 0$) depending on the size of

$$\frac{x_0}{c}\frac{a_0 c}{2b_0}$$

Since we are free to choose x_0 in any application, it is in our interest to choose it so that

$$C'_{p_2} \ll C_p^0$$

More will be said of this in the following section.

We note that *higher* terms in the power series for a and b may be included and a solution for ϕ' obtained in a similar manner. The algebra becomes more tedious, of course.

Results and comparisons with other theoretical and experimental data

We have calculated two examples, a Guderley airfoil and a parabolic arc airfoil, both of 6% thickness ratio, τ, and for Mach numbers near one. These were chosen because they have smooth mean steady pressure distributions (at least for some Mach number range) and because other investigators have obtained results for these airfoils. These two airfoils and their mean, steady pressure distributions are shown in [24]. The Guderley airfoil has a linear mean pressure variation while the parabolic arc has a somewhat more complicated variation including a (theoretical) logarithmic singularity at the leading edge. For $M_\infty = 1$, when $C_{p_{ms}} = 0$ the local Mach number along the chord equals one and if one expanded about that point then $b_0 \equiv 0$, and our procedure would fail in that $\phi' \gg \phi^0$. Hence, one is led to believe that one should choose x_0 as far away from the sonic point, $C_{p_{ms}} \equiv 0$ at $M_\infty \equiv 1$, as possible. To fix this idea more concretely, we first consider the Guderley airfoil.

Guderley airfoil. We have calculated C_p^0 and $C_p^0 + C_p'$ for $M_\infty = 1$. Two different choices of x_0 were used, $x_0 = c/2$ (Figure 4.18) and c (Figure 4.19). Results from Stahara and Spreiter [24] are also shown for reference. As can be seen for $x_0 = c/2$, the 'correction' term, C_p', dominates the basic solution, C_p^0, as $x/c \to 0$. For $x_0 = c$, on the other hand, the correction term is much better behaved, in agreement with our earlier speculation about choosing x_0 as far as possible from the sonic point. Note that if, for example, we choose $x_0 = 0$ this would also work in principle, but now $b_0 < 0$, and a 'subsonic' solution would have to be obtained for ϕ^0.

Parabolic arc airfoil. Similar results have been obtained and are displayed in Figure 4.20 ($x_0 = c/2$) and Figure 4.21 ($x_0 = c_0$). Both of these solutions are well behaved in the sense that $C_p' < C_p^0$, though again the results for $x_0 = c$ appear to be better than those for $x_0 = c/2$. The relatively better behavior of the $x_0 = c/2$ results for the parabolic arc as

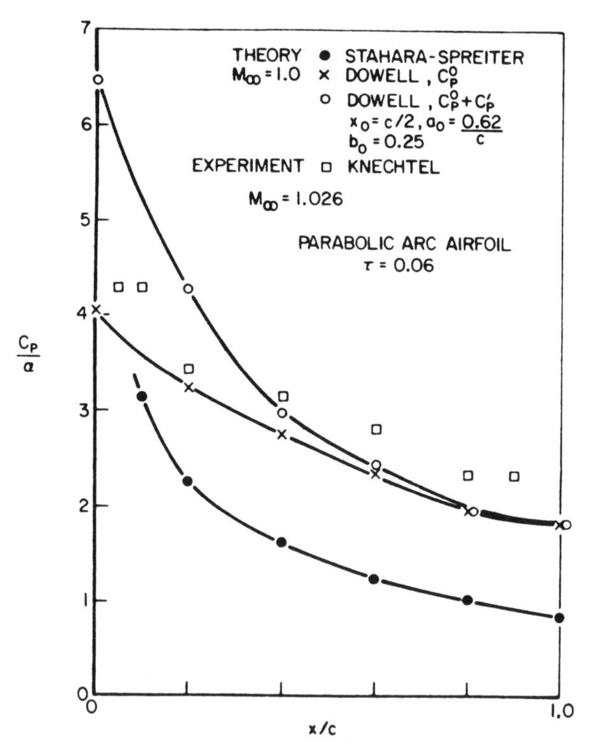

Figure 4.20 Pressure distribution for parabolic arc airfoil at constant angle of attack.

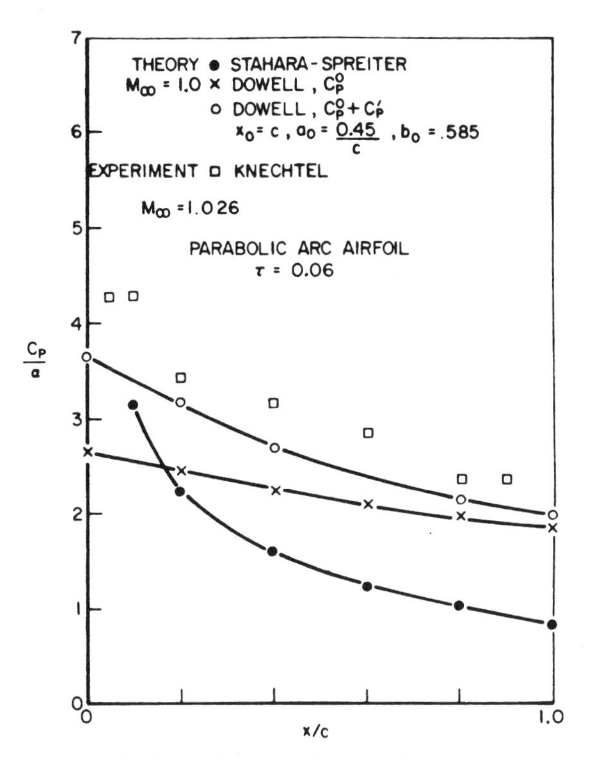

Figure 4.21 Pressure distribution for parabolic arc airfoil at constant angle of attack.

compared with the Guderley airfoil is probably related to the sonic point being further ahead of $x_0 = c/2$ for the former than the latter. See [24]. Also shown in Figures 4.20 and 4.21 are the theoretical results of Stahara–Spreiter [24] and the experimental data of Knechtel [36]. Knechtel indicates the effective Mach number of his experiments should be reduced by approximately 0.03 due to wall interference effects. Also he shows that the measured mean steady pressure distributions at zero angle of attack, $C_{p_{ms}}$, agree well with the theoretical results of Spreiter [24, 37] for $M_\infty \geq 1$. However, for $M_\infty \leq 1$, $C_{p_{ms}}$ deviates from that theoretically predicted; see Figure 4.22 taken from [36]. The change in slope for $C_{p_{ms}}$ near the trailing edge may be expected to be important for computing the lifting case. In Figure 4.23 results are shown for $M_\infty = 0.9$ which dramatically make this point. Shock induced separation of the boundary layer is the probable cause of the difficulty.

Finally, we present a graphical summary of lift curve slope and center

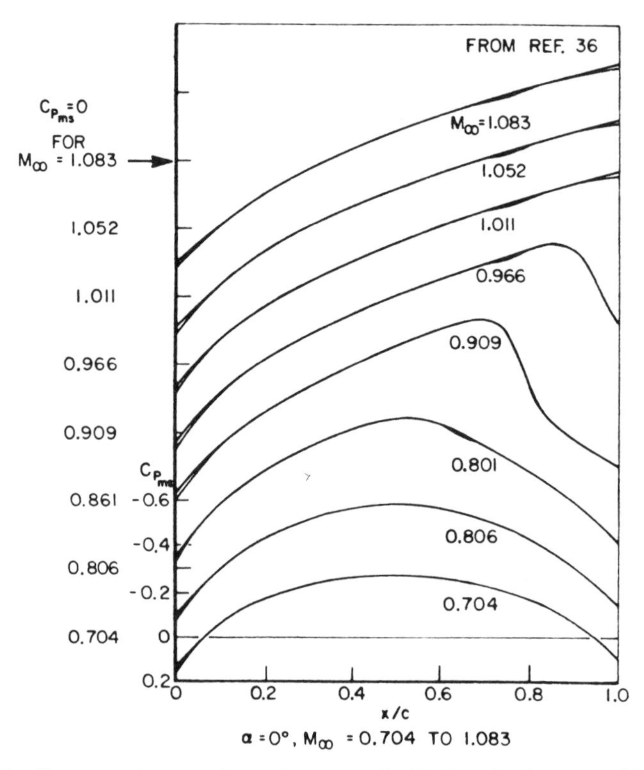

Figure 4.22 Representative experimental pressure distributions for 6-percent-thick circular-arc airfoil with roughness elements near the leading edge.

of pressure for the parabolic arc airfoil comparing results of Knechtel's experimental data and the present analysis. See Figure 4.24.

All things considered the agreement between theory and experiment is rather good; however, it is clear that if $C_{p_{ms}}$ varies in a complicated way one must go beyond the straight line approximation used in obtaining the present results. In principle this can be done; how much effort will be required remains to be determined.

Nonsteady airfoil motion in two-dimensional, 'supersonic' $b_0 > 0$ flow

Solution for ϕ^0. Again taking a Laplace transform with respect to x of (4.5.8) (for $\phi_{yy} \equiv 0$ and $a = a_0$, $b = b_0$) we obtain

$$\phi^{0*}_{zz} - \mu^2 \phi^{0*} = 0 \qquad (4.5.26)$$

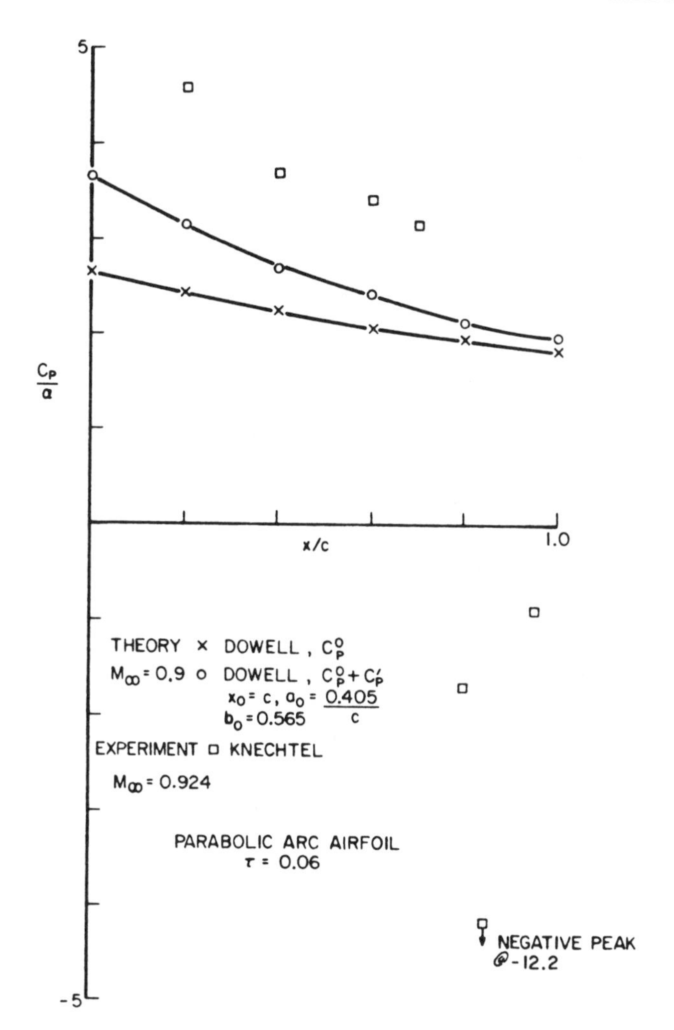

Figure 4.23 Pressure distribution for parabolic arc airfoil at constant angle of attack.

where $\mu \equiv [b_0 p^2 + \tilde{a}_0 p - d]^{\frac{1}{2}}$; b_0—as before

$$\tilde{a}_0 \equiv a_0 + \frac{2U_\infty}{a_\infty^2} i\omega; \qquad d \equiv \left(\frac{\omega}{a_\infty}\right)^2$$

and we have assumed simple harmonic motion in time. Solving (4.5.26) subject to the boundary condition, (4.5.11), and appropriate finiteness

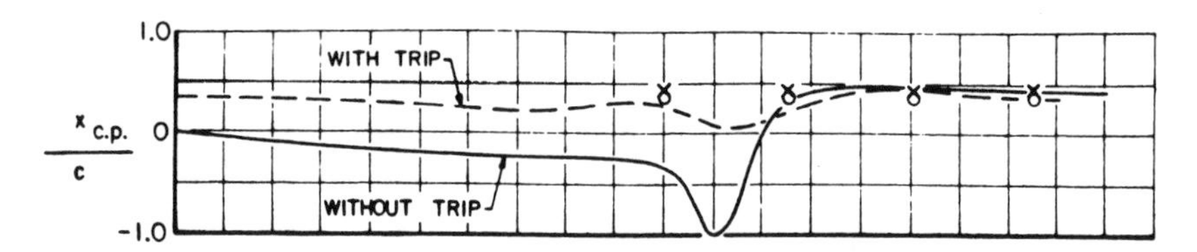

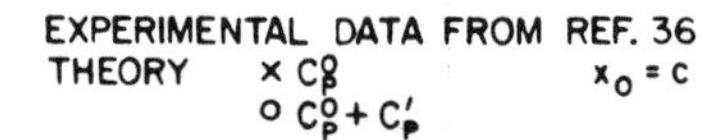

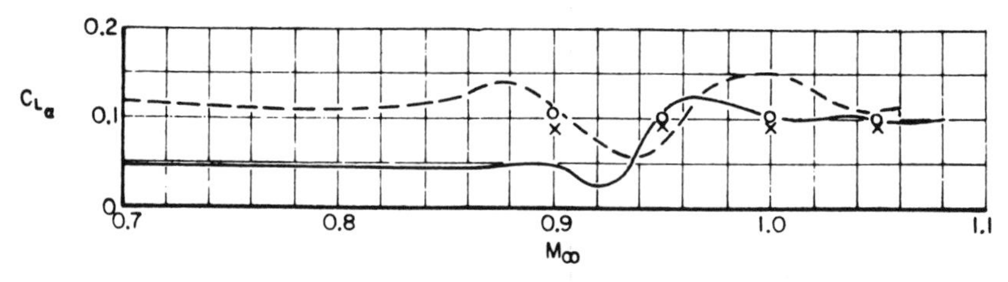

Figure 4.24 Effect of boundary-layer trip on the variation with Mach number of lift-curve slope and center of pressure of the circular-arc airfoil at $\alpha_0 \cong 0°$.

and/or radiation condition at infinity we have (after inversion)

$$\phi^0|_{z=0} = -\int_0^x b_0^{-\frac{1}{2}} \exp\left(\frac{-\tilde{a}_0\xi}{2b_0}\right) I_0\left\{\left[\left(\frac{\tilde{a}_0}{2b_0}\right)^2 + \frac{d}{b_0}\right]^{\frac{1}{2}} \xi\right\} w(x-\xi)\, d\xi \tag{4.5.27}$$

The perturbation pressure on the lower surface is given by

$$p^0 = \rho_\infty[\phi_t^0 + U_\infty \phi_x^0] \tag{4.5.28}$$

which may be evaluated from (4.5.27) directly using Leibnitz' rule

$$C_p^0 = \frac{p^0}{\frac{\rho_\infty U_\infty^2}{2}} = -2b_0^{-\frac{1}{2}}\left\{\exp\left(\frac{-\tilde{a}_0 x}{2b_0}\right) I_0\left[\left(\frac{\tilde{a}_0}{2b_0}\right)^2 + \frac{d}{b}\right]^{\frac{1}{2}} x\right\} \frac{w(0)}{U_\infty}$$
$$+ \int_0^x \exp\left(\frac{-\tilde{a}_0\xi}{2b_0}\right) I_0\left\{\left[\left(\frac{\tilde{a}_0}{2b_0}\right)^2 + \frac{d}{b_0}\right]^{\frac{1}{2}} \xi\right\}$$
$$\cdot \left[i\omega \frac{w(x-\xi)}{U_\infty^2} + \frac{w'(x-\xi)}{U_\infty}\right] d\xi$$

where

$$w'(x) \equiv \frac{dw}{dx} \tag{4.5.29}$$

An alternative form for C_p^0 may be obtained by first interchanging the arguments x and $x-\xi$ in (4.5.27). For $a_0 \equiv 0$, $b_0 \equiv M_\infty^2 - 1$ the above reduces to the classical result. For any a_0 and b_0 and $k \equiv \omega c/U_\infty$ large the results approach those of the classical theory and for $k \to \infty$ approach the 'piston' theory [32]. For the specific case of an airfoil undergoing vertical translation, $w = -h_t$, where h is vertical displacement and h_t is the corresponding velocity, we have the following results,

$$\phi^0|_{z=0} = h_t b_0^{-\frac{1}{2}} \left[\left(\frac{\tilde{a}_0}{2b_0} \right)^2 + \frac{\mathrm{d}}{b_0} \right]^{-\frac{1}{2}} \mathrm{e}^{-e\tilde{x}} \tilde{x} \left\{ I_0(\tilde{x}) + \frac{I_1(\tilde{x})}{e} \right\}$$

where

$$\tilde{x} \equiv \left[\left(\frac{\tilde{a}_0}{2b_0} \right)^2 + \frac{\mathrm{d}}{b_0} \right]^{\frac{1}{2}} x$$

$$e \equiv \frac{\tilde{a}_0}{2b_0} \left[\left(\frac{\tilde{a}_0}{2b_0} \right)^2 + \frac{\mathrm{d}}{b_0} \right]^{-\frac{1}{2}} \tag{4.5.30}$$

In the limit as $b_0 \to 0$, (corresponding to $M_\infty \to 1$ in the classical theory)

$$\left[\left(\frac{\tilde{a}_0}{2b_0} \right)^2 + \frac{\mathrm{d}}{b_0} \right]^{\frac{1}{2}} \to \frac{\tilde{a}_0}{2b_0}; \qquad e \to 1$$

and

$$\phi^0|_{z=0} \to h_t 2 \left(\frac{x}{\tilde{a}_0 \pi} \right)^{\frac{1}{2}} \tag{4.5.31}$$

Using (4.5.30) or (4.5.31) in (4.5.28) gives the perturbation pressure. The latter form is particularly simple

$$\frac{C_p}{ik\dfrac{\bar{h}}{c}\mathrm{e}^{i\omega t}} \equiv \frac{\dfrac{p}{\rho_\infty U_\infty^2}}{ik\dfrac{\bar{h}}{c}\mathrm{e}^{i\omega t}} = (\pi \tilde{a}_0 c)^{-\frac{1}{2}} \left[2(x/c)^{-\frac{1}{2}} + i4k \left(\frac{x}{c} \right)^{\frac{1}{2}} \right] \tag{4.5.32}$$

where

$$h \equiv \bar{h}\mathrm{e}^{i\omega t}; \qquad k \equiv \frac{\omega c}{U_\infty}$$

Solution for ϕ'. Park [38] has computed ϕ' and made comparisons with available experimental and theoretical data. It is well-known, of course, that for sufficiently large k the classical theory itself is accurate transonically [33]. Hence, we also expect the present theory to be more accurate for increasing k.

Results and comparison with other theoretical data

We have calculated a numerical example for the Guderley airfoil for $M = 1$ and $k = 0.5$ in order to compare with the results of Stahara–Spreiter [24]. We have chosen $x_0 = c/2$ for which

$$b_0 = 0.12; \qquad a_0 = 1.2/c$$

For such small b_0, we may use the asymptotic form for $b_0 \to 0$, (4.5.32), and the results are plotted in Figures 4.25 and 4.26 along with the results of

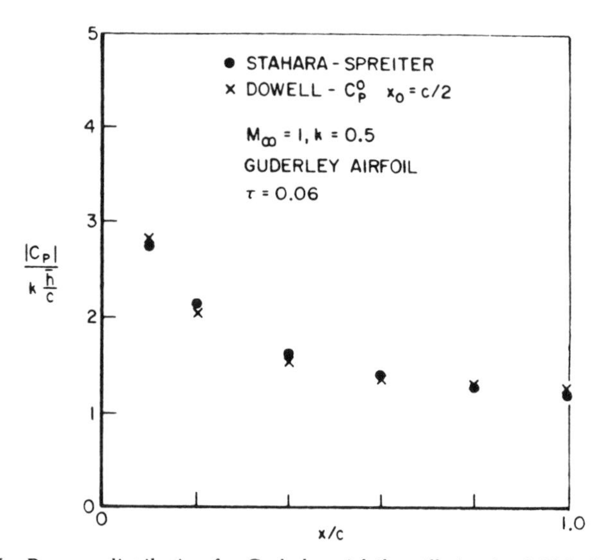

Figure 4.25 Pressure distribution for Guderley airfoil oscillating in rigid body translation.

[24]. As $k \to 0$, the phase angle, Φ, is a constant at 90° and the pressure coefficient amplitude is the same as that of Figure 4.18. Presumably somewhat more accurate results could be obtained by choosing $x_0 = c$ and computing the correction, C_p'. However, the agreement is already good between the present results and those of [24].

As Stahara–Spreiter [24] point out even for k as large as unity there are still substantial quantitative differences between their results (and hence the present results) and those of the classical theory. However, for

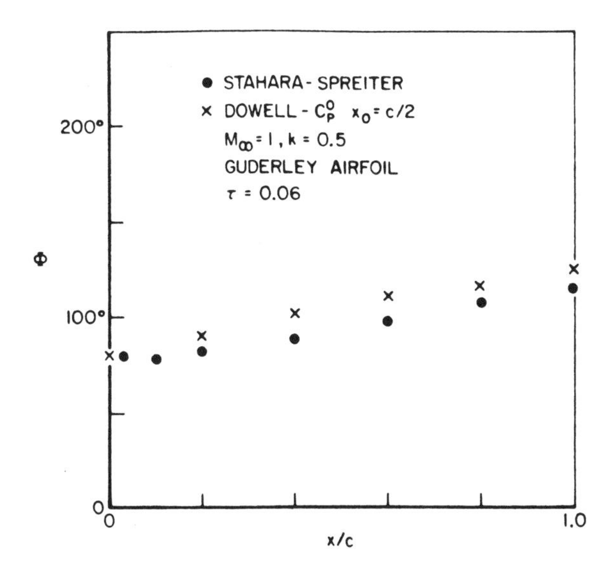

Figure 4.26 Pressure-translation phase angle distribution.

$k \gg 1$, one may expect the present theory and that of [24] to give results which approach those of the classical theory.

Nonsteady airfoil motion in three-dimensional 'supersonic' $(b_0 > 0)$ *flow*

Solution for ϕ^0. We begin with (4.5.1) and take a Fourier transform with respect to y,

$$\phi^\dagger \equiv \int_{-\infty}^{\infty} \phi e^{-i\gamma y}\, dy \tag{4.5.33}$$

and a Laplace transform with respect to x,

$$(\phi^\dagger)^* \equiv \int_0^\infty (\phi^\dagger) e^{-px}\, dx \tag{4.5.34}$$

(4.5.1) becomes

$$\phi_{zz}^{0*\dagger} - \mu^2 \phi^{0*\dagger} = 0 \tag{4.5.35}$$

where

$$\mu \equiv [b_0 p^2 + \tilde{a}_0 p - \tilde{d}]^{\frac{1}{2}}; \qquad b_0, \tilde{a}_0 \quad \text{as before}$$

and

$$\tilde{d} \equiv (\omega/a_\infty)^2 - \gamma^2$$

Solving (4.5.35) subject to the boundary condition, (4.5.11), and appropriate boundary finiteness/radiation conditions at infinity we have (after inversion)

$$\phi^0|_{z=0} = \int_0^x \int_{-\infty}^{\infty} A(x-\xi, y-\eta) w(\xi, \eta)\, d\xi\, d\eta \tag{4.5.36}$$

where

$$A(x, y) = \frac{-\exp\left(\dfrac{-\tilde{a}_0}{2b_0}x\right)}{\pi} r^{-1} \cosh\left\{\left[\left(\frac{\dfrac{\omega}{a_\infty}}{b_0}\right)^2 + \left(\frac{\tilde{a}_0}{2b_0}\right)^2\right]^{\frac{1}{2}} r\right\}$$

for $r^2 > 0$, i.e. $0 < |y| < xb^{-\frac{1}{2}}$

$= 0$ for $r^2 < 0$, i.e. $xb_0^{-\frac{1}{2}} < |y|$

and

$$r^2 \equiv x^2 - b_0^2 y^2 \tag{4.5.37}$$

A is the aerodynamic Green's function required in the Mach Box numerical lifting surface method [32].

For $b_0 \to M_\infty^2 - 1$; $a_0 \to 0$; $\tilde{a}_0 \to 2(i\omega U_\infty/a_\infty^2)$ and A reduces to the classical result. For $b_0 \to 0$, $R_e\tilde{a}_0 > 0$,

$$A \to -\frac{1}{2\pi x} e^{-\tilde{a}_0 y^2/4x} \quad \text{for} \quad x > 0; \quad |y| < \infty \tag{4.5.38}$$

For $b_0 \to 0$, $R_e\tilde{a}_0 < 0$,

$$A \to -\frac{1}{2\pi x} \exp\left(\frac{-\tilde{a}_0 x}{b_0}\right) \quad \text{for} \quad x > 0; \quad |y| < \infty \tag{4.5.39}$$

Non-steady airfoil motion in three-dimensional 'subsonic' ($b_0 < 0$) flow

Solution for ϕ^0. We begin with (4.5.1), assuming simple harmonic motion,

$$-b_0\phi_{xx}^0 - \tilde{a}_0\phi_x^0 + d\phi^0 + \phi_{yy}^0 + \phi_{zz}^0 = 0 \tag{4.5.40}$$

where $\tilde{a}_0$, b_0, d as before.

To put (4.5.40) in canonical form by eliminating the term ϕ_x, we introduce the new dependent variable, Φ

$$\phi^0 \equiv e^{\Omega x}\Phi \tag{4.5.41}$$

where Ω is determined to be

$$\Omega = -\tilde{a}_0/2b_0 \tag{4.5.42}$$

and the equation for Φ is

$$B\Phi_{xx} + \Phi\left[\frac{-a^2}{4B} + \mathrm{d}\right] + \Phi_{yy} + \Phi_{zz} = 0 \tag{4.5.43}$$

and

$$B \equiv -b_0 > 0$$

We further define new independent variables,

$$x' \equiv x, \qquad y' \equiv B^{\frac{1}{2}}y, \qquad z' \equiv B^{\frac{1}{2}}z \tag{4.5.44}$$

then (4.5.43) becomes

$$\Phi_{x'x'} + \Phi_{y'y'} + \Phi_{z'z'} + \tilde{k}^2\Phi = 0 \tag{4.5.45}$$

where

$$\tilde{k}^2 \equiv \left(\mathrm{d} - \frac{a^2}{4B}\right)\Big/ B$$

We are now in a position to use Green's theorem

$$\iiint [\Phi\nabla^2\psi - \psi\nabla^2\Phi]\,\mathrm{d}V = \iint_S \left[\Phi\frac{\partial\psi}{\partial n} - \psi\frac{\partial\Phi}{\partial n}\right]\mathrm{d}S \tag{4.5.46}$$

V volume enclosing fluid
S surface area of volume indented to pass over airfoil surface and wake
n outward normal.

We take Φ to be the solution we seek and choose ψ as

$$\psi \equiv \left(\frac{\mathrm{e}^{-i\tilde{k}r}}{r}\right) \tag{4.5.47}$$

where

$$r \equiv [(x' - x_1')^2 + (y' - y_1')^2 + (z' - z_1')^2]$$

Note that

$$[\nabla^2 + \bar{k}^2]\left(\frac{e^{-i\bar{k}r}}{r}\right) = -4\pi\delta(x' - x_1')\delta(y' - y_1')\delta(z' - z_1') \tag{4.5.48}$$

Thus the LHS of (4.5.7) becomes $-4\pi\Phi(x', y', z')$. On the RHS, there is no contribution from the surface area of sphere at infinity. Thus (4.5.46) becomes

$$4\pi\Phi(x, y, z) = \iint\limits_{\substack{S\\ \text{airfoil}\\ \text{plus wake}}} \left[(\Phi_U - \Phi_L)\frac{\partial}{\partial z_1}\left(\frac{e^{-i\bar{k}r}}{r}\right) - \left(\frac{e^{-i\bar{k}r}}{r}\right)\frac{\partial}{\partial z_1}(\Phi_U - \Phi_L)\right] dx_1\, dy_1 \tag{4.5.49}$$

where

Φ_U, Φ_L upper, lower surface

$\dfrac{\partial}{\partial n} = \dfrac{-\partial}{+\partial z_1}$ on $\begin{matrix}\text{upper}\\ \text{lower}\end{matrix}$ surface

and we have returned to the original independent variables, x, y, z and x_1, y_1, z_1. Since Φ is an odd function of z, z_1,

$$\frac{\partial}{\partial z_1}(\Phi_U - \Phi_L) = 0 \tag{4.5.50}$$

Also

$$\frac{\partial}{\partial z_1}\left(\frac{e^{-i\bar{k}r}}{r}\right) = \frac{\partial}{\partial z}\frac{e^{-i\bar{k}r}}{r}(-1) \tag{4.5.51}$$

Thus (4.5.49) becomes, re-introducing the original dependent variable, ϕ^0,

$$\phi^0(x, y, z) = \frac{-e^{-\Omega x}}{4\pi}\iint \Delta\phi e^{-\Omega x_1}\frac{\partial}{\partial z}\left\{\frac{e^{-i\bar{k}r}}{r}\right\} dx_1\, dy_1 \tag{4.5.52}$$

where

$$\Delta\phi \equiv \phi_U^0 - \phi_L^0$$

Up to this point we have implicitly identified ϕ^0 with the velocity potential. However, *within the approximation*, $a = a_0$, $b = b_0$, $\phi = \phi^0$, $p = p^0$, ϕ and p satisfy the same equation, (4.5.40); hence, we may use

(4.5.54) with ϕ^0 replaced by p^0. Further using Bernoulli's equation, (4.5.5), we may relate ϕ^0 to p^0

$$\phi^0(x, y, z) = -\int_{-\infty}^{x} \frac{p^0(\lambda, y, z)}{\rho_\infty U_\infty} \exp\left[\frac{i\omega(\lambda - x)}{U_\infty}\right] d\lambda \tag{4.5.53}$$

Substituting (4.5.52) into (4.5.53) (where (4.5.52) is now expressed in terms of p^0); introducing a new variable $\xi \equiv \lambda - x_1$ and interchanging the order of integration with respect to ξ and x_1, y_1; gives

$$\phi(x, y, z) = \frac{1}{4\pi} \iint \frac{\Delta p}{\rho_\infty U_\infty}(x_1, y_1) \exp\left[\frac{-i\omega(x - x_1)}{U_\infty}\right] \cdot \left\{\int_{-\infty}^{x - x_1} \exp\left(\frac{[\Omega + i\omega]\xi}{U_\infty}\right) \frac{\partial}{\partial z}\left\{\frac{e^{-i\bar{k}r}}{r}\right\} d\xi\right\} dx_1\, dy_1 \tag{4.5.54}$$

Finally, computing from (4.5.56)

$$w = \left.\frac{\partial \phi}{\partial z}\right|_{z=0}$$

we obtain

$$\frac{w(x, y)}{U_\infty} = \iint \frac{\Delta p}{\rho_\infty U_\infty^2}(x_1, y_1) K(x - x_1, y - y_1)\, dx_1\, dy_1 \tag{4.5.55}$$

where

$$K \equiv \lim_{z \to 0} \frac{\exp\left[\dfrac{-i\omega(x - x_1)}{U_\infty}\right]}{4\pi} \int_{-\infty}^{x - x_1} \exp\left(\left[\Omega + \frac{i\omega}{U_\infty}\right]\xi\right) \frac{\partial^2}{\partial z^2}\left\{\frac{e^{i\bar{k}r}}{r}\right\} d\xi \tag{4.5.56}$$

and

$$r^2 \equiv [\xi^2 + B(y - y_1)^2]$$

The above derivation, though lengthy, is entirely analogous to the classical one. For $a_0 \to 0$, $B \to 1 - M_\infty^2$ we retrieve the known result [32].

It should be noted that in the above derivation we have assumed $\operatorname{Re} \tilde{a}_0 > 0$ and thus $\operatorname{Re} \Omega < 0$. This permits both the radiation and finiteness conditions to be satisfied as $z \to \pm\infty$. For $\operatorname{Re} \tilde{a}_0 < 0$ one may not satisfy both conditions and one must choose between them.

Asymmetric mean flow. In the above derivations we have assumed a mean flow about symmetrical airfoils at zero angle of attack and considered small motions of that configuration. It is of interest to generalize this to a mean flow about asymmetrical airfoils at nonzero angles of attack. First consider the Mach box form of the integral relation between velocity potential and downwash, cf. equation (4.5.36),

$$\phi_U = \iint A_U(x-\xi, y-\eta) w_U(\xi, \eta)\, d\xi\, d\eta \tag{4.5.57}$$

Here we have written the relation as though we knew w_U everywhere on $z = 0^+$. We do not, of course, and thus the need for the Mach Box procedure [32]. Here A_U is that calculated using upper surface parameters, ignoring the lower surface. A similar relation applies for the lower surface with A_U replaced by $-A_L$. Hence, we may compute from (4.5.57) (for lifting motion where $w_U = w_L \equiv w$ on and off the airfoil)

$$\phi_U - \phi_L = \iint A(x-\xi, y-\eta) w(\xi, \eta)\, d\xi\, d\eta \tag{4.5.58}$$

where

$$A \equiv A_U + A_L$$

is the desired aerodynamic influence function. Note that A_U and A_L are the same basic function, but in one the upper surface parameters are used and in the other the lower surface parameters.

Using the Kernel Function approach the situation is somewhat more complicated. Here we have, cf. equation (4.5.55),

$$w_U = \iint K_U(x-\xi, y-\eta) p_U(\xi, \eta)\, d\xi\, d\eta \tag{4.5.59}$$

Note $K_U = 2K_{\Delta p}$ where $K_{\Delta p}$ is the Kernel Function for Δp when the lower surface mean flow parameters are the same as those of the upper surface.

A similar equation may be written for w_L and p_L with K_U replaced by $-K_L$. Again we note $w_L = w_U \equiv w$. These two integral equations must be solved simultaneously for p_U and p_L with given w. Hence, the number of unknowns one must deal with is doubled for different upper and lower surface parameters. This poses a substantial additional burden on the numerics.

There is a possible simplification, however. Define

$$K \equiv \frac{K_U + K_L}{2}; \qquad \Delta K \equiv \frac{K_U - K_L}{2} \tag{4.5.60}$$

If $(\Delta K/K)^2 \ll 1$, then one may simply use K, i.e., the average of the upper and lower surface kernel functions. Formally, one may demonstrate this using perturbation ideas as follows.

Using (4.5.59) (and its counterpart for the lower surface) and (4.5.60) one may compute

$$w_U + w_L \equiv 2w = \iint [K(p_U - p_L) + \Delta K(p_U + p_L)]\, d\xi\, d\eta$$

and

$$w_U - w_L \equiv 0 = \iint [K(p_U + p_L) + \Delta K(p_U - p_L)]\, d\xi\, d\eta \tag{4.5.61}$$

From the second of these equations, the size of the terms may be estimated.

$$\frac{p_U + p_L}{p_U - p_L} \sim 0\left(\frac{\Delta K}{K}\right)$$

Thus in the first of (4.5.61) the two terms on the right hand side are of order

$$K(p_U - p_L) \quad \text{and} \quad \frac{(\Delta K)^2}{K}(p_U - p_L)$$

The second of these terms may be neglected if

$$(\Delta K/K)^2 \ll 1 \tag{4.5.62}$$

and (4.5.61) may be approximated as

$$w(x, y) \approx \iint \frac{K}{2}(x - \xi, y - \eta)\Delta p(\xi, \eta)\, d\xi\, d\eta \tag{4.5.63}$$

where

$$\Delta p \equiv p_U - p_L$$

(4.5.62) would not appear to be an unduly restrictive condition for some applications.

The development in this section is *not* dependent upon the particular method used to compute K_U and/or K_L elsewhere in the text. The crucial assumptions are that (1) the oscillating motion is a small perturbation to the mean flow and (2) the difference between the upper and lower surface Kernel functions is small compared to either.

Concluding remarks

A relatively simple, reasonably accurate and systematic procedure has been developed for transonic flow. A measure of the simplicity of the method is that all numerical results presented herein were computed by hand and analytical forms have been obtained for general 'supersonic' Mach number and airfoil motion for two-dimensional flow. For three-dimensional flow the relevant Green's functions have been determined which may be used in the Kernel Function and Mach Box numerical lifting surface methods.

This approach has recently been extended to include a more accurate form of Bernoulli's equation and airfoil boundary condition. Also numerical examples are now available for two dimensional airfoils in transient motion and three dimensional steady flow over a delta wing. Finally a simple correction for shock induced flow separation has been suggested.*

For a recent, highly readable survey of transonic flow, the reader should consult the paper, by Spreiter and Stahara [40].

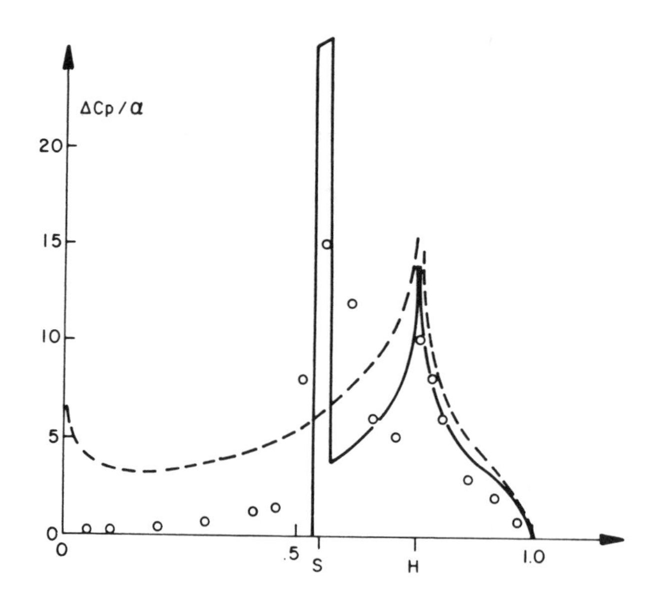

Figure 4.27a

* Dowell [39].

Also important recent advances in finite differences and finite element solutions are discussed in the following papers (all presented at the AIAA Dynamic Specialists Conference, San Diego, March 1977): Chan and Chen [41], Ballhaus and Goorjian [42] and Isogai [43].

In an important, but somewhat, neglected paper Eckhaus [44] gave a transonic flow model including shock waves which considered a constant supersonic Mach number ahead of the shock and a constant subsonic Mach number behind it. An obvious next step is to combine the Eckhaus and Dowell models. M. H. Williams [45], in work recently published, has extended Eckhaus' results by utilizing a somewhat broader theoretical formulation and obtaining more accurate and extensive solutions. He has compared his results to those of Tijdeman and Schippers [46] (experiment) and Ballhaus and Goorjian [42] (finite difference solutions) and obtained good agreement. The comparison with experiment is shown here in Figure 4.27 for a NACA 64 A006 airfoil with a trailing edge quarter chord oscillating flap. The measured steady state shock strength and location for no flap oscillation is used as an input to the theoretical model. Since the flap is downstream of the shock, the theory predicts no

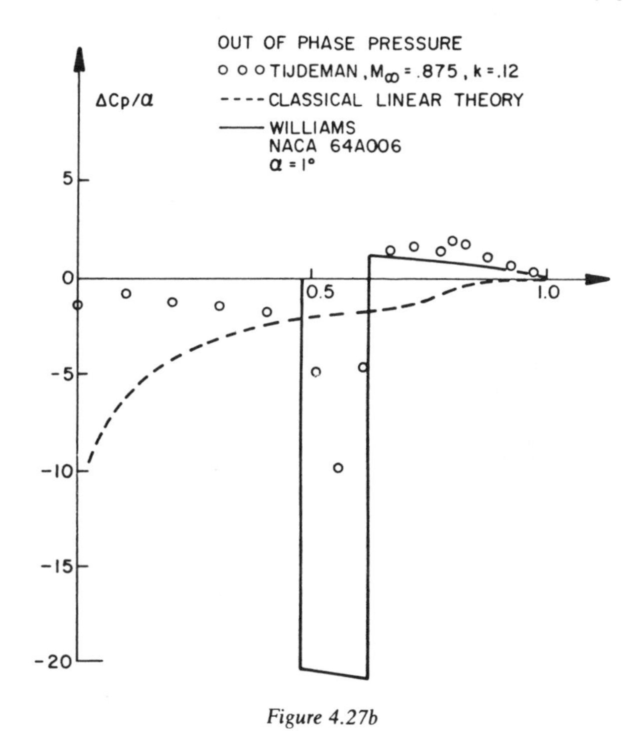

Figure 4.27b

disturbance upstream of the shock. The experiment shows the upstream effect is indeed small. Moreover the agreement on the pressure peaks at the shock and at the flap hinge line is most encouraging. It would appear that the transonic airfoil problem is finally yielding to a combination of analytical and numerical methods. As Tijdeman and others have emphasized, however, the effects of the viscous boundary layer may prove significant for some applications. In particular the poorer agreement between theory and experiment for the imaginary pressure peak at the shock in Figure 4.27 is probably due to the effects of viscosity. The same theoretical model has also been studied by Goldstein, *et al.* for cascades with very interesting results [47]. Rowe, a major contributor to subsonic aerodynamic solution methods, has in the same spirit discussed how the classical boundary conditions and Bernoulli's equation can be modified to partially account for transonic effects as the airfoil critical mach number is approached [48].

For a recent, broad-ranging survey of unsteady fluid dynamics including a discussion of linear potential theory, transonic flow, unsteady boundary layers, unsteady stall, vortex shedding and the Kutta–Joukowski trailing edge condition the paper by McCroskey [49] is recommended. For a discussion of the fundamentals of computational fluid dynamics of unsteady transonic flow, see Chapter 9.

Nomenclature

A	aerodynamic influence function; see equation (4.5.36)
a, b	see definitions following equation (4.5.1)
a_m, b_n	see equation (4.5.5)
$\tilde{a}_0$	see equation (4.5.26)
a_∞	free stream speed of sound
B	$\equiv -b_0$
C_p	$\equiv 2(p-p_\infty)/\rho_\infty U_\infty^2$; pressure coefficient due to airfoil motion
$C_{p_{ms}}$	mean steady pressure coefficient due to airfoil finite thickness at zero angle of attack
C_{L_α}	lift curve slope per degree
c	airfoil chord
d	see equation (4.5.26)
$\tilde{d}$	see equation (4.5.35)
e	see equation (4.5.30)
f	vertical airfoil displacement
h	rigid body translation of airfoil
Im	imaginary part

i	$\equiv(-1)^{\frac{1}{2}}$
K	aerodynamic kernel function; see equation (4.5.55)
k	$\equiv\dfrac{\omega c}{U_\infty}$
$\tilde{k}$	see equation (4.5.45)
L	lift
M	pitching moment about leading edge
M_∞	free stream Mach number
p	perturbation pressure; also Laplace Transform variable
Re	real part
r	see equations (4.5.37), (4.5.47) and (4.5.56)
t	time
U_∞	free stream velocity
w	downwash; see equation (4.5.4)
x, y, z	spatial coordinates
x', x', z'	see equation (4.5.44)
$\tilde{x}$	$\equiv\dfrac{a_0 x}{2b_0}$
$x_{c.p.}$	center of pressure; measured from leading edge
x_0	see equation (4.5.5)
α	angle of attack
γ	ratio of specific heats; also Fourier transform variable
Φ	see equation (4.5.41)
ϕ	velocity potential
ψ	see equation (4.5.47)
ρ_∞	free stream density
Ω	see equation (4.5.42)
ω	frequency of airfoil oscillation
ξ, λ, η	dummy integration variables for x, x, y

Superscripts

0	basic solution
$'$	correction to basic solution
*	Laplace Transform
†	Fourier Transform

Subscripts

U, L	upper, lower surfaces

References for Chapter 4

[1] Liepmann, H. W. and Roshko, A., *Elements of Gasdynamics*, John Wiley, 1957.

[2] Hildebrand, F. B., *Advanced Calculus for Engineers*, Prentice-Hall, Inc., 1961.

[3] van der Vooren, A. I., *Two Dimensional Linearized Theory*, Vol. II, Chapter 2, AGARD Manual on Aeroelasticity.

[4] Lomax, H., Heaslet, M. A., Fuller, F. B. and Sluder, L., 'Two- and Three-dimensional Unsteady Lift Problems in High Speed Flight', *NACA Report 1077*, 1952.

[5] Landahl, M. T. and Stark, V. J. E., 'Numerical Lifting Surface Theory—Problems and Progress', *AIAA Journal* (November 1968) pp. 2049–2060.

[6] Watkins, C. E., *Three Dimensional Supersonic Theory*, Vol. II, Chapter 5, AGARD Manual on Aeroelasticity.

[7] Bateman, H., *Table of Integral Transforms*, McGraw-Hill, 1954.

[8] Many Authors, *Oslo AGARD Symposium Unsteady Aerodynamics for Aeroelastic Analyses of Interfering Surfaces*, Tønsberg, Oslofjorden, Norway (Nov. 1970).

[9] Landahl, M. T. and Ashley, H., *Thickness and Boundary Layer Effects*, Vol. II, Chapter 9, AGARD Manual on Aeroelasticity.

[10] Williams, D. E., *Three-Dimensional Subsonic Theory*, Vol. II, Chapter 3, AGARD Manual on Aeroelasticity.

[11] Albano, E. and Rodden, W. P., 'A Doublet-Lattice Method for Calculating Lift Distributions on Oscillating Surfaces in Subsonic Flows', *AIAA J.* (February 1969) pp. 279–285.

[12] Stratton, J. A., *Electromagnetic Theory*, McGraw-Hill, 1941.

[13] Watkins, C. E., Woolston, D. S. and Cunningham, J. J., 'A Systematic Kernel Function Procedure for Determining Aerodynamic Forces on Oscillating or Steady Finite Wings at Subsonic Speeds', *NASA Technical Report TR*-48, 1959.

[14] Williams, D. E., *Some Mathematical Methods in Three-Dimensional Subsonic Flutter Derivative Theory*, Great Britain Aeronautical Research Council, R&M 3302, 1961.

[15] Cunningham, A. M., Jr., 'Further Developments in the Prediction of Oscillatory Aerodynamics in Mixed Transonic Flow', *AIAA Paper 75–99* (January 1975).

[16] Morino, L., Chen, L. T. and Suciu, E. O., 'Steady and Oscillatory Subsonic and Supersonic Aerodynamics Around Complex Configurations', *AIAA Journal* (March 1975), pp. 368–374.

[17] Rodden, W. P., 'State-of-the-Art in Unsteady Aerodynamics', *AGARD Report No. 650*, 1976.

[18] Ashley, H. and Rodden, W. P., 'Wing-Body Aerodynamic Interaction', *Annual Review of Fluid Mechanics*, Vol. 4, 1972, pp. 431–472.

[19] Theodorsen, T., 'General Theory of Aerodynamic Instability and the Mechanism of Flutter', *NACA Report 496*, 1935.

[20] Abramowitz, M. and Stegun, I. A., *Handbook of Mathematical Functions*, National Bureau of Standards, U.S. Printing Office, 1965.

[21] Edwards, J. V., Ashley, H. and Breakwell, J. B., 'Unsteady Aerodynamics Modeling for Arbitrary Motions', *AIAA Paper 77–451*, AIAA Dynamics Specialist Conference, San Diego, March 1977.

[22] H. Lomax, *Indicial Aerodynamics*, Vol. II, Chapter 6, AGARD Manual on Aeroelasticity, 1960.

[23] Rodden, W. P., Giesing, J. P. and Kálmán, T. P., 'New Developments and Applications of the Subsonic Doublet-Lattice Method for Non-Planar Configurations',

AGARD Symposium on Unsteady Aerodynamics for Aeroelastic Analyses of Interfering Surfaces, Tønsberg, Oslofjorden, Norway (November 3–4) 1970.
[24] Stahara, S. S. and Spreiter, J. R., 'Development of a Nonlinear Unsteady Transonic Flow Theory', *NASA CR-2258* (June 1973).
[25] Rubbert, P. and Landahl, M., 'Solution of the Transonic Airfoil Problem through Parametric Differentiation', *AIAA Journal* (March 1967), pp. 470–479.
[26] Beam, R. M. and Warming, R. F., 'Numerical Calculations of Two-Dimensional, Unsteady Transonic Flows with Circulation', *NASA TN D-7605*, (February 1974).
[27] Ehlers, F. E., 'A Finite Difference Method for the Solution of the Transonic Flow Around Harmonically Oscillating Wings', *NASA CR-2257* (July 1974).
[28] Traci, R. M., Albano, E. D., Farr, J. L., Jr. and Cheng, H. K., 'Small Disturbance Transonic Flow About Oscillating Airfoils', *AFFDL-TR-74-37* (June 1974).
[29] Magnus, R. J. and Yoshihara, H., 'Calculations of Transonic Flow Over an Oscillating Airfoil', *AIAA Paper 75-98* (January 1975).
[30] Cunningham, A. M., Jr., 'Further Developments in the Prediction of Oscillatory Aerodynamics in Mixed Transonic Flow', *AIAA Paper 75-99* (January 1975).
[31] Dowell, E. H. A Simplified Theory of Oscillating Airfoils in Transonic Flow', *Proceedings of Symposium on Unsteady Aerodynamics*, pp. 655–679, University of Arizona (July 1975).
[32] Bisplinghoff, R. L. and Ashley, H., *Principles of Aeroelasticity*, John Wiley and Sons, Inc., New York, 1962.
[33] Landahl, M., *Unsteady Transonic Flow*, Pergamon Press, London, 1961.
[34] Spreiter, J. R., 'Unsteady Transonic Aerodynamics—An Aeronautics Challenge', *Proceedings of Symposium on Unsteady Aerodynamics*, pp. 583–608, University of Arizona (July 1975).
[35] Ballhaus, W. F., Magnus, R. and Yoshihara, H., 'Some Examples of Unsteady Transonic Flows Over Airfoils', *Proceedings of a Symposium on Unsteady Aerodynamics*, pp. 769–792, University of Arizona (July 1975).
[36] Knechtel, E. D., 'Experimental Investigation at Transonic Speeds of Pressure Distributions Over Wedge and Circular-Arc Airfoil Sections and Evaluation of Perforated-Wall Interference', *NASA TN D-15* (August 1959).
[37] Spreiter, J. R. and Alksne, A. Y., 'Thin Airfoil Theory Based on Approximate Solution of the Transonic Flow Equation', *NACA TN 3970*, 1957.
[38] Park, P. H., 'Unsteady Two-Dimensional Flow Using Dowell's Method', *AIAA Journal* (October 1976) pp. 1345–1346. Also see Isogai, K., 'A Method for Predicting Unsteady Aerodynamic Forces on Oscillating Wings with Thickness in Transonic Flow Near Mach Number 1', National Aerospace Laboratory Technical Report NAL-TR-368T, Tokyo, Japan, June 1974. Isogai, using a modified local linearization procedure, obtains aerodynamic forces comparable to Park's and these provide significantly better agreement with transonic flutter experiments on parabolic arc airfoils.
[39] Dowell, E. H., 'A Simplified Theory of Oscillating Airfoils in Transonic Flow: Review and Extension', *AIAA Paper 77-445*, presented at AIAA Dynamic Specialists Conference, San Diego (March 1977).
[40] Spreiter, J. R. and Stahara, S. S., 'Developments in Transonic Steady and Unsteady Flow Theory', *Tenth Congress of the International Council of the Aeronautical Sciences, Paper No. 76-06* (October 1976).
[41] Chan, S. T. K. and Chen, H. C., 'Finite Element Applications to Unsteady Transonic Flow', *AIAA Paper 77-446*.

[42] Ballhaus, W. F. and Goorjian, P. M., 'Computation of Unsteady Transonic Flows by the Indicial Method', *AIAA Paper 77-447*.

[43] Isogai, K., 'Oscillating Airfoils Using the Full Potential Equation', *AIAA Paper 77-448*. Also see NASA TP1120, April 1978.

[44] Eckhaus, W., 'A Theory of Transonic Aileron Buzz, Neglecting Viscous Effects', *Journal of the Aerospace Sciences* (June 1962) pp. 712–718.

[45] Williams, M. H. 'Unsteady Thin Airfoil Theory for Transonic Flow with Embedded Shocks', Princeton University MAE Report No. 1376, May 1978.

[46] Tijdeman, H. and Schippers, P., *Results of Pressure Measurements on an Airfoil with Oscillating Flap in Two-Dimensional High Subsonic and Transonic Flow*, National Aerospace Lab. Report TR 730780, The Netherlands (July 1973).

[47] Goldstein, M. E., Braun, W., and Adamczyk, J. J., 'Unsteady Flow in a Supersonic Cascade with Strong In-Passage Shocks', *J. Fluid Mechanics*, Vol. 83, 3 (1977) pp. 569–604.

[48] Rowe, W. S., Sebastian, J. D., and Redman, M. C., 'Recent Developments in Predicting Unsteady Airloads Caused by Control Surfaces', *J. Aircraft* (December 1976) pp. 955–963.

[49] McCroskey, W. J., 'Some Current Research in Unsteady Fluid Dynamics—The 1976 Freeman Scholar Lecture', *Journal of Fluids Engineering* (March 1977) pp. 8–39.

5

Stall flutter

As the name implies, stall flutter is a phenomenon which occurs with partial or complete separation of the flow from the airfoil periodically during the oscillation. As contrasted with the so-called classical flutter (i.e., flow attached at all times) the mechanism for energy transfer from the airstream to the oscillating airfoil does not rely on elastic and/or aerodynamic coupling between two modes, nor upon a phase lag between a displacement and its aerodynamic reaction. These latter effects are necessary in a linear system to account for an airstream doing positive aerodynamic work on a vibrating wing. The essential feature of stall flutter is the *nonlinear* aerodynamic reaction to the moton of the airfoil/structure. Thus, although coupling and phase lag may alter the results somewhat, the basic instability and its principal features can be explained in terms of nonlinear normal force and moment characteristics.

5.1 Background

Stall flutter of aircraft wings and empennages is associated with very high angles of attack. Large incidence is necessary to induce separation of the flow from the suction surface. This type of operating condition and vibratory response was observed as long ago as World War I at which time stall flutter occurred during sharp pull-up maneuvers in combat. The surfaces were usually monoplane without a great deal of effective external bracing. The cure was to stiffen the structure and avoid the dangerous maneuvers whenever possible.

Electric power transmission cables of circular cross-section, or as modified by bundling or by ice accretion, etc., and structural shapes of various description are classified as bluff bodies. As such they do not require large incidence for flow separation to occur. In fact incidence is chiefly an orientation parameter for these 'airfoils' rather than an indication of the level of steady aerodynamic loading. Again, largely attributable to the nonlinearity in the force and moment as a function of incidence, such structures are prone to stall flutter. These vibrations are sometimes called 'galloping' as in the case of transmission lines. The number and classes of structures that potentially could experience stall flutter are very great, and include such diverse examples as suspension bridges, helicopter rotors and turbomachinery blades. More mundane examples are venetian blind slats and air deflectors or spoilers on automobiles.

The stall flutter of non-airfoil structures is described at greater length in Chapter 6, along with galloping and buffeting. These are all closely related bluff body phenomena from the point of view of vortex method aerodynamics, a subject which is introduced later in the present chapter. The stall flutter of rotorcraft blades is described in greater detail in Chapter 7 where the special kinematic restraints of these rotating structures lead to a unique aeroelastic description. The stall flutter of turbomachinery blades is described more fully in Chapter 8, wherein it is observed that the aeroelastic behavior in stall flutter is distinct from both non-airfoil structures and rotorcraft blades.

When the flow field is measured or visualized during stall flutter oscillations it is observed that free vortices are generated in the vicinity of the separation points. These large vortical structures are shed periodically creating regions of reduced and even reversed velocity in the vicinity of the airfoil. For this reason the aforementioned technique known as the vortex method has been developed recently for the computational modelling of unsteady separation aerodynamics.

It may be shown that the mutual induction, or interaction, of as few as three vortices leads to chaotic behavior. Thus it is confirmed by computation that use of vortex method aerodynamics displays many of the nonlinear aeroelastic phenomena actually observed experimentally in conjunction with stall flutter.

5.2 Analytical formulation

Although analysis of stall flutter based on computational unsteady aerodynamics is becoming feasible, it is nevertheless instructive to couch the problem in analytical terms so as to discriminate clearly the actual mechanism of instability [1]. We will consider two important cases: bending and twisting.

In the case of bending, or plunging displacement of a two-dimensional 'typical section' airfoil, let us assume that the force coefficient, including penetration well into the stall regime, is given by a polynomial approximation in α,

$$-C_n = \sum_{n=0}^{\nu} a_n(\alpha_{ss})\alpha^n \qquad a_0 \cong -C_{nss}(\alpha_{ss}) \tag{5.2.1}$$

where α is the instantaneous departure from the steady state value of angle of attack, α_{ss}, attributable to vibration of the airfoil. This method of expressing the normal force characteristic gives a good local fit with a few terms. However, the coefficients, a_n, depend on the mean angle of attack, α_{ss}. Force has been taken to be positive in the same direction as positive displacement h. (In the usual (static) theory of thin unstalled and uncambered profiles $-C_n = \pi \sin 2\alpha_{ss}$. The a_n could then be obtained by deriving the Maclaurin series expansion of $\pi \sin 2(\alpha_{ss} + \alpha)$ considered as a function of α). In general the $-C_n$ function is an empirically determined function, or characteristic, when stall occurs on a cambered airfoil, but the procedure is still the same. The a_n are in fact given by the slope and higher order derivatives according to

$$a_n = -\frac{1}{n!} \frac{d^n C_n}{d\alpha^n}\bigg|_{\alpha=0} \tag{5.2.2}$$

We next consider a small harmonic bending oscillation

$$h = h_0 \cos \omega t$$

to exist and enquire as to the stability of that motion: Will it amplify or decay?

Under these circumstances, it is possible to interpret the instantaneous angle of attack perturbation to be given by (see Figure 5.1)

$$\alpha = \arctan\left(\tan \alpha_{ss} + \frac{\dot{h}}{V \cos \alpha_{ss}}\right) - \alpha_{ss} \tag{5.2.3}$$

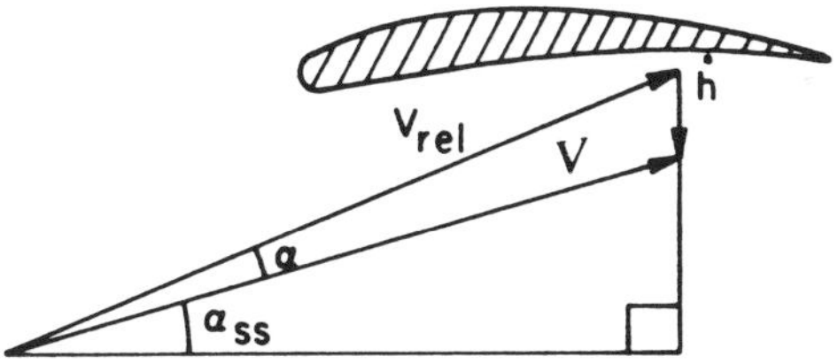

Figure 5.1 Velocity triangle.

with Maclaurin series expansion in powers of $\dot{h}$ as follows

$$\alpha = \cos \alpha_{ss}\left(\frac{\dot{h}}{V}\right) - \tfrac{1}{2}\sin 2\alpha_{ss}\left(\frac{\dot{h}}{V}\right)^2 - \tfrac{1}{3}\cos 3\alpha_{ss}\left(\frac{\dot{h}}{V}\right)^3 + \tfrac{1}{4}\sin 4\alpha_{ss}\left(\frac{\dot{h}}{V}\right)^4 + \cdots \tag{5.2.4}$$

It should be noted that this incidence is relative to a coordinate system fixed to the airfoil. The dynamic pressure also changes periodically with time in this coordinate system according to

$$q_{\text{rel}} = \tfrac{1}{2}\rho V_{\text{rel}}^2 = \tfrac{1}{2}\rho V^2\left[1 + 2\sin\alpha_{ss}\left(\frac{\dot h}{V}\right) + \left(\frac{\dot h}{V}\right)^2\right] \tag{5.2.5}$$

It is assumed for simplicity that the single static characteristic of normal force coefficient versus angle of attack continues to be operative in the dynamic application described above. Thus, the expanded equation for the normal force $N = q(2b)C_n$ is given by

$$N = -\tfrac{1}{2}\rho V^2(2b)\left[1 + 2\sin\alpha_{ss}\left(\frac{\dot h}{V}\right) + \left(\frac{\dot h}{V}\right)^2\right]\sum_{n=0}^{\nu} a_n(\alpha_{ss})\left[\cos\alpha_{ss}\left(\frac{\dot h}{V}\right) - \tfrac{1}{2}\sin 2\alpha_{ss}\left(\frac{\dot h}{V}\right)^2 - \tfrac{1}{3}\cos 3\alpha_{ss}\left(\frac{\dot h}{V}\right)^3 + \tfrac{1}{4}\sin 4\alpha_{ss}\left(\frac{\dot h}{V}\right)^4 + \cdots\right]^n \tag{5.2.6}$$

with

$$\frac{\dot h}{V} = -\frac{\omega h_0}{V}\sin\omega t = -k\frac{h_0}{b}\sin\omega t \tag{5.2.7}$$

A slight concession to the dynamics of stalling may be introduced by the inclusion of a time delay, ψ/ω, in the oscillatory velocity term appearing in the C_n expansion, i.e., within the summation of (5.2.6), but not in the development of q_{rel}. The latter is assumed to respond instantaneously to α or $\dot h$.

5.3 Stability and aerodynamic work

As is common with single degree of freedom systems such as that postulated above, the question of amplification or subsidence of the amplitude of the initial motion can easily be decided on the basis of the work done by this force acting on the displacement. Thus

$$\text{Work/Cycle} = \int_0^T N\dot h \, \mathrm{d}t = \frac{1}{\omega}\int_0^{2\pi} N\dot h \, \mathrm{d}(\omega t) \tag{5.3.1}$$

and since the frequency is effectively the number of cycles per unit time, the power may be expressed as

$$\mathbb{P} = \text{Power} = (\text{Work/Cycle})(\text{Cycles/Second}) = \frac{1}{2\pi}\int_0^{2\pi} N\dot h \, \mathrm{d}(\omega t) \tag{5.3.2}$$

Using the previous expression for N and $\dot h$, it is clear that only even powers of $\sin\omega t$ in the integrand of the power integral will yield nonzero contributions. Also, terms of the form $\sin^n\omega t\cos\omega t$ will integrate to zero for any integer value of n including zero. Restricting the series expansions for $-C_n$ and α to their leading terms such that the power integral displays terms of vibratory amplitude up to the sixth power (i.e., up to h_0^6) results in

$$\mathbb{P}=\tfrac{1}{2}\rho V^3 b[A(\omega h_0/V)^2+B(\omega h_0/V)^4+C(\omega h_0 V)^6+\cdots] \tag{5.3.3a}$$

where

$$\begin{aligned}
A &= -2a_0 \sin \alpha_{ss} - a_1 \cos \alpha_{ss} \cos \psi \\
B &= -\tfrac{1}{4}a_1[-(\cos \alpha_{ss} - \cos 3\alpha_{ss})(1+\tfrac{1}{2}\cos 2\psi) \\
&\quad +(3\cos \alpha_{ss} - \cos 3\alpha_{ss})\cos \psi] \\
&\quad -\tfrac{1}{4}a_2[(\sin \alpha_{ss} + \sin 3\alpha_{ss})(1-\tfrac{3}{2}\cos \psi + \tfrac{1}{2}\cos 2\psi)] \\
&\quad -\tfrac{3}{16}a_3[(3\cos \alpha_{ss} + \cos 3\alpha_{ss})\cos \psi] \qquad (5.3.3b) \\
C &= -\tfrac{1}{16}a_1[(\cos 3\alpha_{ss} - \cos 5\alpha_{ss})(\tfrac{3}{2}+\cos 2\psi) \\
&\quad -\tfrac{1}{16}(3\cos 3\alpha_{ss} - 2\cos 5\alpha_{ss})\cos \psi - \tfrac{1}{3}\cos 3\alpha_{ss}\cos 3\psi] - \cdots
\end{aligned}$$

The cubic dependence on V is a consequence of the dimensions of power, or work per unit time.

5.4 Bending stall flutter

The analytical expression for the aerodynamic power in a sinusoidal bending vibration is too cumbersome for easy physical interpretation. However, for very small amplitudes of motion, as might be triggered by turbulence in the fluid, or other 'noise' in the system, it is clear that the sign of the work flow will be governed by the coefficient of $(\omega h_0/V)^2$. Assuming a small to moderate positive mean incidence, α_{ss}, the coefficient a_0 will be positive. With $\cos \psi$ near unity, a positive power can only occur if a_1 is sufficiently negative, i.e., if the $-C_n$ vs α characteristic has a negative slope at the static operating incidence. More precisely, if $|\psi|< 90°$ and

$$a_1 < -2a_0 \tan \alpha_{ss} \sec \psi \tag{5.4.1}$$

the small amplitude vibration is unstable and the work flow will be such as to feed energy into the vibration and increase its amplitude.

In the previous expression for the power, (5.3.3a),

$$\mathbb{P}/(\tfrac{1}{2}\rho V^3 b) = A(\omega h_0/V)^2 + B(\omega h_0/V)^4 + C(\omega h_0/V)^6 \tag{5.4.2}$$

the coefficients A, B and C are complicated functions of ψ, α_{ss} and the a_n, the coefficients of the power series representation of the normal force characteristic. For example in the highly simplified case of $\alpha_{ss} = \psi = 0$, we obtain

$$A = a_1 = \left.\frac{dC_n}{d\alpha}\right|_{\alpha=0}, \qquad B = \left.\frac{1}{2}\frac{dC_n}{d\alpha}\right|_{\alpha=0} + \left.\frac{1}{8}\frac{d^3C_n}{d\alpha^3}\right|_{\alpha=0}$$

and

$$C = \left.\frac{1}{12}\frac{dC_n}{d\alpha}\right|_{\alpha=0} + \left.\frac{1}{192}\frac{d^5C_n}{d\alpha^5}\right|_{\alpha=0} \tag{5.4.3}$$

In the general case A, B and C individually may be either positive, zero or negative. The several possible cases are of fundamental interest in describing possible bending stall flutter behavior.

I. $A<0$, $B<0$, $C<0$ No flutter is possible.

II. $A>0$, $B>0$, $C>0$ Flutter amplitude grows from zero to very large values.

III. $A>0$, $B<0$, $C<0$ Flutter amplitude grows smoothly from zero to a finite amplitude given by

$$(\omega h_0/V)^2_{\mathrm{III}} = (-|B| + \sqrt{B^2 + 4A\,|C|})/2\,|C|$$

At this amplitude the power once again becomes zero.

IV. $A<0$, $B>0$, $C>0$ No flutter at small amplitudes; if an external 'triggering' disturbance carries the system beyond a certain critical vibratory amplitude

$$(\omega h_0/V)^2_{\mathrm{IV}} = (-B + \sqrt{B^2 + 4\,|A|\,C})/2C$$

the flutter will continue to grow beyond that amplitude up to very large values. At the critical amplitude the power is zero.

V. $A>0$, $B>0$, $C<0$ This is similar to case III except that the finite amplitude, or equilibrium, flutter amplitude

$$(\omega h_0/V)^2_{\mathrm{V}} = (B + \sqrt{B^2 + 4A\,|C|})/2\,|C|$$

might be expected to be somewhat larger.

VI. $A>0$, $B<0$, $C>0$ This is similar to case IV except that the critical vibratory amplitude beyond which flutter may be expected to grow

$$(\omega h_0/V)^2_{\mathrm{VI}} = (|B| + \sqrt{B^2 + 4\,|A|\,C})/2C$$

is perhaps a larger value.

VII. $A>0$, $B<0$, $C>0$ This case has behavior similar to case II if B is very small and similar to case III if C is very small and also very large amplitudes are excluded from consideration.

VIII. $A < 0$, $B > 0$, $C < 0$ This case behavior is similar to case I if B is very small and similar to case IV if C is very small and also very large amplitudes are excluded from consideration.

5.5 Nonlinear mechanics description

A number of these variations of power dependency on amplitude have been sketched in Figure 5.2. Case II is an example of what may be termed 'soft flutter'; given an airstream velocity V, incidence α_{ss} and time delay ψ/ω such as produce values of A, B and C according to case II, the vibratory amplitude of flutter might be expected to grow smoothly from zero.

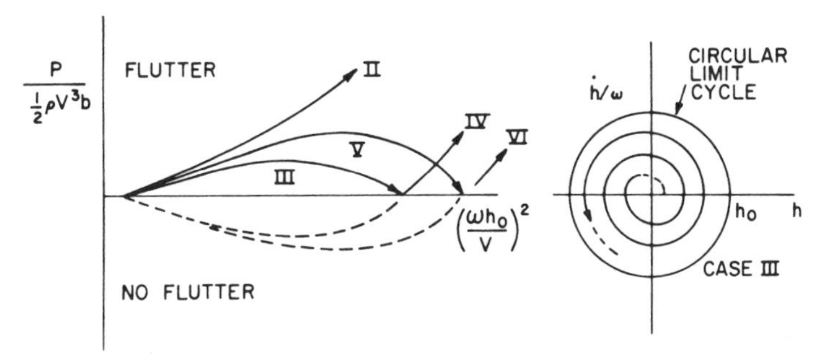

Figure 5.2 Power vs. amplitude.

Cases III and V similarly are examples of soft flutter; in these cases however, the amplitude of vibration reaches a steady value and does not increase further. An equilibrium flutter amplitude is attained after a period of time and maintained thereafter. If, in either of these cases, one were to plot h versus $\dot{h}/\omega$ with time as a parameter, it would be found that the 'trajectory' of the 'characteristic point' would be a spiral around the origin, beginning at the origin at $t = 0$ and asymptotically approaching a circle of radius h_0 for very large time. In the parlance of nonlinear mechanics the circular path is a 'limit cycle' and hence most instances of stall flutter may be termed limit cycle vibrations.

Case IV, or alternatively case VI, describes a type of behavior which may be termed 'hard flutter'. In this situation when flutter appears as a self-sustaining oscillation, the amplitude is immediately a large finite value. Here the motion spirals away from the circular limit cycle to either larger or smaller amplitudes in the phase plane (i.e., the h, $\dot{h}/\omega$ plane). This example is an instance in which the limit cycle is unstable. The slightest perturbation from an initially purely circular path, either to larger or smaller radii, will result in monotonic spiralling away from the limit cycle. The previous example of case III illustrated the case of a stable limit cycle.

The origin of the phase plane is also a degenerate limit cycle in the sense that the limit of a circle is a point in which case only path radii larger than zero have physical meaning. However, the origin may be an unstable limit cycle (soft flutter) or a stable limit cycle (hard flutter).

It is clear from a consideration of cases VII and VIII that more than two limit cycles may obtain; it is a theorem of mechanics that the concentric circles which are limit cycles of a given system are alternately stable and unstable.

5.6 Torsional stall flutter

With pure twisting motion of the profile, the analytical formulation is more complex stemming from the fact that the dynamic angle of incidence is compounded of two effects: the instantaneous angular displacement and the instantaneous linear velocity in a direction normal to the chord. The magnitude of the first effect is a constant independent of frequency and chordal position; the second magnitude is linearly dependent upon the distance along the chord from the elastic axis and upon the frequency of vibration. Both components, of course, vary harmonically with the frequency ω. Thus, assuming a displacement $\theta_0 \cos \omega t$ the 'local' angle of attack becomes

$$\alpha = \theta_0 \cos \omega t + \arctan\left[\tan \alpha_{ss} - \frac{(x - x_0)\omega\theta_0}{V \cos \alpha_{ss}} \sin \omega t\right] - \alpha_{ss} \tag{5.6.1}$$

and the relative dynamic pressure becomes

$$q_{\text{rel}} = \tfrac{1}{2}\rho V_{\text{rel}}^2 = \tfrac{1}{2}\rho V^2\left[1 + 2 \sin \alpha_{ss} \frac{\dot{\theta}(x - x_0)}{V} + \left(\frac{\dot{\theta}(x - x_0)}{V}\right)^2\right] \tag{5.6.2}$$

Since the local incidence varies along the chord in the torsional case, it is not possible to formulate the twisting problem in a simple and analogous manner to the bending case unless a single 'typical' incidence is chosen. From incompressible, potential flow, thin airfoil theory, it is known [2] that the three-quarter chord point is 'most representative' in relating changes in incidence to changes in aerodynamic reaction for an unstalled thin airfoil with parabolic camber. Replacing $x - x_0$ by a *constant*, say eb, for

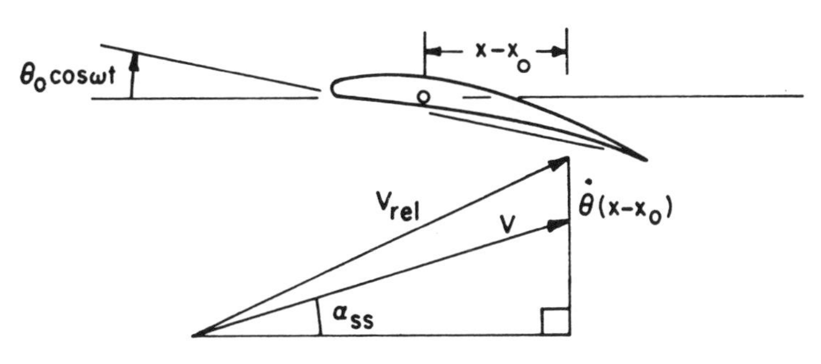

Figure 5.3 Geometry.

simplicity, one has by analogy with bending

$$\alpha = \theta_0 \cos \omega t + \cos \alpha_{ss}(-ek\theta_0) \sin \omega t - \tfrac{1}{2} \sin 2\alpha_{ss}(-ek\theta_0)^2 \cdot \sin^2 \omega t - \tfrac{1}{3} \cos 3\alpha_{ss}(-ek\theta_0)^3 \sin^3 \omega t + \tfrac{1}{4} \cdots \tag{5.6.3}$$

where α is, again, the departure in angle of attack from α_{ss}. The constant e will normally be of order unity for an elastic axis location forward of midchord.

From this point onward, the illustrative analysis involves the substitution of α into an analytical approximation for the aerodynamic moment coefficient

$$C_m = \sum_{n=0}^{\nu} b_n(\alpha_{ss})\alpha^n \tag{5.6.4}$$

In this equation, the b_n may be associated with the slope and higher order derivatives of the characteristic

$$b_n = \frac{1}{n!} \frac{\mathrm{d}^n C_m}{\mathrm{d}\alpha^n}\bigg|_{\alpha=0} \tag{5.6.5}$$

at the mean incidence point, in a manner analogous to the role of the a_n in the normal force coefficient.

The work done by the aerodynamic moment acting on the torsional displacement is given by

$$\text{Work/Cycle} = \int_0^T M\dot{\theta}\, \mathrm{d}t = \frac{1}{\omega} \int_0^{2\pi} M\dot{\theta}\, \mathrm{d}(\omega t) \tag{5.6.6}$$

and hence the work flow, or power, is

$$\mathbb{P}=\frac{1}{2\pi}\int_0^{2\pi} M\dot{\theta}\, d(\omega t) \tag{5.6.7}$$

Using the previously derived expressions contributing to the moment $M=q(2b)^2C_m$ leads to

$$M = \frac{1}{2}\rho V^2(2b)^2 \left[1+2\sin\alpha_{ss}\left(\frac{\dot{\theta}eb}{V}\right)+\left(\frac{\dot{\theta}eb}{V}\right)^2\right]$$

$$\cdot \sum_{m=0}^{\nu} b_n(\alpha_{ss})[\theta_0 \cos\omega t - \cos\alpha_{ss}(ek\theta_o)\sin\omega t$$

$$-\tfrac{1}{2}\sin 2\alpha_{ss}(ek\theta_0)^2\sin^2\omega t+\tfrac{1}{3}\cos 3\alpha_{ss}(ek\theta_0)^3\sin^3\omega t+\cdots]^n \tag{5.6.8}$$

and this expression, in turn inserted into the integrand of (5.6.7), will allow an analytical expression to be derived by quadrature.

At this stage in the development of torsional stall flutter, a key difference emerges more clearly when compared to bending stall flutter; a fundamental component of the moment coefficient appears ($b_1\theta_0\cos\omega t$) which is out of phase with the torsional velocity ($\dot{\theta}=-\omega\theta_0\sin\omega t$). Noting that $\dot{\theta}$ is the second factor in the integrand, it is seen that the final integrated expression for the power will have terms similar in nature to the expression derived for the bending case, and in addition may have terms proportional to

$$b_1\theta_0,\ b_2\theta_0^2,\ b_2\theta_0,\ b_3\theta_0^3,\ b_3\theta_0^2, \text{ etc.}$$

It is not particularly instructive to set out this result in full detail.

However, let us consider briefly the case of very slow oscillations, so that terms proportional to higher powers of the frequency can be ignored. Then

$$\mathbb{P}_1 = -\tfrac{1}{2}\rho V^2(2b)^2\frac{\omega\theta_0}{2\pi}\sum_{n=0}^{\nu} b_n\theta_0^n\int_0^{2\pi}\cos^n(\omega t-\psi)\sin\omega t\, d(\omega t)$$

$$= -\tfrac{1}{2}\rho V^3(4b)k\sin\psi\sum_{n=\text{odd}}^{\nu} b_n\theta_0^{n+1}\frac{1\cdot3\cdot5\cdots n}{2\cdot4\cdot6\cdots(n+1)} \tag{5.6.8}$$

We conclude from this equation that the work flow again will be proportional to a sum of terms in even powers of the vibratory amplitude, but in this instance, the low frequency torsional stall flutter is critically dependent on the time lag ψ/ω between the oscillatory motion and the response of the periodic aerodynamic moment.

Torsional stall flutter is thus seen to be a much more complex phenomenon, with a greater dependence on time lag and exhibiting very strong dependence on the location of the elastic axis. For example, if the elastic axis were artificially moved rearward on an airfoil such as to reduce the effective value of the parameter e to zero, the airfoil flutter behavior would be governed by exactly the same specialization of the analysis as was just termed 'low frequency'. Exactly the same terms would be eliminated from consideration. In qualitative terms one may also conclude that the actual behavior in torsional flutter in the general case (with $e \neq 0$) is some intermediate state between the low frequency behavior (critical dependence on sin ψ) and a type of behavior characteristic of bending stall flutter (critical dependence on the slope of a dynamic characteristic at the mean incidence).

5.7 General comments

An interesting by-product of the nonlinear nature of stall flutter is the ability, in principle, to predict the final equilibrium amplitude of the vibration. This is in contradistinction to classical flutter in which only the stability boundary is usually determined. The condition for constant finite flutter amplitude is that the work, or power flow, again be zero. As we have seen this can be discerned when the power equation is set equal to zero; the resulting quadratic equation is solved for the squared flutter amplitude, either $(h_0/b)^2$ or θ_0^2 as the case may be. Since all the a_n or b_n coefficients are functions of α_{ss}, the two types of flutter are displayed in Figure 5.4 as *presumed* functions of this parameter. Hard flutter displays a sudden jump to finite amplitude as a critical parameter is varied and a lower 'quench' value of that parameter where the vibration suddenly disappears. The two effects conspire to produce the characteristic hysteresis loop indicated by arrows in Figure 5.4.

In summary then, stall flutter is associated with nonlinearity in the aerodynamic characteristic; the phenomenon may occur in a single degree of freedom and the amplitude of vibratory motion will often be limited by the aerodynamic nonlinearities. Although structural material damping has not been considered explicitly, it is clear that since damping is an absorber of energy its presence will serve to limit the flutter amplitudes to smaller values; damping limited amplitudes will obtain when the positive power flow from airstream to airfoil equals the power conversion to heat in the mechanical forms of damping.

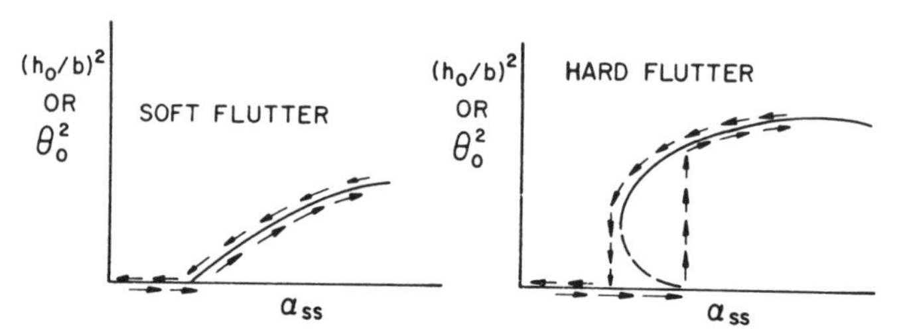

Figure 5.4 Flutter amplitude vs steady state angle of attack.

It is also clear that motion in a third degree of freedom is possible. Oscillatory surging of the airfoil in the chordwise direction can be related to a nonlinear behavior in the drag acting on the profile. However, airfoils are usually very stiff structurally in the chordwise direction and the drag/surging mechanism would normally be of importance only for bluff structural shapes such as bundles of electric power conductors suspended between towers, etc.

Under certain circumstances such as the example noted directly above, stall flutter in more than one degree of freedom may occur. In these cases, the dynamic characteristics of normal force, aerodynamic moment (and drag) become functions of an effective incidence compounded of many sources: plunging velocity, torsional displacement, torsional velocity and surging velocity. The resultant power equation will also contain cross-product terms in the various displacement amplitudes, and hence the equation cannot be used to predict stability or equilibrium flutter amplitudes without additional information concerning the vibration modes.

Perhaps the greatest deficiency in the theory, however, is the fact that even in pure bending motion or pure torsional motion, the dynamic force and moment are in fact frequency dependent: $a_n = a_n(\alpha_{ss}, k)$ and $b_n = b_n(\alpha_{ss}, k)$. And in general $a_0 \neq -C_{nss}$ and $b_0 \neq C_{mss}$. In analogy with classical flutter it may be shown that even this dependence is deficient in that the characteristics in practice may be double valued. That is, for the same value of effective incidence α, the characteristic may have different values depending upon whether α is decreasing or increasing with time. Such a hysteretic characteristic is usually more pronounced at high frequencies of oscillation; an airfoil may have two lift or moment coefficients at a particular angle of attack even in the static case, depending upon how the operating point was approached.

It is for these reasons that practical stall flutter prediction has been at best a semi-empirical process, and often entirely empirical. A model is oscillated in torsion, or bending, in a wind tunnel under controlled conditions with parametric variation of reduced frequency, mean incidence and oscillatory amplitude. Various elastic axis locations also may be studied. Data which are taken may vary from instantaneous normal force and moment down to the actual time-dependent pressure distribution on the profile. Data reduction consists essentially of cross-plotting the various data so that flutter prediction for prototype application is largely a matter of interpolation in model data using dimensionless groups. Specific representative data will be taken up in subsequent chapters where stall flutter applications are studied.

An exception to the previous reliance on experimental data is a theory

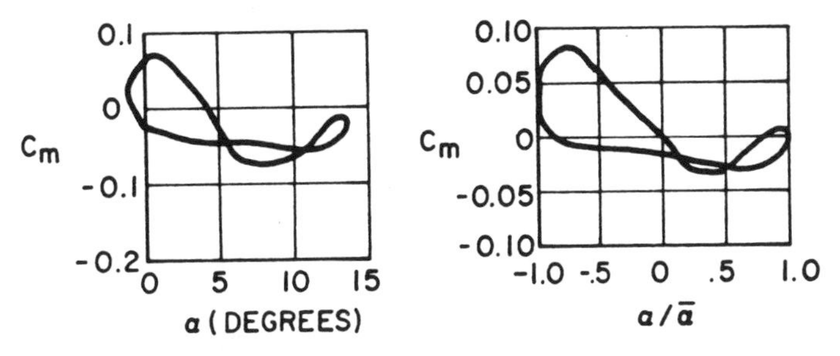

Figure 5.5 Dynamic moment loops.

[3] which postulates that the departure of the normal force and/or aerodynamic moment from the classical (attached flow) values can be modelled by considering a flat plate with separated flow on the suction side. As the plate oscillates harmonically in time, the position of the separation point (from which emanates a free streamline) is also considered to move periodically with the same frequency as the oscillation. The movement of the separation point along the suction surface is between two arbitrarily specified upstream and downstream limits and with an arbitrarily specified phase angle with respect to the oscillatory motion.

Under these circumstances, it is possible to solve the unsteady flow problem (analogous to the classical Theodorsen solution for attached flow) with separation present. In effect the appropriate dynamic force and moment characteristics are generated for each function specifying the separation point movement and airfoil motion. The empirical part of the flutter prediction technique then resides in correlation of the separation point behavior as a function of the airfoil attitude and oscillatory motion. To illustrate the potential of the technique, two moment loops from the reference are shown in Figure 5.5. The one on the left is from an experimental program [4], the one on the right is from [3]. Although the variation of moment with torsional displacement is remarkably similar, it must be emphasized that the particular choice of elastic axis location is different in experiment and theory, and the assumed separation point behavior in the theory was reasonable, but quite arbitrary and unrelated to the unknown separation point behavior in the experiment.

The method of modelling the separation region on the suction surface of the airfoil by a free streamline issuing from the 'separation' point has been generalized subsequently [5]. The method employs simultaneous integral equations and may be applied to subsonic, small perturbation flows of aeroelastic significance. In particular, for cascades of airfoils of interest in axial-flow compressors [6], the method has shown promise of improved stall flutter prediction. A type of stall resulting in a leading edge 'bubble' is also amenable to this type of small perturbation analysis [7] and is more appropriate for sharp leading edges with onset flows that result in reattachment of the separation streamline.

These free-streamline methods are useful when the reattachment point and/or separation point behavior can be predicted beforehand and the mean incidence is not excessive. An example is the thin airfoil with small leading edge radius at moderate incidence where the separation point is 'anchored' at the leading edge and reattachment does not occur.

5.8 Reduced order models

As noted in Chapter 11, Nonlinear Aeroelasticity, reduced order models have been developed to help account for the effect of airfoil vibratory displacement, velocity and acceleration on the associated aerodynamic responses. Since the theoretical underpining for these models is not firmly established for conditions of massive flow separation, the characteristics must be developed by model fitting from experimental data. For this reason these models have also been termed "semi-empirical".

In fact, a low order model is the quasisteady development presented in Section 5.2 for the nonlinear normal force and moment characteristics. The linear quasisteady development in Chapter 3 is another low order model. The steady flow aerodynamics example of that same chapter is of course the model of lowest possible order.

Reduced order modelling for stall flutter and bluff body aeroelasticity has been studied by a number of investigators. Some of these studies are described in Chapter 11 and references to much of the recent literature may be found there. One important and representative study is that by Tang and Dowell [16] in which many of the characteristics attributable to aerodynamic nonlinearities appear. Examples are the asymptotic approach to limit cycles and the development of chaotic pitch displacement and moment coefficient histories for particular values of the advance ratio.

5.9 Computational stalled flow

In recent years the so-called vortex method has begun to be used to model periodically separated flow from bluff bodies [8, 9] as well as streamlined shapes [10] such as airfoils. The vortex method is essentially a computational algorithm which tracks a large collection of discrete vortices in time. Since it is a time-marching procedure, the aerodynamic reactions are obtained with an evolving flow and the aeroelastic response of the structure must evolve in like manner. Hence stability of a specific structure oriented in specific flow cannot be discriminated *ab initio*. The aeroelastic vibration develops in the course of time; hence the method might equally be termed computational fluid elasticity (CFE). The power of the method may be appreciated when it is realized that highly nonlinear aerodynamics (and structure as well) may be modelled and finite amplitudes of the flutter vibration may be predicted. The cost of computation is high since fairly long runs on supercomputers are required for acceptable accuracy.

The vortex method for modelling unsteady separated flow as initiated in [10] and modified in [11] and [12] for oscillating airfoils, is based upon the following fluid dynamic system of equations.

For two-dimensional, viscous, incompressible flow past an infinite linear cascade of airfoils at high Reynolds number, the basic aerodynamic equations that govern the vorticity field derived in [8] are as follows. (For a single airfoil the formulation may be simplified from what is shown here).

Conservation of vorticity in the fluid requires

$$\frac{D\omega}{Dt} = \nu\nabla^2\omega \tag{5.9.1}$$

where the vorticity in the fluid field is

$$\omega = \frac{\partial v}{\partial x} - \frac{\partial u}{\partial y} \tag{5.9.2}$$

Vorticity within the solid is a continuation of the fluid field and represents the motion (vibration) of the solid

$$\omega = 2\Omega_m \tag{5.9.3}$$

The boundary conditions in terms of vorticity can be written as [8]

$$\oint \left(\nu \frac{\partial \omega}{\partial n} \right) \mathrm{d}s = -2R_m \frac{\mathrm{d}\Omega_m}{\mathrm{d}t} \tag{5.9.4}$$

The system of equations governing the vorticity and the system governing the velocity and pressure are equivalent. A stream function ψ can be defined to satisfy the continuity equation

$$u = \frac{-\partial \psi}{\partial y} \quad \text{and} \quad v = \frac{\partial \psi}{\partial x} \tag{5.9.5}$$

Combining (5.9.2) and (5.9.5) results in the Poisson equation

$$\nabla^2 \psi = \omega \tag{5.9.6}$$

The vortex method represents the vorticity field as the sum of a large number (N) of vortex blobs

$$\omega = \sum_{k=1}^{N} \omega_k \tag{5.9.7}$$

and the stream function induced by a collection of vortices is $\sum \psi_k$, where

$$\psi_k = (\Gamma_k/4\pi) \ln |\sin [(2\pi/p)(z - z_k)]|^2 \tag{5.9.8}$$

Here $i = \sqrt{-1}$ and the complex variable notation $z = x + iy$ is used.

The instantaneous coordinates of the mth airfoil surface $[x(t), y(t)]$ under coupled bending-torsion with a frequency of f Hertz are given by

$$x(t) = x_0 - h \sin(2\pi ft + \mu + m\sigma) \sin \beta - y_0 \theta \sin(2\pi ft + m\sigma) \tag{5.9.9a}$$

$$y(t) = y_0 + h \sin(2\pi ft + \mu + m\sigma) \cos \beta + x_0 \theta \sin(2\pi ft + m\sigma) \tag{5.9.9b}$$

where (x_0, y_0) are coordinates for each airfoil without vibration and are measured from its centroid, assumed here for simplicity to coincide with the center of twist. The quantity μ refers to the intrablade phase angle which is the phase difference between the bending and torsional modes. On the other hand, the interblade phase angle, σ, represents the phase shift between neighboring blades. To obtain the corresponding boundary conditions, the nonpenetration condition is imposed as expressed by (4.1.32).

With the definition of the stream function $\partial\psi/\partial s = V_n$ where s and n are local coordinates parallel and normal to the wall, respectively, the incremental value of stream function along each airfoil surface can be determined by

$$d\psi|_s = \int_{s_0}^{s_0+\Delta s} V_n \, ds = \int_{s_0}^{s_0+\Delta s} (\dot{x}_b n_x + \dot{y}_b n_y) \, ds \tag{5.9.10}$$

This equation is used to determine the distribution of the values of the stream function along the boundary points of the airfoils, and then to solve the vorticity-stream function equations. As a consequence of the airfoil motion the values of the stream function are not constant along the boundary of the airfoil. It should also be mentioned that the no-slip condition reflecting the nonzero viscosity of the fluid is satisfied in a weak sense, as discussed in [8].

Computations based on this system of equations have shown [12] that the two-dimensional unsteady flow, as exemplified in a linear cascade of oscillating airfoils, is properly predicted for a range of reduced frequencies at low incidence. Results similar to those derivable analytically by the methods of Section 4.3 in Chapter 4, and also in Chapter 8 for cascades, are confirmed by these computational procedures. With this validation in hand it is possible then to consider larger values of the mean incidence until stall is encountered, and compute the aerodynamic response under intermittent separation, and finally, under complete or 'deep' stall. The rapid change in amplitude and phase for lift due to plunging motion as the mean incidence is increased in steps is shown in the following table, along with streamline pattern at one instant for the highest incidence case, Figure 5.6.

The presence of strong vortices in the flow illustrates an important stability modification mechanism present in stalled flow. These coherent structures are subject to a nonlinear eigenfunction/eigenfrequency interpretation associated entirely with the flow. A completely rigid airfoil (cascade of airfoils) is (are) subject to a flow instability identified as Karman vortex shedding [13] (propagating stall phenomenon). This unsteady periodic behavior has a characteristic frequency and the associated flow pattern is in the guise of an eigenfunction. Thus stall flutter, in

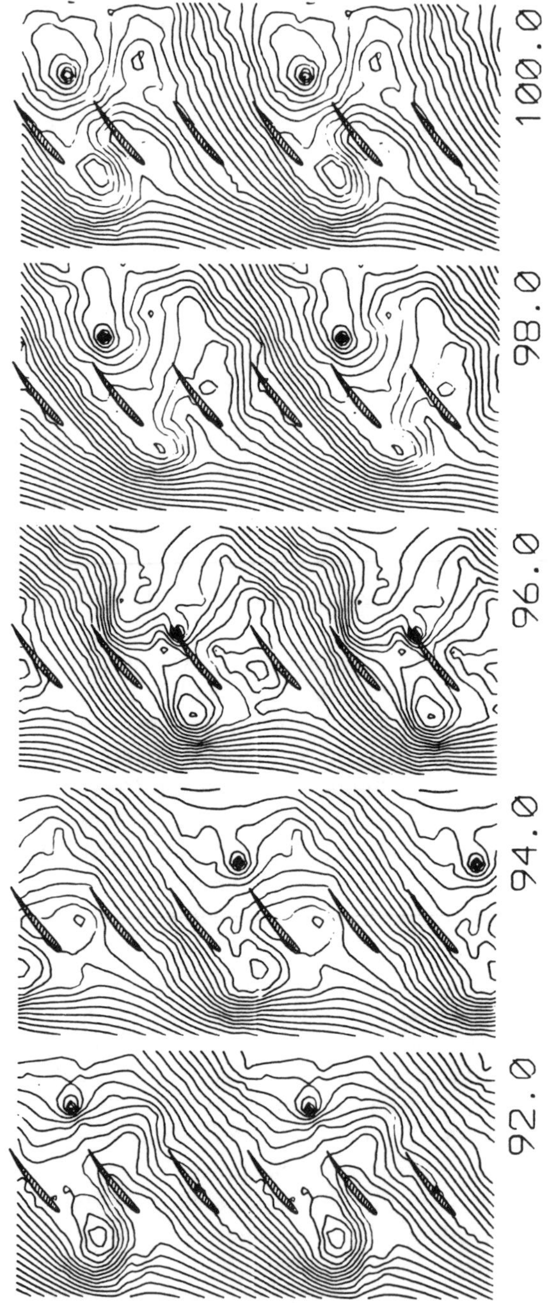

Figure 5.6 **Streamline patterns at several instants for bending vibrations in stall.**

Lift Response vs. Inlet Flow Angle in a Cascade of Unit Solidity (for $k = 0.5$, $\sigma = \pi$, $h/c = 0.05$, $\beta = 0°$).

Incidence	Computational[1]		Analytical	
	Ampl.	Phase	Ampl.	Phase
0°	0.939	−3.28	0.876	−3.16
20°	0.910	−3.19	–	–
40°	0.670	−3.36	–	–
45°	1.105	−2.77	–	–
50°	2.690	−2.49	–	–
55°	2.923	−1.27	–	–

[1] Obtained from spectral analysis of lift data [12].

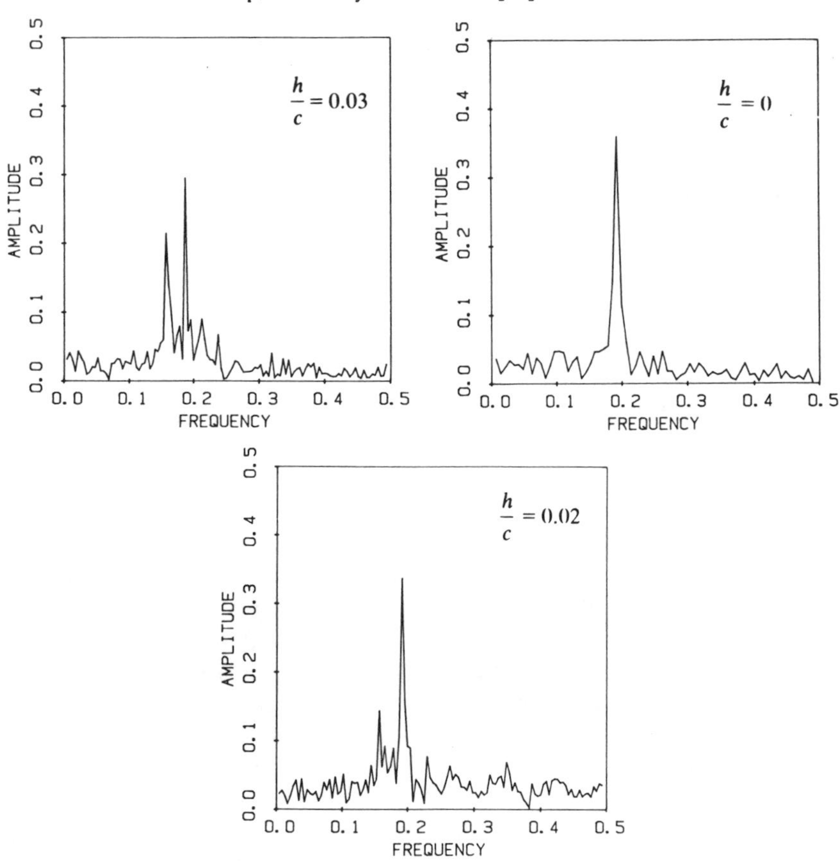

Figure 5.7 Effect of vibration amplitude on lift amplitude and frequency.

a modern interpretation, may be thought of as the aeroelastic coupling of fluid and structure through the vortex shedding and convection mechanism. If an airfoil natural frequency lies close to a natural frequency of the flow instability (either Karman vortex or propagating stall), the vibration of the blade can 'entrain' the stall frequency, resulting in the shift from a forced excitation at the 'stall natural frequency' to a self-excitation at the flutter frequency. This duality of frequencies may be observed in the lift response spectrum during the first few instants of the prescribed motion, Figure 5.7, for several bending amplitudes.

In this figure two distinct frequencies are evident, one associated with the propagating stall that would be present in the absence of any vibration, and the other at the same frequency as the impressed vibration. At a later time the propagating frequency has shifted and is essentially equal to the vibration frequency (which is always taken to include the effect of apparent mass). Frequency synchronization has taken place.

Results of this nature have led to further modelling and computation with the conclusion that stall flutter can be predicted by a computational algorithm in which the airfoil motion is not prescribed beforehand. In [14] and [15] the vortex method aerodynamics subprogram which is executed in parallel and interactively with a structural dynamics subprogram, the entire computation being carried forward in a time marching fashion.

Figure 5.8 from [14] is a computational confirmation of the frequency entrainment phenomenon previously hypothesized to occur for free vibrations. The temporal evolution of the streamline pattern and the accompanying blade vibratory motion for one datum point of Figure 5.8 is shown in Figure 5.9. The propagating stall frequency of a cascade of blades with fixed geometry and onset flow is seen to be relatively unaffected by the presence of flexible blades except in the neighborhood of those blades having a natural frequency near the intrinsic stall frequency. Within the interval of entrainment, however, the stall frequency is physically modified so as to synchronize with the blade natural frequency. Within the entrainment interval stall flutter may be said to occur. In Chapter 6 the synchronization phenomenon as applied to bluff bodies is discussed in greater detail. Further studies are underway to define the interval of synchronization as a function of the governing aeroelastic parameters and to further define the stall flutter behavior within this interval.

The vortex method possesses inherent limitations which are related to the two-dimensionality of the assumed flow and the necessity for a separation criterion embedded in a boundary layer subroutine. These limitations would be removed with the alternative development of Navier-Stokes solvers for full three-dimensional, unsteady, compressible flows. The principal difficulty to be overcome is the provision of an accurate turbulence model that will

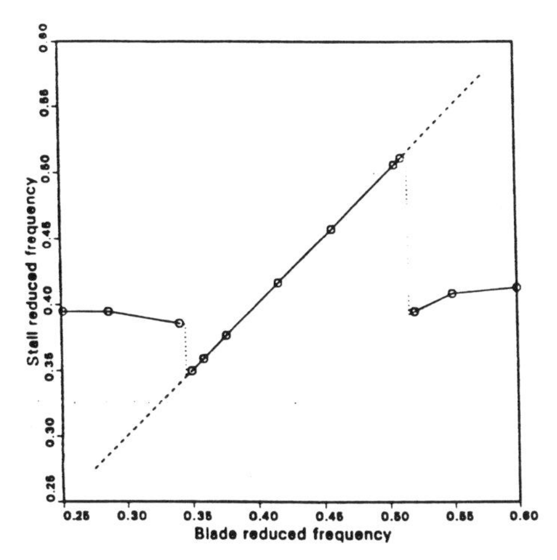

Figure 5.8 *Influence of blade-reduced frequency on the stall-reduced frequency for a cascade in torsional vibration. The plot shows the entrainment of stall frequency on a certain interval of blade frequency.*

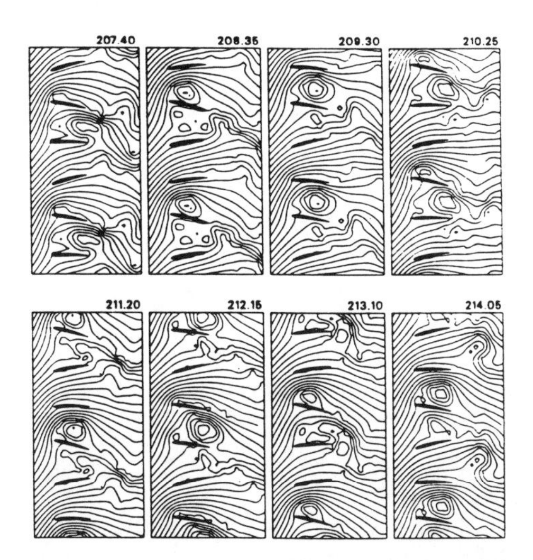

Figure 5.9 *Instantaneous streamline pattern for a cascade in torsional vibration at different time levels over a single period of oscillation (cascade periodicity is three; stagger = 0 deg, inflow angle = 55 deg, blade natural frequency in vacuum = 0.13 Hz, and the corresponding reduced frequency = 0.408).*

result in the necessary resolution of the scale of turbulence for typical cascade geometry. And the much greater number of computations required for this computational model, stemming from the multiplicity of blade passages, makes the forthcoming increase in computational speed a necessary adjunct. Supercomputers, probably involving parallel processing, are a necessity for reliable large scale Navier-Stokes solutions.

References for Chapter 5

[1] Sisto, F., 'Stall Flutter in Cascades', *Journal of the Aeronautical Sciences*, Vol. 20, No. 9 (September 1953) pp. 598–604.

[2] Durand, W. F. (Ed.) *Aerodynamic Theory*, Sec. E., Vol. II, Dover Publications, New York, 1963, p. 50.

[3] Sisto, F. and Perumal, P. V. K., 'Lift and Moment Prediction For An Oscillating Airfoil With a Moving Separation Point', *Trans. ASME, J. Engrg. for Power*, Vol. 96, Series A, No. 4 (October 1974) pp. 372–378.

[4] Gray, Lewis and Liiva, J., Windtunnel Tests of Thin Airfoils Oscillating Near Stall, Vol. II, *USAAVLABS Technical Report* 68–89B, 1969.

[5] Chi, M. R., 'Unsteady Aerodynamics in Stalled Cascade and Stall Flutter Prediction', *ASME Paper* 80–C2/Aero-1, 1980.

[6] Chi, R. M. and Srinavasan, A. V., 'Some Recent Advances in the Understanding and Prediction of Turbomachine Subsonic Stall Flutter', *ASME Paper* 84–GT–151.

[7] Tokel, H. and Sisto, F., 'Dynamic Stall of an Airfoil with Leading Edge Bubble Separation Involving Time-Dependent Reattachment', *ASME Paper* 78–GT–94.

[8] Spalart, P. R., 'Numerical Simulation of Separated Flows', NASA–TM–84238, 1983.

[9] Lewis, R. I. and Porterhouse, D. T. C., 1982, 'A Generalized Numerical Method for Bluff Body Stalling Aerofoil Flow', *ASME Paper* 82–GT–70.

[10] Spalart, P. R., 'Two Recent Extensions of the Vortex Method', *AIAA Paper No.* 84–0343, Reno, 1984.

[11] Speziale, C. G., Sisto, F. and Jonnavithula, S., 'Vortex Simulation of Propagating Stall in a Linear Cascade of Airfoils', *ASME Journal of Fluids Engineering*, Vol. 108, p. 304, 1986.

[12] Sisto, F., Wu, W., Thangam, S. and Jonnavithula, S., 'Computational Aerodynamics of Oscillating Cascades with the Evolution of Stall', *AIAA Journal* Vol. 27, No. 4 (April 1989) pp. 462–471.

[13] Fung, Y. C., *An Introduction to the Theory of Aeroelasticity*, John Wiley and Sons, Inc., New York, N.Y. 1955, p. 65.

[14] Sisto, F., Thangam, S. and Abdelrahim, A., "Computational Prediction of Stall Flutter in Cascaded Airfoils", *AIAA Journal* Vol. 29, No. 7, pp. 1161–1167, 1991.

[15] Sisto, F. Thangam, S. and Abdelrahim, A., "Computational Study of Stall Flutter in Linear Cascades", *ASME Journal of Turbomachinery*, Vol. 115, No. 1, 1993.

[16] Tang, D.M. and Dowell, E.H., "Chaotic Stall Response of Helicopter Rotor in Forward Flight", *Journal of Fluids and Structures*, Vol. 6, No. 6, pp. 311–335, 1992.

[17] Ekaterinaris, J.A. and Platzer, M.F., 'Numerical Investigation of Stail Flutter', ASME Paper No. 94-GT-206, 1994.

6

Aeroelastic problems of civil engineering structures

In recent decades the designs and methods of fabrication of new engineering structures have been such as to render them more flexible and not as highly damped as their counterparts of the past. As a result the sensitivity of such structures to the natural wind has become an object of considerable attention, and many studies in wind engineering have been occasioned. In these not only the character of the wind alone—as loading function—has been taken into account, but also the interaction between wind and structure and consequent structural motion have required consideration. It is the latter group of *aeroelastic* studies that will be examined particularly in this chapter.

In the design of flight vehicles the air flow approaching them is usually considered to be laminar in nature. It is also invariably found desirable to 'streamline' the vehicle surfaces to a refined degree, aiming at the reduction of flow separations, thus alleviating local buffeting effects and lowering drag. On the other hand, civil engineering structures are usually designed to meet non-aerodynamic objectives and consequently must be viewed as aerodynamically bluff objects. When the natural wind, itself of variable velocity with components of turbulence, flows around such structures, highly variable local pressures are engendered, as are turbulent wakes.

A helpful point of view in the analysis of certain interaction phenomena between structures and flow is the two-dimensional one, as frequently employed in aerodynamics to gain fundamental insights. This viewpoint will be taken in the present chapter to help elucidate certain phenomena. From such a view—an alongwind 'slice' of the flow past an object—one of the most common flow events is the vortex that is typically initiated when the flow leaves the vicinity of the body at a sharp downstream corner or point of small radius of curvature.

Smaller vortices shed in the near vicinity of a body characteristically coalesce into larger coherent ones downstream and a typical vortex wake is initiated. Some detailed aspects of such a wake will be discussed at a later point. A first important observation for bluff bodies, however, is that such vortex wakes, in one detailed form or another, are present in all flows for all bluff bodies. In a brief phrase, vortex shedding in some form may be said to take place under all circumstances. Thus vortex shedding may be occurring simultaneously with one or more other aeroelastic phenomena.

It has been common practice to place bluff bodies in laminar flows and measure their two-dimensional lift, drag, and moment coefficients by means of standard low-frequency balances. Given the observations above about shed vortices however, it must be understood that such static coefficients are really time averages of fluctuating values. As will be pointed out in the sequel, certain fluid-structure interaction phenomena can be well modeled analytically using static or steady-state coefficients. Others cannot, as it happens, and important distinctions are therefore in order regarding them. In such cases means must be devised to measure motion-dependent or time-dependent aerodynamic coefficients. In spite of the advent of computational fluid dynamics, few bluff-body aerodynamic results (particularly for moving bodies) have been inferred to date by theory or calculation from first principles; thus certain experiments are still in order.

A useful two-dimensional viewpoint has been handed on from potential flow theory, namely that fluid forces (like lift) on a body in a flow are associated with the flow circulation about the body. Conservation of angular momentum in the entire flow requires then that the total flow picture be irrotational, which means that equal and opposite vorticity must exist somewhere in the flow to balance that about the body. An airfoil at a fixed angle of attack in a flow, for example, must be presumed, under this view, to have left an equal and opposite 'starting vortex' far to its rear that balances its own circulation.

Should the lifting body change its angle of attack it must then shed motion-initiated or attitude-change vorticity in its immediate wake. Thus an oscillating body, for example, creates a motion-related near wake associated with position or attitude change which has important effects upon the pressure distributions over the body.

Viewed globally then, a bluff body in motion within a laminar flow may be expected to manifest some or all of the following wake phenomena:

a) near-wake vortices associated with the natural vortex shedding set up by the flow itself. Such a wake can be associated with either a fixed or a moving body.

b) a far-wake vortex associated with the mean, or steady, forces acting on the body. Such a phenomenon may be considered to be associated only with a fixed or very slowly moving body.

c) near-wake vortices associated with dynamic body positional changes in time. These represent the changes in vorticity accompanying body motion.

In analyzing commonly recognized problems of aeroelasticity the above classifications become useful. Although the problems that arise are associated with specific structures such as towers, tall buildings, and bridges, it has been judged preferable to discuss them below according to phenomenological type rather than structure type. In following out this method of presentation, the phenomena that will be considered are *vortex-shedding*, *galloping*, *divergence*, *flutter*, and *buffeting*. The discussion of each of these in turn will include some commentary on the particular structures that are subject thereto; since many of the phenomena are undesirable from a structural point of view, methods for their prevention will be mentioned where appropriate.

6.1 Vortex shedding

Introduction

The wind harp of the ancient Greeks consisted of a set of strings which vibrated when exposed to breezes, eliciting Aeolian tones. That the oscillations of such strings were mainly across-wind and associated with the shedding of alternating vortices in the wake of the strings was not generally recognized until much later. In 1878 Strouhal [1] published a paper on the production of tones by fluid flow.

It is a wide departure from the more usual topics of the field of aeroelasticity to consider aerodynamically bluff bodies. This is the case for the phenomenon of vortex shedding. Some sense of the complexity thereby introduced has already been given by the discussion of stall flutter in Chapter 5. When a bluff body is placed in a smooth, low-speed flow under ranges of Reynolds number that are typically encountered ($100 \leq R \leq 10{,}000$), perhaps the most outstanding single characteristic observed in the flow is its separation from the body and the initiation of a more or less agitated wake.

Upon detailed examination, this wake, while fluctuating, is seen to possess a good deal of recognizable structure, and, in fact, in certain flow regimes it can be seen to be wholly deterministic, though still quite complex, in character. The outstanding aspect of such character has been the rolling-up of the wake into pairs of alternating vortices which proceed downstream in a quite orderly array. The basic causative agents in the process are the viscous and inertial characteristics of the flow that exist in appropriate proportion to favor the formation of the vortices. In 1908 Bénard [2] and later von Kármán [3] discussed the organized vortex trail.

Almost all bluff bodies in cross flows 'shed' such vortices at regular rhythms that correspond to their respective Strouhal numbers S, the value for each being a constant that is a function of body shape:

$$S = f_s D/U \tag{6.1.1}$$

where f_s is the frequency of the natural vortex-formulation cycle, D is a representative body across-flow dimension, and U is the oncoming uniform flow velocity. Refs. [4, 5] for example, discuss some of the prominent characteristics of vortex shedding from bodies of structural shape. The point of view most commonly adopted in discussing the vortex shedding problem is that the flow is two-dimensional, and this view will be employed in what follows.

For each bluff body the local flow pattern is quite particular to that body, but the vortex trail in its wake eventually acquires a general character that can be said to be roughly similar for most bluff bodies with vortex spacing dependent on the value of S. In the low-speed flow (though with Reynolds number not low enough to characterize the flow as purely viscous) the vortex wake pattern is accompanied by alternating pressures over the body.

Integration over the body of the components of such pressures in lift and drag directions results in alternating lift and drag forces, the former principally at the Strouhal frequency and the latter principally at twice this frequency, a complete Strouhal cycle being considered the complex of events in which the flow conditions at a point in the near wake proceed from a given state to an identical repetition of that state.

It will, in particular, be noted that the vortex wake described so far is a function only of body shape and flow characteristics. For real structures the alternating forces accompanying this phenomenon cause structural deflection at the basic Strouhal rhythms. This deflection may be slight, affecting the established flow pattern only a little, or it may be appreciable, depending on structural support stiffness and—most importantly—on structural natural frequency n_n. If, for example, the rhythmic changes of fluid forces on the body approach a natural resonant frequency of the body, quite large structural deflections may occur, and if they do, this body motion strongly affects the local boundary conditions of the flow. Additional vorticity which bears this influence is then shed into the wake.

Thus—in a somewhat oversimplified description—wake conditions and the associated body forces are engendered that bear the influence of two distinct rhythms: the 'Strouhal rhythm' associated with a fixed body, and a second, 'motion-induced', rhythm associated with body time-dependent displacement, the latter usually occurring near the natural mechanical frequency of the body.

When these two rhythms approach coincidence, which can take place at some particular flow velocity U, body motion, if appreciable, tends to influence the wake more strongly than does the natural Strouhal tendency. Body motion also tends to approach its largest value under these conditions.

In this situation, known as 'lock-in' or synchronization, with the wake controlled by body motion, body natural mechanical rhythm predominates throughout a certain range of velocity U that begins at the inception stage of the phenomenon and extends beyond it some distance. This circumstance has been qualitatively pictured in graphical form by Fung [6] as in Figure 6.1.

Of particular note during lock-in, however, is that, while body response somewhat resembles that of a mechanically resonant system, it—regardless of the low level of mechanical damping that may be present in the structure—never achieves a condition of excessively large amplitudes. Instead, the oscillation amplitude is always observed to be

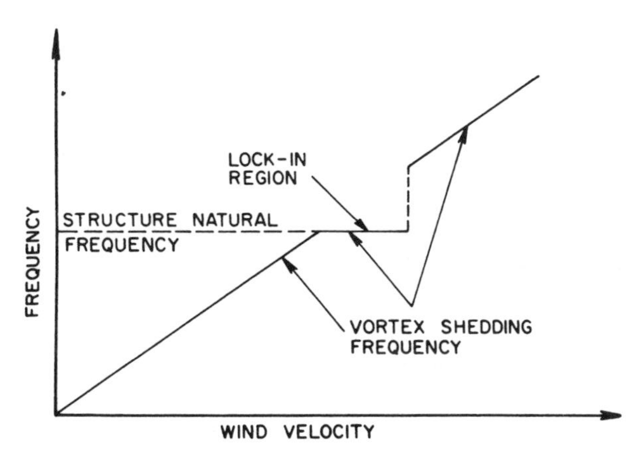

Figure 6.1 Quantitative trend of vortex shedding frequency with wind velocity when lock-in occurs [6].

self-limiting, usually to the order of a fraction of the dimension D. In this respect it is distinctly unlike catastrophic flutter. A complex effect of fluid-engendered damping acts to bring this about.

In light of the apparent success of analytical theory in describing the flutter phenomenon it has long been tempting to the profession to create mathematical fluid-dynamical models that describe the vortex excitation and response of elastically supported bluff bodies. In fact, a vast literature, consisting of several hundred references, presently exists on this topic. Perhaps the size alone of this effort is one indication of its imperfect degree of success. Several models—all empirical—have stood out, however, though each has particular shortcomings; a number of these will be reviewed subsequently.

Certainly the phenomenon of vortex shedding depends strongly, in its detail, upon the particular geometric form of the bluff body that initiates the vortex wake. Hence achievement of great generality in analytical vortex-shedding models may in fact ultimately be illusory. One particular cross-sectional shape, however, has received more treatment than any other. This is the circular section. There are numerous underlying motivations for this concern, but this shape may also be one of the most difficult to model since it permits many nuances of flow-induced response to occur. The curved surface of the section permits flow separation over a range of positions, as against the analogous action from bodies with sharply defined corners or other flow separation points.

Before discussing vortex response models, a general qualitative, phenomenological description of the kind of features that are observed will be given.

Aspects of response to vortex shedding

Some of the generally observed qualitative phenomena associated with the vortex-induced response of a bluff body are:

1. A regular train of alternating vortices produced at the Strouhal frequency is shed behind the body when the latter is fixed in place in a fluid stream.

2. Fixed-body wake characteristics continue but are modified importantly by the changing boundary conditions accompanying body motion, so that the latter changes notably the total effects produced.

3. When the body is free to move it undergoes pressures both near the fixed-body Strouhal rhythm and at the rhythm of its own motion. (The

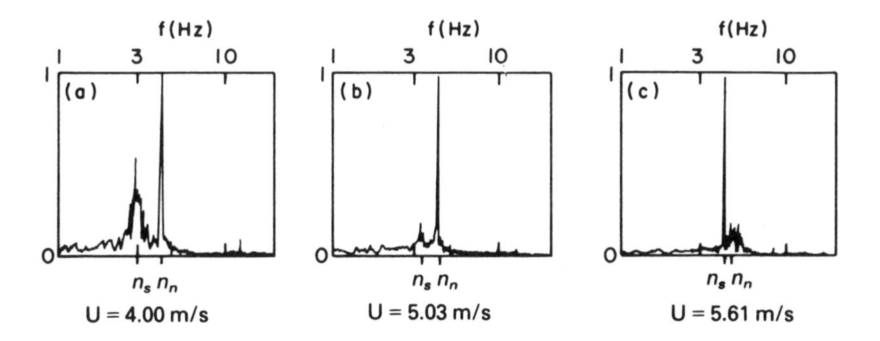

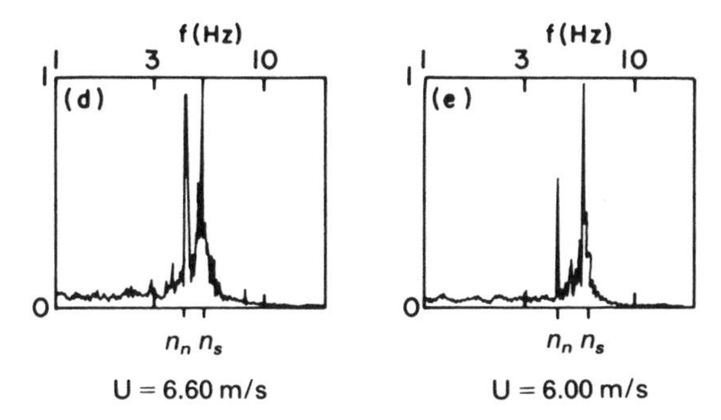

Figure 6.2 Spectra of fluid force on a square cylinder in smooth cross flow at different velocities; n_s = Strouhal frequency, n_n = mechanical frequency [33].

latter occurs quite closely to the natural frequency that obtains in the absence of flow.) (See Figure 6.2.)

4. For certain bodies, at least, other (lower and higher harmonic) rhythms than the Strouhal rhythms also occur in the body pressures. The effects of these are occasionally clearly manifested in body response.

5. Integrated body pressures associated with vortex shedding may excite along- and across-flow motions as well as torsional motion, depending on the degrees of freedom available to the body.

6. The phenomna of lock-in and "resonant" response occur, as suggested in Figures 6.1 and 6.3, for bodies supported elastically in the across-flow direction. Bodies in air flows usually manifest mainly across-flow oscillation, while those in water flows may exhibit both along- and across-flow response.

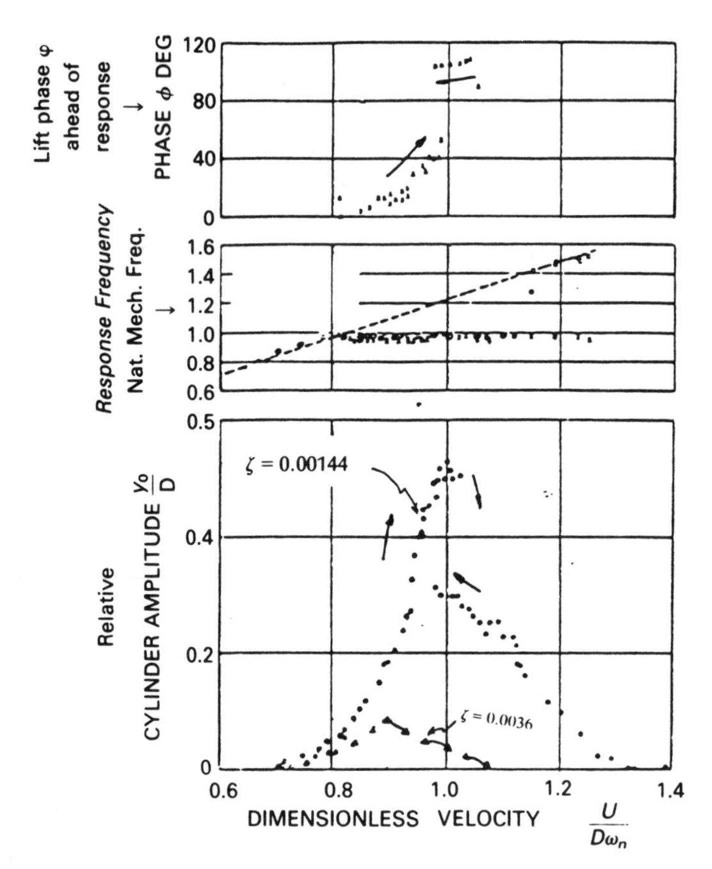

Figure 6.3 Lock-in response amplitude, ratio of response frequency to mechanical frequency, and phase of elastically supported circular cylinder; also amplitude for a higher mechanical damping value [11].

7. The phenomena accompanying both entrance into—and escape from—lock-in are different between the situations of increasing and of decreasing flow velocity U (see Figure 6.3). A "jump phenomenon" thus appears to be involved in the response, suggesting nonlinear properties in the overall phenomenon.

8. The complete mechanico-fluid system acts as a single functional unit insofar as the in- or out-flow of energy associated with vortex-induced oscillation is concerned. For example, the level of mechanical damping strongly affects the nature of vortex-related body forces and the level of body response (see Figures 6.3 and 6.4).

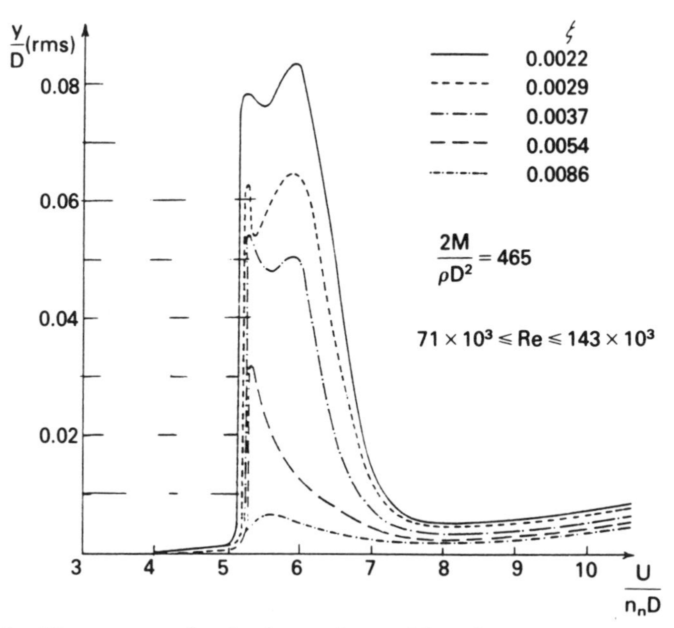

Figure 6.4 The response of a circular-section model stack (Height = 10 diameters) at different values of structural damping (Re subcritical) [30].

9. Body response to vortex-related pressures evidences aspects of self-excitation at low amplitudes, but the response is always self-limiting, evidencing increased fluid damping effects at higher amplitudes.

10. The Strouhal wake detached from a body evidences certain characteristics of an independent oscillator, to which the body is coupled in some manner. The near wake has the greatest effect upon the body.

Empirical models of vortex-induced oscillation

The problem associated with developing response models may be viewed somewhat as follows: Given the observed facts of experiment, how best are analytical models to be created that possess characteristics reflecting those facts? In pursuing the modeling objective it is obvious early-on that while certain salient aspects of the observable facts can be rather simply modeled—at least in first approximation, attainment of appropriate representation of the more subtle facts is a more recondite problem, both as regards correct appreciation of the facts themselves and choice of analytical devices to reflect them.

Among the experimental efforts that provide very useful background data are those of Bishop and Hassan [7, 8], Sarpkaya [9], Staubli [10], Feng [11], and Ferguson [12]. Refs. [8–10] employed mechanically driven systems with controlled amplitudes. Other researches employed flow-excited spring-supported models. A more recent study [114] will be described at a later point.

In what follows attention will be given uniquely to a linear mechanical model located in a smooth fluid stream of velocity U and free to oscillate only in a single across-flow degree of freedom y. Let m be the model mass per unit span normal to the flow, ζ its mechanical damping ratio-to-critical, and $\omega_n = 2\pi n_n$, its natural mechanical circular frequency. Then the equation of motion is

$$m[\ddot{y} + 2\zeta\omega_n\dot{y} + \omega_n^2 y] = F(y, \dot{y}, \ddot{y}, U, t) \tag{6.1.2}$$

where F represents the fluid force. Various models for F will be considered, as well as extensions involving an eventual fluid oscillator.

It is important to reiterate that all models to be discussed are empirical in nature, i.e. not based on first theoretical principles of fluid flow. This situation is dictated by the fact that, for bluff body flows accompanied by fluid separation and turbulence, it has not to date been possible—except possibly through computational fluid mechanics—to duplicate flow circumstances other than by experiment. Thus proposed analytical models have provided forms appropriate to account for general response characteristics that are to be quantified later through parameters determinable by experiment. Hence, with each of the models suggested below, some allusion to necessary experimental support will be made. On the other hand, presentation of full details required in such experiments would go beyond the scope of the treatment to be offered here.

First will be examined some rather simplistic models that acknowledge the observed rhythms of fluid-body forces and body response and, to a limited extent, the levels of body response amplitude.

Uncoupled single-degree-of-freedom models

A very general class of s.d.o.f. response models is provided for by the form:

$$F(y, \dot{y}, \ddot{y}, U, t) = a_2\ddot{y} + a_0 y + \sum_{n=1}^{N} b_{2n-1}\dot{y}^{2n-1} + \sum_{m=1}^{M} c_m \sin(p_m t + \theta_m) \tag{6.1.3}$$

in which in principle the constants a, b, c, p, θ are to be evaluated to match observed response. This model allows for adjustments to natural frequency via coefficients of 'inertia' and 'stiffness' a_2 and a_0, coefficients of nonlinear (positive or negative) damping b_{2n-1}, and driving forces via the sine terms at different frequencies p_m, all unknown coefficients being evaluated for each given velocity U or, for example, reduced frequency $K_n = B\omega_n/U$.

Clearly, aside from the question of whether the model offers enough flexibility to match observed response, there is also the practical question of eventually limiting it sufficiently to the number of parameters that can actually be identified through experimental observables. Various authors have used forms of this model to characterize response considered to be essentially sinusoidal in its time variation.

1. Among the simplest of these is one that drops all terms except a single sine driving term at the Strouhal shedding frequency:

$$F = \tfrac{1}{2}\rho U^2 D C_L \sin \omega_s t \tag{6.1.4}$$

where $\omega_s = 2\pi n_s$ and n_s is the Strouhal forcing frequency:

$$n_s = SU/D \tag{6.1.5}$$

The force F may be restated in the form $F_0 \sin \omega_s t$ where

$$F_0 = \tfrac{1}{2}\rho U^2 D C_L \tag{6.1.6}$$

and C_L is a lift coefficient. An approximate value

$$C_L \cong 0.6 \tag{6.1.7}$$

was suggested in [7] for the fixed circular section. However, as the body of circular section moves, this model proves inadequate, as motion-induced forces also appear.

Sarpkaya [9], examining controlled, forced-amplitude across-flow sinusoidal motion of a circular cylinder over a range of frequencies and amplitudes, obtained both the lift coefficient C_L and its phase with respect to the motion as a function of the reduced velocity U/nD where n was the forced oscillation frequency and D the cylinder diameter. From these data he was able to predict (using F as in (6.1.4)) the dynamic response y of a self-excited cylinder in the lock-in region, where forcing and Strouhal frequencies practically coincide. Of particular interest in the experiments of Sarpkaya was the dramatic sign reversal in C_L (i.e. phase shift) in the region close to $1/S = U/nD = 5$ (a little above 5 for the inertial contribution and a little below 5 for the drag contribution to C_L).

2. Staubli [10], in forced-displacement experiments of similar style to Sarpkaya's, measured the resulting lift coefficient and its phase as functions of dimensionless forced oscillation frequency nD/U, and subjected them to spectral analysis. As a central result he proposed that F be expressed by

$$F = \tfrac{1}{2}\rho U^2 D C_L(t) \tag{6.1.8}$$

with

$$C_L(t) = C_{L_n} \sin(\omega_n t + \theta) + C_{L_s} \sin \omega_s t \tag{6.1.9}$$

where ω_n, ω_s are respectively the radian natural mechanical and fixed-cylinder Strouhal frequencies, these being the ones at which significant components C_{L_n}, C_{L_s} were experimentally present. He then employed this form of C_L to predict the response of the elastically mounted cylinder, with quite good matches to free oscillation results of Feng [11] over the lock-in range.

3. Provisions for fluid modifications to damping and frequency [4] can be included in the model specified by (6.1.8) and (6.1.9) by defining F as

$$F = \tfrac{1}{2}\rho U^2 D\left[C_L(t) + Y_1 \frac{\dot{y}}{U} + Y_2 \frac{y}{D}\right] \tag{6.1.10}$$

thus including additional parameters Y_1, Y_2 to be evaluated by experiment—as, for example, by direct examination of the response of the mechanical system (6.1.2) under fluid flow.

4. A model that extends this idea further to include self-limiting aerodynamic damping characteristics was used by Scanlan [4] (and a very similar one was employed by Vickery and Basu [13]):

$$F = \tfrac{1}{2}\rho U^2 D\left\{C_L(t) + Y_1\left[1 - \varepsilon\left(\frac{y}{D}\right)^2\right]\frac{\dot{y}}{U} + Y_2 \frac{y}{D}\right\} \tag{6.1.11}$$

where Y_1, ε, Y_2 are constants to be identified through experiment. In cases of self-excitation the term in $C_L(t)$ may possibly be omitted, since negative damping effects are provided by the term in ε, which recalls the Van der Pol oscillator.

Models such as 1 to 4 above allow adjustment to some, though not all, of the observable characteristics of the response to vortex shedding. There have therefore been efforts to create models with extended possibilities. These have mainly taken the route of including a second, wake-related, degree of freedom. From one of these a fifth single-degree model has been derived, and this will be discussed at a later point.

Coupled two-degree-of-freedom ('wake oscillator') models

1. Birkhoff [14] proposed, as an approximate model of wake conditions, that the wake in the near vicinity of a circular cylinder be visualized as an inertial lamina extending aft of the cylinder surface a distance L and oscillating at the Strouhal frequency from side to side across the wake centerline through an angle α 'somewhat like the tail of a swimming fish'. This model then served to estimate the value of the Strouhal number as a function of the length L and the cylinder diameter D, which was given as

$$S = \frac{1}{2\sqrt{\pi}}\left(\frac{D}{kL}\right)^{1/2} \tag{6.1.12}$$

with $k = 1.33$ and $L = 1.5D$ for the case of the circular cylinder ($S = 0.2$).

Bishop and Hassan [7] also noted that 'the wake of a cylinder in a flowing fluid behaves as a conventional mechanical oscillator'.

Before discussing typical wake oscillator models proposed in the literature it is perhaps worth while to discuss some general characteristics common to all. The particular choice of a representative wake quantity φ to typify the variable wake is a key distinguishing characteristic of each model. (For example, $\varphi = \alpha$ in the Birkhoff model.) Others are then the assumed mechanisms through which φ is coupled to the body motion y, on the one hand, and the body motion y to the wake, on the other. An overview of the possibilities presented results in the general mathematical model below.

Body motion:

$$m[\ddot{y} + 2\zeta\omega_n\dot{y} + \omega_n^2 y] = F(\varphi, \dot{\varphi}, \ddot{\varphi}, \ldots, U, t) \tag{6.1.13}$$

Wake coupling to body;

$$F(\varphi, \dot{\varphi}, \ddot{\varphi}, \ldots, U, t) = a_2\ddot{\varphi} + a_0\varphi + \sum_{n=1}^{N} a_{2n-1}\dot{\varphi}^{2n-1} \tag{6.1.14}$$

Wake motion:

$$\ddot{\varphi} + b_0\varphi + \sum_{m=1}^{M} b_{2m-1}\dot{\varphi}^{2m-1} = G(y, \dot{y}, \ddot{y}, \ldots, U, t) \tag{6.1.15}$$

Body coupling to wake:

$$\begin{aligned} G(y, \dot{y}, \ddot{y}, \ldots, U, t) &= c_0 y + c_2\ddot{y} + c_3\dddot{y} + c_4\ddddot{y} \\ &+ \ldots + \sum_{r=1}^{R} d_{2r-1}\dot{y}^{2r-1} \end{aligned} \tag{6.1.16}$$

Where all constants a, b, c, d are eventually to be identified from experimental evidence (and/or reasoning about the physical system). Note that no purely time-dependent external driving forces are considered necessary in this model.

The general model presented permits a number of choices for the wake variable and the possibilities of nonlinear wake response, including, in particular, the description of an oscillator of Van der Pol type. Most models couple two second-order oscillator systems only. In the end, all authors have employed empirical approaches, though supported by special reasonings pertinent to their models. These efforts have in general served to identify directly some of the parameters employed and thus evaluate them prior to experiment. Other parameters have been left open to experimental identification. Because of this aspect of the modeling problem there has been some natural tendency toward reducing to a minimum the number of unknown parameters used.

2. Funakawa [15] elaborated on the Birkhoff model, endowing it with certain physical characteristics assumed to be associated with an elliptical oscillator mass representing the 'dead stream' region in the near wake aft of the cylinder. In experiments in which water flowed at uniform velocity U past a solid cylinder of diameter D that was oscillated at the Strouhal frequency across-stream by a mechanism imparting the displacement

$$y = y_0 \sin \omega_s t \tag{6.1.17}$$

flow visualization photographs of the wake configuration were made. These permitted estimates of the nature and action of the near-wake 'dead stream' region at lock-in.

Then, passing to theory for the wake oscillator, this was presumed to act like a kind of horizontal physical pendulum coupled to the cylinder motion and responding according to the equation

$$I\ddot{\alpha} + C\dot{\alpha} + K\left(\alpha + \frac{\dot{y}}{U}\right) = \omega_s^2 I\bar{\alpha} \sin \omega_s t \tag{6.1.18}$$

with C not initially specified, and the following assigned values of the other parameters:

'moment of inertia'

$$I = \tfrac{1}{4}\rho LH(D + L)^2 \qquad \text{(with } H = 1.25D\text{)} \tag{6.1.19}$$

'restoring moment per unit angle':

$$k = \tfrac{1}{2}\rho U^2(2\pi)L(D + L)/2 \tag{6.1.20}$$

$$\bar{\alpha} = 2y_0/(D + L) = 0.625\, y_0/D \qquad \text{(with } L = 2.2D\text{)} \tag{6.1.21}$$

From the solution to (6.1.18) the amplitude of the wake oscillation angle was obtained as a function of I, k, U, c, ω, and y_0 and then associated with assumed Magnus-effect lift and drag forces acting on the cylinder due to the wake oscillator and taken as proportional to the effective wake angular displacement $\alpha + \dot{y}/U$. Thus four force functions $F_1, \ldots, F_4$ were obtained such that, in summary, a) for the lift force:

$$F_L = F_1\dot{y} - F_2 y \tag{6.1.22}$$

and b) for the drag force:

$$F_0 = -F_3\dot{y} - F_4\dot{y}^3 \tag{6.1.23}$$

Using these in the mechanical equation of motion of the spring-supported cylinder in the across-flow direction led finally to the single-degree-of-freedom form:

$$M\ddot{y} - (F_1 - C - F_2)\dot{y} + F_4\dot{y}^3 + (K + F_2)y = 0 \tag{6.1.24}$$

where M, C, K are the mechanical mass, damping and stiffness coefficients, respectively. The critical roles of self-excitation and self-limiting damping of the Van der Pol type are made evident in (6.1.24). Details of the forces $F_1, \ldots, F_4$ are omitted from the present summary.

3. Nakamura [16] espoused the form of the Funakawa–Birkhoff wake oscillator, estimating, from experimental information, the Magnus lift force due to wake angular swing α to be

$$F_L = -0.58\rho U^2 D\alpha \tag{6.1.25}$$

(though remarking that this value was tentative).

Neglecting damping conditions he then obtained the two coupled equations

Cylinder:

$$M\ddot{y} + Ky = F_L \tag{6.1.26}$$

Wake:

$$I\ddot{\alpha} + k\alpha = (-2I\ddot{y})/(D + L) \tag{6.1.27}$$

which he treated analogously to a case of binary flutter, emphasizing the possible 'resonant' character of the fluid-structure system. Finally, the system was further analyzed with the inclusion of a sinusoidal forcing term along with F_L. Nakamura included a list of critical comments on the limitations of this model in the original paper.

4. Tamura and Matsui [17] further developed the Birkhoff–Funa-

kawa wake oscillator concept through the novel introduction of an oscillating change in the length L of the near-wake fluid mass occurring at twice the Strouhal frequency, thus replacing L by a new value:

$$L = \bar{L} - b \sin 2\omega_s t = \bar{L}\left(1 - b^* \frac{\alpha\dot{\alpha}}{\omega_s}\right) \tag{6.1.28}$$

where $\bar{L}$ is the mean length, ω_s is the fixed-body Strouhal circular frequency $\omega_s = 2\pi SU/D$ and

$$b^* = 2b/\alpha_0^2\bar{L} \tag{6.1.29}$$

This device introduced parametric damping.

The net results obtained were

wake;

$$\bar{I}\ddot{\alpha} - \bar{C}\left[1 - \frac{4f^2}{C_{L_0}^2}\alpha^2\right]\dot{\alpha} + \bar{K}\left(\alpha + \frac{\dot{y}}{U}\right) = \frac{-2\bar{I}\ddot{y}}{D + \bar{L}} \tag{6.1.30}$$

cylinder:

$$M\ddot{y} + C\dot{y} + Ky = -\tfrac{1}{2}\rho U^2 D\left[f\left(\alpha + \frac{\dot{y}}{U}\right) + C_D\frac{\dot{y}}{U}\right] \tag{6.1.31}$$

where $\bar{I}$, $\bar{K}$ (mean values) have the same definitions respectively as I, K, with L replaced by $\bar{L}$, and

$$\left.\begin{aligned} &\bar{C} \cong 2\zeta\omega_s\bar{I} \quad (\zeta = 0.038) \\ &f \cong 1.16 \\ &C_{L_0} \cong 0.4 \\ &C_D \cong 1.2 \end{aligned}\right\} \tag{6.1.32}$$

In this model the wake is coupled to the body via $\ddot{y}$ and $\dot{y}$ terms, while the body in turn is coupled to the wake via a term in α. The influence of Van der Pol type damping is manifest in the wake oscillator.

5. Hartlen and Currie [18], while adopting the Birkhoff concept of a wake oscillator, focused their interpretation upon the independent oscillation of the lift coefficient C_L, so that in this context $\varphi = C_L$. Thus, taking into account a number of experimentally observed response results, and arguing qualitatively the need for both self-excitation and self-limiting damping, they postulated the following coupled equations, which they characterized overall as a 'rudimentary' model:

cylinder:

$$M\ddot{y} + C\dot{y} + Ky = \tfrac{1}{2}\rho U^2 D C_L(t) \tag{6.1.33}$$

lift:

$$\ddot{C}_L - a\omega_s \dot{C}_L + \frac{\gamma}{\omega_s}\dot{C}_L^3 + \omega_s^2 C_L = b\dot{y} \tag{6.1.34}$$

where again ω_s is the circular Strouhal frequency corresponding to $S = 0.2$, b is an initially undetermined constant, a is a modified mass ratio

$$a = \rho D^2/8\pi^2 S^2 M$$

and the ratio of a/γ is to be determined from the measured amplitude C_{L_0} of the fluctuating lift coefficient on a stationary cylinder:

$$C_{L_0} = \left(\frac{4a}{3\gamma}\right)^{1/2} \tag{6.1.35}$$

6. Skop and Griffin [19, 20] slightly modified the Hartlen and Currie model, and argued certain relationships among the coefficients, converting (6.1.34) to the form

$$\ddot{C}_L - \omega_s G\left[C_{L_0}^2 - \tfrac{4}{3}\frac{\dot{C}_L^2}{\omega_s^2}\right]\dot{C}_L + \omega_s^2[1 - \tfrac{4}{3}HC_L^2]C_L = F\omega_s \frac{\dot{y}}{D} \tag{6.1.36}$$

where coefficients $C_{L_0}^2$, G, H, F were considered parameters to be evaluated from experiment.

7. Higher order nonlinearities in $\dot{C}_L$ were employed in extensions of the Hartlen–Currie model by Landl [21], Wood [22], and Wood and Parkinson [23]. For example, the governing equation for C_L took the form [21]

$$\ddot{C}_L + (\alpha - \beta C_L^2 + \gamma C_L^4)\dot{C}_L + \omega_s^2 C_L = b\dot{y} \tag{6.1.37}$$

for constants α, β, γ, b to be evaluated.

8. Berger [24], following the style of the Hartlen–Currie model, employed the general form

$$\ddot{C}_L + \sum_{n=1}^{N} b_{2n-1}\dot{C}_L^{2n-1} + \omega_s^2 C_L = \sum_{m=1}^{M} C_{2m-1}\dot{y}^{2m-1} \tag{6.1.38}$$

9. Dowell [25] also employed C_L as the wake variable. Four basic requirements were imposed:

(a) At high frequencies a virtual mass relationship should be preserved between lift and cylinder acceleration, i.e.

$$8C_L = -B_1 \rho D^2 \frac{\ddot{y}}{2} \quad (\omega \rightarrow \infty) \tag{6.1.39}$$

B_1 being a constant.

(b) At low frequencies quasi-steady conditions should hold between C_L and y, i.e., for $\omega \rightarrow 0$

$$C_L = f\left(\frac{\dot{y}}{U}\right) = A_1\left(\frac{\dot{y}}{U}\right) - A_3\left(\frac{\dot{y}}{U}\right)^3 + \cdots \tag{6.1.40}$$

$A_1, A_3, \ldots$ being constants.

(c) For small C_L and $y \equiv 0$ the fluid oscillation should be at the Strouhal frequency ω_s, i.e.

$$\ddot{C}_L + \omega_s^2 C_L = 0 \tag{6.1.41}$$

(d) Characteristics of Van der Pol damping should be included in the response of C_L.

These conditions led to the equation

$$\ddot{C}_L - \varepsilon\left[1 - 4\left(\frac{C_L}{C_{L_0}}\right)^2\right]\omega_s \dot{C}_L + \omega_s^2 C_L \tag{6.1.42}$$
$$= -B_1\left(\frac{D}{U^2}\right)\dddot{y} + \omega_s^2\left[A_1\frac{\dot{y}}{U} - A_3\left(\frac{\dot{y}}{U}\right)^3 + A_5\left(\frac{\dot{y}}{U}\right)^5 - A_7\left(\frac{\dot{y}}{U}\right)^7\right]$$

where ϵ is to be experimentally determined. This model is distinguished from others particularly in its 4th order coupling of y to C_L and its possibilities of describing activity in a broader frequency range (for example, evolving toward the galloping case at low frequency). Dowell gives a complete discussion and comparisons with experiment in the original paper [25].

10. Iwan and Blevins [26, 5] introduced a 'hidden flow variable' to represent the wake. The variable $\varphi = w$ was chosen for this purpose, where $\dot{w}$ represented the 'average transverse component' of fluid velocity within an arbitrary control volume that included the cylinder and two downstream vortex pairs. Thus $\dot{w}$ was defined as

$$\dot{w} = \frac{J_y}{a_0 \rho D^2} \tag{6.1.43}$$

where J_y was the estimated total vertical momentum in the control volume and a_0 was a dimensionless constant.

Omitting the details of the authors' derivations the resulting equations of motion took the form

body:

$$m[\ddot{y} + 2\zeta\omega_n\dot{y} + \omega_n^2 y] = a_2\rho D^2(\ddot{w} - \ddot{y}) + a_1\rho DU(\dot{w} - \dot{y}) \tag{6.1.44}$$

wake:

$$\ddot{w} + c_1\dot{w} + c_3\dot{w}^3 + c_0 w = b_2\ddot{y} + b_1\dot{y} \tag{6.1.45}$$

where the constants a, b, c are to be evaluated either through reasoning about the form of the system or by experimental identification.

11. Billah [112], after an extensive review of the vortex-shedding literature through 1988, suggested the following two-degree-of-freedom model for the circular cylinder:

body:

$$\ddot{y} + 2\zeta\omega_n\dot{y} + \omega_n^2 y + 2\alpha y\phi + 4\beta y^3\phi + (\gamma y + \epsilon y^3)\dot{\phi} = 0 \tag{6.1.46}$$

wake:

$$\ddot{\phi} + f(\phi, \dot{\phi}) + (2\omega_s)^2\phi + \alpha y^2 + \beta y^4 - 2(\gamma y + \epsilon y^3)\dot{y}\dot{\phi} = 0 \tag{6.1.47}$$

where $\alpha, \beta, \gamma, \epsilon$ are constants; the fluid variable ϕ represents the downstream distance from the cylinder to the first wake vortex, and $f(\phi, \dot{\phi})$ is a fluid damping function.

This model contains a number of appropriate features, an obvious one of which is that when the body is fixed, the wake oscillator reverts to one with a natural frequency equal to twice the Strouhal frequency. In fact, the wake oscillator under all conditions imparts a coupling force at frequency $2\omega_s$ to the body.

This latter characteristic of the wake as oscillator is a salient one. In light of it a return to a single-degree model analogous to that of (6.1.11) may be suggested as a simpler alternative. This alternative model has taken the form, for body motion y:

$$\ddot{y} + 2\zeta\omega_n\dot{y} + \omega_n^2 y = F \tag{6.1.48}$$

with

$$F = \frac{\rho U^2 D}{2m}\left[Y_1(K)\frac{\dot{y}}{U} + Y_2(K)\frac{y^2}{D^2}\frac{\dot{y}}{U} + J_1(K)\frac{y}{D} + J_2(K)\frac{y}{D}\cos 2\omega_s t\right] \tag{6.1.49}$$

where

ζ = mechanical damping ratio

ρ = fluid density

U = cross flow velocity

D = cylinder diameter

m = cylinder mass per unit length

$Y_i(K), J_i(K)$ = coefficients to be determined

$K = \frac{D\omega}{U}$ = reduced frequency

ω_n = natural circular frequency

ω_s = Strouhal circular frequency

In dimensionless form, with

$$\eta = \frac{y}{D}, \; s = \frac{Ut}{D}, \; (\;)' = \frac{d}{ds}(\;):$$

(6.1.49) may be rewritten:

$$\eta'' + (2\zeta K_n - RY_1)\eta' - RY_2\eta^2\eta' + (K_n^2 - RJ_1)\eta - RJ_2\eta\cos 2K_s s = 0 \tag{6.1.50}$$

where

$$K_s = \omega_s D/U, \; K_n = \omega_n D/U, \; R = \rho D^2/2m.$$

A point of departure in this model that permits its fitting to experimental results is that the coefficients Y_i and J_i are not fixed, but (as in the classical theory of flutter) are to be considered functions of the reduced frequency K. Goswami [113] and Goswami, Scanlan and Jones [114], reporting experiments like those of Feng [11], performed coefficient evaluation for this model.

In their experiments, done in air, the cylinder oscillation frequency ω_0 proved to coincide practically with the mechanical frequency ω_n. Hence the coefficient J_1 was effectively null. Figure 6.5 portrays, for the light damping ratio $\zeta = 0.15\%$, values of the other three coefficients Y_1, Y_2, J_2 as determined from experiments on aluminum cylinder in an air flow for which the Reynolds number was about 6000. The mass ratio parameter was $R \cong 6.3 \times 10^{-4}$.

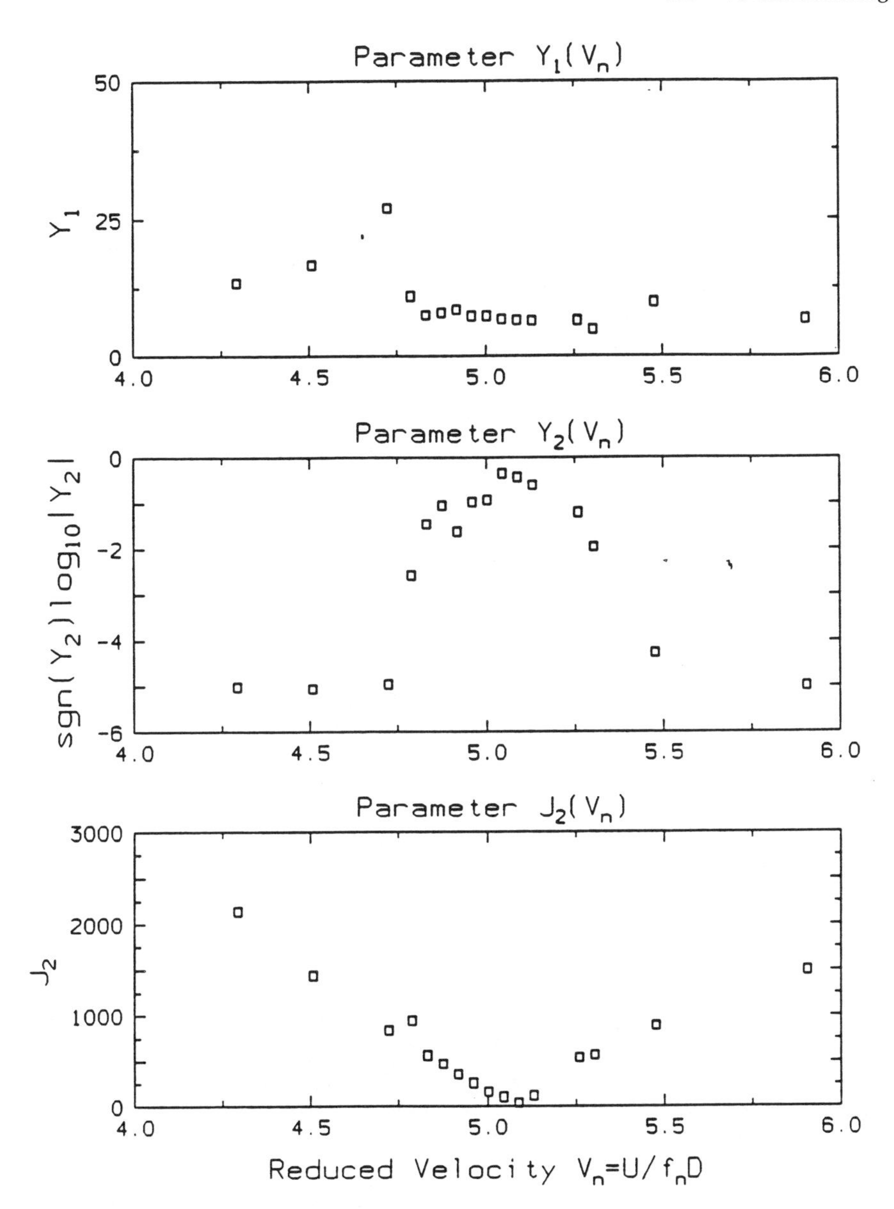

Figure 6.5 Model parameters vs. reduced velocity U/f_nD; $\zeta = 0.15\%$.

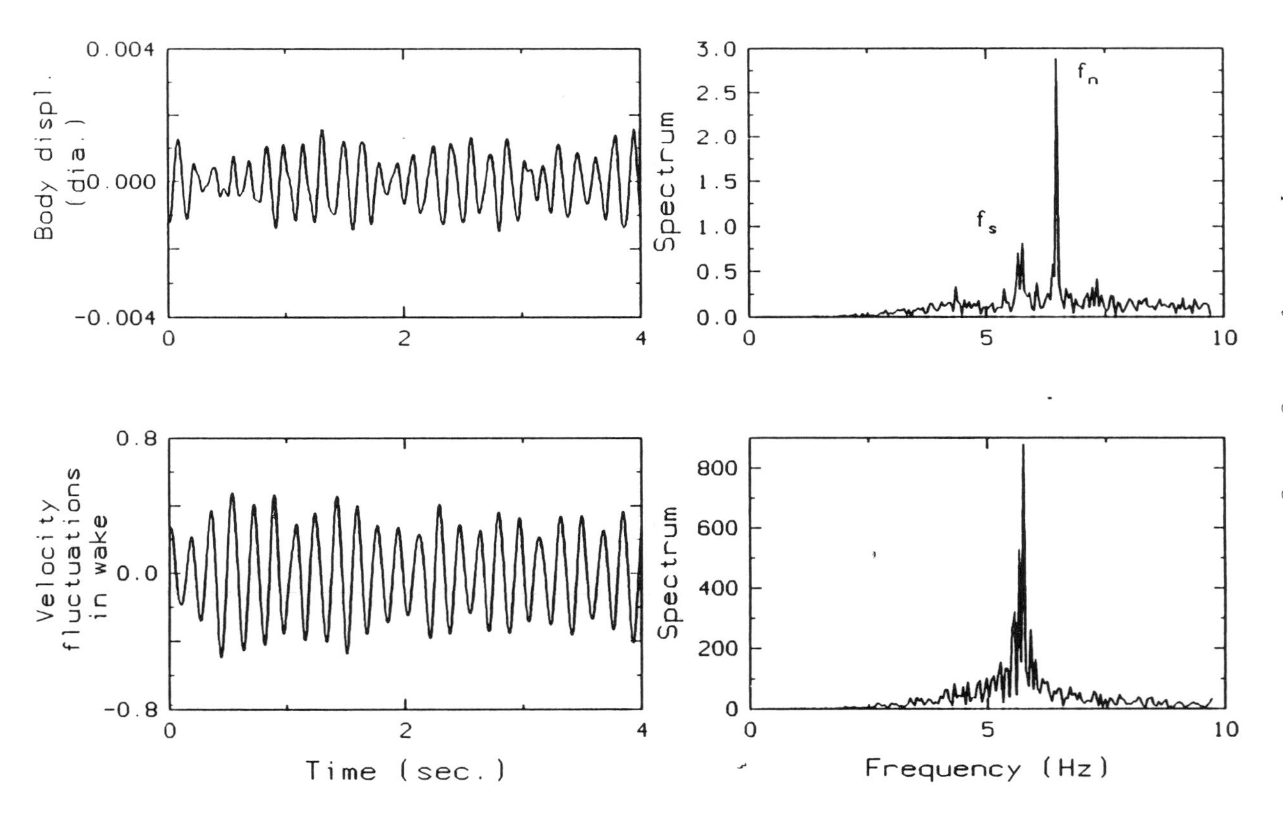

Figure 6.6. Response: $U/f_nD = 4.294$; $\zeta = 0.15\%$.

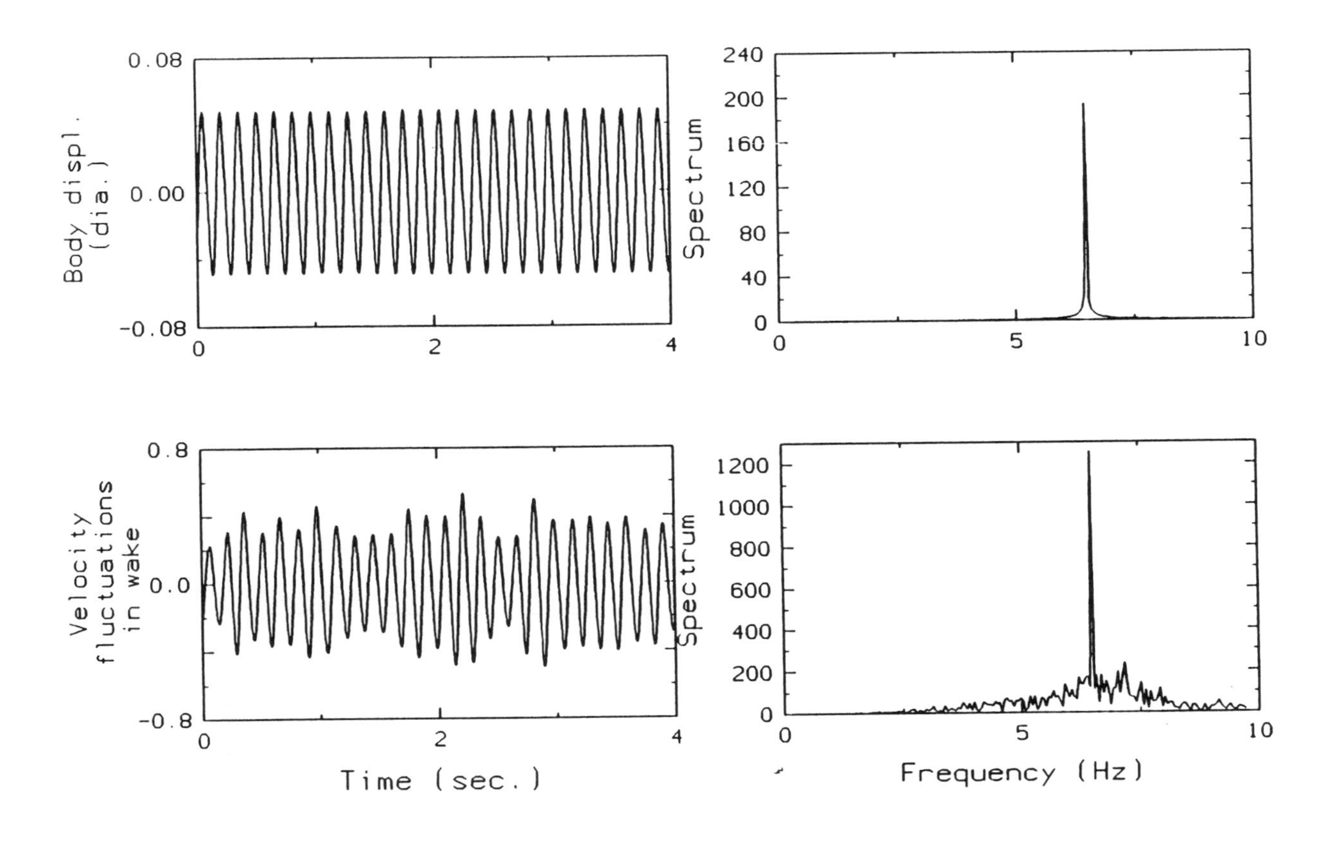

Figure 6.7. *Response:* $U/f_nD = 5.003$; $\zeta = 0.15\%$.

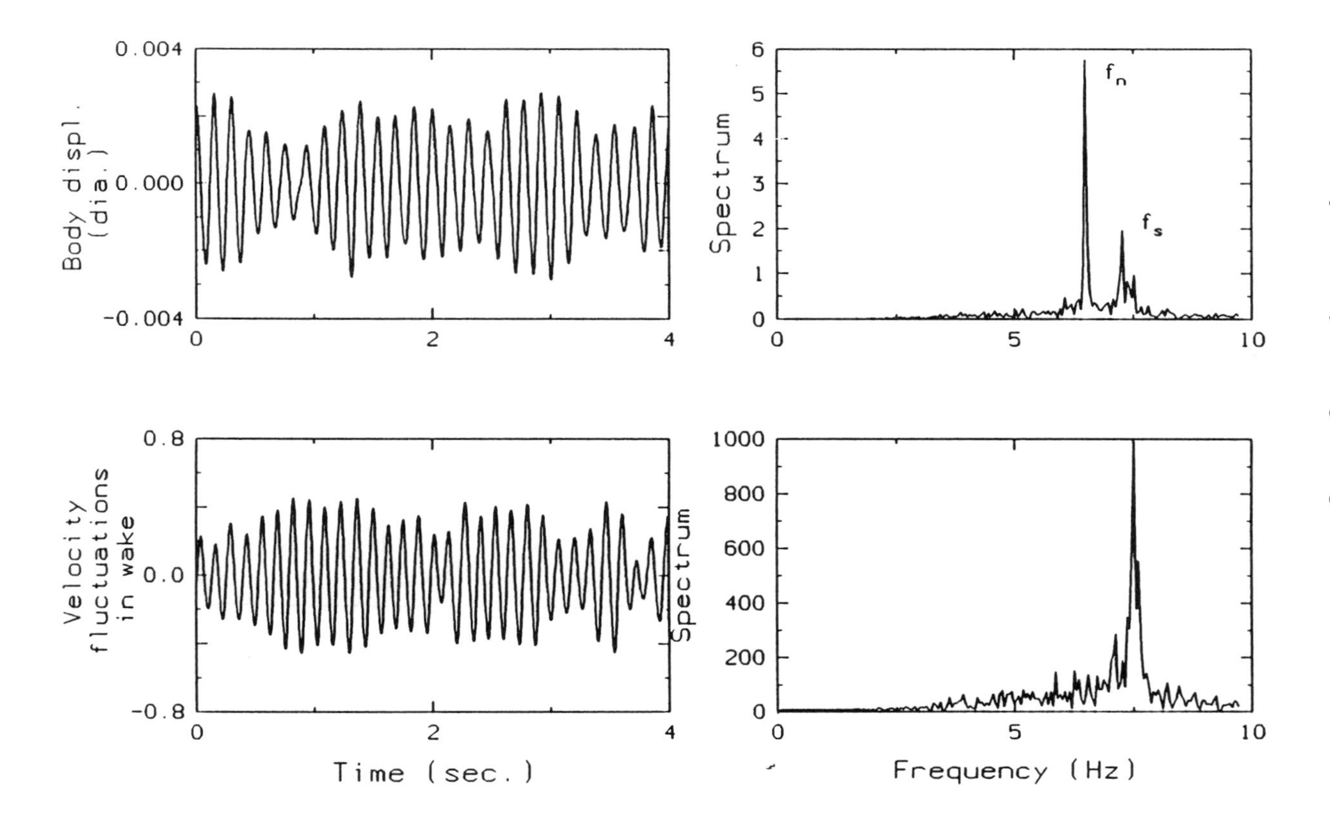

Figure 6.8. *Response:* $U/f_nD = 5.475$; $\zeta = 0.15\%$.

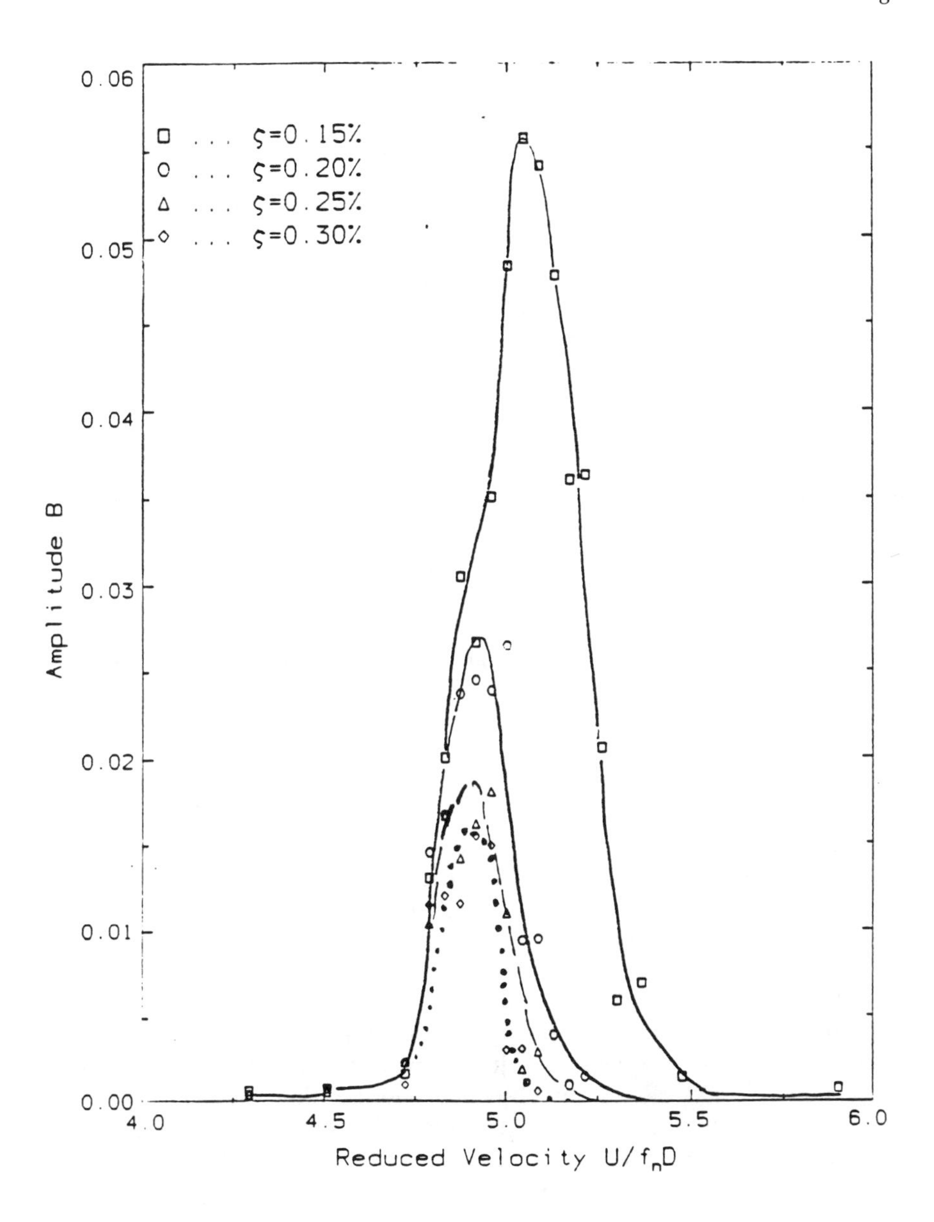

Figure 6.9 Amplitude B vs. reduced velocity $U/f_n D$ for several values of ζ.

The method of coefficient evaluation that was employed will not be discussed in detail here. However, it first was demonstrated that an approximate solution of the form:

$$\eta = A\cos(K_s s - \phi) + B\cos(K_0 s - \psi) \qquad (6.1.51)$$

(cf. eq. (6.1.9) also) can well match experimental results and be compatible with eq. (6.1.50): where A, B, ϕ, ψ are slowly varying parameters and K_0 is the reduced velocity based upon the observed mechanical response frequency ω_0.

Figures 6.6–6.8 demonstrate the experimental time history and power spectral character of the body and near-wake fluid responses for three successive values of reduced velocity U/f_nD, namely $U/f_nD = 4.294, 5.003, 5.475$; i.e. prior to, at, and subsequent to lock-in. In these figures, fluid velocity in the wake was sensed by a hot wire located 2.5 diameters downstream and 1.0 diameter above the cylinder axis. A plot of amplitude B (the principal contribution) for different damping values is given in Figure 6.9. Finally, maximum cylinder amplitude is plotted vs. Scruton number $S_{Sc} = \zeta m/\rho D^2$ in Figure 6.10 and is seen to correspond to a curve of the same character given by Griffin et al [115]. The net model thus achieved has the advantage of relative simplicity, with the possibility of being matched to experiment and thus having utility in engineering application.

Commentary on vortex excitation models

As has been stated, all of the models discussed above are empirical in nature, regardless of the specific dependence of each upon reasoning relative to the case considered. All therefore are open to criticism regarding the degree to which they are ultimately adopted to, and reflect, the true physical situation.

Again, the role that the model is required to serve is an important consideration in evaluating its success or usefulness. The continued dependence of all models on experiment to a greater or lesser degree is not to be considered a negative factor, just as the basic need to measure force coefficients C_L and C_D for any objects exposed to flow is an inescapable practical necessity. Moreover, the potential exists, through experiment, to carry these response models over to may other than circular bluff bodies (cf. [116]). Notable examples are various structural forms exposed to wind or water flows. Here the engineering procedure is first to make a test upon a physical model and then to extrapolate its results to the field, most particularly

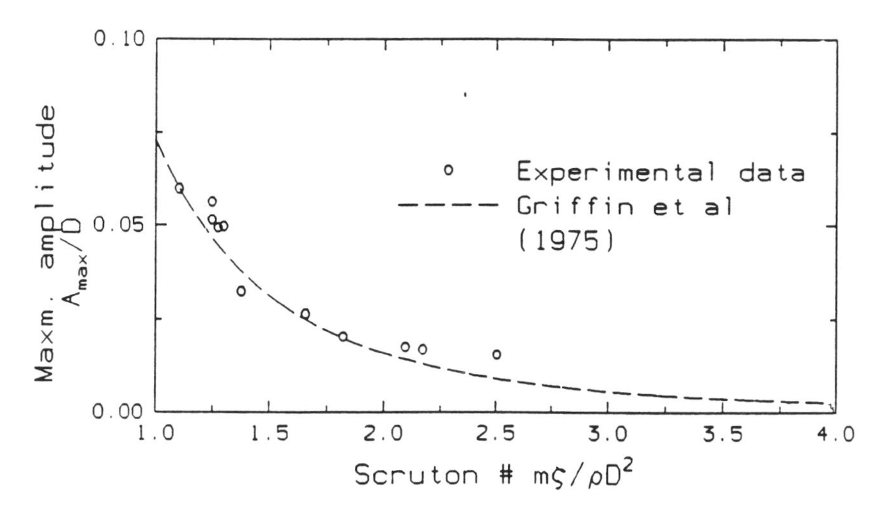

Figure 6.10 Maximum amplitude vs. Scruton number.

at the maximum amplitude accompanying lock-in, which is usually the most critical for engineering design purposes.

The action of real structures under vortex shedding is widely observed. Tall unsupported poles, stacks and towers may oscillate across-wind; cables and lines of a great variety exhibit Aeolian vibrations; certain long, unsupported structural members will vibrate in the wind; and the decks of cable-stayed and suspension bridges may undulate under cross-winds of relatively low velocity. The means of prevention or alleviation are several, but basically they are of two types: mechanical or aerodynamic.

Mechanical approaches have included stiffening, or raising the natural frequency, of a structure, increasing the damping, or installing a device such as a tuned-mass damper. Aerodynamic approaches have sought to spoil the coherence of the flow in order to break up the vortex shedding. As examples, Figures 6.11–6.16 illustrate a tuned-mass damper for suppressing the vortex-induced oscillation of a long structural member (bridge deck hanger), the 'Stockbridge' tuned-mass damper for suppressing the effects, near a support point, of the Aeolian vibration of a power line cable, and several aerodynamic spoiler devices. Refs. [27–32] discuss some of the latter. Figure 6.17a depicts the results of wind-induced vibration in model tests of various fairings in reducing the vortex-induced response of a cable-stayed bridge deck which was in close proximity to a

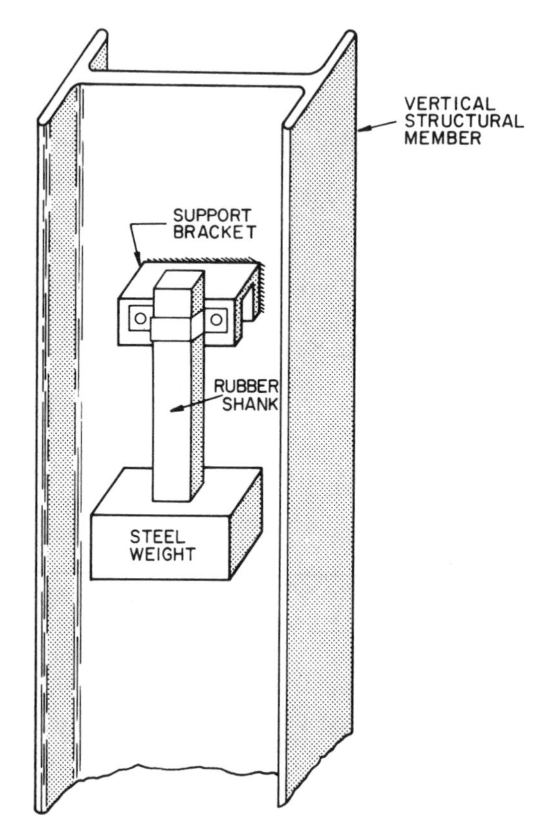

Figure 6.11 Schematic form of tuned-mass damper to suppress vortex-induced oscillations of vertical structural member.

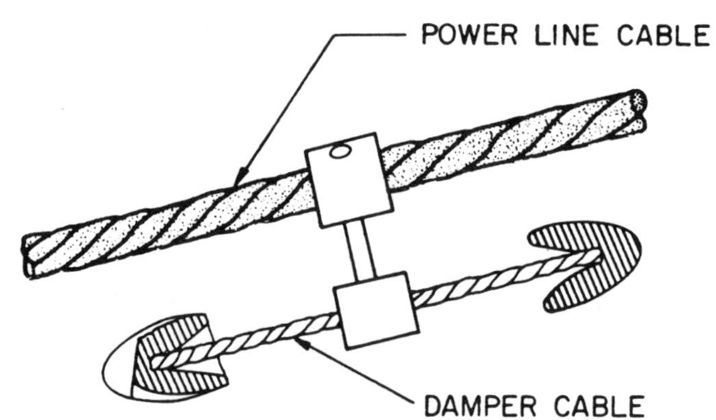

Figure 6.12 Diagrammatic form of the Stockbridge-type power line damper [27].

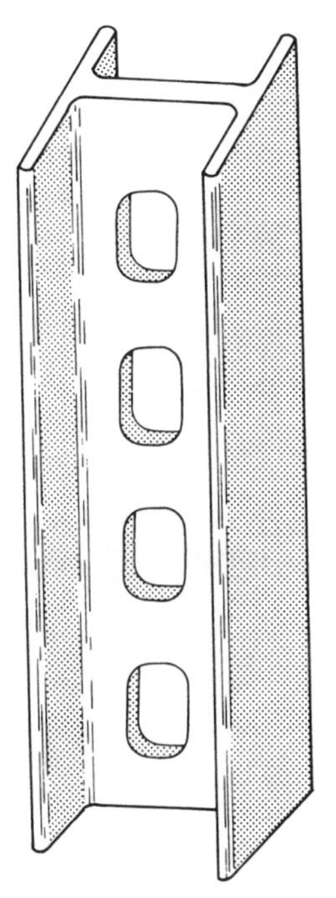

Figure 6.13. Structural member with web holes.

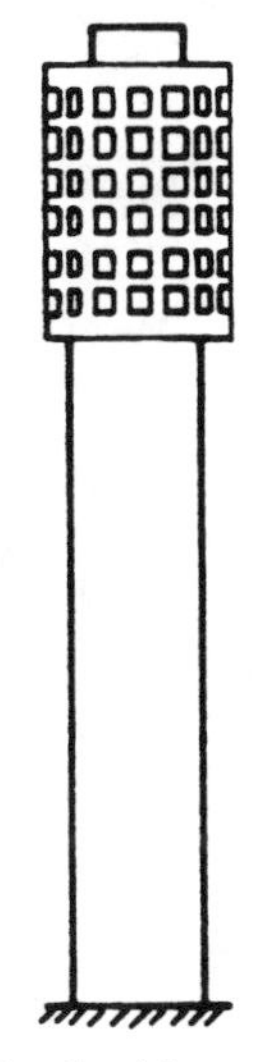

Figure 6.14 Stack with porous shroud [30].

Figure 6.15 Pipeline with fins [100].

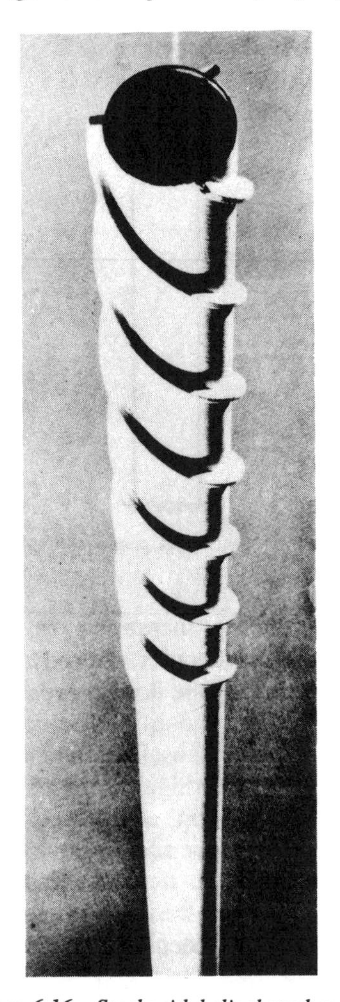

Figure 6.16 Stack with helical strakes [30].

water surface. Figure 6.17b shows the final choice of fairing on this bridge as installed in the field.

Because the literature on vortex shedding is vast, it is inevitable that a number of insightful and valuable references were not cited earlier in this chapter. To compensate in some measure for this lack, Refs. [33] through [68] are included with the 'References for Chapter 6'.

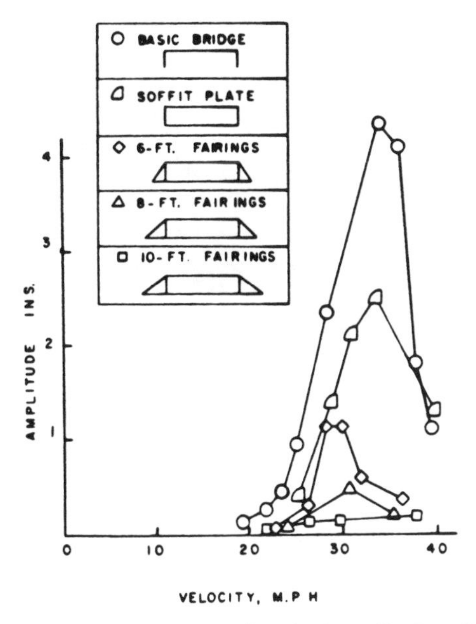

Figure 6.17a Wind tunnel measurements of vertical amplitudes of Long's Creek bridge (bridge height 15 ft.) [27].

6.2 Galloping

This descriptive term reflects the effort to characterize a specific type of large-amplitude oscillation first encountered under practical circumstances, i.e. the mainly vertical-plane vibration of long, ice-laden cable spans caused by cross winds.

Flow about a circular cylinder with perfect symmetry will not trigger galloping, a general prerequisite for which is some asymmetry in the aerodynamic forces associated with the cross flow. Such asymmetry can be caused by the unsymmetrical ice that clings to a cable under certain weather conditions.

At a later historical point than that of the recognition of iced

Figure 6.17b Modified Long's Creek bridge [27].

conductor galloping, another kind of cable instability, caused by asymmetry in the flow itself, was observed, This was wake-initiated galloping, wherein one (leeward) cable lay in the sheared flow near the edge of a wake region behind an upstream (windward) cable. This type of oscillation occurred fairly frequently in certain 'bundled conductor' configurations of high-voltage power lines consisting of groups of several parallel cables.

While vortex shedding proceeds more or less at the Strouhal rhythm during both types of galloping, it happens that the associated relatively rapidly-oscillating fluid forces vary about 'mean' values that are much more slowly varying. Under these circumstances the mean, or quasi-steady, lift and drag values are the prime movers of the body in question. It may then prove possible to create a sufficiently accurate, predictive theory of the main response of the body by incorporating the use of such quasi-steady values. This, in fact, proves to be the case for both of the galloping phenomena mentioned above.

It becomes reasonable and convenient then to define *galloping* more generally as a kind of oscillation undergone by an elastically sprung body, the frequency of whose main motion is low relative to its natural vortex-shedding frequency, and for which a satisfactorily descriptive analytical model can be constructed based on steady-state aerodynamic derivatives.

Across-wind galloping

As noted above, one of the early manifestations of large-amplitude, across-wind oscillation was the galloping of ice-coated power lines under wind. Often a full catenary span would exhibit such a vibration in its fundamental mode, with center amplitudes approximating the cable sag. This phenomenon was examined by Den Hartog [69]. A body having an undistorted circular cross-section cannot gallop but various cross-sectional shapes other than purely circular are gallop-prone. For example, the net configuration in which a fairly heavy cable is suspended from a smaller diameter, strapped-on supporting cable has been observed to gallop under strong, steady winds in the field. Some large-diameter power cables made up of wire strands had a configuration of wire lay that was conducive to galloping [70]. Square and D-section rods (the latter with flat face to the wind) have been demonstrated by tests to be gallop-prone.

The phenomenon will be examined here in terms of the time-averaged or steady aerodynamic properties of the cross section of the galloping object. Consider the cross section shown in Figure 6.18. Let $C_D(\alpha)$ and $C_L(\alpha)$ be the experimentally determined steady drag and lift coefficients, respectively, of the section given as functions of the geometric angle of attack α of the wind.

The effective, or relative, angle of attack of the wind will depend upon the attitude of the fixed body or upon the across-wind velocity of the moving body. If its across-wind displacement is denoted as y (positive downward) the velocity in question is $\dot{y}$. The angle of attack of the relative wind to the body moving across-wind will then be

$$\alpha = \tan^{-1} \frac{\dot{y}}{U} \tag{6.2.1}$$

and the relative wind velocity U_r will have the value

$$U_r = [U^2 + \dot{y}^2]^{1/2} \tag{6.2.2}$$

that is

$$U_r = U \sec \alpha \tag{6.2.3}$$

The average drag force on the section will be

$$D = \tfrac{1}{2}\rho U_r^2 B C_D(\alpha) \tag{6.2.4}$$

and the corresponding lift will be

$$L = \tfrac{1}{2}\rho U_r^2 B C_L(\alpha) \tag{6.2.5}$$

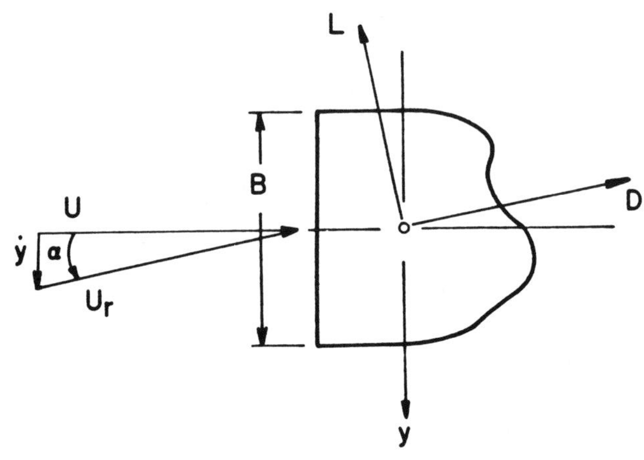

Figure 6.18 Wind and motion components, with resultant lift and drag, on a bluff cross section.

The projection of these components in the across-wind or y direction gives the across-wind force

$$F_y(\alpha) = -D(\alpha)\sin\alpha - L(\alpha)\cos\alpha \tag{6.2.6}$$

$F_y(\alpha)$ may alternately be expressed in the form

$$F_y(\alpha) = \tfrac{1}{2}\rho U^2 B C_{F_y}(\alpha) \tag{6.2.7}$$

where

$$C_{F_y}(\alpha) = -[C_L(\alpha) + C_D(\alpha)\tan\alpha]\sec\alpha \tag{6.2.8}$$

If the section of the galloping body is assumed to be elastically supported so that it can vibrate across wind, the equation of motion of this section can be written

$$m[\ddot{y} + 2\zeta\omega_1\dot{y} + \omega_1^2 y] = F_y \tag{6.2.9}$$

where m is the mass per unit span, ζ is the damping ratio-to-critical of the mechanical system ($\zeta = \delta/2\pi$, where δ is the logarithmic decrement of the system) and ω_1 is the undamped natural circular frequency of the system.

A basic criterion for galloping arises if the situation of incipient motion or initial, small departure from rest is considered; that is, the condition of very small angle of attack α. Then

$$F_y \cong \left.\frac{\partial Fy}{\partial\alpha}\right|_{\alpha=0} \alpha \tag{6.2.10}$$

This leads to examination of $\mathrm{d}C_{F_y}/\mathrm{d}\alpha$, which from (6.2.8) is found to have the following value at $\alpha = 0$:

$$\left.\frac{\mathrm{d}C_{F_y}}{\mathrm{d}\alpha}\right|_{\alpha=0} = -\left(\frac{\mathrm{d}C_L}{\mathrm{d}\alpha} + C_D\right) \tag{6.2.11}$$

Thus, for small motion, the galloping is governed by:

$$m[\ddot{y} + 2\zeta\omega_1\dot{y} + \omega_1^2 y] = -\tfrac{1}{2}\rho U^2 B\left[\frac{\mathrm{d}C_L}{\mathrm{d}\alpha} + C_D\right]\frac{\dot{y}}{U} \tag{6.2.12}$$

The system is then simply a linear oscillator with damping term ($\dot{y}$ term) having the coefficient

$$d = 2m\zeta\omega_1 + \tfrac{1}{2}\rho UB\left[\frac{\mathrm{d}C_L}{\mathrm{d}\alpha} + C_D\right] \tag{6.2.13}$$

According to linear theory the system is then stable if $d > 0$ and marginally or totally unstable if $d \leq 0$.

Since the mechanical damping ζ is always positive, the condition

$$\frac{\mathrm{d}C_L}{\mathrm{d}\alpha} + C_D < 0 \tag{6.2.14}$$

is seen to be necessary for galloping. This is known as the Den Hartog criterion [71]. In many cases examination of this criterion alone suffices to assess the galloping proclivities of a structure. For example, Figure 6.19 [5] illustrates the properties of a section of an octagonal lamp standard that exhibited susceptibility to galloping at certain azimuth angles of the wind because $\mathrm{d}C_L/\mathrm{d}\alpha$ was strongly negative at those angles.

A more complete investigation of the galloping problem reveals, however, that it is in fact strongly nonlinear. Parkinson *et al.* [72–77], Novak [78, 79] and others have investigated these aspects of galloping.

The force F_y, being a function of α as given by (6.2.7), (6.2.8), can, in virtue of (6.2.1), be expressed as a polynomial in $\dot{y}$. This is done, for example, by Novak [78] who expressed F_y in the form:

$$F_y = A_1\left(\frac{\dot{y}}{U}\right) - A_2\left(\frac{\dot{y}}{U}\right)^2\frac{\dot{y}}{|y|} - A_3\left(\frac{\dot{y}}{U}\right)^3 + A_5\left(\frac{\dot{y}}{U}\right)^5 - A_7\left(\frac{\dot{y}}{U}\right)^7 \tag{6.2.15}$$

where the A_i ($i = 1, 2, 3, 5, 7$) are determined by any desired curve-fitting procedure.

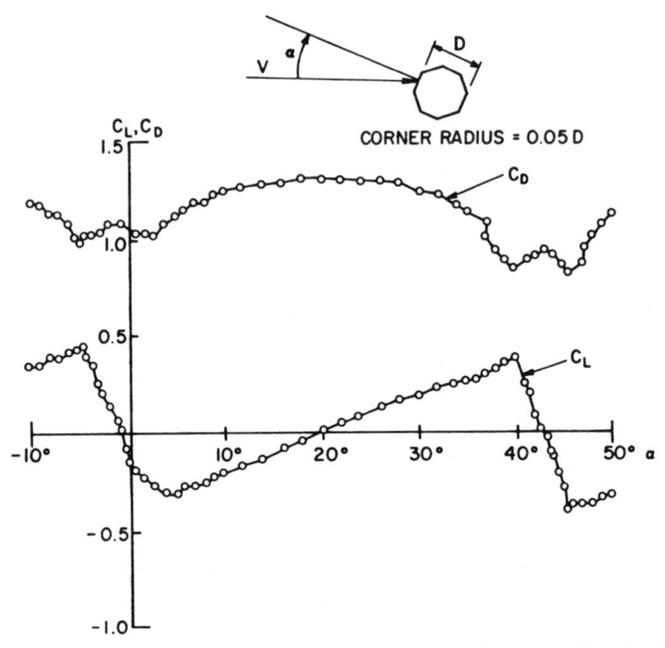

Figure 6.19 Lift and drag coefficients on an octagonal structural section [27].

Next, since the galloping equation (6.2.9) is now nonlinear, special solution methods must be applied. Ref. [78] employs the method of Kryloff and Bogoliuboff [80], postulating as a first approximation to the response the forms:

$$y = a\cos(\omega t + \varphi) \tag{6.2.16a}$$

$$\dot{y} = -a\omega\sin(\omega t + \varphi) \tag{6.2.16b}$$

Where a and φ are considered to be slowly varying functions of time.

Full details of the solution method will not be pursued here, but the theory adduced has proven effective in describing the phenomena observed. From analysis of the type suggested Ref. [79] identifies various type-forms of curves $C_F(\alpha)$ and the trends of the corresponding galloping response amplitude a as function of the reduced velocity parameter $U/B\omega_1$ (see Figure 6.20).

All of the above results were obtained assuming laminar incident flow, which often is adequate as a representation of actual conditions. However, the effects of small-scale turbulence upon galloping can, by affecting the average values of C_L and C_D in certain instances, modify the phenomenon. Finally, [79] notes that there exist situations, for example the condition of a large triggering disturbance of the galloping body, wherein galloping takes place although it can be demonstrated that Den Hartog's criterion (6.2.14) is not satisfied.

In certain cases of elongated sectional dimensions, a more complex form of galloping, involving both across-wind and torsional freedoms, can be demonstrated. Analysis of this borders on the methods of classical flutter analysis, but the necessary aerodynamic information can nonetheless be extracted by average steady-state methods making the analysis 'quasi-steady' in this case also.

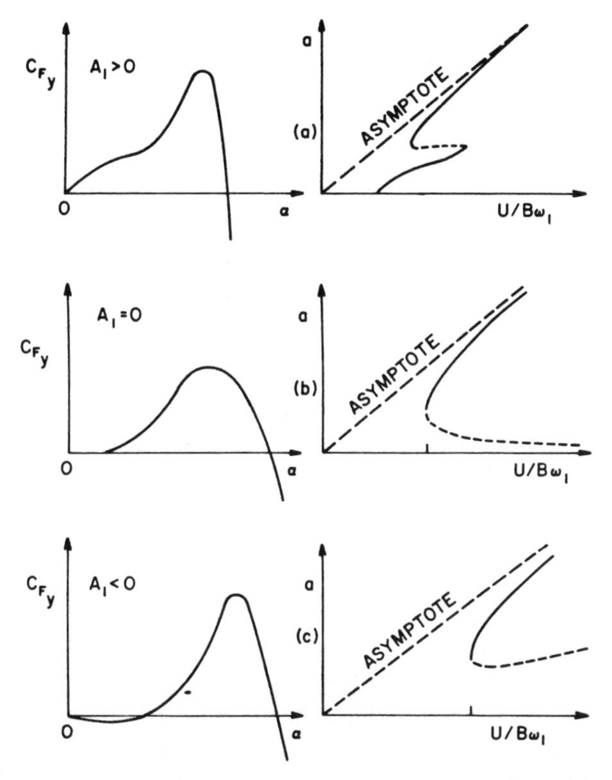

Figure 6.20 Types of across-wind force coefficients and corresponding trends of galloping amplitude [79].

In the case of galloping it is not usually necessary to extend the detailed analysis spanwise into the third dimension, though this has been done [81]. Since the phenomenon is usually unwanted, identification of sectional galloping proclivities and attempts to alleviate them are the main concerns. However, the galloping which occurs on electrical power lines has proven to be elusive and difficult to conquer. The primary reasons for this have been designed into the situation where long, generally unobserved cables are inevitably exposed to the vagaries of the weather. Some anti-galloping dampers, ice-warning and removal devices, the installation of connectors between adjacent cables, and other palliatives have been employed, each with modest success.

Wake galloping

As already noted, objects of circular section cannot develop any average lift in a uniform flow, but, if the flow region about them is sheared, that is, contains an across-flow velocity distribution, it is obvious that net pressure differentials over such objects can develop and lead to across-wind and along-wind forces. In the present section the wake galloping phenomenon, which is called 'subspan galloping' in connection with overhead power lines, will be described as a typical instability resulting from sheared flow.

The phenomenon occurs most prominently with bundled-conductor power lines. These are lines in which, instead of a single cable carrying one phase of the current, a group or 'bundle', consisting of two or more parallel cables, is employed. When parallel cables are so used they are held in the 'bundle' configuration by devices called spacers, which are placed perpendicular to the cables periodically along the span between support towers. The spacers divide a cable-bundle span between towers into subspans.

In the situation described, when wind blows across the span of the cable bundle it frequently occurs that one cable of the bundle is located in the aerodynamic wake of another cable. If it happens to lie in certain regions of that wake, particularly where strongly sheared flow is occurring, the downstream cable may exhibit the unstable oscillatory tendency which is called wake galloping or, in power line parlance, 'subspan oscillation'.

Figure 6.21 depicts the geometry of the situation, where the wake of the windward cylinder (cable) lies between the heavy lines. The quasi-steady aerodynamics of the phenomenon have been explored by Cooper [82, 83], who examined the forces of lift and drag acting on a downstream cylinder in various positions. The results observed are depicted qualitatively in Figure 6.22. Aside from expected velocity and drag defects, as

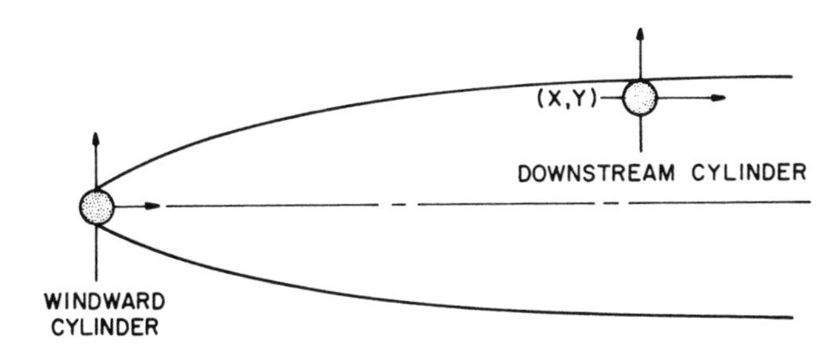

Figure 6.21. Geometry for the wake-galloping phenomenon.

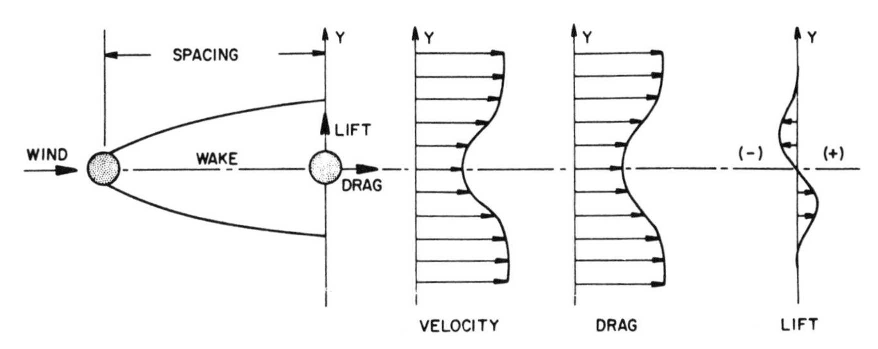

Figure 6.22 Wake properties behind a cylinder [82].

indicated, a striking centering tendency of the lift on the downstream cylinder is to be noted, that is, lift is directed downward when the cylinder is above the wake centerline and upward when it is below that line. The physical reasons for this may stem from the averaging, on the cylinder in the wake, of the inward-directed effects of the alternating vortex flow leaving the windward cylinder.

In any event, when the leeward cylinder is elastically sprung in X- and Y-directions it exhibits oscillatory tendencies when located in certain regions of the wake. Figure 6.23 is an oscillograph trace of the oscillation orbit of a downstream cylinder in this kind of situation.

Figure 6.24 depicts the geometry of the situation for analysis purposes. Equations of motion of the downstream cylinder can be written in the form

$$m\ddot{x} + d_x\dot{x} + k_{xx}x + k_{xy}y = F_x \tag{6.2.17a}$$

$$m\ddot{y} + d_y\dot{y} + k_{yx}x + k_{yy}y = F_y \tag{6.2.17b}$$

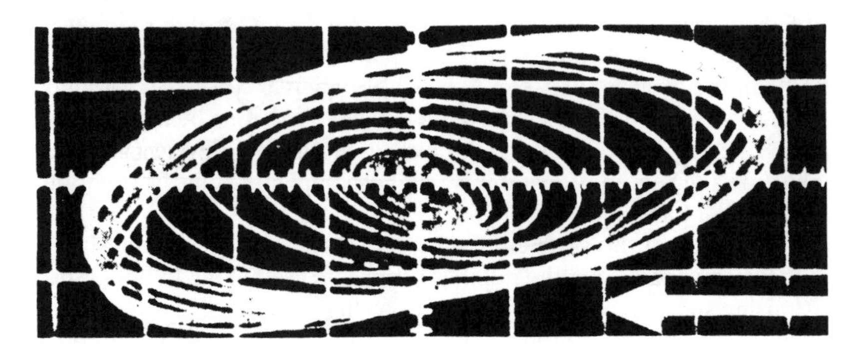

Figure 6.23 *Oscillograph trace of cylinder wake galloping* [83].

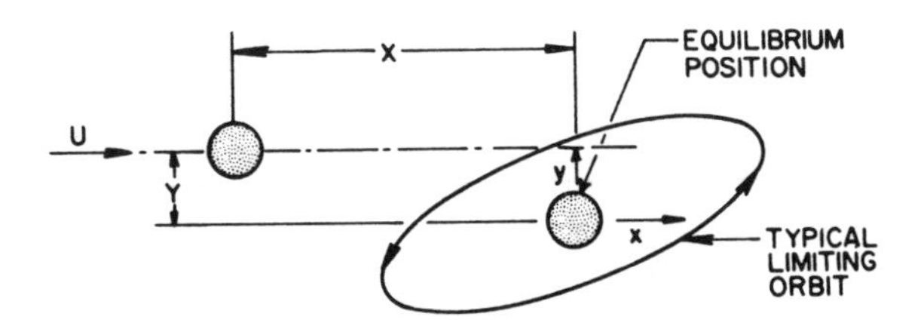

Figure 6.24 *Coordinates for the wake galloping problem* [82].

where m is its mass per unit span, d_x and d_y are damping constants, k_{rs} (r, $s = x$, y) are direct and cross-coupling mechanical spring constants, and F_x, F_y are aerodynamic forces. These are given, from theory and from experimental wake explorations of the type suggested above, in the form [84]:

$$F_x = \tfrac{1}{2}\rho U^2 D\left\{\frac{\partial C_x}{\partial x}x + \frac{\partial C_x}{dy}y + C_y\frac{\dot{y}}{U_w} - 2C_x\frac{\dot{x}}{U_w}\right\} \tag{6.2.18a}$$

$$F_y = \tfrac{1}{2}\rho U^2 D\left\{\frac{\partial C_y}{\partial x}x + \frac{\partial C_y}{\partial y}y - C_y\frac{\dot{y}}{U_w} - 2C_y\frac{\dot{x}}{U_w}\right\} \tag{6.2.18b}$$

where $\frac{1}{2}\rho U^2$ is the free stream dynamic pressure, D is the diameter of the wake cylinder, U_w is the wake velocity at point (x, y) in the wake, and C_x, C_y are aerodynamic drag (x) and lift (y) coefficients referred to free-stream velocity. Equations (6.2.18) appear to have been derived originally by Simpson [85].

Solution of the equations of motion is not unlike that of the flutter problems, with the exception that experimental steady-state data, in real form, suffice for the formulation. The equations being linear, solutions for x and y of the form $e^{\lambda t}$ are seen to be stable when $\lambda = \omega_1 + i\omega_2$ is found to have $\omega_1 < 0$. For $\omega_1 = 0$, boundaries in the wake can then be defined which identify the regions of wake galloping susceptibility. Figure 6.19 which depicts a wake region of instability, suggests typical results [82] obtained experimentally and theoretically.

Alleviation of this problem has been experimentally accomplished for power lines by increasing the number of spacers used along bundled spans, thus raising subspan natural frequencies to higher values; and the twisting of the entire bundle throughout the span, destroying the necessary aerodynamic wake coherence for the phenomenon to take place. In practice, however; the isolated occurrence of subspan oscillation has been difficult to detect and arrest, mainly for several economic reasons

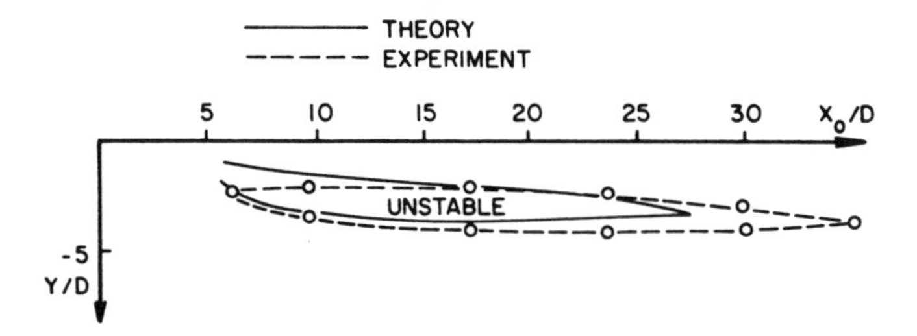

Figure 6.25 Wake galloping: theoretically and experimentally derived definitions of the wake region of instability [82].

attendant on operation of long power line systems. Considerable effort has been directed along the lines of various spacer designs (as dampers, line detuners, etc.) with only modest success.

6.3 Divergence

This is a non-oscillatory phenomenon that, like galloping, can be analytically described through the use of quasi-steady aerodynamic derivatives. It has already been discussed in Section 2.1–2.3, relative to airfoils and lifting surfaces. It is, in general, a rather high-velocity phenomenon and so occurs relatively infrequently in civil structures at the normal speeds of the natural wind. These, except in tornadoes, rarely exceed 125 mph

(1 mph = 0.447 m/sec). Hence the aerodynamics involved remains in the incompressible flow range. Only relatively weak structures risk divergence, but it should nonetheless be considered for all structures of importance.

The phenomenon can be illustrated in the case of a long, torsionally flexible suspension bridge. In terms of a single representative section of the bridge the action of torsional divergence is completely analogous to that described in Section 2.1–2.3 for airfoils and lifting surfaces. Consider the bridge section shown in Figure 6.20.

There, horizontal wind at velocity U is blowing against the structural section, the angle of twist of which is α. The section is restrained by a structural spring of elastic characteristic K_α. If increased twist α implies increased moment M_α acting on the section, there will be, for some high wind velocity U, sufficient torsional moment on the section to overcome the structural restraint, and the section will rotate to destruction (diverge).

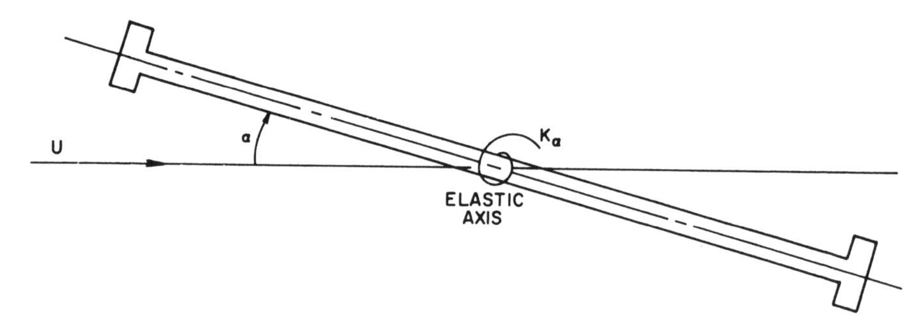

Figure 6.26. Cross section of bridge deck, with torsional restraint.

The phenomenon is not unlike that of structural buckling, and its details are very similar to those already given in Sections 2.1–2.3.

Consider now, to be more explicit, a full bridge span for which the section conditions resemble these in Figure 6.20 but for which the deflection at the representative section is a function of loads both there and elsewhere. To this end let D_T be the matrix of torsional influence coefficients of the deck at various spanwise stations i. An element d_{ij} of D_T will represent the angle of attack at spanwise station i due to a unit torsional moment applied at spanwise station j.

Let C_M be the mean (i.e., time-averaged) aerodynamic moment coefficient about the rotation point of a representative section; then

$$M = \tfrac{1}{2}\rho U^2 B^2 C_M \qquad (6.3.1)$$

is the aerodynamic moment per unit span applied by the wind at the section.* B is a reference length and is often taken to be the full width (chord) of the bridge section.

If the bridge span L is conceptually divided into equal subspans of length ΔL, the moment at section i is

$$M_i = \tfrac{1}{2}\rho U^2 B^2 C_{M_i}\,\Delta L \qquad (6.3.2)$$

and the column of discrete torsional deformations of the bridge $\{\alpha_i\}$ is expressible as

$$\{\alpha_i\} = D_T\{M_i\} \qquad (6.3.3)$$

which will have a definite solution for any assigned wind speed U..

This solution will now be considered. As is well known, the coefficient C_M is found experimentally to be a function of α. See Figure 6.21 for an example of such a coefficient for the experimental bridge deck section shown [86]. Suppose, for example, that C_M is expressible in the form

$$C_M(\alpha) = c_0 + c_1\alpha + c_2\alpha^2 + \cdots + c_n\alpha^n \qquad (6.3.4)$$

that is, the C_M curve analogous to that in Figure 6.21 is fitted by a polynomial in α. Then, if $\frac{1}{2}\rho U^2$ is written as q, M_i takes the form

$$M_i = qB^2\,\Delta L[c_0 + c_1\alpha_i + \cdots + c_n\alpha_i^n] \qquad (6.3.5)$$

and (6.1.3) becomes

$$\{\alpha_i\} = qB^2\,\Delta L D_T\{\{c_0\} + c_1\{\alpha_i\} + \cdots + c_n\{\alpha_i^n\}\} \qquad (6.3.6)$$

This problem may be solved iteratively for an assumed value of q. For example, the value of (α_i) is first estimated by taking $c_1 \cdots c_n$ equal to zero in (6.1.3); then the result (α_i) is employed on the right of (6.1.3) to recalculate (α_i). The process is repeated with the latest value of (α_i) until convergence occurs for a given q. With a new value of q assigned, the iteration scheme is then repeated.

The divergence problem concerns the determination of the highest value of q for which a solution $\{\alpha_i\}$ remains statically stable, that is, the limiting value of q for which convergence of the above-described process will occur. This depends, of course, upon the exact form of the function $C_M(\alpha)$; and equation (6.3.6.) must, in general, be solved approximately.

---

* Strip theory aerodynamics, previously discussed in Sections 2.2 and 3.4, will be employed here.

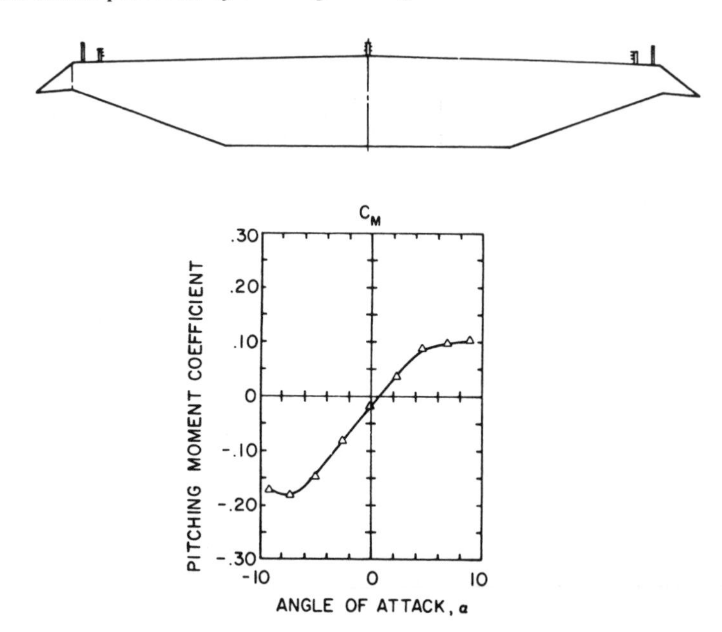

Figure 6.27 Modern bridge deck section and its torsional moment coefficient [86].

A simple form of the problem occurs if the expression (6.1.4) for $C_M(\alpha)$ is linear:

$$C_M(\alpha) = c_0 + c_1\alpha \tag{6.3.7}$$

where c_0 is the value of C_M for $\alpha = 0$ and

$$c_1 = \frac{dC_M}{d\alpha} = C'_m \tag{6.3.8}$$

Letting

$$\frac{1}{p} \equiv qB^2\,\Delta L \tag{6.3.9}$$

the problem reduces to the form

$$[pI - C'_M D_T]\{\alpha_i\} = D_T\{c_0\} \tag{6.3.10}$$

This has infinite (divergent) solutions for α_i when the determinant of the leading matrix vanishes:

$$|pI - C'_M D_T| = 0 \tag{6.3.11}$$

where I here is the identity matrix.

In the further special case that $c_0 = 0$ (moment coefficient is zero for $\alpha = 0$), the classic eigenvalue form for the problem is obtained:

$$[pI - C'_M D_T]\{\alpha_i\} = \{0\} \tag{6.3.12}$$

The largest eigenvalues p from either (6.3.11) or (6.3.12) permit the determination from (6.3.9) of the velocity U at which the divergent instability occurs.

Note that the necessary data, namely the evolution of C_M as a function of α (Figure 6.27) for this problem must be provided by wind tunnel test, for example on a geometrically similar model of a section of the bridge deck.

Many variants on the procedures described above can be devised, but that given serves to illustrate the character of the problem. The phenomenon can, in principle, occur for actual suspension bridge decks, though usually the calculated velocity for divergence of a normally designed bridge is considerably higher than expected wind speeds. Clearly, the most effective means of prevention of torsional divergence—or assuring that it will not occur until very high velocities are reached—are (1) a section moment coefficient with low slope and/or low overall value, and (2) high torsional stiffness of the structure.

6.4 Flutter and buffeting

Basic concepts

Of the several civil structure types exhibiting severe wind-induced oscillations, the suspension bridge has perhaps been the most notable. This is the result of the prominent exposure of bridges to wind and the fact that suspension bridges are quite flexible, with generally low natural frequencies. On several occasions bridges of this type have suffered serious damage or even complete destruction under wind. While the Tacoma Narrows disaster of 1940 occasioned the modern experimental and analytical approaches to problems of bridge aeroelasticity, that occurrence was in fact one of a long line of similar events stretching back 100 years or more into history.

The destructive oscillations of these events are now generally classified as types of flutter, although many appear to be single-degee-of-freedom instabilities in torsion* rather than classical flutter. However, coupled bending-torsion oscillations can and do occur.

Suspension bridge decks fall broadly into two classifications: those with roadways stiffened by open, latticed trusswork, and those—usually with single deck and of more recent design—having a roadway atop a closed, box-like stiffening structure. Figures 6.28 and 6.29 show examples of the two types. While both types are susceptible to flutter, the more streamlined box type will tend to have a higher critical flutter speed and may enter into coupled flutter, whereas the truss-stiffened type is very likely to exhibit single-degree torsional flutter. Very bluff deck sections tend strongly to shed vortices as well as enter single-degree flutter. The

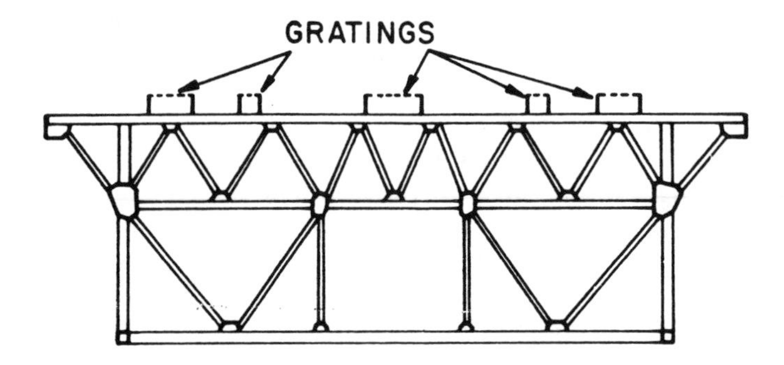

Figure 6.28 Typical cross section of truss-stiffened suspension bridge deck [101].

girder-stiffened original Tacoma Narrows bridge possessed a squat H-like section which was extremely unstable aerodynamically, both in its vortex-shedding and its flutter tendencies.

The collapse of that bridge is often said to be attributable to vortex shedding. In the sense in which this term and the term *flutter* are presently used, some clarification is perhaps in order. In the usual sense of *natural* vortex shedding, a certain rhythm of alternate vortex formation occurs as a result of a fixed, rigidly-supported body being traversed by a smooth flow, with the vortices trailing into its wake. In this case the rhythm is a function only of the geometry of the body and the Reynolds number of the flow. On the other hand, while a fluttering body also produces vortices in its wake, in this case one or more natural structural frequencies influence the rhythm thereof. In fact the strongest of the shed vortices are the direct

* These appear to be a type of flutter, akin to stall flutter, associated with separated flow during some portion of the oscillation cycle. See Chapter 5.

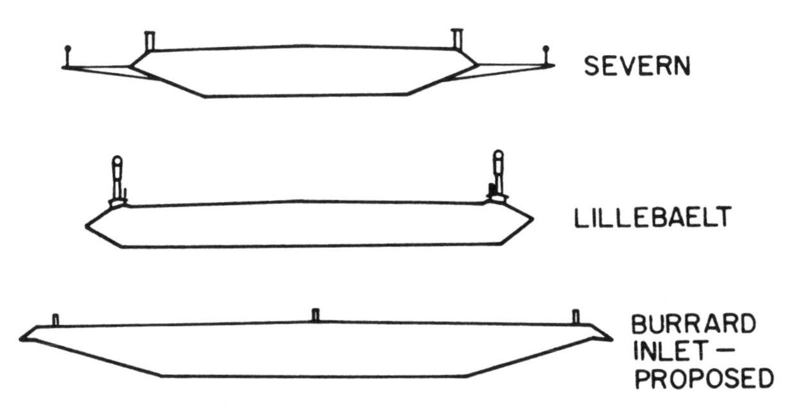

Figure 6.29. Some streamlined deck section forms.

result of that rhythm, which reflects the changing motion (and therefore lift) of the structure itself. Wind speeds at which natural flow-induced vortex-shedding tends to lock in on a structural frequency are usually much lower than flutter speeds; furthermore, increases in wind speeds tend to break the locked-in condition of natural vortex-shedding and leave the structure without much response. However, flutter, once engaged, generally tends to grow more violent as wind speed increases. Thus the Tacoma Narrows flutter, while indeed accompanied by a shed vortex wake, had little or nothing to do with a natural vortex rhythm intrinsic merely to body geometry and the flow alone, but was a type of basically single-degree separated-flow flutter in a low torsion mode of that structure. The main features of the 1940 Tacoma Narrows incident have recently been reviewed by Billah and Scanlan [117].

In recent years a newer type of suspended-span bridge has come into prominence for spans that are usually considerably shorter than those of the greater suspension bridges. This is the so-called cable-stayed bridge, in which the deck is supported not by a catenary cable but by a fan of smaller-diameter cables reaching from the bridge towers to points spaced along the deck. While tending generally to be stiffer structurally than the classical suspension bridge, bridges of this type still require aeroelastic study in their design stages. The structural vibration modes of these bridge types tend in general to be more complex in form than those of catenary suspension bridges.

The study of bridge flutter owes much to the literature on airfoil flutter. The two prove to be enough different in character, however, so that the theoretical aerodynamic formulations for airfoil flutter are not directly transferable to bridges, as will be seen. It is of historical interest that an attempt was made by Bleich [87] to apply airfoil flutter theory to bridges, and that this theory was also used by Selberg [88] as a basis of comparison among bridge stability results. Numerous subsequent studies [89–92] have, however, proven beyond question the dissimilarity between airfoil and bridge deck flutter mechanisms, unless the bridge deck is unusually streamlined.

It has been common practice to study the flutter susceptibility of a suspended-span bridge deck by creating a reduced-scale, geometrically faithful model of a typical section of the deck and suspending it from springs for wind tunnel testing. In view of the fact that this basic device has proved so useful over nearly a half century since its first application in connection with the Tacoma Narrows episode, it serves as an appropriate starting point for the present discussion.

It may be mentioned here however that interpretation of the results from such models has undergone some evolution since their introduction. The earliest conception of the section flutter model was that it effectively duplicated the full bridge. Such models were initially scaled both geometrically and inertially to duplicate the prototype bridge section, and were equipped with spring supports duplicating certain selected vertical and torsional frequencies of the prototype bridge considered to be the lowest of such frequencies susceptible of entering full-bridge flutter. A section model – or any model – with limited dynamical properties and degrees of freedom clearly cannot, because of several shortcomings, duplicate full-bridge action.

The viewpoint on section models has been modified in the more recent literature [93], from which the position emerges that such models are best suited as analog computational devices to provide fundamental aerodynamic stability information. Among the most useful items sought in this connection are the static force derivatives and the motion-related flutter derivatives typical of the particular deck geometry involved. Exploitation of a two-degree-of-freedom deck section model in this sense will first be described.

Let h and α be the vertical and torsional degrees of freedom of a bridge deck section model, referred to its elastic axis, as illustrated in Figure 6.30. (Provisionally, the sway deflection p will not be considered.) The equations of motion of the bridge deck section are mechanically the same as for the typical airfoil section (recall Section 3.2) namely

$$m\ddot{h} + S_\alpha\ddot{\alpha} + c_h\dot{h} + K_h h = L_h \tag{6.4.1a}$$

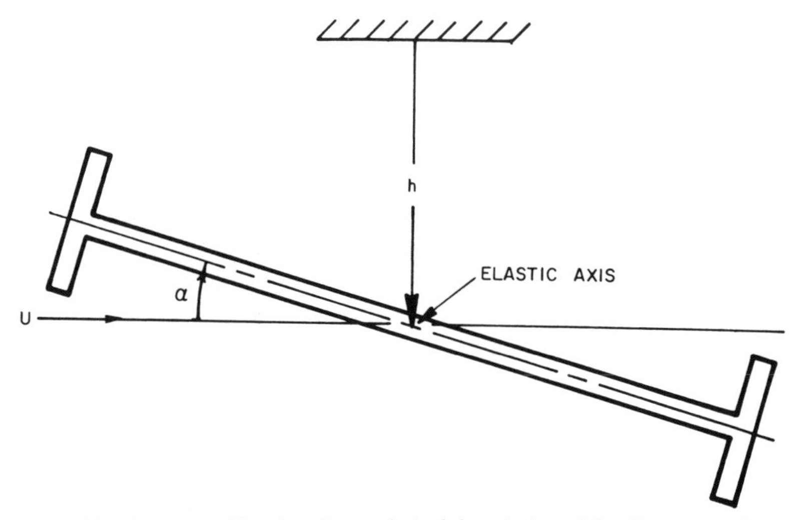

Figure 6.30 Degrees of freedom for analytical description of the divergence, flutter and buffeting problems.

$$S_\alpha \ddot{h} + I\ddot{\alpha} + c_\alpha \dot{\alpha} + K_\alpha \alpha = M_\alpha \tag{6.4.1b}$$

where m, S_α, I are respectively mass, static unbalance about the elastic axis, and mass moment of inertia, all per unit span; c_h and c_α are damping coefficients, K_h and K_α are sectional stiffnesses, and L_h, M_α are respectively aerodynamic lift and moment at the elastic axis. The majority of bridges are symmetric about their centerlines, and have $S_\alpha \equiv 0$.

Various approaches have been employed to obtain experimental results for L_h and M_α. These approaches have centered about two methods: measuring the forces on sinusoidally machine-driven models of bridge deck sections, or inferring the desired results from freely oscillating deck models by system identification techniques. Some results [94] of the U.S. method will be given.

The basic models [94] for the self-excited (flutter) forces and moments on bridge decks were postulated first as linear, thus:

$$L_h = m[H_1 \dot{h} + H_2 \dot{\alpha} + H_3 \alpha] \tag{6.4.2a}$$

$$M_\alpha = I[A_1 \dot{h} + A_2 \dot{\alpha} + A_3 \alpha] \tag{6.4.2b}$$

with H_i, A_i $(i = 1, 2, 3)$ to be determined experimentally (see [94]). In the context of heavy civil engineering structures aerodynamic inertial terms in $\ddot{h}$, $\ddot{\alpha}$ are usually considered negligible. Following thin airfoil practice, at least for laminar flow conditions, terms in the coordinate h were usually neglected also. However, more recent work [104, 118] has included terms in

h, i.e. $H_4 h$ and $A_4 h$, respectively in eqs. (6.4.2a) and (6.4.2b).

It has become the practice to treat H_i, A_i as real in this context, rather than complex as in the aeronautical context. These coefficients are, as in classical flutter, expected to be functions of the reduced frequency $b\omega/U = k$ where ω is the flutter circular frequency, U the wind velocity, and b the half-chord of the surface under study. Usage in the bridge flutter context has replaced k by $K = B\omega/U$, where $B = 2b$ is the full deck width.

It is of course desirable to put the experimentally determined values of the coefficients H_i, A_i into nondimensional forms, which is accomplished in the flutter context by the relations

$$\begin{aligned}
&(a)\ H_1^* = \frac{2mH_1}{\rho B^2\omega} \qquad (e)\ A_1^* = \frac{2IA_1}{\rho B^3\omega}\\
&(b)\ H_2^* = \frac{2mH_2}{\rho B^3\omega} \qquad (f)\ A_2^* = \frac{2IA_2}{\rho B^4\omega}\\
&(c)\ H_3^* = \frac{2mH_3}{\rho B^3\omega^2} \qquad (g)\ A_3^* = \frac{2IA_3}{\rho B^4\omega^2}\\
&(d)\ H_4^* = \frac{2mH_4}{\rho B^2\omega^2} \qquad (h)\ A_4^* = \frac{2IA_4}{\rho B^3\omega^2}
\end{aligned} \tag{6.4.3}$$

where ρ is air density and ω is flutter frequency.

With these definitions of nondimensional aerodynamic coefficients H_i^*, A_i^*, the flutter lift and moment may be written

$$L_h = \frac{1}{2}\rho U^2(B)\left[KH_1^*(K)\frac{\dot{h}}{U} + KH_2^*(K)\frac{B\dot{\alpha}}{U} + K^2H_3^*(K)\alpha + K^2H_4^*\frac{h}{B}\right] \tag{6.4.4a}$$

$$M_\alpha = \frac{1}{2}\rho U^2(B^2)\left[KA_1^*(K)\frac{\dot{h}}{U} + KA_2^*(K)\frac{B\dot{\alpha}}{U} + K^2A_3^*(K)\alpha + K^2A_4^*\frac{h}{B}\right] \tag{6.4.4b}$$

In the context developed here, factors like $K^2H_3^*$ are seen to be analogous to derivatives like $dC_L/d\alpha$; hence, loosely speaking, the coefficients H_i^* and A_i^* may be termed 'flutter derivatives' or 'flutter coefficients'. Since in aeronautical practice it is usual to employ the half-chord b in place of B, it is worth noting that replacement of B by b above throughout would result in a comparable set $\tilde{H}_i^*$, $\tilde{A}_i^*$ bearing the following relation to H_i^*, A_i^*:

$$\begin{array}{llll} \text{(a)} & \tilde{H}_1^* = 4H_1^* & \text{(e)} & \tilde{A}_1^* = 8A_1^* \\ \text{(b)} & \tilde{H}_2^* = 8H_2^* & \text{(f)} & \tilde{A}_2^* = 16A_2^* \\ \text{(c)} & \tilde{H}_3^* = 8H_3^* & \text{(g)} & \tilde{A}_3^* = 16A_3^* \\ \text{(d)} & \tilde{H}_4^* = 4H_4^* & \text{(h)} & \tilde{A}_4^* = 8A_4^* \end{array} \tag{6.4.5}$$

To demonstrate the link of the present formulation to airfoil flutter theory it can then be shown [94] that the set $\tilde{H}_i^*$, $\tilde{A}_i^*$, when written for the classical case of the thin airfoil, has the following values:

$$\begin{array}{ll} \text{(a)} & k\tilde{H}_1^*(k) = -4\pi F(k) \\ \text{(b)} & k\tilde{H}_2^*(k) = -2\pi\left[1 + F(k) + \dfrac{2G(k)}{k}\right] \\ \text{(c)} & k^2\tilde{H}_3^*(k) = -4\pi\left[F(k) - \dfrac{kG(k)}{2}\right] \\ \text{(d)} & k\tilde{H}_4^* = 2\pi[k + 2G] \\ \text{(e)} & k\tilde{A}_1^*(k) = 2\pi F(k) \\ \text{(f)} & k\tilde{A}_2^*(k) = -\pi\left[1 - F(k) - \dfrac{2G(k)}{k}\right] \\ \text{(g)} & k^2\tilde{A}_3^*(k) = 2\pi\left[F(k) - \dfrac{kG(k)}{2} + \dfrac{k^2}{8}\right] \\ \text{(h)} & k\tilde{A}_4^* = -2\pi G \end{array} \tag{6.4.6}$$

where $F(k) + iG(k) = C(k)$ is the well-known Theodorsen circulation function (see Section 4.3). It may be observed in passing that, for the airfoil, the damping derivatives $k\tilde{H}_1^*$ and $k\tilde{A}_2^*$ remain negative for all values of $1/k$.

In contrast to the case of the thin airfoil, all values of H_i^* and A_i^* for bridge decks must be obtained from experiment. Methods for accomplishing this are outlined in [94]. More recent results and methods for flutter derivative identification are discussed in [102] and [119]. References [103, 104, 119] discuss obtaining flutter coefficients under turbulent wind. Examples of the type of results generally obtained are given in Figure 6.31, where coefficients for the airfoil, original Tacoma Narrows bridge, and three truss-stiffened bridge decks are presented as functions of U/NB, where $N = \omega/2\pi$. Figure 6.32 presents analogous results for some box sections, of which 1 and 2 are streamlined to some extent. It is clear from a broad comparison of the thin airfoil results with those of bridge decks that airfoil flutter coefficients are inappropriate for bridge decks, in general. In recent studies [119], however, for very streamlined bridge decks the flutter derivatives more nearly resemble those of the thin airfoil.

Of particularly outstanding difference for bluffer bridge sections is the torsional damping derivative A_2^*, which changes sign for some deck forms with increasing values of $2\pi/K$. This is basic evidence of potential single-degree-of-freedom instability.

The numerical solution for the two-degree-of-freedom flutter problem, that is, of the problem defined by (6.4.1) and (6.4.4), follows the classical method. A value of K is chosen, and the corresponding values of H_i^* and A_i^* for (6.4.4) are determined from results like those of Figures 6.31 and 6.32. Solutions at the flutter condition depend upon the assumption of

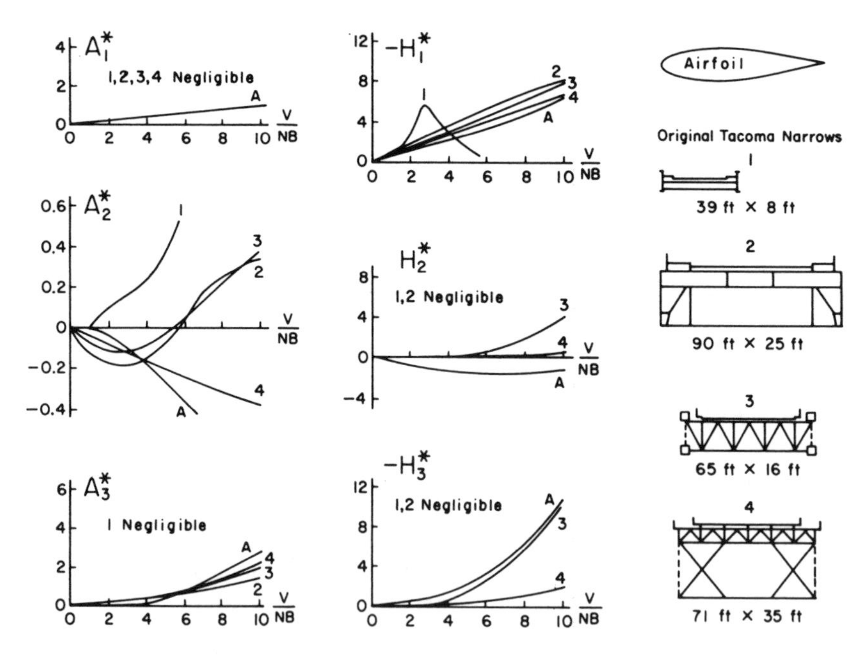

Figure 6.31 Some representative flutter coefficients [94].

responses having the form $h = h_0 e^{\imath\omega t}$. The determinant of coefficients of the equations (6.4.1) and (6.4.4) is then set equal to zero. Defining the unknown flutter frequency in the form

$$X = \omega/\omega_h \tag{6.4.7}$$

where $\omega_h^2 \equiv K_h/m$ and $\omega_\alpha^2 \equiv K_\alpha/I$, one then requires the satisfaction of the following two (real) equations simultaneously.†

$$X^4\left(1+\frac{\rho B^4}{2I}A_3^* - \frac{\rho B^2}{m}\frac{\rho B^4}{4I}A_2^*H_1^* + \frac{\rho B^2}{m}\frac{\rho B^4}{4I}A_1^*H_2^*\right)$$
$$+X^3\left(2\zeta_\alpha\frac{\omega_\alpha}{\omega_h}\frac{\rho B^2}{2m}H_1^* + 2\zeta_h\frac{\rho B^4}{2I}A_2^*\right)$$
$$+X^2\left\{-\frac{\omega_\alpha^2}{\omega_h^2} - 4\zeta_h\zeta_\alpha\frac{\omega_\alpha}{\omega_h} - 1 - \frac{\rho B^4}{2I}A_3^*\right) + 0\cdot X + \left(\frac{w_\alpha}{W_h}\right)^2 = 0 \tag{6.4.8a}$$

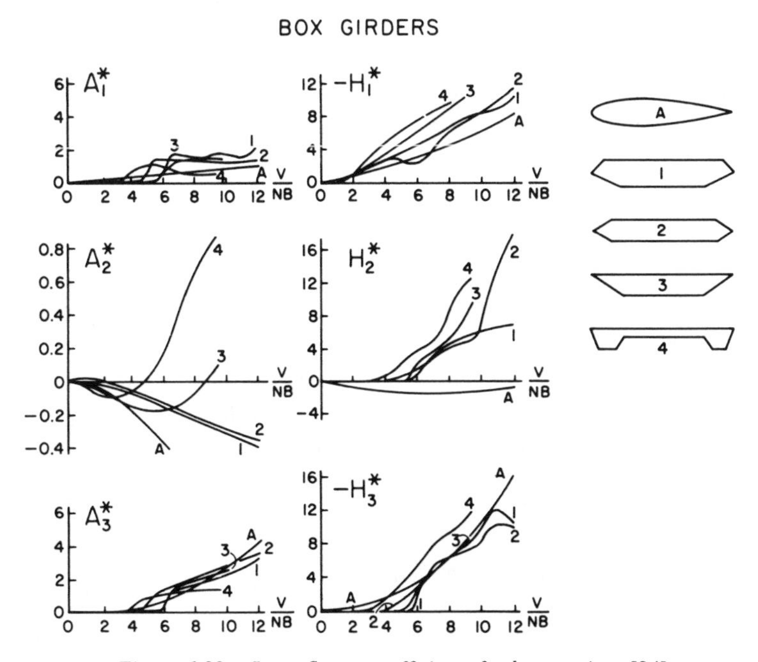

Figure 6.32 Some flutter coefficients for box sections [94].

---

† These results are written for the common case $S_\alpha \equiv 0$, i.e., a symmetric bridge deck.

$$X^3\left(\frac{\rho B^4}{2I}A_2^* + \frac{\rho B^2}{2m}H_1^* + \frac{\rho B^2}{m}\frac{\rho B^4}{4I}H_1^*A_3^* - \frac{\rho B^2}{m}\frac{\rho B^4}{4I}A_1^*H_3^*\right)$$
$$+X^2\left(-2\zeta_\alpha\frac{\omega_\alpha}{\omega_h} - 2\zeta_h - 2\zeta_h\frac{\rho B^4}{2I}A_3^*\right)$$
$$+X\left(-\frac{\rho B^2}{2m}H_1^*\frac{\omega_\alpha^2}{\omega_h^2} - \frac{\rho B^4}{2I}A_2^*\right)$$
$$+\left(2\zeta_h\frac{\omega_\alpha^2}{\omega_h^2} + 2\zeta_\alpha\frac{\omega_\alpha}{\omega_h}\right) = 0 \qquad (6.4.8b)$$

As one typical method of achieving the final result, the problem is pursued to the above point for each of a number of values of K over a chosen range, and the real solutions ω of each of (6.4.8a) and (6.4.8b) are plotted versus K. The point where the solution curves cross yields the desired real flutter frequency ω_c. This, together with the corresponding K value, K_c, permits determination of the flutter velocity U_c:

$$\frac{B\omega_c}{K_c} = U_c \qquad (6.4.9)$$

Some worked examples of this method appear in [95]. More recently, more general computer-based methods have been pursued for multi-degree-of-freedom flutter cases.

In the frequently occurring case of single-degree flutter with a strong torsional component, the above problem is considerably simplified, since then $h \cong 0$ and a single equation in α remains. A more general form of the single-mode problem will be discussed below when the full three-dimensional problem is treated.

Three-dimensional flutter and buffeting

In describing the motion of a bridge deck section, vertical and torsional coordinates h and α respectively were assigned as in Figure 6.30. In order to follow the full displacement of such a section as it participates in three-dimensional structural response, it has been found useful to provide a third coordinate p representing the horizontal translation (or sway) of the deck that takes place in certain bridge vibration modes, particularly of cable-stayed bridges and suspension bridges with decks that are strongly arched in the vertical plane along the direction of their roadway. The forces affecting this coordinate are the wind drag forces, which are generally variable under the natural wind, and flutter-type forces, necessitating the introduction of associated new flutter derivatives P_i^*, as seen presently.

The sectional wind forces per unit span at spanwise station x may then be designated in the form:

lift:

$$L(x, t) = L_{ae} + L_b \tag{6.4.10}$$

moment:

$$M(x, t) = M_{ae} + M_b \tag{6.4.11}$$

drag:

$$D(x, t) = D_{ae} + D_b \tag{6.4.12}$$

where the subscripts 'ae' refer to *aeroelastic* or fluid-structure interactive effects; and 'b' to *buffeting* effects. The linearized forms of the aeroelastic effects, for sinusoidal motions h, α, p, are:

$$L_{ae} = (1/2)\rho U^2 B[KH_1^*(K)(\dot{h}/U) + KH_2^*(K)(B\dot{\alpha}/U) + K^2H_3^*(K)\alpha] + K^2H_4^*(K)\frac{h}{B}] \tag{6.4.13}$$

$$M_{ae} = (1/2)\rho U^2 B^2[KA_1^*(K)(\dot{h}/U) + KA_2^*(K)(B\dot{\alpha}/U) + K^2A_3^*(K)\alpha + K^2A_4^*(K)\frac{h}{B}] \tag{6.4.14}$$

$$D_{ae} = (1/2)\rho U^2 B[KP_1^*(K)(\dot{p}/U) + KP_2^*(K)(B\dot{\alpha}/U) + K^2P_3^*(K)\alpha + K^2P_4^*(K)\frac{p}{B}] \tag{6.4.15}$$

where: ρ = air density;
U = mean cross-wind velocity (horizontal);
B = deck width;
$K = B\omega/U$
ω = oscillation frequency; and

$H_i^* = H_i^*(K)$ experimentally obtained
$A_i^* = A_i^*(K)$ 'flutter derivatives'
$P_i^* = P_i^*(K)$ $(i = 1, 2, 3)$

It will be noted at a later point that the extension of (6.4.13)–(6.4.15) to other than sinusoidal motions can be obtained via use of the *indicial* *force functions* associated specifically with H_i^*, A_i^*, P_i^*. These will be discussed in Section 6.4.4. However, the explicit use of these functions will be avoided in the present formulation by moving eventually to the Fourier-transformed domain associated with the independent frequency parameter K.

Let the buffeting forces be written in the form:

$$L_b(x, t) = (1/2)\rho U^2 B\mathscr{L}(x, t) \tag{6.4.16}$$

$$M_b(x, t) = (1/2)\rho U^2 B^2\mathscr{M}(x, t) \tag{6.4.17}$$

$$D_b(x, t) = (1/2)\rho U^2 B\mathscr{D}(x, t) \tag{6.4.18}$$

where $\mathscr{L}$, $\mathscr{M}$, and $\mathscr{D}$ are dimensionless time-varying forms of these forces, the specific details of which will be reserved to a later point. (It may be assumed that they are determinable by experiment and/or analytical formulation based upon time histories of wind horizontal and vertical gust velocities u, w respectively.)

The analysis from this point will closely follow the development of Ref. [97]. It will now be assumed that a preliminary three-dimensional vibration analysis of the complete bridge structure has been made, yielding definitions of a series of natural vibration modes i, together with their associated circular frequencies ω_i, and their full-bridge generalized masses I_i. Let the dimensionless generalized coordinate associated with mode i be ξ_i, so that the respective components of displacement h, α, p over the bridge deck spanwise section at x are expressible as:

vertical:

$$h(x, t) = \sum_i \xi_i(t) h_i(x) B \tag{6.4.19}$$

twist:

$$\alpha(x, t) = \sum_i \xi_i(t)\alpha_i(x) \tag{6.4.20}$$

sway:

$$p(x, t) = \sum_i \xi_i(t) p_i(x) B \tag{6.4.21}$$

where $h_i(x)$, $\alpha_i(x)$, $p_i(x)$ are respectively the dimensionless components of mode i referred to the deck only, at its local reference point. The equation of motion for the generalized coordinate of the ith mode of the system is then

$$I_i[\ddot{\xi}_i + 2\zeta_i\omega_i\dot{\xi}_i + \omega_i^2\xi_i] = Q_i(t) \qquad (6.4.22)$$

where ζ_i is the mechanical damping ratio of mode i, and $Q_i(t)$ is the generalized force

$$Q_i(t) = \int_{\text{deck}} [L(x,t)h_i(x)B + M(x,t)\alpha_i(x) + D(x,t)p_i(x)B]\,dx \qquad (6.4.23)$$

Note that aerodynamic strip theory is employed in (6.4.23). Alternatives will at present not be resorted to.

It is useful to convert to the dimensionless time variable s defined by

$$s = Ut/B \qquad (6.4.24)$$

whence (6.4.22) becomes

$$I_i[\xi_i'' + 2\zeta_i K_i\xi_i' + K_i^2\xi_i] = (B/U)^2 Q_i(s) \qquad (6.4.25)$$

since

$$\dot{\xi}_i(t) = (\mathrm{d}/\mathrm{d}s)\xi_i(s)(\mathrm{d}s/\mathrm{d}t) = (U/B)\xi_i'(s) \qquad (6.4.26)$$

etc., with

$$K_i = B\omega_i/U. \qquad (6.4.27)$$

Let the Fourier transform of $\xi_i(s)$ be defined by

$$\bar{\xi}_i(K) = \int_0^\infty \xi_i(s)\,\mathrm{e}^{-iKs}\,\mathrm{d}s \qquad (6.4.28)$$

so that the transformed form of (6.4.25) becomes

$$\begin{aligned}
[-K^2 &+ 2i\zeta_i K_1 K + K_i^2]\bar{\xi}_i(K) \\
&= (\rho B^4\ell/2I_i)\int_{\text{deck}} \{[iK^2H_1^*(K)(\bar{h}/B) + iK^2H_2^*(K)\bar{\alpha} \\
&+ K^2H_3^*(K)\bar{\alpha} + K^2H_4^*(K)(\bar{h}/B)]h_i(x) + [iK^2A_1^*(K)(\bar{h}/B) \\
&+ iK^2A_2^*(K)\bar{\alpha} + K^2A_3^*(K)\bar{\alpha} + K^2A_4^*(K)(\bar{h}/b)]\alpha_i(x) \\
&+ [iK^2P_1^*(K)(\bar{p}/B) + iK^2P_2^*(K)\bar{a} + K^2P_3^*(K)\bar{\alpha} \\
&+ K^2P_4^*(K)(\bar{p}/B)]p_i(x) + \bar{\mathcal{L}}(x,K)h_i(x) + \bar{\mathcal{M}}(x,K)\alpha_i(x) \\
&+ \bar{\mathcal{D}}(x,K)p_i(x)\}\,(dx/\ell) \qquad (6.4.29)
\end{aligned}$$

where l is the deck span and, according to (6.4.19)–(6.4.21):

$$\bar{h}(x, K)/B = \sum_i \bar{\xi}_i(K) h_i(x) \tag{6.4.30}$$

$$\bar{\alpha}(x, K) = \sum_i \bar{\xi}_i(K) \alpha_i(x) \tag{6.4.31}$$

$$\bar{p}(x, K)/B = \sum_i \bar{\xi}_i(K) p_i(x) \tag{6.4.32}$$

Equation (6.4.29) contains integrals of the form

$$\int_{\text{deck}} F_i^*(K) r_m(x) s_n(x)\, \mathrm{d}x/l \tag{6.4.33}$$

where $F_i^*(K) = H_i^*(K), A_i^*(K), P_i^*(K) (i = 1, 2, 3, 4)$, and r, s represent h, α, or p, with $r, s = i$ or j. The flutter derivatives F_i^* may be influenced, in particular, by mode shape, wind turbulence, and by displacement amplitude. It is usual to assume, as a first approximation, that the effects upon them of mode shape and (small) displacement amplitude may be ignored. When the effects upon them of wind turbulence are sought, there presently exist some techniques [102–104] for assessing this dependence. In what follows the flutter derivative factors $F_i^*(K)$ will be considered as removable to the outside of integrals of the form (6.4.33).

Defining the dimensionless modal constants by

$$G(r_m, s_n) = \int_{\text{deck}} r_m(x) s_n(x)\, \mathrm{d}x/l \tag{6.4.34}$$

The transformed equations of motion (6.4.29) become

$$\begin{aligned}
[K_i^2 - K^2 + 2i\zeta_i K_i K]\bar{\xi}_i = (\rho B^4 l/2I_i)\Big\{K^2 \sum_j [i(H_1^* G(h_j, h_i) \\
+ H_2^* G(\alpha_j, h_i) + A_1^* G(h_j, \alpha_i) + A_2^* G(\alpha_j, \alpha_i) + P_1^* G(p_j, p_i) \\
+ P_2^* G(\alpha_j, p_i)) + H_3^* G(\alpha_j, h_i) + A_3^* G(\alpha_j, \alpha_i) + P_3^* G(\alpha_j, p_i) \\
+ H_4^* G(h_j, h_i) + A_4^* G(h_j, \alpha_i) + P_4^* G(p_j, p_i)]\bar{\xi}_j \\
+ \int_{\text{deck}} [\bar{\mathscr{L}}(x, K) h_i(x) + \bar{\mathscr{M}}(x, K) \alpha_i(x) + \bar{\mathscr{D}}(x, K) p_i(x)]\, \mathrm{d}x/l\Big\}
\end{aligned} \tag{6.4.35}$$

All coefficients in (6.4.35) are functions of K. If n modes are used, the array of equations takes the matrix form

$$\begin{bmatrix} a_{11}+ib_{11} & a_{12}+ib_{12} & \dots & a_{1n}+ib_{1n} \\ \cdot & \cdot & \dots & \cdot \\ \cdot & \cdot & \dots & \cdot \\ \cdot & \cdot & \dots & \cdot \\ \cdot & \cdot & \dots & \cdot \\ a_{n1}+ib_{n1} & a_{n2}+ib_{n2} & \dots & a_{nn}+ib_{nn} \end{bmatrix} \begin{bmatrix} \bar{\xi}_1 \\ \cdot \\ \cdot \\ \cdot \\ \cdot \\ \bar{\xi}_n \end{bmatrix} = \begin{bmatrix} \bar{Q}_{b1} \\ \cdot \\ \cdot \\ \cdot \\ \cdot \\ \bar{Q}_{bn} \end{bmatrix} \tag{6.4.36}$$

where a_{ij}, b_{ij} are functions of K and

$$\bar{Q}_{bi} = \int_{\text{deck}} [\bar{\mathscr{L}}h_i + \bar{\mathscr{M}}\alpha_i + \bar{\mathscr{D}}p_i]\,\mathrm{d}x/l \tag{6.4.37}$$

Equation (6.4.36) has the symbolic matrix form

$$[A+iB]\{\bar{\xi}\} = \{\bar{Q}_b\} \tag{6.4.38}$$

and the solution form, for any given value of K, is

$$\{\bar{\xi}\} = [A+iB]^{-1}\{\bar{Q}\} = [C+iD]\{\bar{Q}\} \tag{6.4.39}$$

provided $[A+iB]^{-1}$ exists; D and C are formally defined by

$$D = -A^{-1}(BA^{-1}+AB^{-1})^{-1}$$

$$C = -DAB^{-1} \tag{6.4.40}$$

In the event that $A+iB$ is singular, i.e. the determinant of $[A+iB]$ vanishes:

$$|A+iB| = 0 \tag{6.4.41}$$

the system is seen to be unstable and defines the flutter condition. This condition is independent of the form of the buffeting force $\{\bar{Q}_b\}$, according to the present formulation. Prior to development of the explicit forms of the buffeting forces, a discussion of indicial functions will be in order. This follows a brief section dealing with the common case of single-mode flutter.

Single-mode flutter

The formulation to this point considers a multimode response of the bridge system. It has been found, however, that many practical cases of bridge flutter either occur in, or are driven by, a single mode that becomes unstable in its torsional component. (In particular, this means that the value of the associated A_2^* coefficient evolves from negative to positive with increasing values of $U/NB = 2\pi/K$.) In engineering practice it often becomes useful then to consider flutter as occurring when such a single mode i develops net negative damping. From (6.4.35) the following *single-mode flutter criterion* can then be deduced:

The system will be unstable if

$$\frac{K\rho B^4 l}{K_i 4\zeta_i I_i}[H_1^*(K)G(h_i, h_i) + H_2^*(K)G(\alpha_i, h_i) + A_1^*(K)G(h_i, \alpha_i)$$

$$+ A_2^*(K)G(\alpha_i, \alpha_i) + P_1^*(K)G(p_i, p_i) + P_2^*(K)G(\alpha_i p_i)] \geqslant 1 \qquad (6.4.42)$$

with K defined by

$$K^2 =$$

$$= \frac{K_i^2}{1 + (\rho B^4 l/2I_i)[H_3^*(K)G(\alpha_i, h_i) + A_3^*(K)G(\alpha_i, \alpha_i) + P_3^*(K)G(\alpha_i, p_i)}$$

$$+H_4^*(K)G(h_i, h_i) + A_4^*(K)G(\alpha_i, \alpha_i) + P_4^*(K)G(p_i, p_i)] \qquad (6.4.43)$$

A particular observation from this criterion is that favorable (i.e. pronounced negative) values of certain of the flutter derivatives, (notably H_1^* and P_1^*, aerodynamic damping derivatives in vertical and sway motion) together with strong modal action (G values) in these degrees of freedom, may negate possible unfavorable aerodynamic effects in torsion (i.e. positive A_2^* values). This helps account for some practical cases of evident full-bridge stability in the presence of a deck section known to possess torsionally unstable characteristics. The roles of mechanical structure damping ζ_i and generalized inertia I_i are also clear from the criterion (6.4.64), that is, increase of either increases stability tendencies. It may also be remarked that, to the extent that the linear flutter derivatives can reflect the effects of vortex shedding upon aerodynamic damping, the criterion also applies to this type of instability, which, in this sense, joins in a generalized view of what may be termed 'flutter'. Some authors include this general view of flutter in their descriptions of bridge oscillations.

Indicial Function Formulations

In the case of the aeroelastic problem for the thin airfoil, the circulatory lift due to airfoil change of incidence is given by

$$L_{ae} = \frac{1}{2}\rho U^2(B)\frac{dC_L}{d\alpha}\int_{-\infty}^{s} \Phi_L(s-\sigma)\frac{d\alpha(\sigma)}{d\sigma}d\sigma \qquad (6.4.44)$$

where α represents the effective angle of attack $\alpha_{3/4}$ based on the vertical velocity of the airfoil at the 3/4-chord point (see Section 4.3). In the airfoil formulation the function $\Phi_L(s)$ is the Wagner function [120].*

In the case of a bridge deck section in incompressible flow no special role such as that of $\alpha_{3/4}$ can be identified in general, and the Wagner function must be replaced by separate functions $\Phi_{L\alpha}, \phi_{Lh'}$ related independently to pure vertical velocity $\dot{h}$ and to twist α. Note that

$$\dot{h} = \frac{dh}{ds} \times \frac{ds}{dt} = h'\frac{U}{B} \tag{6.4.45}$$

The alternate result for lift per unit span can be expressed by the form

$$\frac{L_{ae}(s)}{\frac{1}{2}\rho U^2 B} = \frac{dC_L}{d\alpha} \int_{-\infty}^{s} \left[\Phi_{L\alpha}(s-\sigma)\frac{d\alpha(\sigma)}{d\sigma} + \Phi_{Lh'}(s-\sigma)\frac{d^2h(\sigma)}{Bd\sigma^2} \right] d\sigma \tag{6.4.46}$$

where $\Phi_{L\alpha}(s), \Phi_{Lh'}(s)$ are appropriate indicial functions for the particular deck shape encountered. Correspondingly, for the moment about the deck section reference point

$$\frac{M(s)}{\frac{1}{2}\rho U^2 B^2} = \frac{dC_M}{d\alpha} \int_{-\infty}^{s} \left[\Phi_{M\alpha}(s-\sigma)\frac{d\alpha(\sigma)}{d\sigma} + \Phi_{Mh'}(s-\sigma)\frac{d^2h(\sigma)}{Bd\sigma^2} \right] d\sigma \tag{6.4.47}$$

Thus, in principle, four new indicial functions of "Wagner type" are required. Figure 6.33 suggests the form of $\Phi_{M\alpha}$ obtained for a particular open-truss bridge deck, contrasting it with the Wagner function.

These indicial functions are related to the flutter derivatives via the Fourier transform. It can be demonstrated that

$$\frac{dC_L}{d\alpha} \int_0^\infty \Phi_{L\alpha} e^{-iKs} ds = K[H_2^*(K) - iH_3^*(K)] \tag{6.4.48}$$

$$\frac{dC_M}{d\alpha} \int_0^\infty \Phi_{M\alpha} e^{-iKs} ds = K[A_2^*(K) - iA_3^*(K)] \tag{6.4.49}$$

$$\frac{dC_L}{d\alpha} \int_0^\infty \Phi_{Lh'} e^{-iKs} ds = -iH_1^*(K) - iH_4^*(K) \tag{6.4.50}$$

* In the bridge deck context, all functions are defined in terms of the variables $K = B\omega/U, s = Ut/B$, in contrast to the usual airfoil context, in which k $= K/2$ and s defined as $Ut/b = 2Ut/B$ are employed.

$$\frac{dC_M}{d\alpha}\int_0^\infty \Phi_{Mh'}e^{-iKs}ds = -iA_1^*(K) - A_4^*(K) \tag{6.4.51}$$

These Φ functions cannot in general be specified theoretically but can be evaluated by inverse Fourier transformation of the experimentally identified flutter derivatives A_i^*, H_i^*.

Another distinct set of indicial functions is associated with gust effects, i.e. the lift and moment associated with gust penetration. Such effects are initiated not by changes of body incidence but by changes in effective angle of attack due to external wind direction changes.

In the case of the thin airfoil (cf. Section 4.3) the lift per unit span occasioned by penetration into a sharp-edged vertical gust is given by

$$\frac{L}{\frac{1}{2}\rho U^2 B} = \frac{dC_L}{d\alpha}\frac{w_0}{U}\psi(s) \tag{6.4.52}$$

where $\psi(s)$ is the Küssner function [121] and w_0 is the constant vertical velocity of the gust. By superposition the lift accompanying thin airfoil penetration into a gust of variable velocity $w(s)$ is given by

$$\begin{aligned} L(s) &= \frac{1}{2}\rho U^2 B\frac{dC_L}{d\alpha}\int_0^s w(\sigma)\psi'(s-\sigma)d\sigma \\ &= \frac{1}{2}\rho U B\frac{dC_L}{d\alpha}\int_0^\infty w'(s-\sigma)\psi(\sigma)d\sigma \end{aligned} \tag{6.4.53}$$

In the case where the gust is sinusoidally variable:

$$w(s) = w_0 e^{iKs} \tag{6.4.54}$$

the corresponding sinusoidal circulatory lift for the airfoil is

$$L(s) = \frac{1}{2}\rho U^2 B\frac{dC_L}{d\alpha}\frac{w_0}{U}\Theta(K)e^{iKs} \tag{6.4.55}$$

where $\Theta(K)$ is the complex Sears function [98]. Clearly from the above

$$\Theta(K) = iK\int_0^\infty \psi(\sigma)e^{-iK\sigma}d\sigma \tag{6.4.56}$$

relates the Sears and Küssner functions.

The Küssner and Sears results for the airfoil are recalled here as background for the analogous results applicable to bridge deck sections. Thus the lift caused by passage of a variable gust $w(s)$ over a bridge deck section is given by

$$L_b(s) = \frac{1}{2}\rho U^2 B^2\frac{dC_L}{d\alpha}\int_0^s \frac{w(\sigma)}{U}\Psi'_{Mh'}(s-\sigma)d\sigma \tag{6.4.57}$$

where $\Psi_{Lh'}(s)$ is the counterpart of the Küssner function. The corresponding moment is

$$M_b(s) = \frac{1}{2}\rho U^2 B^2 \frac{dC_M}{d\alpha} \int_0^s \frac{w(\sigma)}{U} \Psi'_{Mh'}(s-\sigma) d\sigma \tag{6.4.58}$$

where $\Psi_{Mh'}(s)$ is the analogous indicial function. Similar expression can be written for drag. In what follows, drag terms will be omitted.

Force expressions including buffeting

Recalling eqs. (6.4.10) and (6.4.11), but writing them now for general conditions and including buffeting terms, the total lift and moment may be written for a section x along the span:

$$L(x,t) = L_{ae} + L_b \tag{6.4.59}$$

$$M(x,t) = M_{ae} + M_b \tag{6.4.60}$$

where L_{ae}, M_{ae} are given by eqs. (6.4.13 and 6.4.14) and L_b, M_b by eqs. (6.4.57 and 6.4.58).

The wind velocity components affecting two-dimensional buffeting are $U + u(s)$ horizontal and $w(s)$ vertical, where U is the mean wind speed normal to the span and u, w are gust components with zero means. If the wind were very slowly varying, the section lift would be given at fixed angle of attack α_0 by

$$L = \frac{1}{2}\rho(U+u)^2 B C_L(\alpha_0)\psi'(s-\sigma) \tag{6.4.61}$$

Neglecting u^2 compared to U^2, this has the approximate value

$$L = L_0 + L_1 \frac{1}{2}\rho U^2 B C_L(\alpha_0) + \rho U u B C_L(\alpha_0) \tag{6.4.62}$$

where L_0 is the steady component of lift. For rapidly varing gusts $u(s)$, the variable lift is more correctly described by

$$L_1 = \frac{1}{2}\rho U^2 B C_L(\alpha_0) \int_{-\infty}^s \frac{2u(\sigma)}{U} \Phi'_{L\alpha}(s-\sigma) d\sigma \tag{6.4.63}$$

in which $\Phi_{L\alpha}$ is the indicial lift function seen earlier,

$$\Phi'_{L\alpha}(s) = \frac{d\Phi_{L\alpha}}{ds}.$$

The unsteady lift due to vertical gusting $w(s)$ is given analogously by

$$L_2 = \frac{1}{2}\rho U^2 B \left(\frac{dC_L}{d\alpha}\right)_0 \int_{-\infty}^{s} \frac{w(\sigma)}{U} \Psi'_{Lh}(s-\sigma)d\sigma \tag{6.4.64}$$

If, during the process, the deck section is also moving, this initiates the following additional lift components; for α-motion:

$$L_3 = \frac{1}{2}\rho U^2 B \left(\frac{dC_L}{d\alpha}\right)_0 \int_{-\infty}^{s} \alpha(\sigma)\Phi'_{L\alpha}(s-\sigma)d\sigma \tag{6.4.65}$$

for $h-$motion:

$$L_4 = \frac{1}{2}\rho U^2 B \left(\frac{dC_L}{d\alpha}\right)_0 \int_{-\infty}^{s} \frac{h'(\sigma)}{B} \Phi'_{Lh}(s-\sigma)d\sigma \tag{6.4.66}$$

In a completely analogous fashion the following moment components are developed

$$M_1 = \frac{1}{2}\rho U^2 B^2 C_M(\alpha_0) \int_{-\infty}^{s} \frac{2u(\sigma)}{U} \Phi'_{M\alpha}(s-\sigma)d\sigma \tag{6.4.67}$$

$$M_2 = \frac{1}{2}\rho U^2 B^2 \left(\frac{dC_M}{d\alpha}\right)_0 \int_{-\infty}^{s} \frac{w(\sigma)}{U} \Psi'_{Mh}(s-\sigma)d\sigma \tag{6.4.68}$$

$$M_3 = \frac{1}{2}\rho U^2 B^2 \left(\frac{dC_M}{d\alpha}\right)_0 \int_{-\infty}^{s} \alpha(\sigma)\Phi'_{M\alpha}(s-\sigma)d\sigma \tag{6.4.69}$$

$$M_4 = \frac{1}{2}\rho U^2 B^2 \left(\frac{dC_M}{d\alpha}\right)_0 \int_{-\infty}^{s} \frac{h'}{B} \Phi'_{Mh}(s-\sigma)d\sigma \tag{6.4.70}$$

Frequency domain representation

From what has been defined above it follows that the Fourier transforms of the total lift and moment are:

$$\bar{L} = \overline{L_b} + \overline{L_{ae}} \tag{6.4.71}$$

$$\overline{M} = \overline{M_b} + \overline{M_{ae}} \tag{6.7.72}$$

where, with reference to eqs. (6.4.63–6.4.70) (including also the vertical projection of the drag force) we find:

$$\begin{aligned}\overline{L_b} &= \frac{1}{2}\rho U^2 B \Big\{2C_L(\alpha_0)\frac{\bar{u}}{U}\overline{\Phi'_{L\alpha}} + [C'_L(\alpha_0) \\ &+ C_D(\alpha_0)]\frac{\bar{w}}{U}\overline{\Psi'_{Lh}}\Big\}\end{aligned} \tag{6.4.73}$$

$$\overline{M_b} = \frac{1}{2}\rho U^2 B^2 \left\{2C_M(\alpha_0)\frac{\bar{u}}{U}\overline{\Phi'_{M\alpha}} + C'_M(\alpha_0)\frac{\bar{w}}{U}\overline{\Psi'_{Mh}}\right\} \tag{6.4.74}$$

and

$$\overline{L_{ae}} = \frac{1}{2}\rho U^2 BK^2[(iH_1^* + H_4^*)\frac{\bar{h}}{B} + (iH_2^* + H_3^*)\bar{\alpha}] \qquad (6.4.75)$$

$$\overline{M_{ae}} = \frac{1}{2}\rho U^2 BK^2[(iA_1^* + A_4^*)\frac{\bar{h}}{B} + (iA_2^* + A_3^*)\bar{\alpha}] \qquad (6.4.76)$$

Expressions of this kind may now be employed in a frequency-domain response analysis of the full structural system. Full details will not be pursued here. Some points bear emphasis, however. It will be noted that the flutter derivatives play roles of aerodynamic coupling or of stiffness and damping, so that such homogeneous terms can be absorbed with associated (h, α) terms of the structure. The $\bar{u}$ and $\bar{w}$ terms, however, play the roles of active exciters of the system. From eqs. (6.7.75) and (6.4.76) the power spectral densities $S_L(K), S_M(K)$ of buffeting lift and moment, respectively, can be derived. These take the forms

$$\begin{aligned} S_L(K) \;=\; & \left(\frac{\rho UB}{2}\right)^2 \{4C_L^2(\alpha_0)S_u(K)\mathcal{X}_{L\alpha}^2 \\ &+ \; [C_L'(\alpha_0) + C_D(\alpha)]^2 \; S_w(K)\mathcal{X}_{Lh}^2 \\ &+ \; 2C_L(\alpha_0)[C_L'(\alpha_0) + C_D(\alpha_0)] \left[S_{uw}(K)\overline{\Phi'_{L\alpha}\Psi'^*_{Lh}}\right. \\ &+ \; \left. S_{wu}(K)\overline{\Phi'^*_{L\alpha}\Psi'_{Lh}}\right]\} \end{aligned} \qquad (6.4.77)$$

$$\begin{aligned} S_M(K) \;=\; & \left(\frac{\rho UB}{2}\right)^2 \{4C_M^2(\alpha_0)S_u(K)\mathcal{X}_{M\alpha}^2 \\ &+ \; [C_M'(\alpha_0)]^2 \; S_w(K)\mathcal{X}_{Mh}^2 \\ &+ \; 2C_M(\alpha_0)C_M'(\alpha_0) \left[S_{uw}(K)\overline{\Phi'_{M\alpha}\Psi'^*_{Mh}}\right. \\ &+ \; \left. S_{wu}(K)\overline{\Phi'^*_{M\alpha}\Psi'_{Mh}}\right]\} \end{aligned} \qquad (6.4.78)$$

where S_u, S_w, S_{wu} are direct and cross power spectral densities of wind turbulence components u and w.

It is particularly worth noting that certain *aerodynamic admittances* such as

$$\mathcal{X}_{L\alpha}^2(K) = \bar{\Phi}'_{L\alpha}\bar{\Phi}'^*_{L\alpha} = \left(\frac{dC_L}{d\alpha}\right)^{-2} K^{-4}[H_2^{*2} + H_3^{*2}] \qquad (6.4.79)$$

$$\mathcal{X}_{M\alpha}^2(K) = \bar{\Phi}'_{M\alpha}\bar{\Phi}'^*_{M\alpha} = \left(\frac{dC_M}{d\alpha}\right)^{-2} K^{-4}[A_2^{*2} + A_3^{*2}] \qquad (6.4.80)$$

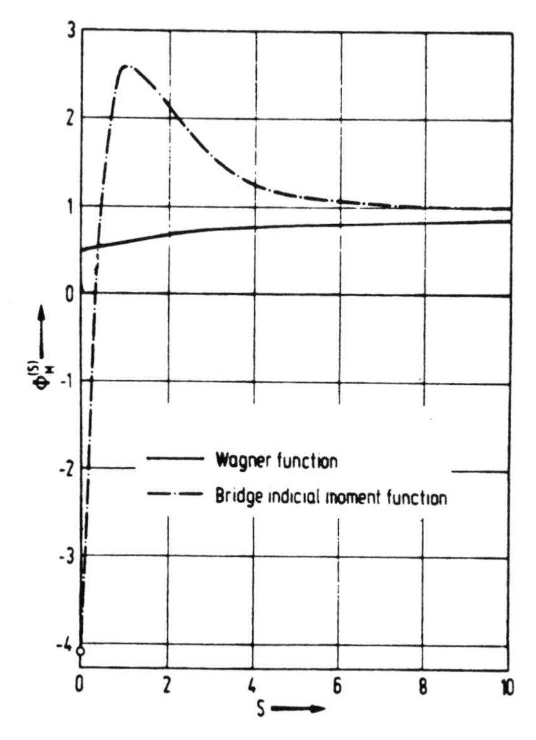

Figure 6.33. Indicial moment function for open-truss bridge deck.

occasioned by this procedure, may be evaluated from knowledge of the flutter derivatives. Others may not. It will be seen, for example, that the admittance $\mathcal{X}_{Lh}^2$ plays a role strictly analogous to that of the square of the absolute value of the Sears function of the airfoil case.

The general approach outlined has found application to numerous practical cases of bridges [122]. The full definition of aerodynamic admittances remains, however, to be explored more completely for such cases.

The three-dimensional flutter and buffeting problems of flexible bridges have been treated here together in an analytical fashion from a specifically adopted viewpoint: the aeroelastic portion is based upon small-amplitiude linear flutter derivatives that are deterministic experimental functions of K; the buffeting forces are considered to be random functions only of time and spatial location, i.e. not interactive with structural motion. In the frequency domain they are also functions of K.

It is undisputed however, that the sectional flutter derivatives are in fact affected to some extent by the state of tutbulence in the oncoming wind flow.

The method proposed here to cope with this situation is that their mean values, for any state of turbulence, be experimentally obtained, as functions of K, and used as such in the analysis. this approach must also account in an appropriate manner for the spanwise effects of turbulence upon the flutter derivatives. These questions will not be pursued here.

References for Chapter 6

[1] Strouhal, V., 'Uber eine besondere Art der Tonerregung', *Annalen der Physik* (Leipzig), Vol. 5, 1878.

[2] Bénard, H., 'Formation de centres de giration à l'arrière d'un obstacle en mouvement, *Comptes Rendus de l'Académie des Sciences*, Paris, Vol. 146, 1908.

[3] von Kármán, Th., 'Über den Mechanismus des Widerstandes den ein bewegter Koerper in einer Fluessigkeit erfaehrt', *Nachricht der Koeniglichen Gesellschaft der Wissenschaft*, Göttingen, 1912.

[4] Simiu, E. and Scanlan, R. H., 1986, *Wind Effects on Structures* (*An Introduction to Wind Engineering*) (2nd Ed.) Wiley, NY, 1986.

[5] Blevins, R. D., *Flow Induced Vibrations*, Van Nostrand Reinhold, NY, 1977.

[6] Fung, Y. C., *The Theory of Aeroelasticity*, Wiley, NY, 1955. (Also reprinted by Dover.)

[7] Bishop, R. E. D. and Hassan, A. Y., 'The Lift and Drag Forces on a Circular Cylinder in a Flowing Fluid', *Proc. Royal Soc.*, Ser. A, Vol. 277, London, 1964, pp. 32–50.

[8] Bishop, R. E. D. and Hassan, A. Y., 'The Lift and Drag Forces on a Circular Cylinder Oscillating in a Flowing Fluid', *Proc. Royal Soc.*, Ser. A, Vol. 277, London, 1964, pp. 51–75.

[9] Sarpkaya, T., 'Fluid Forces on Oscillating Cylinders', ASCE *J. of the Waterway, Port, Coastal and Ocean Div.* (Proc. ASCE), Vol. 97, No. WW4, August 1978, pp. 275–290.

[10] Staubli, T., 'Calculation of the Vibration of an Elastically Mounted Cylinder Using Experimental Data from a Forced Oscillation', *ASME J. of Fluids Engrg.*, Vol. 105, June (1983, pp. 225–229.

[11] Feng, C. C., 'The Measurement of Vortex-Induced Effects in a Flow Past Stationary and Oscillating Circular and D-Section Cylinders', M.S. Thesis, Univ. of British Columbia, 1968.

[12] Ferguson, N., 'The Measurement of Wake and Surface Effects in the Subcritical Flow Past a Circular Cylinder at Rest and in Vortex-Excited Oscillation', M.S. Thesis, Univ. of British Columbia, 1965.

[13] Vickery, B. J. and Basu, R. I., 'Across-Wind Vibrations of Structures of Circular Cross-Section', *J. Wind Engrg. Ind. Aerodyn.*, Vol. 12, 1983, Part 1: 'Development of a Two-Dimensional Model for Two-Dimensional Conditions', pp. 49–73: Part 2: 'Development of a Mathematical Model for Full Scale Application', pp. 75–97.

[14] Birkhoff, G., 'Formation of Vortex Streets', *J. of Appl. Physics*, Vol. 24, No. 1, 1953, pp. 98–103.

[15] Funakawa, M., 'The Vibration of a Cylinder Caused by Wake in a Flow', *Bull. JSME*, Vol. 12, No. 53, 1969, pp. 1003–1010.

[16] Nakamura, Y., 'Vortex Excitation of a Circular Cylinder Treated as a Binary Flutter', *Rep. Res. Inst. Appl. Mech.*, Vol. XVII, No. 59, Kyushu Univ., 1969, pp. 217–234.

[17] Tamura, Y. and Matsui, G., 'Wake-Oscillator Model of Vortex-Induced Oscillation of Circular Cylinder', *Proc. 5th Int. Conf. on Wind Engrg.*, Fort Collins, 1979, pp. 1085–1094.
[18] Hartlen, R. T. and Currie, I. G., 'A Lift-Oscillator Model of Vortex-Induced Vibration', *Proc. ASCE, J. Engrg. Mech. Div.*, Vol. 96, October 1970, pp. 577–591.
[19] Skop, R. A. and Griffin, O. M., 'A Model for the Vortex-Excited Resonant Response of Bluff Cylinders', *J. of Sound & Vibration*, Vol. 27, 1973, pp. 225–233.
[20] Skop, R. A. and Griffin, O. M., 'On a Theory for the Vortex-Excited Oscillations of Flexible Cylindrical Structures', *J. of Sound & Vibration*, Vol. 41, 1975, pp. 263–274.
[21] Landl, R., 'A Mathematical Model for Vortex-Excited Vibrations of Bluff Bodies', *J. of Sound & Vibration*, Vol. 42(2), 1975, pp. 219–234.
[22] Wood, K. N., 'Coupled-Oscillator Models for Vortex-Induced Oscillation of a Circular Cylinder', M.S. Thesis, Univ. of British Columbia, 1976.
[23] Wood, K. N. and Parkinson, G. V., 'A Hysteresis Problem in Vortex-Induced Oscillations', *Proc. Canadian Congress Applied Mechanics*, Vancouver, 1977, pp. 697–698.
[24] Berger, E., 'On a Mechanism of Vortex-Excited Oscillations of a Cylinder', *Preprints, 7th Int'l. Conf. on Wind Engrg.*, Vol. 2, Aachen, Germany, 1987, pp. 169–177.
[25] Dowell, E. H., 'Non-Linear Oscillator Models in Bluff Body Aeroelasticity', *J. of Sound & Vibration*, Vol. 75(2), 1981, pp. 251–264.
[26] Iwan, W. D. and Blevins, R. D., 'A Model for Vortex Induced Oscillation of Structures', *J. of Applied Mechanics, Trans. ASME*, September 1974, pp. 581–586.
[27] Scanlan, R. H. and Wardlaw, R. L., 'Reduction of Flow-Induced Structural Vibrations', Section 2 of *Isolation of Mechanical Vibration, Impact and Noise*, AMD Vol. 1, 1973, American Society of Mechanical Engineers, New York.
[28] Scruton, C., 'A note on a device for the suppression of the vortex-excited oscillations of flexible structures of circular or near circular section, with special reference to its application to tall stacks', *NPL Aero Report* 1012, National Physical Lab., U.K., 1963.
[29] Walshe, D. E. and Wooton, L. R., 'Preventing wind-induced oscillations of structures of circular section', *Proc. Inst. of Civil Engineers*, Vol. 47, September 1970, pp. 1–24.
[30] Wooton, L. R. and Scruton, C., 'Aerodynamic Stability', *Modern Design of Wind-Sensitive Structures*, Seminar, Construction Research and Information Association, U.K., 1970, pp. 65–81.
[31] Wardlaw, R. L. and Cooper, K. R., 'Mechanisms and Alleviation of Wind-Induced Structural Vibrations', *Proceedings, 2nd Symposium on Applications of Solid Mechanics*, McMaster Univ., Hamilton, Ontario, Canada, June 1974, pp. 369–399.
[32] Wardlaw, R. L., 'Approaches to the Suppression of Wind-Induced Vibrations of Structures', *Practical Experiences with Flow-Induced Vibrations*, (Eds. E. Naudascher and D. Rockwell) Proc. IAHR/IUTAM Symposium, Karlsruhe, Germany 1979 (Springer-Verlag., N.Y.), Paper F6, pp. 650–672.
1[33] Otsuki, Y., Washizu, K., Tomizawa, H. and Ohya, A., 'A Note on the Aeroelastic Instability of a Prismatic Bar with Square Section', *J. of Sound & Vibration*, Vol. 34(2), 1974, pp. 233–245.
[34] Griffin, O. M. Skop, R. A. and Koopmann, G. H., 'The Vortex-Excited Resonant Vibrations of Circular Cylinders', *J. of Sound & Vibration*, Vol. 31, 1973, pp. 235–249.
[35] Bearman, P. W., 'Vortex Shedding from Oscillating Bluff Bodies', *Annual Rev. of Fluid Mech.*, Vol. 16, 1984, pp. 195–222.
[36] Bearman, P. W. and Graham, J. M. R., 'Vortex Shedding from Bluff Bodies in

Oscillatory Flow', A report on Euromech 119, *J. of Fluid Mechanics*, Vol. 99, 1980, pp. 225–245.

[37] Bearman, P. W. and Currie, I. C., 'Pressure Fluctuation Measurements on an Oscillating Circular Cylinder', *J. of Fluid Mechanics*, Vol. 91, 1979, pp. 661–667.

[38] Berger, E., 'Zwei fundamentale Aspekte wirbelerregter Schwingungen', Hermann-Fottinger-Institut fur Thermo- und Fluiddynamik, Berlin, 1984.

[39] Berger, E. and Wille, R., 'Periodic Flow Phenomena', *Ann. Rev. of Fluid Mechanics*, Vol. 4, 1972, pp.313–340.

[40] Currie, I. G., Hartlen, R. T. and Martin, W. W., 'The Response of Circular Cylinders, to Vortex Shedding', *IUTAM Conf. on Flow-Induced Structural Vibrations*, 1974 (Ed., E. Naudascher) Springer, Berlin, pp. 128–142.

[41] Ferguson, N. and Parkinson, G. V., 'Surface and Wake Flow Phenomena of the Vortex-Excited Oscillation of a Circular Cylinder', *ASME J. of Engrg. Indus.* 1967, pp. 831–838.

[42] Griffin, O. M., 'Vortex-Excited Cross-Flow Vibrations of a Single Cylindrical Tube', *ASME J. of Pressure Vessel Tech.*, Vol. 102, 1980, pp. 158–166.

[43] Griffin, O. M. and Koopman, G. H., 'The Vortex-Excited Lift and Reaction Forces on Resonantly Vibrating Cylinders', *J. of Sound & Vibration*, Vol. 54(3) 1977, pp. 435–448.

[44] Hall, S. A. and Iwan, W. D., 'Oscillations of a Self-Excited, Nonlinear System', *J. of Applied Mechanics*, Vol. 51, 1984, pp. 892–898.

[45] Iwan, W. D. and Botelho, D. L. R., 'Vortex-Induced Oscillation of Structures in Water', *ASCE J. of Waterways, Port, Coastal & Ocean Engrg.*, Vol. 111(2), 1985, pp. 289–303.

[46] Jones, N. P., 'Flow-Induced Vibrations of Long Structures', Ph.D. Thesis, 1986, California Institute of Technology.

[47] Jones, G. W., Cincotta, J. J. and Walker, R. W., 'Aerodynamic Forces on a Stationary and Oscillating Circular Cylinder at High Reynolds Number', NASA Rep. TR-R-300, 1969.

[48] Mair, W. A. and Maull, D. J., 'Bluff Bodies in Vortex Shedding', A report on Euromech 17, *J. of Fluid Mechanics*, Vol. 45, 1971, pp. 209–224.

[49] Marris, A. W., 'A Review on Vortex Streets, Periodic Wakes, and Induced Vibration Phenomena', *ASME J. of Basic Engrg.*, Vol. 88, 1964, pp. 185–196.

[50] Moeller, M. J., 'Measurement of Unsteady Forces on Circular Cylinder in Cross-Flow at Subcritical Reynolds Number', Ph.D. Thesis, 1982, Dept. of Ocean *Engrg*, MIT.

[51] Morkovin, M. V., 'Flow around Circular Cylinders: A Kaleidoscope of Challenging Fluid Phenomena', *ASME Symp. on Fully Separated Flows*, 1964, Philadelphia, pp. 102–108.

[52] Ongoren, A., 'Unsteady Structure and Control of Near-Wakes', Ph.D. Thesis, 1986, Lehigh University.

[53] Parkinson, G. V., 'Nonlinear Oscillator Modelling of Flow-Induced Vibration', Paper H3, *Practical Experiences with Flow-Induced Vibrations* (E. Naudascher and D. Rockwell, Eds.), IAHR–IUTAM Symposium, Karlsruhe, 1979, Springer-Verlag, Berlin, 1980, pp. 786–797.

[54] Parkinson, G. V., 'Mathematical Models of Flow-Induced Vibrations', *Flow-Induced Structural Vibrations* (Ed., E. Naudascher), Springer-Verlag, Berlin, 1974, pp. 81–127.

[55] Parkinson, G. V. Feng, G. and Ferguson, N., 'Mechanisms of Vortex-Excited Oscillations of Bluff Cylinders', *Proceedings, Symposium on Wind Effects on Buildings*

and Structures, 1966, Loughborough Univ. of Tech., U.K.

[56] Poore, A. B., Doedel, E. J. and Cermak, J. E., 'Dynamics of the Iwan-Blevins Wake Oscillator Model', *International J. of Nonlinear Mech.*, Vol. 21, No. 4, 1986, pp. 291–302.

[57] Roshko, A., 'On the Development of Turbulent Wakes from Vortex Streets', NACA Technical Note 2913, 1953.

[58] Sarpkaya, T. and Isaacson, M., *Mechanics of Wave Forces on Offshore Structures*, Van Nostrand-Reinhold, N.Y., 1981.

[59] Scruton, C. and Flint, A. R., 'Wind-Excited Oscillations of Structures', *Proceedings, Institution of Civil Engineers*, Vol. 27, 1964, pp. 673–702.

[60] Stansby, P. K., 'The Locking-on of Vortex Shedding Due to the Cross-Stream Vibration of Circular Cylinders in Uniform and Shear Flows', *J. of Fluid Mechanics*, Vol. 74, 1976, pp. 641–665.

[61] Toebes, G. H., 'Fluidelastic features of Flow around Cylinders', *Proceedings, Internat. Research Seminar on Wind Effects on Buildings and Structures*, Ottawa, Canada, 1967, Vol. 2, 1968, pp. 323–334.

[62] Toebes, G. H. and Ramamurthy, A. S., 'Fluidelastic Forces on Circular Cylinders', *ASCE J. Engg. Mechanics*, Vol. 93, 1967, pp. 51–75.

[63] Toebes, G. H., 'The Unsteady Flow and Wake near an Oscillating Cylinder', *ASME J. Basic Engineering*, Vol. 91, 1968, pp. 493–502.

[64] Toebes, G. H., 'Fluidelastic Features of Flow Around Cylinders', *Proceedings, Internat. Research Seminar on Wind Effects on Buildings and Structures*, Ottawa, Canada, 1967, Vol. 2, 1968, pp. 323–334.

[65] Zdravkovich, M. M., 'Review and Classification of Various Aerodynamic and Hydrodynamic Means for Suppressing Vortex Shedding', *J. of Wind Engrg. & Industrial Aerodynamics.*, Vol. 7, 1981, pp. 145–189.

[66] Zdravkovich, M. M., 'Modification of Vortex Shedding in the Synchronization Range', *ASME J. of Fluid Engrg.*, Vol. 104, 1982, pp. 513–517.

[67] Wille, R., 'Karman Vortex Streets', *Advances in Applied Mechanics*, Vol. VI, Academic Press, New York, 1960, pp. 273–287.

[68] Williamson, C. H. K. and Roshko, A., 'Vortex Formation in the Wake of an Oscillating Cylinder', *J. Fluids and Structures*, Vol. 2, 1988, pp. 355–381.

[69] Den Hartog, J. P., 'Transmission Line Vibration Due to Sleet', *Transactions, AIEE*, 1932, pp. 1074–1076.

[70] Richards, D. J. W., 'Aerodynamic Properties of the Severn Crossing Conductor', *Wind Effects on Buildings and Structures*, National Physical Laboratory, U.K., 1965, Vol. II, pp. 687–765.

[71] Den Hartog, J. P., *Mechanical Vibrations*, McGraw-Hill, New York, 4th Ed., 1956.

[72] Parkinson, G. V. and Brooks, N. P. H., 'On the Aeroelastic Instability of Bluff Cylinders', *Trans. ASME J. Appl. Mech.*, Vol. 83, 1961, pp. 252–258.

[73] Parkinson, G. V. and Smith, J. D., 'An Aeroelastic Oscillator with Two Stable Limit Cycles', *Trans. ASME J. Appl. Mech.*, Vol. 84, 1962, pp. 444–445.

[74] Parkinson, G. V., 'Aeroelastic Galloping in One Degree of Freedom', *Proc. Sympos. on Wind Effects on Bldgs. & Struct.*, Teddington, England, 1963, pp. 581–609.

[75] Parkinson, G. V. and Smith, J. D., 'The Square Prism as an Aeroelastic Nonlinear Oscillator', *Quart. J. of Mech. and Appl. Math.* Oxford Press, London, Vol. XVII Part 2, 1964, pp. 225–239.

[76] Parkinson, G. V. and Santosham, T. V., 'Cylinders of Rectangular Section as Aeroelastic Nonlinear Oscillators', Paper *Vibrations Conf.*, ASME, Boston, Mass. March 1967.

[77] Parkinson, G. V., 'Mathematical Models of Flow-Induced Vibrations of Bluff Bodies', *Flow-Induced Structural Vibrations*, (Ed., E. Naudascher), Springer-Verlag, 1974, pp. 81–127.
[78] Novak, M., 'Aeroelastic Galloping of Prismatic Bodies', *Proc. ASCE J. EMD*, Vol. 9, No. EM1, Feb. 1969, pp. 115–142.
[79] Novak, M., 'Galloping Oscillations of Prismatic Structures', *Proc. ASCE J. EMD*, Vol. 98, No. EM1, Feb. 1972, pp. 27–46.
[80] Kryloff, N. and Bogoliuboff, N., 'Introduction to Nonlinear Mechanics', (*Annals of Math Studies*, No. 11) Translator S. Lefschetz, Princeton Univ. Press, Princeton, New Jersey, 1947.
[81] Richardson, A. S., Martuccelli, J. R. and Price, W. S., 'Research Study on Galloping of Electric Power Transmission Lines', *Proceedings, Symposium on Wind Effects on Buildings and Structures*, Vol. II, Paper 7, Teddington, England, 1965.
[82] Wardlaw, R. L., Cooper, K. R. and Scanlan, R. H., 'Observations on the Problem of Subspan Oscillation of Bundled Power Conductors', *DME/NAE Quarterly Bulletin*, 1973(1), National Research Council, April 1973, Ottawa, Ont. Canada.
[83] Cooper, K. R., 'A Wind Tunnel Investigation of Twin-Bundled Power Conductors', *Report LTR-LA-96 NAE*, National Research Council, May 1972, Ottawa, Ont. Canada.
[84] Scanlan, R. H., 'A Wind Tunnel Investigation into the Aerodynamic Stability of Bundled Power Line Conductors for Hydro-Quebec', *Report LTR-LA-121 NAE* (Part VI), National Research Council, Sept. 1972, Ottawa, Ont. Canada.
[85] Simpson, A., 'On the Flutter of a Smooth Cylinder in a Wake', *The Aeronautical Quarterly*, Feb. 1971, pp. 25–41.
[86] Wardlaw, R. L., 'Static Force Measurements of Six Deck Sections for the Proposed New Burrard Inlet Crossing', *Report LTR-LA-53*, NAE, National Research Council, June 1970, Ottawa, Ont., Canada.
[87] Bleich, F., 'Dynamic Instability of Truss-Stiffened Suspension Bridges under Wind Action', *Proceedings ASCE*, Vol. 75, No. 3, March 1949, pp. 413–416, and Vol. 75, No. 6, June 1949, pp. 855–865.
[88] Selberg, A., 'Oscillation and Aerodynamic Stability of Suspension Bridges', *Acta Polytechnica Scandinavica*, Civil Engrg. and Construction Series No. 13, 1961.
[89] Selberg, A. and Hjorth-Hansen, E., 'The Fate of Flat-Plate Aerodynamics in the World of Bridge Decks', *Proc. Theodorsen Colloq., Det Kongelige Norske Videnskabers Selskab*, Trondheim, 1976, pp. 101–113.
[90] Scanlan, R. H. and Budlong, K. S., 'Flutter and Aerodynamic Response Considerations for Bluff Objects in a Smooth Flow', *Flow Induced Structural Vibrations* (Ed., E. Naudascher) Springer-Verlag, N.Y., August 1974, pp. 339–354.
[91] Scanlan, R. H., Beliveau, J.-G. and Budlong, K. S., 'Indicial Aerodynamic Functions for Bridge Decks', *Proceedings ASCE, J Engrg. Mech. Div.*, August 1974, pp. 657–672.
[92] Scanlan, R. H., 'On the State of Stability Considerations for Suspended-Span Bridges Under Wind', Paper F1, *Practical Experiences with Flow-Induced Vibrations* (E. Naudaschev and D. Rockwell, Eds.) Springer-Verlag, Berlin, New York 1980, pp. 595–618.
[93] Scanlan, R. H., 'Interpreting Aeroelastic Models of Cable-Stayed Bridges', *J. Engrg. Mech., ASCE*, Vol. 113, No. 4, April 1987, pp. 555–575.
[94] Scanlan, R. H. and Tomko, J. J., 'Airfoil and Bridge Deck Flutter Derivatives', *ASCE J. of Engrg. Mech. Div.* Vol. 97, No. EM6, December 1971, pp. 1717–1737.
[95] Scanlan, R. H., 'State-of-the-Art Methods for Calculating Flutter, Vortex-Induced,

and Buffeting Response of Bridge Structures', *Report No. FHWA/RD*-80/050, April 1981, Federal Highway Administration Offices of Research and Development, Structures and Applied Mechanics Division, Washington, D.C. 20590.

[96] Scanlan, R. H., 'The Action of Flexible Bridges under Wind, I: Flutter Theory; II: Buffeting Theory', *J. Sound and Vibration*, Vol. 60, No. 2, 1978, pp. 187–199 and 201–211.

[97] Scanlan, R. H., 'On Flutter and Buffeting Mechanisms in Long-Span Bridges', *Probabilistic Engineering Mechanics* Vol. 3, No. 1, March 1988, pp. 22–27. Computational Mechanics Publications, Southampton, England.

[98] Sears, W. R., 'Some Aspects of Non-Stationary Airfoil Theory and its Practical Application', *J. Aeron. Sci.* Vol. 8, 1941, pp. 104–108.

[99] King, J. P. C. and Davenport, A. G., 'The Determination of Dynamic Wind Loads on Long Span Bridges', *Proceedings, Fifth U.S. National Conference on Wind Engineering*, Lubbock, Texas, Nov. 1985, pp. 4A–25 to 4A–32.

[100] Baird, R. C., 'Wind-Induced Vibration of a Pipe-Line Suspension Bridge and its Cure', *Transactions Amer. Soc. of Mech. Engrs.*, Vol. 77, August 1955, pp. 797–804.

[101] Okubo, T., Okauchi, I. and Murakami, E., 'Aerodynamic Response of a Large-Scale Bridge Model against Natural Wind', *Reliability Approach in Structural Engineering*, Maruzen Co. Ltd., Tokyo, 1975, pp. 315–328.

[102] Huston, D. R., 'Flutter Derivatives Extracted from Fourteen Generic Deck Sections', *Bridges and Transmission Line Structures* (L. Tall, Ed.), pp. 281–291, Proc. ASCE Structures Congress, Orlando, Florida, August 1987.

[103] Scanlan, R. H. and Lin, W.-H., 'Effects of Turbulence on Bridge Flutter Derivatives', *J. Engrg. Mech. Div. ASCE*, Vol. 104, No. EM4, 1978, pp. 719–733.

[104] Huston, D. R., 'The Effects of Upstream Gusting on the Aeroelastic Behavior of Long Suspended-Span Bridges', Doctoral Dissertation, Princeton University, May 1986.

[105] Lin, Y. K. and Ariaratnam, S. T., 'Stability of Bridge Motion in Turbulent Winds', *J. Struct. Mech.* Vol. 8, No. 1, 1980, pp. 1–15.

[106] Scanlan, R. H., 'Role of Indicial Functions in Buffeting Analysis of Bridges', *J. Struct. Div. ASCE* Vol. 110, No. 7, July 1984, pp. 1433–1446.

[107] Bucher, C. G. and Lin, Y. K., 'Stochastic Stability of Bridges Considering Coupled Modes', *Reports* CAS 87-7, 87-9, Florida Atlantic University, Sept., Oct. 1987.

[108] Davenport, A. G., 'The Response of Slender, Line-Like Structures to a Gusty Wind', *Proceedings Institution of Civil Engineers*, Paper 6610, London, 1962, pp. 389–407.

[109] Davenport, A. G., 'The Application of Statistical Concepts to the Wind Loading of Structures', *Proc. Inst. Civ. Eng.*, Vol. 19, 1961, London, pp. 449–472.

[110] Liepmann, H. W., 'On the Application of Statistical Concepts to the Buffeting Problem', *J. Aeron. Sci.*, Vol. 19, No. 12, December 1952, pp. 793–800, 822.

[111] Scanlan, R. H. and Gade, R. H., 'Motion of Suspended Bridge Spans under Gusty Wind', *Proc. ASCE, J. of the Struct. Div.* Vol. 103, 1977, pp. 1867–1883.

[112] Billah, K.Y.R., 'A Study of Vortex-Induced Vibration', Doctoral dissertation, Princeton University, 1989.

[113] Goswami, I., 'Vortex-Induced Vibrations of Circular Cylinders', Doctoral dissertation, Johns Hopkins University, 1991.

[114] Goswami, I., Scanlan, R.H., and Jones, N.P., 'Vortex-Induced Vibration of a Circular Cylinder in Air: I. New Experimental data'; II. A new Model', it ASCE Jnl. of Engineering Mechanics, Vol. 119, No. 11, Nov. 1993, pp. 2270–2302.

[115] Griffin, O.M., Skop, R.A., and Ramberg, S.E., 'The Resonant Vortex-Excited Vibrations of Structures and Cable Systems', Offshore Technology Conference Paper OTC-2319, Houston, TX 1975.

[116] Ehsan, F. and Scanlan, R.H., 'Vortex-Induced Vibrations of Flexible Bridges', *ASCE Jnl. Engineering Mechanics*, 116 (6) June 1990.
[117] Billah, K.Y.R. and Scanlan, R.H., 'Resonance, Tacoma Narrows bridge failure, and undergraduate physics textbooks', *Amer. Jnl. Physics*, 59 (2), Feb. 1991, pp. 118–124.
[118] Scanlan, R.H., 'Problematics in the Formulation of Wind Force Models for Bridge Decks', To appear in *ASCR Jnl. Engineering Mechanics*, Vol. 119, No. 7, July 1993, pp. 1353–1375.
[119] Sarkar, P., 'New Identification Methods Applied to the Response of Flexible Bridges to Wind', Doctoral dissertation, Johns Hopkins University, June 1992.
[120] Wagner, H. Über die Entstehung des dynamischen Auftriebes von Tragflügeln', *Z. angew. Math. u. Mech.*, 5, 1925, pp. 17–35.
[121] Küssner, H.G., 'Zusammenfassender Bericht über den instationären Auftrieb von Flügeln,', *Luftfahrt-Forschung*, 13 p. 410, 1936.
[122] Scanlan, R.H. and Jones, N.P. 'Aeroelastic Analysis of Cable-Stayed Bridges', *ASCE Jnl. Struct. Engineering*, 116 (2), Feb. 1990, pp. 279–297.

7

Aeroelastic response of rotorcraft

In this chapter we will examine a number of aeroelastic phenomena associated with helicopters and other rotor or propeller driven aircraft. Certain areas have been selected for treatment to illustrate some significant stability problems which are associated with the design of helicopters. The approach to be followed employs simplified modelling of various problems such that physical insight into the nature of the phenomena can be obtained. In general a complete and precise formulation of many of the problem areas discussed is highly complex and the reader is referred to the literature for these more detailed formulations.

A basic introduction to the mechanics and aerodynamics of helicopters may be found in [1] and [2]. Extensive reviews of helicopter aeroelasticity may be found in [3] and [4]. Ref. [4] provides an excellent discussion of the considerations necessary in modelling helicopter aeroelasticity and illustrates the complexity of a general formulation as well as the care required to obtain a complete and precise analytical model.

Helicopter rotors in use may be broadly classified in four types, semi-articulated or teetering, fully-articulated, hingeless and bearingless. This classification is based on the manner in which the blades are mechanically connected to the rotor hub. The teetering rotor is typically a two-bladed rotor with the blades connected together and attached to the shaft by a pin which allows the two-blade assembly to rotate such that tips of the blades may freely move up and down with respect to the plane of rotation (flapping motion). In the fully-articulated rotor, each blade is individually attached to the hub through two perpendicular hinges allowing rigid motion of the blade in two directions, out of the plane of rotation (flapping motion) and in the plane of rotation (lag motion). The third type is the hingeless rotor in which the rotor blade is a cantilever beam, but with soft flexures near the root, simulating hinges. Fourth, bearingless rotors further replace the pitch bearing by the softness in torsion of the root of the blade. Thus, pitch changes are introduced through torsional deformations.

Because of their greater flexibility, elastic deformations of hingeless and bearingless rotors are significant in the analysis of the dynamics of the vehicle. Bending out of the plane of rotation is referred to as flap bending and in-plane as lag bending. These three rotor configurations are shown schematically in Figure 7.1. Rotation of the blade about its long axis is controlled by a pitch change mechanism suitably connected to the pilot's stick. For further details see [1] for articulated rotors and [5] for hingeless rotors. Other variations in rotor hub geometry are found such as the gimballed rotor described in [6]. We will concentrate our discussion on the aeroelastic behavior of fully-articulated and hingeless rotors. However, it is important to realize that for the aeroelastic analysis of rotors the precise details of the hub and blade geometry must be carefully modelled.

Problems in helicopter aeroelasticity may be classified by the degrees-of-freedom which are significantly coupled. Typically, the dynamics of a single blade are of interest although coupling among blades can be present through the elasticity of the blade pitch control system or the aerodynamic wake [7]. The degrees-of-freedom of a single blade include rigid body motion in the case of the articulated system as well as elastic motion. Elastic motions of interest include bending in two directions and twisting or torsion. These elastic deformations are coupled in general. In addition to individual blade aeroelastic problems, the blade degrees-of-freedom can couple with the rigid body degrees-of-freedom of the fuselage in flight as well as the elastic deformations of the fuselage [8–10] or with the fuselage/landing gear system on the ground [10]. In fact, a complete aeroelastic model of the helicopter typically involves a dynamic model with a large number of degrees-of-freedom. We do not propose to examine these very complex models, but rather will consider simple formulations of certain significant stability problems which will give some insight into the importance of aeroelasticity in helicopter design. Avoiding resonances is also of considerable significance, but is not discussed here. First, aeroelastic problems associated with an individual blade are described and then those associated with blade/body coupling are examined. Finally, we will consider phenomena associated with the dynamics of the wake.

7.1 Blade dynamics

Classical flutter and divergence of a rotor blade involving coupling of flap bending and torsion have not been particularly significant due to the fact that, in the past, rotor blades have been designed with their elastic axis, aerodynamic center, and center of mass coincident at the quarter chord.

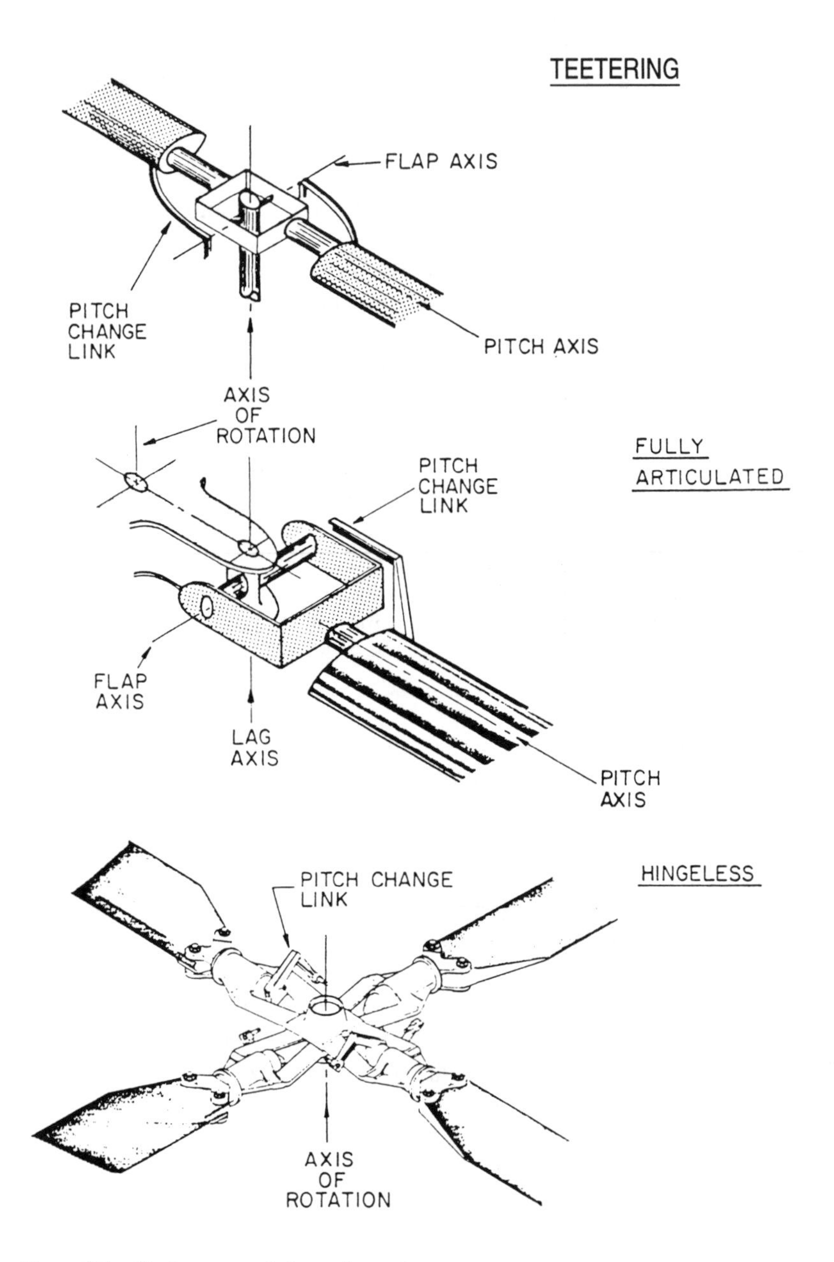

Figure 7.1 Various rotor hub configurations.

In addition the blades are torsionally stiff (a typical torsion frequency of a modern rotor blade is about 5–8 per revolution) which minimizes coupling between elastic flap bending and torsion. It is important to note that torsional stiffness is used here in the sense that control system flexibility is included as well as blade flexibility. Rotor systems with low torsional stiffness [11] have experienced flutter problems, and on hingeless and bearingless rotors, the blade section center of mass and elastic axis position can be moved from the quarter chord to provide a favorable effect on the overall flight stability [12] which may mean that these classical problem areas will have to be reviewed more carefully in the future. Sweep has also been employed on rotor blades [13] and this couples flap bending with torsion. However, we will not consider flutter and divergence here, but will instead concentrate on problems more frequently encountered in practice. Further discussion of classical bending-torsion flutter and divergence of rotor blades may be found in [3] and [14].

Articulated, rigid blade motion

In orter to introduce the nature of rotor blade motion we first develop the equations of motion for the flapping and lagging of a fully articulated blade assuming that the blade is rigid. Consider a single blade which has only a flapping hinge located on the axis of rotation as shown in Figure 7.2 The blade flapping angle is denoted by β_s and the blade rotational speed by Ω. We proceed to derive the equation of motion of the blade about the flapping axis. We assume that the rotor is in a hovering state with no translational velocity. It is most convenient to use a Newtonian approach to this problem. Since the flapping pin is at rest in space we may write the equation of motion for the blade as follows [15]

$$\dot{\bar{H}}_P + \bar{\Omega}_B \times \bar{H}_P = \int_0^R \bar{r} \times \mathrm{d}\bar{F}_A \tag{7.1.1}$$

A blade-body axis system denoted by the subscript $_B$ is employed and $\bar{H}_P$ is the moment of momentum of the blade with respect to the flapping pin. $\mathrm{d}\bar{F}_A$ is the aerodynamic force acting on the blade at the radial station $\bar{r}$. The gravity force on the blade is neglected owing to the comparatively high rotational velocity. Figure 7.2 also shows the coordinate system and variables involved. The blade is modelled as a very slender rod, and the body axes are principal axes such that the inertia characteristics of the blade are

$$I_B \cong I_y \cong I_z; \qquad I_x \cong 0$$

Therefore

$$\bar{H} = (I_B q_B)\bar{j}_B + (I_B r_B)\bar{k}_B \tag{7.1.2}$$

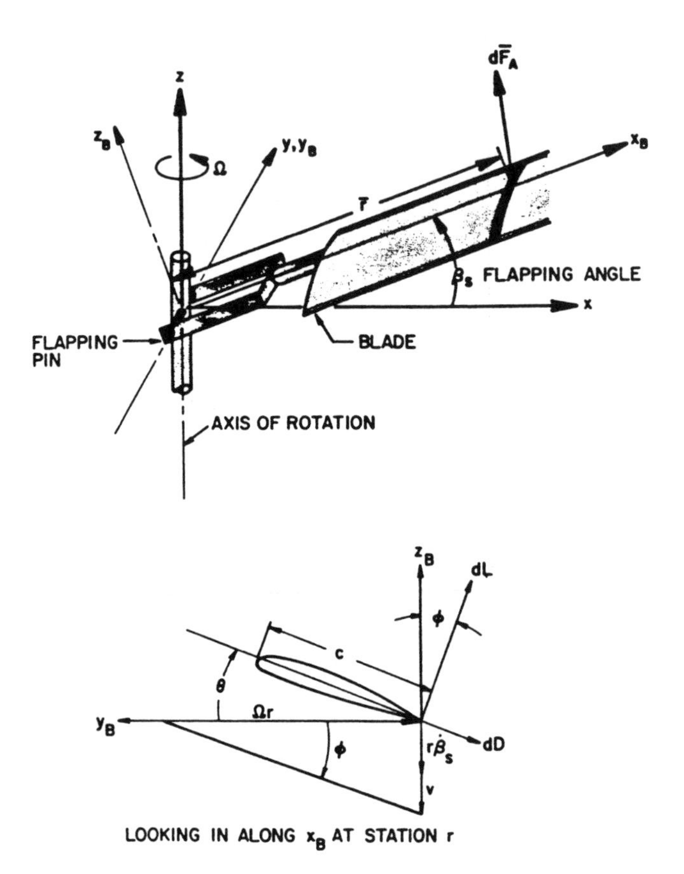

Figure 7.2 Coordinate systems and aerodynamics for blade flapping analysis.

where

$$\bar{\Omega} = p_B \bar{i}_B + q_B \bar{j}_B + r_B \bar{k}_B \tag{7.1.3}$$

The equation of motion, (7.1.1), becomes

$$I_B[\dot{q}_B - p_B r_B]\bar{j}_B + I_B[\dot{r}_B + p_B q_B]\bar{k}_B = \int_0^R \bar{r} \times \mathrm{d}\bar{F}_A \tag{7.1.4}$$

Now we must express the angular body rates in terms of the variables of interest in the problem, Ω the angular velocity, and β_s the flap angle. The angular velocity must be resolved into the blade axis system by rotation through β_s, and then the flapping velocity $\dot{\beta}_s$ added.

$$\begin{Bmatrix} p_B \\ q_B \\ r_B \end{Bmatrix} = \begin{bmatrix} \cos\beta_s & 0 & \sin\beta_s \\ 0 & 1 & 0 \\ -\sin\beta_s & 0 & \cos\beta_s \end{bmatrix} \begin{Bmatrix} 0 \\ 0 \\ \Omega \end{Bmatrix} + \begin{Bmatrix} 0 \\ -\dot{\beta}_s \\ 0 \end{Bmatrix}$$

That is

$$\begin{aligned} p_B &= \Omega \sin\beta_s \\ q_B &= -\bar{\beta}_s \\ r_B &= \Omega \cos\beta_s \end{aligned} \tag{7.1.5}$$

Substitution of (7.1.5) into (7.1.4) gives

$$I_B[-\ddot{\beta}_s - \Omega^2 \cos\beta_s \sin\beta_s]\bar{j}_B + [-2\Omega \sin\beta_s \dot{\beta}_s]\bar{k}_B = \int_0^R \bar{r} \times \mathrm{d}\bar{F}_A \tag{7.1.6}$$

The first term on the left hand side is the angular acceleration of the blade about the y_B axis and the second term is the angular acceleration of the blade about the z_B axis. The second term can be considered an inertial torque about the z_B axis (i.e., in the lag direction), which arises as a result of out-of-plane (flapping) motion of the blade. The aerodynamic force on the blade element is comprised of the lift and drag and is formulated from strip theory (usually called blade-element theory) [1, 2]. Also, see the discussion in Section 3.4. Three-dimensional effects are obtained by including the induced velocity which, for our purposes, may be calculated by momentum theory [1]. Thus from Figure 7.2

$$\mathrm{d}\bar{F}_A = \mathrm{d}L\bar{k}_B + (-\mathrm{d}D - \phi\, \mathrm{d}L)\bar{j}_B \tag{7.1.7}$$

where the inflow angle ϕ is assumed to be small and is made up of the effect of induced velocity (downwash) and the induced angle due to flapping velocity. Therefore

$$\mathrm{d}L = \frac{1}{2}\rho(\Omega r)^2 c\, \mathrm{d}r a(\theta - \phi)$$

$$\mathrm{d}D = \frac{1}{2}\rho(\Omega r)^2 c\, \mathrm{d}r\delta$$

$$\phi = \frac{r\dot{\beta}_s + v}{\Omega r}$$

Define

$$x \equiv \frac{r}{R}\,; \quad \lambda \equiv -\frac{v}{\Omega R}\,; \quad \gamma \equiv \frac{\rho a c R^4}{I_B}, \qquad \text{the Lock number} \tag{7.1.8}$$

The blade chord, c, and pitch angle, θ, are taken to be independent of x, for simplicity, although rotor blades are usually twisted. The blade section drag coefficient is denoted by δ and is also assumed to be independent of the radial station. Thus

$$\begin{aligned} \mathrm{d}L &= \frac{I_B\gamma\Omega^2}{R}\left[\theta - \frac{\dot{\beta}_s}{\Omega} + \frac{\lambda}{x}\right] x^2\,\mathrm{d}x \\ \mathrm{d}D &= \frac{I_B\gamma\Omega^2}{R}\left(\frac{\delta}{a}\right) x^2\,\mathrm{d}x \end{aligned} \tag{7.1.9}$$

and

$$\bar{r} = xR\bar{\imath}_B$$

The total rotor thrust is found by integrating the lift along the radius, averaging over one revolution, and multiplying by the number of blades to give [1]

$$\frac{2C_T}{a\sigma} = \frac{\theta}{3} + \frac{\lambda}{2} \tag{7.1.10}$$

where

$$\sigma = \frac{bc}{\pi R}$$

and b is the number of blades. The thrust coefficient is

$$C_T = \frac{T}{\rho\pi R^2(\Omega R)^2}$$

Momentum theory results in the following expression for the induced velocity

$$\lambda = -\sqrt{\frac{C_T}{2}}$$

so that the integral on the right-hand side of equation (7.1.6) becomes

$$\int_0^R \bar{r}\times \mathrm{d}\bar{F}_A = -\frac{I_B\gamma\Omega^2}{8}\left[\theta + \frac{4\lambda}{3} - \frac{\dot{\beta}_s}{\Omega}\right]\bar{\jmath}_B$$

$$+\frac{I_B\gamma^2}{8}\left[-\frac{\delta}{a} + \frac{\dot{\beta}_s}{\Omega}\left(\frac{\dot{\beta}_s}{\Omega} - \theta\right) + \frac{4}{3}\left(\theta - 2\frac{\dot{\beta}_s}{\Omega}\right)\lambda + 2\lambda^2\right]\bar{k}_B \tag{7.1.11}$$

The $\bar{j}_B$ components contribute to the flapping equation of motion which may be expressed from equations (7.1.6) and (7.1.11) as

$$\ddot{\beta}_s + \frac{\gamma\Omega}{8}\dot{\beta}_s + \Omega^2 \cos\beta_s \sin\beta_s = \frac{\gamma\Omega^2}{8}\left[\theta + \frac{4\lambda}{3}\right] \tag{7.1.12}$$

The $\bar{k}_B$ component of equation (7.1.11) is the aerodynamic torque about the z_B axis or in the lag direction. There is a steady component and a component proportional to flapping velocity. Each of these components is important either for loads or for stability of the inplane motion.

If we assume that the flapping motion is small as is typical of rotor blade motion then the flapping equation becomes linear.

$$\ddot{\beta}_s + \frac{\gamma\Omega}{8}\dot{\beta}_s + \Omega^2\beta_s = \frac{\gamma\Omega^2}{8}\left[\theta + \frac{4\lambda}{3}\right] \tag{7.1.13}$$

The linearized blade flapping equation may be recognized as a second order system with a natural frequency equal to the rotor angular velocity and a damping ratio equal to $\gamma/16$ which arises from the aerodynamic moment about the flapping pin. This motion is well damped as γ is between 5 and 15 for typical rotor blades. It is good that the system is well damped since the aerodynamic inputs characteristically occur in forward flight at Ω, and thus the blade flapping motion is forced at resonance.

The spring or displacement term can be interpreted as arising from the centrifugal force [1]. This same stiffening effect will appear in the flexible blade analysis and will increase the natural frequency as rotational speed is increased.

If the more general case of flapping in forward flight is considered, then the equation of motion for flapping (7.1.12) will contain periodic coefficients which can lead to instabilities [16]. However, the flight speed at which such instabilities occur is well beyond the performance range of conventional helicopters, unless they have positive pitch-flap coupling.

Now we include the lag degree-of-freedom to obtain a complete description of rigid motion of a fully-articulated rotor blade. The complete development of this two-degree-of-freedom problem is quite lengthy and will not be reproduced here [17].

Following the approach given above, assuming that the flap angle and lag angle are small and that the lag hinge and flap hinge are coincident and located a small distance e (hinge offset) from the axis of rotation as shown in Figure

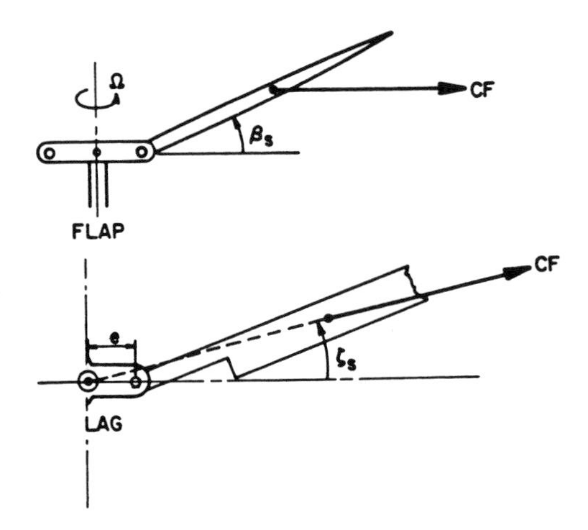

Figure 7.3 Direction of centrifugal force for flap and lag motion.

7.3, and further accounting for the effect of lag velocity on the aerodynamics forces acting on the blade, the lift is given by

$$\mathrm{d}L = \frac{1}{2}\rho[(\Omega + \dot{\zeta}_s r]^2 c\,\mathrm{d}ra\left[\theta - \frac{(r\dot{\beta}_s + v)}{(\Omega + \dot{\zeta}_s)r}\right] \tag{7.1.14}$$

where the effect of the small distance e on the aerodynamics is neglected. The lag angle is defined as positive in the direction of rotor rotation. Care must be taken in formulating the inertial terms since we have noted above that a term like $\dot{\beta}_s \sin\beta_s$ is of significance in the equations of motion, and thus the small angle assumption must not be made until after the expressions for the acceleration have been obtained. Rotating by the flap angle first and then by the lag angle, the angular rates in the blade body axis system are given by

$$\begin{Bmatrix} p_b \\ q_B \\ r_b \end{Bmatrix} = \begin{bmatrix} \cos\zeta_s & \sin\zeta_S & 0 \\ -\sin\zeta_s & \cos\zeta_s & 0 \\ 0 & 0 & 1 \end{bmatrix} \begin{bmatrix} cos\beta_s & 0 & \sin\beta_s \\ 0 & 1 & 9 \\ -\sin\beta_s & 0 & \cos\beta_s \end{bmatrix} \begin{Bmatrix} 0 \\ 0 \\ \Omega \end{Bmatrix}$$
$$+ \begin{bmatrix} \cos\zeta s & \sin\zeta_s & 0 \\ -\sin\zeta_s & \cos\zeta_s & 0 \\ 0 & 0 & 1 \end{bmatrix} \begin{Bmatrix} 0 \\ -\dot{\beta}_s \\ 0 \end{Bmatrix} + \begin{Bmatrix} 0 \\ 0 \\ \dot{\zeta}_s \end{Bmatrix} \tag{7.1.15}$$

We must also account for the fact that the hinge point of the blade is no longer at rest but is accelerating [15]. Since the hinge point is located at a distance e

from the axis of rotation, the equation of motion, (7.1.1), must be modified to read

$$\dot{\bar{H}}_P + \bar{\Omega}_B \times \bar{H}_P = \int_0^R \bar{r} \times \mathrm{d}\bar{F}_A + \bar{E} \times M_B \bar{a}_P \tag{7.1.16}$$

where $\bar{a}_P$ is the acceleration of the hinge point

$$\bar{a}_P = \bar{\Omega} \times (\bar{\Omega} \times \bar{E}) + \dot{\bar{\Omega}} \times \bar{E} \tag{7.1.17}$$

$\bar{E}$ is the offset distance and M_B is the blade mass.

Accounting for all of these factors, and assuming that the flapping and lagging motion amplitudes are small, the equations of motion for this two-degree-of-freedom system may be expressed [17, 18] as

$$-\ddot{\beta}_s - \Omega^2 \left(1 + \frac{3}{2}\bar{e}\right) \beta_s - 2\beta_s \dot{\zeta}_s \Omega = -\frac{\gamma \Omega^2}{8} \left[\theta + \frac{4}{3}\lambda - \frac{\dot{\beta}_s}{\Omega} + \left(2\theta + \frac{4}{3}\lambda\right) \frac{\dot{\zeta}_s}{\Omega}\right] - 2\beta_s \dot{\beta}_s \Omega + \ddot{\zeta}_s + \frac{3}{2}\bar{e}\Omega^2 \zeta_s = \frac{\gamma \Omega^2}{8} \left[-\left(\theta + \frac{8}{3}\lambda\right) \frac{\dot{\beta}_s}{\Omega} - \left(2\frac{\delta}{a} - \frac{4}{3}\lambda\theta\right) \frac{\dot{\zeta}_s}{\Omega} - -\frac{\delta}{a} + \frac{4}{3}\lambda\theta + 2\lambda^2\right] \tag{7.1.18}$$

where

$$\bar{e} = \frac{e}{R}.$$

It has been assumed that the blade has a uniform mass distribution. These results can be displayed more conveniently by nondimensionalizing time by rotor angular velocity Ω and also expressing the variables as the sum of a constant equilibrium part and a perturbation

$$\begin{aligned} \beta_s &= \beta_0 + \beta \\ \zeta_s &= \zeta_0 + \zeta \end{aligned} \tag{7.1.19}$$

Retaining only linear terms, the equilibrium equations are

$$\begin{aligned} \beta_0 &= \frac{\gamma}{8(1 + \frac{3}{2}\bar{e})} \left[\theta + \frac{4\lambda}{3}\right] \\ \zeta_0 &= \frac{\gamma}{12\bar{e}} \left[-\frac{\delta}{a} + \frac{4}{3}\lambda\theta + 2\lambda^2\right] = -\frac{1}{3}\frac{\gamma}{\bar{e}} \left(\frac{2C_q}{a\sigma}\right) \end{aligned} \tag{7.1.20}$$

The steady value of the flapping, β_0, is referred to as the coning angle. The steady value of the lag angle, ζ_0, is proportional to the rotor torque coefficient, C_q [1].

The perturbation equations are

$$\ddot{\beta} + \frac{\gamma}{8} + \left(1 + \frac{3}{2}\bar{e}\right)\beta + \left[2\beta_0 - \frac{\gamma}{8}\left(2\theta + \frac{4}{3}\lambda\right)\right]\dot{\zeta} = 0$$

$$\left[-2\beta_0 + \frac{\gamma}{8}\left(\theta + \frac{8}{3}\lambda\right)\right]\dot{\beta} + \ddot{\zeta} + \frac{\gamma}{8}\left(2\frac{\delta}{a} - \frac{4}{3}\lambda\theta\right)\dot{\zeta} + \frac{3}{2}\bar{e}\zeta = 0 \tag{7.1.21}$$

These equations describe the coupled flap-lag motion of a rotor blade. A number of features can be noted. The effect of the blade angular velocity on the lag frequency is much weaker than on flap frequency. The uncoupled natural frequency in flap expressed as a fraction of the blade angular velocity is

$$\frac{\omega_\beta}{\Omega} = \sqrt{1 + \frac{3}{2}\bar{e}} \tag{7.1.22}$$

and the uncoupled frequency in lag is

$$\frac{\omega_\zeta}{\Omega} = \sqrt{\frac{3}{2}\bar{e}} \tag{7.1.23}$$

For a typical hinge offset of $\bar{e} = 0.05$, the rigid flap frequency is

$$\frac{\omega_\beta}{\Omega} = 1.04$$

and the rigid lag frequency is

$$\frac{\omega_\zeta}{\Omega} = 0.27$$

The flap natural frequency is thus somewhat higher than the rotational speed and the lag frequency is roughly one-quarter of the rotational speed. This difference is due to the weaker effect of the restoring moment due to centrifugal force in the lag direction as indicated in Figure 7.3.

The uncoupled lag damping arises primarily from the blade drag and is equal to

$$D_L \equiv 2\frac{\delta}{a}\left(\frac{\gamma}{8}\right) \tag{7.1.24}$$

The lift curve slope of the blade, a, is the order of 6 per radian and the drag coefficient, δ, is the order of 0.015 giving a physical lag damping which is

0.005 times the flap damping or characteristically negligible. The damping ratio of the uncoupled lag motion for a Lock number of 8 is

$$\zeta_L = 0.009$$

This low value of aerodynamic damping indicates that structural damping will be of significance in estimating the lag damping. Any coupling between these equations which reduces the lag damping tends to result in an instability. Equations (7.1.21) can be rewritten

$$\ddot{\beta} + \frac{\gamma}{8}\dot{\beta} + \left(1 + \frac{3}{2}\bar{e}\right)\beta + \left(\beta_0 - \frac{\gamma}{8}\theta\right)\dot{\zeta} = 0$$

$$\frac{-\gamma}{8}\theta\dot{\beta} + \ddot{\zeta} + \frac{\gamma}{8}\left[2\frac{\delta}{a}\right]\dot{\zeta} + \frac{3}{2}\,\bar{e}\zeta = 0 \tag{7.1.25}$$

where the equilibrium relationship for β_0 has been introduced (7.1.20) with the effect of hinge offset on coning neglected. It can be shown that the coupling present in this two-degree-of-freedom system arising from inertial and aerodynamic forces will not lead to an instability. However, with the hinge offset or with minor features of the hub geometry this can lead to instability. The equilibrium lag angle is proportional to rotor torque (equation 7.1.20)) and consequently varies over a wide range from high power flight to autorotation as a result of the weak centrifugal stiffening. Thus the simple pitch link geometry shown in Figure 7.4 will produce a pitch change with lag depending upon the equilibrium lag angle. The blade pitch angle variation with lag angle can be expressed as

$$\Delta\theta = \theta_\zeta\zeta$$

This expression is inserted into equations (7.1.18). Retaining only the linear homogeneous terms, the perturbation equations are

$$\ddot{\beta} + \frac{\gamma}{8}\dot{\beta} + \left(1 + \frac{3}{2}\bar{e}\right)\beta + \left(\beta_0 - \frac{\gamma}{8}\theta\right)\dot{\zeta} - \frac{\gamma}{8}\theta_\zeta\zeta = 0$$

$$\frac{-\gamma}{8}\theta\dot{\beta} + \ddot{\zeta} + \frac{\gamma}{8}\left[2\frac{\delta}{a}\right]\dot{\zeta} + \frac{3}{2}\bar{e}\zeta - \frac{\gamma}{6}\lambda\theta_\zeta\zeta = 0 \tag{7.1.26}$$

We can now sketch a root locus for the effect of θ_ζ on the dynamics of this system. Expressing the equations of motion in operational notation, the root

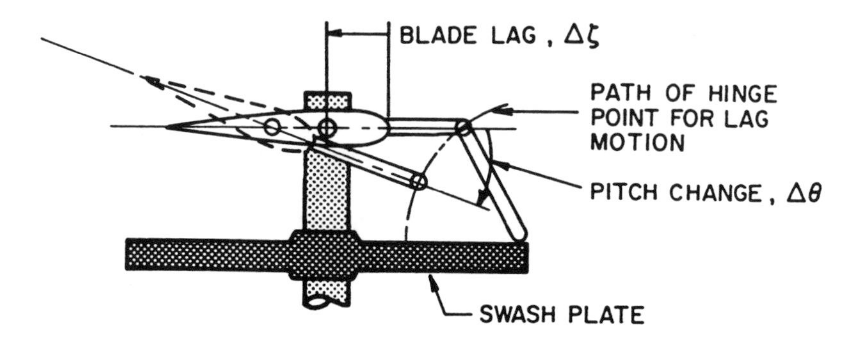

Figure 7.4 Pitch-lag coupling due to pitch link geometry. Articulated rotor.

locus equation for variations in θ_ζ is

$$\frac{-\theta_\zeta \frac{\gamma\lambda}{6}\left[s^2 + \frac{3}{4}\frac{\beta_0}{\lambda}s + 1 + \frac{3}{2}\bar{e}\right]}{\left[s^2 + \frac{\gamma}{8}s + \left(1 + \frac{3}{2}\bar{e}\right)\right]\left[s^2 + \frac{\gamma}{8}\left[2\frac{\delta}{a}\right]s + \frac{3}{2}\bar{e}\right] + \frac{\gamma}{8}\theta\left(\beta_0 - \frac{\gamma}{8}\theta\right)s^2} = -1 \tag{7.1.27}$$

The root locus shown in Figure 7.5 illustrates the effect of this geometric coupling, indicating that the critical case where instability occurs corresponds to the 180° locus (θ_ζ is positive). Recall that λ is negative. Thus, if forward lag produces an increase in pitch, an instability is likely to occur for a soft-inplane rotor. The effect is also proportional to thrust coefficient indicating that the instability is more likely to occur as the thrust is increased [18, 19]. Increasing thrust also increases the steady-state lag angle, hence increasing the geometric coupling for the geometry shown. In general, this instability tends to be of a rather mild nature, but it has destroyed tail rotors. Mechanical dampers are often installed about the lag axes for reasons to be discussed and these also provide additional lag damping and thus alleviate the instability.

This example serves to illustrate that great care must be taken in the geometric design of the articulated rotor hub to avoid undesirable couplings and possible instabilities. We now turn to the elastic hingeless blade.

Elastic motion of hingeless blades

The dynamics of a single hingeless blade will now be examined. Again we will use a simplified analysis which yields the essential features of the dynamic motion, and the reader is referred to the literature for a more detailed approach. In general, the flap and lag elastic deformations (as referred to a shaft axis

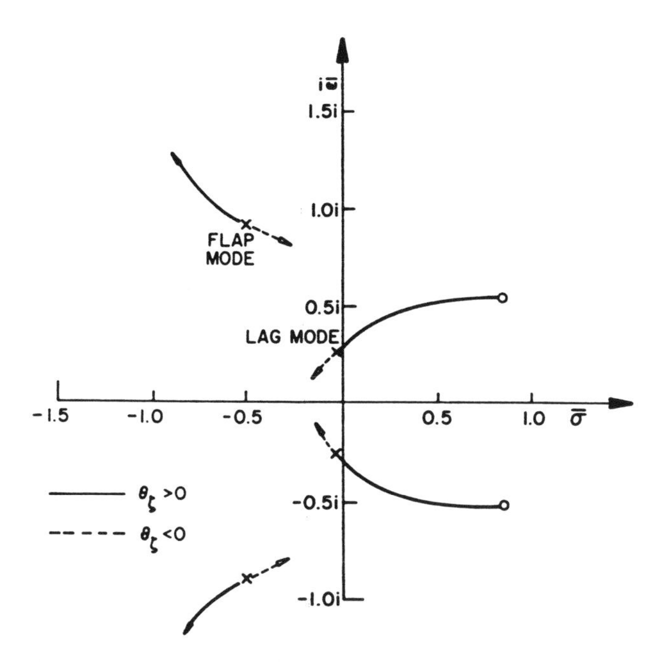

Figure 7.5 Effect of pitch-lag coupling on flap-lag stability.

system) are coupled as a result of the fact that the principal elastic axes of the blade will be inclined with respect to the shaft due to pitch angle. In fact, the term flexible "blade", as used here, includes the hub as well as the blade itself. Hub is used to refer to the portion of the blade structure inboard of the radial location where the pitch change takes place. The rotation of the blade principal elastic axes with blade pitch will depend upon the relative stiffness of the hub and the blade. It can be seen physically that, if the hub is soft in comparison to the blade, then the principal axes of this flexible system tend to remain fixed as the pitch of the blade is changed. However, if the hub is stiff and the blade is soft, the principal elastic axes rotate in a 1:1 relationship with blade pitch. An additional source of elastic coupling between flap and lag deflections arises from the built-in blade twist. A third source of elastic coupling between flap and lag arises from inclusion of torsion as a degree-of-freedom. For the typical rotor blade with a high torsional frequency, the effect of torsional flexibility on flap-lag coupling can be obtained through a quasistatic approximation to the torsional motion. That is, for a first-order estimate, the torsional inertia and damping can be neglected; and the coupling effects of torsional flexibility can be expressed in terms of a geometric coupling similar in form to the hub geometry effects described in connection with the fully-articulated rotor. A detailed analysis of the flap-lag-torsion motion of a hingeless rotor blade may be found in [20] and [21], and the complete equations of motion for elastic bending and torsion of rotor blades may be found in [22].

We now proceed to examine the flap-lag motion of a hingeless rotor blade from a simplified viewpoint.

If it is assumed that the rotor blade is untwisted, has zero pitch, and is torsionally rigid; the natural frequencies of the rotating blade can be expressed in terms of its mode shapes, ϕ, and derivatives with respect to radial distance ϕ' and ϕ'' as [23–25]

$$\omega_\beta^2 = \frac{\int_0^R EI_\beta(\phi''_\beta)^2\,\mathrm{d}r + \Omega^2\int_0^R(\phi'_\beta)^2\left(\int_r^R mn\,\mathrm{d}n\right)\mathrm{d}r}{\int_0^R m\phi_\beta^2\,\mathrm{d}r}$$

$$\omega_\zeta^2 = \frac{\int_0^R EI_\zeta(\phi''_\zeta)^2\,\mathrm{d}r + \Omega^2\left\{\int_0^R(\phi'_\zeta)^2\left(\int_r^R mn\,\mathrm{d}n\right)\mathrm{d}r - \int_0^R m\phi_\zeta^2\,\mathrm{d}r\right\}}{\int_0^R m\phi_\zeta^2\,\mathrm{d}r} \tag{7.1.28}$$

m is the running mass of the blade and EI is the stiffness. The first term in each of these expressions gives the nonrotating natural frequency and the second term gives the effect of centrifugal stiffening due to rotation. The coefficient of the square of the angular velocity Ω in the expression for flapping frequency is usually referred to as the Southwell coefficient. Note that the effect of the centrifugal stiffening is considerably weaker in the lag direction than in the flap direction as would be expected from previous discussion of the articulated rotor.

Denoting the Southwell coefficient by K_s

$$K_s = \frac{\int_0^R(\phi'_\beta)^2\left(\int_r^R mn\,\mathrm{d}n\right)\mathrm{d}r}{\int_0^R m\phi_\beta^2\,\mathrm{d}r} \tag{7.1.29}$$

and the nonrotating frequencies by

$$\omega_{\beta_0}^2 = \frac{\int_0^R EI_\beta(\phi''_\beta)^2\,\mathrm{d}r}{\int_0^R m\phi_\beta^2\,\mathrm{d}r}$$

$$\omega_{\zeta_0}^2 = \frac{\int_0^R EI_\zeta(\phi''_\zeta)^2\,\mathrm{d}r}{\int_0^R m\phi_\zeta^2\,\mathrm{d}r} \tag{7.1.30}$$

If the flap and lag mode shapes are assumed to be the same, the rotating frequencies can be written as

$$\begin{aligned} \omega_\beta^2 &= \omega_{\beta_0}^2 + K_s \Omega^2 \\ \omega_\zeta^2 &= \omega_{\zeta_0}^2 + (K_s - 1)\Omega^2 \end{aligned} \tag{7.1.31}$$

It is interesting to note that, if the mode shape is assumed to be that of a rigid articulated blade with hinge offset, $\bar{e}$, i.e.,

$$\begin{aligned} \phi &= 0 \quad 0 < x < \bar{e} \\ \phi &= (x - \bar{e}) \quad \bar{e} < x < 1 \end{aligned} \tag{7.1.32}$$

for a uniform mass distribution and small $\bar{e}$, then from (7.1.29)

$$K_s \equiv 1 + \frac{3}{2}\bar{e} \tag{7.1.33}$$

Thus the natural frequencies are from (7.1.30), (7.1.31) and (7.1.33)

$$\begin{aligned} \omega_\beta^2 &= \Omega^2 \left(1 + \frac{3}{2}\bar{e}\right) \\ \omega_\zeta^2 &= \Omega^2 \left(\frac{3}{2}\bar{e}\right) \end{aligned} \tag{7.1.34}$$

reducing to the results for the rigid blade. For typical blade mass and stiffness distributions the Southwell coefficient is of the order of 1.2 [24].

A simplified model for the elastic rotor blade follows. The elastic blade is modelled as a rigid blade with hinge offset $\bar{e}$ and two orthogonal springs (K_β and K_ζ) located at the hinge to represent the flap and lag stiffness characteristics. The natural frequencies for this model of the blade are

$$\omega_\beta^2 = \frac{K_\beta}{I_B} + \left(1 + \frac{3}{2}\bar{e}\right)\Omega^2$$

$$\omega_\zeta^2 = \frac{K_\zeta}{I_B} + \left(\frac{3}{2}\bar{e}\right)\Omega^2$$

The spring constants K_β and K_ζ can be chosen to match the nonrotating frequencies of the actual elastic blade and the offset is chosen to match the Southwell coefficient and in this way the dependence of frequency on rotor angular velocity is matched. Owing to the fact that the Southwell coefficient

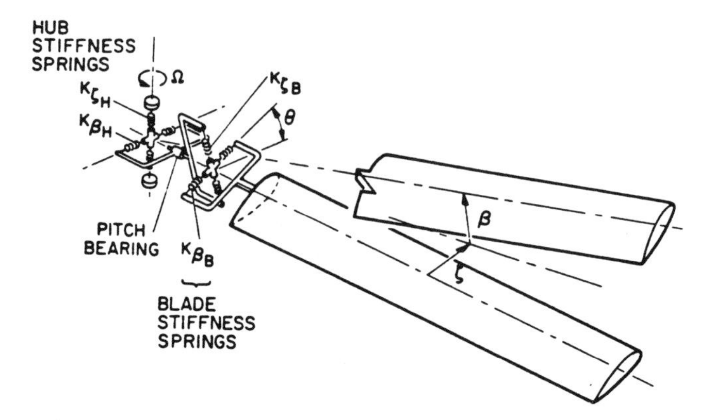

Figure 7.6 Spring model for elastic blade and hub.

is close to unity (i.e., the equivalent offset, $\bar{e}$, is amall), in many investigations the dependence of the Southwell coefficient on $\bar{e}$ is neglected [17] giving

$$\begin{aligned}\omega_\beta^2 &= \frac{K_\beta}{I_B} + \Omega^2 \\ \omega_\zeta^2 &= \frac{K_\zeta}{I_B}\end{aligned} \tag{7.1.36}$$

Thus, with this approximation there is no centrifugal stiffening in the lag direction. We will use this approximation in the analysis which follows. Recall that these frequencies are assumed to be uncoupled and therefore are defined with respect to the blade axes. Thus, they will appear coupled in a shaft oriented axis system. In order to include the effect of hub flexibility in the analysis, the hub (the portion of the blade system which does not rotate with pitch) is modelled by a second pair of orthogonal springs which are oriented parallel and perpendicular to the shaft and do not rotate when the blade pitch is changed [17]. These spring constants are denoted K_{β_H} and K_{ζ_H}. The springs representing blade stiffness (K_{β_B} and K_{ζ_B}) are also located at the root since offset has been neglected. However, this pair of springs rotate with the blade as pitch is changed. Figure 7.6 shows the geometry.

This model for the hub and blade gives rise to elastic coupling between flap and lag motion. Essentially, a mode shape $\phi = x$ is being employed to describe the elastic deflection of the blade in both directions such that the aerodynamic and inertial coupling terms developed for the articulated blade model (equation (7.1.21) apply directly to this approximate model of the hingeless blade. The equations of motion for flap-lag dynamics are therefore

$$\ddot{\beta} + \frac{\gamma}{8}\dot{\beta} + p^2\beta - \left\{\frac{\gamma}{8}\left(2\theta + \frac{4}{3}\lambda\right) - 2\beta_0\right\}\dot{\zeta} + z^2\zeta = 0$$

$$\left[-2\beta_0 + \frac{\gamma}{8}\left(\theta + \frac{8}{3}\lambda\right)\right]\dot{\beta} + z^2\beta + \ddot{\zeta} + \frac{\gamma}{8}\left(2\frac{\delta}{a} - \frac{4}{3}\lambda\theta\right)\dot{\zeta} + q^2\zeta = 0 \qquad (7.1.37)$$

where the difference between these equations of motion and those presented for the articulated blade (7.1.21) arise from the terms p, q, and z. p and q are the ratios of the oncoupled natural frequencies (i.e., those at zero pitch) to the rotor rpm, and z is the elastic coupling effect. For the spring model described above these terms can be expressed as [17]

$$p^2 = 1 + \frac{1}{\Delta}(\bar{\omega}_\beta^2 + R(\bar{\omega}_\beta^2)\sin^2\theta)$$

$$q^2 = \frac{1}{\Delta}(\bar{\omega}_\zeta^2 - R(\bar{\omega}_\zeta^2)\sin^2\theta)$$

$$z^2 = \frac{R}{2\Delta}(\bar{\omega}_\zeta^2 - \bar{\omega}_\beta^2)\sin 2\theta$$

$$\Delta = 1 + R(1 - R)\frac{(\bar{\omega}_\zeta^2 - \bar{\omega}_\beta^2)^2}{\bar{\omega}_\zeta^2\bar{\omega}_\beta^2}\sin 2\theta \qquad (7.1.38)$$

$$\bar{\omega}_\beta^2 = \frac{K_\beta}{I_B\Omega^2}\bar{\omega}_\zeta^2 = \frac{K_\zeta}{I_B\Omega^2}$$

$$K_\beta = \frac{K_{\beta_B}K_{\beta_B}}{K_{\beta_B} + K_{\beta_B}} \qquad K_\zeta = \frac{K_{\zeta_B}K_{\zeta_H}}{K_{\zeta_B} + K_{\zeta_H}} \qquad R = \frac{\bar{\omega}_\zeta^2\frac{K_\beta}{K_{\beta_B}} - \bar{\omega}_\beta^2\frac{K_\zeta}{K_{\zeta_B}}}{\bar{\omega}_\zeta^2 - \bar{\omega}_\beta^2}$$

R is referred to as the elastic coupling parameter. The physical significance of this parameter can be understood by examining the relationship between the rotation of the principal axes of the blade-hub system, η, and the blade pitch angle, θ, [26]

$$\tan 2\eta = \frac{R\sin 2\theta}{R\cos 2\theta + (1 - R)} \qquad (7.1.39)$$

It can be seen from this expression that if $R = 0$ the principal axes remain fixed as blade pitch is changed and consequently there is no elastic coupling. The flap and lag natural frequencies are

$$p^2 = \bar{\omega}_\beta^2 + 1$$

$$q^2 = \bar{\omega}_\zeta^2$$

where $\bar{\omega}_\beta^2$ and $\bar{\omega}_\zeta^2$ are the dimensionless nonrotating frequencies. This is the case in which the hub is flexible and the blade is rigid. At the other limit $R = 1$, equation (7.1.39) indicates that the principal axes rotate in a 1:1 relationship with the blade pitch ($\eta = \theta$). In this case elastic coupling is present, and expressions for the natural frequencies (7.1.38) simply represent the fact that, as the blade is rotated through 90° pitch, the nonrotating frequencies must interchange. In addition to the case $R = 0$ where the elastic coupling between flap and lag vanishes, another interesting case exists in which no elastic coupling is present. This is the case referred to as matched stiffness, i.e., when the nonrotating frequencies of the blade are equal in both directions ($\bar{\omega}_\zeta = \bar{\omega}_\beta$). Various advantages accrue from this particular design choice as will be discussed below.

In principle, the designer has at his disposal the selection of the nonrotating frequencies of the blade. Consider some of the options in this regard. For simplicity, only the behavior of the rotor at zero pitch is examined. One choice is the matter of the hub stiffness relative to the blade stiffness which has an important impact on the flap-lag behavior of the rotor through the parameter R as will be discussed below. The flap frequency is largely chosen on the basis of the desired helicopter stability and control characteristics [5, 12]. Since the rotor blade is, in general, a long slender member, the flap frequency will tend to be relatively near to the rotor rpm. Typical ratios of flap frequency to blade angular velocity for hingeless rotor helicopters are of the order of $p =$ 1.05–1.15 [25] although at least one helicopter has flown with a flap frequency ratio of 1.4 [27]. The second major design decision is the choice of the lag frequency. Characteristically, the nonrotating lag frequency will tend to be considerably higher than the flap frequency owing to the larger dimensions of the blade and hub in the chordwise direction compared to the flapwise direction. As mentioned above, lag hinges are provided on articulated rotors to relieve lag stresses arising from flapping. Owing to the fact that the flap frequency is only slightly larger than once per revolution on a typical hingeless blade there will be considerable flap bending of the rotor blades. In fact, the amplitude of the vertical displacement of the blade tip on a hingeless blade will be quite similar to the flapping amplitude of the fully articulated rotor. The relationship between amplitude of tip motion of the hingeless blade and the flapping amplitude of the articulated blade is given by [28]

$$|\beta_H| = \frac{|\beta_A|}{\left\{1 + \left(\frac{8}{\gamma}(p^2 - 1)\right)^2\right\}^{1/2}}$$

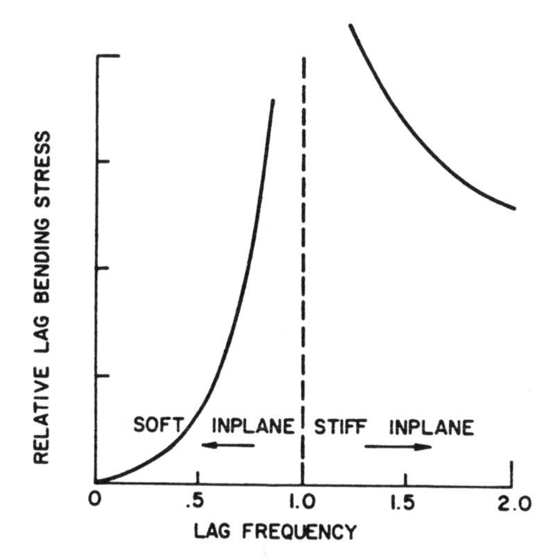

Figure 7.7 Dependance of lag bending stress on lag frequency.

Therefore, the inplane forces due to flap bending will cause the significant root stresses on a hingeless rotor. The dependence of these stresses on the selection of lag frequency can be seen by assuming that the flap and lag bending are loosely coupled ($z = 0$). The lag bending amplitude arising from sinusoidal flap bending at one per rev can be expressed from the equations (7.1.37), neglecting the lag damping, as

$$|\frac{\zeta}{\beta}| = \frac{\left[2\beta_0 - \frac{\gamma}{8}\left(\theta + \frac{8}{3}\lambda\right)\right]}{(q^2 - 1)} \tag{7.1.40}$$

The lag bending moment at the blade root, $K_\zeta \zeta$, thus varies as $q^2/(q^2 - 1)$ as shown in Figure 7.7. It can be seen that if the lag frequency is selected above one per rev, large root bending stresses occur. The bending moment is reduced by choosing a lag frequency well below one per rev. A lag frequency below one per rev incidentally would be characteristic of a matched stiffness blade. For example, if

$$p^2 = 1.2 = \bar{\omega}_\beta^2 + 1$$

and

$$\bar{\omega}_\zeta^2 = \bar{\omega}_\beta^2$$

then

$$\bar{\omega}_\zeta = 0.45$$

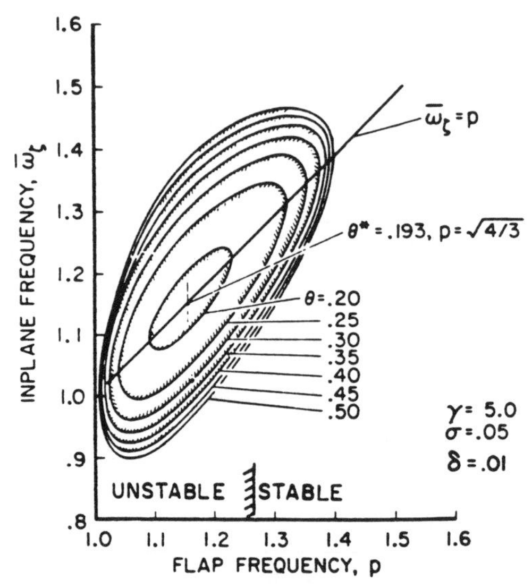

Figure 7.8 Flap lag stability boundaries. $R = 0$, *no elastic coupling* [17].

Rotor blades are usually characterized by their lag frequency as soft inplane ($\bar{\omega}_\zeta < 1$) or stiff inplane ($\bar{\omega}_\zeta > 1$). Thus, rotor blade lag stresses can be reduced by choosing a soft inplane blade design and it should be kept in mind in the discussion that follows that there is a significant variation in the root bending stress with lag frequency. In the following, the influence of lag frequency on the dynamics of a hingeless blade is examined. Also it may be noted at this point that in contrast to the articulated rotor, in which large mechanical motion in lag allows mechanical lag dampers to be effective, this is usually more difficult with the hingeless rotor. Nevertheless, hingeless rotor helicopters have been equipped with lag dampers [10, 29].

Note also that if the lag frequency is selected such that the operating condition of the rotor it is less than one per rev, then resonance in the lag mode will be encountered as the rotor is run up to operating speed.

Flap-lag stability characteristics as predicted by the equations of motion given by equations (7.1.37) are now examined. First consider the case in which the hub is considerably more flexible than the blade ($R = 0$). In Figure 7.8 the stability boundaries given by equations (7.1.37) are shown as a function of flap and lag frequency and blade ptich angle for a typical rotor blade. This figure was obtained by determining the conditions under which Routh's

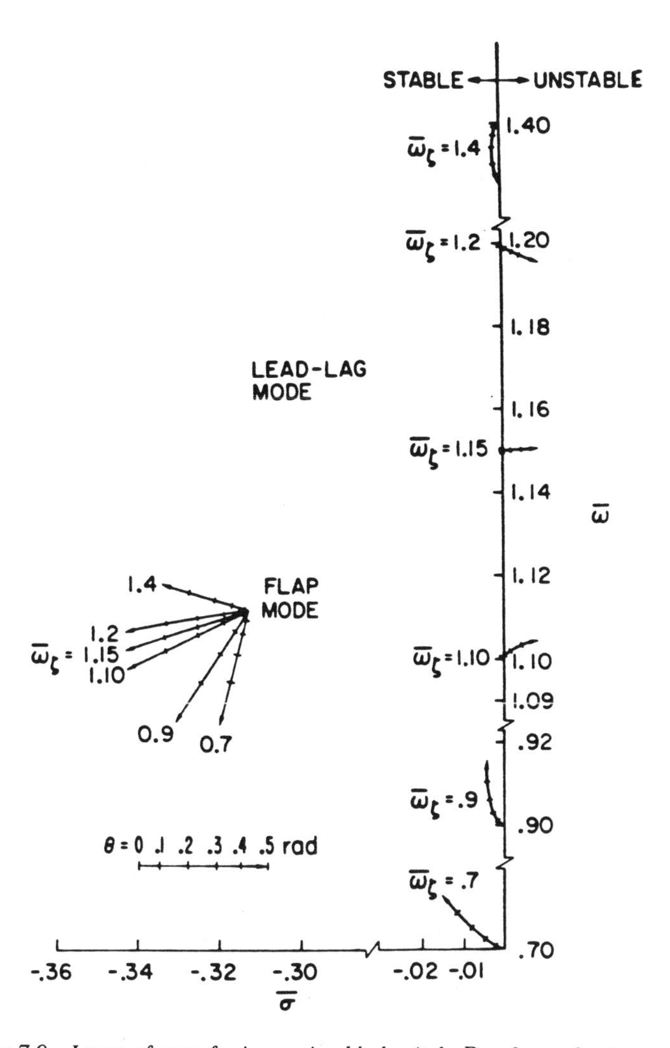

Figure 7.9 Locus of roots for increasing blade pitch. $R = 0$, no elastic coupling [17].

discriminant equals zero. It can be seen that an approximately elliptical region of instability occurs which increases in extent as blade pitch is increased. It is centered around a lag bending frequency of 1.15 and flap frequency of 1.15 indicating that, in this particular case, flap-lag instability is more likely to be a problem for stiff inplane rotors. It can be seen that the stiff inplane blade $(1.1 < \bar{\omega}_\zeta < 1.2)$ is destabilized with increasing pitch. Figure 7.10 shows the effect of various ratios of hub stiffness to blade stiffness (different values of R) indicating the importance of careful modelling of the blade and hub in the study of flap-lag stability. This theory has been correlated with experiment in [30]. At large pitch angles where the blade encounters stall, wider ranges of instability occur as shown in [30]. This increase in the region of instability is primarily a result of the loss in flap damping owing to reduction in blade lift curve slope, a.

Various other configuration details have an impact on the flap-lag stability such as precone (the inclination of the blade feathering or pitch change axis with respect to a plane perpendicular to the hub). Precone is usually employed to relieve the root bending stresses that arise from the steady flap bending moment due to average blade lift. The blade may also have droop and sweep [21] (the inclinations of the blade axis with respect to the pitch change axis in the flap and lag directions respectively) which will also have an impact on the flap-lag stability. The presence of kinematic pitch-lag coupling will have important effects on hingeless blade stability which depend strongly on the lag stiffness and the elastic coupling parameter R [17]. Reference [31] provides a closed-form damping expression with a physical exploration of the effect of each parameter.

If torsional flexibility is included, elastic coupling between pitch, lag and flap will exist. This can be most readily understood by extending the simple spring model of blade flexibility to include a torsion spring. Consider a blade hub system as shown in Figure 7.11 with a flap angle β and a lag angle ζ. Owing to the root spring orientation there will be torques exerted about the torsion axis which depend on the respective stiffnesses in the two directions. Representing the torsional stiffness of the blade and control system by K_θ, the equation for torsional equilibrium is (neglecting torsional inertia and damping)

$$K_\theta \theta = (K_\beta - K_\zeta)\beta\zeta \tag{7.1.41}$$

Linearizing about the blade equilibrium position, β_0, ζ_0,

$$\Delta\theta = \frac{1}{K_\theta}\left[(K_\beta - K_\zeta)\beta_0\Delta\zeta + (K_\beta - K_\zeta)\zeta_0\Delta\beta\right] \tag{7.1.42}$$

That is, torsional flexibility results in both pitch-lag coupling

$$\theta_\zeta = \left(\frac{K_\beta - K_\zeta}{K_\theta}\right)\beta_0 \tag{7.1.43}$$

and pitch-flap coupling

$$\theta_\beta = \left(\frac{K_\beta - K_\zeta}{K_\theta}\right)\zeta_0 \tag{7.1.44}$$

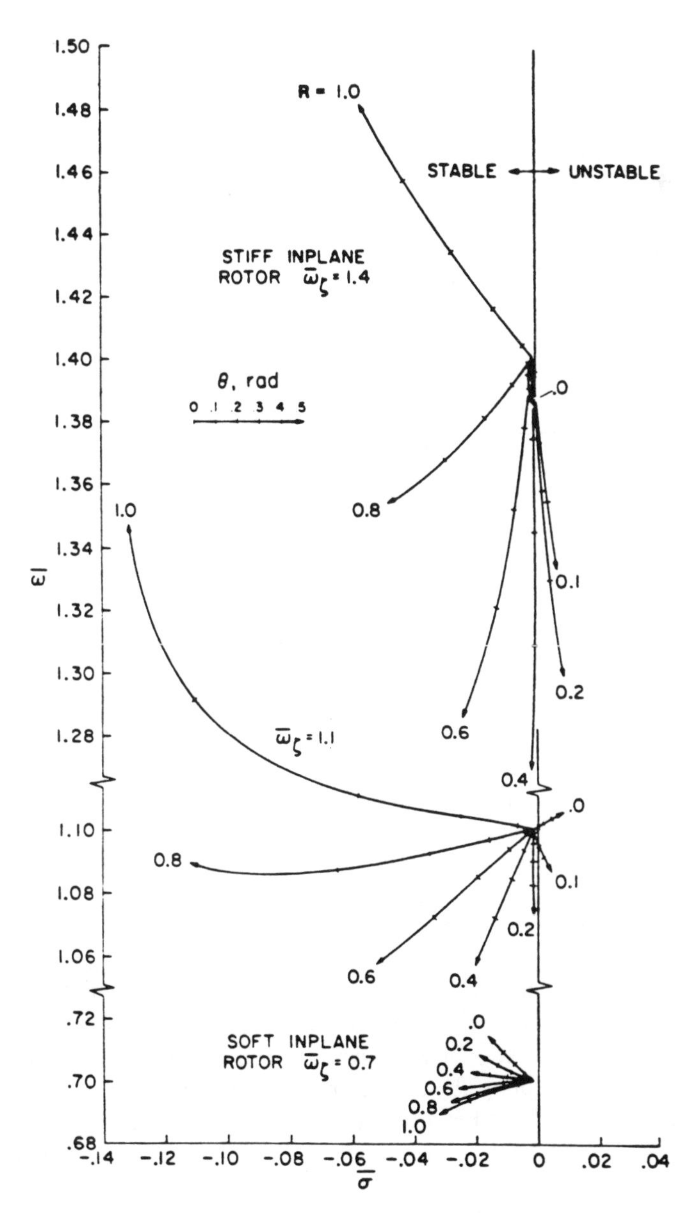

Figure 7.10 *Locus of roots for increasing blade pitch with various levels of elastic coupling* [17].

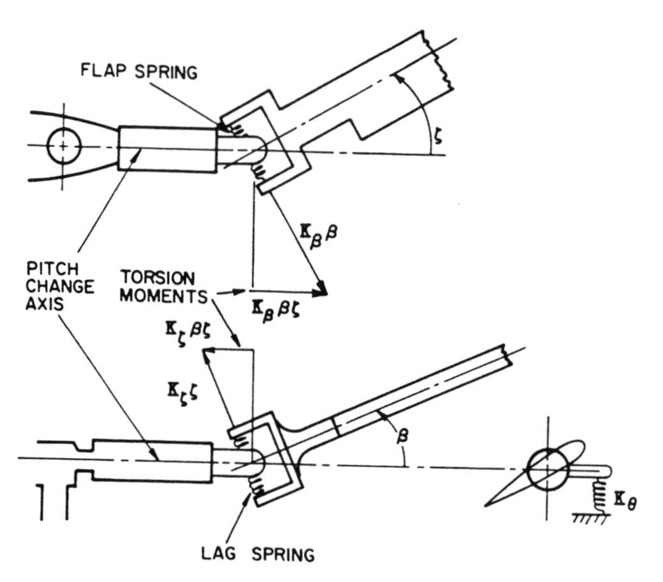

Figure 7.11 Simplified blade model for flap-lag-torsion coupling.

These couplings depend upon the relative stiffness of the blade in the flap and lag directions and the equilibrium values of the flap deflection and the lag deflection. A matched stiffness blade ($K_\beta = K_\zeta$) eliminates these couplings and is perhaps the primary reason for interest in a matched stiffness blade. For typical blade frequencies, K_ζ is larger than K_β and therefore θ_ζ tends to be negative and θ_β positive. θ_β is equivalent to what is usually referred to as a δ_3 hinge on an articulated blade. In powered flight ζ_0 is negative (7.1.20), and the sign of the effect is equivalent to negative δ_3 [32]. this pitch change arising from flapping is statically destabilizing in the sense that an upward flapping produces an increase in pitch. If this term becomes sufficiently large, flapping divergence can occur. In autorotation, this coupling would change sign as the equilibrium lag angle is positive. The characteristically negative value of pitch-lag coupling θ_ζ tends to produce a stabilizing effect in most cases as may be seen from the articulated rotor example. Negative values of θ_ζ can be destabilizing for a stiff inplane rotor with small values of R [17]. Precone, that is rotation of the pitch change axis in the flap direction, has a significant effect on the pitch-lag coupling. The coning angle β_0, in equation (7.1.43) refers only to the elastic deflection of the blade. Consequently, with perfect precone, that is, when the precone angle is equal to the equilibrium

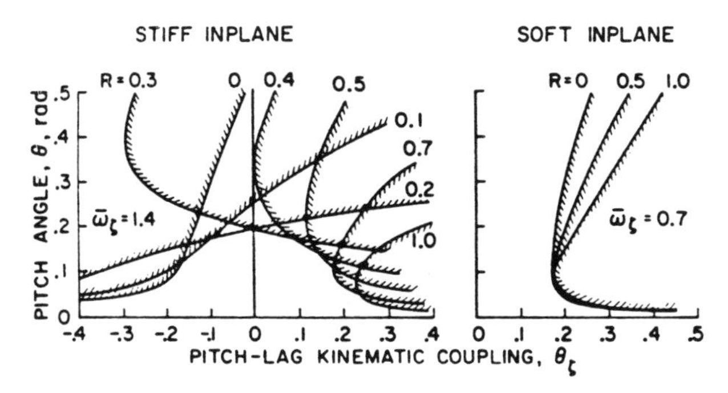

Figure 7.12 Flap-lag stability boundaries as a function of pitch-lag coupling and elastic coupling [17].

steady flap angle given by equation (7.1.20), the elastic deflection is zero and the pitch-lag coupling is zero. For excessive precone (i.e., if the rotor is operated well below its design thrust), β_0 is negative and a destabilizing pitch-lag coupling occurs. It should be noted that hub flexibility will also have an important impact on these kinematic couplings since it will determine the deflection of the pitch change axis. A precise formulation of flap-lag-torsion coupling as well as further discussion of its influence on blade stability can be found in [12]. Comparisons of theory and experiment can be found in [33].

In summary, a soft-inplane rotor blade tends to be less susceptible to isolated blade instabilities while the stiff-inplane blade tends to exhibit instabilities along with a conserably more complex behavior with changes in parameters. Figure 7.12 contrasts the effect of pitch-lag coupling on these two rotor blade types illustrating the complexity of the stability boundaries for the stiff inplane case in contrast to the soft inplane case which is quite similar to the articulated rotor.

7.2 Stall flutter

A single degree-of-freedom instability encountered by helicopter blades which also occurs in gas turbines is referred to as stall flutter. The reader should consult Chapter 5 for a discussion of stall flutter on a nonrotating airfoil. Stall flutter is primarily associated with high speed flight and maneuvering of a helicopter and arises from the fact that stalling of the rotor blade is encountered at various locations on the rotor disc. For a rotor blade, stall flutter does not constitute a destructive instability, but rather produces a limit

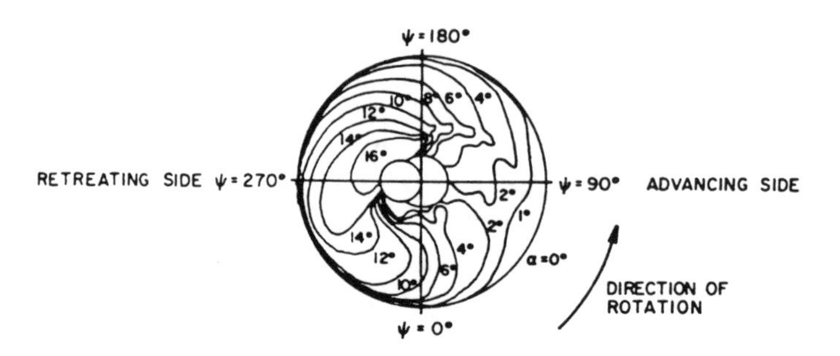

Figure 7.13 Angle of attack distribution of helicopter rotor at 140 knots (advance ratio = 0.33) [34].

cycle behavior owing to the varying aerodynamic conditions encountered by the blade as it rotates in forward flight.

Consider the aerodynamic conditions existing on a rotor blade in high speed flight. On the advancing side of the rotor disc, the dynamic pressure at a blade section depends on the sum of the translation velocity of the helicopter and the rotational velocity, while on the retreating side of the disc it depends on the difference between these two velocities. Consequently, if the rolling moment produced by the rotor is equal to zero, as required for equilibrium flight, the angle-of-attack of the blade is considerably smaller on the advancing side than on the retreating side. A typical angle-of-attack distribution at an advance ratio of 0.33 (140 kts) is shown in Figure 7.13. This resulting is produced by a combination of flapping or flap bending motion and the pilot's control input.

Note that on the advancing side the angle-of-attack is small and varies comparatively slowly with azimuth angle. On the retreating side the angle-of-attack is large and changes rapidly with azimuth angle. Consequently, prediction of the airload on the blade requires a model for the aerodynamics of the blade element which includes unsteady effects both in the potential flow region as well as in the stalled region. The source of the stall flutter instability is related to the unsteady aerodynamic characteristics of an airfoil under stalled conditions. Since the stalled region is only encountered by the blade over a portion of the rotor disc, however, if an instability occurs as a result of the aerodynamics at stall, it will not give rise to continuing unstable motion since a short time later the blade element will be at a low angle-of-attack, well below stall.

Owing to the complexity of the flow field around a stalled airfoil, we must have recourse to experimental data in order to determine the unsteady aerodynamic characteristics of an airfoil oscillating at high angle-of-attack.

Experimental data have become available in recent years on typical helicopter airfoil sections [34–36], which make it possible to characterize the aerodynamics of an airfoil oscillating about stall. In addition a number of investigations have been conducted which give insight into the nature and complexity of the aerodynamic flow field under stalled conditions [37–39].

For a simplified treatment of stall flutter, it is assumed that the blade motion can be adequately described by a model involving only the blade torsional degree-of-freedom. The influence of flapping or heave motion of the section is neglected such that $\theta = \alpha$. The equation of motion for this single degree-of-freedom system is therefore

$$\ddot{\alpha} + \omega_\theta^2 \alpha = \left(\frac{\rho(\Omega R)^2 c^2}{2 I_\theta} \right) C_M(\dot{\alpha} \qquad (7.2.1)$$

where aerodynamic strip theory analysis is employed. Since the aerodynamic damping is a complex function of the angular velocity it is convenient to express equation (7.2.1) as an energy equation by multiplying by $\mathrm{d}\alpha$ and integrating over one cycle to obtain

$$\Delta \left\{ \frac{\dot{\alpha}^2}{2} + \omega_\theta^2 \frac{\alpha^2}{2} \right\} = \frac{\rho(\Omega R)^2 c^2}{2 I_\theta} \int C_M(\dot{\alpha}) \, \mathrm{d}\alpha \qquad (7.2.2)$$

The left hand side of equation (7.2.2) expresses the change in energy over one cycle which is produced by the dependence of aerodynamic pitching moment on angle-of-attack rate as given by the right hand side. Figure 7.14 shows the time history of the pitching moment and normal force coefficients as a function of angle-of-attack for an airfoil oscillating at a reduced frequency typical of one per rev motion at three mean angles-of-attack. The arrows on the figure denote the direction of change of C_N and C_M. Note the large hysteresis loop which occurs in the normal force in the dynamic case when the mean angle-of-attack is near stall. In the potential flow region the effects are rather small and are predicted by Theodorsen's method (see Chapter 4). Proper representation of the unsteady lift behavior does have an important bearing on the prediction of rotor performance, but will not be discussed further. The pitching moment characteristics are of primary interest here.

The pitching moment is well behaved in the potential flow region, and well above stall, resulting in small elliptically shaped loops over one cycle. In the vicinity of lift stall, two interesting effects occur; the average pitching moment increases markedly in a nose down sense, a phenomenon that is referred to as moment stall [35], and the moment time history looks like a figure eight.

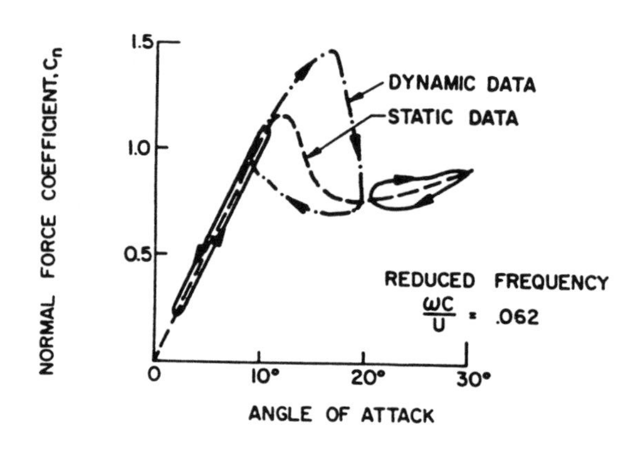

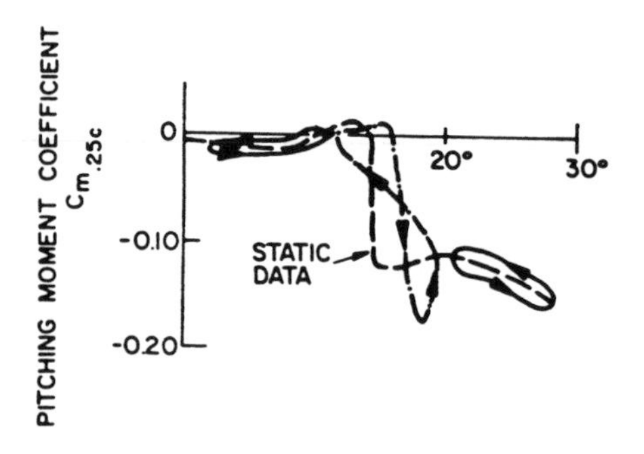

Figure 7.14 Typical oscillating airfoil data [35].

The change in energy over one cycle given by equation (7.2.2) is proportional to

$$\int C_M(\dot{\alpha})\mathrm{d}\alpha$$

The value of this integral is equal to the area enclosed by the loop, and its sign is given by the direction in which the loop is traversed. If the loop is traversed in a counter clockwise direction, then this integral will have a negative value indicating that energy is being removed from the structure or that there is positive damping. Thus the low and high angle-of-attack traces indicate positive damping. Around the angle-of-attack at which moment stall occurs statically however, the figure-eight-like behavior indicates that there is essentially no net dissipation of energy over a cycle or possibly that

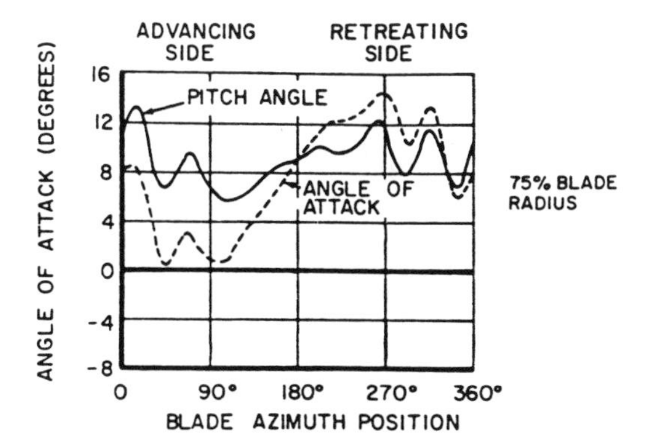

Figure 7.15 Typical time history of blade motion for blade encountering stall flutter [42].

energy is being fed into the structure (the integral on the right hand side of equation (7.2.2) is positive). This pitching moment characteristic gives rise to the phenomenon referred to as stall flutter. To actually encounter stall flutter, this behavior must occur over some appreciable span of the blade [40]. Reference [40] also discusses the importance of the rate of change of angle-of-attack with time on the dynamics of this process and concludes that delay in the development of dynamic stall depending upon $\dot{\alpha}$ is responsible for stall occurring over a significant radius of the blade with consequent effects on the rotor loads and vibrations. Of course, the rotor blade only encounters this instability over a small azimuth range and consequently the complete motion is essentially a limit cycle. The loss in damping at stall coupled with the marked change in the average pitching moment gives rise to large torsional motion with perhaps 2 or 3 cycles of the torsion excited before being damped by the low angle-of-attack aerodynamics. A typical time history of blade torsional motion and angle-of-attack when stall flutter is encountered is shown in Figure 7.15. The dominant effect of the occurrence of stall flutter on a helicopter is to give rise to a marked increase in the vibratory loads in the blade pitch control system [40]. Reference [40] discusses approximate methods for incorporating unsteady stall aerodynamics into the rotor blade equations of motion. The most significant assumption in the analysis of rotor blade stall flutter relates to the applicability of two-dimensional data on airfoils oscillating sinusoidally to a highly three-dimensional flow field in which the motions are nonsinusoidal.

Further understanding of the aerodynamics of stall may make it possible to design airfoil sections which would minimize the occurrence of stall flutter and the associated control loads. Blade section design, however, has many constraints owing to the wide range in aerodynamic conditions encountered in one revolution and the aerodynamic phenomena described appear to be characteristic of airfoils oscillating about an angle-of-attack beyond the onset of static stall. In more recent work, analytic methods have been developed to model dynamic stall [41] and [42]; and CFD codes have been formulated to predict dynamic stall [43].

7.3 Rotor-body coupling

Another important topic is the aeroelastic instability of helicopters associated with coupling of blade motion and body motion. This problem is of considerable significance in articulated and hingeless rotor helicopter design, and was first encountered on autogiros. This violent instability was at first attributed to rotor blade flutter until a theory was developed during the period 1942 to 1947 showing it to be a new phenomenon. The instability is called *ground resonance* and was first analyzed and explained by Coleman [44] who modelled the essential features of the instability for articulated rotor helicopters. The name ground "resonance" is somewhate confusing since, in fact, the dynamic system of the helicopter and blades is unstable. The instability occurs at a particular rotor angular velocity and therefore it appears in some sense like a resonance, but it is not. Further, the ground enters the problem owing to the mechanical support provided the helicopter fuselage by the landing gear. The particularly interesting result obtained by Coleman is that the instability can be predicted neglecting the rotor aerodynamics; that is, ground resonance is purely a mechanical instability, the energy source being the rotor angular velocity. In the discussion below, Coleman's development is followed. Then there is qualitative discussion of the more complex formulation of this problem as applied to hingeless rotors. For an articulated rotor, the aerodynamics tend to be unimportant and only the lag degree-of-freedom nees to be included. For hingeless rotors, the flapping degree-of-freedom is important in the analysis and aerodynamic forces play a significant role [10]. The addition of the flapping degrees-of-freedom leads to a similar instability in flight referred to as air resonance.

Following Coleman's analysis we consider a simplified model of a helicopter resting on the ground. The degrees-of-freedom assumed are: pitch and roll of the rotor shaft or pylon which arise from the landing gear oleo strut flexibility and the lag degree-of-freedom of each rotor blade. Discussion is restricted to the case in which the rotor has three or more rotor blades and thus has polar symmetry. The two-bladed rotor is a somewhat more complex problem and a few remarks on this special case are made at the end of this section. A four-bladed rotor system is used as the example since the approach is most easily visualized in this case. The generalization to three or more blades is described at the end of this section.

Consider the helicopter shown in Figure 7.16. The system has six degress-of-freedom, the lag motion of each of the four rotor blades and the two pylon deflections. Each rotor blade is modelled as an articulated blade with hinge offset $\bar{e}$. A spring is included at the root since a centering spring may be employed about the lag hinge.

A coordinate system is chosen which is fixed in space in order to allow the simplest mathematical treatment of the asymmetric stiffness and inertia characteristics associated with pitch and roll motion of the fuselage on the landing gear. If a rotating coordinate system is employed, then the differential equations describing the dynamics would involve periodic coefficients with attendant problems in unravelling the solution. In fact, it is this difference in the form of the equations of motion in fixed and rotating coordinate systems which gives rise to difficulties in analyzing the two-bladed rotor system with asymmetric pylon characteristics. The two-bladed rotor lacks polar symmetry; and, therefore, a fixed coordinate system approach will give rise to periodic coefficients from the rotor, while a rotating coordinate system analysis will give rise to periodic coefficients from the asymmetric pylon characteristics. Thus, periodic coefficients can not be eliminated in the two-bladed case unless the pylon frequencies are equal. For a rotor with three or more blades, the use of a fixed coordinate system allows treatment of asymmetric pylon characteristics without encountering the problem of solving equations with periodic coefficients.

First we consider the equations of motion describing blade lag dynamics in a fixed coordinate system to illustrate the influence of coordinate system motion. All of our previous examples have used a coordinate system rotating with the blade. Simplification of this problem can be effected by defining new coordinates to describe the rotor lag motion. These new coordinates are linear combination of the lag motion of the individual blades. They usually are referred to as multi-blade coordinates [45] and are defined for a four-bladed rotor as

$$\begin{aligned}
\gamma_0 &= \frac{\zeta_1 + \zeta_2 + \zeta_3 + \zeta_4}{4} \\
\gamma_1 &= \frac{\zeta_1 - \zeta_3}{2} \\
\gamma_2 &= \frac{\zeta_2 - \zeta_4}{2} \\
\gamma_3 &= \frac{(\zeta_1 + \zeta_3) - (\zeta_2 + \zeta_4)}{4}
\end{aligned} \tag{7.3.1}$$

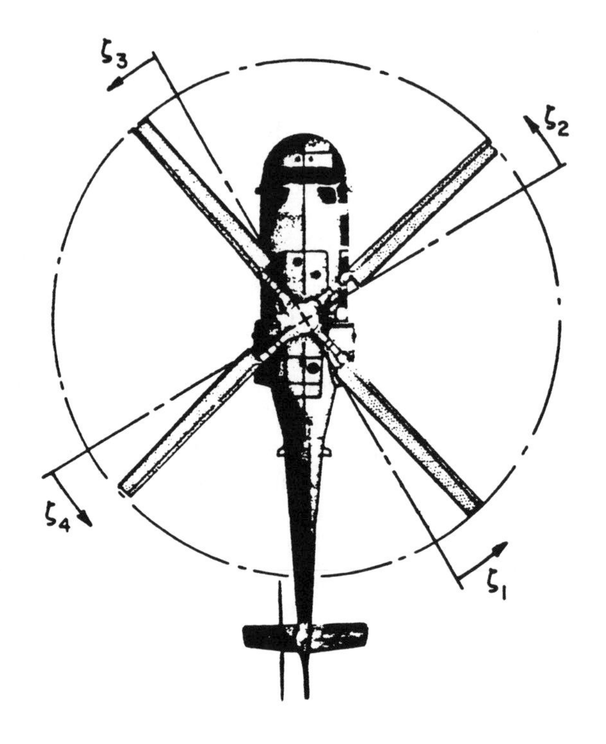

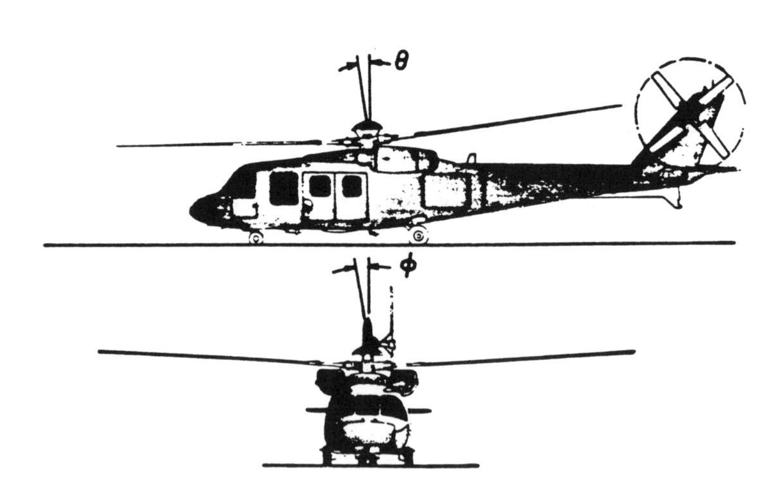

Figure 7.16 Mechanical degrees of freedom for ground resonance analysis.

The new coordinates, γ_1 and γ_2, describe the motion of the center of mass of the rotor system with respect to the axis of rotation and thus are responsible for coupling of rotor motion to pylon motion, while γ_0 and γ_3 describe motions of the rotor in which the center of mass of the rotor system remains on the axis of rotation. If $\gamma_2 = \gamma_2 = 0$, then motions corresponding to γ_0 and γ_3 are such that opposite blades move as though rigidly attached together with a vertical pin at the root. These motion variables, γ_0 and γ_3, will be uncoupled for the dynamic problem of interest; and, consequently, the system is reduced to four degrees-of-freedom by introducing these coordinates as will be shown.

Now the equations of motion for γ_1 and γ_2 are developed in a moving coordinate system and then transformed to a stationary coordiante system. With the hub fixed, the lag motion of each blade without aerodynamics is, as shown earlier,

$$\ddot{\zeta}_i + \Omega^2(\bar{\omega}_\zeta^2)\zeta_i = 0 \quad i = (1, 2, 3, 4) \tag{7.3.2}$$

The natural frequency, $\bar{\omega}_\zeta^2$ arises from a mechanical spring on the hinge and the offset or centrifugal stiffening effect and is given by equation (7.1.35). The equations of motion for γ_1 and γ_2 are from 7.3.1) and (7.3.2)

$$\begin{aligned} \ddot{\gamma}_1 + \Omega^2(\bar{\omega}_\zeta^2)\gamma_1 &= 0 \\ \ddot{\gamma}_2 + \Omega^2(\bar{\omega}_\zeta^2)\gamma_2 &= 0 \end{aligned} \tag{7.3.3}$$

These equations may be thought of as describing the motion of the center of mass of the rotor system in two directions with respect to the coordinate system rotating at the rotor angular velocity Ω. Resolving to a fixed coordinate system as shown in Figure 7.17,

$$\begin{aligned} \gamma_1 &= \Gamma_1 \cos \Omega t + \Gamma_2 \sin \Omega t \\ \gamma_2 &= -\Gamma_1 \sin \Omega t + \Gamma_2 \cos \Omega t \end{aligned} \tag{7.3.4}$$

Differentiating and substitution (7.3.4) into (7.3.3) we obtain the equations

$$\begin{aligned} &[\ddot{\Gamma}_1 + 2\Omega\dot{\Gamma}_2 + \{\Omega^2(\bar{\omega}_\zeta^2 - 1)\}\Gamma_1 \ \cos \Omega t \\ &\quad +[\ddot{\Gamma}_2 - 2\Omega\dot{\Gamma}_1 + \{\Omega^2(\bar{\omega}_\zeta^2 - 1)\}\Gamma_2 \ \sin \Omega t = 0 \\ &[\ddot{\Gamma}_1 + 2\Omega\dot{\Gamma}_2 + \{\Omega^2(\bar{\omega}_\zeta^2 - 1))\}\Gamma_1] \sin \Omega t \\ &\quad -[\ddot{\Gamma}_2 - 2\Omega\dot{\Gamma}_1 + \{\Omega^2(\bar{\omega}_\zeta^2 - 1)\}\Gamma_2] \cos \Omega t = 0 \end{aligned} \tag{7.3.5}$$

The second equation appears similar to the first with the coefficients of the sine and cosine terms reversed. Although the variables have been transformed

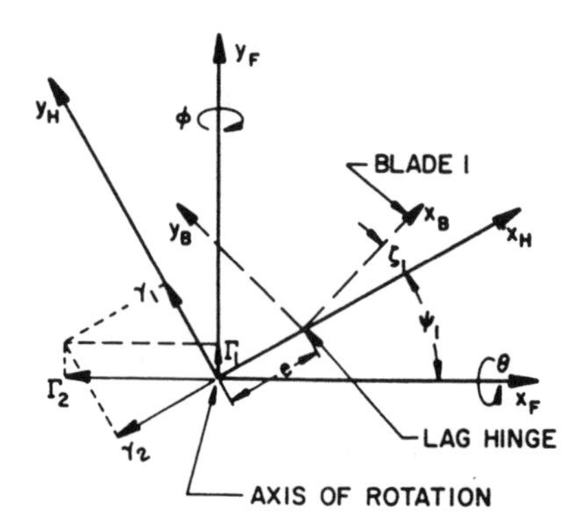

Figure 7.17 Axis systems and coordinates for ground resonance analysis.

to a fixed system, the equations of motion are still referred to a moving frame which accounts for the presence of the cosine and sine terms. To complete the transformation, multiply the first equation by $\cos t$ and add to the second multiplied by $\sin t$ to obtain one fixed-axis equation. Multiplying the first equation by $\sin t$ and subtracting from the second multiplied by $\cos t$ yields the second equation. The two equations of motion are

$$\begin{aligned} &\ddot{\Gamma}_1 + 2\Omega\dot{\Gamma}_2 + \{\Omega^2(\bar{\omega}_\zeta^2 - 1)\}\Gamma_1 = 0 \\ &\ddot{\Gamma}_2 - 2\Omega\dot{\Gamma}_1 + \{\Omega^2(\bar{\omega}_\zeta^2 - 1)\}\Gamma_2 = 0 \end{aligned} \tag{7.3.6}$$

These are the equations of motion for the new lag coordinates (or CM motion) in the nonrotating coordinate system. Note that the variables are coupled due to the effects of rotation. The characteristic equation for the dynamics of the lag motion is now obtained from equations (7.3.6) as

$$\{s^2 + \Omega^2(\bar{\omega}_\zeta^2 - 1)\}^2 + 4\Omega^2 s^2 = 0 \tag{7.3.7}$$

The roots of this characteristic equation are

$$\begin{aligned} s_{1,2} &= \pm i\Omega(\bar{\omega}_\zeta + 1) \\ s_{3,4} &= \pm i\Omega(\bar{\omega}_\zeta - 1) \end{aligned} \tag{7.3.8}$$

Thus, the coordinate transformation has resulted in natural frequencies in the fixed coordinate system which are equal to the natural frequencies in the rotating system given by equations (7.3.3) ($\bar{\omega}_\zeta\Omega$) plus or minus the rotational speed (the angular velocity of the coordinate system). This is a basic characteristic of natural frequencies when calculated in rotating and fixed coordinate systems which must be kept in mind in analyzing rotating systems. At this point we consider one other aspect of the dynamics of this type of system which is helpful in visualizing the motion. Consider the eigenvectors describing the amplitude and phase of the two variables in transient motion. These ratios are obtained from the equations of motion and the characteristic roots.

$$\left|\frac{\Gamma_1}{\Gamma_2}\right| = \left.\frac{-2\Omega s}{s^2 + \Omega^2(\bar{\omega}_\zeta^2 - 1)}\right|_{s_{1,2,3,4}}$$

Therefore

$$\left|\frac{\Gamma_1}{\Gamma_2}\right| = \pm i\Big|_{s_{1,2}}$$
$$\left|\frac{\Gamma_1}{\Gamma_2}\right| = \mp i\Big|_{s_{3,4}} \tag{7.3.9}$$

The upper sign corresponds to the upper sign in the roots (7.3.8).

In either of these characteristic motions, Γ_1 and Γ_2 are of equal amplitude and Γ_1 either leads or lags Γ_2 by 90°. Thus the transient motion of the rotor system center of mass is a circular motion. This symmetry which occurs in many rotating systems permits an elegant formulation using complex coordinates [28, 44]. The two variables Γ_1 and Γ_2 can be combined into one single complex variable, as will be discussed below. Further, since the transient motion is circular, these modes are referred to as whirling modes and the whirling may be described as advancing or regressing depending upon whether the mode of motion corresponds to transient motion in the direction of rotor rotation or against the direction of rotation. Consider the root

$$s_1 = +i\Omega(\bar{\omega}_\zeta + 1)$$

corresponding to a counter clockwise rotation of the variables Γ_1 and Γ_2 in Figure 7.17. From the eigenvectors (7.3.9), we see that Γ_1 leads Γ_2 by 90°. Thus Γ_1 reaches a maximum and then Γ_2 reaches a maximum and so the oscillation proceeds in the direction of rotation and is an advancing mode. Similarly, $s_2 = -\Omega(\bar{\omega}_\zeta + 1)$, corresponds to the two vectors rotating in a clockwise direction, but now Γ_1 lags Γ_2 and so this is also an advancing mode. Hence, the mode with frequency $(\bar{\omega}_\zeta)\Omega$ is an advancing mode. Following a similar argument for the mode $(\bar{\omega}_\zeta - 1)\Omega$ we find that it is a regressing mode, when $\bar{\omega}_\zeta$ is greater than 1. One must be careful of this terminology, since in a rotating coordinate system modes are also described as advancing and regressing modes, but because of the change in coordinate system angular velocity, modes may be regressing in

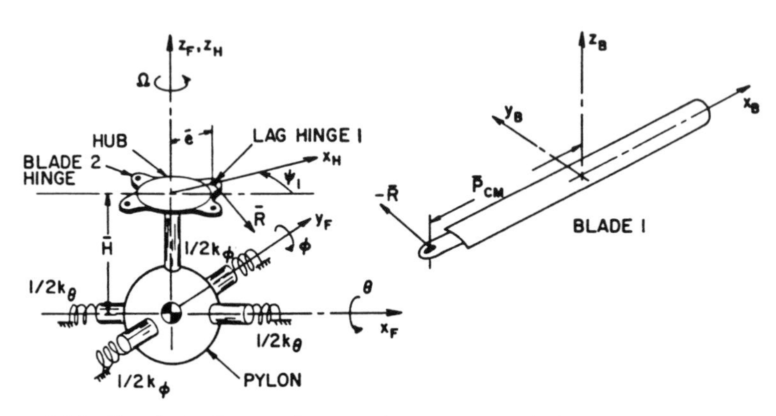

Figure 7.18 Free body diagram for ground resonance analysis.

this system and advancing in the stationary system. From a geometric point of view, there are two whirling modes corresponding to the four characteristic roots. Use of complex coordinates helps to visualize the direction of rotation of the modes simply [28].

Thus, the transient motion of the center of mass of the rotor system may be described in terms of two circular or whirling modes. When viewed in the fixed frame of reference, one is an advancing whirl (in the same direction as the rotation of the rotor) and one is a regressing whirl at low rotor angular velocity ($\omega_\zeta > \Omega$) and a slow advancing whirl at larger angular velocities ($\omega_\zeta < \Omega$). Recall that the frequencies as seen in the rotating frame are simply equal to $\pm\omega_\zeta$ and one is an advancing mode while the other is always a regressing mode.

Now the effect of the pylon motion is added. It is assumed that the pylon is sufficiently long and the angular deflections are sufficiently small such that the hub motion lies in a horizontal plane. The equations of motion are developed using a Newtonian approach. First, a single blade is considered; and then the effects of the other blades are added. It is most convenient to derive the equations with the pylon or fuselage motion referred to a fixed axis system and the lag angle referred to a moving axis system and then to transform the lag angle to a fixed coordinate system. This will illustrate the manner in which periodic coefficients enter the equations. The equations of motion for the blade and fuselage system may be written from the free body diagram shown in Figure 7.18, as

$$\bar{H}_b + \bar{\Omega}_B \times \bar{H}_B = \bar{P}_{CM} \times \bar{R}$$

$$M_B \bar{a}_{CM} = -\bar{R} \qquad (7.3.10)$$
$$\bar{H}_F = \bar{H} \times \bar{R}$$

where $\bar{R}$ is the reaction force at the hinge, $\bar{H}_B$ is the moment of momentum of the blade about its center of mass and $\bar{H}_F$ is the moment of momentum of the fuselage about its center of mass which is assumed fixed in space. $\bar{\Omega}_B$ is the angular velocity of the blade, $\bar{P}_{CM}$ is the distance from the hinge to the blade center of mass and $\bar{H}$ is the height of the rotor hub above the CM. The acceleration of the blade center of mass in terms of the acceleration at the hinge point is

$$\bar{a}_{CM} = \bar{a}_E + \bar{\Omega}_B \times (\bar{\Omega}_B \times \bar{P}_{CM}) + \dot{\Omega}_B \times \bar{P}_{CM} \qquad (7.3.11)$$

and the acceleration of the hinge in terms of the acceleration of the hub, $\bar{a}_0$, and the rotational velocity of the hub is

$$\bar{a}_E = \bar{a}_0 + \bar{\Omega} \times (\bar{\Omega} \times \bar{e}) \qquad (7.3.12)$$

The angular velocity of the hub is assumed constant. Three sets of unit vectors are defined. The subscript $_B$ refers to the set of unit vectors fixed to the blade, the subscript $_H$ to the set fixed in the hub and the subscript $_F$ refers to a set fixed in space. The lag angle is assumed to be small, so that the relationships among these unit vectors for blade number 1 (Figure 7.17) are

$$\begin{Bmatrix} \bar{i}_F \\ \bar{j}_F \\ \bar{k}_F \end{Bmatrix} = \begin{bmatrix} \cos\psi_1 & -\sin\psi_1 & 0 \\ \sin\psi_1 & \cos\psi_1 & 0 \\ 0 & 0 & 1 \end{bmatrix} \begin{Bmatrix} \bar{i}_H \\ \bar{j}_H \\ \bar{k}_H \end{Bmatrix} \qquad (7.3.13)$$

$$\begin{Bmatrix} \bar{i}_H \\ \bar{j}_H \\ \bar{k}_H \end{Bmatrix} = \begin{bmatrix} 1 & -\zeta_1 & 0 \\ \zeta_1 & 1 & 0 \\ 0 & 0 & 1 \end{bmatrix} \begin{Bmatrix} \bar{i}_B \\ \bar{j}_B \\ \bar{k}_B \end{Bmatrix} \qquad (7.3.14)$$

The various quantities involved in the equations of motion are

$$\begin{aligned} \bar{H}_b &= I_{CM}(\Omega + \dot{\zeta}_1)\bar{k}_B \\ \bar{H}_F &= I_x \dot{\theta} \bar{i}_F + I_y \dot{\phi} \bar{j}_F \\ \bar{\Omega} &= \Omega \bar{k}_H \\ \bar{\Omega}_B &= (\Omega + \dot{\zeta}_1)\bar{k}_B \\ \bar{H} &= h\bar{k}_F \\ \bar{e} &= e\bar{i}_H \\ \bar{P}_{CM} &= r_{CM}\bar{i}_B \\ \bar{a} &= \ddot{x}_H \bar{i}_F + \ddot{y}_H \bar{j}_F = h\ddot{\phi}\bar{i}_F - h\ddot{\theta}\bar{j}_F \end{aligned} \qquad (7.3.15)$$

Substituting equations (7.3.11)-(7.3.15) into the equations of motion (7.3.10) for the blade and body motion, noting that $\bar{R} = -M_B\bar{a}_{CM}$, we obtain one blade equation of motion and two body equations of motion

$$\begin{aligned}
\ddot{\zeta}_1 + \frac{er_{CM}M_B}{I_B}\Omega^2\zeta_1 &= \frac{M_B r_{CM} h}{I_B}(\ddot{\theta}\cos\psi_1 + \ddot{\phi}\sin\psi_1 \\
(I_y + M_B h^2)\ddot{\phi} &= M_B h[e\Omega^2 + r_{CM}(\Omega + \dot{\zeta}_1)^2]\cos\psi \\
&\quad + M_B r_{CM} h(\ddot{\zeta}_1 - \Omega^2\zeta_1)\sin\psi_1 \qquad (7.3.16) \\
(I_x + M_B h^2)\ddot{\theta} &= -M_B h[e\Omega^2 + rCM(\Omega + \dot{\zeta}_1)^2]\sin\psi_1 \\
&\quad + M_B r_{CM} h(\ddot{\zeta}_1 - \Omega^2\zeta_1)\cos\psi_1
\end{aligned}$$

The subscript 1 has been added to note that only one blade has been considered.

The equations of motion for the other three blades are identical to blade one with the azimuth angle suitably shifted, i.e., the equation of motion of blade 2 in terms of the azimuth angle of blade 1 is

$$\ddot{\zeta}_2 + \frac{er_{CM}M_B}{I_B}\Omega^2\zeta_2 = \frac{M_B r_{CM} h}{I_B}\left(\ddot{\theta}\cos\left(\psi_1 + \frac{\pi}{2}\right) + \ddot{\phi}\sin\left(\psi_1 + \frac{\pi}{2}\right)\right) \qquad (7.3.17)$$

The equations of motion for the new coordinates, γ_0, γ_1, etc., are formulated by linear combinations of the blade equations and are

$$\begin{aligned}
\ddot{\gamma}_0 + \frac{er_{CM}M_B}{I_B}\Omega^2\gamma_0 &= 0 \\
\ddot{\gamma}_1 + \frac{er_{CM}M_B}{I_B}\Omega^2\gamma_1 &= \frac{M_B r_{CM} h}{I_B}(\ddot{\theta}\cos\psi_1 + \ddot{\phi}\sin\psi_1) \\
&\qquad (7.3.18) \\
\ddot{\gamma}_2 + \frac{er_{CM}M_B}{I_B}\Omega^2\gamma_2 &= \frac{M_B r_{CM} h}{I_B}(-\ddot{\theta}\sin\psi_1 + \ddot{\phi}\cos\psi_1) \\
\ddot{\gamma}_3 + \frac{er_{CM}M_B}{I_B}\Omega^2\gamma_3 &= 0
\end{aligned}$$

We thus see as discussed earlier that γ_0 and γ_3 are not coupled to the hub motion and thus do not need to be considered further. Note that the equations ofmotion for γ_1 and γ_2 have periodic coefficients since $\psi_1 = \Omega t$. The influence of the other three blades must be added to the fuselage equations. The first equation becomes

$$(I_y + 4M_B h^2)\ddot{\phi} =$$

$$\sum_{i=1}^{v} M_B h(e\Omega^2 + r_{CM}(\Omega + \dot{\zeta}_i)^2)\cos\left(\psi_1 + \frac{(i-1)\pi}{2}\right)$$
$$+M_B r_{CM} h(\ddot{\zeta}_i - \Omega^2\zeta_i)\sin\left(\psi_1 + \frac{(i-1)\pi}{2}\right) \tag{7.3.19}$$

and a similar form is obtained for the other fuselage equation. Using trigonometric identities and the definitions of the multi-blade coordinates (7.3.1) the two fuselage equations become, retaining only linear terms,

$$\begin{aligned}(I_y + 4M_B h^2)\ddot{\phi} &= 2M_B r_{CM} h\{(\ddot{\gamma}_1 - 2\Omega\dot{\gamma}_2 - \Omega^2\gamma_1)\sin\psi_1 \\ &\quad +(\ddot{\gamma}_2 + 2\Omega\dot{\gamma}_1 - \Omega^2\gamma_2)\cos\psi_1\} \\ (I_x + 4M_B h^2)\ddot{\theta} &= 2M_B r_{CM} h\{(\ddot{\gamma}_1 - 2\Omega\dot{\gamma}_2 - \Omega^2\gamma_1)\cos\psi_1 \\ &\quad -(\ddot{\gamma}_2 + 2\Omega\gamma_1 - \Omega^2\gamma_2)\sin\psi_1\}\end{aligned} \tag{7.3.20}$$

Again we see that the coordinates γ_0 and γ_3 do not appear.

These equations involve periodic coefficients. The periodic coefficients are a consequence of defining the lag motion in a rotating system and the fuselage motion in a fixed system as noted earlier. The periodic coefficients can eliminated by transforming the lag motion to fixed coordinates as described above. This transformation involves the relationships given by equations (7.3.4). A centering spring about the lag hinge is incorporated in the lag equations such that the lag frequency is given by

$$\omega_\zeta^2 = \frac{K_\zeta}{I_B} + \frac{e r_{CM} M_B}{I_B}\Omega^2$$

This is equivalent to equation (7.1.35) without the assumption of a uniform blade mass distribution. Employing equations (7.3.4) the blade equations (7.3.18) become

$$\begin{aligned}\ddot{\Gamma}_1 + (\omega_\zeta^2 - \Omega^2)\Gamma_1 + 2\Omega\dot{\Gamma}_2 &= \frac{M_B r_{CM} h}{I_B}\ddot{\theta} \\ -2\Omega\dot{\Gamma}_1 + \ddot{\Gamma}_2 + (\omega_\zeta^2 - \Omega^2)\Gamma_2 &= \frac{M_B r_{CM} h}{I_B}\ddot{\phi}\end{aligned} \tag{7.3.21}$$

The fuselage equations, including the effects of the supporting springs k_ϕ and k_θ, are

$$\begin{aligned}(I_y + 4M_B h^2)\ddot{\phi} + k_\phi\phi &= 2M_B r_{CM} h\ddot{\Gamma}_2 \\ (I_x + 4M_B h^2)\ddot{\theta} + k_\theta\theta &= 2M_B r_{CM} h\ddot{\Gamma}_1\end{aligned} \tag{7.3.22}$$

Equations (7.3.22) can be placed in the form given in [43] by converting the pylon rotations θ and ϕ to linear hub translations x and y. From Figure 7.17, dropping the subscripts $_F$ on x and y,

$$x = h\phi, \bar{x} = \frac{x}{R}$$

$$y = -h\theta, \bar{y} = \frac{y}{R}$$

A uniform mass blade is assumed such that

$$I_B = M_B \frac{R^2}{3}$$

Define effective fuselage mass and spring constants by

$$M_{Fx} = \frac{I_y}{h^2} \; k_x = \frac{k_\phi}{h^2}$$

$$M_{Fy} = \frac{I_x}{h^2} \; k_y = \frac{k_\theta}{h^2}$$

These definitions eliminate the parameter h from equations (7.3.21) and (7.3.22). Equations (7.3.21) and (7.3.22) become

$$\begin{aligned}
&\ddot{\Gamma}_1 + (\omega_\zeta^2 - \Omega^2)\Gamma_1 + 2\Omega\dot{\Gamma}_2 = -\frac{3}{2}\ddot{y} \\
&-2\Omega\dot{\Gamma}_1 + \ddot{\Gamma}_2 + (\omega_\zeta^2 - \Omega^2)\Gamma_2 = \frac{3}{2}\ddot{x} \\
&(M_{Fx} + 4M_B)\ddot{x} + k_x\bar{x} = M_B\ddot{\Gamma}_2 \\
&(M_{Fy} + 4M_B)\ddot{y} + k_y\bar{y} = -M_B\ddot{\Gamma}_1
\end{aligned} \tag{7.3.23}$$

Note that the periodic coefficients will not be eliminated if we attempt to transform the body motion into rotating coordinates except in the special case where fuselage inertias and springs are identical about both axes. The procedure followed above for four blades will produce identical results for three or more blades. A generalized description of this procedure may be found in [6].

If the rotor has two blades, the only way to eliminate the periodic coefficients is to convert the pylon motion to rotating coordinates. Only in the special case of equal inertia and stiffness will the periodic coefficients be eliminated [43]. In general, if there is polar symmetry in one frame of reference and a lack of symmetry in the other frame, expressing the equation of motion in this latter frame will eliminate the necessity of dealing with periodic coefficients. With a two bladed rotor, the rotor lacks polar symmetry. If the support system also lacks polar symmetry, the periodic coefficients cannot be eliminated, and Floquet theory must be employed to analyze the stability of the system. The simplest case (equal support stiffness) of the two-bladed rotor is analyzed in the rotating system and the three or more bladed rotor in the fixed frame. Recall from the previous discussion that this will give quite a different picture of the variation of system natural frequencies with rpm. For simplicity, only the multibladed rotor with pylon symmetry is discussed which may be treated in either reference frame. The pylon characteristics are assumed to be

$$M_{Fx} = M_{Fy} = M_F$$

$$k_x = k_y = k_F$$

The important parameter governing the coupling between the blade motion and the fuselage motion is the ratio of the total blade mass to the total system mass defined by μ,

$$\mu = \frac{4M_B}{M_F + 4M_B}$$

It is convenient to nondimensionalize the time by the support frequency with the blade mass concentrated at the hub

$$\omega_F^2 = \frac{k_F}{M_F + 4M_B}$$

since rotor angular velocity is considered to be the variable parameter. The frequencies nondimensionalized in this fashion are denoted by $\hat{\omega}_\zeta$ and $\hat{\Omega}$. Introducing these definitions, equations (7.3.23) become,

$$\begin{aligned}
&\ddot{\Gamma}_1 + (\hat{\omega}_\zeta^2 - \hat{\Omega}^2)\Gamma_1 + 2\hat{\Omega}\dot{\Gamma}_2 + \frac{3}{2}\ddot{\bar{y}} = 0 \\
&-2\hat{\Omega}\dot{\Gamma}_1 + \ddot{\Gamma}_2 + (\hat{\omega}_\zeta^2 - \hat{\Omega}^2)\Gamma_2 - \frac{3}{2}\ddot{\bar{x}} = 0 \\
&\frac{\mu}{4}\ddot{\Gamma}_1 + \ddot{\bar{y}} + \bar{y} = 0 \\
&= \frac{\mu}{4}\ddot{\Gamma}_2 + \ddot{\bar{x}} + \bar{x} = 0
\end{aligned} \tag{7.3.24}$$

The stability of the system defined by equations (7.3.24) is examined as a function of the various physical parameters of the problem. First, consider the limiting case in which the blade mass is zero ($\mu = 0$). This eliminates the coupling between the fuselage motion and the blade motion. The natural frequencies of the system are composed of the uncoupled blade dynamics and the fuselage dynamics. The roots of the characteristic equation are therefore

$$\pm i(\hat{\Omega} + \hat{\omega}_\varsigma), \quad \pm i(\hat{\Omega} - \hat{\omega}_\varsigma), \quad \pm i, \quad \pm i$$

The latter two pairs correspond to the fuselage motion and the former to the blade motion. The modes of motion are whirling or circular modes owing to polar symmetry. Figure 7.19 shows the whirling modes, i.e., only the four frequencies with signs that correspond to the direction of whirling, positive being an advancing mode. The frequencies are shown as a function of rotor angular velocity. In the numerical example shown, a centering spring (K_ς) is included such that $\hat{\omega} = 0.3$ when the rotor is not rotating ($\hat{\Omega} = 0$) and the hinge offset $\bar{e} = 0.05$. ($\hat{\Omega} - \hat{\omega}_\varsigma$) is a regressing mode when negative, ($\hat{\Omega} + \hat{\omega}_\varsigma$) is an advancing mode. The fuselage modes ($\pm i$) are advancing and regressing modes respectively. These four whirling modes constitute the dynamics of the system in the limiting case of no hub mass. For comparison purposes, if the system is analyzed in the rotating frame, the result will be equivalent to subtracting the angular velocity $\hat{\Omega}$ from the frequencies shown in Figure 7.19 resulting in the diagram shown in Figure 7.20. Thus, the appearance of the figure depends upon the coordinate system. For two bladed rotors, one is likely to see a graph similar to Figure 7.20; while, for multibladed rotor analysis, one usually sees the fixed coordinate plot shown in Figure 7.19. It may be noted that at two rotor angular velocities ($\hat{\Omega} = 0.65$ and $\hat{\Omega} = 1.51$) the frequency of one of the blade modes is equal to a pylon mode. It would be expected that the coupling effects due to blade mass are most significant in these regions.

Now the influence of the mass ratio μ on the dynamics of the system is examined. It would be most convenient if root locus techniques could be used. This is not possible directly with equations (7.3.24) since μ will not appear linearly in the characteristic equation. Through introduction of complex coordinates, root locus techniques can be employed [28]. Define

$$\bar{z} = \bar{x} + i\bar{y}$$

$$\delta = \Gamma_2 - i\Gamma_1$$

This coordinate change reduces the four equations (7.3.24) to two equations owing to the symmetry properties of these equations, i.e.,

$$\ddot{\delta} - 2i\hat{\Omega}\dot{\delta} + (\hat{\omega}_\varsigma^2 - \hat{\Omega}^2)\delta - \frac{3}{2}\ddot{\bar{z}} = 0 \tag{7.3.25}$$

$$-\frac{\mu}{4}\ddot{\delta} + \ddot{\bar{z}} + \bar{z} = 0$$

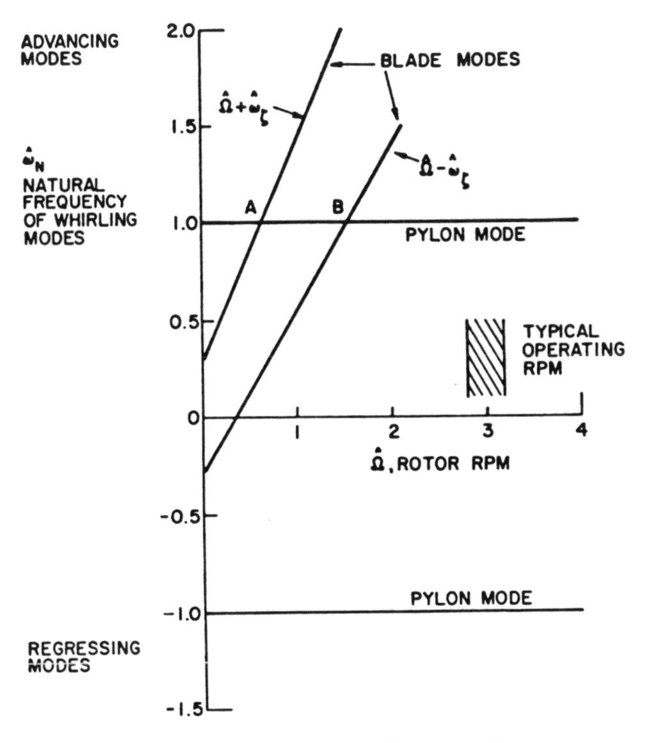

Figure 7.19 Uncoupled whirling modes ($\mu = 0$). Fixed coordinate system.

We now have a fourth order system in place of an eighth order system (Equations 7.3.24), and the roots of this system are the whirling modes only, i.e.,

$$i(\hat{\Omega} + \hat{\omega}_\zeta)$$

$$i(\hat{\Omega} - \hat{\omega}_\zeta)$$

$$\pm i$$

The characteristic equation of this system can now be written as

$$(s^2 + 1)(s^2 - 2i\hat{\Omega}s + (\hat{\omega}_\zeta^2 - \hat{\Omega}^2)) - \frac{3}{8}\mu s^4 = 0$$

Now μ appears as a linear parameter and the characteristic equation can be written as

$$\frac{\frac{3}{8}\mu s^4}{(s^2 + 1)(s^2 - 2i\hat{\Omega}s + (\hat{\omega}_\zeta^2 - \hat{\Omega}^2))} = 1 \tag{7.3.26}$$

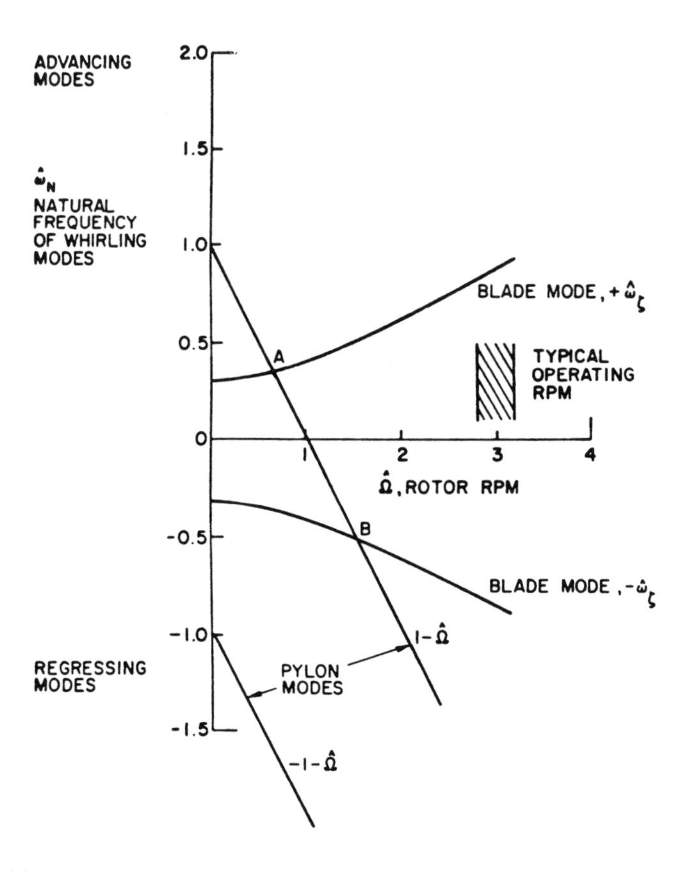

Figure 7.20 Uncoupled whirling modes (μ = 00. Rotating coordinate system.

Equation (7.3.26) is in root locus form with μ as the variable parameter and a zero degree locus is indicated. The usual root locus rules apply to equations with complex coefficients as well as to those with real coefficients. Figure 7.21 shows root loci for increasing μ for two values of $\hat{\Omega}$ (0.2 and 1.3) and indicates that the influence of μ on the dynamics is quite different depending upon the sign of $(\hat{\Omega} - \hat{\omega}_\zeta)$. When $(\hat{\Omega} - \hat{\omega})$ is negative it can be seen from Figure 7.21 that the coupling effect of increasing μ is to separate the system frequencies. However, when $(\hat{\Omega} - \hat{\omega}_\zeta)$ is positive, the two intermediate ferquencies come together; and, if μ is sufficiently large, instability occurs. The most critical case occurs at intersection B of Figure 7.19 (i.e., when the regressing mode frequency $(\hat{\Omega} - \hat{\omega}_\zeta) = 1$ such that the two intermediate frequencies are equal).

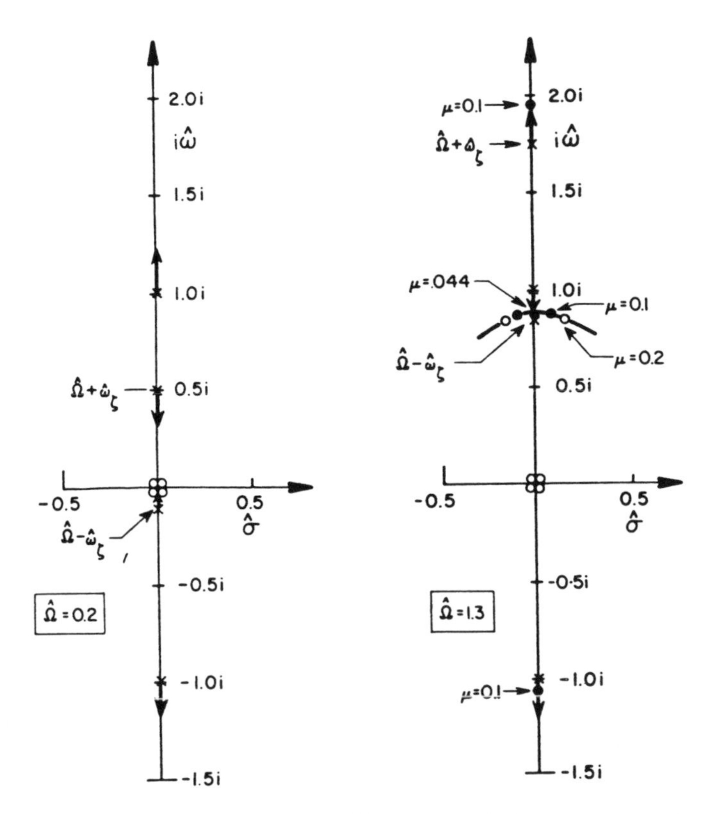

Figure 7.21 Root locus for increasing blade mass ratio (μ) for two operating conditions.

At this operating condition, any value of μ causes instability. Intersection A is not critical because of the large separation of the two intermediate frequencies ($\hat{\Omega}+\hat{\omega}_\zeta$and$\hat{\Omega}-\hat{\omega}_\zeta$). In the typical case, for an articulated rotor $\hat{\omega}_\zeta$ is the order of one; and $\hat{\Omega}$ at the operating condition ($\hat{\Omega}_{\mathrm{OP}}$) is the order of three, so that intersection B occurs below operating rpm. That is

$$\hat{\Omega}_{\mathrm{CR}} = 1 + \hat{\omega}_\zeta \equiv 2$$

Thus, to completely eliminate the possibility of this instability, which is called ground resonance, one must have $\hat{\Omega}_{\mathrm{CR}} > \hat{\Omega}_{\mathrm{OP}} = 3$, and a very large offset is required since $\hat{\omega}_\zeta > 2$ and therefore, $\bar{\omega}_\zeta = 0.67$ which corresponds to a hinge offset of 0.3 (7.1.35) without a centering spring. Note that this ratio $\hat{\Omega}_{\mathrm{OP}}$ is largely determined by considerations other than rotor stability, such as the rotor operating rpm and the shock absorbing character of the landing gear. Since this large hinge offset is not practical, a centering spring may be

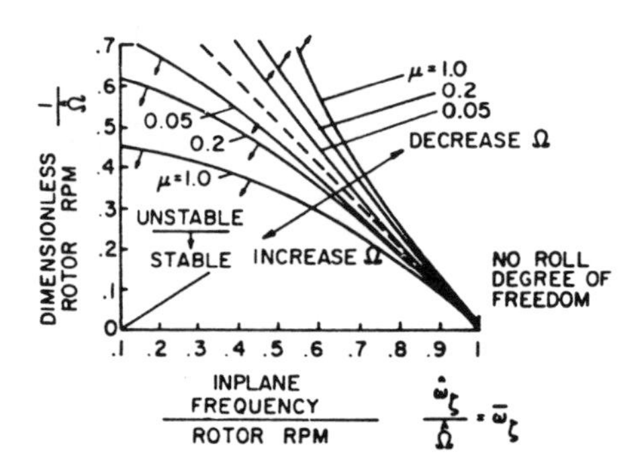

Figure 7.22 Extent of unstable region for various mass ratios [10].

employed to increase $\bar{\omega}_\zeta$; however this will increase the root bending moment, the reason the lag hinge was installed.

The unstable region extends below and above this intersection to an extent depending upon the mass ratio as well as the other geometric parameters. Various criteria can be found in the literature as to the size of the unstable region as a function of mass ratio. A typical graph of the frequencies as well as extent of the unstable region as a function of μ is shown in Figure 7.22 taken from [10]. This result applies to the case where the pylon has only one degree-of-freedom in contrast tot he example here where the pylon has two degrees-of-freedom.

Now we examine the influence of mechanical damping on the stability of the system. Damping in the rotation (lag damping) and nonrotating (pylon damping) parts of the system is considered separately. Consider first the influence of damping on the pylon. This would lead to terms $C_F\dot{\phi}$ and $C_F\dot{\theta}$ in equations (7.3.22). Adding this in complex form to the second of equations (7.2.25) and expressing the characteristic equation in root locus form

$$\frac{\hat{C}_F s[s^2 - 2i\hat{\Omega}s + (\hat{\omega}_\zeta^2 - \hat{\Omega}^2)]}{(s^2+1)(s^2 - 2i\hat{\Omega}s + (\hat{\omega}_\zeta^2 - \hat{\Omega}^2)) - \frac{3}{8}\mu s^4} = -1 \qquad (7.3.27)$$

where

$$\hat{C}_F = \frac{C_F\omega_F}{k_F h^2}$$

This root locus has two zeros at the uncoupled lag modes. Figure 7.23 shows the influence of increasing damping for two cases. In the first, μ is small so that the basic system is neutrally stable. Adding only fixed axis damping destabilizes the system. In the second case where μ is large enough such that the basic system is unstable no amount of damping will stabilize the system.

Now consider adding dapming to the lag motion of the blades. It must be noted that this damping will be in the rotating coordinate system (about the blade hinge) and so to directly add damping terms to the equations of motion, the rotating frame equations must be used. The damping then appears as $C_R\dot{\gamma}_1$ and $C_R\dot{\gamma}_2$. If the transformations are followed, this will ultimately result (in the rotating frame with complex notation) in the damping appearing in the first of equations (7.3.25) as

$$\hat{C}_R(\dot{\delta} - i\hat{\Omega}\delta)$$

The $i\hat{\Omega}$ term appears because rotating coordinate system damping is expressed with respect to a fixed frame. Adding this damping to the first of the two equations and expressing the characteristic equation in root locus form as

$$\frac{\hat{C}_R(s - i\hat{\Omega})(s^2 + 1)}{(s^2 + 1)(s^2 - 2i\hat{\Omega}s + (\hat{\omega}_\zeta^2 - \hat{\Omega}^2)) - \frac{3}{8}\mu s^4} = -1 \tag{7.3.28}$$

the root locus shown in Figure 7.24 is obtained. Again it is interesting to note that adding damping only in the rotating frame results in destabilizing one of the fuselage modes when the system is initially neutrally stable (small μ). For large μ the situation is similar to the fixed axis damping case. These rather surprising effects of damping in a rotating system indicate that damping must be handled with considerable care. Owing to the order of the system, it is rather difficult to obtain physical insight into the source of these effects. A *combination* of damping in the rotating frame (blades) and stationary frame (pylon) is required to stabilize the system; although, as can be seen from the root locus sketches, there will always be one zero near to the fuselage, blade lag mode making it difficult to provide a large amount of damping in one of the modes. There would, of course, generally be damping in the pylon frequencies. Particularly on articulated rotors, blade lag dampers are added since, as noted, this region of instability must be traversed in increasing the rotor speed to operating rpm. Ref. [44] presents boundaries showing the damping required to eliminate the instability region for articulated rotor helicopters.

The treatment of more general problems, including the blade flapping degree-of-freedom and discussion of its importance in the hingeless rotor case, may be found in [8], [9], and [10]. The two-bladed rotor is treated in [44].

Flapping motion of articulated rotors with small hinge offset does not produce appreciable hub moments and consequently there is only weak coupling between the flapping motion and the pylon motion. The hingeless rotor, however, produces large hub moments and consequently the flapping motion is

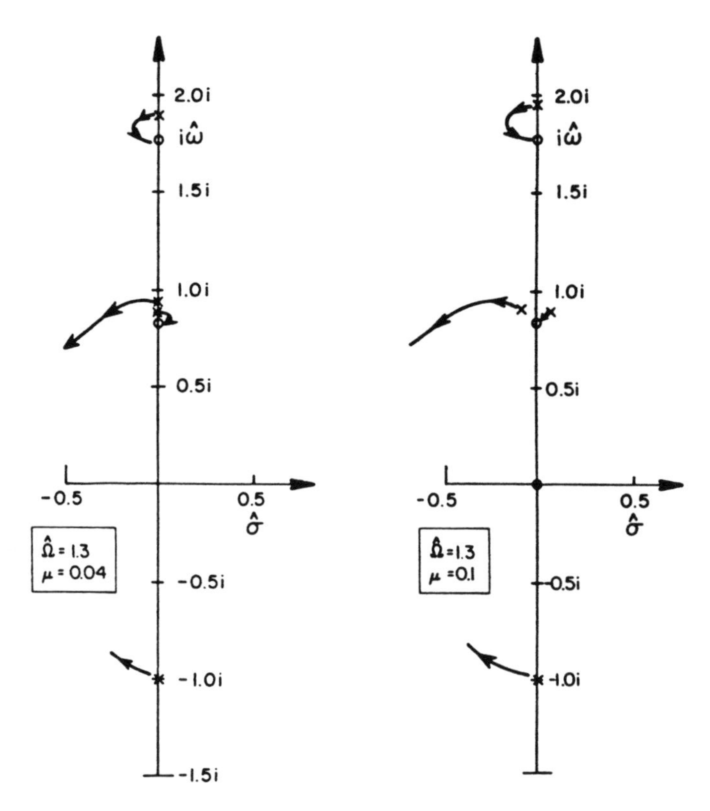

Figure 7.23 Root locus for increasing pylon (fixed axis) damping on ground resonance stability.

coupled into the pylong, lag dynamics [10]. Figure 7.25 shows the influence of the flapping frequency on the stability boundaries for $\mu = 0.1$.

There are now three frequencies involved in the problem, the pylon frequency, ω_F, the lag frequency, $\hat{\Omega} - \hat{\omega}_\zeta$ and the flap frequency p ($p - 1$ in the stationary frame). In addition to the destructive instability which occurs when the coupled pylon frequency is equal to the lag frequency $(\hat{\Omega} - \hat{\omega}_\zeta)$, a mild instability occurs when the coupled flap frequency is equal to the lag frequency $(\hat{\Omega} - \hat{\omega}_\zeta)$ as shown by Figure 7.26. Note that pylon and flap frequencies are significantly changed by the coupling. The ground resonance problem now becomes quite complex and difficult to generalize. The reader is referred to [10] for further details. It may be noted that an increasingly detailed model of the rotor blades must be employed. Since both flap and lag degrees-of-

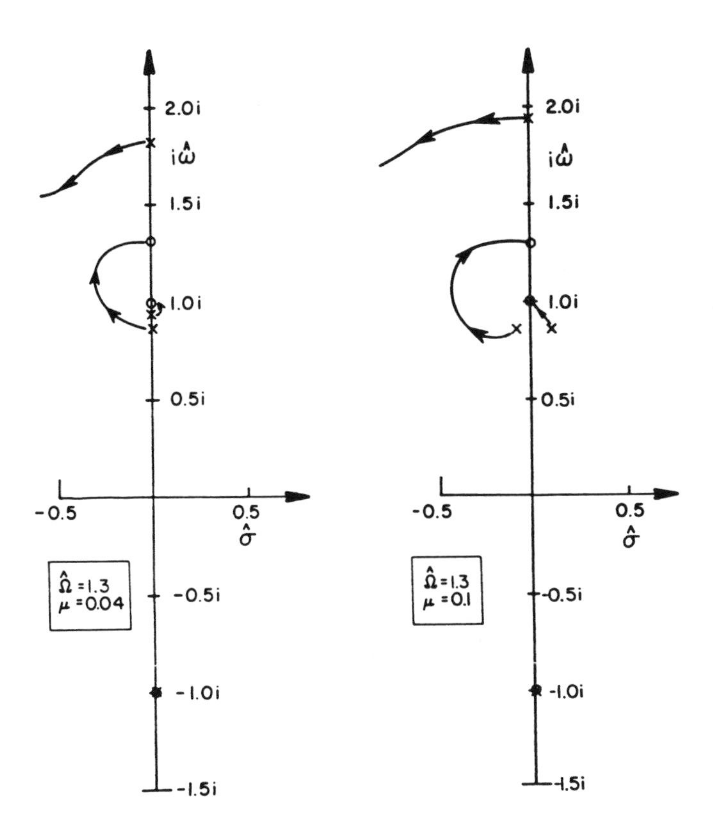

Figure 7.24 *Root locus for increasing lag (rotating axis) damping on ground resonance stability.*

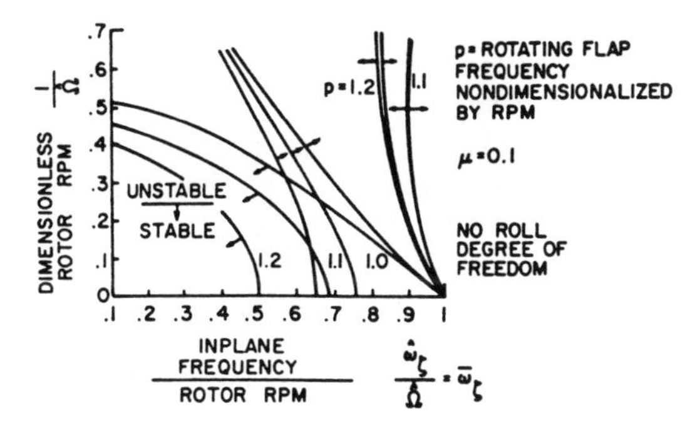

Figure 7.25 *Influence of flap frequency on ground resonance stability boundaries* [10].

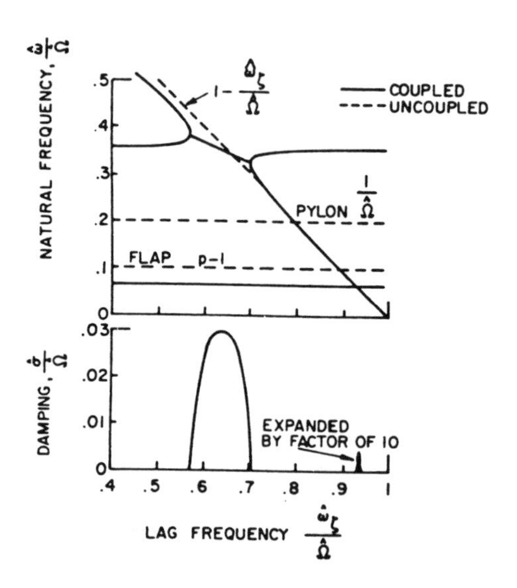

Figure 7.26 Frequency and damping of rotor-body system. Flap, lag and pylon pitch degrees of freedom. No aerodynamics [10].

freedom are involved, it is important to model the coupling between these motions which occurs as a result of hub and blade geometry. Aerodynamic forces and structural damping are also significant.

Air resonance refers to the form this dynamic problem takes with the landing gear restraint absent, that is, with the vehicle in the air. Coupling of flapping motion, body motion and lag motion is involved. Fuselage inertia and damping characteristics can have a significant impact on the stability. The air resonance problem is clearly asymmetrical and characteristically the roll axis is more critical owing to its low inertia and small aerodynamic damping [8]. Furthermore, air resonance generally involves the unsteady flow field [46].

Since the primary source of damping in this physical system arises from flap bending, it is possible that the nature of the flight control system can have an impact on air resonance stability as shown in [47]. Essentially an attitude feedback from the body to cyclic pitch tends to maintain the rotor in a horizontal plane thus effectively removing the aerodynamic damping from the flapping/body dynamics.

There are also other indications that the flight control system feedbacks have an impact on rotor system stability [48]; however, this problem does not appear to be well understood.

Another problem associated with propeller and prop/rotor driven aircraft which involves a blade motion-support coupling is whirl flutter which has been experienced on conventional aircraft [49] as well as on V/STOL aircraft [6]. This instability in the case of the conventional aircraft can be explained by considering only the wing as flexible (i.e., the propeller blades may be assumed to be rigid). For the tilt prop/rotor aricraft, where blade flexibility is important, the primary source of the instability is the same as in the rigid propeller case. It is a result of the aerodynamics characteristics of propellers and prop rotors at high inflows typical of cruising flight. It can be shown that the source of the whirl flutter instability is primarily associated with the fact that an angle-of-attack change on a propeller produces a yawing moment, and a sideslip angle produces a pitching moment. Further, the magnitude of this moment change grows with the square of the tangent of the inflow angle [6] and results in a rapid onset of the instability.

For the prop/rotor, a complex model with a large number of degrees-of-freedom is required to predict the dynamics of the system accurately [6]. The whirl flutter instability can occur on articulated rotors as well as hingeless rotors although for somewhat different physical reasons [50]. Here inplane force dependence on angular rate produces unstable damping moments acting on the support. The hingeless rotor which produces significant hub moments is similar to the rigid propeller. Young [51] has shown by a simplified analysis that, under certain circumstances, the occurrence of this instability can be minimized by suitable selection of the flapping frequency. Ref. [6] contains an excellent discussion of these various problem areas.

A typical predicted variation of damping with flight speed for a tilt-prop-rotor aircraft is shown in Figure 7.27. As mentioned earlier, it is important in modelling this dynamic system to insure that the structural details of the hub, blade and pitch control system are precisely modelled. Ref. [52] indicates the impact that relatively small modelling details can have on the flutter speed, as well as describing in detail the modelling requirements for prop-rotor whirl flutter.

Aeroelastic analysis of two-bladed rotors requires special considerations since the two blades are connected together. The reader is referred to [53] and [54] for the analysis of two-bladed rotors.

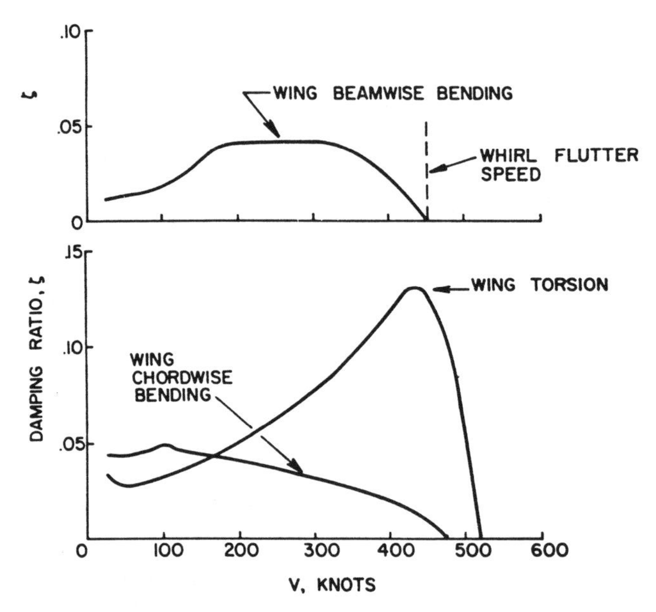

Figure 7.27 Damping of wing bending modes as a function of airspeed. Tilt prop/rotor aircraft with gimballed rotor [6].

7.4 Unsteady aerodynamics

Thus far in this chapter, we have utilized only the simplest quasi-steady, blade-element aerodynamics. However, it is well-known that the unsteady dynamics of the rotor flow field can have a profound effect on rotorcraft aeroelastic response. This effect is routinely included in rotor vibration and performance codes through vortex-lattice wake modeling [55]. However, in rotor aeroelastic computations, which involve eigenvalue computations, the vortex wake is usually frozen in time. When not frozen in time, the eigenvalue computation can take hours of computing time [56]. Therefore, more approximate methods have been developed to include the important effects of wake dynamics in rotorcraft aeroelasticity. Some of the more useful ones are explained below.

Dynamic Inflow

Amer [57] analyzed the problem of rotor damping in roll and correlated the predicted data with flight test measurements in hover and forward flight. Amer observes that a discrepancy in damping is "due primarily to changes in induced velocity which occur during rolling (or pitching) because of changes in the distribution of thrust around the rotor disk". This observation more or less forms the stimulus for most of the subsequent dynamic inflow work. It fell to Sissingh [58], however, to explain this discrepancy by quantifying Amer's observation with the inclusion of variable inflow or, more precisely, of changes in induced velocities caused by transient changes in rotor disk loading. Starting from Glauert's classical momentum theory postulate, he gives the formula

$$k(\delta\lambda/\lambda) = \delta C_T/C_T \tag{7.4.1}$$

where $k = 2$ in hovering and $k = 1$ in forward flight with $V > 40$ mph ($\mu \ll \lambda$). For transitional flight conditions when induced flow λ cannot be neglected in comparison to μ, Sissingh suggests an "appropriate" value for $k(1 < k < 2)$ on an *ad hoc* basis.

It is easily seen that equation (7.4.1) follows from the classical results

$$\lambda = -\sqrt{C_T/2}, \mu = 0 \tag{7.4.2}$$

$$\lambda = -C_T/2\mu, \mu \gg \lambda \tag{7.4.3}$$

Sissingh was probably the first to initiate a systematic exposition that established a relation between instantaneous perturbations (or transients) in thrust δT, and perturbations in induced flow, $\delta\lambda$. The induced flow λ is an involved function of both radius, r, and spatial azimuth position, ψ. To arrive at a tractable model, he uses first harmonic inflow and lift distributions, without radial variation,

$$\lambda = \lambda_o + \lambda_s \sin\psi + \lambda_c \cos\psi \tag{7.4.4}$$

Here, λ_0 is the uniform inflow, while λ_s and λ_c are side-to-side and fore-to-aft inflow variations. His analysis convincingly shows that the inclusion of induced velocity perturbations, as typified in equation (7.4.4), improves correlation of predicted damping values with those of the flight test data of Amer. (Sissingh's distribution has been used by several other investigators [59–61].) As seen from this equation, the distribution has two disadvantages. First, it neglects the effects of radial variation completely. Second, it exhibits a discontinuity at $r = 0$. As a means of improving the inflow distribution to account for radial variation to some degree and to avoid discontinuity, Peters [62] approximates dynamic inflow perturbations in induced flow by a truncated Fourier series with a prescribed radial distribution. The dynamic flow ν is perturbed with respect to the steady inflow λ such that the total inflow is

$$-\lambda = \bar{\lambda} + \nu \tag{7.4.5}$$

and dynamic inflow is

$$\nu = \nu_o + \nu_s \frac{r}{R} \sin\psi + \nu_c \frac{r}{R} \cos\psi \tag{7.4.6}$$

Similar to the development of [58], the inflow takes the form

$$\begin{Bmatrix} \nu_o \\ \nu_s \\ \nu_c \end{Bmatrix} = \frac{1}{V} \begin{bmatrix} 1/2 & 0 & 0 \\ 0 & -2 & 0 \\ 0 & 0 & -2 \end{bmatrix} \begin{Bmatrix} \delta C_T \\ \delta C_L \\ \delta C_M \end{Bmatrix} \tag{7.4.7}$$

where δC_T, δC_L, and δC_M are perturbations in thrust, roll-moment, and pitch-moment coefficients and where the mass-flow parameter, V, is obtained from momentum theory as

$$V = \frac{\mu^2 + \bar{\lambda}(\bar{\lambda} + \bar{\nu})}{\sqrt{\mu^2 + \bar{\lambda}^2}} \tag{7.4.8}$$

where $\bar{\nu}$ is the part of $\bar{\lambda}$ due to thrust (the remainder is due to climb) and μ is the ratio of forward speed to tip speed.

The preceding development from equation (7.4.7) implies that perturbations in disk loading $(\delta C_T, \delta C_L, \delta C_M)$ create instantaneous perturbations in inflow (ν_0, ν_s, ν_c). In other words, the feedback between changes in disk loading and inflow takes place without time lag. However, in transient downwash dynamics, a large mass of air is involved; and it is natural to expect that mass effects will have an influence on the complete build up of inflow perturbations due to disk-loading perturbations and *vice versa*. That is, the feedback will have some form of time delay due to mass effects. This aspect of the problem was investigated by Carpenter and Fridovich [63] during the early 1950s. The inclusion of the mass effects forms an integral part of the development of unsteady inflow models as an extension of the quasisteady inflow treated in the preceding paragraph. Substantial data-correlation experience with the quasisteady momentum model clearly demonstrates that *unsteady* wake effects (not quasisteady alone) play a dominant role in hover, in transitional flight and at low collective pitch [61, 62]. We will bypass the mathematical details [64] and include the rate terms, $[M]$, in the quasisteady equation:

$$\frac{1}{\Omega}[M] \begin{Bmatrix} \dot{\nu}_o \\ \dot{\nu}_s \\ \dot{\nu}_c \end{Bmatrix} + [L]^{-1} \begin{Bmatrix} \nu_o \\ \nu_s \\ \nu_c \end{Bmatrix} = \begin{Bmatrix} \delta C_T \\ \delta C_L \\ \delta C_M \end{Bmatrix} \tag{7.4.9}$$

where $[L]$ is the matrix of influence coefficients in equation (7.4.7); or, symbolically

$$\frac{1}{\Omega}[M]\{\dot{U}\} + [L]^{-1}\{U\} = \{\delta F\} \tag{7.4.10}$$

When premultiplied by $[L]$, equation (10b) takes the form

$$[\tau]\{\dot{U}\} + \{U\} = [L]\{\delta F\} \tag{7.4.11}$$

where $[\tau] = [L][M]/\Omega$.

In equation (7.4.11) $[\tau]$ and $[L]$ have the physical significance of time constants and gains, respectively. The elements of $[\tau]$ can also be treated as filter constants. This means, unsteady inflow can be simulated by passing the quastisteady inflow through a low-pass filter.

We now turn to the problem of evaluating these rate (or apparent mass) terms. This problem has been the subject matter of extensive studies. In [63], apparent mass terms are identified in terms of reaction forces (or moments) of an impermeable disk which undergos instantaneously acceleration (or rotation) in still air. The problem of finding reactions on an impermeable disk basically leads to the solution of a potential flow problem in terms of elliptic integrals. The values for the apparent mass of air m_A and apparent inertia of air I_A are: [64]:

$$m_A = \frac{8}{3}\rho R^3 \text{ and } I_A = \frac{16}{45}\rho R^5 \tag{7.4.12}$$

In other words, these values represent 64 per cent of the mass and 57 per cent of the rotary inertia of a sphere of air or radius R; and we have a diagonal $[M]$ matrix with

$$m_{11} = m_A/\rho\pi R^3 = 8/(3\pi) \tag{7.4.13}$$

$$m_{22} = m_{33} = -I_A/\rho\pi R^5 = -16/(45\pi) \tag{7.4.14}$$

which give time constants of 0.4244/V for δC_T and 0.2264/V for δC_L or δC_M. Given the complexity of the actual apparent mass terms of a lifting rotor, it would seem that the methodology adopted to arrive at the time constants is at best a crude approximation. Surprisingly, tests of Hohenemser et al. [59], and more recent analytical studies of Pitt and Peters [64] arrive at time constants or mass terms which are within a few percent of those given by equation (7.4.12). From the symmetry of the flow problem in hover, it is clearly seen that M is a diagonal matrix with $m_{22} = m_{33}$. Therefore, we have

$$[M] = \begin{bmatrix} \frac{8}{3\pi} & 0 & 0 \\ 0 & \frac{-16}{45\pi} & 0 \\ 0 & 0 & \frac{16}{45\pi} \end{bmatrix} \tag{7.4.15}$$

Equation (7.4.9) with $[M]$ and $[L]^{-1}$ from equations (7.4.15) and(7.4.7) forms the theory of dynamic inflow in hover. Numerous correlations with experimental data have shown the model accurate and crucial in rotor aeroelastic modelling. This includes frequency response [59, 62], control derivatives [62], and air and ground resonance [65]. Figures 7.28 and 7.29, taken from [65], show measured and computed frequencies of a ground-resonance model versus rotor speed. Figure 7.28 has no dynamic inflow modelling, and Figure 7.29 includes modelling of the type of equation (7.4.9). We are particularly interested in the range $300 < \Omega < 1000$ in which ground resonance can occur. The modes, labelled on the basis of theoretical eigenvectors are: regressing inplane (ζ_R), regressing flapping (β_R), roll (ϕ), pitch (θ). Note that, in Figure 7.28, the regressing inplane mode shows good frequency correlation; but all other modes are significantly off of the correct frequency. In Figure 7.29, a new mode appears (due to the added inflow degrees of freedom). It is labelled λ and is a mode dominated by inflow motions. The new results show excellent correlation of all frequencies with the exception of a roll-pitch coupling for $200 < \Omega < 400$. The results show that the regressing flap mode becomes critically damped at $\Omega = 750$, and that the measured modes are crucially impacted by the dynamic inflow.

One of the interesting aspects of the dynamic inflow theory refers to the formulation of equilvalent Lock number and drag coefficient (γ^* and C_d^*) [66]. This formulation reveals that there is an intrinsic correlation between downwash dynamics and unsteady airfoil aerodynamics. After all, any three-dimensional, unsteady vorticity theory automatically includes induced flow theory as a local approximation to transient downwash dynamics. Further, dynamic inflow decreases lift and increases profile drag. Therefore, we should expect an equivalent γ (or γ^*) that is lower than γ, and an equivalent C_d (or C_d^*) that is higher than C_d. Thus, the $\gamma^* - C_d^*$ concept leads to one of the simplest methods of crudely accounting for dynamic inflow in conventional "no-inflow"- programs. One simply must change γ to γ^* and C_d to C_d^*. Furthermore, the concept brings out the physics of dynamic inflow in a simple and visible manner.

In quasi-steady inflow theory, apparent mass effects are neglected. Therefore, the inflow differential equations reduce to algebraic equations without increasing the system dimension. If we stipulate the condition of axial flow (e.g., $\mu = 0$) in the quasisteady formulation, we may obtain γ^* and C_d^*, directly as detailed in [66].

$$\gamma^* = \gamma / \left(1 + \frac{a\sigma}{8V}\right) \tag{7.4.16}$$

and

$$(C_d/a)^* = \frac{C_d}{a}\left(1 + \frac{a\sigma}{8V}\right) + \frac{a\sigma}{8V}(\bar{\theta} - \phi)^2 \tag{7.4.17}$$

where $(\bar{\theta} - \phi)$ can be approximated by $6C_T/a\sigma$.

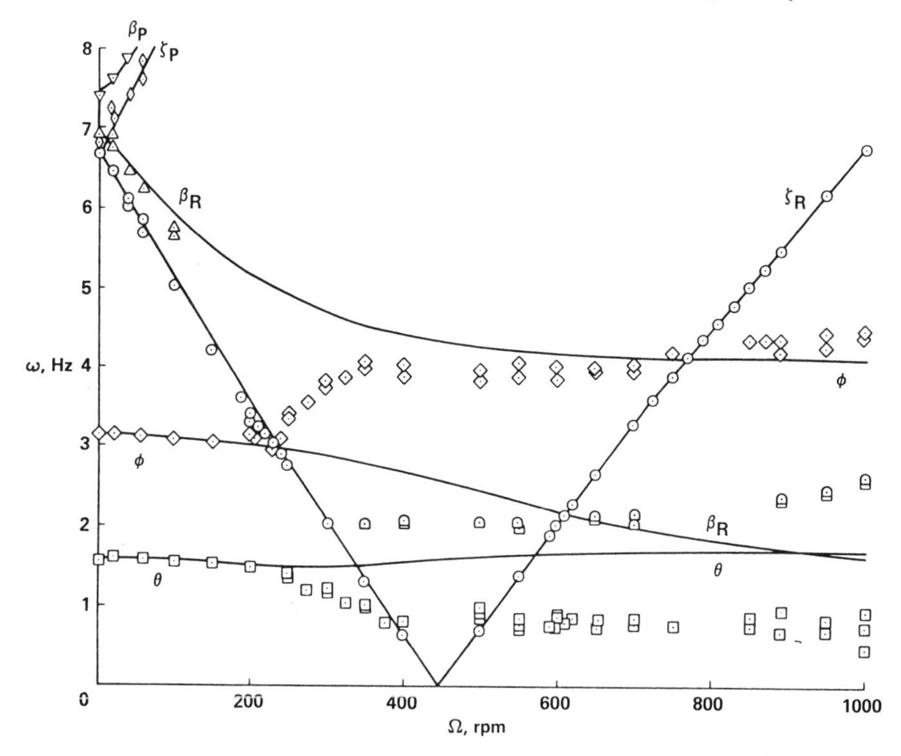

Figure 7.28 Influence of unsteady aerodynamics on hingeless rotor ground resonance. Comparison of measured modal frequencies, and calculations without dynamic inflow.

In forward flight, the model of equation (7.4.9) can be used but with alterations to the $[L]$ matrix. In [64], a general (L) matrix is defined based on potential flow theory. A new parameter, X, is introduced which is defined as the tangent of one-half of the wake skew angle. In hover or axial flight, $X = \tan(0) = 0$ and, in edgewise flight, $X = \tan(\pi/4) = 1$. Thus, S varies from zero to one as we transition from axial flight to edgewise flight. The corresponding $[L]$ matrix is

$$[L] = \frac{1}{V}\begin{bmatrix} \frac{1}{2} & 0 & \frac{15\pi}{64}X \\ 0 & -2(1+X^2) & 0 \\ \frac{15\pi}{64}X & 0 & -2(1-X^2) \end{bmatrix} \tag{7.4.18}$$

In axial flow, $X = 0$, and this matrix reduces to the momentum-theory values in equation (7.4.7). In edgewise flow, significant couplings develop between thrust and the fore-to-aft gradient in flow (and between pitch moment and

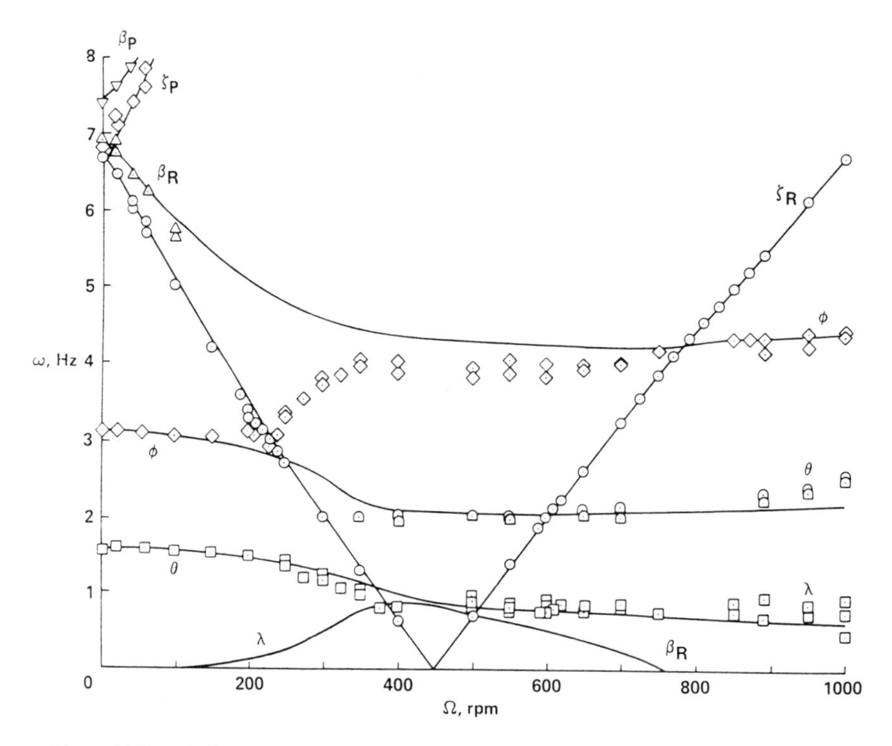

Figure 7.29 Influence of unsteady aerodynamics on hingeless rotor ground resonance. Comparison of measured modal frequencies, and calculations including dynamic inflow.

uniform flow). Equation (7.4.18) has been proven accurate by exhaustive correlations with forward-flight data [67].

Frequency domain

The theory of dynamic inflow treats low-frequency inflow effects in the range of 0/rev to 2/rev. For high-frequency effects, investigators traditionally have relied upon frequency-domain aerodynamics such as the Theodorsen function [68]. However, the wake model for Theodorsen theory (a flat two-dimensional wake) is simply inadequate for rotor work. A very useful alternative is Loewy theory [69]. The model for Loewy theory is similar to that for Theodorsen theory except that the wake is assumed to return in an infinite number of layers spaced apart by a fixed distance, h semi-chords. The resultant lift-deficiency function (the Loewy function) takes the following form

$$C'(k) = \frac{1}{1 + A(k)} \tag{7.4.19}$$

$$A(k) = \frac{Y_0(k) + iJ_0(k)(1 + 2W)}{J_1(k)(1 + 2W) - iY_1(k)} \tag{7.4.20}$$

$$W = [\exp(kh + 2\pi i\omega/Q) - 1]^{-1} \tag{7.4.21}$$

where k is the rotating-system reduced frequency, Y_n and J_n are Bessel Functions, ω is the frequency per revolution as seen in the non-rotating system, and Q is the number of blades.

W is called the wake-spacing function. For infinite wake spacing ($h \to \infty$), $W \to 0$; and consequently, equation (7.4.20) reduces to Theodorsen theory. For finite wake spacing, W becomes largest when ω is an integer multiple of Q, and the resulting $C'(k)$ is low. Thus, lift is lost when the shed vorticity of successive layers is aligned. For small k, the near-wake approximation for $A(k)$ is [71]

$$A(k) \approx \pi k \left(\frac{1}{2} + W\right) \tag{7.4.22}$$

The above theory has a connection to dynamic inflow theory (although the latter is in the time-domain). In particular, when ω is an integer multiple of Q and k is small, we can write

$$A(k) \approx \pi k \left(\frac{1}{kh}\right) = \frac{\pi}{h} \tag{7.4.23}$$

Now, if we notice that $a = 2\pi$ for Loewy theory and that $h = 4V/\sigma$, where ($\sigma = 2bQ/\pi R$ (where b = semi-chord). Then

$$A = \frac{\sigma a}{8V}, \quad C'(k) = \frac{1}{1 + \frac{\sigma a}{8V}} \tag{7.4.24}$$

which is the quasi-steady approximation for dynamic inflow, γ^*/γ.

Thus, $C'(k)$ tends to oscillate between $\frac{1}{1+\sigma a/8V}$ and 1 (as ω is varied). The lowest points are at integer multiple of Q, and the highest points are at odd multiples of $Q/2$.

An alternative theory for $C'(k)$ is given by Miller [70], which neglects the W in the denominator of $A(k)$. Both theories have the samen near-wake approximation.

In order to apply a lift deficiency function to rotor problems, one must also account for the effect of $C'(k)$ on lift and drag. This can be accomplished through the C_D^* approach.

$$a^* = aC'(k) = \frac{a}{1 + A(k)} \tag{7.4.25}$$

$$C_D^* = C_D + \frac{A(k)}{1 + A(k)}(\bar{\theta} - \phi)^2 \tag{7.4.26}$$

Despite the elegance and power of lift-deficiency functions, their use in rotor problems has been limited by several shortcomings. First, the theories are limited to a two-dimensional approximation in axial flow. Second, they are in the frequency domain, which is inconvenient for periodic-coefficient eigen-analysis (although finite-state approximations can be obtained) [72]. Third, Loewy theory exhibits a singularity as ω and k approach zero simultaneously. Due to these drawbacks, investigators often use dynamic inflow for the low-frequency effects (since it is a three-dimensional, time-domain theory) but neglect wake effects in other frequency ranges.

Finite-state wake modeling

More recently, a complete three-dimensional wake model has been developed that includes dynamic inflow and the Loewy function implicitly [73]. In this theory, the induced flow on the rotor disk is expressed as an expansion in a Fourier series (azimuthally) and in special polynomials (radially) with power of $\bar{r} = r/R$.

$$\nu(\bar{r}, \psi, t) = \sum_{m=0}^{\infty} \sum_{n=m+1,m+3}^{\infty} \hat{\phi}_n(\bar{r})[\alpha_n^m \cos m\psi + \beta_n^m \sin m\psi] \tag{7.4.27}$$

where

$$\hat{\phi}_n^m(\bar{r}) = \frac{1}{2}\sqrt{\pi(2n+1)} \sum_{q=m,m+2,\ldots}^{n-1} \bar{r}^q \frac{(-1)^{(q-m)/2}(n+q)!!}{(q-m)!!(q+m)!!(n-q-1)!!}$$

and α_n^m and β_n^m are the expansion coefficients. Thus, α_1^0, α_2^1, and β_2^1 take on the role of ν_o, ν_c and ν_s in dynamic inflow; and the higher-order terms allow a more detailed inflow model to any order desired. When the above expansion is combined with an acceleration-potential for the three-dimensional flow field, differential equations are formed that are similar in character to dynamic inflow.

$$\frac{1}{\Omega}\{\dot{\alpha}_n^m\} + [L^c]^{-1}\{\alpha_n^m\} = \{\tau_n^{mc}\} \tag{7.4.29}$$

$$\frac{1}{\Omega}\{\dot{\beta}_n^m\} + [L^s]^{-1}\{\beta_m^n\} = \{\tau_n^{ms}\} \tag{7.4.30}$$

The $[L]$ matrices are influence coefficients that depend on X (tangent of one-half of the wake skew angle). They are partitioned by the harmonic numbers m, r (m = inflow harmonic, r = pressure harmonic) and (within each mr partition) there is a row-column pair (j, n) for the inflow and pressure shape function, respectively.

$$\begin{aligned}
[L_{jn}^{om}]^c &= \frac{1}{V} X^m \Gamma_{jn}^{0m} \\
\left[L_{jn}^{rm}\right]^c &= \frac{1}{V}\left[X^{|m-r|} + (-1)^\ell X^{|m+l|}\right]\Gamma_{jn}^{rm} \\
\left[L_{jn}^{rm}\right]^c &= \frac{1}{V}\left[X^{|m-r|} - (-1)^\ell X^{|m+r|}\right]\Gamma_{jn}^{rm}
\end{aligned} \tag{7.4.31}$$

where $\ell = \min(r,m)$ and

$$\begin{aligned}
\Gamma_{jn}^{rm} &= \frac{(-1)^{\frac{n+j-2r}{2}} 4\sqrt{(2n+1)(2j+1)}}{\pi(j+n)(j+n+2)[(j-n)^2-1]} \quad \text{for} \quad r+m \text{ even} \\
\Gamma_{jn}^{rm} &= \frac{sgn(r-m)}{\sqrt{(2n+1)(2j+1)}} \quad \text{for} \quad r+m \text{ odd}, j = n \pm 1 \\
\Gamma_{jn}^{rm} &= 0 \quad \text{for} \quad r+m \text{ odd}, j \neq n \pm 1
\end{aligned} \tag{7.4.32}$$

The right hand sides of equations (7.4.29–30) are generalized forces obtained from the integral over each blade of the circulatory lift per unit length (and then summed over all the blades, the q-th blade being at ψ_q).

$$\begin{aligned}
\tau_n^{oc} &= \frac{1}{2\pi}\sum_{q=1}^{Q}\left[\int_0^1 \frac{L_q}{\rho\Omega^2 R^3}\hat{\phi}_n^0(\bar{r})d\bar{r}\right] \\
\tau_n^{mc} &= \frac{1}{\pi}\sum_{q=1}^{Q}\left[\int_0^1 \frac{L_q}{\rho\Omega^2 R^3}\hat{\phi}_n^m(\bar{r})d\bar{r}\right]\cos(m\psi_q) \\
\tau_n^{ms} &= \frac{1}{\pi}\sum_{q=1}^{Q}\left[\int_0^1 \frac{L_q}{\rho\Omega^2 R^3}\hat{\phi}_n^m(\bar{r})d\bar{r}\right]\sin(m\psi_q)
\end{aligned} \tag{7.4.33}$$

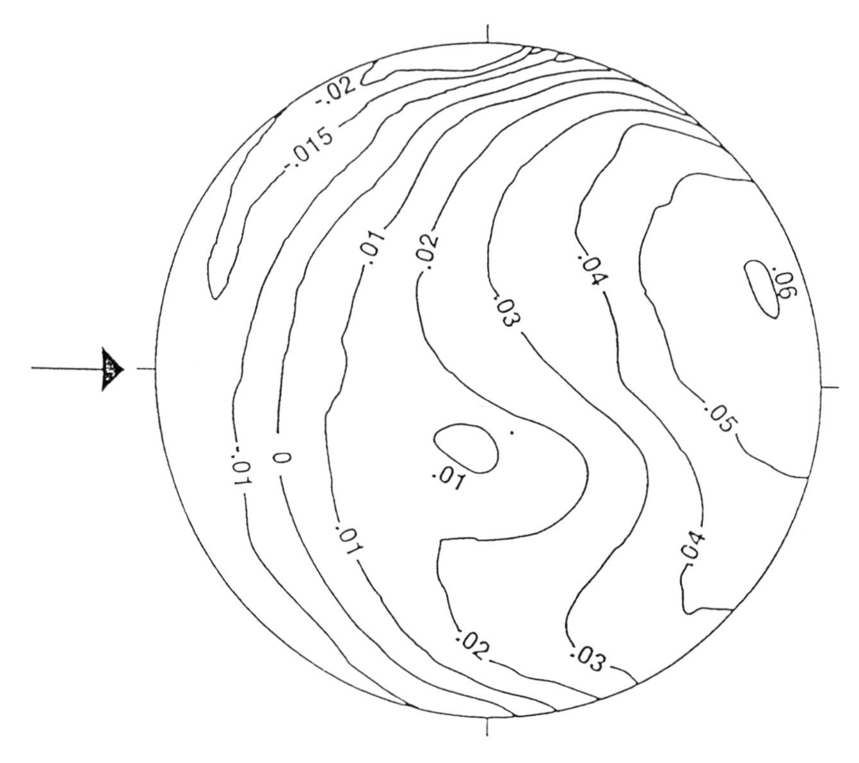

Figure 7.30 Experimental induced inflow distribution. Tapered blade, μ = 0.15.

It is interesting to note that τ_1^{0c}, τ_2^{1c}, and τ_2^{1s} are proportional to C_T, C_M, and C_L (respectively). Thus, when only these three are present, we recover dynamic inflow. The higher expansion terms are taken to the same order as we take velocity expansions.

The L_q terms (lift per unit length of q-th blade) can be inserted in equation (7.4.33) from any lifting theory. When they are taken from blade element theory (and when the radial direction is neglected), one can prove that the system of equations (7.4.29–30) reduces to the Loewy theory for $X = 0$ (axial flow) [73]. It should be pointed out that these equations are perturbation equations with ν and L_q taken as perturbations with respect to a steady state. A complete, nonlinear theory is available [74] but is not presented here.

Figures 7.30 and 7.31 show measured and calculated induced flows on a rotor at $\mu = 0.15$, from [74]. The major features of the flow-field are

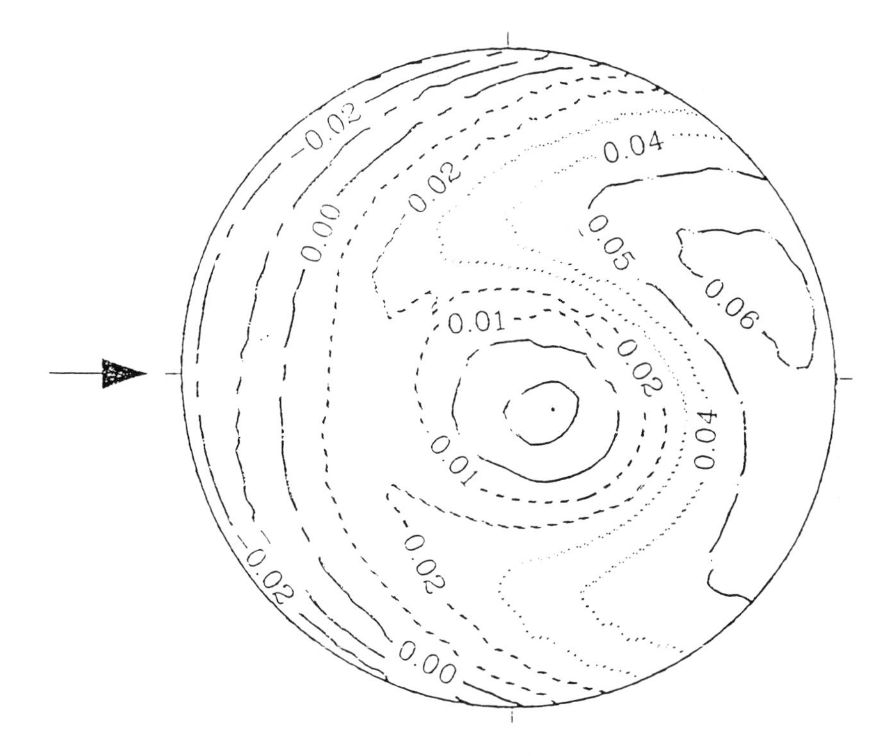

Figure 7.31 Theoretical induced inflow distribution, tapered blades with fuselage, $\mu = 0.15, C_T = 0.0064, \alpha = 3^\circ, M = 4, S = 33$.

captured by the model. In [33], the new wake model is used to greatly improve correlation of stability computations with experimental data.

Summary

In recent years, rotor aeroelasticity has relied more heavily on unsteady aerodynamic modeling to improve predictive capabilities. The major modeling tools are dynamic inflow, lift-deficiency functions, and finite-state modeling. The last of these includes the other two.

1. Gessow, A. and Myers, G.C., Jr., *Aerodynamics of the Helicopter*, The Macmillan Company, New York, 1952.
2. Bramwell, A.R.S., *Helicopter Dynamics*, John Wiley & SOns, New York, 1976.
3. Loewy, R.G., "Review of Rotary Wing V/STOL Dynamics and Aeroelastic Problems," *Journal of American Helicopter Society*, Vol. 14, No. 3 (July 1969).
4. Friedmann, P., "Recent Developments in Rotary Wing Aeroelasticity," Paper No. 11, *Second European Rotorcraft and Powered Lift Aircraft Forum* (september 20–22, 1977).
5. Hohenemser, K.H., "Hingeless Rotorcraft Flight Dynamics," *Agardograph* 197, 1974.
6. Johnson, W., "Dynamics of Tilting Proprotor Aircraft in Cruise Flight," NASA TN D-7677 (May 1974).
7. Manjunath, A., Nagabhushanam, J., Gaonkar, G., Peters, D., and Su, A., "Flap-Lag Damping in Hover and Forward Flight with a Three-Dimensional Wake," Proceedings of the 48th Annual Forum of the American Helicopter Society, Washington, June 3–5, 1992.
8. Burkham, J.E. and Miao, W.L., "Exploration of Aeroelastic Stability Boundaries with a Soft-in-Plane Hingeless-Rotor Model," *Journal of the American Helicopter Society*, Vol. 17, No. 4 (October 1972).
9. Donham, R.E., *et al.*, "Ground and Air Resonance Characteristics of a Soft In-Plane Rigid-Rotor System," *Journal of the American Helicopter Society*, Vol. 14, No. 4 (October 1969).
10. Ormiston, R.A., "Aeromechanical Stability of Soft Inplane Hingeless Rotor Helicopters," Paper No. 25, *Third European Rotorcraft and Powered Lift Aircraft Forum*, Aix-en-Provence, France (September 7–9, 1977).
11. Goland, L. and Perlmutter, A.A., "A Comparison of the Calculated and Observed Flutter Characteristics of a Helicopter Rotor Blade," *Journal of the Aeronautical Sciences*, Vol. 24, No. 4 (April 1957).
12. Reichert, G. and Huber, H., "Influence of Elastic Coupling Effects on the Handling Qualities of a Hingeless Rotor Helicopter," *Agard Conference Proceedings No. 121 Advanced Rotorcraft* (February 1973).
13. Lentine, F.P, *et al.*, "Research in Manuverability of the XH-51A Compund Helicopter," USA AVLABS TR 68–23 (June 1968).
14. Ham, N.D., "Helicopter Blade Flutter," AGARD Report 607 (January 1973).
15. Shames, I.H., *Engineering Mechanics*, Vol. II: *Dynamics*, Prentice-Hall, Inc., Englewood Cliffs, N.J., 1958.
16. Peters, D.A. and Hohenemser, K.H., "Application of the Floquet Transition Matrix to Problems of Lifting Rotor Stability," *Journal of the American Helicopter Society*, Vol. 16, No. 2 (April 1971).
17. Ormiston, R.A. and Hodges, D.H., "Linear Flap-Lag Dynamics of Hingeless Helicopter Rotor Blades in Hover," *Journal of the American Helicopater Society*, Vol. 17, No. 2 (April 1972).
18. Chou, P.C., "Pitch Lag Instability of Helicopter Rotors," *Institute of Aeronautical Sciences Preprint* 805, 1958.
19. Bennett, R.M. and Curtiss, H.C., Jr., "An Experimental Investigation of Helicopter Stability Characteristics Near Hovering Flight Using a Dynamically Similar Model," *Princeton University Department of Aeronautical Engineering Report* 517 (July 1960).

20. Hansford, R.E. and Simons, I.A., "Torsion-Flap-Lag Coupling on Helicopter Rotor Blades," *Journal of the American Helicopter Society*, Vol. 18, No. 4 (October 1973).
21. Hodges, D.H. and Ormiston, R.A., "Stability of Elastic Bending and Torsion of Uniform Cantilever Rotor Blades in Hover with Variable Structural Coupling," NASA TN D-8192 (April 1976).
22. Hodges, D.H. and Dowell, E.H., "Non-Linear Equations of Motion for the Elastic Bending and Torsion of Twisted Nonuniform Rotor Blades," NASA TN D-7818 (July 1974).
23. Flax, A.H. and Goland, L., "Dynamic Effects in Rotor Blade Bending," *Journal of the Aeronautical Sciences*, Vol. 18, No. 12 (December 1951).
24. Yntema, R.T., "Simplified Procedures and Charts for Rapid Estimation of Bending Frequencies of Rotating Beams," NACA TN 3459 (June 1955).
25. Young, M.I., "A Simplified Theory of Hingeless Rotors with Application to Tandem Helicopters," *Proceedings of the 18th Annual National Forum. American Helicopter Society*, (May 1962) pp. 38–45.
26. Curtiss, H.C., Jr., "Sensitivity of Hingeless Rotor Blade Flap-Lag Stability in Hover to Analytical Modelling Assumptions," *Princeton University Department of Aerospace and Mechanical Sciences Report 1236* (January 1975).
27. Halley, D.H., "ABC Helicopter Stability, Control and Vibration Evaluation on the Princeton Dynamic Model Track," *American Helicopter Society Preprint* 744 (May 1973).
28. Curtiss, H.C., Jr., "Complex Coordinates in Near Hovering Rotor Dynamics," *Journal of Aircraft*, Vol. 10, No. 8 (May 1973).
29. Berrington, D.K., "Design and Development of the Westland Sea Lynx," *Journal of American Helicopter Society*, Vol. 19, No. 1 (January 1974).
30. Ormiston, R.A. and Bousman, W.G., "A Study of Stall-Induced Flap-Lag Instability of Hingeless Rotors," *Journal of the American Helicopter Society*, Vol. 20, No. 1 (January 1975).
31. Peters, David A., "An Approximate Closed-Form Solution for Lead-Lag Damping of Rotor Blades in Hover," NASA TM X-62,425 (April 1975).
32. Gaffey, T.M., "The Effect of Positive Pitch-Flap Coupling (Negative δ_3) on Rotor Blade Motion Stability and Flapping," *Journal of the American Helicopter Society*, Vol. 14, No. 2 (April 1969).
33. de Andrade, Donizeti an Peters, David, "On a Finite-State Inflow Application to Flap-Lag-Torsion Damping in Hover," Proceedings of the Eighteenth European Rotorcraft Forum, Avignon, France, September 15–18, 1992, Paper No. 2.
34. Ham, N.D. and Garelick, M.S., "Dynamic Stall Considerations in Helicopter Rotors," *Journal of the American Helicopter Society*, Vol. 13, No. 2 (April 1968).
35. Liiva, J. and Davenport, F.J., "Dynamic Stall of Airfoil Sections for High-Speed Rotors," *Journal of the American Helicopter Society*, Vol. 14, No. 2 (April 1969).
36. Martin, J.M. *et al.*, "An Experimental Analysis of Dynamic Stall on an Oscillating Airfoil," *Journal of the American Helicopter Society*, Vol. 19, No. 1 (January 1974).
37. McCroskey, W.J. and Fisher, R.K., Jr., "Detailed Aerodynamic Measurements on a Model Rotor in the Blade Stall Regime," *Journal of the American Helicopter Society*, Vol. 17, No. 1 (January 1972).
38. Johnson, W. and Ham, N.D., "On the Mechanism of Dynamic Stall," *Journal of the American Helicopter Society*, Vol. 17, No. 4 (October 1972).
39. Ericsson, L.E. and Reding, J.P., "Dynamic Stall of Helicopter Blades," *Journal of the American Helicopter Society*, Vol. 17, No. 4 (October 1972).

40. Tarzanin, F.J., Jr., "Prediction of Control Loads Due to Blade Stall," *Journal of the American Helicopter Society*, Vol. 17, No. 2 (April 1972).
41. Leishman, J.G. and Beddoes, T.S., "A Semi-Empirical Moel for Dynamic Stall," *Journal of the American Helicopter Society*, Vol. 34, No. 4 (July 1989).
42. Peleau, B. and Petot, D., "Aeroelastic Prediction of Rotor Loads in Forward Flight," *Vertica*, Vol. 13, No. 2 (1989) pp. 107–118.
43. Narramore, J.C., Sankar, L.N., and Vermeland, R., "An Evaluation of a Navier-Stokes Code for Calculation of Retreating Blade Stall on Helicopter Rotor," Proceedings of the 44th Annual Forum of the American Helicopter Society, Washington (June 160018) pp. 797–808.
44. Coleman, R.P. and Feingold, A.M., "Theory of Self-Excited Mechanical Oscillations of Helicopter Rotors with Hinged Blades," NACA TN 3844 (February 1957).
45. Hohenemser, K.H. and Yin, S.K., "Some Applications of the Method of Multi-Blade Coordinates," *Journal of the American Helicopter Society*, Vol. 17, No. 3 (July 1972).
46. Gaonkar, G.H., Mitra, A.K., Reddy, T.S.R., and Peters, D.A., "Sensitivity of Helicopter Aeromechanical Stability to Dynamic Inflow,"' *Vertica*, Vol. 6 (1982) pp. 59–75.
47. Lytwyn, R.T., *et al.*, "Airborne and Ground Resonance of Hingeless Rotors," *Journal of the American Helicopter Society*, Vol. 16, No. 2 (April 1971).
48. Briczinski, S. and Cooper, D.E., "Flight Investigation of Rotor/Vehicle State Feedback," NASA CR-132546 (1974).
49. Reed, W.H., III, "Review of Propeller-Rotor Whirl Flutter," NASA TR 4-264 (July 1968).
50. Hall, E.W., Jr., "Prop-Rotor Stability at High Advance Ratios," *Journal of the American Helicopter Society*, Vol. 11, No. 2 (April 1966).
51. Young, M.I. and Lytwyn, R.T., "The Influence of Blade Flapping Restraint on the Dynamic Stability of Low Disc Loading Propeller-Rotors," *Journal of the American Helicopter Society*, Vol. 12, No. 4 (October 1967).
52. Johnson, W., "Analytical Modelling Requirements for Tilting Prop Rotor Aircraft Dynamics," NASA TN D-8013 (July 1975).
53. Shamie, J. and Friedmann, P., "Aeroelastic Stability of Complete Rotors with Application to a Teetering Rotor in Forward Flight," *American Helicopter Society Preprint* No. 1031 (May 1976).
54. Kawakami, N., "Dynamics of an Elastic Seesaw Rotor," *Journal of Aircraft*, Vol. 14, No. 3 (March 1977).
55. Landgrebe, A.J., "An Analytical Method for Predicting Rotor Wake Geometry,"' *Journal of the American Helicopter Society*, Vol. 14, No. 4 (October 1969).
56. Kwon, Oh Joon, Hodges, D.H., and Sankar, L.N., "Stability of Hingeless Rotors in Hover Using Three-Dimensional Unsteady Aerodynamics," Proceedings of the 45th Annual National Forum of the American Helicopter Society, Boston, 1989, and *Journal of the American Helicopter Society*, Vol. 36, No. 2 (April 1991) pp. 21–31.
57. Amer, K.B., "Theory of Helicopter Damping in Pitch or Roll and Comparison with Flight Measurements," NASA, TN 2136 (October 1948).
58. Sissingh, G.J., "The Effect of Induced Velocity Variation on Helicopter Rotor Damping Pitch or Roll," Aeronautical Research Council (Great Britain), A.R.C. Technical Report G.P. No. 101 (14,757), (1952).
59. Hohenemser, K.H., "Hingeless Rotorcraft Flight Dynamics," (AGARD-AG-197 (1974).

60. Curtiss, H.C., Jr. and Shupe,N.K., "A Stability and Control Theory for Hingeless Rotors," Annual National Forum of the American Helicopter Society, Washington, D.C. (May 1971).
61. Ormiston, R.A. and Peters, D.A., "Hingeless Helicopter Rotor Response with Non-Uniform Inflow and Elastic Blade Banding," *Journal of Aircraft*, Vol. 9, No. 10 (October 1972) pp. 730–736.
62. Peters, D.A., "Hingeless Rotor Frequency Response with Unsteady Inflow," presented at the AHS/NASA Ames Specialists Meeting on Rotorcraft Dynamics, NASA SP-362 (February 1974).
63. Carpenter, P.J. and Fridovich, B., "Effect of a Rapid Blade Pitch Increase on the Thrust and Induced Velocity Response of a Full Scale Helicopter Rotor," NASA, TN 3044 (November 1953).
64. Pitt, D.M. and Peters, D.A., "Theoretical Prediction of Dynamic Inflow Derivatives," *Vertica*, Vol. 5, No. 1 (March 1981) pp. 21–34.
65. Johnson, W., "Influence of Unsteady Aerodynamics on Hingeless Rotor Ground Resonance," *Journal of Aircraft*, Vol. 29, No. 9 (August 1982) pp. 668–673.
66. Gaonkar, G.H., Mitra, A.K., Reddy, T.S.R., and Peters, D.A., "Sensitivity of Helicopter Aeromechanical Stability to Dynamic Inflow," *Vertica*, Vol. 6, No. 1 (1982) pp. 59–75.
67. Gaonkar, G.H. and Peters, D.A., "Effectiveness of Current Dynamic-Inflow Models in Hover and Forward Flight," *Journal of the American Helicopter Society*, Vol. 31, No. 2 (April 1986) pp. 47–57.
68. Theodorsen, T., "General Theory of Aerodynamic Instabilities and the Mechanism of Flutter," NACA TR 496 (1949).
69. Loewy, Robert G., "A Two-Dimensional Approach to the Unsteady Aerodynamics of Rotary Wings," *Journal of the Aerospace Sciences*, Vol. 24, No. 2 (February 1957) pp. 82–98.
70. Miller, R.H., "Rotor Blade Harmonic Air Loading," *AIAA Journal*, Vol. 2, No. 7 (July 1964) pp. 1254–1269.
71. Johnson, Wayne, *Helicopter Theory*, Princeton University Press, Princeton (1980) pp. 484–492.
72. Friedmann, P.P. and Venkatesan, C., "Finite State Modeling of Unsteady Aerodynamics and Its Application to a Rotor Dynamics Problem," Eleventh European Rotorcraft Forum, London, Paper No. 77 (September 1985).
73. Peters, David A., Boyd, David Doug, and He, Cheng Jian, "Finite-State Induced-Flow Model for Rotors in Hover and Forward Flight," *Journal of the American Helicopter Society*, Vol. 34, No. 4 (October 1989) pp. 5-17.
74. Peters, David A. and He, Cheng, Jian, "Correlation of Measured Induced Velocities with a Finite-State Wake Model," *Journal of the American Helicopter Society*, Vol. 36, No. 3 (July 1991) pp. 59–70.

8

Aeroelasticity in turbomachines

The advent of the jet engine and the high performance axial-flow compressor toward the end of World War II focussed attention on certain aeroelastic problems in turbomachines.

The concern for very light weight in the aircraft propulsion application, and the desire to achieve the highest possible isentropic efficiency by minimizing parasitic losses led inevitably to axial-flow compressors with cantilever airfoils of high aspect ratio. Very early in their development history these machines were found to experience severe vibration of the rotor blades at part speed operation; diagnosis revealed that these were in fact stall flutter (see Chapter 5) oscillations. The seriousness of the problem was underlined by the fact that the engine operating regime was more precisely termed the 'part corrected speed' condition, and that in addition to passing through this regime at ground start up, the regime could be reentered during high flight speed conditions at low altitude. In either flight condition destructive behavior of the turbojet engine could not be tolerated.

In retrospect it is probable that flutter had occurred previously in some axial flow compressors of more robust construction and in the latter stages of low pressure axial-flow steam turbines as well. Subsequently a variety of significant forced and self-excited vibration phenomena have been detected and studied in axial-flow turbomachinery blades.

In 1987 and 1988 two volumes of the AGARD *Manual on Aeroelasticity in Turbomachines* [1, 2] were published with 22 chapters in all. The sometimes disparate topics contributed by nineteen different authors and/or co-authors form a detailed and extensive reference base related to the subject material of the present Chapter 8. The reader is urged to refer to the AGARD compendium for in-depth development and discussion of many of the topics to be introduced here, and for related topics (such as the role of experimentation) not properly included here.

8.1 Aeroelastic environment in turbomachines

Consider an airfoil or blade in an axial flow turbine or compressor which is running at some constant rotational speed. For reasons of steady aero-dynamic and structural performance the blade has certain geometric properties defined by its length, root and tip fixation, possible mechanical attachment to other blades and by the chord, camber, thickness, stagger and profile shape which are functions of the radial coordinate. Furthermore, the blade may be constructed in such a manner that the line of centroids and the line of shear centers are neither radial nor straight, but are defined by schedules of axial and tangential coordinates as functions of radius. In fact, in certain cases, it may not be possible to define the elastic axis (i.e., the line of shear centers). The possibility of a built-up sheet metal and spar construction, a laid-up plastic laminate construction, movable or articulated fixations and/or supplemental damping devices attached to the blade would complicate the picture even further.

The blade under consideration, which may now be assumed to be completely defined from a geometrical and kinematical point of view, is capable of deforming* in an infinite variety of ways depending upon the loading to which it is subjected. In general, the elastic axis (if such can be defined) will assume some position given by axial and tangential coordinates which will be continuous functions of the radius (flapwise and chordwise bending). About this axis a certain schedule of twisting deformations may occur (defined, say by the angular displacement of a straight line between leading and trailing edges). Finally, a schedule of plate type bending deformations may occur as functions of radius and the chordwise coordinate. (Radial extensions summoned by centrifugal forces may further complicate the situation).

Although divergence is not a significant problem in turbomachines, the alternative static aeroelastic problem defined above, possibly resulting in measurable untwist and uncamber of the blades, can have important consequences with respect to the steady performance and with respect to the occurrence of blade stall and surge.

One has now to distinguish between steady and oscillatory phenomena. If the flow through the machine is completely steady in time

* Deformations are reckoned relative to a steadily rotating coordinate system in the case of a rotor blade.

and there are no mechanical disturbances affecting the blade through its connections to other parts of the machine, the blade will assume some deformed position as described above (and as compared to its manufactured shape) which is also steady in time. This shape or position will depend on the elastic and structural properties of the blade and upon the steady aerodynamic and centrifugal loading. (The centrifugal contribution naturally does not apply to a stator vane.)

Consider the ultimate situation, however, where disturbances may exist in the airstream, or may be transmitted through mechanical attachments from other parts of the structure. Due to the unsteadiness of the aerodynamic and/or the external forces the blade will assume a series of time-dependent positions. If there is a certain repetitive nature with time of the displacements relative to the equilibrium position, the blade is said to be executing vibrations, the term being taken to include those cases where the amplitude of the time-dependent displacements is either increasing, decreasing or remaining constant as time progresses.

It is the prediction and control of these vibrations with which the turbomachine aeroelastician is concerned. In doing so it is found that a portion of the difficulty is involved in the fact that once the blade is vibrating the aerodynamic forces are no longer a function only of the airstream characteristics and the blade's angular position and velocity in the disturbance field, but depend in general upon the blade's vibratory position, velocity and acceleration as well. There is a strong interaction between the blade's time-dependent motion and the time-dependent aerodynamic forces which it experiences. It is appropriate at this point to note that in certain cases the disturbances may be exceedingly small, serving only to 'trigger' the unsteady motion, and that the vibrationmay be sustained or amplified purely by the aforementioned interdependence between the harmonic variation with time of the blade's position and the harmonic variation with time of the aerodynamic forces (the flutter condition).

A further complication is that a blade cannot be considered as an isolated structure. There exist aerodynamic and possibly structural coupling between neighboring blades which dictate a modal description of the entire vibrating bladed-disk assembly. Thus an interblade phase angle, σ, is defined and found to play a crucial role in turbomachine aeroelasticity. Nonuniformities among nominally 'identical' blades in a row, or stage, are found to be extremely important in turbomachine aeroelasticity; stemming from manufacturing and assembly tolerances every blade row is 'mistuned' to a certain extent.

8.2 The compressor performance map

The axial flow compressor, and its aeroelastic problems, are typical; the other major important turbomachine variant being the axial flow turbine (gas or steam). In the compressor the angle of attack of each rotor airfoil at each radius r is compounded of the tangential velocity of the airfoil section due to rotor rotation and the through flow velocity as modified in direction by the upstream stator row. Denoting the axial component by V_x and the angular velocity by Ω as in Figure 8.1, it is clear that the angle of attack will increase inversely with the ratio $\phi = V_x/(r\Omega)$. In the compressor, an increase in angle of attack (or an increase in 'loading') results in more work being done on the fluid and a greater stagnation pressure increment Δp_0 being imparted to it. Hence the general aspects of the single 'stage' (i.e., pair of fixed and moving blade rows) characteristics in Figure 8.2 are not without rational explanation. Note that the massflow through the stage equals the integral over the flow annulus of the product of V_x and fluid density.

When the various parameters are expressed in dimensionless terms, and the complete multistage compressor is compounded of a number of stages, the overall compressor 'map', or graphical representation of multistage characteristics, appears as in Figure 8.3, where $\dot{m}$ is massflow, γ and R are the ratio of specified heats and gas constant, respectively; T_0 is stagnation temperature and A is a reference flow area in the compressor. Conventionally the constants γ and R are omitted where the identity of

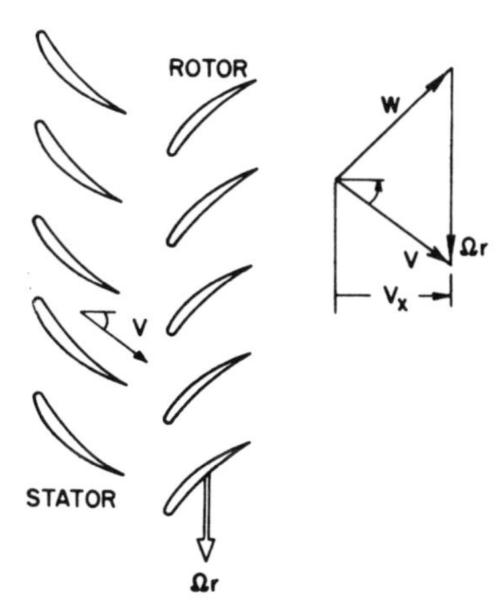

Figure 8.1 Velocity triangle in an axial compressor.

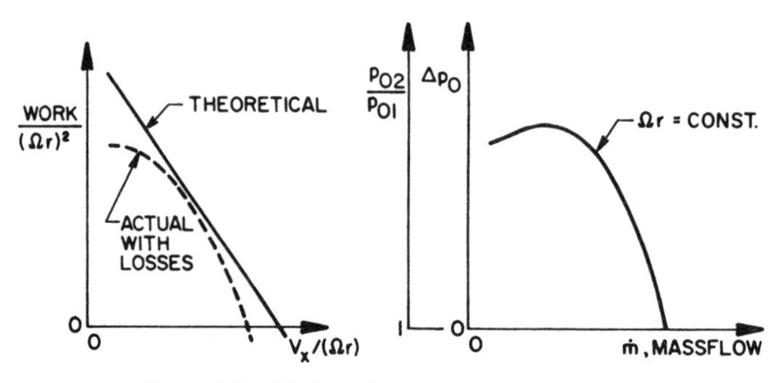

Figure 8.2 Work and pressure ratio relationships.

the working fluid is understood (e.g., air). The quantity A is a scaling parameter relating the absolute massflow of geometrically similar machines and is also conventionally omitted. The tangential velocity of the rotor blade tip, Ωr_{tip}, is conventionally replaced by the rotational speed in rpm. The latter omission and replacement are justified when discussing a particular compressor.

An important property of the compressor map is the fact that to each point there corresponds theoretically a unique value for angle of attack (or incidence) at any reference airfoil section in the compressor. For example, taking a station near the tip of the first rotor blade as a reference, contours of incidence may be superposed on the map coordinates. In Figure 8.4 such angle contours have been shown for a specific machine. As defined here, α_i is the angle between the relative approach velocity W and the

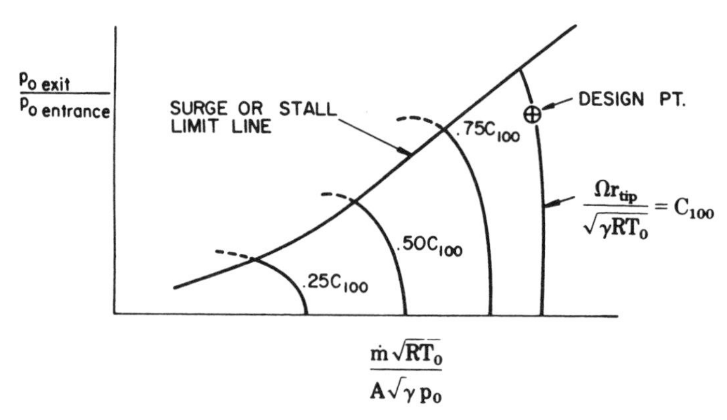

Figure 8.3 Compressor map.

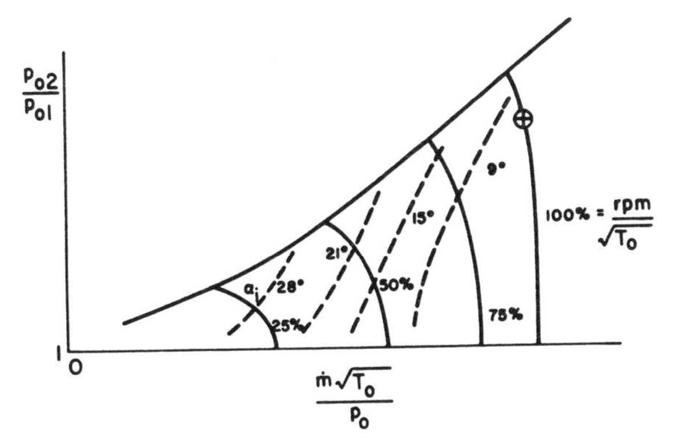

Figure 8.4 Map showing incidence as a parameter.

chord of the airfoil. Here again axial velocity V_x (or massflow) is seen to display an inverse variation with respect to angle of attack as a line of constant rotational speed is traversed. The basic reason that such incidence contours can be established is that the two parameters which locate a point on the map, $\dot{m}\sqrt{T_0}/p_0$ and $\Omega r/\sqrt{T_0}$, are effectively a Mach number in the latter case and a unique function of Mach number in the former case. Thus the 'Mach number triangles' are established which yield the same 'angle of attack' as the velocity triangles to which they are similar, Figure 8.5.

As a matter for later reference, contours of $V/(b\omega)$ for a particular

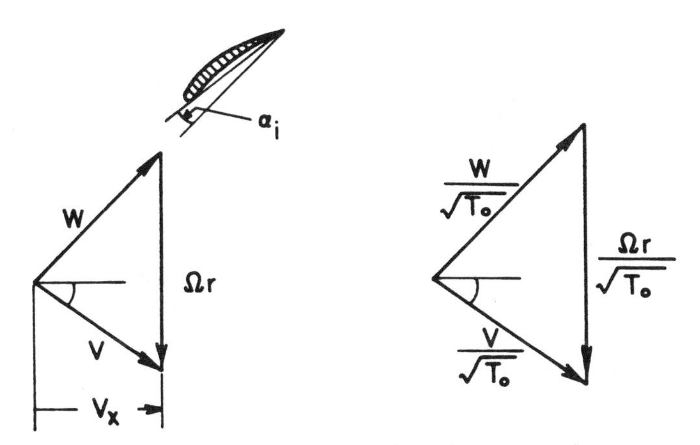

Figure 8.5 Velocity and Mach number correspondence.

stator airfoil, or else $W/(b\omega)$ for a particular rotor airfoil, can be superimposed on the same map, provided the environmental stagnation temperature, T_0, is specified. These contours are roughly parallel, though not exactly, to the constant rotational speed lines. The natural frequency of vibration, ω, tends to be constant for a rotor blade at a given rotational speed; and of course a stator blade's frequency does not depend directly on rotation. However, upon viewing the velocity triangles in Figure 8.5, it is clear that if Ωr is kept constant and the *direction* of V is kept constant, the size of W may increase or decrease as V_x (or massflow) is changed. In fact, if the angle between V and W is initially close to 90 degrees, a not uncommon situation, the change in the magnitude of W will be minimal. For computing the stator parameter, $V/(b\omega)$, the direction of *W leaving* a rotor is considered to be virtually constant, and the corresponding changes in V (length and direction) as V_x is varied lead to similar conclusions with regard to angle of attack and magnitude of V experienced by the following stator. The values of $W/(b\omega)$ increase with increasing values of Ωr_{tip}, since the changes in W (or V) will dominate the somewhat smaller changes in the appropriate frequency ω, at least in the first few stages of the compressor. Compressibility phenomena, when they become significant will sometimes alter these general conclusions.

8.3 Blade mode shapes and materials of construction

Flutter and vibration of turbomachinery blades occur with a wide variety of these beam-like structures and their degrees of end restraint. Rotor blades in use vary from cantilever with perfect root fixity all the way to a single pinned attachment such that the blade behaves in bending like a pendulum 'flying out' and being maintained in a more or less radial orientation by the centrifugal (rather than a gravity) field. Stator vanes may be cantilevered from the outer housing or may be attached at both ends, with degrees of fixity ranging from 'encastred' to 'pinned'.

The natural modes and frequencies of these blades, or blade-disc systems when the blades are attached to their neighbors in the same row or the discs are not effectively rigid, are obtainable by standard methods of structural dynamics. Usually twisting and two directions of bending are incorporated in a beam-type finite element analysis. If plate-type deformations are significant, the beam representation must be replaced by more sophisticated plate or shell elements which recognize static twist and variable thickness.

In predicting the first several natural modes and frequencies of rotor blades it is essential to take into account the effect of rotor rotational speed. Although the description is not analytically precise in all respects, the effect of rotational speed can be approximately described by stating $\omega_n^2 = \omega_{0n}^2 + \kappa_n \Omega^2$ where ω_{0n} is the static (nonrotating) frequency of the rotor blade and κ_n is a proportionality constant for any particular blade in the nth mode. The effect is most pronounced in the natural modes which exhibit predominantly bending displacements; the modes associated with the two gravest frequencies are usually of this type, and it is here that the effects is most important. The effective κ_n may be negative under some circumstances.

Materials of construction are conventionally aluminum alloys, steel or stainless steel (high nickel and/or chromium content). However, in recent applications Titanium and later Beryllium have become significant. In all these examples, considering flutter or else forced vibration in air as the surrounding fluid, the mass ratios are such that the critical mode and frequency may be taken to be one or a combination of the modes calculated, or measured, in a vacuum.

More recently there has been a reconsideration of using blades and vanes made of laminated materials such as glass cloth, graphite or metal oxide fibers laid up in polymeric or metal matrix materials and molded under pressure to final airfoil contours. Determining the modes and frequencies of these composite beams is more exacting. However, once determined, these data may be used in the same manner as with conventional metal blades. It should also be noted that aeroelastic programs related to turbomachinery often make a great deal of practical use out of mode and frequency data determined experimentally from prototype and development hardware.

A major consideration in all material and mode of construction studies is the determination of mechanical damping characteristics. Briefly stated the damping may be categorized as material and structural. The former is taken to describe a volume-distributed property in which the rate of energy dissipation into heat (and thus removed from the mechanical system) is locally proportional to a small power of the amplitude of the local cyclical strain. The proportionality constant is determined by many factors, including the type of material, state of mean or steady strain, temperature and other minor determinants.

The structural damping will usually be related to interfacial effects, for example in the blade attachment to the disk or drum, and will depend on normal pressure across the interface, coefficient of friction between the surfaces, mode shape of vibration, and modification of these determinants by previous fretting or wear. Detailed knowledge about damping is usually not known with precision, and damping information is usually determined and used in 'lumped' or averaged fashion. Comparative calculations may be used to predict such gross damping parameters for a new configuration, basing the prediction on the known information for an existing and somewhat similar configuration. By this statement it is not meant to imply that this is a satisfactory state of affairs. More precise damping prediction capabilities would be very welcome in modern aeroelastic studies of turbomachines, and some studies of this nature are reported in Refs. [1] and [2].

The aeroelastic input is central to the analysis of fatigue and fracture of turbomachinery blades. The question of crack initiation, crack propagation and destructive failure cannot be addressed without due attention being given to the type of excitation (forced or self-excited) and the parametric dependencies of the nonsteady aerodynamic forces. This may be appreciated when it is noted that the modal shape functions, frequencies and structural damping of a blade change with the crack growth of the specimen. This concatenation of aeroelasticity and blade failure prediction is presently an active area of development.

8.4 Nonsteady potential flow in cascades

Unwrapping an annulus of differential height dr from the blade row flow passage of an axial turbomachine results in a two-dimensional representation of a cascade of airfoils and the flow about them. The airfoils are identical in shape, equally spaced, mutually congruent and infinite in number.

When a cascade is considered, as opposed to a single airfoil, the fact that the flexible blades may be vibrating means that the relative pitch and stagger may be functions of time and also position in the cascade. The steady flow, instead of being a uniform stream, will now undergo turning; large velocity gradients may occur in the vicinity of the blades and in the passages between them. These complications imply that the blade thickness and steady lift distribution must be taken into account for more complete fidelity in formulating the nonsteady aerodynamic reactions. See chapters by Whitehead and Verdon in Ref. [1].

A fundamental complication which occurs is the necessity for treating the wakes of shed vorticity from *all* the blades in the cascade.

Assume the flow is incompressible. Standard methods of analyzing steady cascade performance provide the steady vorticity distribution common to all the blades, $\gamma_s(x)$, and its dependence on W_1 and β_1. As a simple example of cascading effects consider only this steady lift distribution on each blade in the cascade and compute the disturbance velocity produced at the reference blade by a vibration of all the blades in the cascade.

In what follows the imaginary index j for geometry and the imaginary index i for time variation (i.e., complex exponential) cannot be 'mixed'. That is $ij \neq -1$. Furthermore, it is convenient to replace the coordinate normal to the chord, z, by y and the upwash on the reference airfoil w_a by v. The velocities induced by an infinite column of vortices of equal

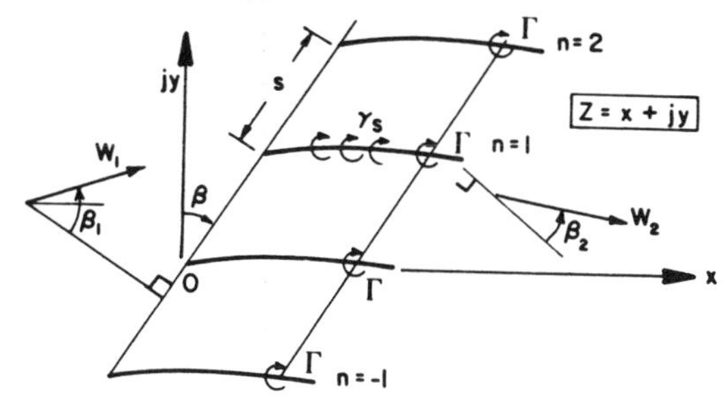

Figure 8.6 Cascade camberlines modelled by vortex sheets.

strength, Γ, are given by

$$\delta[u(z) - jv(z)] = \frac{j\Gamma}{2\pi} \sum_{n=-\infty}^{\infty} \frac{1}{Z - \zeta_n} \tag{8.4.1}$$

where the location ζ_n of the nth vortex

$$\zeta_n = \xi + jnse^{-j\beta} + jY_n(\xi_n, t) + X_n(t) \tag{8.4.2}$$

indicates small deviations from uniform spacing s, ($Y_n \ll s$, $X_n \ll c$). The point Z is on the zeroth or reference blade

$$Z = x + jY(x, t) + X(t) \tag{8.4.3}$$

and the location of the vortices will ultimately be congruent points on different blades so that

$$\xi_n = \xi + ns \sin\beta \tag{8.4.4}$$

(The subscript naught, indicating the zeroth blade, is conventionally omitted.) Finally, harmonic time dependence with time lag $-\tau$ between the motions of adjacent blades* is indicated by

* The so-called 'periodicity assumption' of unsteady cascade aerodynamics lends order, in principle and often in practice, to the processes of cascade aeroelasticity. The mode of every blade is assumed to be identical, with the same amplitude and frequency but with the indicated blade-to-blade phase shift. Such a blade row, would be termed 'perfectly tuned'. Absent this assumption, the cascade representing a rotor of n blades could have n distinct components (type of mode, modal amplitude, frequency).

$$Y_n(\xi_n, t) = e^{in\omega\tau} Y(\xi, t) \tag{8.4.5}$$

With these provisions the Cauchy kernel in (8.4.1) may be written

$$\frac{1}{Z-\zeta_n} = \frac{1}{x-\xi-jns\,e^{-j\beta}+j[Y(x,t)-Y_n(\xi_n,t)]+X(t)-X_n(t)} \tag{8.4.6}$$

and summing (8.4.5) over all blades

$$\sum_{n=-\infty}^{n=\infty} \frac{1}{Z-\zeta_n} = \frac{1}{x-\xi+j[Y(x,t)-Y(\xi,t)]} + \sum_{n=-\infty}^{\infty}{}' \frac{1}{Z-\zeta_n} \tag{8.4.7}$$

where the primed summation indicates $n = 0$ is excluded. The first term on the RHS of (8.4.7) is a self-induced effect of the zeroth foil. The part $Y(x, t) - Y(\xi, t)$ is conventionally ignored in the thin-airfoil theory; it is small compared to $x - \xi$ and vanishes with $x - \xi$. Hence the first term supplies the single airfoil or self-induced part of the steady state solution. Expanding the remaining term yields

$$\begin{aligned}\sum{}' \frac{1}{Z-\zeta_n} \cong \sum{}' \frac{1}{x-\xi-jns\,e^{-j\beta}} &+ j\sum{}' \frac{Y_n(\xi_n,t)-Y(x,t)}{(x-\xi-jns\,e^{-j\beta})^2} \\ &+ \sum{}' \frac{X_n(t)-X(t)}{(x-\xi-jns\,e^{-j\beta})^2} + \cdots\end{aligned} \tag{8.4.8}$$

where the last two summations on the RHS of (8.4.8) are the *time-dependent portions*. The corresponding unsteady induced velocities from (8.4.1) may be expressed as follows using the preceding results

$$\begin{aligned}\delta[\tilde{u}(x') - j\tilde{v}(x')] \simeq -\frac{\gamma_s(\xi')\,\delta\xi'}{2\pi c} P^2 \Bigg\{ &\sum{}' \frac{e^{in\omega\tau} Y(\xi',t) - Y(x',t)}{(\chi - jn\pi)^2} \\ &+ \frac{1}{j} \sum{}' \frac{e^{in\omega\tau} X(t) - X(t)}{(\chi - jn\pi)^2} \Bigg\}\end{aligned} \tag{8.4.9}$$

where the primed variables are dimensionless w.r.t. the chord,

$$P = \pi e^{j\beta} c/s \tag{8.4.10a}$$

$$\chi = P(x' - \xi') \tag{8.4.10b}$$

and $\tilde{u}$, $\tilde{v}$ are the time dependent parts of u, v. The local chordwise distributed vortex strength $\gamma_s(\xi)\,d\xi$ has replaced Γ the discrete vortex strength in the last step, (8.4.9). With the notation

$$q = 1 - \omega\tau/\pi \tag{8.4.11}$$

the summations may be established in closed form. For example, when the blades move perpendicular to their chordlines with the same amplitude all along the chord (pure bending) the displacement function is a constant

$$Y = -\bar{h}\,\mathrm{e}^{i\omega t} = -h \tag{8.4.12}$$

and, upon integrating over the chord in (8.4.9), one obtains

$$\tilde{u}(x') - j\tilde{v}(x') = \frac{P^2}{2\pi c}\int_0^1 \gamma_s(\xi')\sum{}' \frac{\mathrm{e}^{in\omega\tau}h(t) - h(t)}{(\chi - jn\pi)^2}\,\mathrm{d}\xi'$$

or

$$\tilde{u} = -\frac{h}{2\pi c}\int_0^1 \gamma_s(\xi')[F - iI]\,\mathrm{d}\xi' \tag{8.4.13a}$$

$$\tilde{v} = \frac{h}{2\pi c}\int_0^1 \gamma_s(\xi')[G + iH]\,\mathrm{d}\xi' \tag{8.4.13b}$$

where

$$F + iG = P^2\,\frac{q\sinh\chi\sinh q\chi - \cosh\chi\cosh q\chi + 1}{\sinh^2\chi} \tag{8.4.13c}$$

$$H + iI = P^2\,\frac{q\sinh\chi\cosh q\chi - \cosh\chi\sinh q\chi}{\sinh^2\chi} \tag{8.4.13d}$$

Similar disturbance velocity fields can be derived for torsional motion, pure chordwise motion, etc. Another separate set of disturbance fields may be generated to take account of the blade thickness effects by augmenting the steady vorticity distribution $\gamma(x)$ by, say $-\mathrm{j}\varepsilon(x)$, the steady source distribution, in the above development.

The net input to the computation of oscillatory aerodynamic coefficients is then obtained by adding the $\tilde{v}$ of all the effects so described to the LHS of the integral equation which follows

$$\overbrace{v_1(x) + v_2(x) + v_3(x)}^{\text{on } y=0,\ 0<x<c} = \frac{1}{2\pi}\int_0^c [\gamma_1(\xi) + \gamma_2(\xi) + \gamma_3(\xi)]K(\xi - x)\,\mathrm{d}\xi$$

$$+\frac{1}{2\pi}\int_c^\infty [\gamma_1(\xi) + \gamma_2(\xi) + \gamma_3(\xi)]K(\xi - x)\,\mathrm{d}\xi \tag{8.4.14}$$

In this formulation v_1 may be identified with the unsteady upwash, if any, convected as a gust with the mean flow and v_2 is the unsteady upwash attributable to vibratory displacement of all the blades in the cascade, where each blade is represented by steady vortex and source/sink distribution. It is v_2 that was described for one special component (pure bending) in the derivation of $\tilde{v}$ leading to (8.4.13b).

The component v_3 may be identified with the unsteady upwash relative to the zeroth airfoil occasioned by its harmonic vibration.

Since we are dealing here with a linear problem each of the subscripted sub-problems may be solved separately and independently of the others. It is also important to note that since the vortex distributions γ_1, γ_2 and γ_3 representing the lift distributions on the cascade chordlines are unsteady they must give rise to distributions of free vortices in the wake of each airfoil of the cascade. In other words vortex wakes emanate from the trailing edge of each airfoil and are convected downstream: at a point with fixed coordinates in the wake, the strength of the vortex element instantaneously occupying that point will vary with time. Hence, the integral equation will in general contain a term that is an integral over the wake $(c < \xi < \infty)$ to account for the additional induced velocities from the infinite number of semi-infinite vortex wakes. The kernel $\frac{1}{2}\pi K(\xi - x)$ accounts in every case for the velocity induced at $(x, 0)$ by a vortex element at the point $(\xi, 0)$ on the chord or wake of the reference, or zeroth, airfoil plus an element of equal strength located at the congruent point $(\xi + ns \sin\beta, ns\cos\beta)$ of every other profile of the cascade or its wake. The form of K may in fact be derived by returning to the previous derivation for $\tilde{v}$ in (8.4.9) and (8.4.13b) and extracting the terms

$$\overbrace{\frac{1}{\xi - x}}^{\text{isolated airfoil}} + \overbrace{\sum_{n=-\infty}^{\infty}{}' \frac{1}{\xi - x + jns\, e^{-j\beta}}}^{\text{cascade effect}} \tag{8.4.15}$$

In this expression the signs have been changed to imply calculations of positive v (rather than $-jv$) and with each term it is now necessary to associate a strength $\gamma_r(\xi)\exp(in\omega\tau)$ $(r = 1, 2 \text{ or } 3)$ since the inducing vortexes now pulsate rather than being steady in time. The kernel now appears as

$$\frac{1}{2\pi}K(\xi - x) = \frac{1}{2\pi}\sum_{n=-\infty}^{\infty}\frac{e^{in\omega\tau}}{\xi - x + jns\exp(-j\beta)} \tag{8.4.16}$$

which may be summed in closed form to yield

$$\frac{1}{2\pi}K(\xi - x) = \frac{e^{j\beta}}{2s} \cdot \frac{\cosh[(1-\sigma/\pi)\pi\exp(j\beta)(\xi - x)/s] + ij\sinh[(1-\sigma/\pi)\pi\exp(j\beta)(\xi - x)/s]}{\sinh[\pi\exp(j\beta)(\xi - x)/s]} \tag{8.4.17}$$

where $\sigma = \omega\tau$ is known as the interblade phase angle, an assumed constant.

The term for $n = 0$ in the summation (8.4.16) is

$$\frac{1}{2\pi}K_0(\xi - x) = \frac{1}{2\pi}\frac{1}{\xi - x} \tag{8.4.18}$$

which is the kernel for the isolated airfoil. Hence the added complexity of solving the cascaded airfoil problem is attributed to the additional terms giving the more complicated kernel displayed in (8.4.17).

In contradistinction to the isolated airfoil case, solutions of the unsteady aerodynamics integral equation cannot be solved in closed form, or in terms of tabulated functions, for arbitrary geometry (β and s/c) and arbitrary interblade phase angle, σ. In fact, as noted previously, the thickness distribution of the profiles and the steady lift distribution become important when cascades of small space/chord ratio are considered to vibrate with nonzero interblade phasing. Consequently, solutions to the equation are always obtained numerically. It is found that the new parameters β, s/c and σ are strong determinants of the unsteady aerodynamic reactions. A tabular comparison of the effect of these variables on the lift due to bending taken from the data in [3] appears below. In this chart, the central stencil gives the *lift coefficient* for the reference values of $s/c = 1.0, \beta = 45°, \sigma = 0.4\pi$. Other values in the matrix give the coefficient resulting from changing one and only one of the governing parameters.

The effects of thickness and steady lift cannot be easily displayed, and are conventionally determined numerically for each application. See Chapter III in [1].

8.5 Compressible flow

The linearized problem of unsteady cascade flow in a compressible fluid may be conveniently formulated in terms of the acceleration potential, $-p/\rho$, where p is the perturbation pressure, i.e., the small unsteady component of fluid pressure. Using the acceleration potential as the

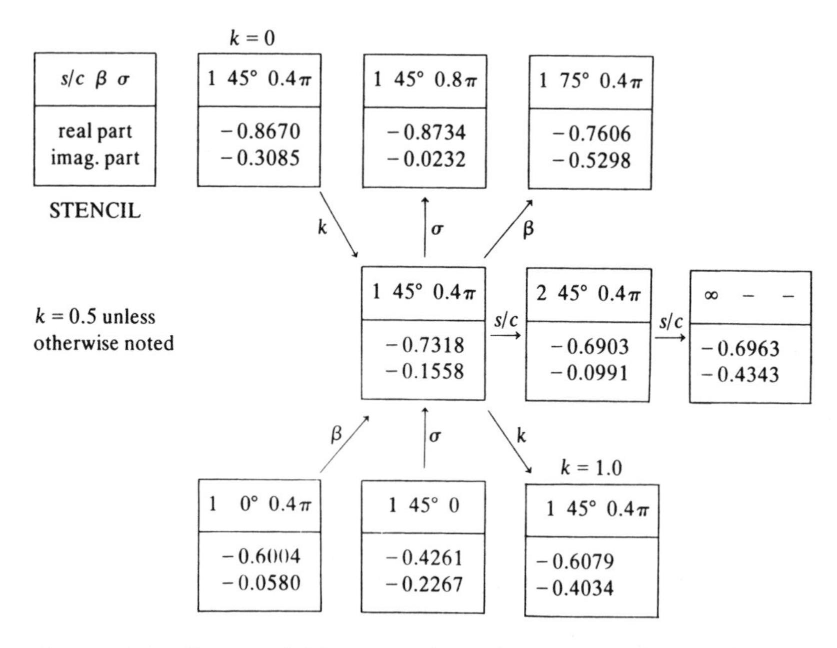

primary dependent variable, a number of compact solutions have been obtained for the flat plate cascade at zero incidence. The most reliable in subsonic flow is that due to Smith [4], and in supersonic flow the solutions of Verdon [5] and Adamczyk [6] are representative.

Supersonic flow relative to the blades of a turbomachine is of practical importance in steam turbines and near the tips of transonic compressor blades. In these cases the axial component of the velocity remains subsonic; hence analytic solutions in this flow regime (the so-called subsonic leading edge locus) are of the most interest. It may be that in future applications the axial component will be supersonic. In this event the theory actually becomes simpler so that the present concentration on subsonic values of M_{axial} represents the most difficult problem. Currently efforts are underway to account for such complicating effects as changing back-pressure on the stage, flow turning, shock waves, etc.

To illustrate the effect of varying the Mach number from incompressible on up to supersonic, a particular unsteady aerodynamic coefficient has been graphed in Figure 8.7 as a function of the relative Mach number. It is seen that the variation of the coefficient in the subsonic regime is not great except in the immediate neighborhood of the so-called 'resonant' Mach number, or the Mach number at which 'aerodynamic resonance' occurs.

It is possible to generalize the situation with respect to compressibility

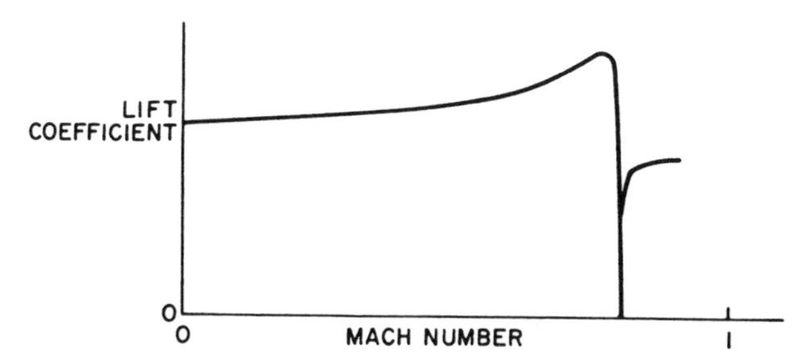

Figure 8.7 The aerodynamic resonance phenomenon.

by indicating that the small disturbance approximations are retained, but the velocities, velocity potential, acceleration potential, or pressure (in every case the disturbance component of these quantities) no longer satisfy the Laplace equation, but rather an equation of the following type.

$$(1 - M^2)\phi_{xx} + \phi_{yy} - \frac{1}{a^2}\phi_{tt} - 2\frac{M}{a}\phi_{xt} = 0 \tag{8.5.1}$$

Here M is the relative Mach number and a is the sound speed. Note that the presence of time derivatives make this partial differential equation hyperbolic whatever the magnitude of M, a situation quite different from the steady flow equation.

Although the above equation is appropriate to either subsonic or supersonic flow, the resonance phenomenon occurs in the regime of subsonic axial component of the relative velocity when geometric and flow conditions satisfy a certain relationship.

Equating the time of propagation of a disturbance along the cascade to the time for an integral number of oscillations to take place plus the time lag associated with the interblade phase angle, σ, yields

$$\frac{s}{V_p^+} = \frac{2\pi\nu}{\omega} - \frac{\sigma}{\omega} \tag{8.5.2a}$$

$$\frac{s}{V_p^-} = \frac{2\pi\nu}{\omega} + \frac{\sigma}{\omega} \tag{8.5.2b}$$

where $V_p^{\pm}$, the velocity of propagation, has two distinct values associated with the two directions along the cascade, see Figure 8.8.

$$V_p^{\pm} = a[\sqrt{1 - M^2\cos^2\beta} \pm M\sin\beta] \tag{8.5.3}$$

These expressions can be reduced to the equation

$$\frac{\omega s}{a} = (2\nu\pi \pm \sigma)(\sqrt{1 - M^2\cos^2\beta} \mp M\sin\beta) \tag{8.5.4}$$

where ν may be any positive integer, and with the upper set of signs may also be zero.

Equation (8.5.4) may be graphed and potential acoustic resonances discerned by plotting the characteristics of a given stage on the same sheet for possible coincidence. (It is convenient to take β as the parameter with axes $\omega s/a$ and M.) Acoustic resonances of the variety described above may be dangerous because they account for the vanishing, or near

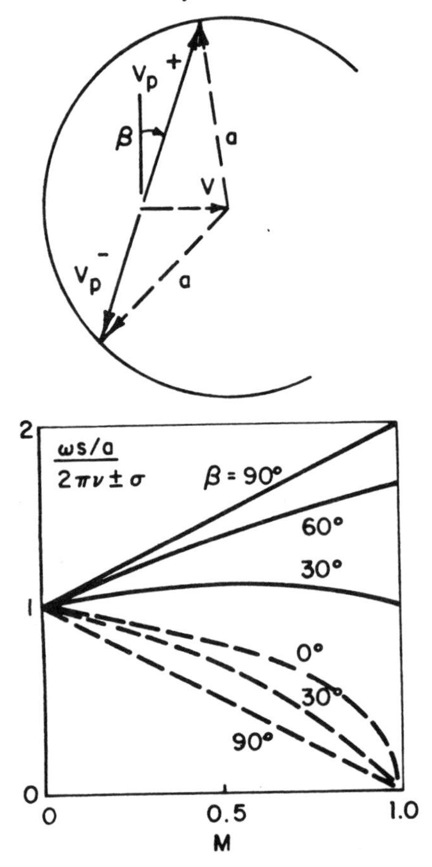

Figure 8.8 Resonant values of governing parameters.

vanishing, of all nonsteady aerodynamic reactions including therefore the important aerodynamic damping. Although it is difficult to establish with certainty, several cases of large vibratory stresses have been presumably correlated with the acoustic resonance formulation. It should be recognized that the effects of blade thickness and nonconstant Mach number throughout the field are such as to render the foregoing formulation somewhat approximate.

The foregoing development may also be based more rigorously on the theoretically derived integral equation relating the harmonically varying downwash on the blade to the resulting harmonically varying pressure difference across the blade's thickness. Symbolically

$$\bar{v}_a(x) = \int_0^c K(\xi - x)\,\Delta\bar{p}_a(\xi)\,\mathrm{d}\xi \tag{8.5.5}$$

and the acoustic resonance manifests itself by a singularity appearing in the kernel K for special values of k, s, c, τ and β of which K is a function. Under this circumstance the downwash v_a can only remain finite, as it must physically, by a vanishing of Δp_a as noted above. The previous development shows why the compressible flow solutions have received such an impetus from, and are so closely related to, the acoustic properties of compressor and fan cascades.

Thus the field of aeroacoustics, as exemplified in the text of Goldstein [7], and the field of turbomachine aeroelasticity are in a synergistic relationship. This is discussed more fully in [1].

The acoustic resonance phenomenon, as just described results from standing waves in blade-fixed coordinates, albeit with impressed throughflow velocity, of the fluid occupying the interblade passages.

8.6 Periodically stalled flow in turbomachines

Rotating, or propagating, stall are terms which describe a phenomenon of circumferentially asymmetric flow in axial compressors. Such a flow usually appears at rotationally part-speed conditions and manifests itself as one or more regions of reduced (or even reversed) throughflow which rotate about the compressor axis at a speed somewhat less than rotor speed, albeit in the same direction.

A major distinction between propagating stall and surge is that in the former case the integrated massflow over the entire annulus remains steady with time whereas in the latter case this is not true. The absolute propagation rate can be brought to zero or even made slightly negative by choosing pathological compressor design parameters.

If the instigation of this phenomenon can be attributed to a single blade row (as it obviously must in a single-stage compressor) then insofar as this blade row is concerned, it represents a periodic stalling and unstalling of each blade in the row. Later or preceding blade rows (i.e., half-stages) may or may not experience individual blade stall periodically, depending on the magnitude of the flow fluctuation at that stage, as well as the cascade stall limits in that stage.

The regions of stalled flow may extend across the flow annulus (full span) or may be confined either to the root or tip regions of the blades (partial-span stall). The number of such regions which may exist in the annulus at any one time varies from perhaps 1 to 10 with greater numbers possible in special types of apparatus.

The periodic loading and unloading of the blades may prove to be extremely harmful if a resonant condition of vibration obtains. Unfortunately the frequency of excitation cannot be accurately predicted at the present time so that avoidance of resonance is extremely difficult.

The results of various theories concerning propagating stall are all moderately successful in predicting the propagational speed. However the number of stall patches which occur (i.e., the circumferential wavelength of the disturbance) seems to be analytically unpredictable so that the frequency of excitation remains uncertain. Furthermore, the identification of the particular stage which is controlling the propagating stall, in the sense noted above, is often uncertain or impossible.

This situation with regard to propagating stall has recently been impacted by a CFD approach using the vortex method of description. In Chapter 5 the vortex method was applied to the analysis of stall flutter. The earlier application, however, was to propagating (or rotating) stall, i.e. for the flow instability which can occur with completely rigid blades. The vortex method is presently being intensively developed for propagating stall prediction and preliminary results [9] indicate that success in the long sought objective of wavelength prediction is at hand. Improvements that are required for more useful results are in the boundary layer subroutine (executed at each time step for each blade) and in the enlargement of computing capacity to handle the number of blades in realistic annular cascades. A further improvement that is desirable is in the vortex merging algorithm. The vortex method is a time-marching CFD routine in which the location of a large number of individual vortices are tracked on the computational domain. New vortices are created at each time step to satisfy the boundary conditions and separation criteria. Hence, to limit the total number of vortices in the field after many time steps, it is necessary to merge individual vortices, preferably downstream of the cascade. Many merging criteria may be considered, related to the strength and position of the candidate vortices, and this subject is under intensive study.

Although the precise classification of vibratory phenomena of an aeromechanical nature is often somewhat difficult in turbomachines because of the complication due to cascading and multistaging, it is nevertheless necessary to make such distinctions as are implied by an attempt at classification. The manifestation of stall flutter in turbomachines is a good example of what is meant. When a given blade row, or cascade, approaches the stalling incidence in some sense (i.e., stalling defined by rapid increase of relative total pressure loss, or defined by rapid increase in deviation angle, or defined by the appearance of flow separation from the suction surface of the blades, etc.) it is found experimentally that a variety of phenomena may exist. Thus the region of reduced throughflow may partially coalesce into discrete patches which propagate relative to the cascade giving rise to the type of flow instability previously discussed under rotating stall. There is no dependence on blade flexibility.

Under certain other overall operating conditions it is found that in the absence of, or even coexistent with, the previous manifestation, the blades vibrate somewhat sporadically at or near their individual natural frequencies. There is no immediately obvious correlation between the motions of adjacent blades, and the amplitudes of vibrations change with time in an apparently random manner. (We exclude here all vibration attributable to resonance with the propagating stall frequency, should the propagation phenomenon also be present.) This behavior is termed stall flutter or stalling flutter and the motion is often in the fundamental bending mode. Another term is random vibration. Since the phenomenon may be explained on the basis of nonlinear mechanics, (see the chapter on Stall Flutter) the sporadicity of the vibration can be attributed provisionally to the fact that the excitation has not been strong enough to cause 'entrainment of frequency', a characteristic of many nonlinear systems. Hence, each blade vibrates, *on the average*, as if the adjacent blades were not also vibrating. However, a careful analysis demonstrates that the instantaneous amplitude of a particular blade is effected somewhat by the 'instantaneous phase difference' between its motion and the motion of the adjacent blade(s). One must also speak of 'instantaneous' frequency since a frequency modulation is also apparent. As a general statement it must be said that the frequency, amplitude and phase of adjacent blades are functionally linked in some complicated aeromechanical manner which results in modulations of all three qualities as functions of time. While the frequency modulation will normally be small (perhaps less than 1 or 2 percent) the amplitude and the phase modulations can be quite large. Here the term phase difference has been used rather loosely to describe the relationship between two motions of slightly different frequency. Since this aerodynamic coupling would also depend on the instantaneous amplitude of the adjacent blade(s), it is not surprising that the vibration gives a certain appearance of randomness. On the linear theory for identically tuned blades one would not expect to find sporadic behavior as described above. However, it is just precisely the failure to satisfy these two conditions that accounts for the observed motion; the average blade system consists of an assembly of slightly detuned blades (nonidentical frequencies) and furthermore the oscillation mechanism is nonlinear.

Application of vortex method aerodynamics to a cascade of elastically supported blades recently has demonstrated [12], in a computational sense, the features of randomness and sporadicity as described above.

When the relative magnitudes of the nonsteady aerodynamic forces increase it may be expected that entrainment of frequency will occur. In certain nonlinear systems it can be shown that the 'normalized' frequency interval $(\omega - \omega_0)/\omega$ (where ω is the impressed frequency and ω_0 is the frequency of self-excitation) within which one observes entrainment, is proportional to h/h_0, where h is the amplitude of the impressed motion and h_0 is the amplitude of the self-excited oscillation. In case of entrainment one would expect to find a common phase difference between the motion of adjacent blades which implies also motion with a common flutter frequency. This latter phenomenon is also termed stalling flutter, although the term limit-cycle vibration is sometimes used to emphasize the constant-amplitude nature of the motion, which is often in the fundamental torsional mode.

Finally it should be noted that the distinction between blade instability (flutter) and flow instability (rotating stall) is not always perfectly distinct. When the sporadic stall flutter occurs it is clear that there is no steady tangentially propagating feature of the instability. Similarly, when propagating stall occurs with little or no vibration (stiff blades away from resonance) it is apparent that the instability is not associated with vibratory motion of the blade. However, the limit cycle type of behavior can be looked upon (due to the simultaneously observed constant interblade phase relationship) as the propagation of a disturbance along the cascade. Furthermore, the vorticity shed downstream of the blade row would have every aspect of a propagating stall region. For instance if the interblade phasing was 180° the apparent stall region would be on one blade pitch in tangential extent and each would be separated by one blade pitch of unstalled throughflow. The tangential wavelength is two blade pitches. Because of the large number of such regions, and the small tangential extent of each, this situation is still properly termed stall flutter since the blades are controlling and the blade amplitudes are constant. At the other extreme when one or two stall patches appear in the annulus it is obvious that the flow instability is controlling and then the phase relationships between adjacent blades' motions may appear to be rather sporadic. At any rate, in the middle ground between these extremes it is probable that a strong interaction between flow stability and blade stability exists and the two phenomena cannot be easily separated.

Another distinction may be attempted to assist in understanding the operative phenonena. When a *single airfoil* is subjected to an increasing angle of attack an instability of the fluid may arise, related to the Karman vortex frequency or the extension of this concept to a distributed frequency spectrum. If the frequency of this fluid instability coincides with the natural frequency of the blade in any mode, the phenomenon is termed buffetting. If the dynamic moment coefficient (or force coefficient) attains a negative slope a self-excited vibration known as stall flutter occurs. The two phenomena may merge when airfoil vibration exerts some influence on the vortex shedding frequency. Stall-flutter is usually observed in the torsional mode and buffetting in the bending mode, but this distinction is not always possible. These concepts cannot be carried over directly to the cascade where steady bending amplitudes of the limit cycle variety have been observed. The explanation rests on the additional degrees of freedom present in the cascaded configuration.

8.7 Stall flutter in turbomachines

On account of the foregoing complications and the very recent emergence of quantitative CFD-based theories noted in Chapter 5 it is not surprising that past prediction for turbomachines has rested almost entirely on correlation of experimental data. The single most important parameter governing stalling is the incidence, and the reduced frequency has been seen in all aeroelastic formulations to exert a profound influence. Hence it is not surprising that these variables have been used to correlate the data.

Typically stall flutter will occur at part-speed operation and will be confined to those rotor stages operating at higher than average incidence. With luck the region of flutter will be above the operating line on a compressor map and extend up to the surge line. Under less fortunate circumstances the operating line will penetrate the flutter region. The flutter boundary will have the appearance shown on Figure 8.9. Contours of constant flutter stress (or tip amplitude) will run more or less parallel to, and within, the boundary. Traditional parameters for this typical experimental correlation are reduced velocity, $W/b\omega$, (the inverse of reduced frequency) and incidence, at some characteristic radius such as 75% or 80% of the blade span for a cantilever blade. The curve is typical of data obtained in turbomachines or cascades; essentially a new correlation is required for each major change of any aerodynamic variable (Mach number, stagger, blade contour, etc.). The mode shape will usually be first torsion. The single contour shown in the previous figure is for that level of cyclic stress (or strain) in the blade material that is arbitrarily taken to represent some distinct and repeatable measurement attributable to the flutter vibration and discernible above the 'noise' in the strain measuring system. A typical number might be a stress of 10 000 kPa used to define the flutter boundary. However, small changes in relative airspeed, W, may increase the flutter stress substantially, or, in the case of 'hard' flutter, a small increase in incidence might have a similar effect. Hence, in keeping with the nonlinear behavior described in Chapter 5, the contours of constant flutter stress may be quite closely spaced in some regions of the correlation diagram.

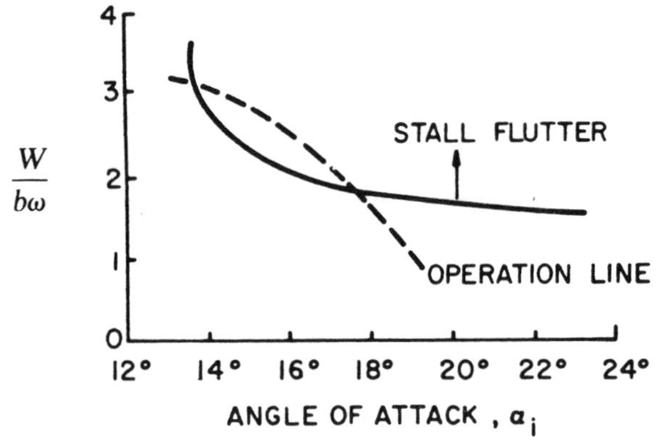

Figure 8.9 Experimental stall flutter correlation.

Naturally, when considering three-dimensional effects it is the net energy passing from airstream to airfoil that determines whether flutter will occur, or not. The stalled tip of a rotor blade, for example, must extract more energy from the airstream than is put back into the airstream by airfoil sections at smaller radii and that is dissipated from the system by damping.

This total description of stall flutter in turbomachine rotor blades is consistent with the appearance of the stall flutter boundary as it appears on the following typical compressor performance map (Figure 8.10), the vibrations are usually confined to the first two or three stages. This figure may be viewed in conjunction with the performance map on Figure 8.4 which shows typical angles of attack for a rotor blade tip in the first stage of a compressor. Keeping in mind that the mass flow parameter $\dot{m}\sqrt{T_0}/p_0$ is virtually proportional to the throughflow velocity in the first few stages of a compressor, it is clear that any typical operating line as shown on the

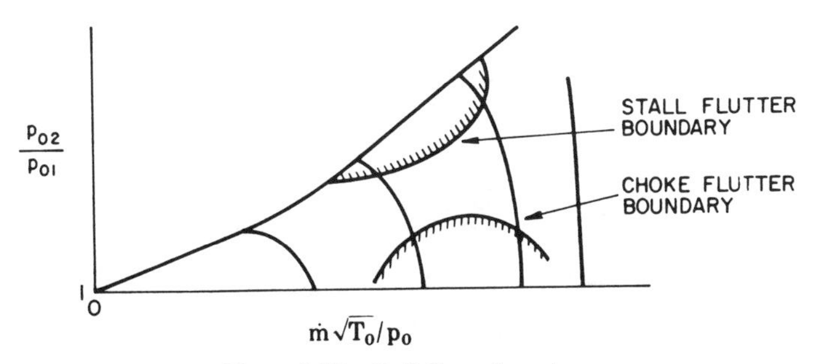

Figure 8.10 Stall flutter boundary.

compressor map will traverse the flutter boundary somewhat as the dotted line on Figure 8.9.

This explains the general shape and location of the region of occurrence of stall flutter. Experimental determinations confirm increasing stresses as the region is penetrated from below and the specific behavior is a function of the aeroelastic properties of the individual machine, consistent with the broad principles enunciated here.

8.8 Choking flutter

In the middle stages of a multistage compressor it may be possible to discern another region on the compressor map wherein so-called choking flutter will appear. This will normally occur at part-speed operation and will be confined to those rotor stages operating at lower than average incidence (probably negative values are encountered). The region of flutter will normally lie below the usual operating line on a compressor map, but individual stages may encounter this type of instability without greatly affecting operating line; this is particularly true when the design setting angle of a particular row of rotor blades had been arbitrarily changed from the average of adjacent stages through inadvertence or by a sequence of aerodynamic redesigns.

The physical manifestation of choking flutter is usually discriminated by a plot of a stage's operating line on coordinates of relative Mach number vs. incidence, as in Figure 8.11. On these same coordinates the choke boundaries are shown; a coincidence or intersection of these graphs indicate the possible presence of choking flutter and is usually confined to a very small range of incidence values. The mechanism of choking flutter is not fully understood. It is related to compressibility phenomena in the fluid and separation of the flow is probably also involved. The graph

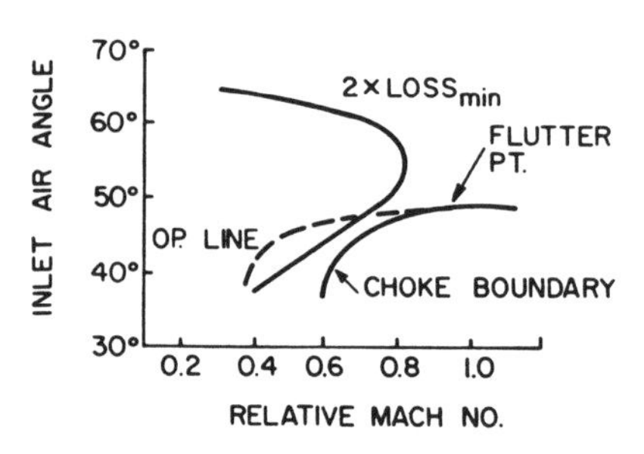

Figure 8.11 Choking flutter correlation.

labelled '$2 \times \text{Loss}_{\text{min}}$' is a locus of constant aerothermodynamic loss coefficient (closely related to the drag coefficient of an airfoil); the interior of the nose-shaped region representing low values of loss, or efficient operation of the compressor stage. The curve labelled 'choke boundary' represents the combination of relative Mach number and flow angle at which the minimum flow area between adjacent blades (the throat) is passing the flow with the local sonic velocity. Presumably separation of the flow at the nose of each airfoil on the pressure surface, and the *relative* motion between adjacent blades as they vibrate, conspire to change the effective throat location in a time dependent manner. These oscillatory changes effect the pressure distribution on each blade in such a fashion (including a phase angle) as to pump energy from the airstream into the vibration and thus sustain the presumed motion.

Experimental results [13, 14] bear out the general description of choking flutter described above. The analytically-based predictions [15, 16] lend further credence to the mechanism, although the aerodynamic formulation is confined to quasisteady time dependence. Ultimately a satisfactory explanation and prediction technique will be attained with a time marching computational capability using the compressible Navier–Stokes equations.

Choking flutter occurs in practice with sufficient frequency and destructive potential as to be an important area for current research efforts as noted above.

8.9 Aeroelastic eigenvalues

Traditionally the analytical prediction of flutter has been conducted by computation of the aeroelastic eigenvalues for the particular system under investigation. In turbomachines the eigenvalue determinations have been conducted in the frequency domain, and the unsteady aerodynamics, excluding separated or choked flow, have been based on the solutions of the small disturbance equations as described in [3–6] and others, and as reviewed effectively in the AGARD Manual [1]. A representative sample analysis for the steady loading effect in an infinite cascade was introduced in Section 8.4. In the literature a large number of additional effects are treated, including compressibility, finite flow deflection, three-dimensionality, finite shock strength and shock movement, section thickness and turbine-type geometry.

In every case, however, the initial formulation of the eigenvalue problem for an N-bladed annular cascade results in a system of mN equations, where m is the number of degrees of freedom (or else structural modes) assigned to each blade. Since the disc on which the turbomachine blades are attached will not be completely rigid, these modeshapes will be 'system' modes in which nodal circles and diameters may be discerned on the disc proper (and its extension into the flow annulus).

$$-\omega^2[M_n]\{q_n\}+[M_n\omega_n^2(1+ig_n)]\{q_n\}=\pi b^3\omega^2[F]\{q_n\} \tag{8.9.1}$$

In (8.9.1) the aeroelastic equation has been specialized for one degree of freedom per blade ($m=1$); hence n ranges from 0 to $N-1$. This equation, adapted from Crawley's Chapter 19 in [2], assumes harmonic time dependence at frequency ω and the nth blade has its individual mass, M_n, natural frequency, ω_n, and structural damping coefficient, g_n. The development leading to (8.9.1) parallels that for (3.7.32) for a single foil. (The principal result of considering $m>1$ is to replace each matrix element by a submatrix and enlarge the displacement vector, $\{q_n\}$).

When the blades on the disc are structurally uncoupled (rigid disc and no interconnecting shrouds or lacing wires) the square matrices on the LHS are diagonal and the equations are coupled only through the aerodynamic force matrix

$$[F]=\begin{bmatrix} F_0 & F_{N-1} & F_{N-2} & \cdots & F_1 \\ F_1 & F_0 & F_{N-1} & \cdots & F_2 \\ \cdot & \cdot & \cdot & & \cdot \\ \cdot & \cdot & \cdot & & \cdot \\ F_{N-1} & F_{N-2} & F_{N-3} & \cdots & F_0 \end{bmatrix}. \tag{8.9.2}$$

The matrix is completely populated and each element is an aerodynamic influence coefficient: the force effect on the row-identified blade due to the motion of the column-identified blade. These are the terms derivable from the previously described analytical theories under the assumption of constant interblade phase angle, σ, and harmonic displacement given by

$$q_n = \mathrm{Re}\,[\bar{q}_n \exp{(i\omega t)}] \tag{8.9.3}$$

although Fourier decomposition of the aerodynamic force is necessary to obtain the form implied by (8.9.1) and (8.9.2).

For a 'tuned' stage the mass, natural frequency and damping coefficient for every blade are the same so that the N equations are identical ($[F]$ is circulant) and the complex eigenvalues

$$\omega = \omega_R + i\omega_I \tag{8.9.4}$$

may be obtained from any one of the individual blade equations. Since there are N possible tuned values of σ, there are N possible $[F]$ matrices and N corresponding eigenvalues. The particular eigenvalue that obtains in practice will be for those values of airspeed W (embedded in F) and g_n that just produce $\omega_I = 0$. That is, the typical V, g plot is replaced by a family of contours with σ as the parameter. The critical flutter speed is then obtained by minimizing W with respect to σ, see [10]. In this sense the aeroelastic behavior of tuned cascades is a straightforward extension of the single airfoil procedure, to include an additional parameter, the interblade phase angle.

One of the most intensive recent efforts in turbomachine aeroelastic studies has been in the area of 'mistuned' blade rows. When the mass and/or stiffness of all airfoils are not identical, or the coupling through the discs or shrouds is not uniform, then structural mistuning is present. Analogous aerodynamic mistuning results from nonuniform blade spacing, setting angle and section profile. Such mistuned stages are inevitably manufactured, subject in degree to inspection and tolerance acceptance procedures at assembly. The general effect of mistuning is to reduce the symmetry and cyclical nature of the matrices in the flutter equation, (8.9.1). The character of the eigenvalue plots and the eigenfunctions become more varied. Thus, at flutter, all blades are found to vibrate with the same frequency; the relative blade amplitudes and phase angles are constant with respect to time, but not with respect to location in the blade row. For each of the eigensolutions, however, there may be associated a 'tuned' interblade phase angle [2]. The most significant effect of mistuning is to change the value of ω_I. If the shift of the least stable eigenvalue is in the direction of increased stability, the proclivity to flutter is reduced and it is for this reason that mistuning is considered to be a powerful design tool for improving aeroelastic stability in cascaded airfoils. Figures 8.12a and 8.12b adapted from [11] show the effect of mistuning on the position of the eigenvalues (actually $i\omega$ rather than ω) for a 14-blade cascade.

It is demonstrated that a necessary but not sufficient condition for aeroelastic stability is that the blades be self-damped; the effect of a blade's motion upon itself must be to contribute positive aerodynamic damping. The unsteady interactions amongst or between blades in the cascade are destabilizing for at least one possible σ. This blade-to-blade destabilizing interference is reduced by mistuning and is hence desirable. Mistuning, however, can never produce stability when self-damping is

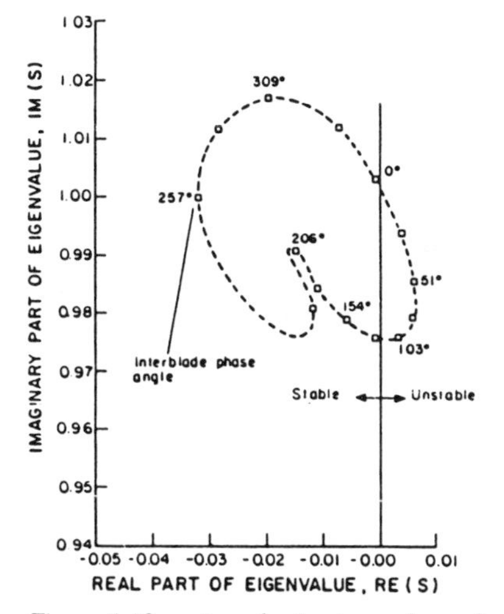

Figure 8.12a Aeroelastic eigenvalues of a 14-bladed tuned rotor.

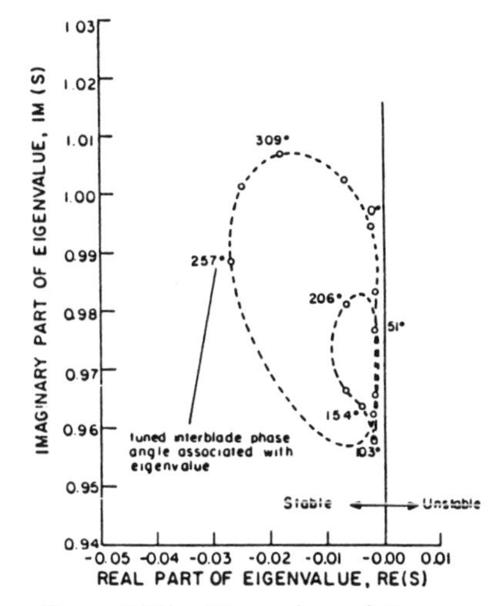

Figure 8.12b Eigenvalues of the same rotor with 'optimal' mistuning.

negative. With nonzero structural damping, blades of larger (blade to air) mass ratio are relatively more stable.

The effect of kinematic coupling, (e.g. the presence of some bending displacements in a predominantly torsional mode) may be quite important in determining stability whereas dynamic coupling (e.g. through the aerodynamic reactions) is usually not strong enough to be of significance. The effect of mean loading is speculated as being a possible source of flutter near stall, and stability trends with reduced velocity are discussed qualitatively in [2], noting both structural and aerodynamic implications of the reduced frequency parameter.

Optimal mistuning as an intentional manufacturing procedure at assembly is an important concept, although it must be tempered with the knowledge that, under forced aerodynamic resonance, so-called 'rogue' blades may be identified which will vibrate at dangerously high amplitude. More research on mistuning may be expected to yield increasingly practical results for the turbomachine aeroelastician to apply beneficially, see [17, 18] and [19].

8.10 Recent trends

A number of supersonic flutter regimes have been encountered in practice, see Regions III, IV and V in Figure 8.13. Only Region III flutter,

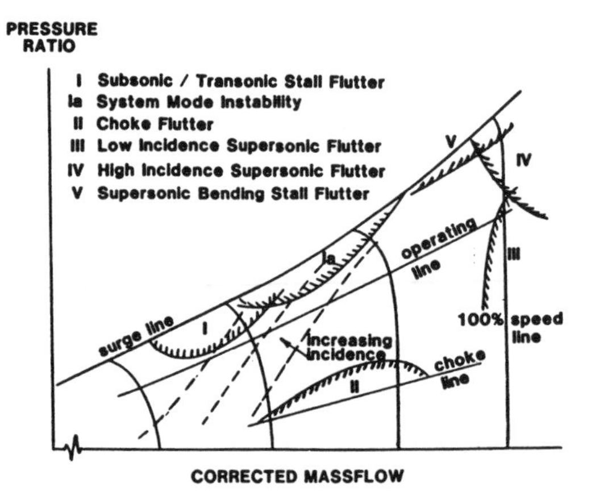

Figure 8.13 Axial compressor or fan characteristic map showing principal types of flutter and regions of occurrence.

in either pitching or plunging, will usually be encountered along a normal operating line, and then only at corrected overspeed conditions. Supersonic aerodynamic theories have been developed to explain and confirm Region III flutter. Low incidence formulations were reported by a number of investigators, with greatest interest being attached to the onset flows having a subsonic axial component. The survey papers by Platzer [21–24] give an excellent summary of the aerodynamics literature and experience up to 1982 including summaries of relevant papers by authors in the former Soviet Union.

Regions IV and V in Figure 8.13 are at higher compressor pressure ratio, above the normal equilibrium operating line, and, in Region V, may involve stalling at supersonic blade relative Mach number. Unsteady aerodynamic analyses appropriate to this regime have been presented [25, 26]. For the first time account was taken of the effect of shock waves which may appear when the surface Mach number exceeds unity. Flutter observed in these regions has been mostly flexural, although not exclusively. In Region V stalling of the flow has been implicated since the region is in the neighborhood of the surge or stall limit line. Hence Region V is provisionally termed 'supersonic bending stall flutter' and it is assumed that there is a detached bow shock at each blade passage entrance; i.e., the passage is unstarted. By contrast, the flutter mechanism in Region IV is thought to involve an in-passage shock wave whose oscillatory movement is essential for the instability mechanism.

A counterclockwise continuation around Figure 8.13 returns one to Region I, delineated earlier in Figure 8.10 and which, it now appears, should be divided into more than one subregion. The so-called system mode instability seems to be associated with the upper end of this region, and although the blade loading is high, flutter may not involve flow separation as an essential part of the mechanism. Instead it has been hypothesized [27] that even with a subsonic onset flow the surface Mach number can exceed unity locally and oscillating shocks may help explain the appearance of negative aerodynamic damping. It seems that these instability mechanisms (separation, oscillating shocks) may both appear in this general region of the fan or compressor map, although not both at the same time in a particular machine. Thus the non-aerodynamic factors, which are not revealed by the map parameters and are discussed in Section 8.1, may determine which, if any, of these flutter types will manifest itself in any particular instance. The clarification of this matter is still required so that Region I is now provisionally labelled Subsonic/Transonic Stall Flutter and System Mode Instability. Region II, discussed in Section 8.8 and of relatively lesser importance, is associated with choking of the passage and is labelled Choke Flutter. As such the role of oscillatory shock waves is again indicated to be important. Hence for relatively low negative incidence and high enough subsonic relative Mach numbers, appropriate to a middle stage of a multistage compressor, the mechanism of choke flutter has many similarities to the transonic stall flutter of Region I. In addition, some authors [28] add a second sub-region at a larger negative incidence and lower relative Mach number, and term it negative incidence stall flutter. The choke flutter mechanism is still controversial; it may involve the type of machine (fan, compressor or turbine), type of stage (front, middle, or rear) and structural details (shrouded vs unshrouded, disc vs drum, etc.).

Three-dimensional unsteady cascade flow was first formulated in the 1970s [29, 30]. In order to apply two-dimensional theory to the aeroelastic problems of real blade systems one must either use a representative section analysis or else apply the strip hypothesis; the aerodynamics at one radius is uncoupled from the aerodynamics at any other radius. In particular, it is known that at 'aerodynamic resonance' the strip theory breaks down and the acoustic modes are strongly coupled radially.

Along with aerodynamic advances the structural description of the bladed-disc assembly [31, 32], has received a great impetus, and the importance of forward and backward travelling waves has been firmly established. Within a particular number of nodal diameters, coupling between modes has been shown to be significant [33] and the role of the 'twin modes' (i.e. sin $n\phi$ and cos $n\phi$) in determining propagation has been clarified. Ford and Foord [34] have used the twin mode concept in both analysis and flutter measurement. Furthermore, the number of nodal diameters affects the fundamental natural frequencies slightly so that they cluster together. Coupling of modes with closely spaced frequencies by aerodynamic means therefore becomes appreciable and the resulting flutter mode may contain significant content from two or three modes with consecutive numbers of diametral nodes.

A great concentration of studies recently has been in the area of Computational Fluid Dynamics (CFD) coupled with a Finite Element Method (FEM) description of the blade and disk structure. Typically these sets of governing equations are solved interactively in a time marching fashion to yield the developing flutter amplitudes. Stability limits are not determined directly per se. For nonlinear systems the limit cycle amplitudes are predicted while for linear systems the temporal growth of amplitude identifies those values of the operating variables that lie within the instability boundary.

Usually in these models only spanwise displacements in plunging, pitching and surging are allowed, leading to beam-type finite elements for representing a tapered, twisted blade of variable cross-section [35][36]. Consequently, when plate- or shell-type elements are necessitated by airfoil thicknesses on the order of 4 or 5 percent, the chordwise deformations cannot be neglected and full three-dimensional FEM packages must be utilized. Essentially the camber schedule of the blade profiles change with time in these cases.

The FEM-based structural analysis is also essential for static aeroelastic studies in the nascent field of compliant blade performance modification. The compliance of the blade in an annular cascade represents a *passive* means of controlling the aerothermodynamic performance of the turbomachine by aeroelastic tailoring. This topic comes under the overarching subject of aeroservoelasticity, the application of automatic control theory to fundamental aeroelastic problems. In the blading of turbomachinery the enhanced compliance, and its chordwise distribution, are introduced intentionally by design. The resulting configuration must be checked for freedom from dynamic aeroelastic instability, or flutter, over the entire operating range of the compressor map such as that appearing in Figure 8.4. It may be remarked that the concept of performance "map" will have to be extended to include the parametric dependence of performance on a representative value of a new dimensionless quantity: the ratio of the dynamic pressure of the fluid to the Young's modulus of the structure. In effect the augmentation of compliance introduces variable geometry into the turbomachine blading.

The small compliance, or conversely great rigidity, of conventional blades is responsible for only slight amounts of untwist and uncambering. In the design and development of traditional turbomachines these effects, in turn, have been reflected in very slight corrections to the aerothermodynamic performance as compared to assuming complete rigidity of the airfoils. This situation will be changed with the application of static aeroservoelasticity to the design of turbomachines with compliant blades.

Recent applications of unsteady Navier-Stokes codes to cascaded airfoils appear in references [37][38] and [39]. These early studies using Navier-Stokes solvers for unsteady flows with moving boundaries are chiefly of interest because of the promise that they indicate for ultimate reliability of the method for computational prediction. At present the needed confidence and accuracy are not being obtained because of the inadequacy of the turbulence model and the extreme requirements on computer capacity alluded to above.

Subjects receiving attention very recently that have not been treated fully include such topics as finite shock motion, variable shock strength, thick and highly cambered blades in a compressible flow, and the effects of curvilinear wakes and vorticity transport. These and other large amplitude and therefore nonlinear perturbations, which prevent the linear superposition implicit in classical modal analysis, have certain implications relative to the traditional solutions of the aeroelastic eigenvalue problem. The field of aeroelasticity in turbomachines continues to be under active investigation, driven by the needs of aircraft powerplant, gas turbine and steam turbine designers.

References for Chapter 8

[1] Platzer, M. F. and Carta, F. O. (Eds.), *AGARD Manual on Aeroelasticity in Axial-Flow Turbomachines*, 'Volume 1, Unsteady Aerodynamics', AGARDograph No. 298, 1987.

[2] *Ibid* 'Volume 2, Structural Dynamics and Aeroelasticity', AGARDograph No. 298, 1988.

[3] Whitehead, D. S., 'Force and Moment Coefficients for Vibrating Aerofoils in Cascade', *ARC R&M* 3254, London, 1960.

[4] Smith, S. M., 'Discrete Frequency Sound Generation in Axial Flow Turbomachines', *ARC R&M* 3709, London, 1972.

[5] Verdon, J. M. and Caspar, J. R., 'A Linearized Unsteady Aerodynamic Analysis for Transonic Cascades', *J. of Fluid Mechanics*, Vol. 149 (1984), pp. 403–429.
[6] Adamczyk, J. J. and Goldstein, M. E., 'Unsteady Flow in a Supersonic Cascade with Subsonic Leading Edge Locus', *AIAA Journal*, Vol. 16, No. 12 (1978), pp. 1248–1254.
[7] Goldstein, M. E., *Aeroacoustics*, McGraw-Hill Publishing Company, New York, 1976.
[8] Spalart, P. R., 'Two Recent Extensions of the Vortex Method', AIAA Paper No. 84–0343, Reno, 1984.
[9] Speziale, C. G., Sisto, F. and Jonnavithula, S., 'Vortex Simulation of Propagating Stall in a Linear Cascade of Airfoils', *ASME Journal of Fluids Engineering*, Vol. 108, No. 3 (1986), pp. 304–312.
[10] Lane, F., 'System Mode Shapes in the Flutter of Compressor Blade Rows', *Journal of the Aeronautical Sciences*, Vol. 23, No. 1 (1956), pp. 54–66.
[11] Crawley, E. F. and Hall, K. C., 'Optimization and Mechanism of Mistuning in Cascades', *Journal of Engineering for Gas Turbines and Power*, Vol. 107, No. 2 (1985), pp. 418–426.
[12] Sisto, F., Wu, W., Thangam, S. and Jonnavithula, S., 'Computational Aerodynamics of Oscillating Cascades with the Evolution of Stall', AIAA Journal, Vol. 27, No. 4 (1989), pp. 462–471.
[13] Tanida, Y. and Saito, Y., 'On Choking Flutter', *Journal of Fluid Mechanics*, Vol. 82 (1977), pp. 179–191.
[14] Jutras, R. R., Stallone, M. J. and Bankhead, H. R., 'Experimental Investigation of Flutter in Mid-Stage Compressor Designs', AIAA Paper 80-0786 1980, pp. 729–740.
[15] Micklow, J. and Jeffers, J., 'Semi-Actuator Disc Theory for Compressor Choke Flutter', NASA Contractor Report 3426, 1981.
[16] Tang, Z. M. and Zhou, S., 'Numerical Prediction of Choking Flutter of Axial Compressor Blades', AIAA Paper 83-0006, Reno, 1983.
[17] Bendiksen, O. O. and Valero, N. A., 'Localization of Natural Modes of Vibration in Bladed Disks', ASME Paper 87-GT-47, Anaheim, California, June 1987.
[18] Kaza, K. R. and Kielb, R. E., 'Flutter and Response of a Mistuned Cascade in Incompressible Flow', *AIAA Journal*, Vol. 20, No. 8 (August 1982), pp. 1120–1127.
[19] Srinavasan, A. V., 'Influence of Mistuning on Blade Torsional Flutter', NASA CR-165137, August, 1980.
[20] Snyder, L. E. and Commerford, G. L., 'Supersonic Unstalled Flutter in Fan Rotors; Analytical and Experimental Results', ASME Trans., *Journal of Engineering for Power*, Vol. 96, Series A, No. 4 (1974), pp. 379–386.
[21] Platzer, M. F., 'Transonic Blade Flutter: A Survey', *Shock and Vibration Digest*, Vol. 7, No. 7 (1975), pp. 97–106.
[22] Platzer, M. F., 'Unsteady Flows in Turbomachines—A Review of Current Developments', In AGARD-CP-227 *Unsteady Aerodynamics*, Ottawa, 1977.
[23] Platzer, M. F., 'Transonic Blade Flutter: A Survey of New Developments', *Shock and Vibration Digest*, Vol. 10, No. 9 (1978), pp. 11–20.
[24] Platzer, M. F., 'Transonic Blade Flutter: A Survey of New Developments', *Shock and Vibration Digest*, Vol. 14, No. 7 (1982), pp. 3–8.
[25] Adamczyk, J. J., 'Analysis of Supersonic Stall Bending Flutter in Axial-Flow Compressor by Actuator Disc Theory', NASA. Tech. Paper 1345, 1978.
[26] Adamczyk, J. J., Stevens, W. and Jutras, R., 'Supersonic Stall Flutter of High-Speed Fans', *Trans. ASME Journal of Engineering for Power*, Vol. 104, No. 3 (1982), pp. 675–682.

[27] Stargardter, H. 'Subsonic/Transonic Stall Flutter Study', Final Report, NASA CR-165256, PWA 5517-31, 1979.

[28] Fleeter, S., 'Aeroelasticity Research for Turbomachine Applications', *Journal of Aircraft*, Vol. 16, No. 5 (1979), pp. 320–326.

[29] Namba, M., 1972, 'Lifting Surface Theory for a Rotating Subsonic or Transonic Blade Row', *Aeronautical Research Council, R&M* 3740, London, 1972.

[30] Salaün, P., 'Pressions Aerodynamiques Instationnaires sur Une Grille Annulaire en Eccoulement Subsonique', Publication ONERA No. 158, 1974.

[31] Ewins, D. J., 'Vibration Characteristics of Bladed Disc Assemblies', *Journal of Mechanical Engineering Sciences*, Vol. 15, No. 3, 1973, pp. 165–186.

[32] Srinivasan, A. V. (Ed.), 'Structural Dynamic Aspects of Bladed Disk Assemblies', *Proc. ASME Winter Annual Meeting*, New York, 1976.

[33] Chi, R. M. and Srinivasan, A. V., 'Some Recent Advances in the Understanding and Prediction of Turbomachine Subsonic Stall Flutter', ASME Paper 84-GT-151, 1984.

[34] Ford, R. A. J. and Foord, C. A., 'An Analysis of Aeroengine Fan Flutter using Twin Orthogonal Vibration Modes', ASME Paper 79-GT-126, 1979.

[35] Sisto, F. and Chang, A.T., 'A Finite Element for Vibration Analysis of Twisted Blades Based on Beam Theory', *AIAA Journal*, Vol. 22, No. 11, pp. 1646–1651, 1984.

[36] Sisto, F. and Chang, A.T., 'Influence of Rotation and Pretwist on Cantilever Fan Blade Flutter', Proc. 7th International Symposium on Airbreathing Engines, Beijing, Sept. 1985.

[37] Rai, M.M., 'Navier-Stokes Simulations of Rotor-Stator Interaction Using Patched and Overlaid Grids', *AIAA Journal for Propulsion and Power*, Vol. 3, No. 9, pp. 387–396, 1987.

[38] Schroeder, L.M. and Fleeter, S., 'Viscous Aerodynamic Analysis of an Oscillating Flat Plate Airfoil with Locally Analytic Solution', *AIAA* Paper 88–0130, January 1988.

[39] Clarkson, J.D., Ekaterinaris, J.A. and Platzer, M.F., 'Computational Investigation of Airfoil Stall Flutter', Sixth International Symposium on Unsteady Aerodynamics and Aeroacoustics and Aeroelasticity of Turbomachines and Propellers', Notre Dame, IN, Sept. 1991.

9

*Unsteady transonic aerodynamics and aeroelasticity**

Summary

In recent years substantial progress has been made in the development of an improved understanding of unsteady aerodynamics and aeroelasticity in the transonic flow regime. This flow regime is often the most critical for aeroelastic phenomena yet it has proven the most difficult to master in terms of basic understanding of physical phenomena and the development of predictive mathematical models. The difficulty is primarily a result of the nonlinearities which may be important in transonic flow. The emerging mathematical models have relied principally on finite difference solutions to the governing nonlinear partial differential equations of fluid mechanics. Here are addressed fundamental questions of current interest which will provide the reader with a basis for understanding the recent and current literature in the field.

Four principal questions are discussed.

(1) Under what conditions are the aerodynamic forces essentially linear functions of the airfoil motion?
(2) Are there viable alternative methods to finite difference procedures for solving the relevant fluid dynamical equations?
(3) Under those conditions when the aerodynamic forces are *nonlinear* functions of the airfoil motion, when is the significance of the multiple (nonunique) solutions that are sometimes observed?
(4) What are effective, efficient computational procedures for using

---

* An earlier version of this chapter has appeared in *Recent Advances in Aerodynamics*, edited by A. Krothapelli and C. A. Smith, Springer-Verlag, New York, 1986. Reprinted by permission.

unsteady transonic aerodynamic computer codes in aeroelastic (e.g., flutter) analyses?

Nomenclature

C_L, C_M lift, moment coefficients
$C_{L\alpha}, C_{M\alpha}$ lift, moment curve slopes
C_p pressure coefficient
c airfoil chord
K $= (\gamma + 1) M_\infty^2 \tau / \beta^3$
k $= \omega c / U_\infty$; reduced frequency
M Mach number
s $= (\beta^2 t U_\infty / c) / M_\infty^2$
t time
x, y spatialcoordinates in freestream and vertical directions
x_s shock location
Δx_s shock displacement normalized by the airfoil chord
α_0, α_1 mean angle of attack; dynamic angle of attack in degrees
β $= (1 - M_\infty^2)^{1/2}$
γ ratio of specific heats
ν $= k M_\infty^2 / \beta^2$
$\phi^{(0)}, \phi^{(1)}$ velocity potentials of steady flow and unsteady airfoil motion respectively
ϕ phase angle
τ thickness ratio of airfoil
ω frequency
Δ gradient operator

Subscripts

∞ freestream
L local; also lift
M moment
max maximum
0, 1 mean, dynamic
TE trailing edge

Superscripts

c shock first forms
tc shock reaches the trailing edge

Section 5

M numberof structural modes
NF number of reduced frequencies needed for a flutter analysis
NR number of response levels for a nonlinear flutter analysis
P number of parameters
T_A computational time for aerodynamic code to reach a steady state lift value for a prescribed airfoil motion
T_F computational time for a simultaneous fluid-structural calculation to complete a transient
T_{AF} computational time for aerodynamic code to determine aerodynamic forces for one reduced frequency

Section 6

$A(\ ), A_L, A_M$ indicial response functions
a distance of elastic axis from mid-chord: percent semichord, positive downstream
b semichord length
c full chord length
C_L^N nonlinear lift coefficient
C_M^N nonlinear moment coefficient about mid-chord
C_{Me}^N nonlinear moment coefficient about elastic axis
$\bar{D}_L, \bar{D}_M$ components of describing function

F	output of describing function
G	structural transfer function
H	nonlinear aerodynamic transfer function
$\hat{H}$	aerodynamic describing function
h	plunging displacement of elastic axis (positive down)
h_c	plunging displacement of mid-chord (positive down)
I_α	moment of inertia per unit span about elastic axis
k	$= c\omega/u$, reduced frequency
L	lift force
M	moment force about mid-chord (positive nose-up)
m	mass per unit span
M_∞	Mach number of uniform airflow
R	$= \omega_h/\omega_\alpha$, uncoupled frequency ratio
r_α	dimensionless radius of gyration about elastic axis (based on semichord); $r_\alpha^2 = r_{cg}^2 + (x_{cg} - a)^2$.
r_{cg}	dimensionless radius of gyration about center of gravity (based on semichord)
S_α	static unbalance
s	dimensionless variable of Laplace operator; $s = ik$ for harmonic oscillation
t	time
U	dimensionless airspeed $u/c\omega_\alpha\sqrt{\mu}$.
u	dimensional airspeed
x_{cg}	distance of center of gravity from mid-chord; percent semichord, positive downstream
α	pitching displacement
ϕ	effective induced angle-of-attack; see (6.1)
ϕ_1	amplitude of ϕ oscillation
μ	mass ratio $m/\pi\rho b^2$
ω_h, ω_α	uncoupled circular frequency of the airfoil in plunging and in pitch, respectively
ρ	air density
τ	dimensionless time ut/c

Superscripts

T	transpose of matrix
$\hat{}$	quantity associated with describing function
$'$	$= d/dt$
$\bar{}$	quantity in the subsidiary domain of Laplace Operator

9.1 Introduction

The four questions cited in the summary are chosen to provide the framework of this chapter. This selection was made for several reasons.

- They are fundamental questions that are expected to be of lasting significance.
- Answers to these questions have important consequences for aeroelastic applications of unsteady transonic aerodynamics.
- Recent work has led to at least partial answers.

The four questions are addressed in Sections 9.2, 9.3, 9.4, and 9.5–9.6, respectively. Each section may be read relatively independently of the others, and the reader may wish to take advantage of that option.

9.2 Linear/nonlinear behavior in unsteady transonic aerodynamics

Motivation and general background

The aeroelastician uses linear dynamic system theory for most aeroelastic analyses. The motivation for doing so is clear. Extensive experience, understanding, and effective computational/experimental procedures have been developed for linear systems. By contrast, although nonlinear methods of analysis and experimentation are available, the results are far more expensive to obtain and also more difficult to interpret. Hence linear models, where applicable, are very powerful, relatively simple, and extremely valuable. Thus, it is highly important to determine the domain of validity of any linear model.

Here our concern is with possible aerodynamic nonlinearities in transonic flow. Of course, aerodynamic nonlinearities may arise in other flow regimes; however, it is in transonic flow where they tend to be most important. Indeed, it is often observed that the transonic flow regime is inherently nonlinear in the governing field equations. However, at any Mach number for any airfoil, if the angle of attack is sufficiently small, the aerodynamic forces and shock motion will be linear in the angle of attack. Moreover, as the frequency of the angle of attack motion increases, the range of angle of attack over which linear behavior persists increases. It is our purpose here to study when linear or nonlinear behavior occurs using as our principal analytical method the low frequency, transonic small disturbance (LTRAN2) procedure of Ballhaus and Goorjian [1, 2]. Any other present or future nonlinear aerodynamic method could (and should) be used for similar purposes.

It will be helpful to discuss first the shock and its motion, which is sometimes a source of confusion. A consequence of any consistent linearization of steady transonic small disturbance aerodynamic theory in the dynamic angle of attack is that a concentrated force or pressure pulse (sometimes called a shock doublet) will appear at the location of the steady state shock [58, 59]. The strength of the pressure pulse is equal to the steady state shock pressure jump and its width is proportional to the dynamic angle of attack. By contrast, elsewhere on the airfoil chord (away from the shock doublet whose center is at the steady state shock location), the pressure magnitudes (in a transonic linear theory) are proportional to the dynamic angle of attack and become smaller in proportion as the dynamic angle of attack is smaller. Of course this latter behavior is also true in classical theory. The most important (although not the only) distinction between classical, linear theory and transonic, linear theory is the presence of the shock (and its motion) in the latter that creates the concentrated shock force doublet. LTRAN2 and some other transonic computer codes include both the shock and its motion while classical aerodynamic theory includes neither. Some inconsistent transonic methods include the shock's presence, but not its motion.

The behavior described above is seen in a nonlinear dynamic theory as well, when the dynamic angle of attack becomes small. Consider Figure 1 which was obtained using LTRAN2. It shows the chordwise differential (lower surface minus upper) pressure distribution for an NACA 64A006 airfoil at $M_\infty = 0.86$ for several angles of attack. Here, for simplicity, the reduced frequency is set to zero, so there is no distinction (numerically) between steady and dynamic angle of attack. As may be seen for small angles of attack, say $\alpha = 0.125$ deg, 0.25 deg, the pressure distribution has a shock doublet centered at the mean (zero angle of attack) shock location, $x_s/c = 0.584$. The width of the shock

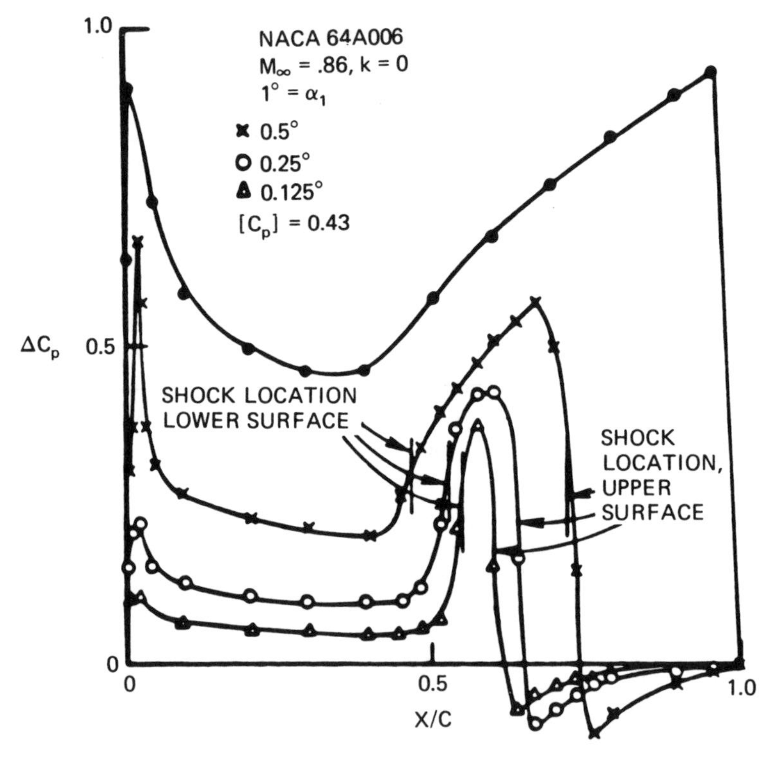

Figure 9.1 Differential pressure distribution.

doublet is indicated by the vertical lines, the forward one is at the lower surface shock location and the rearward one at the upper surface location. The shock doublet width is proportional to α for the smaller α; however, as α increases to 1 deg, the lower surface shock disappears while the upper surface shock moves to the trailing edge and remains there.

Also, for the smaller α the shock doublet magnitude is essentially equal to the pressure jump through the shock at $\alpha = 0$ deg, i.e., 0.43. Away from the shock doublet, the pressures are proportional to α for small α.

Finally, note a matter of practical importance. For small α as the shock doublet width narrows, any finite difference scheme nonlinear in α will have a resolution problem as $\alpha \to 0$. By contrast a method a priori linearized in α avoids this difficulty as it computes the shock motion explicitly, e.g., see [58, 59] and [19]. Also see the discussion of [50] and [51] for a critical assessment of theory and experiment. The experimental study of Davis and Malcolm [12] is particularly relevant here as it provides confirmation of the above in broad terms.

NACA 64A006 airfoil at $M_\infty = 0.86$ pitching about its leading edge

The following principal issues were studied [15]: effect of dynamic angle of attack at various reduced frequencies on dynamic forces and shock motion; boundary for linear/nonlinear behavior; effect of reduced frequency and dynamic amplitude on aerodynamic transfer functions; effect of dynamic angle of attack on steady state forces and shock displacement; and effect of steady-state angle of attack on dynamic forces and shock motion. For the sake of brevity, only the first two issues will be considered here.

Effect of dynamic angle of attack on dynamic forces and shock motion

It is desirable to assess at what dynamic amplitude nonlinear effects become important in order to determine the relative linear vs nonlinear behavior of lift, pitching moment, and shock motion. Note that the total lift (moment, shock motion) is characterized by $C_L = C_{L_0} + C_{L_1}$, where C_{L_0} is defined to be the lift due to the mean angle of attack, α_0, and C_{L_1} that due to the dynamic angle of attack, α_1 for given α_0. In classical linear theory (but not transonic linear theory), C_{L_1} is independent of α_0.

In Figure 9.2 lift, pitching moment, and shock displacement amplitudes are shown as a function of dynamic amplitude, α_1, for a reduced frequency of $k = 0.2$. Lift and moment coefficient have their usual definitions and the moment is about the midchord. The shock displacement is normalized by the airfoil chord. Phases are also presented for lift and pitching moment. The shock motion phase was also computed; however, it tended to be less accurately determined. Since it is not needed for our present purposes, it is not shown.

It is seen that lift tends to remain linear to higher dynamic amplitudes than moment, which, in turn, tends to remain linear to higher amplitudes than shock motion. Moreover, as will be seen, the larger the reduced frequency, the greater the range of linear behavior. Phase information generally, although not universally, is a more sensitive indicator of departure from linearity than lift, moment, or shock amplitude information. In a strictly linear theory, of course, the phase is independent of the dynamic angle of attack.

It is noted that no measurable higher harmonic content was found in

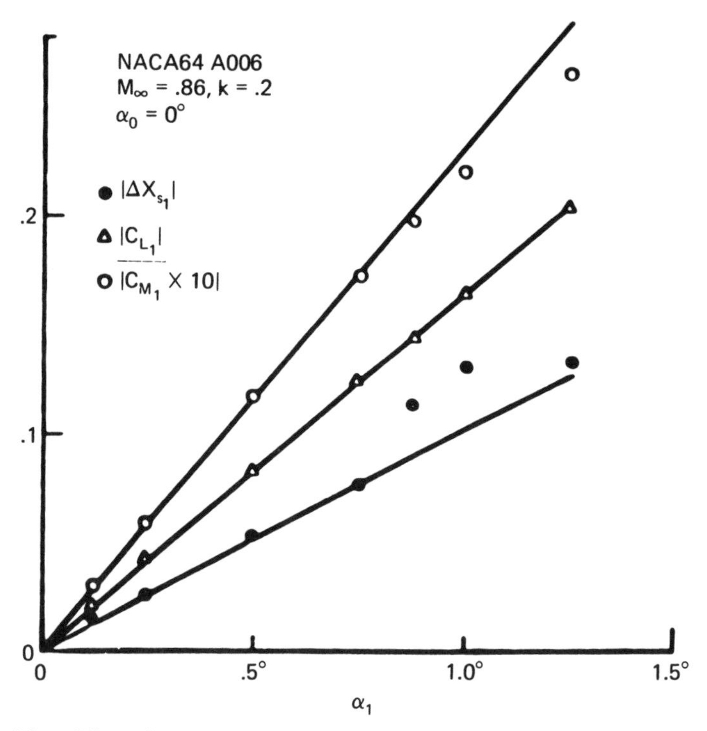

Figure 9.2a Effect of dynamic angle of attack on dynamic forces and shock motion: amplitudes.

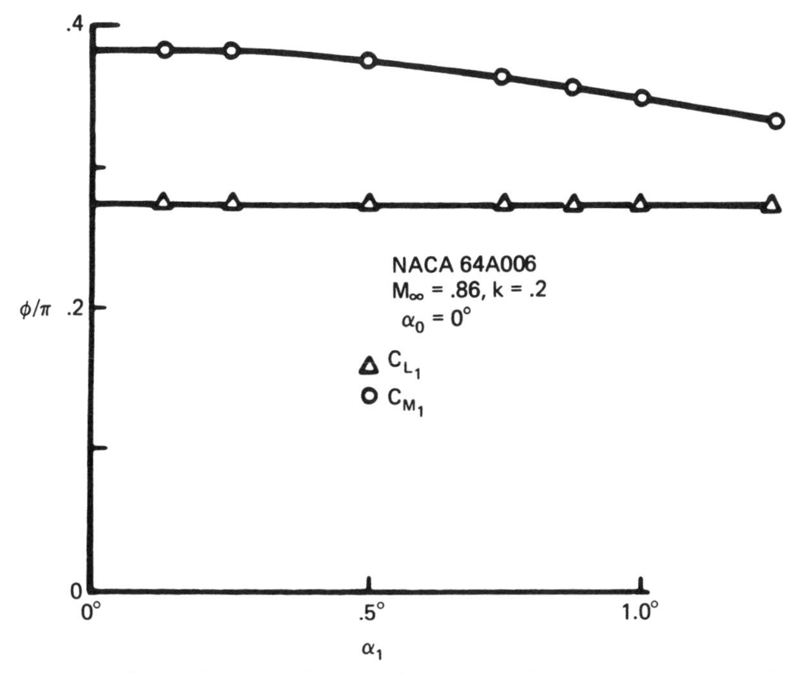

Figure 9.2b Effect of dynamic angle of attack on dynamic forces and shock motion: phases.

any of the numerical results. The results were virtually sinusoidal signals for lift, moment, and shock motion; hence, determination of magnitude and phase was readily done by any one of several conventional methods. The exception was shock motion phase, which is difficult to determine accurately by any method because of the relatively coarse finite difference mesh resolution of the shock.

Boundary for linear/nonlinear behavior

It is highly desirable to provide a criterion by which the aeroelastician may assess when a linear dynamical theory may be used.

Figure 9.3 has been constructed from Figure 9.2 and other similar results by identifying the k, α_1 combinations for which the pitching moment deviates by 5% in amplitude or phase from linearity. As expected, at higher k the pitching moment remains linear to larger α_1.

Although Figure 9.3 provides very useful information, it requires a nonlinear dynamical theory to construct it. A question thus arises: Is

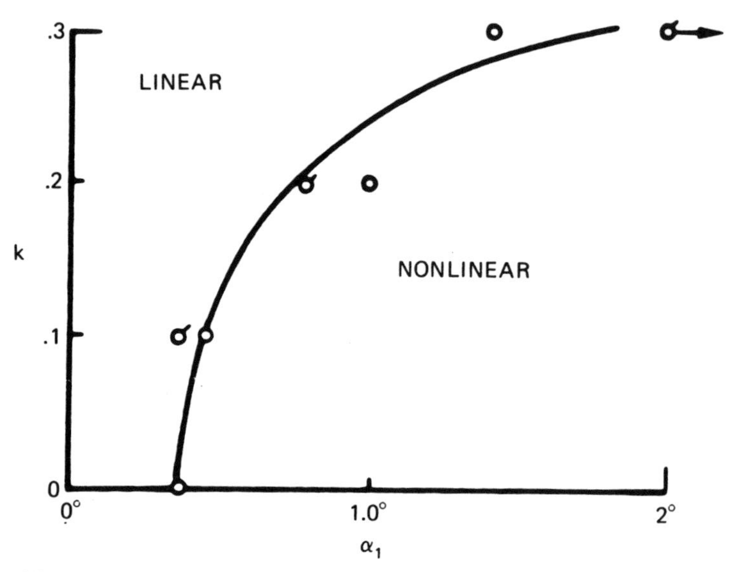

Figure 9.3 Boundary for linear/nonlinear behavior in terms of reduced frequency and dynamic angle of attack.

there a similar, but perhaps more conservative, criterion that may be used with a linear dynamical theory? The answer is provided by the shock motion. In Figure 9.4 a similar boundary to that shown in Figure 9.3 is constructed (again from information such as that provided by Figure 9.2) based on shock motion rather than pitching moment. It is observed in Figure 9.2 that for shock displacement amplitudes of less than 5% the shock motion (as well as lift and pitching moment) behave in a linear fashion. Hence, a 5% shock motion boundary is shown in Figure 9.4. Note that this boundary could be constructed from a linear dynamical theory. A second boundary (less conservative) based on the first detectable deviation of shock motion from linearity is also shown. Finally, the boundary from Figure 9.3 is shown for reference. These results are consistent with those of Ballhaus and Goorjian [3, 4] who also suggested that shock motions of less than 5% chord correspond to linear behavior.

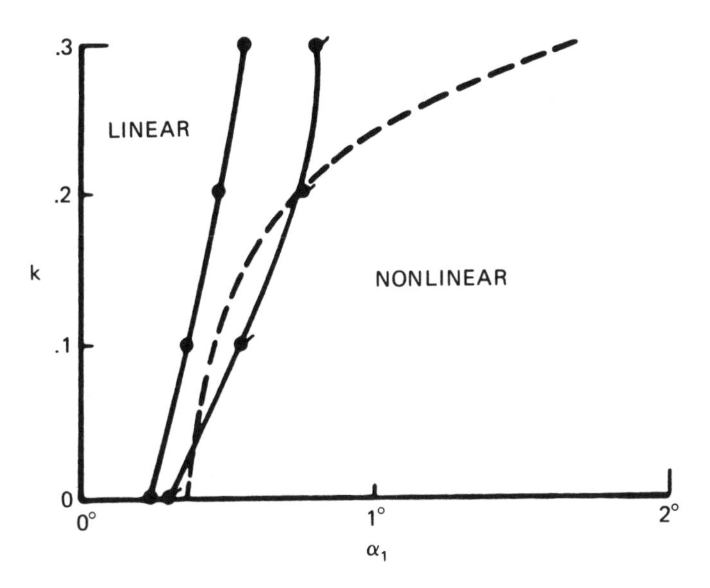

Figure 9.4 Conservative boundary for linear/nonlinear behavior based on shock motion amplitude.

Thus, it is concluded that a simple criterion for departure from nonlinearity based on shock motion may be used. It can be evaluated by a linear dynamical theory in principle (which enhances its practical utility), although the present results were obtained using a nonlinear, dynamical theory.

A brief digression is in order to explain why the shock motion criterion is extremely useful to the aeroelastician. After the flutter mode is determined from a conventional flutter analysis using linearized but transonic aerodynamics, one may compute the amplitude of the flutter motion which will correspond to a 5% shock motion using the linear transonic aerodynamic model employed in the flutter analysis. This will give the aeroelastician the limit on amplitude for which the linear, flutter calculation is valid. This is very useful information.

Mach number trends

Similarity law

Here the effects of Mach number are studied systematically for the NACA 64A006 airfoil. We note that a similarity rule holds for low frequency, transonic flow, which gives the following results for any family airfoils:

$$C_p = \frac{\tau}{\beta}\bar{C}_p(x/c,\, s;\, K,\, \nu,\, \alpha/\tau) \tag{9.2.1}$$

where $\bar{C}_p$ is a universal function of its arguments and

$$\beta \equiv (1 - M_\infty^2)^{1/2}, \qquad \nu \equiv kM_\infty^2/\beta^2$$

$$K \equiv \frac{(\gamma+1)M_\infty^2}{\beta^3}, \qquad s \equiv \frac{\beta^2 tU_\infty/c}{M_\infty^2}$$

where $\tau \equiv$ thickness ratio of airfoil, and $\alpha \equiv$ angle of attack.

Equation (9.2.1) may be further specialized for the case $\alpha \rightarrow 0$, by expanding in a Taylor series, i.e.,

$$C_p = \frac{\tau}{\beta}\bar{C}_p(x/c,\, s;\, K) + \frac{\alpha}{\beta}\,\mathrm{Re}\{e^{i\nu s}\bar{C}_{p1}(x/c;\, K,\, \nu)\} \tag{9.2.2}$$

This is the similarity law for dynamic linearization in α, i.e., $\alpha = \alpha_1$. Zero mean angle of attack is assumed for simplicity, $\alpha_0 = 0$, although the result is readily extended. From (9.2.2) it is seen that similarity for the harmonic component requires only that K and ν be the same for two different flows.

Finally, it is noted that since the shock is simply a discontinuity surface of ϕ_x, it satisfies a similarity law expressed by

$$x_s = x_s(\beta y/c,\, s;\, K,\, \alpha/\tau,\, \nu) \tag{9.2.3}$$

For the limit, $\alpha \rightarrow 0$

$$x_s = x_{s_0}(\beta y/c;\, K) + \alpha/\tau\,\mathrm{Re}\{e^{i\nu s}x_{s_1}(\beta y/c;\, K,\, \nu)\} \tag{9.2.4}$$

The similarity law given by (9.2.1) was known to Miles (1959). Equations (9.2.2)–(9.2.4) are extensions of his results due to M. H. Williams.

Using the similarity rules, the results for the 64A006 airfoil may be used to obtain results for any other airfoil of the same family, in particular, the 64A010.

It may be inferred from (9.2.4) that the 5% shock motion criterion has the functional form (for a given family of airfoils)

$$\alpha/\tau = F(\nu, K) \tag{9.2.5}$$

It is interesting to note that Fung *et al.* [19] proposed a criterion for the validity of linearization of the form

$$\frac{\alpha}{\tau}/K \ll 0.1 \tag{9.2.6}$$

Equation (9.2.6) is clearly a special case of (9.2.5). Using (9.2.5), the data of Figures 9.3, 9.4 (and 9.5, subsequently) may be reinterpreted in terms of similarity variables, and thereby generalized.

Boundary for linear/nonlinear behavior

Using results such as those shown in Figure 9.2 and invoking the 5% shock displacement criterion, a linear/nonlinear boundary may be con-

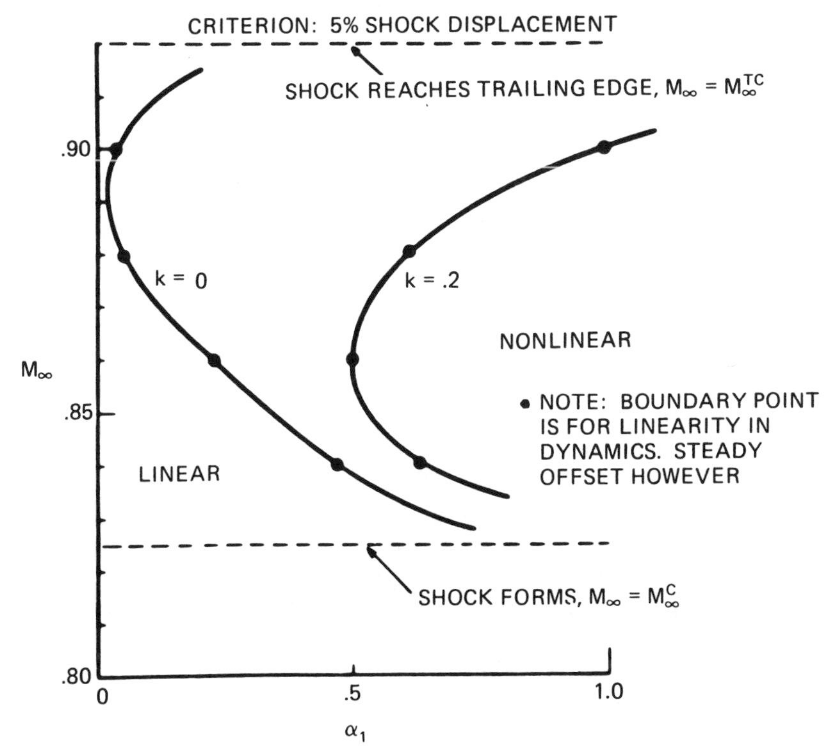

Figure 9.5 Effect of reduced frequency on boundary for linear/nonlinear behavior: Mach number vs. dynamic angle of attack.

structed in terms of Mach number vs amplitude of airfoil oscillation. Of course, as the shock reaches very near the trailing edge, the 5% criterion would need to be modified. Results are shown in Figure 9.5 for $k = 0$ and 0.2. Note that for steady flow ($k \equiv 0$) the angle of attack must be very small when $M_\infty = 0.88$ and 0.9 for linear behavior to occur. However, as we have seen before, the 5% shock displacement criterion is conservative. That is, lift and moment tend to remain linear in α to higher α than this criterion would suggest. Nevertheless, the trend should not change using any other reasonable criterion. By contrast, for $k = 0.2$ the linear region is much enlarged.

For $M_\infty < M_\infty^c$ or $M_\infty > M_\infty^{tc}$ the linear region is for all practical purposes unbounded. In practice, in this region other physical effects, e.g., viscosity, are likely to come into play before inviscid, small disturbance, transonic theory nonlinearities become important. M_∞^c is the Mach number when the shock first forms and M_∞^{tc} that when it reaches the airfoil trailing edge.

Aerodynamic transfer functions

In the linear region it is of interest to display aerodynamic transfer functions vs. Mach number. Perhaps the most familiar of these is lift curve slope, C_{L_1}/α_1. Its amplitude is shown in Figure 9.6a from LTRAN2 for $k = 0$. Also shown are results from full potential theory, classical subsonic theory, and local linearization. The latter is shown for $M_\infty > M_\infty^{tc}$, i.e., the shock is at the trailing edge. It uses the local trailing edge supersonic Mach number in classical (supersonic) theory. One concludes that for $M_\infty < M_\infty^c$, classical theory gives reasonable results, and for $M_\infty > M_\infty^{tc}$ local linearization gives reasonable results. For $M_\infty^c < M_\infty < M_\infty^{tc}$, LTRAN2 gives markedly different results although it likely fails for $M_\infty = 0.88, 0.90$. Note the difference between transonic small disturbance theory (LTRAN2), which falls well off scale at $M_\infty = 0.88$ and 0.9, and full potential theory [6].

It should be noted that the full potential results shown in Figure 9.6a were obtained using a nonconservative finite difference scheme. Full potential results obtained using a quasiconservative finite difference scheme (for technical reasons results were only obtained for $M_\infty < 0.87$) are essentially identical to those of transonic small disturbance theory using a conservative finite difference scheme (LTRAN2). Hence, the difference shown in Figure 9.6 should be attributed to the distinction between conservative and nonconservative finite differences and not to the distinction between small disturbance and full potential theory. To the extent that the nonconservative finite difference method may be said to have some form of numerical (as opposed to physical) viscosity, the differences may be attributed to the qualitative distinction between inviscid and viscous flow.

In Figure 9.6b results are shown for $k = 0.2$. For reference, the LTRAN2 results for $k = 0$ are also shown. Again it is seen that the classical subsonic theory and local linearization theory give reasonable results (better than for $k = 0$) for $M_\infty < M_\infty^c$ and $M_\infty > M_\infty^{tc}$, respectively. Moreover, LTRAN2 appears to give reasonable results over the entire Mach number range, although there is no better theory to validate it. Note that from $M_\infty = 0.9$ to 0.92 there is a somewhat abrupt change.

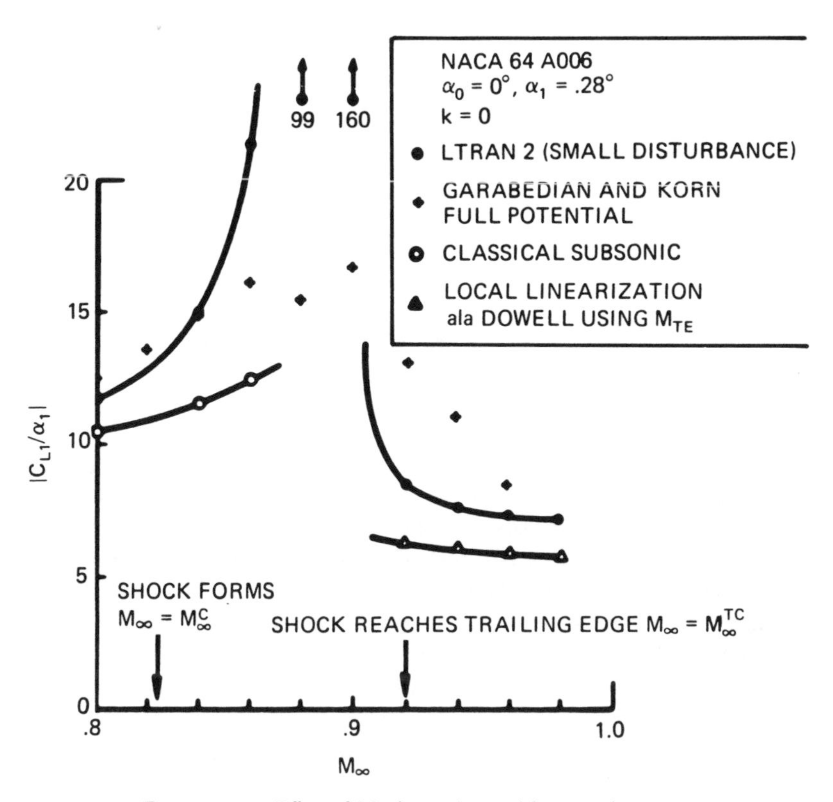

Figure. 9.6a Effect of Mach number on lift curve slope.

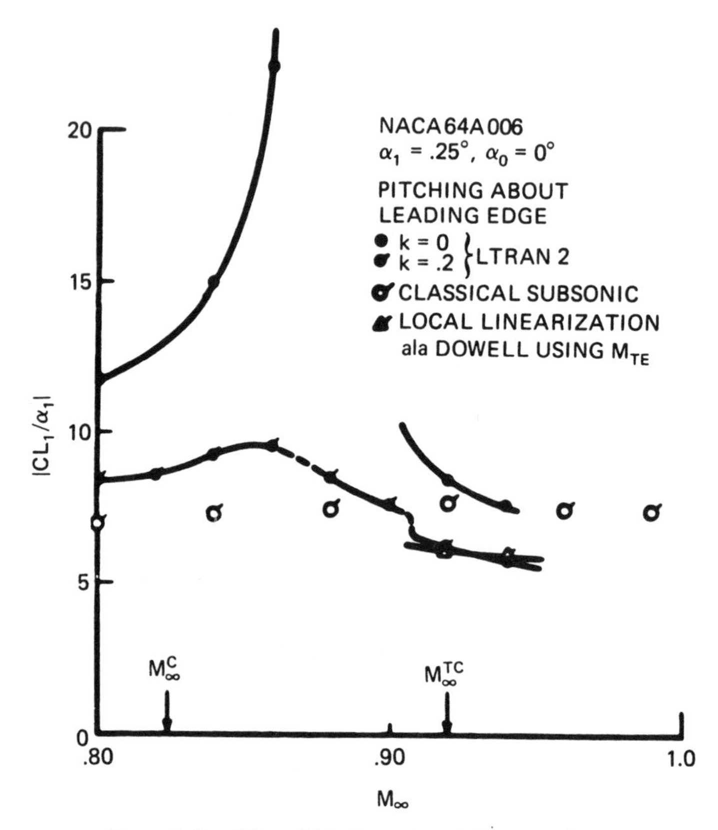

Figure 9.6b Effect of Mach number on lift curve slope.

Conclusions

For $M_\infty < M_\infty^c$, where no shock exists, the aerodynamic forces are linear over a substantial range of angle of attack. This is also true for $M_\infty > M_\infty^{tc}$, i.e., where the shock has moved to the trailing edge. For $M_\infty^c < M_\infty < M_\infty^{tc}$ a boundary of linear/nonlinear behavior may be constructed, which shows the angle of attack must be quite small for linear behavior to occur for steady flow. However, the region of linear behavior increases substantially for unsteady flow.

In the range $M_\infty^c < M_\infty < M_\infty^{tc}$, transonic small disturbance theory (LTRAN2) and full potential theory appear to fail for steady flow for some narrow band of M_∞ where they substantially overestimate the shock displacement and, hence, the aerodynamic forces. This is tentatively attributed to the absence of viscosity in the theories.

Classical subsonic theory and local linearization are useful approximate tools for unsteady flow provided their limitations are recognized.

Aerodynamic transfer functions are expected to retain their utility even when nonlinear dynamic effects are important. This is for several reasons, including:

(1) Nonlinear effects diminish with increasing frequency.
(2) At high frequencies, classical linear theory is expected to be reasonably accurate and indeed most inviscid theories will approach classical theory as the frequency becomes larger [58, 59].
(3) The preceding suggests that several theories may be used to provide a composite aerodynamic representation in the frequency domain. For example, one might use BGK for $k = 0$, LTRAN2 for $k = 0.05$–0.2, Williams for $k = 0.2$–1.0, and classical theory (which Williams' theory smoothly approaches) for $k > 1.0$.

A similarity law for low frequency transonic small disturbance theory is available that reduces the number of aerodynamic computations required and generalizes results for one airfoil to an entire family.

Although two-dimensional flows have been discussed here, the general concepts and approach should be useful for three-dimensional flows. In particular, one expects the effect of three-dimensionality to increase the region of linear behavior for transonic flows. For example, the accuracies of transonic small disturbance theory, local linearization, and classical theory should be enhanced by three-dimensional effects.

No transonic method of aerodynamic analysis can be expected to give useful information to the aeroelastician unless the mean steady flow it predicts and uses is accurate. Hence, it is highly desirable to be able to input directly the best steady flow information which is available including that from experiment. The latter would include implicitly viscosity effects on the mean steady flow; in particular it would place the mean shock in the correct position.

The reader may wish to consult the lucid survey article by Tijdeman and Seebass [51] which provides a context in which to evaluate the present results and conclusions. Also Nixon and colleagues have discussed extensively how the transonic, linear theory may be used in aeroelastic analyses. For example, see [40, 41]. Finally see the subsequent discussion in Section 9.5.

9.3 Viable alternative solution procedures to finite difference methods

Although continuing advances in computer technology will lead to diminishing costs, economics alone will dictate for the next decade a substantial effort to improve the efficiency of finite difference methods and/or consider less expensive alternative solution techniques. Here, the latter is discussed drawing largely on the recent work of Hounjet [25] and Cockey [11]. Both of these authors have used integral equation methods (IEM), although from rather different points of view. Prior work by Hounjet [24] was based on the Williams-Eckhaus model [58, 59], which also is the point of departure for Cockey. The motivation for considering IEM and a concise description of earlier work is well covered by Hounjet [25], Morino [36], Morino and Tseng [37], Albano and Rodden [1], Nixon [39], Voss [55], Williams [50], and Liu [31].

Both Hounjet and Cockey adopt a transonic small disturbance equation approximation and the associated velocity potential is divided into steady, $\phi^{(0)}$, and unsteady (due to airfoil motion), $\phi^{(1)}$, parts. By assuming (infinitesimally) small airfoil (harmonic) motion, the governing field equation for $\phi^{(1)}$ is linear with variable coefficients that depend on $\phi^{(0)}$, *viz.*,

$$(1-M_\infty^2)\phi_{xx}^{(1)}+\phi_{yy}^{(1)}-2ikM_\infty^2\phi_x^{(1)}+k^2M_\infty^2\phi^{(1)} = [(M_L^2-M_\infty^2)\phi_x^{(1)}]_x \tag{9.3.1}$$

where M_∞ is the freestream Mach number, k is the reduced frequency, $k=\omega c/2u$, in which c denotes the airfoil chord. c is used to make lengths dimensionless. M_L is the local Mach number:

$$M_L^2 = M_\infty^2+[3-(2-\gamma)M_\infty^2]M_\infty^2\phi_x^{(0)} \tag{9.3.2}$$

Subscripts on $\phi^{(1)}$ denote spatial derivatives.

From this point the approaches of Hounjet and Cockey follow different paths. Note that by setting the right hand side of (9.3.1) to zero, one retrieves classical aerodynamic theory.

Hounjet

By using the Green's function of classical aerodynamic theory [the LHS of (9.3.3)], one obtains an integral equation for the unknown $\phi^{(1)}$. It is

$$\phi^{(1)}(x, y) = \int_0^\infty \Delta\phi^{(1)}(u) \frac{\partial}{\partial y}[E(x-u, y; k, M)]\,\mathrm{d}u$$
$$+ \int_{-\infty}^{\infty}\int_{-\infty}^{\infty} m(u, v)E(x-u, y-v; k, M)\,\mathrm{d}u\,\mathrm{d}v \qquad (9.3.3)$$

On the right hand side of (9.3.3), the first term is an integral along the airfoil chord (and wake), while the second integral is over the *entire* (but see below) flow field. E represents an elementary source solution that satisfies the radiation condition and the following equation:

$$(1 - M_\infty^2)E_{xx} + E_{yy} - 2ikM_\infty^2 E_x + k^2 M_\infty^2 E = e^{i\bar{k}x}\delta(x)\delta(y) \qquad (9.3.4)$$

where

$$\bar{k} \equiv kM_\infty^2/\beta^2$$

m is given by

$$m(x, y) = [3 - (2-\gamma)M_\infty^2]M_\infty^2[\phi_x^{(0)}\phi_x^{(1)}] \qquad (9.3.5)$$

In classical (integral equation) aerodynamic theory, of course, the second integral of (9.2.3) is not present because $\phi_x^{(0)}$ is zero. In transonic IEM the second term may be neglected everywhere in the flow field where $\phi_x^{(0)} = 0$, i.e., the steady flow field is sensibly uniform. Hence, only a relatively small part of the total flow field will contribute to the second integral term and this simplifies the subsequent calculation very substantially. This is the key point which allows the possibility of an efficient computer code. Then the numerical solution of equation (9.3.3) proceeds as described by Hounjet [25].

Numerical results obtained by Hounjet show two principal features:

(1) The accuracy (for a linearization in the dynamic airfoil motion) is the same as that obtained by finite difference procedures.
(2) The computer time is approximately one quarter of that of LTRAN2 (a popular finite difference code).

Recently, Hounjet has extended this approach to three-dimensional flow fields.

Cockey

First rewrite (9.2.1) as

$$(1 - M_\infty^2)\phi_{xx}^{(1)} + \phi_{yy}^{(1)} - 2ikM_\infty^2\phi_x^{(1)} + k^2M_\infty^2\phi^{(1)}$$
$$-(M_L^2 - M_\infty^2)\phi_{xx}^{(1)} - 2M_L\frac{\mathrm{d}M_L}{\mathrm{d}x}\phi_x^{(1)} = 0 \qquad (9.3.6)$$

If one could determine the Green's function for the LHS of (9.3.6), then an integral equation for $\phi^{(1)}$ will only have an integral along the airfoil chord, shock, and wake and none in the flow field per se (except possibly along the shock). Unfortunately, obtaining this Green's function is difficult because the last two forms of the LHS of (9.3.6) have variable coefficients.

Thus, Cockey modifies (9.3.6) by a local linearization approximation to M_L and $\mathrm{d}M_L/\mathrm{d}x$ for the purpose of obtaining the Green's function (approximately). The subsequent calculation follows standard techniques except that integrals extend along the shock as well as the airfoil chord (and wake).

Numerical results obtained by Cockey show two principal features:

(1) The accuracy is substantially less than that obtained by finite difference procedures, even though the shock and its movement is taken into account.
(2) However, the computer cost is no greater than that associated with classical aerodynamic theory.

(1) is, of course, a disappointing result. However, the successful incorporation of the shock into the Cockey model and Hounjet's substantial success suggest a possible way of advantageously combining the features of methods of Hounjet and Cockey.

A possible synthesis of the Hounjet and Cockey methods

Consider again (9.3.1). Recognizing that in the Cockey method, M_L, $\mathrm{d}M_L/\mathrm{d}x$ are approximated for the purpose of obtaining a Green's function (9.3.1) is rewritten as follows:

$$
\begin{aligned}
&(1-M_\infty^2)\phi_{xx}^{(1)}+\phi_{yy}^{(1)}-2ikM_\infty^2\phi_x^{(1)}+k^2M_\infty^2\phi^{(1)}\\
&\qquad-(M_{LA}^2-M_\infty^2)\phi_{xx}^{(1)}-2M_{LA}\frac{\mathrm{d}M_{LA}}{\mathrm{d}x}\phi_x^{(1)}\\
&\quad=(-M_{LA}^2+M_L^2)\phi_{xx}^{(1)}+2\left(M_L\frac{\mathrm{d}M_L}{\mathrm{d}x}-M_{LA}\frac{\mathrm{d}M_{LA}}{\mathrm{d}x}\right)\phi_x^{(1)}
\end{aligned}
\tag{9.3.7}
$$

Setting the RHS of (9.3.7) to zero, we retrieve the Cockey model. However, if the RHS is retained, then Hounjet's method could be used to solve the resulting integral equation where now Cockey's approximate Green's function is used corresponding to the LHS of (9.3.7) rather than the classical Green's function (as Hounjet has used) corresponding to the

LHS of (9.3.1). Presumably the advantage of using the hybrid approach [i.e., (9.3.7) rather than (9.3.1)] is that the RHS of (9.3.7) is usually smaller than the RHS of (9.3.1) and thus the region in the flow field which is included in Hounjet's approach may be smaller and thus the resulting computer code will be more efficient.

We note finally that this approach (Hounjet's original method or the hybrid approach suggested here) is reminiscent of Lighthill's theory of jet noise [21] except that here, of course, the RHS in any of its several possible forms is known exactly.

9.4 Nonuniqueness, transient decay times, and mean values for unsteady oscillations in transonic flow

Early work

Kerlick and Nixon [29] have made the important point that, when using a finite difference, time marching computer code to investigate the lift on an oscillating airfoil in transonic flow, it is necessary to carry the solution sufficiently far forward in time that an essentially steady state solution is obtained. Moreover, they offer a method for estimating the transient time before the steady state is reached. They note that, if one stops the time marching solution before the transient is complete and the steady state is reached, then one may reach the incorrect conclusion that a change in the mean lift has occurred due to the oscillating motion of the airfoil when in fact no such change has occurred.

Nevertheless, what is perhaps surprising is that for a narrow Mach number range the time for the transient to decay and a steady state to be reached is extraordinarily long. Moreover, for a very narrow range of Mach number a non-zero mean value of lift can occur for an airfoil of symmetrical profile oscillating about a zero angle of attack.

Recent work

In Kerlick and Nixon [29] a NACA 64A006 airfoil was studied using the LTRAN2 computer code for a freestream Mach number of 0.875 with the airfoil oscillating at a peak angle of attack of 0.25° and with various reduced frequencies, k. It was shown that typically up to six cycles of airfoil oscillation must be considered for the mean lift to be less than 1% of the corresponding oscillatory lift peak value. Dowell *et al.* [14, 15] (also using the LTRAN2 computer code) examined various airfoils, 64A006,

64A010, MBB-A3, at various Mach numbers and reduced frequencies. The calculations in [14, 15] were carried forward in time through six cycles of airfoil oscillation. As was found by Kerlick and Nixon, the results of [14, 15] for the (symmetric) 64A006 airfoil showed the mean or average lift with the airfoil oscillating to be essentially unchanged from its value for no airfoil oscillation (i.e., zero) for most Mach numbers studied after six cycles of oscillation. However, at $M_\infty = 0.88$ and particularly at $M_\infty = 0.9$ this was not the case. Hence, the present author incorrectly concluded that a change in mean or average lift had occurred. The correct conclusion is that at $M_\infty = 0.9$, 0.88, many cycles of oscillation (>40) are required for the mean lift to decay to essentially zero.

The results presented in Figure 9.7 are for $M_\infty = 0.9$, $k = 0.2$ and various peak oscillatory angle of attack [14, 15]. They are for the lift coefficient; similar results (not shown for brevity) were obtained for moment (about mid-chord) coefficient and shock displacement (normalized by airfoil chord). Both mean steady values and first harmonic are shown after six cycles of oscillation. As can be seen the change in mean lift

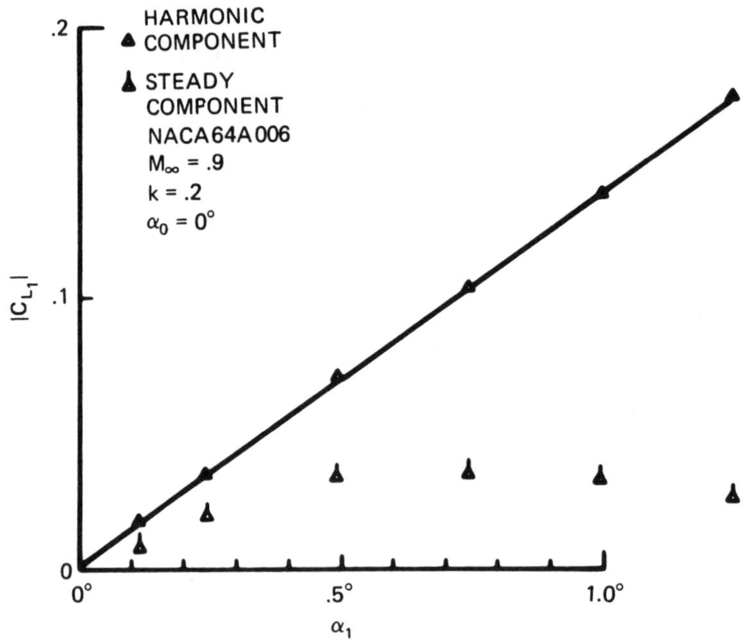

Figure 9.7 Steady state and first harmonic lift components vs. dynamic angle of attack. NACA 64A006, $M_\infty = 0.9$, $k = 0.2$, $\alpha_0 = 0°$.

at this Mach number can be as much as 50% of the peak oscillatory lift. For $M = 0.88$ the mean or average components are never more than 10% of the first harmonic oscillatory components, and, hence, these results are not displayed.

Inspired by the note of Kerlick and Nixon, the case of $k = 0.2$ and a peak oscillating angle of 0.5° was carried forward in time through many more cycles of airfoil oscillation for several M_∞. A typical result is shown in Figure 9.8 for $M_\infty = 0.9$ in terms of lift vs. angle of attack through 20 cycles of oscillation. Time varies along the (hysteresis) curve. As may be seen after twenty cycles, the average lift is still distinctly different from zero and, moreover, is a substantial fraction of the oscillatory lift. The average lift is defined as

$$C_{L_{AVG}} = \frac{C_L^+ + C_L^-}{2}$$

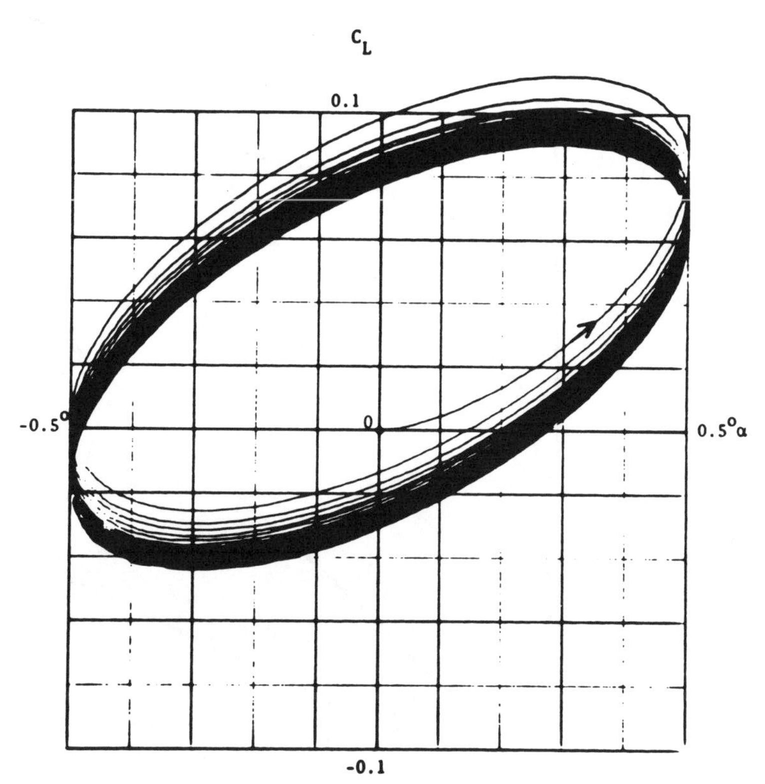

Figure 9.8 NACA 64A006 airfoil, $M_\infty = 0.9$, $k = 0.2$. Hysteresis curve of lift vs. dynamic angle of attack.

and the oscillatory lift as

$$C_{L_{OSC}} = \frac{C_L^+ - C_L^-}{2}$$

where C_L^+, C_L^- are peak values, positive and negative respectively, of any two adjacent peaks.

The essential results can be summarized compactly in Figure 9.9 where $C_{L_{AVG}}$ is plotted against $1/N$, where N is the cycle number of the angle of attack oscillation. The results are plotted against $1/N$ so that the limit as $N \to \infty$ or $1/N \to 0$ may be more readily examined. As may be seen for $M_\infty = 0.86$ and 0.875, the average lift rapidly declines and is essentially zero for $N > 10$ or $1/N < 0.1$. (However it should be noted that the oscillatory lift, $C_{L_{OSC}}$, converges much more rapidly than the average lift. For the sake of brevity, $C_{L_{OSC}}$ is not shown.) By contrast for $M_\infty = 0.9$ and 0.885 the average lift persists in measurable values through a much larger

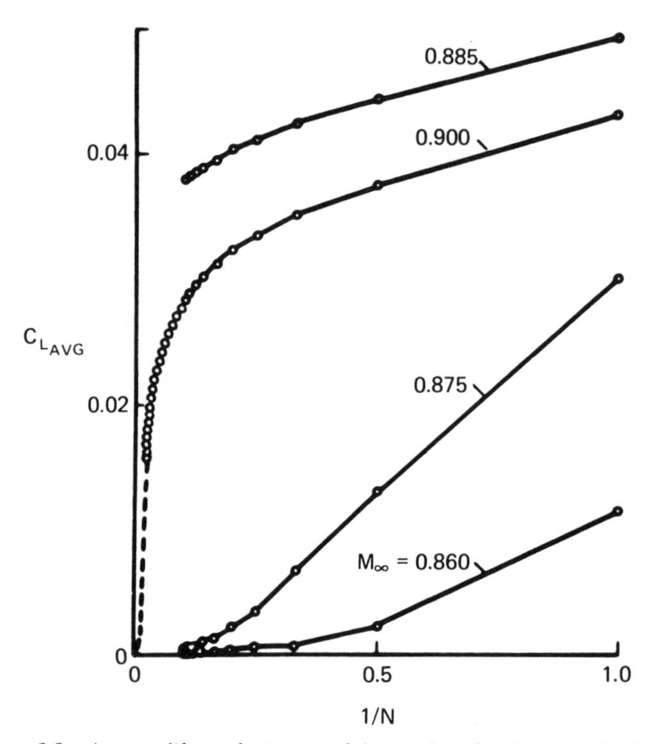

Figure 9.9 Average lift vs. the inverse of the number of cycles of oscillations.

number of cycles. For example, for $M_\infty = 0.9$ at $N = 40$ or $1/N = 0.025$, there is a clear trend toward zero average lift as $N \to \infty$ or $1/N \to 0$, but at $N = 6$ this would be difficult to perceive. Of course, these results have important practical consequences. A run for $N = 40$ takes 1672 sec of CPU time on an IBM 3033.

Hence, at this point the present author concluded that while a non-zero average lift is mathematically possible for a nonlinear aerodynamic system responding to an oscillating angle of attack, no such lift was observed using LTRAN2 for the range of parameters studied. However, at some Mach numbers the time for the average lift to decay to zero is extraordinarily long.

Next the present author greatly benefitted from a discussion with Dr Peter Goorjian [22]. He had carried out calculations [2] at $M_\infty = 0.89$ which indicated that a non-zero average lift did occur. Dr. Goorjian's results are more fully discussed in [15], where an extended version of the

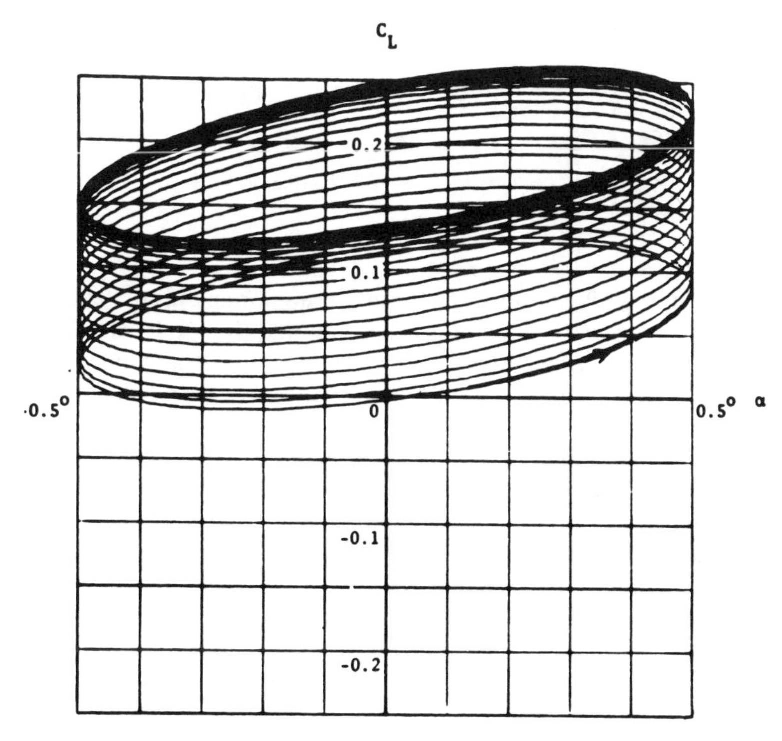

Figure 9.10a NACA 64A006 *airfoil,* $M_\infty = 0.89$, $k = 0.2$, $N = 1$–20 *cycles. Hysteresis curve of lift vs. dynamic angle of attack.*

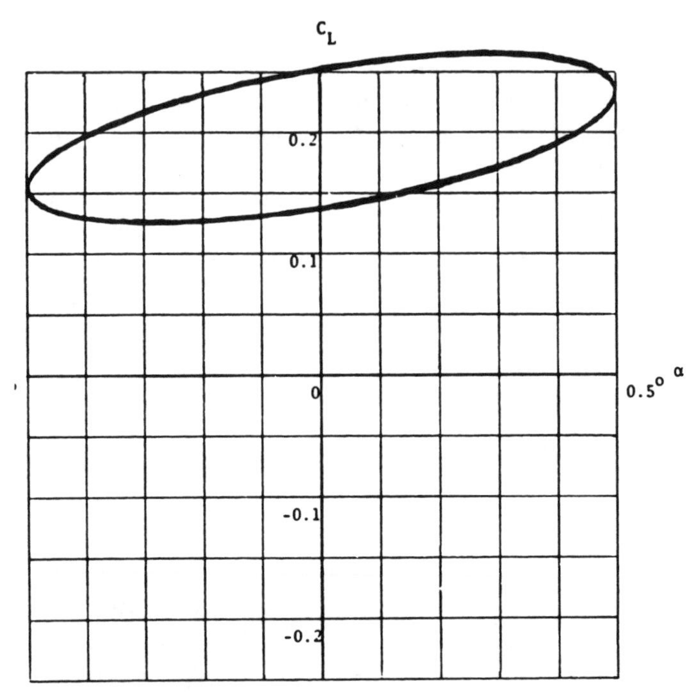

Figure 9.10b NACA 64A006 *airfoil*, $M_\infty = 0.89$, $k = 0.2$, $N = 21$–40 *cycles. Hysteresis curve of lift vs. dynamic angle of attack.*

present account is given. Hence, the present author and his colleague, Dr. Ueda, also carried out calculations at this M_∞ and the results are shown in Figure 9.10 (analogous to Figure 9.8) and Figure 9.3 (analogous to Figure 9.9). These results clearly suggest that a non-zero average lift does occur at $M_\infty = 0.89$.

As explained by Goorjian, if one starts the airfoil oscillation with an initial negative angle of attack rather than a positive one, the mean lift is correspondingly negative (with the same magnitude) rather than positive. Indeed, the entire hysteresis curve is a double mirror image with C_L and α both undergoing sign inversions. Moreover, if the oscillating airfoil is now rendered motionless at exactly zero angle of attack, the mean lift persists at some finite, nonzero value. This implies that three solutions for lift occur at this single (zero) angle of attack. This result is fully consistent with that of Steinhoff and Jameson [48] who obtain nonunique (multi-valued) steady flow solutions by direct calculation for the full potential equations.

More recently, Salas *et al.* [44] have studied the full potential equations and the Euler equations for steady flow. For similar conditions he has found nonunique solutions for the full potential equations, but not for the Euler equations. However, it is very difficult to prove by numerical calculations, the absence of nonunique solutions for all possible conditions of interest.

Questions still remain of course.

- Why does this nonzero average lift occur only over a narrow range of Mach number? Note that the Mach number, $M_\infty \simeq 0.89$, at which nonunique solutions are observed corresponds to the Mach number for which the range of linear behavior is smallest as $k \to 0$. See Figure 9.11.
- At what level of mathematical modeling, if any, do nonunique solutions no longer occur?
- Is the result physically significant? In particular, what would be the counterpart, if any, for a viscous fluid model?

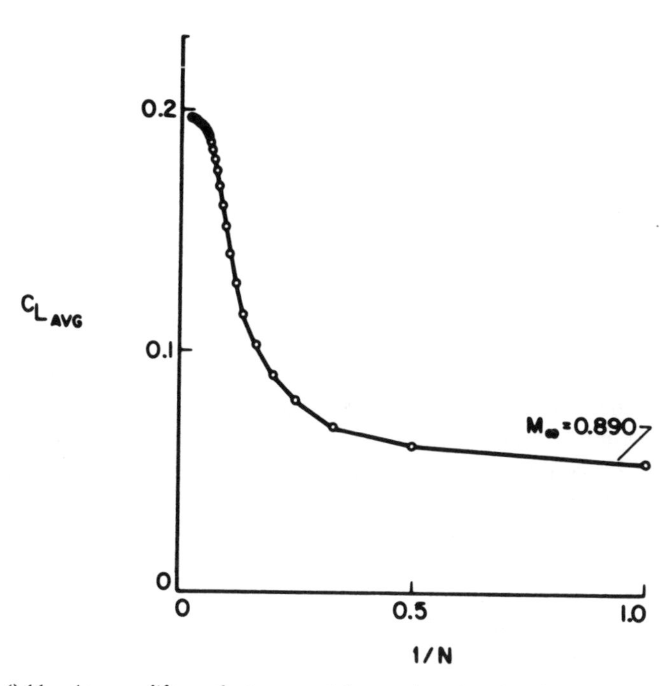

Figure 9.11 Average lift vs. the inverse of the number of cycles of oscillations. $M_\infty = 0.89$.

Studies of Williams et al. and Salas et al.

Other studies, which have provided at least partial answers to these questions and added to our knowledge concerning the nonuniqueness of the potential flow solutions, include those of Williams, Bland, and Edwards [57] and Salas and Gumbert [45]. Williams *et al.* have studied the stability (instability) of the multiple static equilibrium solutions predicted by potential flow theory (zero lift, positive and negative lift) and shown that, as expected, the zero lift solution is unstable. Furthermore, they have investigated the long transient decay times associated with the approach to the two stable, non-zero lift, static equilibria. The long transient times are not totally unexpected, since, even in the classical fully linear aerodynamic theory, the decay times are long as the freestream Mach number nears one. Here it is the average local Mach number that is near one.

They also have combined a potential flow model with a simplified (lag-entrainment) boundary layer theory to see if viscous effects remove the multiple equilibria and restore the stability of the zero lift solution. Apparently they do not.

Finally, they have noted that the small-disturbance potential flow equations are a consistent physical model and should represent a proper limit of the Euler equations as the thickness of the airfoil tends toward zero. Unfortunately present finite difference codes have difficulty solving the Euler equations in this limit.

Salas and Gumbert [45] have computed steady flow results from potential flow theory for several (relatively thick) airfoils over a range of Mach number. They show that the predicted nonuniquenees is common if the angle of attack is sufficiently large. However, the nonuniqueness only occurs over a limited range of Mach number at zero angle of attack.

For one airfoil, NACA 0012, at two Mach numbers $M_\infty = 0.67$ and 0.75, Euler solutions are also provided by Salas and Gumbert and compared to those from potential flow theory. See Figures 6 and 7 of their paper. The linearity of lift with angle of attack as predicted by the Euler code is typical for thick airfoils and is remarkable. It is argued by Salas and Gumbert that the deviation of the results of potential theory from those of the Euler code is an indication of the failure of the potential theory. This may well be; however, such an explanation still does not address the predictions of nonuniqueness by the potential flow model for thin airfoils at nearly zero angle of attack.

Fortunately (or otherwise depending on one's point of view), most airfoils of practical interest are of a certain thickness where the Euler codes apparently give reliable answers and where nonuniqueness has not been observed. Moreover, and perhaps more importantly, recently published results by Williams [20] show that for three-dimensional flows over wings of moderate to small aspect ratio, the nonuniqueness observed in the two-dimensional flows over airfoils is absent.

Aileron buzz (flutter) and the Steger–Bailey calculation

No discussion of unsteady transonic aerodynamics and aeroelasticity would be complete without a summary of this pioneering study by Steger and Bailey [47]. They used an implicit finite difference computer code to simulate the fluid dynamic behavior of a two-layer algebraic eddy viscosity model of the Navier–Stokes equations for the flow about a NACA 65-213 airfoil with an oscillating control surface (aileron). Using a simple (one-degree-of-freedom with inertia only) dynamic model of the control surface itself, they determined the stability (flutter) boundary of the control surface and also limit cycle behavior amplitude and frequency during the flutter oscillation.

In earlier experimental work, Erickson and Stephenson [18] had concluded that aileron buzz is a one-degree-of-freedom flutter in which shock-wave motion causes a phase shift in the response of the control surface hinge moment to the aileron motion. They observed oscillatory aileron motion for various combinations of Mach number, geometry and mechanical damping.

For the specifics of the fluid dynamic modeling and computational method, Steger and Bailey [47] should be consulted. Here we focus on the aeroelastic results.

Representative results are shown in Figures 7 and 10 of their paper. In Figure 7 the transient oscillation of the aileron is shown for $M_\infty = 0.82$, a mean airfoil angle of attack of $\alpha = -1.0°$, and an initial control surface angle of δ_a $(t = 0) = 4°$. As can be seen, after several cycles of oscillation the aileron approaches an apparent steady state limit cycle oscillation. On the other hand, for δ_a $(t = 0) \cong 0°$, the aileron oscillation decayed to zero after a transient moition and no steady state oscillation was observed. As the Mach number was increased to $M_\infty = 0.83$, however, a steady state oscillation was observed even for δ_a $(t = 0) \cong 0°$. At lower Mach numbers, $M_\infty = 0.76$ and 0.79, no steady state oscillation was observed even for δ_a $(t = 0) = 4°$.

Such results were used to construct a stability boundary as shown in Figure 10 of their paper, which was compared to that found experimentally. The comparison is reasonably good. Moreover, the (limit cycle) amplitude, and frequency of the flutter oscillation predicted by the calculation was in reasonable agreement with that measured, as well. Of course, the demonstrated sensitivity of the limit cycle motion to initial conditions gives some reason for caution in comparing calculated and experimental results.

Steger and Bailey [47] also carried out inviscid calculations. For $\alpha = -1.0°$ they found flutter occurred at $M_\infty = 0.84$, but not at 0.83. Recall that with viscous effects included flutter occurred at $M_\infty = 0.83$ and 0.82 (but not at 0.79). Hence, there is some difference between the two calculated results, inviscid and viscous, with respect to the predicted onset of flutter (the stability boundary). More important, however, the iviscid calculation failed to predict a steady state oscillation, instead only showing an exponentially, diverging or decaying oscillation. Thus, the dominant nonlinear effects leading to a finite limit cycle amplitude appear to be due to viscosity. As Steger and Bailey concluded: "From these data we find that while inviscid unsteady shock wave motion is the driving force of transonic aileron buzz, the viscosity is nevertheless crucial and can both sustain and moderate the flap motion."

They continued: "Finally for the aileron *held fixed* at a high Mach number, $M_\infty = 0.85$, we find that the viscous flow [itself] does not reach a steady state [here they really mean, static equilibrium] but buffets [reaches an oscillatory steady (or chaotic?) state]... if the aileron is then released, it no longer oscillates in a simple sinusoidal motion. Viscous effects appear to be much more dominant [at this Mach number] and change the frequency and amplitude of aileron motion. The motion appears to repeat every fourth oscillation [the motion is periodic with two (three?) dominant frequencies in the ratio of 4 (to 2?) to 1]...".

Altogether these results are most fascinating and one can only hope that more such will be forthcoming.

9.5 Effective, efficient computational approaches for determining aeroelastic response using unsteady transonic aerodynamic codes

Various computational approaches and their relative merits

The basic issue is whether

- one should run an aerodynamic code for prescribed airfoil motions and store the resultant aerodynamic data prior to carrying out a flutter analysis (Option I) or whether

- one should run an aerodynamic code in conjunction with a structural dynamics code to determine simultaneously the time history of the airfoil and aerodynamic forces (Option II)?

There are, of course, two types of aerodynamic computer codes available, (1) those which calculate in the time domain [2, 3, 8–10, 23, 26, 43], and (2) those which calculate in the frequency domain [17, 25, 58, 59]. The use of a code of the type (2), of course, precludes the pursuit of option II. However, the use of a computer code of type (1) permits the use of option I or II.

Below, each type of computer code will be considered in terms of how it may be used most effectively, and for a type (1) code the relative merits of options I and II will be discussed. Until otherwise noted, it will be assumed that we are anticipating a linear flutter analysis that only seeks to determine the conditions for the onset of flutter. A type (2) aerodynamic code implicitly assumes this to be the case. Again, as will be discussed later, use of a type (1) aerodynamic code will permit either a linear or a nonlinear flutter analysis to be conducted.

Type (1) aerodynamic code-time domain

Assume that flutter solutions are to be found for P parameter combinations (e.g., dynamic pressure values, tip tank masses, control surface stiffnesses) and M structural modes. Moreover, assume that T_A is the computational time it takes for the aerodynamic code to reach a steady state lift value for a prescribed airfoil motion and T_F is the computational time it takes for a simultaneous fluid – structural time marching calculation to complete a transient. The relative sizes of T_A and T_F depend on airfoil profile and Mach number (for T_A) and structural damping, (T_F). Generally, $T_F > T_A$, but there are exceptions. For simplicity, think of the Mach number as fixed, since calculations at several M_∞ will simply increase all computations by the same factor.

Consider now the relative merits of options I and II.

Option I: Generate and store aerodynamic data prior to aeroelastic calculation.

The total computational time to generate the aerodynamic forces is

$$M * T_A \tag{9.5.1}$$

which is independent of P. There is some additional time required for flutter solutions *per se*, but it is assumed this is negligible compared to the time required to generate the aerodynamic forces.

Option II: Generate aerodynamic data and structural data simultaneously.

The total computational time will be

$$P * T_F \tag{9.5.2}$$

which, of course, is independent of M. Clearly, which of the two options is most attractive depends on whether $M * T_A \gtreqless P * T_F$. For option II to be more attractive, the number of modes, M, should be somewhat larger than the number of parameters, P. Thus option II will be more attractive in a design verification study while option I will tend to be more attractive in a preliminary design phase.

Type (2) aerodynamic code-frequency domain

Here only option I has been used to date; however, see the discussion below. Let NF be the number of reduced frequencies needed for the flutter analysis. Let T_{AF} be the time for the aerodynamic code to determine the aerodynamic forces for one frequency. Assume that an aerodynamic influence coefficient approach is used so that the number of modes does not influence the computational time.*

Option I

The computational time to generate the aerodynamic forces is

$$NF * T_{AF} \tag{9.5.3}$$

Compare (9.5.3) to the computational time associated with time domain aerodynamics [see previous discussion and (9.5.1)]:

$$NF * T_{AF} \gtreqless M * T_A$$

If T_{AF} and T_A are comparable (one might expect competition would tend to make them so), then the method of choice as between (9.5.1) and (9.5.3) will depend on the number of frequencies, NF, compared to the number of modes, M, needed in the aeroelastic calculations.

* If a relaxation scheme is used rather than direct inversion of the aerodynamic matrix, then another set of issues arises.

Option IA

Although this approach has not been pursued to date, one could take the frequency domain aerodynamic forces, curve fit them with Padé approximants or comparable representation [13], use these to deduce a differential equation aerodynamic force representation [52, 34], and then do a time marching flutter solution. The computational time would still be approximately

$$NF * T_{AF} \tag{9.5.3a}$$

The comparable computational time using a direct time-marching aerodynamic code was (see previous discussion)

$$P * T_F \tag{9.5.2a}$$

Assuming $T_{AF} = T_F$, one concludes that, if the number of parameters is large compared to the number of reduced frequencies needed, then the frequency domain aerodynamic method tends to be the method of choice over the time domain method which uses option II.

Summary comparison

By comparing the estimates, (9.5.1), (9.5.2), (9.5.3), (9.5.3a), one may make an initial judgement as to the method of choice in a given situation.

Nonlinear flutter analysis

General considerations

If a nonlinear flutter analysis is needed, then only the time domain aerodynamic method is available, type (1).

Of course, the aeroelastic calculation may be done in either the time or frequency domain. However, there is still a trade-off between options I and II. Now, for option I, a further multiplicative factor must be used which is the number of airfoil response levels, NR, which are of interest. For linear flutter analysis only one response level is of interest (strictly speaking an infinitesimal response level which approaches zero). Thus the computational times to compare are

Option I:

$$NR * M * T_A$$

Option II:

$$P * T_F.$$

It should be noted, moreover, that for nonlinear flutter analysis the number of parameters, P, will tend to be somewhat larger than for linear flutter analysis.

The relative attractiveness of the two options is as before but with a bias shift toward option II, because of the factor NR appearing in option I. The use of option I in a nonlinear flutter analysis does in fact lead to a flutter analysis in the frequency domain, and the methodology by which that is done is described below in Section 6. Of course, this methodology reduces to the classical frequency domain flutter solution method when the airfoil response levels are small. The solution procedures for option II as currently practiced are straightforward and will not be elaborated upon further here. It is also worthy of note that

- a frequency domain nonlinear flutter analysis usually introduces approximations beyond those of a time domain analysis, but
- fortunately, linear flutter analysis will often suffice and hence the whole set of questions is frequently moot.

9.6 Nonlinear flutter analysis in the frequency domain and comparison with time marching solutions

The nonlinear effects of transonic aerodynamic forces on the flutter boundary of a typical section airfoil are discussed. The amplitude dependence on flow velocity is obtained by utilizing a novel variation of the describing function method that takes into account the first fundamental harmonic of the nonlinear oscillatory motion. By using an aerodynamic describing function, traditional frequency domain flutter analysis methods may still be used while including (approximately) the effects of aerodynamic nonlinearities. Results from such a flutter analysis are compared with those of brute force and periodic shooting time marching solutions. The aerodynamic forces are computed by the LTRAN2 aerodynamic code for a NACA 64A006 airfoil at $M_\infty = 0.86$.

Motivation and background

Recent developments in computational aerodynamics have led to renewed interest in the prediction of flutter boundaries of an airfoil in the transonic flow regime [3, 4, 61, 28]. Until recently, flutter calculations have either assumed the transonic aerodynamic forces could be approximated as linear functions of the airfoil motion so that traditional linear flutter analysis methods could be used or, alternatively, taken a brute force approach by structural and aerodynamic equations. The latter method does, of course, fully account for aerodynamic nonlinearities.

Ballhaus and Goorjian [3, 4] calculated the aeroelastic response of a NACA 64A006 airfoil with a single-degree-of-freedom control surface by simultaneously integrating numerically in time the structural equation of motion and also the aerodynamic equations. They used their own code, LTRAN2, for unsteady transonic flow. The indicial method, whereby an aerodynamic impulse function is first calculated by the aerodynamic code and then used in the flutter calculation via a convolution integral, was also studied. The indicial method assumes linearity of the aerodynamic forces with respect to airfoil motion. The flutter of the same airfoil but with two-degrees-of-freedom was analyzed by Yang *et al.* [63] with aerodynamic forces obtained by three different methods. These forces were obtained by the time integration method, the indicial method (both of these employed the LTRAN2 code), and the harmonic analysis method in the frequency domain using the ULTRANS2 code. The latter method also assumes linearity of the aerodynamic forces. In general, all three methods agree well for the range of parameters studied by Yang. After the flutter boundary was obtained, the response was confirmed near the flutter boundary by simultaneous time integration of the governing structural and aerodynamic equations. Isogai [28] studied the transonic behavior of the NACA 64A010 airfoil by using his own USTS transonic aerodynamic code, which can be applied to supercritical Mach numbers for reduced frequencies, $0 < k < 1.0$. (By contrast, the aerodynamic methods used by Yang were limited to small k.) Isogai used the time integration method for evaluating the aerodynamic forces, but then converted them to linearized harmonic forces for the flutter calculations. See Yang *et al.* [63] for further discussion of previous work on flutter calculations including that of other investigators who have used brute force, simultaneous numerical integration of the structural and aerodynamic equations.

A discussion of when the aerodynamic forces may be treated as linear in the airfoil motion is given in Dowell *et al.* [14] and previously in the present chapter. The analysis of Yang *et al.* [63] and Isogai [28] described above assumed linear characteristics for the aerodynamic forces in the flutter calculations. Linearity can be ensured if the amplitude of the airfoil oscillation is sufficiently small [14], even though the governing fluid equations are inherently nonlinear for transonic flow fields. Yang *et al.*

[63] fixed an amplitude of pitching motion at 0.01 radian (0.574°), whereas Isogai [28] used 0.1 in degrees for the computation of the aerodynamic forces. Dowell *et al.* [14] pointed out that increasing the value of the reduced frequency increases the range of amplitude of oscillation for which linear behavior exists in transonic flow. However, the aerodynamic forces often begin to deviate from linear behavior for amplitudes of relatively small value such as 1.0° in pitching motion. Such amplitudes may be attained due to the disturbances an aircraft wing encounters during its flight. It is of importance, therefore, to clarify the aerodynamic nonlinear effect on a flutter boundary, especially when the nonlinear effect may create an aeroelastic softening system, i.e., the flutter speed decreases as the amplitude of oscillation increases. Such softening behavior may cause a dangerous unconservative estimation of a flutter boundary by linear analysis.

Here we study the nonlinear effect of transonic aerodynamic forces on a flutter boundary by utilizing a novel variation of the describing function method [27], which takes into account the first fundamental harmonic of the nonlinear oscillatory motion. By using an aerodynamic describing function, traditional frequency domain flutter analysis methods may still be used while including (approximately) the effects of aerodynamic nonlinearities. Brute force time marching calculations are also presented for comparison purposes.

The method used to calculate the describing functions is briefly this. A step change in angle of attack is specified and the transient aerodynamic force time history (calculated numerically by an appropriate aerodynamic code) is identified as a nonlinear impulse function. The Fourier transform of this impulse function (which in general depends on the step input level or amplitude) is taken as the aerodynamic describing function (nonlinear transfer function). Calculations have shown that this describing function agrees very well with the one determined by using a harmonic angle of attack input to the aerodynamic code. The latter method of calculation is, of course, much more expensive and time consuming for the range of frequencies needed in flutter analysis. The LTRAN2 computer code is used for determining the aerodynamic forces. However, any other nonlinear code could be used in a similar fashion.

Typical airfoil section

A typical airfoil section subjected to transonic flow is considered as shown in Figure 9.12. Since it can be assumed that the structural deformation is linearly dependent on the aerodynamic load for wings of ordinary modern

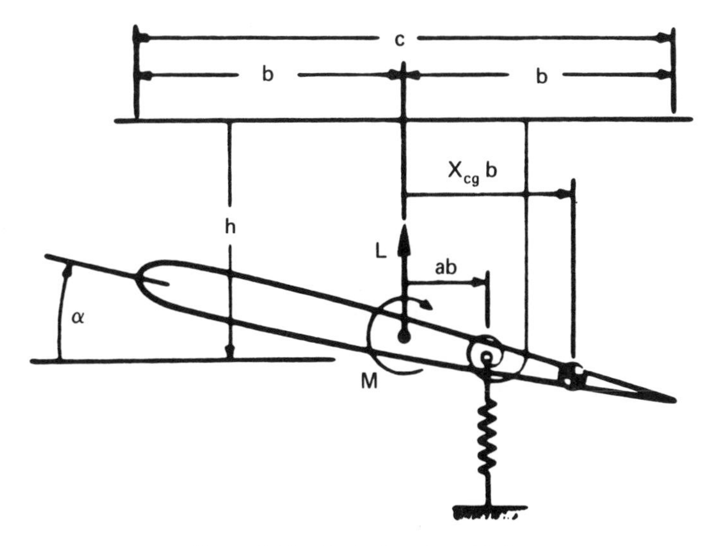

Figure 9.12 Typical section airfoil.

aircraft during its normal flight, a linear structural transfer function is used. The aerodynamic force, however, may depend in a nonlinear manner on the structural deformation in the transonic flow range [14]. In order to include the nonlinear effect of large (r) amplitudes of motion on the aerodynamic forces and, hence, on a flutter boundary, we use a nonlinear aerodynamic transfer function by employing the describing function method.

Aerodynamic describing function

Here we give a summary of the relevant standard describing function results and place the present method in context.

If we assume that the frequency of motion is relatively low, the aerodynamic forces due to the airfoil motion can be approximated as a function of the effective induced angle-of-attack, which is given by

$$\phi = \alpha + \frac{\dot{h}_c}{u} \tag{9.6.1}$$

This quasi-steady approximation is compatible with the low frequency assumption in the LTRAN2 transonic unsteady aerodynamic code which we use in the present flutter calculation. Taking into account the nonlinear effects of the amplitudes of motion, we assume the aerodynamic forces take the form:

$$L = \tfrac{1}{2}\rho u^2 c C_L^N(\phi, \dot{\phi}), \tag{9.6.2}$$

$$M = \tfrac{1}{2}\rho u^2 c^2 C_M^N(\phi, \dot{\phi}), \tag{9.6.3}$$

where C_L^N, C_M^N are functionals of ϕ, $\dot{\phi}$, i.e., they may, in principle, include the complete time history of ϕ and $\dot{\phi}$.

For general periodic time dependent motion, the effective angle-of-attack ϕ can be expanded in a Fourier series as

$$\phi = \tfrac{1}{2}\phi_0 + \sum_n^N [\phi_{I,n} \cos(nk\tau) + \phi_{R,n} \sin(nk\tau)] \tag{9.6.4}$$

According to the describing function method, only the first harmonic of ϕ is taken as an input to the aerodynamic force transfer function, i.e.,

$$\phi = \phi_1 \sin k\tau \tag{9.6.5}$$

This input motion generates aerodynamic forces through the nonlinear fluid element, call it H, which, in general, includes higher harmonics. Thus,

$$C_L^N(\phi, \dot{\phi}) = \tfrac{1}{2}C_{L_{1,0}}(\phi_1) + \sum_n^N [C_{L_{I,n}}(\phi_1)\cos(nk\tau) + C_{L_{R,n}}(\phi_1)\sin(nk\tau)], \tag{9.6.6}$$

$$C_M^N(\phi, \dot{\phi}) = \tfrac{1}{2}C_{M_{1,0}}(\phi_1) + \sum_n^N [C_{M_{I,n}}(\phi_1)\cos(nk\tau) + C_{M_{r,n}}(\phi_1)\sin(nk\tau)] \tag{9.6.7}$$

The describing function method, however, replaces the nonlinear element H by another nonlinear element H with the property that it operates on any sinusoidal input, (9.6.5), by passing its fundamental frequency in exactly the same manner as H; however, whatever the input frequency k, H generates no higher harmonics. This replacement allows us to write

$$\hat{C}_L^N(\phi, \dot{\phi}) = C_{L_{I,1}}(\phi_1)\cos(k\tau) + C_{L_{R,1}}(\phi_1)\sin(k\tau) = D_{L_R}(\phi_1)\phi + D_{L_I}(\phi_1)\dot{\phi}/k, \tag{9.6.8}$$

$$\hat{C}_M^N(\phi, \dot{\phi}) = C_{M_{I,1}}(\phi_1)\cos(k\tau) + C_{M_{R,1}}(\phi_1)\sin(k\tau) = D_{M_R}(\phi_1)\phi + D_{M_I}(\phi_1)\dot{\phi}/k \tag{9.6.9}$$

where

$$D_{L_R} = (1/\pi\phi_1)\int_0^{2\pi} C_L^N(\phi, \dot{\phi})\sin(k\tau)\,\mathrm{d}(k\tau), \tag{9.6.10}$$

$$D_{L_I} = (1/\pi\phi_1)\int_0^{2\pi} C_L^N(\phi, \dot{\phi})\cos(k\tau)\,\mathrm{d}(k\tau), \tag{9.6.11}$$

$$D_{M_R} = (1/\pi\phi_1)\int_0^{2\pi} C_M^N(\phi, \dot{\phi})\sin(k\tau)\,\mathrm{d}(k\tau), \tag{9.6.12}$$

$$D_{M_I} = (1/\pi\phi_1)\int_0^{2\pi} C_M^N(\phi, \dot{\phi})\cos(k\tau)\,\mathrm{d}(k\tau) \tag{9.6.13}$$

Using complex notation for (9.6.8) and (9.6.9) yields a more compact result, i.e.,

$$\hat{C}_L^N(\phi, \dot{\phi}) = D_{L_1}(\phi_1)\phi, \tag{9.6.14}$$

$$\hat{C}_M^N(\phi, \dot{\phi}) = D_{M_1}(\phi_1)\phi \tag{9.6.15}$$

where

$$D_{L_1}(\phi_1) = D_{L_R} + iD_{L_I} \tag{9.6.16}$$

$$D_{M_1}(\phi_1) = D_{M_R} + iD_{M_I} \tag{9.6.17}$$

In (9.6.14) and (9.6.15), the coefficients, $\hat{C}_L^N$ and C_M^N, also have complex values whose real parts correspond to (9.6.8) and (9.6.9), respectively.

If the amplitude ϕ_1 is fixed, (9.6.14) and (9.6.15) take a form identical to that for a linear system. This implies the applicability of the same stability analysis as that for a linear system.

The coefficients defined in (9.6.10)–(9.6.13) to construct the describing function can be computed by a time integration code for transonic flow. It is also possible to evaluate them from wind-tunnel experimental data measured on a harmonically (or impulsively) excited airfoil. In the present study, we utilize an *extended nonlinear* version of the indicial method [3, 4] to calculate the aerodynamic coefficients.

Since the describing function assumes the same form as a linear transfer function when the amplitude is fixed, we can regard a typical such element, $\hat{H}_\phi$, which relates any representative aerodynamic forces, F, to airfoil motion, ϕ, as a linear system with respect to variations in frequency.

$$F(\tau) = \hat{H}_\phi(ik, \phi_1)\phi_1 e^{ik\tau} \tag{9.6.18}$$

This relation corresponds in the subsidiary (effectively frequency) domain of the Laplace operator to the following:

$$\bar{F}(s) = \hat{H}(s, \phi_1)\frac{\phi_1}{s - ik} \tag{9.6.19}$$

If we put $k = 0$, then (9.6.19) represents an indicial response relationship.

$$\bar{\bar{A}}(s, \phi_1) = \hat{H}_\phi(s, \phi_1)\frac{\phi_1}{s} \tag{9.6.20}$$

By using (9.6.20) we can obtain the describing function, $H_\phi(ik, \phi_1)$, from the indicial response to a step input with amplitude ϕ_1. Furthermore, if we neglect the effect of higher harmonics, an assumption already made in the describing function method, then $\hat{H}\phi(ik, \phi_1)$ can be approximated by using the indicial response $\bar{A}(s, \phi_1)$ of the element H as

$$\hat{H}_\phi(ik, \phi_1) = \bar{A}(ik, \phi_1)ik/\phi_1 \tag{9.6.21}$$

From a linear system, starting from (9.6.18) one may proceed through (9.6.19), (9.6.20), to (9.6.21) and *vice versa* by standard mathematical methods. However, as the careful reader will note this is not strictly possible for a nonlinear system, i.e., (9.6.19) and (9.6.20) follow from (9.6.18) only by analogy to linear system results. Indeed we may take (9.6.18) and (9.6.21) [or (9.6.20)] as two independent definitions of $\hat{H}_\phi(ik, \phi_1)$, the describing function that will be used in the flutter analysis. However, by numerical example we will show that, in fact, the two definitions lead to similar results. This is fortunate, because the less obvious definition, (9.6.21), is far easier to use in practice for generating aerodynamic forces to employ in flutter calculations.

Working form of the aeroelastic system equations

The governing structural equations of the system are given in nondimensional form by

$$\frac{\pi\mu}{2}\left(\frac{h}{c}\right)'' + \frac{\pi\mu}{2}\left(\frac{S_\alpha}{mc}\right)\alpha'' + \frac{\pi\mu}{2}\left(\frac{c\omega_h}{u}\right)^2\left(\frac{h}{c}\right) = -C_L^N, \tag{9.6.22}$$

$$\frac{\pi\mu}{2}\left(\frac{S_\alpha}{mc}\right)\left(\frac{h}{c}\right)'' + \frac{\pi\mu}{2}\left(\frac{I_\alpha}{mc^2}\right)\alpha'' + \frac{\pi\mu}{2}\left(\frac{I_\alpha}{mc^2}\right)\left(\frac{c\omega_\alpha}{u}\right)^2\alpha = C_{Me}^N \tag{9.6.23}$$

From (9.6.22) and (9.6.23) the structural transfer function for the state vector, $[(h/c)\alpha]^T$, is

$$G^{-1}(s, U) = \begin{bmatrix} \frac{\pi}{2}\{\mu s^2 + R^2/U^2\} & \frac{\pi\mu}{4}(x_{cg} - a)s^2 \\ \frac{\pi\mu}{4}(x_{cg} - a)s^2 & \frac{\pi}{8} r_\alpha^2\{\mu s^2 + 1/U^2\} \end{bmatrix} \tag{9.6.24}$$

As to the aerodynamic describing function, we first assume the indicial response $A(r, \phi_1)$ to a step change in ϕ in (9.6.21) can be expressed in the following form for the lift and moment forces;

$$A_L(\tau, \phi_1) = a_0^L(\phi_1) + \sum_{i=1}^{N} a_i^L(\phi_1)e^{b_i^L\tau}, \tag{9.6.25}$$

$$A_M(\tau, \phi_1) = a_0^M(\phi_1) + \sum_{i=1}^{N} a_i^M(\phi_1)e^{b_i^M\tau} \tag{9.6.26}$$

where a_0^L and a_0^M are chosen to be identical to the steady state values for $\phi = \phi_1$, since every b_i is chosen to be a negative real number. The coefficients, a_i^L, a_i^M are determined by the least square method for fixed values of the b_is. The b_i are selected to be in the vicinity of the negative of the k values for which the imaginary parts of the aerodynamic transfer function have extrema. This procedure for selecting the b_i is discussed in detail in Dowell [13].

After determining the coefficients in (9.6.25) and (9.6.26), the indicial response functions can be written in the subsidiary domain of the Laplace operator (frequency domain) as,

$$\bar{A}_L(s, \phi_1) = \frac{a_0^L}{s} + \sum_{i=1}^{N} \frac{a_i^L}{s - b_i^L}, \tag{9.6.27}$$

$$\bar{A}_M(s, \phi_1) = \frac{a_0^M}{s} + \sum_{i=1}^{N} \frac{a_i^M}{s - b_i^M} \tag{9.6.28}$$

Then from (9.6.21), the aerodynamic describing function is obtained for the state variable ϕ as

$$\hat{H}_\phi(ik, \phi_1) = [\bar{D}_L(ik, \phi_1)\ D_M(ik, \phi_1)]^T \tag{9.6.29}$$

where

$$\bar{D}_L(ik, \phi_1) = ik\bar{A}_L(ik, \phi_1)/\phi_1, \tag{9.6.30}$$

$$\bar{D}_M(ik, \phi_1) = ik\bar{A}_M(ik, \phi_1)/\phi_1 \tag{9.6.31}$$

In order to construct the aerodynamic describing functions so that they are compatible with the structural transfer function, we must transform ϕ and C_M^N to those variables used in the structural equations of motion. The relationships for the state vectors and the moment coefficients are as follows:

$$\bar{\phi}(ik) = \begin{pmatrix} ik & 1 - \frac{a}{2} ik \end{pmatrix} \begin{pmatrix} \bar{h}/c \\ \bar{\alpha} \end{pmatrix}, \tag{9.6.32}$$

$$C_{Me}^N = C_M^N + \frac{a}{2} C_L^N \tag{9.6.33}$$

As the aerodynamic describing function $\hat{H}(ik, \phi_1)$ is defined by

$$\begin{pmatrix} -\bar{C}_L^N(ik, \phi_1) \\ -\bar{C}_{Me}^N(ik, \phi_1) \end{pmatrix} = [\hat{H}(ik, \phi_1] \begin{pmatrix} \bar{h}/c \\ \bar{\alpha} \end{pmatrix} \tag{9.6.34}$$

it becomes, using (9.6.30)–(9.6.34)

$$[\hat{H}(ik, \phi_1)] = \begin{bmatrix} A_{11} & A_{12} \\ A_{21} & A_{22} \end{bmatrix} \tag{9.6.35}$$

where

$$A_{11} = -\bar{D}_{L_1}(ik, \phi_1) \cdot ik$$

$$A_{12} = -\bar{D}_{L_1}(ik, \phi_1)\left(1 - \frac{a}{2} ik\right),$$

$$A_{21} = -\bar{D}_{L_1}(ik, \phi_1)\frac{a}{2} ik + \bar{D}_{M_1}(ik, \phi_1) \cdot ik,$$

$$A_{22} = \bar{D}_{L_1}(ik, \phi_1)\frac{a}{2}\left(1 - \frac{a}{2} ik\right) + \bar{D}_{M_1}(ik, \phi_1)\left(1 - \frac{a}{2} ik\right)$$

Using the structural transfer function (9.6.24) and the aerodynamic describing function (9.6.35), a self-sustained oscillation of the system shown in Figure 9.2 is characterized by the equation

$$|G^{-1}(ik, U) - \hat{H}(ik, \phi_1)| = 0 \tag{9.6.36}$$

Equation (9.6.36) corresponds to the so-called flutter determinant if the system is linear. For the present nonlinear system (9.6.36) allows one to determine the amplitude of the flutter motion as a function of some system parameter, say airspeed, U.

Extension of the describing function

In the earlier discussion of aerodynamic describing functions, we assumed the aerodynamic forces can be given as functions of a single variable, ϕ. More rigorously, however, the upwash of the mean camber line of an airfoil is given by

$$-\frac{w}{u} = \phi + \frac{c\dot{\alpha}}{u}(x/c - 0.5) \tag{9.6.37}$$

where the airfoil is located on $0 \leq x \leq c$. The second term in (9.6.37) includes the effect of the angular velocity of the airfoil motion, $\dot{\alpha}$. If we take into account this aerodynamic effect in the flutter analysis, the aerodynamic describing function must be determined separately for h- and α-motion, or alternatively for ϕ and α. Moreover, this procedure makes the describing function method less obvious as to its theoretical basis. Neglecting the effect of $\dot{\alpha}$, however, we encountered fictitious instabilities at high frequencies as well as a decrease of flutter speeds. To eliminate this artifact, an improvement was made to the aerodynamic describing function by adding the component that is derived from the second term of (9.6.37). This is based on the assumption that the $\dot{\alpha}$ effect compared to ϕ is generally secondary for the aerodynamic forces at low reduced frequencies, k, where nonlinear transonic aerodynamic effects are most significant. Thus, the components of the describing function, (9.6.35), are redefined by

$$A_{12} = A_{12} - \bar{D}_{L_2}(ik) \cdot ik$$

$$A_{22} = A_{22} + \bar{D}_{L_2}(ik)\frac{a}{2} \cdot ik + \bar{D}_{M_2}(ik) \cdot ik \tag{9.6.38}$$

where A_{12} and A_{22} on the right hand side of (9.6.38) are those of (9.6.35). $\bar{D}_{L_2}$ and $\bar{D}_{M_2}$ are the components obtained from an indicial response of the second term of (9.6.37). For brevity, details are omitted.

Results and discussion

Aerodynamic results

All the original aerodynamic data in the following were computed by the LTRAN2 [2] code for a NACA 64A006 airfoil with Mach number set to $M_\infty = 0.86$ where, characteristically, transonic effects can be observed. The zero angle of attack, steady state shock stands roughly at mid-chord

[14]. Results were also calculated for a NACA 64A010 airfoil, but these are omitted here.

In this calculation a 113×97 finite difference mesh was employed. The non-dimensional time increment, $\Delta\tau$, for integration was selected at $\pi/12$ to obtain indicial responses. The lift and moment forces at every five time increments were used to evaluate the coefficients in (9.6.25) and (9.6.26). In order to compare the aerodynamic forces with those obtained by the present extended nonlinear indicial method, the time integration for the airfoil undergoing harmonic excitation of pitching motion about mid-chord axis was also performed with 120 time steps per cycle at various reduced frequencies. The latter results [using (9.6.18)] were compared to those obtained by the indicial method using the Laplace transform versions of (9.6.25) and (9.6.26), which derive from (9.6.21). In general, good agreement was obtained.

The indicial response for lift to step functions of different amplitudes, ϕ_1, are shown in Figure 9.13. Generally, in this type of indicial response, a spike should appear near $\tau = 0$. However, the low frequency theory of LTRAN 2 can not follow a piston-wave-type pressure change because of its infinite propagation rate [3, 4]. Hence, in the present calculations the lift coefficient increases gradually from zero at $\tau = 0$.

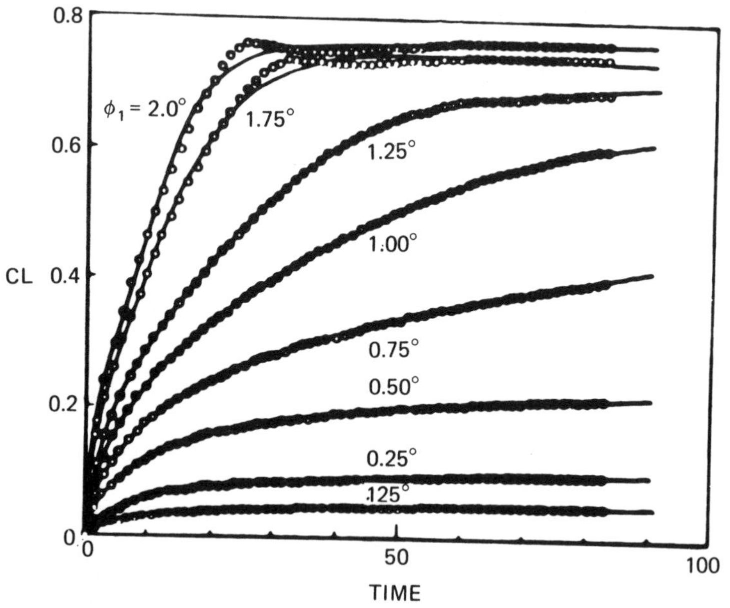

Figure 9.13 Indicial responses computed by LTRAN2 and exponential curve fit: lift.

In the results of Figure 9.13, the curves prescribed by (9.6.25) are shown after the coefficients were determined by the least squares method using 64 time data points for the indicial response at each amplitude. The b_i^Ls and b_i^Ms, are selected at six values ($N=6$), which are -0.01875, -0.0375, -0.075, -0.15, -0.3, and -0.6. These results obtained from (9.6.25) are in excellent agreement with the indicial responses computed by the LTRAN2 code, especially for the lower amplitude values of step

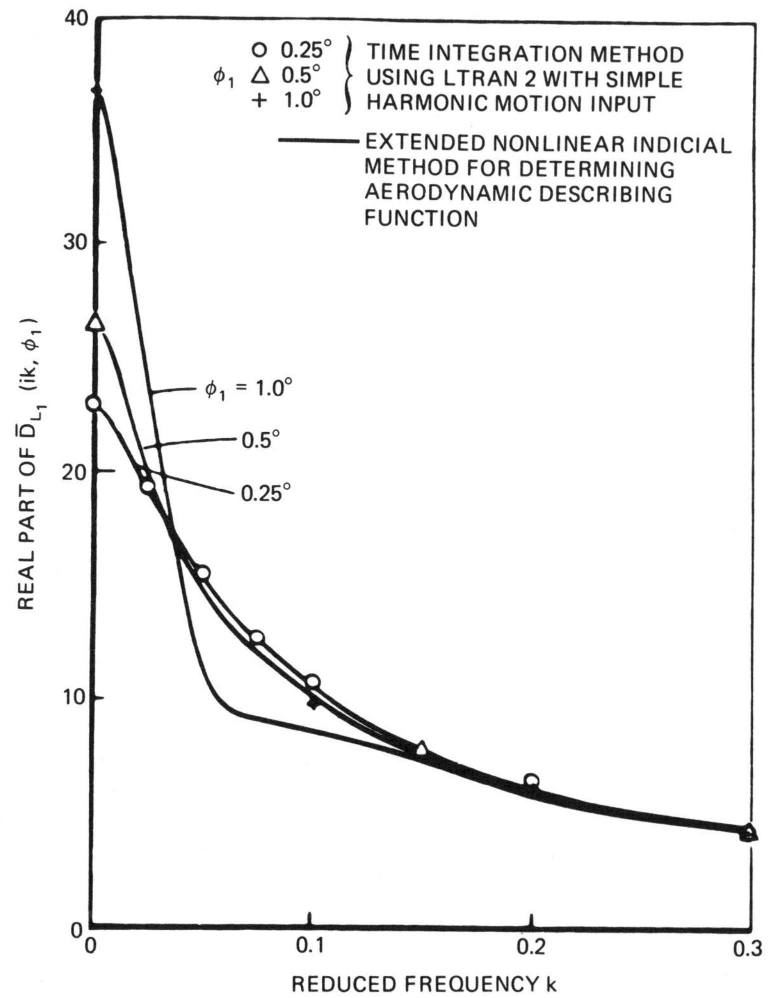

Figure 9.14 Comparison of extended nonlinear indicial method with time integration method (NACA 64A006, $M_\infty = 0.86$).

inputs. Using the coefficients a_i^L, a_i^M, and (9.6.21) we can obtain the elements of the aerodynamic functions, $\bar{D}_L\ (ik, \phi_1)$ and $\bar{D}_M\ (ik, \phi_1)$. The real parts of the former are shown in Figure 9.14. They are plotted for reduced frequencies up to 0.3. Although the describing function for higher frequencies can be calculated by (9.6.30) and (9.6.31), they are no longer meaningful at those frequencies because of the low frequency limitation [2] in LTRAN2.

In Figure 9.14, the describing functions thus obtained are also compared with the results of the time integration method for simple harmonic motion inputs that use (9.6.10)–(9.6.13) [see also (9.6.18)]. The agreement between the two methods is generally satisfactory. However, it was seen that the agreement is better for smaller amplitude than larger ones, for lift than for moment, and for real part than for imaginary part [53, 54]. It should be emphasized that the extended nonlinear indicial method has substantially greater simplicity and efficiency in determining the aerodynamic describing function, as compared to the time integration method for simple harmonic motion inputs.

Flutter results

Some flutter calculations have been done using these aerodynamic describing functions for typical section airfoils. First, the parameters of a typical section airfoil were chosen to compare with the results by Yang *et al.* [63]. A comparison is made in Ueda and Dowell [53]. The flutter boundary calculated by the present method for an amplitude, ϕ_1, between 0.5° and 1.0° agrees well with that obtained from the linear indicial method by Yang *et al.* [63].

To investigate the amplitude effect on the flutter boundary, a typical section airfoil corresponding to case B in Isogai [28] was considered next, although the results cannot be compared directly with those in Isogai [28] due to the use of a different airfoil profile. The results for the flutter speed as well as for the reduced frequency, bending/torsion amplitude ratio and phase, are shown in Fig. 15. Those without the $\dot{\alpha}$ effect are depicted by dashed curves. In this case, the effect of angular velocity on flutter boundaries was very small [53, 54].

As the aerodynamic describing function method invokes several assumptions, a fully nonlinear time marching solution was computed to verify the above results. The numerical integration scheme adopted for structural equations is the state transition matrix method which Edwards *et al.* [16] recommended after examining seven different integrators for a similar calculation (12 in [16]). Three time marching calculations with different speed parameters have been carried out for the case shown in Figure 9.15. The initial state vector was determined from the flutter solution of the describing function method with $\phi_1 = 0.5°$, namely, $U_F = 0.1798$, $k_F = 0.2126$, $|\bar{h}/c/\bar{\alpha}| = 5.446$, and $\phi_{h\alpha} = 4.24°$. If we choose the

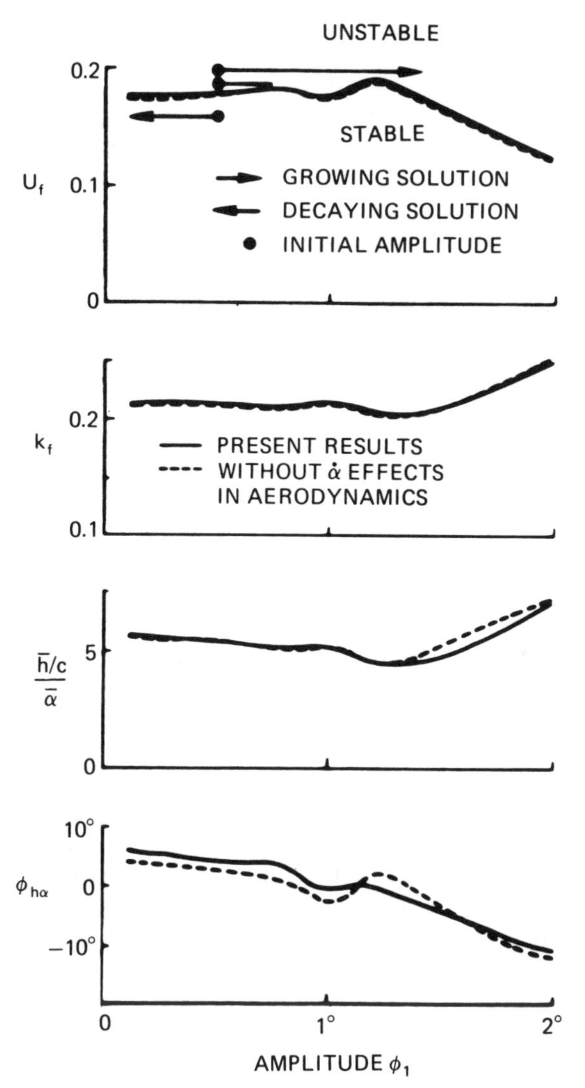

Figure 9.15 **Flutter parameters vs. amplitudes** (*NACA* 64A006, $M_\infty = 0.86$, $a = -0.3$, $x_{cg} = 0.24$, $r_{cg}^2 = 0.24$, $R = 0.2$, $\mu = 60$).

initial time, $\tau = 0$, as the instant when $\dot{\alpha} = 0$, the flutter solution gives the initial state vector for the time marching as $\mathbf{x} = (0.03164, 0.00583, -0.00049, 0)^T$. The time increment for integration was selected as $\Delta\tau = 0.25$, which, considering the flutter reduced frequency, corresponds roughly to 120 steps per cycle. Although the initial state vector is determined for the describing function flutter motion, the time marching is started from a steady-state initial condition of the airfoil at a static angle of attack for the aerodynamic calculations. For example, the initial effective induced angle of attack becomes 0.305 degrees for this case. It should be noted that the second term of (9.6.37) vanishes at the initial upwash since the starting time is set at the instant when the angular velocity becomes zero. The time marching is continued up to $\tau = 250$, which contains one thousand time steps. The variations of the amplitude ϕ_1 of these solutions are shown in Figure 9.15 and time histories in Figure 9.16. At $U = 0.16$, the airfoil shows decaying motion, whereas the oscillation is growing at $U = 0.2$. At $U = 0.19$, the oscillation is almost neutrally stable although it is slightly growing. The changes of the peak values in the effective angle of attack of these oscillations are illustrated in Figure 9.15. The solid line shows the flutter boundary (limit cycle curve) obtained by the describing function method. In this case, the flutter boundary is nearly horizontal at small amplitudes. As can be seen from the figure, the results from the describing function method agree well with those of the time marching solution. Furthermore, the last cycle of the time marching solution at $U = 0.19$ gives the values of $\phi_1 = 0.723°$, $k = 0.212$, $|\bar{h}/c/\bar{\alpha}| = 5.18$, and $\phi h_\alpha \leqslant 5°$. The agreement of these values with the results in Figure 9.15 is excellent. It should be noted that the damping in the time marching solutions is attributed to the aerodynamic forces, since we use no structural damping nor artificial damping due to numerical integration schemes. It is known that the transition matrix integrator gives exactly neutral solutions for free structural vibration irrespective of its time step size.

Since the nonlinear effect is most important at relatively low reduced frequencies (see Figure 9.14), the center of gravity was next placed at $x_{cg} = -0.25$ and the frequency ratio at $R = 0.1$ in order to obtain a distinctly nonlinear effect. The results are shown in Figure 9.17. On those portions of the curve where amplitude increases with airspeed, a stable nonlinear limit cycle is predicted. On those portions of the curve where amplitude increases with decreases in airspeed, an unstable limit cycle occurs.

Further time marching calculations have been performed to confirm the limit cycle. The initial state vector to start time integrations is varied

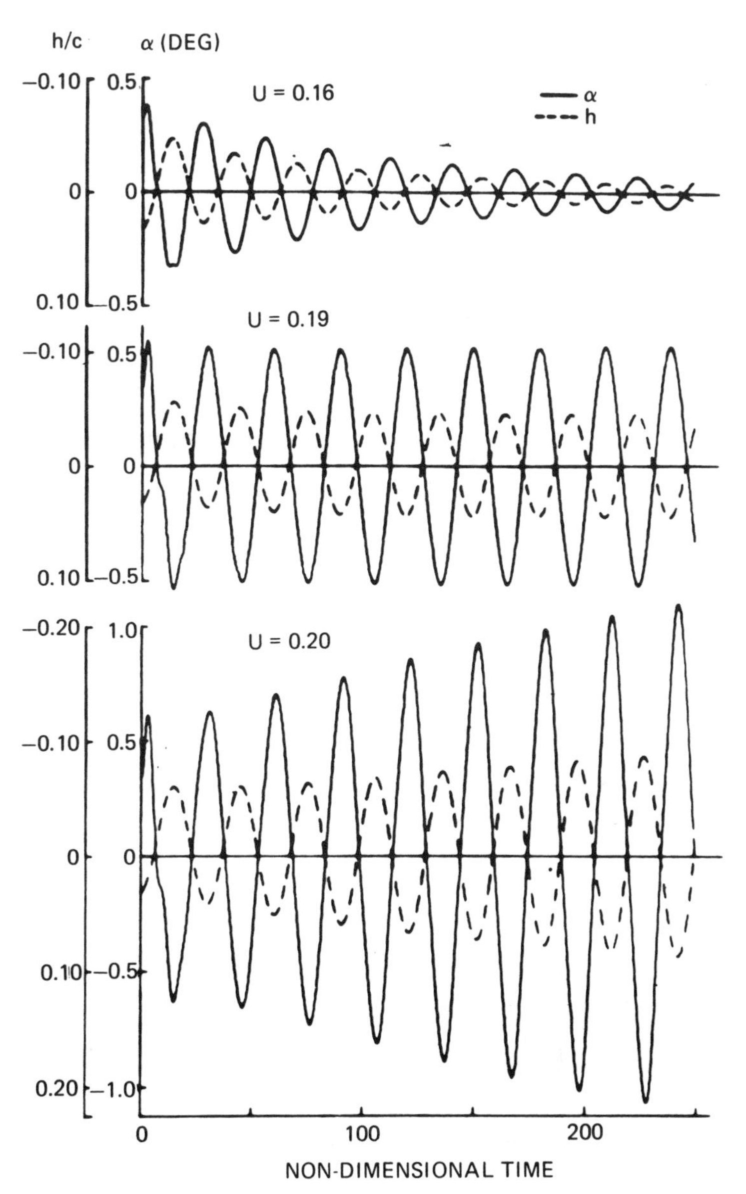

Figure 9.16 Time history of time marching solutions.

proportionally to that of the flutter solution of the describing function method with the amplitude, $\phi_1 = -0.25°$, which gives $\mathbf{x} = (0.04353, 0.00562, -0.005726, 0)^T$ for $\phi_1 = 0.25°$. This state vector yields the effective induced angle of attack, $-0.0063°$ at $\tau = 0$. Since the reduced frequencies of flutter are about 0.1 for small amplitudes, the time step size of integration was chosen as $\Delta\tau = 0.5$.

As the solid curve of the describing function in Figure 9.17 anti-

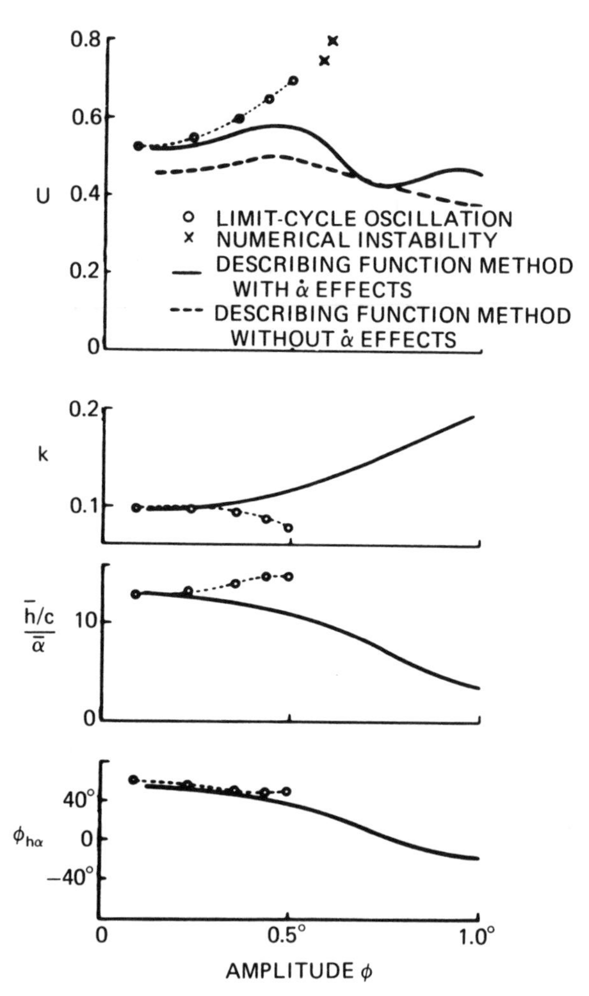

Figure 9.17 Limit cycle oscillations (*NACA* 64*A*006, $M_\infty = 0.86$, $a = -0.3$, $x_{cg} = 0.2$, $r_{cg}^2 = 0.24$, $R = 0.2$, $\mu = 60$).

cipates, the limit cycle flutter is also shown by the time marching solutions for small amplitudes. The amplitudes, reduced frequencies, amplitude ratio, and phase angles of these limit cycle oscillations were also calculated from the time history of the solutions and are depicted with open circles in the figure. Convergence to the limit cycle is determined by changing the initial amplitude. For example, the time history with two different initial amplitudes at $U = 0.6$ is shown in Figure 9.18. Both oscillations converge to the same limit cycle with the amplitude $\phi_1 = 0.355°$ beyond $\tau = 400$.

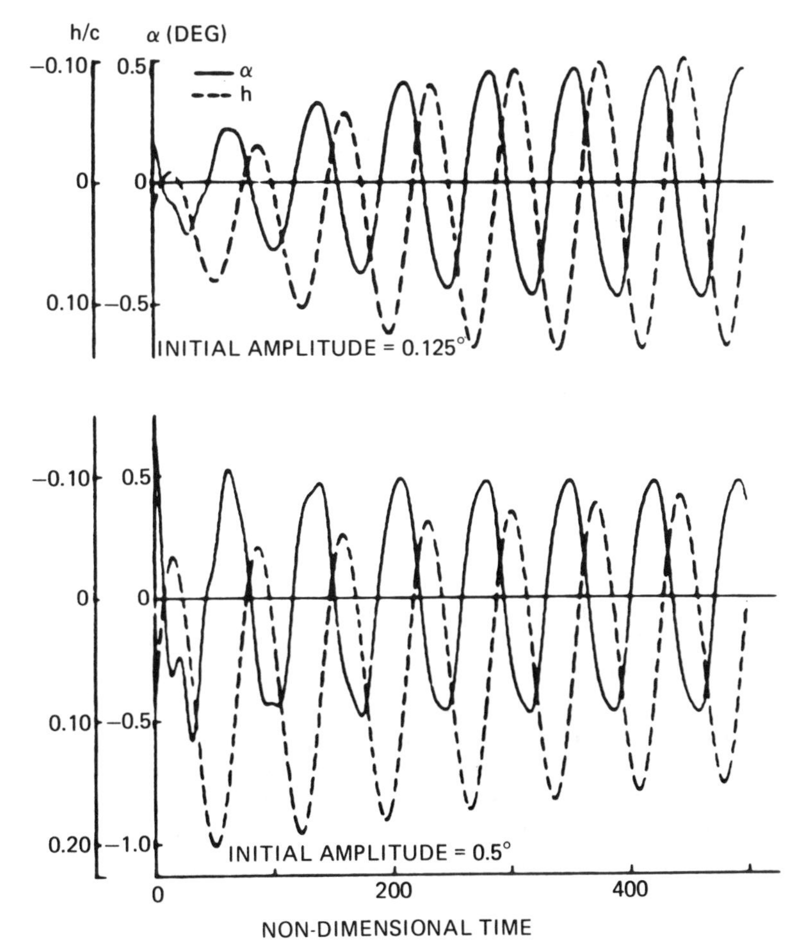

Figure 9.18. Time history of limit cycle flutter, ($U = 0.6$, $\phi_1 = 0.355°$, $k_F = 0.089$, $|\bar{h}/c/\bar{\alpha}| = 14.4$, $\phi h_\alpha = 49.7$).

However, the average displacement for the h-motion at least up to $\tau = 500$ is different for the two initial conditions. This slow convergence for the average displacement of the neutral position may be attributed to the small frequency ratio, R, which implies weak stiffness of the structure against the h-motion. It should be noted that the average translational displacement of an airfoil has no effect on the aerodynamic forces.

The results of the time marching solution can be compared with those obtained by the describing function method in Figure 9.17. For small amplitudes, the agreement of the results is satisfactory. The reduced frequency of the limit cycle by the time marching solution, however, decreases as the flutter amplitude increases, while the describing function method predicts monotonically increasing reduced frequencies. Furthermore, the time marching solution shows stable limit cycle flutter up to the speed of $U = 0.7$, where as the describing function method predicts no stable limit cycle solution above $U = 0.58$. At $U = 0.75$ and 0.8, the time marching solutions with the initial amplitude of $\phi_{\mathrm{I}} = 0.5°$ are terminated by a numerical instability of the aerodynamic calculations. The peak values of ϕ just before the occurrence of the instability are plotted with the symbol x. This kind of difficulty was frequently encountered when an initial amplitude of more than 0.5 degrees was used in the time marching calculations. Possibly for this reason we failed to detect the divergent unstable limit cycle flutter with larger amplitudes which is predicted by the describing function method.

Study by Peters and Ventura

The work of Peters and Ventura [42] has added very substantially to our understanding of the apparent differences in the results from the time-marching and describing function methods. They have used yet a different flutter solution method, the periodic shooting technique. Briefly it is this. In the periodic shooting method, one seeks a periodic solution by choosing initial conditions and guessing the period of the motion. A time marching solution is then carried out over the assumed period. If at the end of the assumed period, the system conditions are equal to the assumed initial conditions, then a periodic solution has been found. Usually, this will not be the case and an iteration procedure must be used to converge to a set of initial conditions and period that correspond to a steady state motion. This method has two advantages over the other approaches.

- Relative to a conventional time marching solution, which can only determine stable periodic motions, periodic shooting determines both stable *and* unstable periodic motions.

- Relative to a describing function approach, there is no assumption of simple harmonic motion, and the solutions are essentially exact.

The unexpected result from the periodic shooting solutions was that this method essentially reproduced the results previously obtained by the describing function method and differed in some respects from those results found by a conventional time marching solution. Consider again Figure 17.

In the plot of U vs. ϕ_1 the solid line is the result obtained by *both* the describing function and periodic shooting solutions. They both show that above a certain flow velocity, $U = 0.58$, no periodic motions are possible.

How then does one explain the apparent periodic motions found by the conventional time marching solutions for some $U > 0.58$? The answer is now known from the work of Peters and Ventura [42]. As the time marching solution is calculated from some initial conditions for, say, $U = 0.6$, the motion is oscillatory and appears to reach a periodic solution for $\phi_1 = 0.35$ after a hundred pseudo-cycles of oscillation or so. At this U and ϕ_1, one is near a true periodic motion which occurs for a slightly smaller U (see Figure 9.17) and the growth rate of the oscillation is small. However, if the calculation is continued over several hundred pseudo-cycles, ϕ_1 does increase and as it moves well beyond 0.35 the rate of growth becomes more rapid and it is clear that for $U = 0.6$, in fact, the motion becomes unbounded. This result is important for

- the caution it provides with respect to carrying out and interpreting conventional time marching solutions

and

- its possible implications for conducting and interpreting wind tunnel experiments or flight tests.

Conclusions

An extended nonlinear indicial approach to modeling nonlinear aerodynamic forces for aeroelastic analyses has been developed. The basic approach is based upon describing function ideas.

The flutter boundaries obtained by the describing function method are generally verified by periodic shooting and time marching solutions for sufficiently small amplitude flutter motion. Hence, the former, less costly method is useful for determining the significance of initial departures from linear behavior. More specific conclusions are listed below.

- Generally the accuracy of the describing function method decreases as the amplitude of the motion increases. The describing function method, however, is a powerful tool to predict the characteristics of transonic flutter, since it generally requires a very small amount of computational time for the aerodynamic forces compared to time marching solutions, particularly if a parameter study is to be undertaken.
- The stable nonlinear limit cycle flutter predicted by the describing function method, is also observed in the time marching solutions.
- The periodic shooting method has confirmed the results for the stable *and* unstable flutter limit cycle oscillations found by the describing function method. This work has also emphasized the care that must be taken when performing conventional time integration flutter solutions.

For an alternative suggestion for achieving the same goals, the reader is directed to Taylor *et al.* [49]. Also, the recent work of Bland and Edwards [7] should be cited. They deal with the important (static nonlinear aerodynamic) effects of airfoil shape and thickness (though not nonlinear dynamic effects per se) in a manner similar to that of the present paper. Also see the work of Batina [5] and Williams [56]. The latter is in the context of panel flutter rather than airfoil flutter. In general the effects of airfoil shape and thickness (static nonlinearities) are more important for flutter than nonlinear dynamics effects *per se* [7, 14, 15]. Hence, dynamically linear aeroelastic analyses will continue to play a dominant role even in the transonic flow regime. However see the recent interesting work by Kousen and Benediksen [64].

9.7 Concluding remarks

Some present answers

The chapter began with four questions. The answers to these will undoubtedly be refined in future years. However, based on present knowledge, partial answers may be formulated as follows:

(1) For sufficiently small airfoil motions (leading to sufficiently small shock motions, $\approx 5\%$ of airfoil chord or less), the aerodynamic forces will be linear functions of the airfoil motion [2, 14, 15, 19, 33, 46, 50, 51].

(2) At least one viable alternative solution technique to finite difference methods is available, the field panel method of Hounjet [25].

The physical phenomena associated with unsteady transonic flow

An excellent article with this title has been published by Dennis Mabey [66]. A brief summary of his discussion is offered here along with a strong recommendation that the reader who finds this topic of interest consult the original article. For the reader who masters the material of this chapter, reading Mabey's article will richly repay one's investment.

Transonic flow is of surpassing importance to the study of aeroelasticity for at least two reasons. The transonic flow regime is often the most critical for design to prevent flutter and, secondly, this flow regime has most of the complications one encounters in other flow regimes plus others as well. Shock waves, separated flow, static and dynamic flow nonlinearities, all of these may be present.

- When does transonic flow occur? For streamlined aerodynamic shapes, at small angles of attack, it occurs when the free stream Mach number is near one. However, as Mabey points out, for bluff bodies (including streamlined bodies at high angles of attack) the local flow about the body may exhibit significant transonic effects even when the free stream flow is subsonic on the one hand or distinctly supersonic, even hypersonic, on the other.

Even so, it is the transonic flow around streamlined bodies at small angles of attack that is our primary interest here. Consider the flow over a relatively small range of Mach numbers, typically .8 – 1.0, where the shock first forms at the lowest of these Mach numbers and then strengthens and moves toward the trailing edge as the Mach number increases. Because of viscous effects, the flow may separate behind the shock when the latter reaches sufficient strength.

The shock reduces substantially the range of angle of attack over which the relationship between aerodynamic forces and moments change in a linear fashion with changes in angle of attack. Recall Fig. 9.5 for a boundary map dividing linear from nonlinear behavior in terms of Mach number and angle of attack. Note that if the airfoil oscillates, increasing the frequency of oscillation expands the domain of (dynamically) linear behavior.

For thicker airfoils, an even more striking nonlinear effect may occur. There is a (narrow) range of (transonic) Mach number where the fluid flow may oscillate (e.g. the shocks oscillate back and forth) even when the airfoil is perfectly stationary. This is an example of a dynamically unstable flow exhibiting a limit cycle oscillation. Some numerical simulations of the Navier-Stokes equation of a viscous fluid including the effects of shock wave/boundary layer separation appear to describe these results at least qualitatively.

Such oscillations may give rise to what has been called buffeting when the oscillations of the flow provide a forced excitation of the wing structure. Of course, under some conditions, the structural motion may lead to a significant change in the flow oscillations and an aeroelastic feedback mechanism will then be established.

The type of buffeting is strongly dependent on the wing aspect ratio and three-dimensional nature of the flow. A classification for the difficulty of these flows has been developed by Mabey. "Difficulty" refers to difficulty in describing, understanding and computing the flow field. Vortices emanating from the wing leading edges further complicate matters. However, low aspect ratio wings whose flow may be dominated by these leading edge vortices, are actually the easiest to model theoretically. Further details of the flow complexity for a swept wing in transonic flow are given in Ref. 66.

Aileron (or more generally, control surface) buzz is a topic that has vexed and perplexed aeroelasticians for some years. As distinct from buffet, aeroelastic feedback between the flow and aileron (control surface) oscillation appears essential to our understanding of this phenomenon, i.e. aileron buzz is a particular type of flutter.

A strictly inviscid flow model predicts that there is a Mach number above which this type of flutter may occur. Typically at this Mach number, there is a shock wave on the airfoil not too far upstream of the aileron hinge line. As the aileron oscillates, the shock also oscillates, and for an inviscid flow these oscillations grow exponentially and are unbounded. Recall the previous discussion of the work of Steger and Bailey [47]. If viscosity is included in the theoretical model (as it is surely present in the physical model), the flutter oscillations are bounded, i.e. a limit cycle exists. Interestingly, however, the viscous effects are not entirely favorable. It turns out that at lower Mach numbers than those predicted by an inviscid theory, flutter may still occur if the aileron is given a sufficiently large disturbance, but not if the disturbance is small. Thus the limit cycle exhibits a form of hysteresis with Mach number when viscous effects are included. At higher Mach numbers this type of flutter is absent altogether.

Further refinements to that approach are expected, possibly including the work of Cockey [11] as discussed here-in and also including the extension to three-dimensional flow fields. The relative merits of finite difference vs. finite element vs. field panel method remains a subject for future study and, no doubt, vigorous debate.

(3) The nonuniqueness that has been observed under some conditions in transonic small disturbance and full potential flow solutions is inherent in the governing field equations themselves and not a numerical artifact of some solution method [15, 22, 45, 48]. To date, corresponding nonuniqueness in solutions to the Euler equations has not been observed.

However, it is not clear at this time whether or not such nonuniqueness in the potential flow equation solutions may have its less (or equally) dramatic counterpart in the solutions to the Euler equations under some conditions. This remains an important topic for future study. Moreover, the physical significance, if any, of such solutions remains to be clarified.

(4) While the option of simultaneous time integration of the fluid dynamical and structural dynamical equations of motion to determine aeroelastic response will be attractive for some applications, flutter analysis in the frequency domain will continue to be an important, and at times more attractive, option as well. Methods for generating the frequency domain aerodynamic forces are now available from

- aerodynamic methods that presuppose infinitesimal, harmonic dynamic airfoil motions [17, 25, 58, 59],
- impulse-transfer function ideas [3, 4, 40, 41, 53, 54] which allow the generation of frequency domain data from a single time history record determined by a time marching aerodynamic code. This approach can be extended (approximately) to large airfoil motions where a nonlinear relationship exists between airfoil motions and aerodynamic forces by using ideas based on the describing function method [53, 54].

(5) Simple order of magnitude estimates of the relative computational times for aeroelastic analyses can be made and these have been discussed in the text. Dynamically linear flutter analysis will continue to play a dominant role even in the transonic flow regime [7, 16, 53, 54, 63].

Future work

A long list of worthy research topics could be given. Here we focus on a few that have as their common theme improved physical modeling and understanding of the fluid dynamic and aeroelastic phenomena of interest.

(1) There is a clear need to understand better the apparent qualitative difference between the solutions of the potential flow equations and those of the Euler equations under those parameter conditions where the former exhibit nonunique solutions. Until this difference is both understood and resolved, it throws into question the whole approach of combining potential flow solution with boundary layer corrections to correct for viscosity under those conditions where nonunique solutions are observed.

(2) A complementary issue is how can more effective (efficient) solution methods be devised for the Euler equations for unsteady flows. Subsequent to the pioneering work of [32], little has been done.

One possible approach which has several prospective advantages is to assume that the flow may be treated as a nonlinear mean steady flow plus a small (infinitesimal) linear dynamic perturbation. This technique has been exploited effectively by [17, 19, 25, 58, 59] and others for the potential flow model. In one of its limiting forms this approach has been used by Lighthill [30], Williams *et al.* [60], and others for modeling boundary layer effects in panel flutter (non-lifting) and lifting surface aerodynamics using the Euler equations.

Such an approach will lead to a set of time dependent, linear, partial differential equations with variable (spatially dependent) coefficients. These will depend in turn on the solution of the time-independent, nonlinear, partial differential, Euler equations for the mean steady flow. The expected advantages of this approach include the following

(3) Existing and future steady flow codes may be exploited to maximum advantage to provide input data for the dynamic perturbation equations and their solution. (Also the finite difference grid developed for the mean steady flow may be retained for the dynamic perturbation flow if a finite difference solution method is used for the latter.)

(4) Solution methods for linear equations with variable coefficients may be brought to bear, although these will require extensions and generalizations. See the work of Hounjet (1981b) and Williams et al. (1977).

(5) Solving the linear, dynamic equations will

- in all likelihood meet the requirements of the aeroelastician (within the context of the Euler equations).
- permit inclusion of viscous effects in the mean steady flow modeling and thus indirectly and partially (but not directly and completely) in the dynamic perturbation equations [30, 60].
- allow one to examine the dynamic stability of the mean, steady flow itself and thus contribute to a better understanding of the prospect for nonunique solutions of the Euler equations.

Hall and Crawley have shown recently some of the advantages of such an approach for internal flows in turbomachinery [65].

References for Chapter 9

[1] Albano, E. and Rodden, W. P., 'A Doublet Lattice Method for Calculating Lift Distribution on Oscillating Wings in Subsonic Flows', *AIAA Journal*, Vol. 7 (1969), pp. 279–85.

[2] Ballhaus, W. F. and Goorjian, P. M., 'Implicit Finite Difference Computations of Unsteady Transonic Flows About Airfoils', *AIAA Journal*, Vol. 15 (1977), pp. 1728–35.

[3] Ballhaus, W. F. and Goorjian, P. M., 'Efficient Solution of Unsteady Transonic Flows About Airfoils', *Paper 14, AGARD Conference Proceedings No. 226, Unsteady Airload in Separated and Transonic Flows*, 1978a.

[4] Ballhaus, W. F. and Goorjian, P. M., 'Computation of Unsteady Transonic Flows by the Indicial Method', *AIAA Journal*, Vol. 16, No. 2 (1978b), pp. 117–24.

[5] Batina, J. T., 'Effects of Airfoil Shape Thickness, Camber, and Angle of Attack on Calculated Transonic Unsteady Airloads', NASA TM 86320 (1985).

[6] Bauer, R., Garabedian, P. and Korn, D., 'Supercritical Wing Sections', *Lecture Notes in Economics and Mathematical Systems*, Vol. 66, Springer-Verlag, 1972.

[7] Bland, S. R. and Edwards, J. W., 'Airfoil Shape and Thickness Effects on Transonic Airloads and Flutter', AIAA SDM Conference in Lake Tahoe, CA, May 1983, *AIAA Paper No. 83-0959.*

[8] Borland, C. J. and Rizzetta, D. P., 'Transonic Unsteady Aerodynamics for Aeroelastic Applications, I: Technical Development Summary', *AFWAL TR 80-3107, I*, June 1982a.

[9] Borland, C. J. and Rizzetta, D. P., 'Nonlinear Transonic Flutter Analysis', *AIAA Journal*, Vol. 20, No. 11 (1982b), pp. 1606–15.

[10] Borland, C. J., Rizzetta, D. P. and Yoshihara, H., 'Numerical Solution of Three-Dimensional Unsteady Transonic Flow Over Swept Wings', *AIAA Journal*, Vol. 20, No. 3 (1982), pp. 340–37.

[11] Cockey, W. D., 'Panel Method for Perturbations of Transonic Flows with Finite Shocks', Ph.D. thesis, Princeton University, June 1983.

[12] Davis, S. S. and Malcolm, G., 'Experiment in Unsteady Transonic Flows', *Proceedings of the AIAA/ASME/ASCE 20th Structures, Structural Dynamics and Materials Conference*, St. Louis, MO, April 1979.

[13] Dowell, E. H., 'A Simple Method for Converting Frequency-Domain Aerodynamics to the Time Domain', *NASA TM 81844*, 1980.

[14] Dowell, E. H., Bland, S. R. and Williams, M. H., 'Linear/Nonlinear Behavior in Unsteady Transonic Aerodynamics', *AIAA Journal*, Vol. 21 (1983), pp. 38–46.

[15] Dowell, E. H., Ueda, T. and Goorjian, P. M., 'Transient Decay Times and Mean Values of Unsteady Oscillations in Transonic Flow', *AIAA Journal*, Vol. 21, No. 12, (1983), pp. 1762–64.

[16] Edwards, J. W., *et al.*, 'Time-Marching Transonic Flutter Solutions Including Angle-of-Attack Effects', *AIAA Paper 82-0685*, presented at the 23rd SDM Conference, New Orleans, LA, May 1982.

[17] Ehlers, F. E. and Weatherill, W. H., 'A Harmonic Analysis Method for Unsteady Transonic Flow and its Application to the Flutter of Airfoils', *NASA CR-3537*, 1982.

[18] Erickson, A. L. and Stephenson, J. D., 'A Suggested Method of Analyzing Transonic Flutter of Control Surfaces Based on Available Experimental Evidence', NACA RM A7F30 (1947).

[19] Fung, K. Y., Yu, N. J. and Seebass, R., 'Small Unsteady Perturbations in Transonic Flows', *AIAA Journal*, Vol. 16 (1978), pp. 815–22.

[20] Gibbons, M., Whitlow, W., Jr. and Williams, M. H., 'Noninsentropic Unsteady Three Dimensional Small Disturbance Potential Theory', AIAA Paper 86-0863, May 1986.

[21] Goldstein, M. E., *Aeroacoustics*, New York: McGraw-Hill, 1976.

[22] Goorjian, P. M., Personal communication, NASA Ames Research Center.

[23] Hessenius, K. A. and Goorjian, P. M., 'Validation of LTRAN2-HI by Comparison with Unsteady Transonic Experiment', *AIAA Journal*, Vol. 20, No. 5 (1982), pp. 731–32.

[24] Hounjet, M. H. L., 'A Transonic Panel Method to Determine Loads on Oscillating Airfoils with Shocks', *AIAA Journal*, Vol. 19 (1981a), pp. 559–66.

[25] Hounjet, M. H. L., 'A Field Panel Method for the Calculation of Inviscid Transonic Flow About Thin Oscillating Airfoils with Shocks', *NLR MP 81043 U*, National Aerospace Laboratory, Netherlands. Presented at the International Symposium on Aeroelasticity, Nuremburg, October 5–7, 1981b, Germany.

[26] Houwink, R. and van der Vooren, J., 'Improved Version of LTRAN2 for Unsteady Transonic Flow Computations', *AIAA Journal*, Vol. 18, No. 8 (1980), pp. 1008–10.

[27] Hsu, J. C. and Meyer, A. U., *Modern Control Principles and Applications*, New York: McGraw-Hill, 1968.

[28] Isogai, K., 'Numerical Study of Transonic Flutter of a Two-Dimensional Airfoil', *Technical Report of National Aerospace Laboratory*, Japan, NAL-TR-617T, 1980.

[29] Kerlick, G. D. and Nixon, D., 'Mean Values of Unsteady Oscillations in Transonic Flow Calculations', *AIAA Journal*, Vol. 19, No. 11 (1981), pp. 1496–98.

[30] Lighthill, M. J., 'On Boundary Layers and Upstream Influence, II; Supersonic Flows Without Separation', *Proceedings of the Royal Society*, *A217* (1953), 478–507.

[31] Liu, D. D., 'A Lifting Surface Theory Based on an Unsteady Linearized Transonic Flow Model', *AIAA Paper 78-501*, 1978.

[32] Magnus, R. J. and Yoshihara, H., 'Calculation of Transonic Flow Over an Oscillating Airfoil', *AIAA Paper 75-98*, Jan. 1975.

[33] McCroskey, W. J., 'Unsteady Airfoils', *Annual Review of Fluid Mechanics*, Vol. 14 (1982), pp. 285–311.

[34] McIntosh, S., Personal communication.

[35] Miles, J. W., *The Potential Theory of Unsteady Supersonic Flow*, Cambridge: Cambridge University Press, 1959, pp. 4–13.

[36] Morino, L., 'A General Theory of Unsteady Compressible Potential Aerodynamics', NASA CR-2464, December 1974.

[37] Morino, L. and Tseng, K. 'Time-Domain Green's Function Method for Three-Dimensional Nonlinear Subsonic Flows', AIAA Paper 78-1204, 1978.

[38] Murman, E. M., 'Analysis of Embedded Shock Waves Calculated by Relaxation Methods', *Proceedings of the AIAA CFD Conference*, July 1973, pp. 27–40.

[39] Nixon, D., 'Calculation of Unsteady Transonic Flows Using the Integral Equation Method', *AIAA Paper 78-13*, January 1978.

[40] Nixon, D., 'On the Derivation of Universal Indicial Functions', *AIAA Paper 81-0328*, January 1981.

[41] Nixon, D. and Kerlick, G. D., 'Calculation of Unsteady Transonic Pressure Distributions by Indicial Methods', *Nielsen Engineering and Research Paper 117*, 1980.

[42] Peters, D. A. and Ventura, L., 'Applications of Various Solutions Techniques to the Calculation of Transonic Flutter Boundaries', in *Fluid-Structure Interaction and Aerodynamic Damping*, edited by E. H. Dowell and M. K. Au-Yang, ASME, (1985), pp. 29–49.

[43] Rizzetta, D. P. and Borland, C. J., 'Unsteady Transonic Flow Over Wings Including Inviscid/Viscous Interactions', *AIAA Journal*, Vol. 21, No. 3 (1983), pp. 363–71.

[44] Salas, M. D., Jameson, A. and Melnik, R. E., 'A Comparative Study of the Nonuniqueness Problem of the Potential Equation', presented at the 6th AIAA CFD Conference, Danvers, Mass., July 13–15, 1983.

[45] Salas, M. D., and Gumbert, C. R., 'Breakdown of the Conservative Potential Equation', *AIAA Paper 85-0367*, Jan. 1985.

[46] Seebass, R., 'Advances in the Understanding and Computation of Unsteady Transonic Flows', *Recent Advances in Aerodynamics*, edited by A. Krothapalli and C. Smith, 1984.

[47] Steger, J. L. and Bailey, H. E., 'Calculations of Transonic Buzz', *AIAA Journal*, Vol. 18, (1980), pp. 249–255.
[48] Steinhoff, J. and Jameson, A., 'Multiple Solutions of the Transonic Potential Flow Equations', *AIAA Paper No. 81-1019*, AIAA Computational Fluid Dynamics Conference, Palo Alto, CA, June 1981.
[49] Taylor, R. F., Bogner, F. K., and Stanley, E. C., 'A Stability Prediction Method for Nonlinear Aeroelasticity', *AIAA Paper 80-0797*, presented at the AIAA/ASME/AHS/ASCE Conference, Seattle, WA, May 12–14, 1980.
[50] Tijdeman, H., 'Investigation of the Transonic Flow Around Oscillating Airfoils', Ph.D. thesis, Delft University, 1977.
[51] Tijdeman, H. and Seebass, R., 'Transonic Flow Past Oscillating Airfoils', *Annual Review of Fluid Mechanics*, Vol. 12 (1980), pp. 181–222.
[52] Tran, C. T. and Petot, M., 'Semi-empirical Model for the Dynamic Stall of Airfoils in View of the Application of Responses of a Helicopter Blade in Forward Flight', *Paper 48*, *Proceedings*, 6th European Rotorcraft and Powered-lift Aircraft Forum, Bristol, England, 1980.
[53] Ueda, T. and Dowell, E. H., 'Flutter Analysis Using Nonlinear Aerodynamic Forces', *AIAA Paper 82-0728*, 1982. Also see *J. Aircraft*, Vol. 21 (1984), pp. 101–109.
[54] Ueda, T. and Dowell, E. H., 'Describing Function Flutter Analysis for Transonic Flow: Extension and Comparison with Marching Analysis', *AIAA Paper 83-0958*, 1983.
[55] Voss, R., 'Time-Linearized Calculation of Two-Dimensional Unsteady Transonic Flow at Small Disturbances', DFVLR FB-81-01, 1981.
[56] Williams, M. H., 'The Effect of a Normal Shock on the Aeroelastic Stability of a Panel', *J. Applied Mechanics*, Vol. 50, (1983), pp. 275–282.
[57] Williams, M. H., Bland, S. R., and Edwards, J. W., 'Flow Instabilities in Transonic Small-Disturbance Theory', *AIAA Journal*, Vol. 23, (1985), 1491–1496.
[58] Williams, M. H., 'Unsteady Thin Airfoil Theory for Transonic Flows with Embedded Shocks', Department of Mechanical and Aerospace Engineering, *Report No. 1376*, Princeton University, May 1978; also *AIAA Journal*, Vol. 18 (1980), pp. 615–24; also see 'Unsteady Airloads in Supercritical Transonic Flows', *Proceedings of the AIAA/ASME/ASCE 20th Structures, Structural Dynamics and Materials Conference*, St. Louis, MO, April 1979b.
[59] Williams, M. H., 'The Linearization of Transonic Flows Containing Shocks', *AIAA Journal*, Vol. 17 (1979a), pp. 394–97.
[60] Williams, M. H., *et al.*, 'Aerodynamic Effects of Inviscid Parallel Shear Flows', *AIAA Journal*, Vol. 15, No. 8 (1977), pp. 1159–66.
[61] Yang, T. Y., Guruswamy, P. and Striz, A. G., 'Aeroelastic Response Analysis of Two-Dimensional, Single and Two Degree of Freedom Airfoils in Low Frequency, Small-Disturbance Unsteady Transonic Flow', *AFFDL-TR-79-3077*, 1979.
[62] Yang, T. Y., Guruswamy, P., and Striz, A. G., 'Application of Transonic Codes to Flutter Analysis of Conventional and Supercritical Airfoils', AIAA SDM Conference in Atlanta, GA, *AIAA Paper 81-0609*, 1981.
[63] Yang, T. Y., Guruswamy, P., Striz, A. G., and Olsen, J. J., 'Flutter Analysis of a NACA 64A006 Airfoil in Small Disturbance Transonic Flow', *J. Aircraft*, Vol. 17, No. 4 (1980), pp. 225–32.
[64] Kousen, K. A. and Bendiksen, O., 'Limit Cycle Phenomena in Computational Transonic Aeroelasticity', AIAA SDM Conference in Mobile, ALA, AIAA Paper 89-1185, 1989.

[65] Hall, K. C. and Crawley, E. F., 'Calculation of Unsteady Flows in Turbomachinery, Using the Linearized Euler Equations', Presented at the Fourth Symposium on Unsteady Aerodynamics and Aeroelasticity of Turbomachines and Propellers, Aachen, West Germany, September 6-10, 1987.

[66]. Mabey, D., "The Physical Phenomena Associated with Unsteady Transonic Flow," in "Unsteady Transonic Flow," Edited by David Nixon, American Institute of Aeronautics and Astronautics, Washington, D.C., 1989, Chapter 1, pp. 1-55.

10

Experimental aeroelasticity

Much of this text has been devoted to mathematical modeling of physical phenomena in the field of aeroelasticity. Yet one of the most challenging and important aspects of the subject is the conduct of physical experiments. Experiments are useful for many purposes, e.g. to assess the accuracy and validity of theoretical models, to study phenomena beyond the current reach of theory, and/or to verify the safety and integrity of aeroelastic systems through wind tunnel tests or flight tests. A thorough exposition of this topic would require a volume in itself. Here a few of the fundamental aspects of experimental aeroelasticity are discussed. The focus is on aeroelastic tests per se rather than structural dynamic tests or unsteady aerodynamic measurements. However the latter will be touched on as well insofar as they are relevant to our principal topic.

For another authoritative treatment of this subject the discussion by Ricketts [1] is highly recommended.

Before an aeroelastic experiment is conducted, it is usual to make measurements of the natural modes and frequencies of the structural model. Hence our discussion begins there.

10.1 Review of structural dynamics experiments

In the jargon of the practitioners, these are referred to as ground vibration tests or GVT. The basic requirements are a means for exciting (forcing) the structure into its resonant, natural modes and also a means for measuring the response of the structure. For excitation systems a variety of means have been used including mechanical forces, electromagnetic forces and acoustical excitation. The choice depends primarily upon the level of force excitation required and the range of frequency to be covered. For lightly damped systems excited in a resonant mode, the force level needed can be estimated as the mass of the structure times the frequency squared times (twice) the critical damping ratio times the amplitude of response required, i.e. the inertial and stiffness terms nearly cancel and the exciting force is balanced by structural dissipation or damping. The response amplitude required is typically determined by the sensitivity of available response measurement instrumentation or perhaps the need for the response to be in a certain linear (or less usually, nonlinear) range of response.

In practical terms mechanical excitation systems are used for low frequencies (say 1-100 Hz), electromagnetic exciters (shakers) are used for moderate frequencies (say 10-1000 Hz) and acoustic excitation at high frequencies (say 100-10000 Hz).

The response measurement systems may be either mechanical (strain gauges or accelerometers), electromagnetic (some electromagnetic devices may be used as either exciters or response measurement devices) or, more recently piezoelectric devices* that may be used to either serve as exciter or responder.

The basic measurement technique is to excite the system at its resonant frequencies (usually having theoretical calculations as a guide) with the excitation and response devices placed at locations on the structure expected to have large response. Multiple exciters are used to distinguish between symmetric and anti-symmetric natural modes or to excite modes with complex shapes. In principle, a continuous distribution of excitation with a distribution of force amplitude proportional to the expected (mass weighted) natural mode (and therefore orthogonal to all other natural modes) is optimum. Rarely can so many exciters be used in practice to approach this ideal.

If a pure frequency excitation is used, then a transient decay time history or a half-power frequency response plot may be used to estimate modal natural frequency and damping (e.g. see Thomson** or any standard text on vibration theory).

Also a random excitation over the range of relevant frequencies may be used to identify multiple modes with one excitation. This is used only when test time is limited. Another possibility is pulse excitation in the time domain and the use of Fast Fourier Transform theory to extract multiple nodes information. Commercial hardware and software is now widely available to perform the latter measurement***. However for precise work the old fashioned methods may still be preferred.

Of course, if any significant nonlinearities are present, nonlinear theory must be used to guide the form of excitation, measurement and data interpretation. The range of possibilities is too extensive to be easily summarized. But the presence of higher harmonics in the response measurement is often a key observation that suggests nonlinearities are present and important.

10.2 Wind tunnel experiments

With the presence of flow the environment for excitation and measurement of response is more complex, but the basic devices for creating, measuring and interpreting the responses remain the same as for the simpler structural dynamics experiments. Sometimes the aerodynamic turbulence in the tunnel is used to provide a random excitation and, of course, for self-excited instabilities (e.g. flutter) no special requirements are (in principle) necessary to excite the aeroelastic system. Nevertheless, where practicable, it is desirable to have an

* See Crawley [2].
** See Thomson [3].
*** See, e.g., ZONIX, HEWLETT PACKARD and other manufacturers' catalogs and equipment manuals.

excitation system available of the conventional sort, e.g. mechanical, electromagnetic, acoustic, or perhaps, piezoelectric. Such excitations allow one to conduct sub-critical (i.e. below the flutter boundary or stable) experiments. One of the principal challenges in flutter testing is to be able to extrapolate to flutter (critical) conditions from sub-critical measurements.

Sub-critical futter testing

By monitoring the change of modal damping with change in flow dynamic pressure, for example, one may try to anticipate the value of dynamic pressure for which the modal damping will become zero and then negative. Because of the sometimes complicated and rapid variation of damping with dynamic pressure and the necessity to monitor several potentially critical modes, it is often difficult to extrapolate to this flutter condition. Indeed extrapolation techniques for this purpose remain an active area of research*.

For certain types of flutter, monitoring the changes in modal frequencies may also be a useful guide to help predict the onset of flutter.

Approaching the flutter boundary

For low speed (incompressible flow) flutter tests, the flutter boundary is normally approached by increasing the flow velocity in suitable increments. For high speed (compressible flow) flutter tests, the Mach number is normally fixed, and the flutter boundary is approached by increasing the wind tunnel stagnation pressure, and hence dynamic pressure, in suitable increments. Then the Mach number is changed and the process repeated. At very high Mach numbers, a blow-down (transient flow) wind tunnel may be the only flow facility available. However, a continuous flow, closed return tunnel is to be preferred when available in order to assure well defined flow conditions and give adequate time for accurate response measurements.

Safety devices

Normally some provision will be made for suppression of the flutter response by a rapidly applied restraint to the flutter model, in order to protect the model from damage due to flutter.

* See Matsuzaki [4].

Research tests vs. clearance tests

Research tests are normally conducted to provide experimental data for comparison with theory and hence usually rather detailed data sets are obtained over a wide range of flow and/or structural parameters. Clearance tests are designed simply to show a particular flutter model is unlikely to encounter an instability over a range of anticipated operational conditions.

*Scaling laws**

By expressing the aeroelastic equations of motion in non-dimensional form or by simply using dimensional analyses, it is possible to relate the behavior of the small scale models typically tested in wind tunnels to that of full-scale aircraft in flight. Often not all relevant non-dimensional parameters can be matched between tunnel scale and flight scale due to the limitations of modal fabrication and wind tunnel flow conditions. Selecting an appropriate set of scaling parameters is a matter of intelligent application of theory (i.e. matching those non-dimensional parameters that are most important and sensitive as predicted by analysis) and judgement based upon experience. Normally modal frequency ratios, reduced frequency, Mach number and a non-dimensional ratio of dynamic pressure to model stiffness are matched. Frequently fluid/structural mass ratio is not.

Wind tunnel tests are extraordinarily valuable and often fill in gaps in our knowledge where theory is unavailable or unreliable.

10.3 Flight experiments

Virtually all previous comments for wind tunnel tests apply to flight tests as well. However the need for safety is now paramount and the challenges of providing a well defined excitation force are considerably higher. Also the test procedure is necessarily different.

Approaching the flutter boundary

Normally the flutter boundary has been estimated from a suitable combination of analysis and wind tunnel experiment prior to the flight test and presented in terms of altitude (corresponding to a certain static or dynamic pressure) vs Mach number. Usually the Mach number at which flutter will occur increases with increasing altitude. Hence flutter testing normally begins at high altitude (this also provides more margin for emergency procedures including the pilot leaving the aircraft). At fixed altitude the Mach number is increased in small increments

* See Dugundji and Calligeros for a particularly valuable discussion [5].

TABLE 10.I.

FLIGHT FLUTTER TESTING

EXAMPLES OF RECENT FLIGHT FLUTTER TEST PROGRAMS

- X-29 FORWARD SWEPT WING DEMONSTRATOR
  - THREE DIFFERENT FLIGHT CONTROL SYSTEMS TO TEST
  - NEW STRUCTURE
  - 218+ TEST POINTS TO CLEAR THE FLIGHT ENVELOPE
  - EXCITATION: TURBULENCE AND ROTARY INERTIA SHAKERS FOR THE FLAPERONS

- F-18 HARV AIRPLANE (TO BE TESTED SUMMER 1991)
  - MODIFICATION: TURNING VANES -- STRUCTURE AND FLIGHT CONTROL LAWS
  - EXCITATION: COMMANDS TO THE FLIGHT CONTROL SURFACES
  - TWO DIFFERENT FLIGHT CONTROL SYSTEMS TO TEST
  - ESTIMATE OVER 80 TEST POINTS TO CLEAR FLIGHT ENVELOPE
    - ANGLE OF ATTACK RANGE: 0° TO 70°
    - MACH NUMBER: MULTIPLE POINTS UP TO 0.7 MACH

- SCHWEIZER 1-36 DEEP-STALL SAILPLANE
  - MODIFICATION: HORIZONTAL STABILIZER MODIFIED TO PIVOT TO 70° FOR CONTROLLABILITY RESEARCH WITH WING COMPLETELY STALLED
  - GROUND TEST PERFORMED PRIOR TO FLIGHT TEST DUE TO NONLINEAR STRUCTURAL DYNAMICS BEHAVIOR OF TAIL
  - EXCITATION: TURBULENCE -- DATA ACQUIRED AT CONSTANT SPEED DURING CONTINUOUS DESCENT IN ALTITUDE BANDS OF ±1000 FEET ABOUT TEST ALTITUDE
  - STABILITY ANALYSIS: CLEARED IN REAL TIME BY MONITORING STRIP CHARTS

- AFTI/F-16 AEROSERVOELASTIC AND FLUTTER TEST
  - MODIFICATIONS: DIGITAL FLIGHT CONTROL SYSTEM AND CANARDS
  - EXCITATION: TURBULENCE AND STICK RAPS
  - STABILITY ANALYSIS: RECURSIVE IDENTIFICATION ALGORITHM USED TO SEPARATE CLOSELY SPACED MODES

FLIGHT FLUTTER TESTING

FLUTTER EXCITATION METHODS

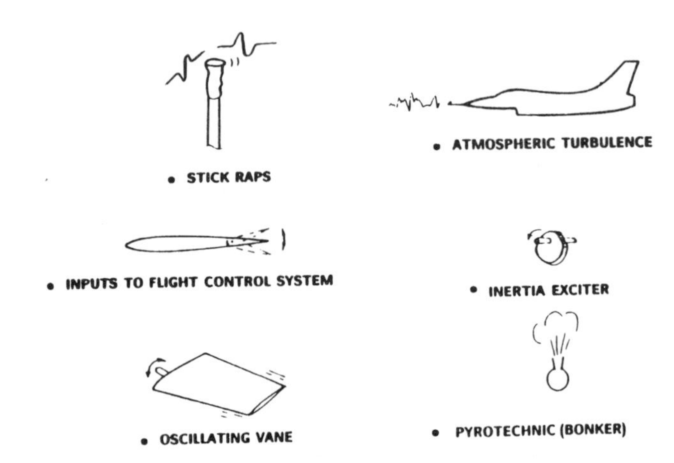

Figure 10.1.

FLIGHT FLUTTER TESTING

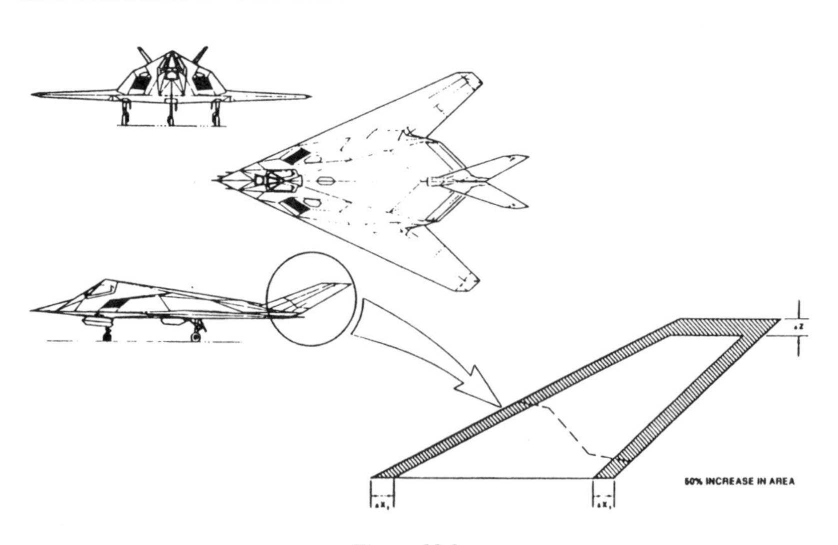

Figure 10.2.

TABLE 10.II.

F–14 FLUTTER PREVENT PROGRAM RELATIVE COSTS

• ANALYSIS	29%	
• WIND TUNNEL	27%	71% (Wind tunnel, GVT, flight flutter test)
• GVT	19%	
• FLIGHT FLUTTER TEST	25%	

R & D COST 0.5%

TABLE 10.III.

AIRCRAFT FLUTTER CLEARANCE PROCESS

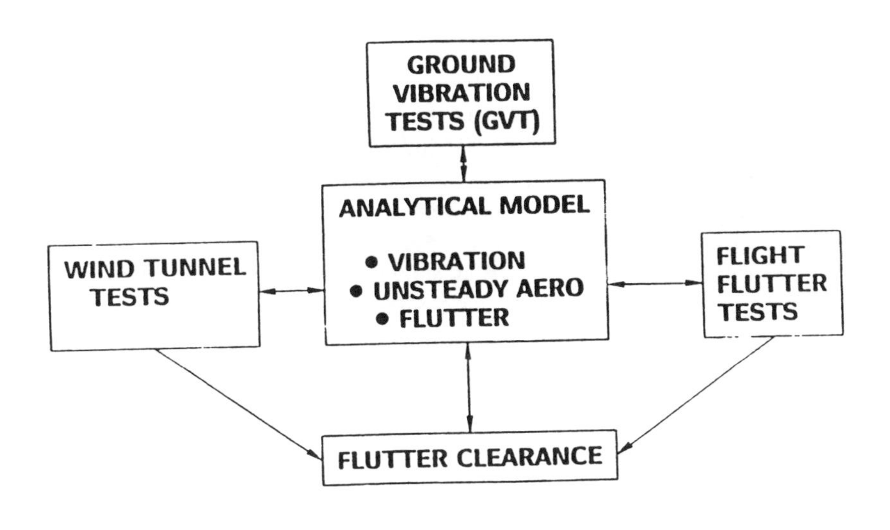

until flutter occurs or the maximum Mach number capability of the aircraft is reached.

When is flight flutter testing required?

For new aircraft, for substantial modifications of existing aircraft and for new uses of an existing aircraft, flutter testing is usually required.

Excitation

Several excitation methods have been proposed and used. None are clearly superior. Use of existing hardware, e.g. control stick raps, electronic inputs to the control system, or atmospheric turbulence, obviously minimize cost. On the other hand, add-on devices, such as oscillating vanes, inertial mass oscillations, or pyrotechnic devices presumably give greater control and range to the excitation. The rotating slotted cylinder device proposed by Reed* shows promise of being a good compromise between cost and performance. Examples of excitation systems that have been used in practice are shown in Fig. 10.1** .

Examples of recent flight flutter test programs

To remind the reader of the danger inherent in such tests, Fig. 10.2 shows the loss of a substantial portion of the tail surface from the recent flutter testing of the F-117A Stealth fighter. Other examples of recent programs are described in Table 10.I.

10.4 The role of experimentation and theory in design

In designing a new aircraft with acceptable aeroelastic behavior, a synergistic combination of theory, wind tunnel tests and flight tests is normally employed. Here a brief overview is presented of how this is usually done.

One measure of the relative importance of each of these synergistic elements is their cost. Baird*** has estimated these for the F-14 aircraft. See Table 10.II.

In Table 10.III a flow chart is shown that indicates the interaction among these elements. Note that each element normally influences another. For example, analysis and wind tunnel tests help define the flight flutter test program. Conversely any anomalies determined during flight test will almost assuredly lead to additional analysis and wind tunnel tests.

---

* See Reed [6].

** After Reed [6]. All Figures and Tables in this Chapter are drawn from [6].

*** Baird [7].

Finally, although the emphasis here has been on flutter experiments, gust response experiments or static aeroelastic behavior may be the subject of tests as well. The techniques employed are similar to those for flutter, with pilot and aircraft safety usually not such a critical concern as with flutter tests.

References for Chapter 10

[1] Ricketts, R., "Experimental Aeroelasticity," AIAA Paper 90-0978, 1990.

[2] Crawley, E. F. and Lazarus, K. B., "Induced Strain Actuation of Isotropic and Anisotropic Plates," *AIAA Journal*, Vol. 29, No. 6 (June 1991) pp. 944-951.

[3] Thomson, W. T., *Theory of Vibration with Applications*, Third Edition, Prentice-Hall, 1988.

[4] Matsusaki, Y. and Ando, Y., "Estimation of Flutter Boundary from Random Responses due to Turbulence of Subcritical Speeds," *J. of Aircraft*, Vol. 18, No. 10 (October 1981) pp. 862-868.

[5] Dugundji, J. and Calligeros, J. M., "Similarity Laws for Aerothermoelastic Testing," *J. of the Aerospace Sciences*, Vol. 29, No. 8 (August 1962) pp. 935-950.

[6] Reed, W. H. III, "Flight Flutter Testing: Equipment and Techniques," Presented at the FAA Southwest Region Annual Designer Conference, Ft. Worth, Texas, November 5 and 6, 1991.

[7] Baird, E. F. J., personal communication with W. H. Reed III.

11

Nonlinear aeroelasticity

Abstract

The study of nonlinear aeroelasticity has flourished in the last decade in recognition of improved methods of modeling nonlinear phenomena and also the important role that nonlinearities sometimes play in various physical settings. Here a discussion is presented of the physical domain of nonlinear aeroelasticity, the mathematical consequences of nonlinearity for modeling physical phenomena, representative results, and finally some thoughts about the future of the subject. In this rapidly advancing field, the present discussion necessarily is a glimpse in time and a projection for the future. As such, it is hoped that it will prove useful to the reader.

11.1 Introduction

Some years ago, Theodore Von Karman wrote a seminal article entitled, "The Engineer Grapples with Nonlinearities," [1]. It is fair to say that engineers are still grappling, but on the whole more successfully. The superior computer capability of today relative to that available to Von Karman and his contemporaries is one reason, of course. But it is also true that as the linear theory and practice of aeroelasticity have become more widely assimilated, the research frontiers have naturally moved toward the effects of nonlinearity. The physical importance of nonlinearities has been known for some years dating from the time of Von Karman and before, e.g. nonlinearities associated with transonic flow, large deflection of elastic structures, etc.

In this chapter the following topics will be covered. See Fig. 11.1. The study of the subject will be motivated by considering the physical domain of nonlinear aeroelasticity. Then the consequences of nonlinearity for constructing mathematical models will be discussed. Next representative results will be presented and, finally, some thoughts for the future will be outlined.

A recent book by Ilgamov and the author [2] has considered some studies of nonlinear aeroelasticity. Here we focus on very recent results and their implications for possible future trends. The reader will be referred to Ref. 2 for additional detail on some of the subjects covered here. Also see Ref. 3 that treats the nonlinear aeroelasticity of plates and shells and the earlier book by Bolotin [4] that discusses some of the topics of Ref. 3. Of course, there are several book length treatments of linear aeroelasticity starting with the well known works of Scanlan and Rosenbaum [5], Fung [6], Bisplinghoff, Ashley and Halfman [7] and continuing on to the present day [8-15]. Finally, the present volume also includes a chapter on nonlinear, unsteady transonic aerodynamics and aeroelasticity and another chapter on stall flutter.

TOPICS

- The physical domain of nonlinear aeroelasticity
- The mathematical consequences of nonlinearity
- Representative results
- Thoughts about the future

Figure 11.1.

DYNAMICS OF NONLINEAR AND NONCONSERVATIVE SYSTEMS

- Flutter of plates and shells
  - Strain-displacement structural nonlinearities
  - Chaotic structural-fluid motion

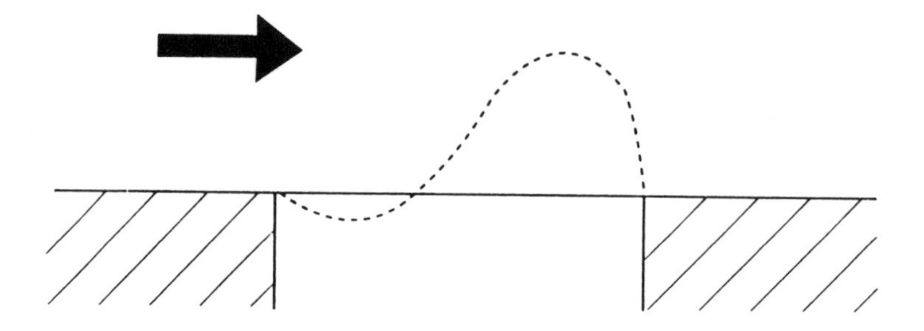

Figure 11.2.

FLUTTER OF AIRFOILS IN TRANSONIC FLOW

- Fluid shock induced nonlinearity
- Nonunique flow behavior

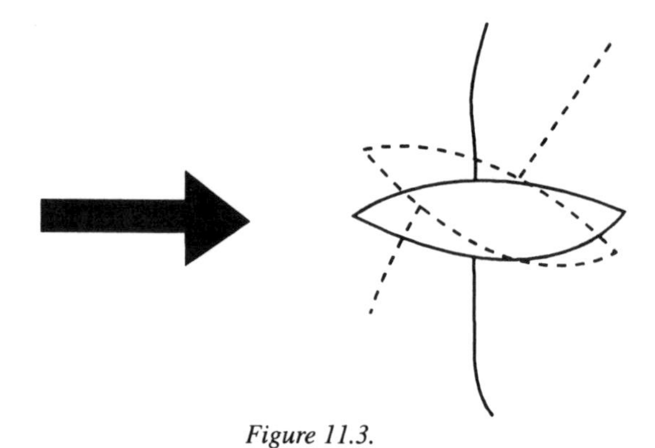

Figure 11.3.

11.2 The physical domain of nonlinear aeroelasticity

The physical phenomena of interest are richly varied and here only a representative selection will be discussed. Of these, the flutter of plates and shells has received perhaps the longest, most sustained attention. See Fig. 11.2. Here a flat (or curved) elastic plate embedded in an otherwise rigid surface has a high speed flow (supersonic flow is of greatest practical interest) over its surface. This models the individual skin panels of high performance aircraft and missiles. The dominant nonlinearity is the geometric structural nonlinearity in the relationship between longitudinal strain and transverse displacement of the plate or shell. For a certain transverse displacement of the plate, the length of the plate is extended, thereby inducing a tension in the plane of the plate. This tension is proportional to the (integral of the) square of the slope of the plate and it induces a nonlinear stiffness comparable to the linear bending stiffness when the plate deflection is on the order of the plate thickness. Thus aerodynamic nonlinear effects, which become significant only when the plate deflection is on the order of the plate length, are normally negligible. In recent years it has been determined that the nonlinear flutter motion may develop chaotic oscillations under some circumstances, e.g. the flutter of buckled plates.

The flutter of airfoils in transonic flow is a critical design consideration for most high performance aircraft. See Fig. 11.3. Both airframes and engines are susceptible to flutter. For many years, the development of theoretical models was inhibited by the widely held view that such models were necessarily nonlinear. This was because of the well known breakdown of linear aerodynamic theory for two-dimensional, steady (time-independent) flow about an airfoil as the Mach number approaches unity [16]. The physical manifestation of this nonlinearity is the development of shock waves (discontinuity of pressure) on the airfoil for transonic flow conditions. Indeed the development of shock waves may be taken as a definition of transonic flow. It is only with the introduction of finite difference (and finite element or finite volume) mathematical methods in conjunction with modern computational facilities (Computational Fluid Dynamics or CFD) that the effective and common calculation of transonic flows around airfoils has become possible. One of the pleasant ironies for the aeroelastician has been that such calculations have revealed that a linear dynamic model perturbed from a nonlinear static or steady flow model which includes shock waves is sufficiently accurate for describing many flutter phenomena. Nevertheless under some special circumstances, e.g. aerodynamic stall, full dynamic nonlinearities may be significant.

Bluff body oscillations in a flowing fluid arise in many interesting physical contexts. See Fig. 11.4. Smoke stacks, tall buildings, or missiles on a launch pad subjected to a transverse wind are some typical examples. The dominant nonlinearity is a result of flow separation from the surface of the bluff body, usually near its shoulders. The flow itself may be periodic (a Karman vortex street may develop) or turbulent. Interaction between the flow and the motion of the elastic structure gives rise to a fascinating array of nonlinear aeroelastic motions including some that exhibit limit cycles and chaos.

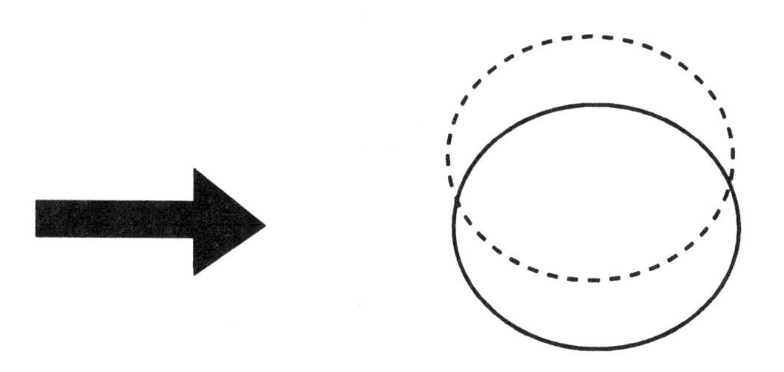

Figure 11.4.

Theoretical models based upon the first principles of fluid mechanics have yet to be developed in a definitive treatment. However phenomenological fluid models have been developed that mimic the essential features of the flow. Typically these postulate a nonlinear fluid oscillator with inertia, damping and stiffness. A Van der Pol equation with linear inertia and stiffness terms and a nonlinear damping term has been suggested and studied by several investigators. See Ref. 2, for example.

The aeroelastic behavior of hingeless (cantilevered) helicopter rotor blades is most complex. See Fig. 11.5. The geometric structural nonlinearities and stall aerodynamics are dominant although now the deformations are typically of the order of the blade chord or span. The possible combinations of structural motions (bending in the lift and drag directions and twisting) and flow field conditions (hover and high speed forward flight) are several. An authoritative account of present knowledge has been given recently by Ormiston, Warmbrodt, Hodges and Peters [17]. The reader is referred to their paper for an excellent summary of this important subject. Also see Chapter 7 of the present volume. It will not be further discussed here.

"Survey of ARMY/NASA Rotorcraft Aeroelastic Stability Research,"
Ormiston, Warmbrodt, Hodges and Peters,
NASA TM101026 and
USAASCOM TR 88-A-005

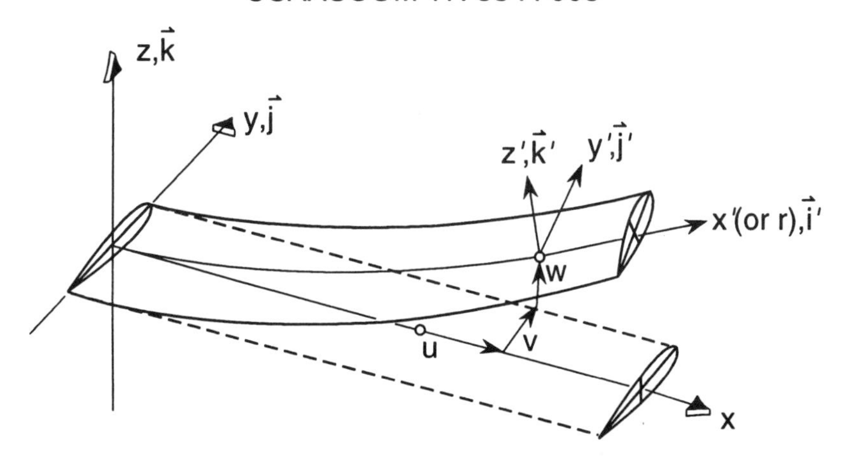

- Large bending and twisting of a helicopter rotor blade

Figure 11.5.

The observation by Kramer some years ago that dolphins appear to have an anomalously low drag has led to a number of investigations of compliant walls to reduce drag. See Fig. 11.6.

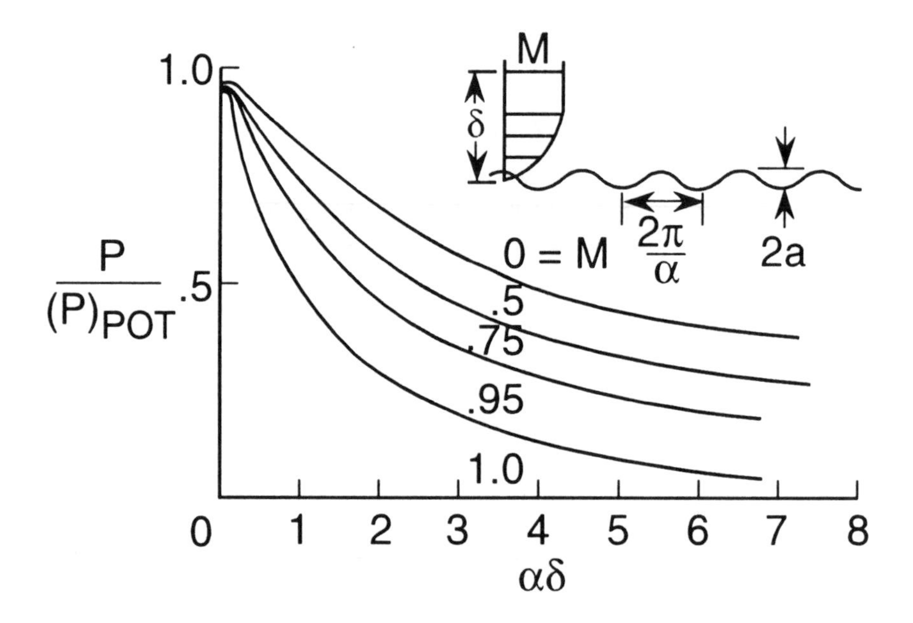

Figure 11.6.

Originally it was speculated that a compliant wall, properly designed, might be used to delay the onset of the (linear) hydrodynamic instability of the laminar boundary layer. Early analyses were pessimistic in this respect, although more recent analyses have questioned these previous results. The modeling of the full nonlinear phenomena including the interaction of a laminar or turbulent boundary layer and a compliant (elastic or viscoelastic) wall has yet to be done in a definitive treatment. For a summary of the state-of-the-art, see Ref. 2. Again, no further discussion will be undertaken here.

11.3 The mathematical consequences of nonlinearity

There are several classical texts on nonlinear dynamics, e.g. Minorsky [18], Stokes [19], and Cunningham [20], as well as more recent ones that reflect the work on chaos, e.g. Guckenheimer and Holmes [21], Thompson and Stewart [22], Moon [23] and the tutorial paper by Dowell [24]. The reader is directed to those sources for a more thorough discussion. Here a few fundamental issues will be addressed.

<u>All real physical phenomena are</u> nonlinear and have an infinite number of degrees of freedom and are governed by an infinite number of parameters. However, <u>some physical phenomena may be usefully described by</u>

- linear (vs nonlinear)

and/or

- finite (vs infinite) degrees of freedom

and/or

- finite (vs infinite) number of parameter <u>models</u>.

Thus the <u>complexity of a model</u> may be determined by whether the model is <u>linear</u> or <u>nonlinear</u> and the <u>number</u> of <u>degrees of freedom</u> and the <u>number</u> of <u>parameters</u>. Obviously, the fewer the number of needed degrees of freedom and the number of parameters, the less complex the model. It is also clear that linear models are simpler than nonlinear ones. However it may be worthwhile to note a fundamental blurring of the distinction between parameters and degrees of freedom that occurs when the model is nonlinear. For a linear system the long time, steady state solution is independent of the initial conditions, though it does depend on model parameters. However for nonlinear models the long term, steady state solution <u>may</u> depend upon initial conditions. Thus in a nonlinear model, initial conditions become possible parameters. In the mathematical literature, the number of parameters is called the <u>co-dimension</u> and (twice) the number of degrees of freedom is called the <u>dimension</u> of the dynamical system.

In making the distinction between linear and nonlinear models, it is useful to consider three classes of models.

- Fully Linear Models

Both the static and dynamic behavior of the physical system are described by linear models.

- Dynamically Linear Models

The static behavior is described by a nonlinear model, but the dynamic behavior is described by a linear model. Note that the parameters of the linear dynamic model will generally depend upon the static (equilibrium) model. Also note that there may be several different static equilibria.

- Fully Nonlinear Models

Both the static and dynamic behavior are described by nonlinear models. Note that there may be several distinct possible dynamic equilibria. The one observed will depend, in general, upon the system parameters and initial conditions.

11.4 Representative results

Flutter of airfoils in transonic flow

By definition, transonic flow is that regime of flow near a Mach number of unity when a shock wave forms on the airfoil. Such shock waves have pressure discontinuities across them and thereby may significantly alter the aerodynamic pressure distribution (loading) on the airfoil, and thus its flutter behavior.

In Fig. 11.3, a sketch of an airfoil is shown with its shock waves. For a symmetrical airfoil at zero angle of attack, the shocks on the upper and lower surfaces will be symmetrically placed. When the airfoil is rotated (pitched) to a positive angle of attack, the upper surface shock wave moves aft and the lower surface shock wave moves forward. This gives rise to a substantial net lift force in the chordwise region between the shock waves. For small angles of attack, the width of this region (distance between the upper and lower surface shocks) is proportional to angle of attack and the net pressure difference is simply the pressure drop through the shock. This "shock doublet", as Holt Ashley has called it, is the most striking feature of aerodynamic loading in transonic flow.

As the airfoil oscillates in pitch, the shock waves oscillate also with some phase lag depending on the frequency of oscillation.

A number of investigators around the world have developed computer codes (e.g. Ref. 25-32) for the finite difference solution of the governing fluid equations. Most current aeroelastic solutions are based upon transonic, small disturbance, potential flow mathematical models. However some investigators have combined potential flow models with boundary layer models and others have considered the Euler or even Navier-Stokes equations.

The practical reason for these studies is because of the critical importance of the transonic flow regime for flutter of high performance aircraft. Typically the minimum dynamic pressure for which flutter will occur is in the transonic range. See Figure 11.7 for a schematic showing a flutter boundary in terms of dynamic pressure, q, vs Mach number of the free stream flow, M_∞. The difference between experiment and classical theory (that ignores the shock waves) is representative and the reason for much of the research effort.

The results of this research have been quite encouraging in one respect. Only the static shock nonlinearity is important as long as the flow does not separate (remains attached). Thus a linear dynamic analysis for flutter usually is still adequate. The key is to determine accurately the steady flow over the airfoil including the static shock strength and location. This requires a nonlinear steady (static) aerodynamic analysis. Then consideration of a linear dynamic perturbation about this steady (static) flow field will be sufficient. While the aerodynamic model is somewhat more complicated than the classical one, which is linear in both a static and dynamic sense, the flutter analysis per se need be no more complicated than for subsonic or supersonic flow.

A summary of the various types of aerodynamic (and hence aeroelastic) models, the motivation for using them, and the criteria for the validity and utility of their use is shown in Table 11.I.

Despite the general encouragement regarding the validity of linear, dynamic models for transonic flutter analysis, some questions of both a fundamental and pragmatic character remain. These are listed in Table 11.II. Since these questions and the best available present answers are discussed at length in Ref. 2, only a summary will be given here. The significance of these questions derives from several considerations. They are fundamental questions which are expected to be of lasting significance. Moreover, answers to these questions have important consequences for aeroelastic applications of unsteady transonic aerodynamics. Finally, recent work has led to, at least, partial answers. These answers are given in Table 11.III.

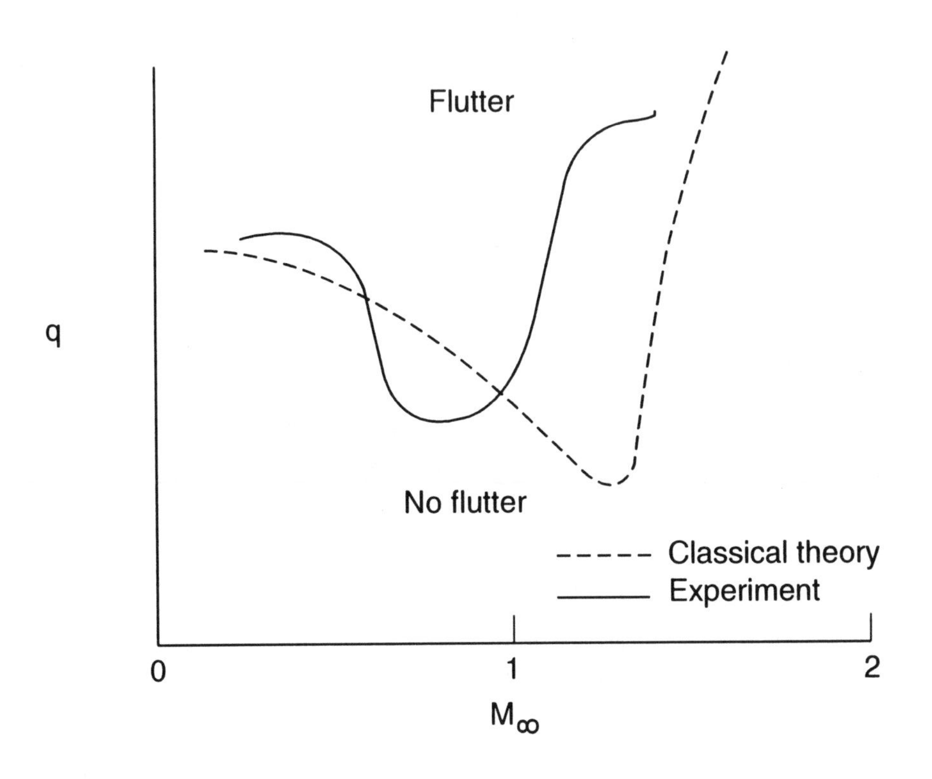

Figure 11.7. Airfoil flutter boundary in transonic flow.

TABLE 11.I. ALTERNATIVE AEROELASTIC ANALYSES FOR TRANSONIC FLOW.

TYPE OF ANALYSIS	MOTIVATION	CRITERIA FOR USE
FULLY LINEAR	INEXPENSIVE COMPUTATION	TORSIONAL DEFORMATION/ ANGLE OF ATTACK ARE SMALL; NO SHOCK WAVES
STATICALLY NONLINEAR; DYNAMICALLY LINEAR	DETERMINE EFFECTS OF STATIC WING DEFORMATION AND/OR WING PROFILE ON SHOCK WAVES	ONLY FLUTTER ONSET IS DETERMINED; NONLINEARITIES MUST BE HARD; I.E. NO EFFECT OF DISTURBANCE SIZE ON FLUTTER ONSET
FULLY NONLINEAR	DETERMINE FLUTTER LIMIT CYCLE AND EFFECT OF DISTURBANCE SIZE ON FLUTTER ONSET	WHEN ALL ELSE FAILS AND FOR FATIGUE PREDICTION

TABLE 11.II FOUR PRINCIPAL QUESTIONS

- UNDER WHAT CONDITIONS ARE THE AERODYNAMIC FORCES ESSENTIALLY LINEAR FUNCTIONS OF THE AIRFOIL MOTION?

- ARE THERE VIABLE ALTERNATIVE METHODS TO FINITE DIFFERENCE PROCEDURES FOR SOLVING THE RELEVANT FLUID DYNAMICAL EQUATIONS?

- UNDER THOSE CONDITIONS WHEN THE AERODYNAMIC FORCES ARE NONLINEAR FUNCTIONS OF THE AIRFOIL MOTION, WHAT IS THE SIGNIFICANCE OF THE MULTIPLE (NONUNIQUE) SOLUTIONS WHICH ARE SOMETIMES OBSERVED?

- WHAT ARE EFFECTIVE, EFFICIENT COMPUTATIONAL PROCEDURES FOR USING UNSTEADY TRANSONIC AERODYNAMIC COMPUTER CODES IN AEROELASTIC (E.G. FLUTTER) ANALYSES?

For a discussion of the significant advances which have been made in finite difference solution methods and consequent improved physical understanding of transonic flows, the reader is referred to Ref. 31. For a discussion of the important effects of viscosity and flow separation, go on to the next section.

Flutter of airfoils at high angles of attack

What is meant by high angle of attack? Here this means the angle of attack is sufficiently high that the flow separates from the airfoil and hence stall (loss of lift) occurs. If the airfoil is oscillating, then separation may occur during only a portion of the oscillatory motion.

TABLE 11.III. SOME PRESENT ANSWERS

- FOR SUFFICIENTLY SMALL AIRFOIL MOTIONS, E.G. THE SHOCK WAVE MOTION IS SMALL COMPARED TO THE AIRFOIL CHORD, THE AERODYNAMIC FORCES WILL BE LINEAR FUNCTIONS OF THE AIRFOIL MOTION.

- AT LEAST ONE VIABLE ALTERNATIVE SOLUTION TECHNIQUE TO FINITE DIFFERENCE METHODS IS AVAILABLE, THE FIELD PANEL METHOD OF HOUNJET.

- THE NONUNIQUENESS WHICH HAS BEEN OBSERVED UNDER SOME CONDITIONS IN TRANSONIC SMALL DISTURBANCE AND FULL POTENTIAL FL,OW SOLUTIONS IS INHERENT IN THE GOVERNING FIELD EQUATIONS THEMSELVES AND NOT A NUMERICAL ARTIFACT OF SOME SOLUTION METHOD.

- WHILE THE OPTION OF SIMULTANEOUS TIME INTEGRATION OF THE FLUID DYNAMICAL AND STRUCTURAL DYNAMICAL EQUATIONS OF MOTION TO DETERMINE AEROELASTIC RESPONSE WILL BE ATTRACTIVE FOR SOME APPLICATIONS, FLUTTER ANALYSIS IN THE FREQUENCYDOMAIN WILL CONTINUE TO BE AN IMPORTANT, AND AT TIMES MORE ATTRACTIVE, OPTION AS WELL. LINEAR FLUTTER ANALYSIS WILL CONTINUE TO PLAY A DOMINANT ROLE EVEN IN THE TRANSONIC FLOW REGIME.

Currently there are two approaches to modeling this physical phenomenon. One approach is to postulate a phenomenological mathematical model that contains some of the basic elements observed in the physical experiments. Several of these models are little more than curve fits to experimental data. The second approach is a numerical solution to the fundamental equations of fluid mechanics, e.g. Ref. 32.

Examples of both approaches will be discussed. However, the first, while phenomenologically based, is suggestive of another approach based upon first principles of fluid mechanics, e.g. the Navier-Stokes equations. In this third approach, based directly upon the Euler or Navier-Stokes equations, a method for reducing substantially the size of the finite difference representation of the fluid mechanical model is developed, while retaining a sufficiently accurate description of the physics of the fluid. These are so called "reduced order aerodynamic models".

A Class of Phenomenological Models. For definiteness the subsequent discussion will be in the context of the well known two-degree-of-freedom in heave and pitch (translation and rotation), typical section model. See Fig. 11.8.

In one class of phenomenological models, it is postulated that the fluid behaves as an oscillator with inertia, damping and stiffness and that the fluid is driven by the motion of a body within the fluid. See Ref. 33-35, for example. Such models have been developed independently for bluff body oscillations [33] (to be discussed later in this chapter) and for rotorcraft applications [34]. Here the latter, somewhat simpler, models are discussed after Ref. 35.

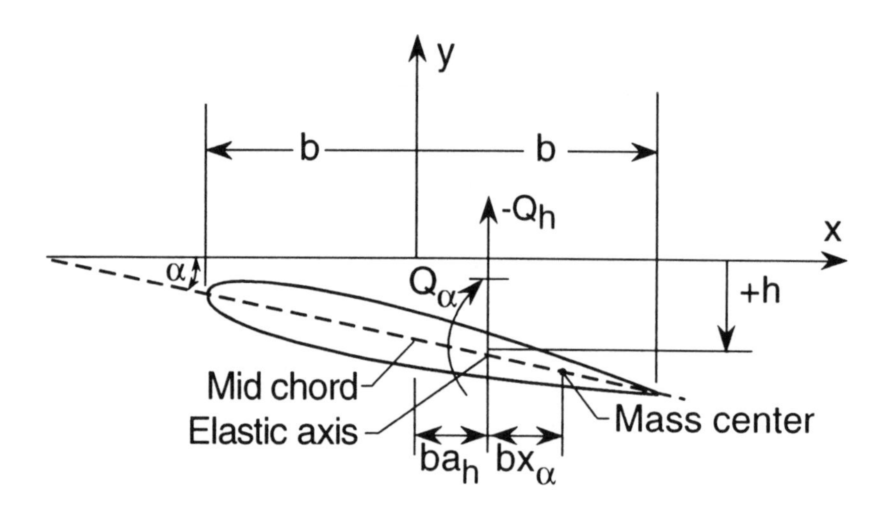

Figure 11.8.

As an example, consider the lift on an oscillating body. The lift coefficient, C_L, is postulated to obey an equation of the following form.

$$\ddot{C}_L + 2\,\zeta_A\,\omega_A\,\dot{C}_L + \omega_A^2\,C_L = F_L\,(\alpha,\,\dot{\alpha},\,\ddot{\alpha},\ldots,M_\infty,Re) \tag{1}$$

Here $\dot{}$ denotes time derivative, ζ_A is a critical damping ratio for the fluid oscillator, ω_A is a natural frequency of the oscillator, α is the angle of attack of the airfoil due to its motion and F_L is some function of α and its time derivatives (usually taken to be a polynomial), the Mach number (M_∞), and the Reynolds number (Re). F_L also depends upon the airfoil shape and planform, of course. More generally α would be replaced by the non-dimensional downwash of the airfoil or wing incorporating the various possible motions. Note that F_L is a nonlinear function of α, etc., in general.

In its purest form, such models assume the damping, ζ_A, and natural frequency, ω_A, are only functions of the fluid, and thus analogous equations hold for other aerodynamic generalized forces such as moment coefficient, C_M. Thus in (1), C_L would be replaced by C_M, and F_L by another function, F_M.

ζ_A , ω_A and the form of F_L, F_M, etc., are determined from experimental data, i.e. physical wind tunnel experiments or numerical (CFD) experiments using the Euler equations or Navier-Stokes equations. For a more extensive discussion of these matters, see Ref. 34 and 35.

Once Equation (1) is fully specified, it and its counterpart for (C_M) may be combined, for example, with the two degree of freedom structural equations of motion for translation and rotation (heave and pitch) of the typical section aeroelastic model to determine the time histories of the motion variables. The well known typical section structural equations of motion are as follows.

$$\begin{aligned} m\ddot{h} + S_\alpha\ddot{\alpha} + K_h h &= -L \\ I_\alpha\ddot{\alpha} + S_\alpha\ddot{h} + K_\alpha\alpha &= M \end{aligned} \tag{2}$$

where h,α are the heaving and pitching motion variables, m, S_α, I_α the mass, static unbalance and moment of inertia, K_h, K_α the translational and torsional spring constants, and L,M the aerodynamic lift and moment.

A representative result [35] is shown in Fig. 11.9. As discussed before, one of the mathematical (and physical) consequences of nonlinearity is that the long time oscillation (limit cycle) of the system may depend on initial conditions. That is, several long time motions may be possible, and the one actually achieved may depend upon how the motion is started (initial conditions). Thus, in Fig. 11.9 a map is displayed showing various regions in terms of non-dimensional flow velocity, V*, vs. initial (t = 0) angle of attack $\alpha_0 \equiv \alpha(t = 0)$. Results for classical linear theory are shown as well as for the semi-empirical phenomenological theory.

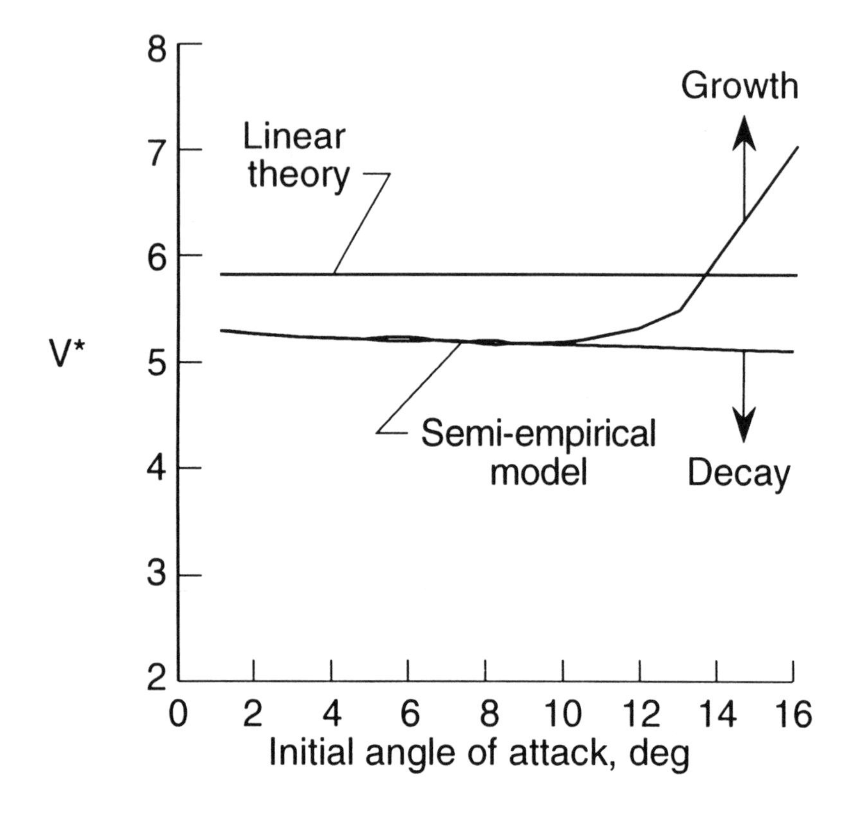

Figure 11.9.

First, consider the results of classical linear theory. Below a certain critical value of V* (roughly V* ≅ 5.8), independent of α_0, the long time solution is a decaying oscillation. Above this same critical V*, also independent of α_0, the long time solution is a growing oscillation. This result is well known, of course.

Next consider the results from the nonlinear, semi-empirical model. Below a certain V*, which depends upon α_0, the long time oscillations decay. Above another certain V*, which also depends on α_0, the long term oscillations grow. For small α_o, these two values of V* are virtually the same. However for sufficiently large α_o, say $\alpha_o > 10°$, the two values of V* are distinctly different. Moreover in the range between these two values of V*, a yet different type of long time oscillation is possible, namely a limit cycle oscillation of finite amplitude. The maximum or peak amplitude of the limit cycle oscillation in terms of the pitch motion variable, α_{max}, is shown in Fig. 11.10 as a function of V* for a given (large) initial angle of attack, i.e. $\alpha_o = 15°$. Also shown are (much more limited) results using a finite difference solver with the Navier-Stokes equations in combination with the typical section structural model. It is noted that there is approximately a thousandfold difference in computation times between the semi-empirical model and the Navier-Stokes solver. In both cases, it is seen that below some value of V*, the long term oscillations decay; above some (higher) V*, the long term oscillations grow; and in a range between these two values of V*, a finite amplitude limit cycle motion exists. However there are some quantitative differences between the two results.

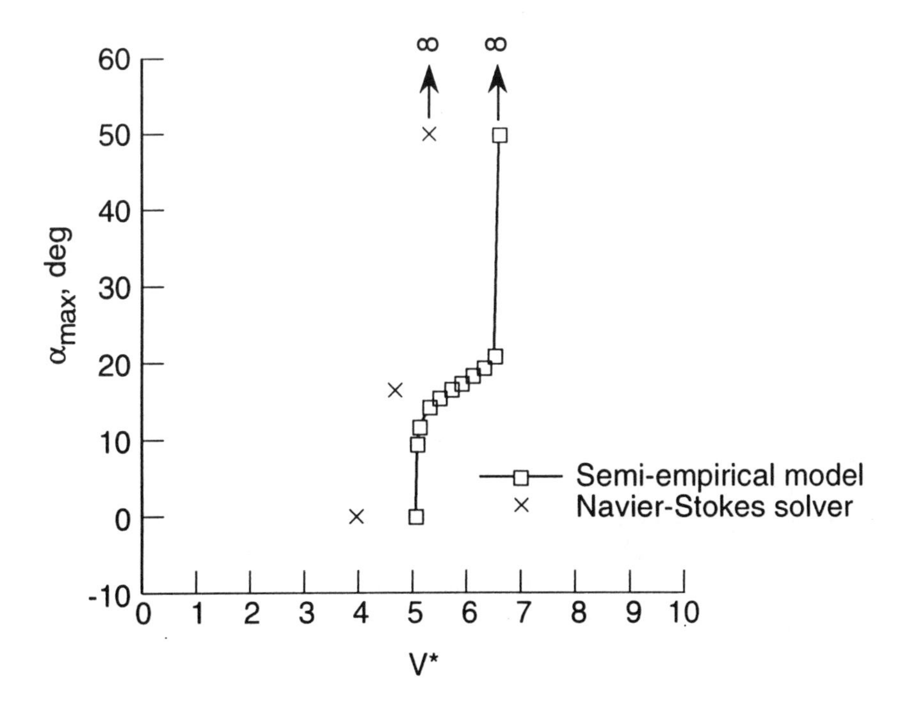

Figure 11.10.

This naturally raises the question, can the results of the semi-empirical model be improved? Referring back to equation (1), a possible improvement suggests itself. Why not include the higher modes of the fluid, i.e. additional oscillators with additional ζ_A and ω_A? The question arises, how to determine these from the first principles of fluid mechanics? That is the subject of the following discussion of a "reduced order aerodynamic model".

Eigenvalues of the Euler Equations and/or Navier-Stokes Equations. The basic idea is quite simple, but the implementation is complex. Hence, only recently have results been obtained, see Ref. 36 and 37. The premise is that by determining the eigenvalues and eigenvectors of the fluid mechanical equations for small dynamic oscillations (perturbations) about a mean steady flow, one may determine the modal structure of the fluid. From the eigenvectors, a linear coordinate transformation may be made from the primitive fluid variables, density, pressure, velocity, etc., to modal coordinates. It is hypothesized, that the number of modal coordinates needed to describe phenomena of interest may be much smaller than the number of primitive fluid variables. Hence one may obtain a "reduced order aerodynamic model".

This general approach is well known for linear, elastic structural models and some efforts have been made to extend it to nonlinear structural models. For example, the work on nonlinear flutter analysis of plates and shells that will be discussed later uses this approach.

However, for a fluid this approach has several additional complications. The eigenvalue problem is for a (non self-adjoint) non-conservative problem, the primitive model from which the eigenvalues are extracted is very large (order of tens of thousands of degrees of freedom), and the reduced order model itself is still fairly large (order of tens or hundreds degrees of freedom). Nevertheless, recently substantial and encouraging progress has been made.

Moreover, there is an added bonus from the results. By examining the eigenvalues of the fluid, one may address directly the numerical stability of the (spatial) finite difference models commonly employed in CFD work and, in particular, examine the effects of artificial viscosity terms on the numerical and physical instabilities of the model.

For many years, artificial (i.e. non-physical) viscosity terms have been added to finite difference spatial CFD codes to ensure numerical stability of the solutions when time marching the code. This has concerned thoughtful investigators, particularly in the context of aeroelastic applications. One of the principal objects of an aeroelastic analysis is to determine the physical stability characteristics of the fluid-structural system. If artificial viscosity is added to suppress numerical instabilities, what is its effect on physical instabilities?

The results of Fig. 11.11-11.13 are suggestive. In Fig. 11.11, the (smallest) eigenvalues of a standard Euler code [32,36,37] are shown for no artificial viscosity added. Some eigenvalues have positive real parts indicating the mathematical model is unstable. When artificial viscosity is added (typical of that used in practice), see Fig. 11.12, all eigenvalues have negative real parts, indicating the mathematical model is stable. Note however the very substantial change in the eigenvalue distribution (constellation) between Fig. 11.11 and 11.12. This naturally raises the question, can we reconcile the amount of artificial viscosity used and the effect on model stability? For a more thorough discussion of these matters, see Ref. 36 and 37. Here a summary is given.

There are two types of artificial viscosity typically added to CFI models. One is explicit, ε_E, and the other implicit, ε_I. ε_E and ε_I are measures of the amount of these types of artificial viscosity. According to the eigenvalue analysis any positive ε_I is sufficient to stabilize the model, no matter how small, and there is no effect of ε_E on stability. However, in practice, it is known that both ε_E and ε_I are required to stabilize time marching solutions of CFD codes. How can this apparent difference be resolved? It turns out that as the time step, Δt, tends to zero, the numerical stability criterion determined from time marching the solutions does indeed correspond to that determined from eigenvalue analysis. See Fig. 11.13.

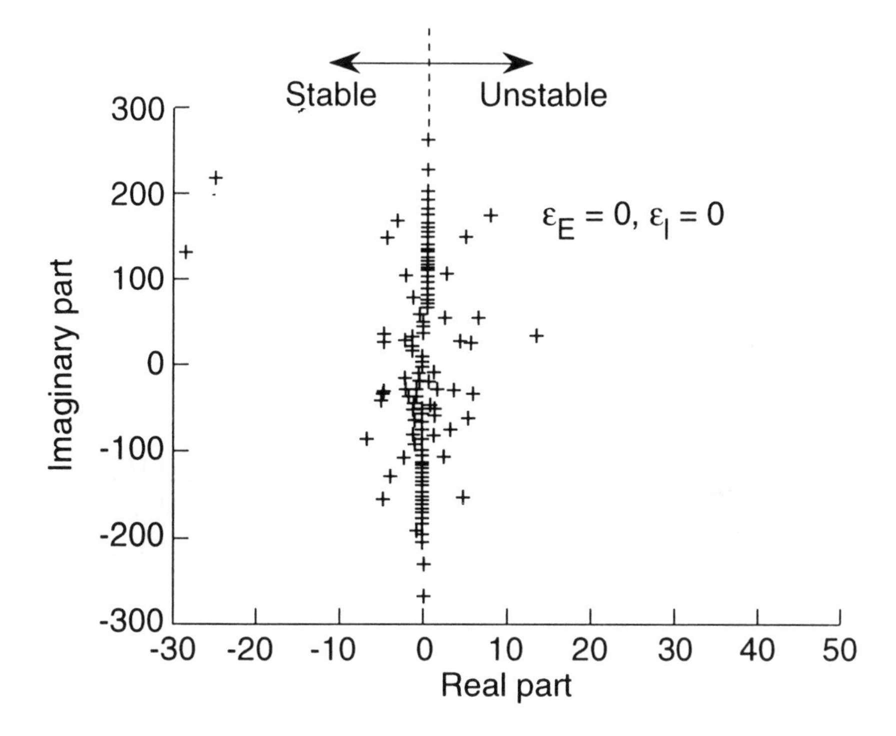

Figure 11.11.

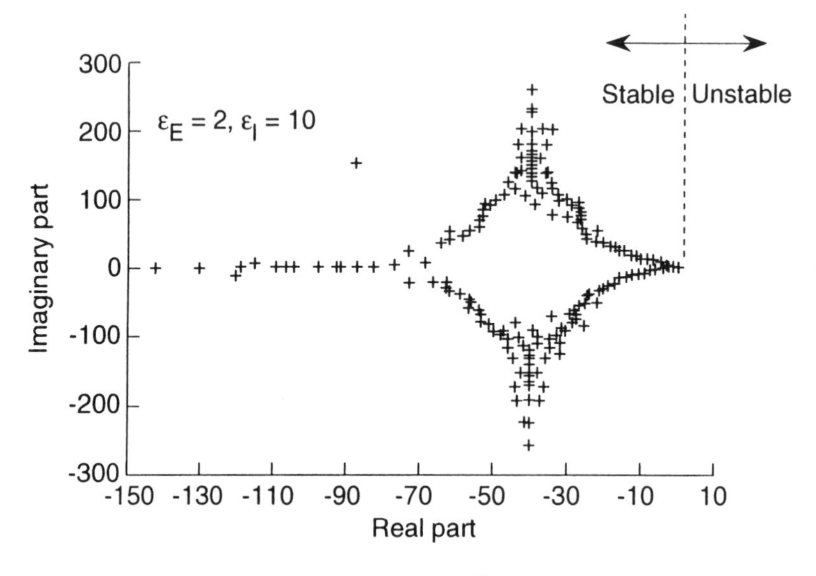

Figure 11.12.

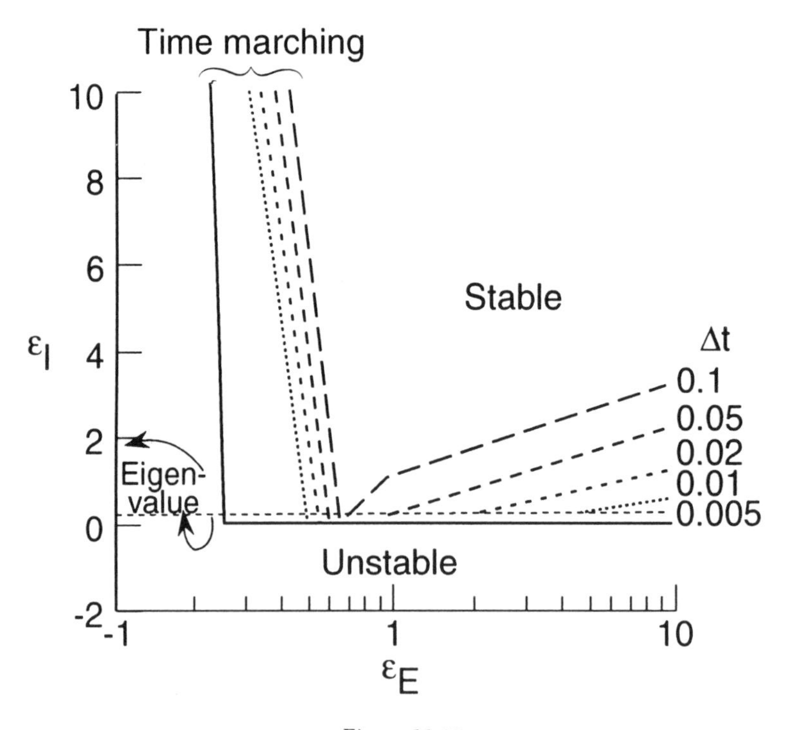

Figure 11.13.

This gives then a basis for assessing the effects of artificial viscosity on physical instabilities and also gives reassurance that an eigenvalue-eigenvector analysis may be directly used for flutter analysis of the combined fluid-structural system as well as used to construct computationally efficient reduced order aerodynamic models.

It is interesting to note that the work of Mahajan, et al., was motivated by interest in the nonlinear aeroelasticity of turbomachinery. The study of this subject is still in its early stages and requires the extension of isolated airfoil results to multiple airfoils, e.g. to airfoil cascades in two-dimensions or airfoils attached to a (rotating) disk in three-dimensions. For further background on turbomachinery aeroelasticity, see the chapters by Sisto in this volume and the chapter by Verdon in Ref. 31.

Flutter of an airfoil with free-play structural nonlinearities

Here the work of Brase and Eversman [38] is used to illustrate several interesting points about the effect of structural nonlinearities on airfoil flutter. These authors also used the typical section model with classical, incompressible, potential flow (Theodorsen) aerodynamics. The nonlinearity was in the structural stiffness (spring) of the pitch or rotation degree of freedom in the form of free-play. See Fig. 11.14, that plots spring restoring force versus the amplitude of rotation. The solid line shows a typical linear spring characteristic and the dashed lines show typical nonlinear characteristics for free-play and friction respectively. Here the free-play nonlinearity will be discussed. In this case there is a range of rotational amplitude, α, for which the restoring force is zero. Outside this range a linear relationship between restoring force and spring rotational amplitude exists. As will be seen, the rotational amplitude which demarks the change in spring stiffness characteristic defines the nonlinear scale of the phenomena of interest.

Fig. 11.15 displays generic time histories for heave (translation), h, and pitch (rotation), α, for various flow speeds. The other parameters are typical. Note in particular that the limits of the free-play regime are $= \pm .2°$ and the initial angle of pitch is α at $t = 0 \equiv \alpha_0 = .1°$. As can be seen for V = 17, 33 and 50 ft/sec there is a decaying oscillation indicating stability or no flutter. For V = 167, 175, 183.3 ft/sec there is a limit cycle oscillation or bounded amplitude flutter. Note that the amplitude increases with V and the lower bound of the α oscillation is at the free-play limit, $\alpha_{free\text{-}play} = +.2°$ or $2\pi/18 \times 10^{-2}$ radians. At V = 187.5 ft/sec there is a bounded flutter oscillation jump phenomenon from one side of the free-play region to the other. Although there is too short a time history to be certain, this is almost assuredly a chaotic (non-periodic) oscillation analogous to that found in the forced or aeroelastic oscillation of a buckled beam or plate. (See the discussion in a later section of the present chapter). For

V = 188.3 ft/sec (and probably higher V), the flutter motion is an exponentially growing oscillation, similar to that found in linear models.

L.O. Brase and W. Eversman, "Application of Transient Aerodynamics to the Structural Nonlinear Flutter Problem", Journal of Aircraft, Vol. 25, No. 11, November, 1988.

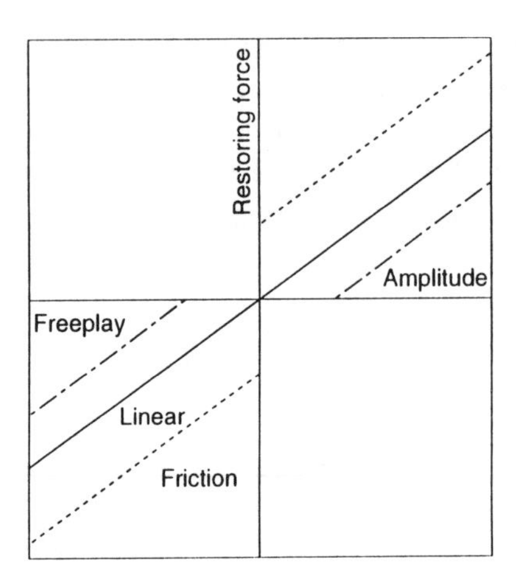

Figure 11.14.

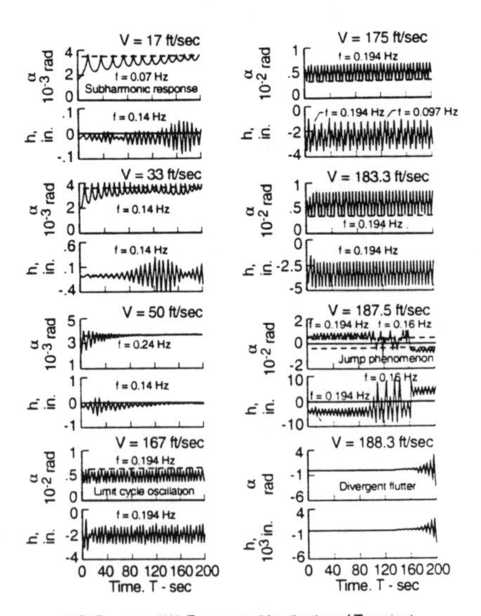

L.O. Brase and W. Eversman, "Application of Transient Aerodynamics to the Structural Nonlinear Flutter Problem," Journal of Aircraft, Vol. 25, No. 11, November, 1988

Figure 11.15.

Finally, in Fig. 11.16, a summary of the interesting results of Brase and Eversman is shown in terms of the velocity, V, at which divergent (exponentially growing) oscillations occur versus the initial pitch rotation, α_o, for various free-play angles, $\alpha_{free\text{-}play}$. Note that if one non-dimensionalizes α_o by $\alpha_{free\text{-}play}$ the results of the three curves fall on top of one another. By examining the governing equations one may show this to be true. Any small differences must be attributable to numerical inaccuracies. Note this is also true for the results of Fig. 11.15, i.e. these curves may be made universal by non-dimensionalizing α by $\alpha_{free\text{-}play}$, α_o by $\alpha_{free\text{-}play}$ and h by $S\alpha\ \alpha_{free\text{-}play}/m$.

As can be seen from Fig. 11.16, when the initial disturbance, α_o, is small the flow velocity for divergent flutter oscillations is high. However when the initial disturbance, α_o, is roughly 3.4 times $\alpha_{free\text{-}play}$ there is a dramatic reduction in this velocity. This again emphasizes the importance of the initial disturbance for the behavior of nonlinear systems.

For a very informative discussion of physical phenomena and practical problems of airfoil and wing nonlinear aeroelasticity, see the articles by Mabey and Cunningham [31].

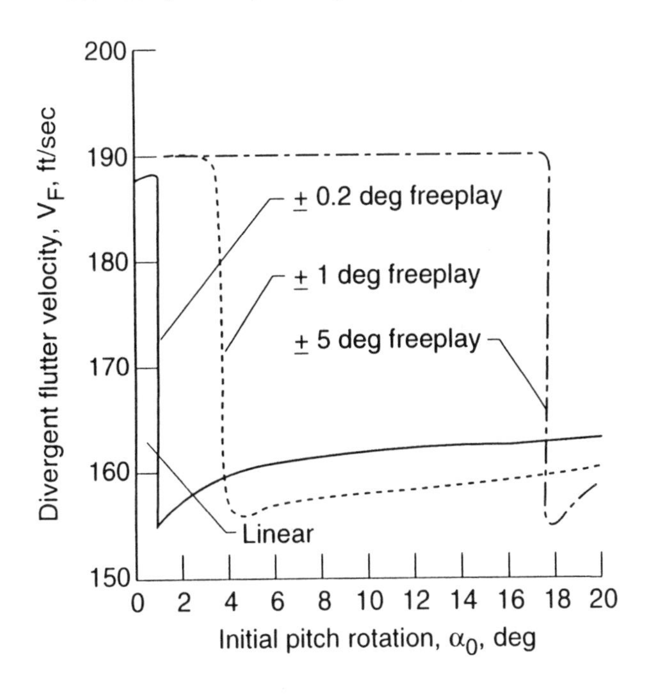

Figure 11.16.

Nonlinear fluid oscillator models in bluff body aeroelasticity

In considering the (separated) flow about a bluff body (mounted on an elastic spring support), one can again model the fluid by a phenomenological representation, or by the Navier-Stokes equations, or possibly by a reduced order aerodynamic model. Recall Fig. 11.4. There is now a good deal of literature on the first of these alternatives [2,33], very little on the second, and none as yet on the third. However, one may expect the latter two to receive more attention in the future. Here the first will be discussed further to illustrate the possibilities of these phenomenological models. This discussion, in turn, may suggest what is in store when first principle models are more extensively studied.

A simple single-degree-of-freedom linear structural oscillator model will be assumed

$$M\left[\ddot{y} + 2\,\zeta_o\omega_o\dot{y} + \omega_o^2\,y\right] = q\,D\,C_L \tag{3}$$

M is the structural mass per unit span, y is the heave (translation) displacement, $\cdot$ denotes time derivative, ζ_0 is the critical structural damping ratio, ω_0 is the structural natural frequency, q is the fluid dynamic pressure, D is a representative length scale of the cross-section of the structure (e.g. diameter), and C_L the lift coefficient.

The fluid is also represented by an oscillator, but in this case a nonlinear oscillator.

$$\begin{aligned}&\ddot{C}_L - \varepsilon\left[1 - 4\,(C_L/C_{Lo})^2\right]\,\omega_A\,\dot{C}_L + \omega_A^2\,C_L \\ &= -B_1\frac{D}{V^2}\,\ddddot{y} + \omega_A^2\left[A_1\frac{\dot{y}}{V} - A_3\left(\frac{\dot{y}}{V}\right)^3\right] \equiv F_L\end{aligned} \tag{4}$$

Compare Equation (4) with Equation (1). They are similar except that Equation (1) has a linear, viscous damping term,

$$2\,\zeta_A\,\omega_A\,\dot{C}_L$$

while Equation (4) has a nonlinear damping term

$$-\,e\,[1 - 4\,(CL/CLo)2]\,wA$$

Note this damping term and indeed the left hand side of Equation (4) is a form of Van der Pol's Equation for the dependent variable C_L. This choice of oscillator model was made by Hartlen and Currie [39] based upon the physical experimental data of Bishop and Hassan [40]. The latter suggested that the results of their measurements of lift on an oscillating cylinder with prescribed motion could be represented by the response of a Van der Pol oscillator. Hartlen and Currie used such a mathematical model for the lift and investigated its consequences when combined with a linear structural oscillator model. A number of authors have subsequently studied such models and some of the key results are summarized in Ref. 2.

A final comment about the nonlinear damping term is appropriate. As is well known this term gives negative damping for small C_L/C_{Lo}. Hence a limit cycle for the fluid alone is possible (even with no structural motion). Thus this fluid oscillator can model, albeit crudely, the fluid oscillations due to alternate vortex shedding from the shoulders of the bluff body.

Flutter of plates and shells

For this class of aeroelastic phenomena the principal nonlinearity is the in-plane stretching due to out-of-plane bending of the plate or shell. For simplicity, a one dimensional flat plate under a high supersonic flow (where piston theory aerodynamics is sufficient) will be considered. Recall Fig. 11.2. The focus of the discussion will be the flutter oscillations of a buckled plate. The variety of possible motions is rich indeed, including chaotic motions and a form of aeroelastic turbulence. For a detailed discussion, the reader is referred to Ref. 2, 3 and 41. Only a few highlights will be discussed here.

There are two dominant parameters (co-dimension two). They are

$$\lambda \equiv \frac{\rho_\infty U_\infty^2 a^3}{M D}$$

and

$$Rx \equiv N_X^E a^2/D$$

where
- ρ_∞ - flow density
- U_∞ - flow velocity
- a - plate length
- M - flow Mach number
- D - plate bending stiffness
- N_X^E - externally applied in-plane load; positive in tension

A parameter (or co-dimension) map can be constructed in terms of λ vs R_X that marks the several regions where various types of oscillations may occur. See Fig. 11.17. Of greatest interest here is the region of large, negative R_X (compressive in-plane load) and moderate λ where chaotic oscillations may occur. Chaotic oscillations may be studied by various descriptors, e.g.

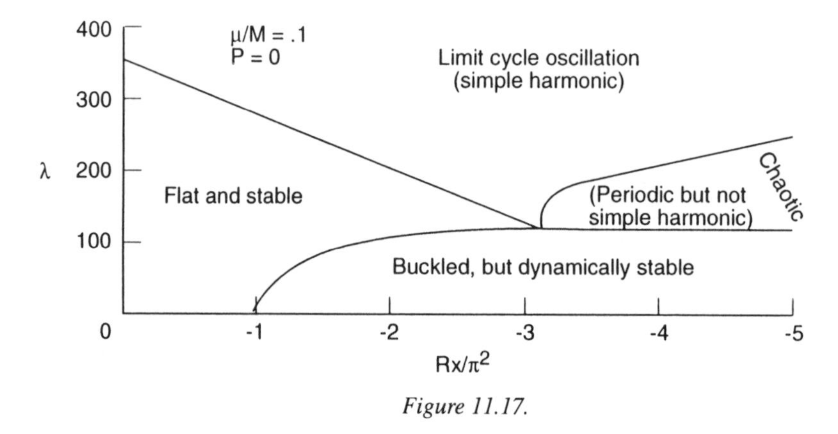

Figure 11.17.

time histories, phase plane portraits, power spectral densities, and Poincare maps. Whatever the descriptor used, such oscillations are endemic to nonlinear systems and have to all appearances the characteristics of random oscillations (e.g. broad band power spectra), even though all system parameters and inputs are entirely deterministic.

For the region of the parameter map where higher order periodic motion and then chaos occurs, a certain scenario for the beginning and evolution of chaotic motion takes place. As one enters the edges of this region, changes in the period of the limit cycle motion occur including sometimes abrupt changes through a period doubling sequence. Further, in the phase plane the periodic motion will trace a closed curve. As the parameters are changed to enter the chaotic region, the phase plane portrait diffuses and only a shadow of the closed curve of the nearby periodic motion is still observable. Also the sharp power spectral peaks of periodic motion broaden and the power spectra takes on a smoother distribution. Finally the Poincare maps (that are stroboscopic snapshots in time of the phase plane) also diffuse as the chaotic region is penetrated more deeply.

One further observation about the motion that occurs well within the chaotic region is appropriate. With deeper penetration into the chaotic region, more and higher structural modes become significant participants in the chaotic motion. There is a very interesting consequence of this higher mode participation. If the spatial-temporal cross-correlation of the plate deflection is examined, it is seen that the spatial correlation diminishes as the chaotic regime is penetrated further (R_x becomes more negative) and the higher modes are more prominent. This diminished spatial correlation due to multi-mode participation combined with loss of temporal correlation due to chaotic motion in each structural (aeroelastic) mode leads to a form of motion that may be appropriately called "turbulence". See Ref. 41 for further discussion of this phenomenon and representative results.

11.5 The future

It is always interesting, but also possibly hazardous to one's reputation for foresight, to speculate as to what the future may bring. A few predictions will be offered in the spirit of stimulating further thought and work by the reader. These are listed in Table 11.IV. It is hoped that this chapter has provided a basis for the reader to consider and improve upon the assertions of Table 11.IV.

TABLE 11.IV. THE FUTURE

- CALCULATION OF EIGENVALUES AND EIGENVECTORS OF FLUID FLOWS WILL BECOME STANDARD
- REDUCED ORDER AERODYNAMIC MODELS WILL BE WIDELY EMPLOYED
- NONUNIQUE SOLUTIONS TO FLOW FIELDS WILL BE RECOGNIZED AS PHYSICALLY MEANINGFUL
- AEROELASTIC TURBULENCE WILL PROVIDE A PARADIGM FOR UNDERSTANDING OTHER FORMS OF TURBULENCE
- EXPLOITATION OF NONLINEAR EFFECTS WILL LEAD TO IMPROVED VEHICLE PERFORMANCE
- HELICOPTER AND TURBOMACHINERY AEROELASTICITY AND COMPLIANT WALL DRAG REDUCTION WILL STILL PROVIDE FORMIDABLE CHALLENGES

While pondering the future, it is well to recall that although much of our knowledge is of recent origin, there were several early pioneers who began the study of nonlinear aeroelasticity many years ago. Some of these are acknowledged in the references. Among the earliest and most notable studies is that of Sisto [42] who considered stall flutter in cascades from a nonlinear perspective four decades ago.

References for Chapter 11

1. Von Karman, Theodore, "The Engineer Grapples with Nonlinear Problems," *Bulletin of the American Mathematical Society*, Vol. 46, 1940.
2. Dowell, E. H. and Ilgamov, M., "Studies in Nonlinear Aeroelasticity," Springer-Verlag, 1988.
3. Dowell, E. H., "Aeroelasticity of Plates and Shells," Noordhoff International Publishing, Leyden, 1975.
4. Bolotin, V. V., "Nonconservative Problems of the Elastic Theory of Stability," Pergamon Press, 1963.
5. Scanlan, R. H. and Rosenbaum, R., "Introduction to the Study of Aircraft Vibration and Flutter," The Macmillan Company, New York, N.Y., 1951. Also available in Dover Edition.
6. Fung, Y. C., "An Introduction to the Theory of Aeroelasticity," John Wiley and Sons, Inc., New York, N.Y., 1955. Also available in Dover Edition.
7. Bisplinghoff, R. L., Ashley, H. and Halfman, R. L., "Aeroelasticity," Addison-Wesley Publishing Company, Cambridge, Mass., 1955.
8. Bisplinghoff, R. L. and Ashley, H., "Principles of Aeroelasticity," John Wiley and Sons, Inc., New York, N.Y., 1962. Also available in Dover Edition.
9. AGARD Manual on Aeroelasticity, Vols. I-VII, Beginning 1959 with continual updating.
10. Ashley, H., Dugundji, J. and Rainey, A. G., "Notebook for Aeroelasticity," AIAA Professional Seminar Series, 1969.
11. Simiu, E. and Scanlan, R. H., Wind Effects on Structures: An Introduction to Wind Engineering, John Wiley and Sons, 1978.
12. Johnson, W., "Helicopter Theory," Princeton University Press, 1980.
13. Petre, A., "Theory of Aeroelasticity," Vol. I "Statics," Vol. II "Dynamics," In Romanian. Publishing House of the Academy of the Socialist Republic of Romania, Bucharest, 1966.
14. Forsching, H. W., "Fundamentals of Aeroelasticity," In German. Springer-Verlag, Berlin, 1974.
15. Dowell, E. H., Curtiss, Jr., H. D., Scanlan, R. H., and Sisto, F., A Modern Course in Aeroelasticity, Second Revised and Enlarged Edition, Kluwer Academic Publishers, Dordrecht, The Netherlands, 1989.
16. Liepmann, H. W., and Roshko, A., Elements of Gasdynamics, John Wiley, 1957.
17. Ormiston, R. A., Warmbrodt, W. G., Hodges, D. H., and Peters, D. A., "Survey of Army/NASA Rotorcraft Aeroelastic Stability Research," NASA TM 101026 and USAAVSCOM TR 88-A-005, October 1988.

18. Minorsky, N., Introduction to Nonlinear Mechanics, J. W. Edwards Publisher, Inc., Ann Arbor, Mich., 1947.
19. Stoker, J. J., Nonlinear Vibrations, Interscience Publishers, Inc., New York, 1950.
20. Cunningham, W. J., Introduction to Nonlinear Analysis, McGraw Hill Book Company, Inc., New York, 1959.
21. Guckenheimer, J. and Holmes, P. J., Nonlinear Oscillations, Dynamical Systems and Bifurcation of Vector Fields, Springer-Verlag, New York, 1983.
22. Thompson, J. and Stewart, H., Nonlinear Dynamics and Chaos, Wiley, London, 1986.
23. Moon, F. C., Chaotic Vibrations, Wiley, New York, 1987.
24. Dowell, E. H., "Chaotic Oscillations in Mechanical Systems," *Computational Mechanics*, Vol. 3, (1988) pp. 199-216.
25. Isogai, K., "Numerical Study of Transonic Flutter of a Two-Dimensional Airfoil," Technical Report of National Aerospace Laboratory, Japan, NAL-TR-617T, 1980.
26. Hounjet, M. H. L., "A Transonic Panel Method to Determine Loads on Oscillating Airfoils with Shocks," *AIAA Journal*, Vol. 19 (1981) pp. 559-66.
27. Seebass, R., "Advances in the Understanding and Computation of Unsteady Transonic Flows," Recent Advances in Aerodynamics, edited by A. Krothapalli and C. Smith, 1984.
28. Ballhaus, W. F. and Goorjian, P. M., "Efficient Solution of Unsteady Transonic Flows About Airfoils," Paper 14, AGARD Conference Proceedings No. 226, Unsteady Airload in Separated and Transonic Flows, 1978.
29. Borland, C. J. and Rizzetta, D. P., "Nonlinear Transonic Flutter Analysis," *AIAA Journal*, Vol. 20, No. 11 (1982), pp. 1606-15.
30. Batina, J. T., "Effects of Airfoil Shape Thickness, Camber, and Angle of Attack on Calculated Transonic Unsteady Airloads," NASA TM 86320 (1985).
31. Nixon, D., Editor, Unsteady Transonic Aerodynamics, *Progress in Astronautics and Aeronautics*, Volume 120, AIAA, 1989.
32. Sankar, N. L. and Tang, W., "Numerical Solution of Unsteady Viscous Flow Past Rotor Sections," AIAA Paper 85-0129, 1985.
33. Dowell, E. H., "Nonlinear Oscillator Models in Bluff Body Aeroelasticity," *J. Sound Vibration*, Vol. 75, No. 2 (1981), pp. 251-264.
34. Tran, C. T. and Petot, D., "Semi-empirical Model for the Dynamic Stall of Airfoils in View of the Application to the Calculation of Responses of a

Helicopter Blade in Forward Flight," Sixth European Rotorcraft and Powered Lift Forum, Bristol, England, September 1980.

35. Mahajan, A. J. and Kaza, K. R. V., "Semi-empirical Model for Prediction of Unsteady Forces on an Airfoil with Application to Flutter," submitted to *J. Fluids and Structures.*.
36. Mahajan, A. J., Dowell, E. H. and Bliss, D. B., "Eigensystem Analysis Procedure for a Navier-Stokes Solver with Application to Flows over Airfoils," *J. Comput. Phys.*, Vol. 97, No. 2 (1991), pp. 398-413.
37. Mahajan, A. J., Dowell, E. H., and Bliss, D. B., "On the Role of Artificial Viscosity in the Navier-Stokes Solvers," *AIAA J.*, Vol. 29, No. 4 (1991), pp. 555-559.
38. Brase, L. O., and Eversman, W., "Application of Transient Aerodynamics to the Structural Nonlinear Flutter Problem," *J. Aircraft,* Vol. 25, No. 11 (1988), pp. 1060-1068.
39. Hartlen, R. T., and Currie, I. G., "Lift Oscillator Model of Vortex-induced Vibration," ASCE EM5, (1970), pp. 577-591.
40. Bishop, R. E. D. and Hassan, A. Y., "The Lift and Drag Forces on an Oscillating Cylinder," *Proceedings of the Royal Society*, A277, (1964), pp. 32-75.
41. Dowell, E. H. and Virgin, L. N., "On Spatial Chaos, Asymptotic Modal Analysis and Turbulence,"*J. Applied Mechanics,* Vol. 57, No. 4 (1990), pp. 1094-1097
42. Sisto, F., "Staff-Flutter in Cascades," *J. Aeronautical Sciences*, Vol. 20, No. 9 (1953), pp. 598-604..

12

Aeroelastic control

The active control of aeroelastic systems, sometimes known as aeroservoelasticity, has as its objective the modification of the aeroelastic behavior of the system by the introduction of deliberate control forces. Aeroelastic control is in fact an intersection of aeroelasticity and of controlled structures technology (Figure 12.1). Aeroelasticity examines the interaction of aerodynamics, structures and dynamics, while controlled structures technology examines the interaction of control systems and structural dynamics.

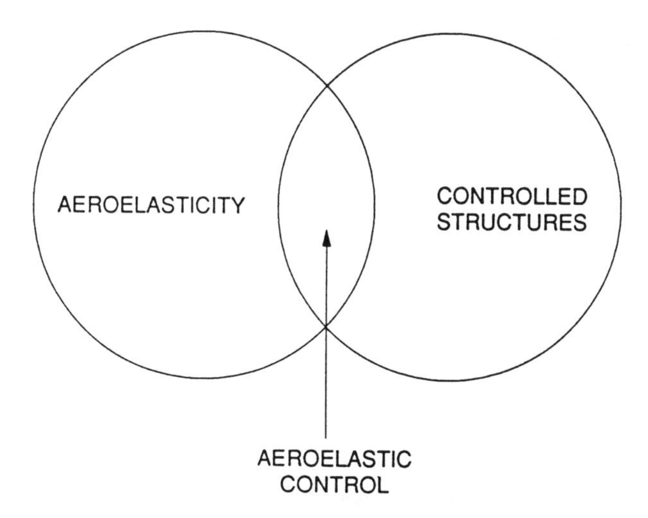

Figure 12.1: The relationship of aeroelastic control to aeroelasticity and controlled structures.

Aeroelastic control can be more challenging than the conventional controlled structures problem, in that the dynamics of the system change dramatically with the flight conditions. Similarly, aeroelastic control can be more difficult than traditional aeroelastic problems, in that the control system introduces a second potential source of instability. Despite its challenges, aeroelastic control holds the promise of significant improvements in performance: reducing the ambient vibration level, increasing the maneuver responsiveness, and stabilizing an otherwise unstable system. As a reflection of its importance, there is a growing literature on this topic. References [1-6] are especially recommended to the student.

Any aeroelastic plant can be compensated by using feedforward or feedback control. This chapter will focus on the feedback control of stationary wings. However, control can also be applied to other aeroelastic phenomena such as panel flutter, and in rotorcraft and turbomachine systems.

The objective of this chapter is to present a unified approach to aeroelastic control, which will lead to an understanding of its methods, applications, and limitations. The approach is to present an overall view of the methodology. A more detailed description will only be given of those necessary tools which are neither presented in other sections of this book nor normally covered in texts on control theory. This chapter *plus* the remainder of this text *and* a good text on modern control constitute the necessary primer on aeroelastic control.

The chapter begins with a review of the principal issues of aeroelastic control. The next section discusses the modeling of an aeroelastic system necessary for control design. In Section 12.3 the open loop dynamics of a typical section are developed, and the physical parameters which influence compensator design are identified. The final section of the chapter illustrates several possible approaches to compensation, and the fundamental advantages and limitations of each.

12.1 Objectives and elements of aeroelastic control

The design and analysis of a controlled aeroelastic system requires a clear understanding of the objectives and the elements of the problem. When a loop is closed around any plant, all aspects of the problem shown in Figure 12.2 must be considered. The complete system consists of: the aeroelastic plant and its accompanying sensors and actuators; the compensator which acts between the outputs of the measurements made by the sensors and the inputs to the actuators; the exogenous inputs to the system, the reference command input and disturbance input; and the performance output of the system, i.e. the performance objective function or metric.

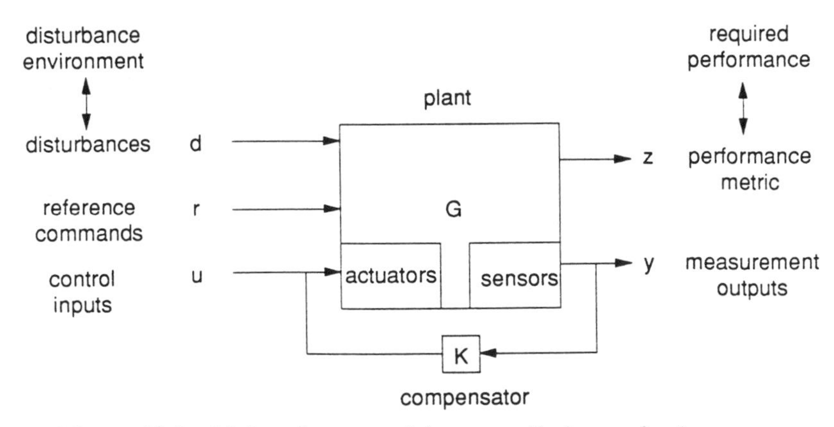

Figure 12.2: Major elements of the controlled aeroelastic system.

In the absence of compensation, the transfer function between reference input or disturbance environment to performance metric produces the reference or open loop performance of the system. Feedforward and feedback closed loop control are used when this open loop performance does not meet the required performance called for in the system specification.

The design of the compensator then depends on the objective function being used in aeroelastic control. The first and perhaps most frequently discussed objective of aeroelastic control is flutter suppression. Operationally, the objective is to enlarge the operational flutter envelope of the aircraft. This is done by stabilizing an aeroelastic system which would otherwise have encountered an instability or flutter. It can be thought of in a Laplace domain representation as moving poles by active control from the right half plane to the stable, or left half plane. The stability or instability of the aeroelastic system can be determined by examining the transfer function from any actuator to any state measurement, since the stability is established by the plant poles which are common to all transfer functions. From a control perspective, the problem of flutter suppression is synonymous with that of stabilization of an unstable plant. Stabilization is primarily a homogeneous modification to the system. Of course once stabilized, the system must also have good disturbance rejection characteristics.

The second possible objective of aeroelastic control is gust alleviation, which operationally smoothes the ride by minimizing the response to unsteady dynamic air loads. This can be for the benefit of the passenger, in which case it is called ride comfort, or for structural load alleviation. The specific objective of ride comfort is to minimize the acceleration response at the payload. The primary objective of load alleviation is to minimize the unsteady structural loads. Load alleviation can be used for increasing static load margin or extending fatigue life of an aircraft. In both of these cases, the important disturbance inputs to the system are the unsteady nonhomogeneous aerodynamic forces and moments. The performance metrics are the wing strains or fuselage accelerations. From a control perspective, the objective is to keep the aeroelastic system well-regulated and to increase its disturbance rejection properties.

The third possible objective of aeroelastic control is maneuver enhancement. Operationally, the goal is to improve the ability of the aircraft to produce sudden changes in lift and moment required for maneuver. The transfer functions important in this case are from the commanded reference inputs to the lift and moments produced on the lifting surface. From a control perspective, the objective is to make the wing a good servo-command tracking system.

Aside from the wing itself, the next most visible parts of an aeroelastic control system are the actuators. Potential actuators include conventional aerodynamic surfaces and less conventional structural actuation devices which directly produce bending and torsional moments on the lifting surface.

Generally, but not universally, lifting surfaces are fitted with some combination of leading edge and/or trailing edge flaps or ailerons. In principle, these can be made with sufficiently high bandwidth and authority to influence the aeroelastic problem. These surfaces are usually present and have the advantage of relatively high authority in terms of the aerodynamic loads they can produce. Significant lags between command and the generation of aerodynamic forces are the main disadvantage of control surfaces as actuator. These delays include the actuator lags (between command and control surface motion) due to the hydraulic and electromechanical systems. Cascaded with the actuator delays are aerodynamic lags (between control surface motion and the generation of lift and moment) due to the unsteady convective vorticity. Control surfaces also produce a non-collocated combination of lift and moment. The unsteady moment and lift generated are in no sense acting at the point of the hinge motion. Control surface effectiveness can also be reduced as the aircraft approaches reversal.

An alternative approach, which is a subject of current research, introduces direct structural control of wing bending or torsional moments. Such control can be introduced by the distributed placement of strain actuation such as piezoelectrics, electrostrictives, or shape memory alloys, or by the introduction of discrete actuator devices within a built-up wing structure. The advantages of these actuators include: they can be designed with high bandwidth relative to the aeroelastic frequencies of interest, they are not subject to reversal or reduced effectiveness, and their influence is independent of flight Mach number, dynamic pressure and angle of attack. The lags between the command input and actual appearance of force or moment are significantly less than those associated with unsteady aerodynamic control surfaces. Furthermore, the applied forces can be accompanied by collocated measurements of the motion. The disadvantages of such devices are that they are not part of the standard aircraft design and will add weight and complexity to aircraft. The authority of such actuators must also be carefully compared with that of lifting surface devices.

The measurements realistically available to the aeroelastic control designer are wing accelerations and strain in the wing. Accelerations are not states of the system, but can be related to states through the dynamics of the system. If discrete actuation devices are used, the motion and rate of these devices are also available, in principle.

The aeroelastic plant is that set of dynamics which relates the inputs to the outputs. It consists of four principal parts: the structural dynamic model, the aerodynamic model, and the dynamic models of the actuators and of the sensors. Typical structural dynamic models for aeroelastic control problems are similar to those used in aeroelastic analysis, often retaining the three to five lowest modes of the system for the purposes of performance as well as an additional number of modes in the cross-over frequency range of the system to insure closed loop stability. The cross-over frequency is that frequency below which closed loop control is effectively exerted and above which the authority rapidly diminishes.

Appended to the structural model is an aerodynamic model which must capture the complex frequency, Mach number and dynamic pressure dependence of the aerodynamics. The actuators and sensors contain additional electrical, mechanical, and aerodynamic lags which must also be included in the plant model.

The final aspect of the closed loop system is the compensator itself, which operates on measurement outputs to produce actuator inputs. The goals of compensation are to insure stability of the closed loop system, improve performance, and be robust to uncertainty or variation in the plant. The important classes of compensation are classical compensation and optimal compensation. Classical compensation entails either static output feedback or dynamic feedback and has the advantage of allowing the designer more direct insight into the performance and stability trade off and design. It is more readily applied to single input/single output (SISO) problems, but can be extrapolated to low order multiple input/multiple output (MIMO) problems. Optimal compensation in the form of state feedback or dynamic compensation can be more easily extended to MIMO problems, but provides less direct insight into the trade off between performance and robustness.

12.2 Modeling for aeroelastic control

Building an aeroelastic model of sufficient accuracy for use in designing closed loop controllers requires considerable knowledge of structural dynamics, aerodynamics and dynamics of other control system components. Figure 12.3 shows the elements of the model and the logical flow of tasks which lead to its development. Each element contributes states to the full order model. The full system includes models of the plant dynamics (structural dynamics and unsteady aerodynamics), actuator and sensor dynamics, disturbance dynamics, and dynamics associated with electronic and mechanical components.

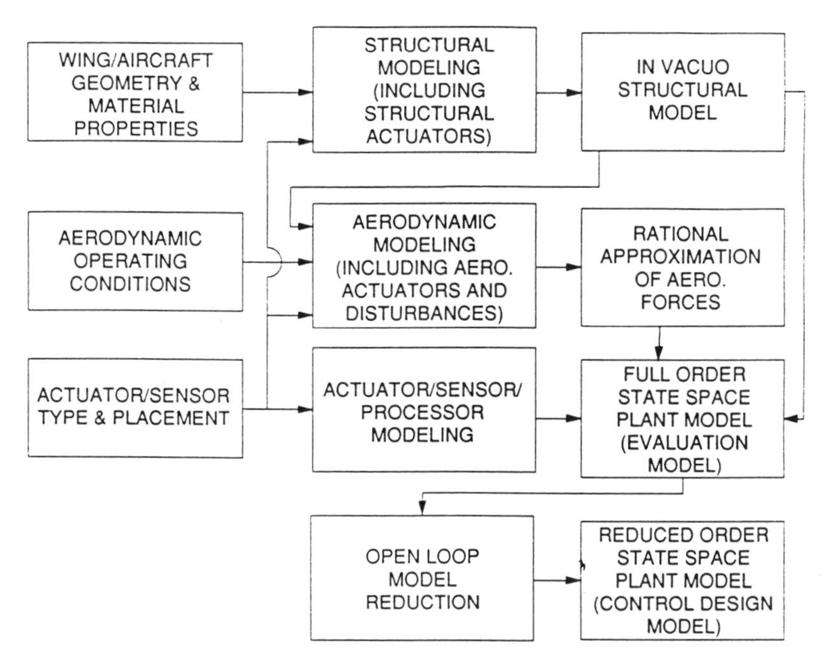

Figure 12.3: The process of deriving a model for aeroelastic control.

In the notation of Bisplinghoff, Ashley and Halfman [7], the aeroelastic control model can be written

$$S(q) - A(q) - I(q) = C(q) + \tilde{Q}_w \tag{12.2.1}$$

where the terms represent the structural, aerodynamic, inertial, and control operators and the exogenous inputs respectively. All but the last depend on the generalized coordinates q. The new aspects of the aeroelastic control problem are the appearance of the control operator, and the fact that the exogenous or nonhomogeneous term is now

$$\tilde{Q}_w = \tilde{Q}_g + \tilde{Q}_r \tag{12.2.2}$$

where $\tilde{Q}_g$ are the true gust disturbances and $\tilde{Q}_r$ are related to the reference commands, present when the controlled lifting surface is part of a servo system. The control problem is also accompanied by the output relations

$$y = \lfloor C_{yq} \; C_{y\dot{q}} \rfloor \begin{Bmatrix} q \\ \dot{q} \end{Bmatrix} + D_{yr} r \tag{12.2.3}$$

$$z = \lfloor C_{zq} \; C_{z\dot{q}} \rfloor \begin{Bmatrix} q \\ \dot{q} \end{Bmatrix} + D_{zr} r \tag{12.2.4}$$

where y is the output available for closed loop control, r is the reference input and z is the measure of the performance variable or metric. The output and performance variables can be distinct both spatially and temporally. For example, in a gust alleviation problem, the output variable y might be a tip accelerometer output, and the performance variable z the root bending strain.

The information in equations (12.2.1) through (12.2.4) can be represented in terms of the standard problem of Figure 12.4. The exogenous inputs w drive the forces $\tilde{Q}_w$ which act on the system. The control inputs u drive the control forces $C(q)$. Implicit in Figure 12.4 are four transfer functions: from disturbance to performance G_{zw}, disturbance to output G_{yw}, input to output G_{yu} and input to performance G_{zu}.

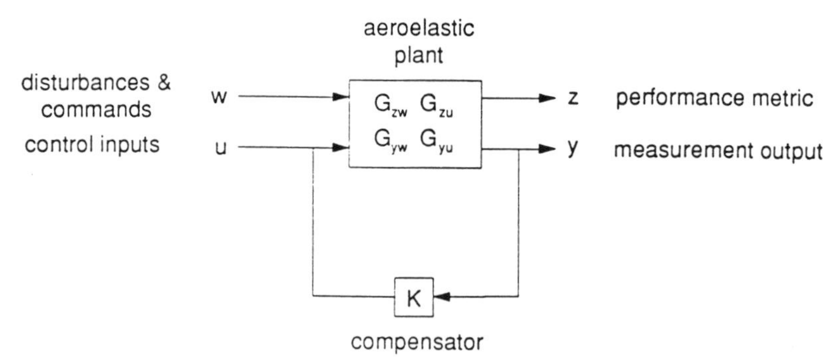

Figure 12.4: The four system transfer functions.

The requirements for a structural dynamics model are: (1) that it contain enough states and information to accurately represent these four transfer functions in the bandwidth of interest; and (2) that its order be as small as possible, in order to not unnecessarily complicate the control design. These two requirements are usually in conflict, as will be discussed below.

The remainder of this section is organized along the lines of the three main rows of Figure 12.3: the structural (and inertial) modeling; the aerodynamic modeling; and the other model elements, including the actuators, sensors, and control system. The section concludes with a discussion of the reduction of the resulting high order model to one appropriate for control system design.

Structural modeling

The structural operators of Equation (12.2.1) are modeled by the *in vacuo* open loop dynamics of the lifting surface. In order to isolate the structural dynamics, the aerodynamic operator is moved to the right hand side

$$S(q)-I(q)=\tilde{Q}=\tilde{Q}_{AH}+\tilde{Q}_{A\delta}+\tilde{Q}_{\Lambda}+\tilde{Q}_{g}+\tilde{Q}_{r} \tag{12.2.5}$$

where the terms on the right hand side represent the homogeneous aerodynamics $\tilde{Q}_{AH}$, the aerodynamic forces due to control surface motion $\tilde{Q}_{A\delta}$, the forces due to structural actuators $\tilde{Q}_{\Lambda}$, the gust disturbance forces $\tilde{Q}_{g}$ and the terms associated with reference command inputs $\tilde{Q}_{r}$.

The actual starting point of the model is a discretized structural equation

$$m\ddot{q}+kq=\tilde{Q} \tag{12.2.6}$$

derived from a finite element, Ritz or other numerical scheme. In Equation (12.2.6), m and k include: the mass and stiffness of the structure; the structural actuators, and any stores or attachments; and the mass, but not the stiffness, associated with articulating control surfaces.

The normal modes and frequencies of Equation (12.2.6) are determined, and the coordinates expanded using the modes as a basis

$$q=\Phi\xi \tag{12.2.7}$$

where Φ is a matrix of the individual eigenvectors ϕ_j, which can include rigid body translations and small angle rotations. Substituting Equation (12.2.7) into (12.2.6), premultiplying by Φ^T, and invoking orthogonality yields the *in vacuo* open loop modal equations

$$\begin{bmatrix}\ddots & & \\ & M_i & \\ & & \ddots\end{bmatrix}\{\ddot{\xi}\}+\begin{bmatrix}\ddots & & \\ & M_i\omega_i^2 & \\ & & \ddots\end{bmatrix}\{\xi\}=\{\Xi\} \tag{12.2.8}$$

where the modal mass and force are

$$M_i = \phi_i^T m \phi_i$$
$$\Xi_i = \phi_i^T \tilde{Q}$$
$$(12.2.9)$$

Structural damping can be included as modal linear viscous damping

$$\begin{bmatrix} \ddots & & \\ & M_i & \\ & & \ddots \end{bmatrix}\{\ddot{\xi}\} + \begin{bmatrix} \ddots & & \\ & 2\zeta_i M_i \omega_i & \\ & & \ddots \end{bmatrix}\{\dot{\xi}\} + \begin{bmatrix} \ddots & & \\ & M_i \omega_i^2 & \\ & & \ddots \end{bmatrix}\{\xi\} = \{\Xi\} \tag{12.2.10}$$

The dimension of this model is equal to the number of assumed degrees of freedom (N_q). In general, this is too large to conveniently use in the subsequent modeling steps. Therefore a first truncation is made to yield a model of order N_S, which retains the dynamics beyond the bandwidth of interest, but truncates dynamics which are not physically meaningful or are well outside the bandwidth of interest. This truncation yields the structural modal equation

$$M_{\xi\xi}\ddot{\xi} + C_{\xi\xi}\dot{\xi} + K_{\xi\xi}\xi = \Phi_\xi^T \tilde{Q} \tag{12.2.11}$$

where $M_{\xi\xi}$, $C_{\xi\xi}$ and $K_{\xi\xi}$ are of order N_S x N_S . The rectangular matrix Φ_ξ retains those modes which have not been truncated and is of order N_q x N_S.

Since the aerodynamic actuators are part of the dynamic system and are affected by the aerodynamics (which will be modeled next), they must be appended to the model at this time. The motion of the control surfaces can be expressed as additional shape functions, so that the generalized coordinates can be written in terms of the normal modal coordinates and control surface deflections

$$q = \begin{bmatrix} \Phi_\xi & \Phi_\delta \end{bmatrix} \begin{Bmatrix} \xi \\ \delta \end{Bmatrix} \tag{12.2.12}$$

where Φ_δ is the N_q x N_δ matrix of vectors corresponding to the N_δ deflections of the control surfaces δ, defined relative to the adjacent lifting surface. Note that Φ_δ is not orthogonal to Φ_ξ, so that substitution of Equation (12.2.12) into (12.2.6) will yield

$$\begin{bmatrix} M_{\xi\xi} & M_{\xi\delta} \\ M_{\xi\delta}^T & M_{\delta\delta} \end{bmatrix} \begin{Bmatrix} \ddot{\xi} \\ \ddot{\delta} \end{Bmatrix} + \begin{bmatrix} C_{\xi\xi} & 0 \\ 0 & 0 \end{bmatrix} \begin{Bmatrix} \dot{\xi} \\ \dot{\delta} \end{Bmatrix} + \begin{bmatrix} K_{\xi\xi} & 0 \\ 0 & 0 \end{bmatrix} \begin{Bmatrix} \xi \\ \delta \end{Bmatrix}$$
$$= \begin{bmatrix} \Phi_\xi^T \\ \Phi_\delta^T \end{bmatrix} \left\{ \tilde{Q}_{AH} + \tilde{Q}_{A\delta} + \tilde{Q}_\Lambda + \tilde{Q}_g + \tilde{Q}_r \right\} + \begin{bmatrix} 0 \\ M_\delta \end{bmatrix} \qquad (12.2.13)$$

The choice of δ as a relative coordinate produces a coupled mass matrix, but uncoupled damping and stiffness matrices, and has isolated the hinge moments M_δ (of dimension N_δ x 1) due to the actuators at the hinges of the control surfaces.

Aerodynamic modeling

The second row of Figure 12.3 represents the process of aerodynamic modeling. Modeling of the aerodynamics provides the terms in Equation (12.2.5) identified as $\tilde{Q}_{AH}$, $\tilde{Q}_{A\delta}$, and $\tilde{Q}_g$. The homogeneous aerodynamics $\tilde{Q}_{AH}$ depend on the lifting surface motion due to flexibility and rigid body vehicle motion. The control surface induced forces $\tilde{Q}_{A\delta}$ depend on the control surface deflection δ relative to the adjacent lifting surface. The gust forces $\tilde{Q}_g$ depend on the gust velocities at a reference point, typically the quarter chord.

The traditional form in which the aerodynamic forces are calculated contains an implicit or explicit irrational dependence on reduced frequency, as in (4.2.15) or (4.3.65). However, to be compatible with most control design approaches, the aerodynamics must be expressed as an explicit rational function of the states of the system. As implied by Figure 12.3, this transformation can be accomplished by employing a two step process:

1. Identify the structural modes for which aerodynamic effects will be included. Then calculate the unsteady aerodynamic forces due to these modal motions, the control surface motions and gust inputs. Use an aerodynamic model appropriate for the Mach number of interest, at distinct values of the reduced frequency k, and over the frequency range of interest.
2. Fit the aerodynamic forces with a rational function approximation in the frequency domain, and transform the rational function approximation to the time domain.

This process produces aerodynamic forces which depend on displacements, velocities and accelerations, as well as terms which depend on augmented states which incorporate the physics of the aerodynamic lags.

The first step begins by defining the coordinates for which aerodynamic influences will be calculated. Assuming that aerodynamics are important for all of the N_S structural modes and the N_δ control surface deflections of (12.2.12), the coordinate of these modes can be defined as

$$\xi_A = \begin{Bmatrix} \xi \\ \delta \end{Bmatrix} \tag{12.2.14}$$

where ξ_A is N_A x 1. The modal forces in (12.2.13) due to the aerodynamics are then

$$\begin{bmatrix} \Phi_\xi^T \\ \Phi_\delta^T \end{bmatrix}\left[\tilde{Q}_{AH} + \tilde{Q}_{A\delta} + \tilde{Q}_g\right] = \left[\Phi_{\xi A}^T\right]\left[\tilde{Q}_{\xi A} + \tilde{Q}_g\right] = \bar{q}Q_{\xi A}\xi_A(s) + \bar{q}Q_g\alpha_g(s) \tag{12.2.15}$$

where Φ_ξ are the structural modes and Φ_δ are the control surface modes for which aerodynamic forces will be calculated, $\bar{q}$ is the dynamic pressure, $Q_{\xi A}$ is the matrix of modal aerodynamic force coefficients due to structural modal motion and control surface deflections, Q_g is the corresponding matrix of coefficients due to gusts, and α_g is the nondimensional gust velocity

$$\alpha_g = \begin{Bmatrix} \dfrac{w_g}{U} & \dfrac{v_g}{U} \end{Bmatrix}^T \tag{12.2.16}$$

of order N_g x 1.

The elements of the influence matrices $Q_{\xi A}$ and Q_g of (12.2.15) are calculated by

$$Q_{ij}(s) = \iint_{area} \frac{\Delta P_j(x,y,s)}{\bar{q}}\phi_i(x,y)\,dA \tag{12.2.17}$$

where ΔP_j is the pressure difference due to the motion of the *j'th* structural mode, control surface deflection , and gust input as appropriate. These influence matrices are calculated using the methods outlined in Chapter 4. For example, techniques such as the doublet lattice or kernel function methods are employed for subsonic flow. Appropriate interpolations must be used to transform forces between the structural nodes and the aerodynamic control points. The results of these calculations are the actual values of the coefficient matrices Q_{ij}, evaluated at a number of discrete reduced frequencies $Q_{ij}(ik_n)$.

Step two is to fit these exact values $Q_{ij}(ik_n)$ using a rational function approximation. The most common form of the approximation is

$$\hat{Q}_{ij} = (A_0)_{ij} + (A_1)_{ij}(ik) + (A_2)_{ij}(ik)^2 + \sum_{\ell=1}^{N_L} (A_{\ell+2}) \frac{(ik)}{(ik) + \bar{a}_\ell} \tag{12.2.18}$$

where $\hat{Q}_{ij}$ is the estimate of the exact calculated values and N_L is the order of the denominator polynomial used to fit the lags. Physically, A_o, A_1 and A_2 capture the dependence of the unsteady aerodynamics on displacement, velocity and acceleration. The partial fraction sum captures the lag in the build up of aerodynamic forces associated with the circulatory effects. Therefore the poles in the partial fraction are commonly referred to as the lag poles. Note that since there is an (ik) in the numerator of the last term in (12.2.18), the lags actually are referenced to the velocity of motion. Various other forms of (12.2.18) can used for the fit, as will be discussed below. However, all forms of the fit include the physics captured in (12.2.18).

The transformation to the time domain is accomplished by first transforming from the Fourier to Laplace domain. Allowing

$$ik = \frac{i\omega b}{U} = \frac{sb}{U} \tag{12.2.19}$$

and substituting into Equation (12.2.18), yields

$$\hat{Q}_{ij}(s) = (\bar{A}_o)_{ij} + (\bar{A}_1)_{ij} s + (\bar{A}_2)_{ij} s^2 + \sum_{\ell=1}^{N_\ell} (\bar{A}_{\ell+2})_{ij} \frac{s}{s + \bar{a}_\ell} \tag{12.2.20}$$

where

$$\bar{a}_\ell = \frac{U}{b} a_\ell \qquad \bar{A}_o = A_o$$

$$\bar{A}_1 = \frac{b}{U} A_1 \qquad \bar{A}_2 = \left(\frac{b}{U}\right)^2 A_2 \quad \bar{A}_{\ell+2} = A_{\ell+2} \tag{12.2.21}$$

The impact of these steps is to transform a result which is rigorously only valid for oscillatory motion into one which can be applied to arbitrary motion. Mathematically, the concept of analytical continuation is cited as justification for this transformation. Physically, the approximation made by this transformation holds up well for lightly damped systems, and systems on the verge of instability.

Three common forms of the rational function approximation are used (see Tiffany and Adams [3] for an excellent discussion of this and extended forms, as well as the issue of the numerical fitting). In the most common of the three forms, called simply the Least Squares Method, Equation (12.2.20), which is written for each element of the influence matrix, is actually evaluated for the matrix as a whole.

$$\hat{Q} = \overline{A}_o + \overline{A}_1 s + \overline{A}_2 s^2 + \sum_{\ell=1}^{N_\ell} (\overline{A}_{\ell+2}) \frac{s}{s+\overline{a}_\ell} \tag{12.2.22}$$

This has the effect of introducing the same lag dynamics for all $N_A + N_g$ coordinates. The matrix of (12.2.22) is of dimension $N_A \times (N_A + N_g)$. Substituting (12.2.22) into (12.2.15) gives for the aerodynamic modal forces

$$\left[\Phi_{\xi A}^T\right]\left[\tilde{Q}_{\xi A} + \tilde{Q}_g\right] = \tag{12.2.23}$$

$$\overline{q}\left\{\overline{A}_{o_A} + \overline{A}_{1_A} s + \overline{A}_{2_A} s^2 + \sum_{\ell=1}^{N_\ell} (\overline{A}_{\ell+2})_A \frac{s}{s+\overline{a}_\ell}\right\}\xi_A$$

$$+\overline{q}\left\{\overline{A}_{og} + \overline{A}_{1g} s + \overline{A}_{2g} s^2 + \sum_{\ell=1}^{N_\ell} (\overline{A}_{\ell+2})_g \frac{s}{s+\overline{a}_\ell}\right\}\alpha_g$$

The $\overline{A}_{2g}$ term can usually be eliminated with no loss of physical generality [4]. Augmented aerodynamic states can be defined by

$$x_{A\ell} = \frac{s}{s+\overline{a}_\ell}\xi_A \tag{12.2.24}$$

$$x_{g\ell} = \frac{s}{s+\overline{a}_\ell}\alpha_g$$

Substitution into (12.2.23) and transformation to the time domain yields the sought after expression for the modal unsteady aerodynamic forces

$$\left[\Phi_{\xi A}^{T}\right]\left[\tilde{Q}_{\xi A}+\tilde{Q}_{g}\right]= \tag{12.2.25}$$

$$\bar{q}\left\{\bar{A}_{o_A}\xi_A+\bar{A}_{1_A}\dot{\xi}_A+\bar{A}_{2_A}\ddot{\xi}_A+\sum_{\ell=1}^{N_\ell}(\bar{A}_{\ell+2})_A x_{A\ell}\right\}$$

$$+\bar{q}\left\{\bar{A}_{og}\alpha_g+\bar{A}_{1g}\dot{\alpha}_g+\bar{A}_{2g}\ddot{\alpha}_g+\sum_{\ell=1}^{N_\ell}(\bar{A}_{\ell+2})_g x_{g\ell}\right\}$$

$$\begin{aligned}\dot{x}_{A\ell}+\bar{a}_\ell x_{A\ell}&=\dot{\xi}_A\\ \dot{x}_{g\ell}+\bar{a}_\ell x_{g\ell}&=\dot{\alpha}_g\end{aligned} \tag{12.2.26}$$

where the total number of augmented states is $\left(N_A+N_g\right)\times N_\ell$. In a slight refinement of this method, the number of augmented aerodynamic states can be reduced to $N_A\times N_\ell$ [2]. Finally, this modal forcing term can be incorporated into the structural dynamic equation of motion (12.2.13). The final form of these equations will be shown below in (12.2.47).

In another variation, known as the Modified Matrix Padé formulation, the form of Equation (12.2.20) is assumed to hold for each column of the matrix of influence coefficients, such that

$$\hat{Q}_j=(\bar{A}_o)_j+(\bar{A}_i)_j s+(\bar{A}_2)_j s^2+\sum_{\ell=1}^{N_\ell}(A_{\ell+2})_j\frac{s}{s+\bar{a}_\ell} \tag{12.2.27}$$

Since the columns are associated with the aerodynamic forces due to the modal and gust coordinates, this allows separate fitting of the poles for the influence of each structural mode, control surface motion, and gust. In principle, different numbers of poles can be fit for each column j. The total number of augmented aerodynamic states is

$$\sum_{j=1}^{N_A+N_g}N_{\ell_j} \tag{12.2.28}$$

If all structural and gust modes are fit with the same number of lags, this reduces to $(N_A+N_g)\times N_\ell$, the same as the simple Least Squares Method.

In a more direct fitting approach, Karpel [4] suggested an alternative relation to (12.2.20) of the form

$$\hat{Q}=\bar{A}_o+\bar{A}_1 s+\bar{A}_2 s^2+\bar{D}\left[sI-\bar{R}\right]^{-1}\bar{E}s \tag{12.2.29}$$

where the total number of aerodynamic lag states is set by the choice of the dimension of $\bar{R}$. This approach has the potential of substantially reducing the number of states necessary to represent the aerodynamic lags, with accuracy equivalent to the two previous methods.

For any of the rational function approximation forms, an important step is the actual process of fitting the approximation to the data points. In the simplest approach, the pole locations are set by judgment and inspection of the operators over the reduced frequency range of interest. Then the square error between the calculated values and functional fit is minimized.

$$\varepsilon_{ij} = \sum_{n} [\hat{Q}_{ij}(ik_n) - Q_{ij}(ik_n)]^2 \tag{12.2.30}$$

When the poles are fixed, minimization leads to a set of linear equations.

$$\frac{\partial \varepsilon_{ij}}{\partial (A_m)_{ij}} = 0 \quad m = 1,\ldots,N_\ell + 2 \tag{12.2.31}$$

This fit is often constrained. For example, by matching the zero frequency asymptote exactly and/or by matching the data exactly at the expected flutter frequency. In more advanced solutions, the value of the poles and the order of the rational function approximation are varied to find truly optimal fits. These improvements lead to more computationally intensive solution routines, but generally reduce the order of the model necessary for a given accuracy.

Other model elements

Other elements, listed in the third row of Figure 12.3, which must be modeled within the bandwidth of interest include: the structural actuators, reference commands, sensors and the performance variables, as well as lags in actuators, sensors, and processing.

Structural actuators are those which directly deform the wing, such as internal hydraulic actuators or piezoelectric strain actuators. They influence only the structural normal modes, so that the appropriate term of the right hand side of (12.2.13) is

$$\begin{bmatrix} \Phi_\xi^T \\ \Phi_\delta^T \end{bmatrix} \tilde{Q}_\Lambda = \begin{bmatrix} \Phi_\xi^T \beta_\Lambda \\ 0 \end{bmatrix} u_{\Lambda s} \tag{12.2.32}$$

where $\tilde{Q}_{\Lambda}$ is the vector of physical forces and $u_{\Lambda s}$ is the (potentially smaller) vector of structural control inputs. The mass and stiffness of structural actuators should be included in the original finite element model (12.2.6).

The influence of the control inputs on the control surfaces appears in the vector of hinge moments

$$M_{\delta} = \beta_{\delta} u_{\delta s} \tag{12.2.33}$$

where M_{δ} is the vector of physical torques in (12.2.13), and $u_{\delta s}$ is a (potentially smaller) vector of structural control inputs. The mass of control surfaces should be included in the original finite element model (12.2.6).

The output of structural sensors which depend on the states may be represented as

$$y_{sx} = \begin{bmatrix} C_{yq} \, C_{y\delta} \, C_{y\dot{q}} \, C_{y\dot{\delta}} \end{bmatrix} \begin{Bmatrix} q \\ \delta \\ \dot{q} \\ \dot{\delta} \end{Bmatrix} = \begin{bmatrix} C_{yq}\Phi_{\xi} & C_{y\delta} & C_{y\dot{q}}\Phi_{\xi} & C_{y\dot{\delta}} \end{bmatrix} \begin{Bmatrix} \xi \\ \delta \\ \dot{\xi} \\ \dot{\delta} \end{Bmatrix} \tag{12.2.34}$$

Likewise, the performance metric which depends on the states is

$$z_{x} = \begin{bmatrix} C_{zq} \, C_{z\delta} \, C_{z\dot{q}} \, C_{z\dot{\delta}} \end{bmatrix} \begin{Bmatrix} q \\ \delta \\ \dot{q} \\ \dot{\delta} \end{Bmatrix} = \begin{bmatrix} C_{zq}\Phi_{\xi} & C_{z\delta} & C_{z\dot{q}}\Phi_{\xi} & C_{z\dot{\delta}} \end{bmatrix} \begin{Bmatrix} \xi \\ \delta \\ \dot{\xi} \\ \dot{\delta} \end{Bmatrix} \tag{12.2.35}$$

It is chosen to reflect the actual objective of the design. For example, if load alleviation is an objective, a performance output matrix which reflects root strain is appropriate.

Reference commands to the system are usually in the form of displacements and velocities. Since the objective of such a servo is to cause the measured outputs to track the commands, these reference inputs to the system often take the form of reference signals in the perceived measurement and performance outputs of the system.

$$\begin{aligned} y_r &= D_{yr} r \\ z_r &= D_{zr} r \end{aligned} \tag{12.2.36}$$

These are added to the actual measurements and performance metrics as shown in Figure 12.5. In the event that there are dynamics associated with the servo tracking loop, as is the case when integral trim is added, the reference command can also appear in the input to the state equations in such a way as to drive the augmented dynamics associated with the servo.

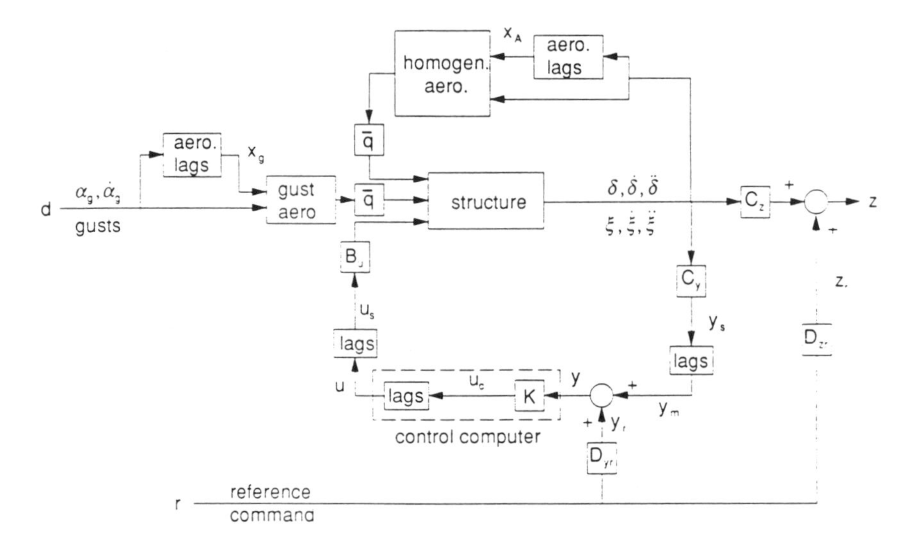

Figure 12.5: Detailed block diagram of the controlled aeroelastic system.

Actuator-sensor dynamics: Many physical devices can have lags associated with computational delays, and dynamics associated with the device itself or its signal conditioning circuitry. In the event the actuator has dynamics in the bandwidth of interest, these must be added to the system model. In such a case, it can be assumed that the control input to the plant, i.e. the signal which runs on the wires from the control computer in Figure 12.5,

$$u = \begin{Bmatrix} u_\Lambda \\ u_\delta \end{Bmatrix} \tag{12.2.37}$$

drives actuator dynamic models which capture the actual device lags

$$\begin{aligned} \dot{x}_{\Lambda_i} &= -A_{\Lambda_i} x_{\Lambda_i} + A_{\Lambda_i} u_{\Lambda_i} \qquad & i = 1,\ldots,N_\Lambda \\ \dot{x}_{\delta_i} &= -A_{\delta_i} x_{\delta_i} + A_{\delta_i} u_{\delta_i} \qquad & i = 1,\ldots,N_\delta \end{aligned} \tag{12.2.38}$$

here modeled as simple first order poles. Of course, higher order actuator dynamics are possible, but these can be reduced to an appropriate first order state space model of the form of (12.2.38). The actual structural inputs now equal these lagged variables

$$u_s = \begin{Bmatrix} u_{\Lambda s} \\ u_{\delta s} \end{Bmatrix} = \begin{Bmatrix} x_\Lambda \\ x_\delta \end{Bmatrix} \tag{12.2.39}$$

instead of $u_s = u$, which is the case with no actuator dynamics.

For a case when sensors have dynamics, and/or when accelerometers are present, the structural outputs y_s can be partitioned

$$y_s = \begin{Bmatrix} y_{sx} \\ y_{s\ddot{q}} \end{Bmatrix} = \begin{bmatrix} C_{yq} \, C_{y\dot{q}} & 0 \\ 0 \quad 0 & C_{y\ddot{q}} \end{bmatrix} \begin{Bmatrix} q \\ \dot{q} \\ \ddot{q} \end{Bmatrix} + \begin{bmatrix} C_{y\delta} \, C_{y\dot{\delta}} & 0 \\ 0 \quad 0 & C_{y\ddot{\delta}} \end{bmatrix} \begin{Bmatrix} \delta \\ \dot{\delta} \\ \ddot{\delta} \end{Bmatrix} \tag{12.2.40}$$

where y_{sx} depends on the structural states and $y_{s\ddot{q}}$ depends on the structural accelerations. The sensor dynamics are modeled by letting the state measurements drive first order dynamics

$$\dot{x}_{y_i} = -A_{y_i} x_{y_i} + A_{y_i} y_{sx\,i} \qquad i = 1,\ldots,N_{y_{sx}} \tag{12.2.41}$$

Acceleration is not a state, and therefore does not appear in the state vector. However, acceleration is often an output. If the accelerometer has dynamics, then the measured structural acceleration drives these dynamics.

$$\dot{x}_{\ddot{y}_i} = -A_{\ddot{y}_i} x_{\ddot{y}_i} + A_{\ddot{y}_i} y_{s\ddot{q}_i} \qquad i = 1,\ldots,N_{y_{s\ddot{q}}} \tag{12.2.42}$$

In order to calculate the driving term of this equation, $\ddot{\xi}$ and $\ddot{\delta}$ are expressed in terms of $\xi, \dot{\xi}, \delta, \dot{\delta}$ and the force inputs using the equations of motion. For example, in the open loop case, in the absence of aerodynamic or control forces

$$\ddot{\xi} = -M_{\xi\xi}^{-1} K_{\xi\xi} \xi - M_{\xi\xi}^{-1} C_{\xi\xi} \dot{\xi} + M_{\xi\xi}^{-1} \Xi \tag{12.2.43}$$

In the presence of aerodynamics and closed loop control, the full closed loop state equation should be used to evaluate the accelerations. This is notationally more complex, but conceptually identical to (12.2.43).

In the case of lags in the sensors and accelerometer measurements, the system outputs as shown in Figure 12.5 are now

$$y=\begin{Bmatrix} x_y \\ x_{\ddot{y}} \end{Bmatrix}+y_r=y_m+y_r \tag{12.2.44}$$

where y_m is the vector of measurement outputs. In the case of no sensor dynamics, then $y=y_s+y_r$.

Computational lags: Most modern compensation is performed by digital computers, which introduce both a time delay and a discrete nature of the output. If the state equations are transformed to a discrete time domain representation [8], these lags can be treated exactly. A more common approach is to derive compensators in the continuous time domain, and model the time delay with an approximation. This lag associated with the time delay actually occurs within the processor, but can be represented by lags at either the outputs or the inputs. Since there are usually fewer inputs to the plant, introducing the computational lags there will minimize the additional states required.

In the frequency domain, the influence of a discrete time delay can be modeled by approximations of increasing order and accuracy. The simplest first order frequency domain approximation of a time delay is

$$u=\frac{1-{Ts}/{2}}{1+{Ts}/{2}}u_c=\left(\frac{2}{1+{Ts}/{2}}-1\right)u_c \tag{12.2.45}$$

where T is the time delay. This can be written in state space form

$$\begin{aligned} \dot{x}_d &= -a_d x_d + 2a_d u_c \\ u &= x_d - u_c \end{aligned} \tag{12.2.46}$$

Higher order approximations of a time delay can be similarly transformed to a continuous time state space representation. Note by reference to Figure 12.5 that u_c is the input that a compensator is tricked into thinking it makes, but when physically implemented in hardware, the output of the digital computer u is the input which actually will be used to drive the system.

Model summary

In order to demonstrate a complete plant model, the equations for a typical case will be summarized. For the case of aerodynamic lags on all modes, dynamics in all sensors and actuators (but no accelerometer outputs), a uniform computational delay and reference commands which correspond to measured outputs, the second order equations of motion are

$$\begin{bmatrix} M_{\xi\xi} & M_{\xi\delta} \\ M_{\xi\delta}^T & M_{\delta\delta} \end{bmatrix} \begin{Bmatrix} \ddot{\xi} \\ \ddot{\delta} \end{Bmatrix} + \begin{bmatrix} C_{\xi\xi} & 0 \\ 0 & 0 \end{bmatrix} \begin{Bmatrix} \dot{\xi} \\ \dot{\delta} \end{Bmatrix} + \begin{bmatrix} K_{\xi\xi} & 0 \\ 0 & 0 \end{bmatrix} \begin{Bmatrix} \xi \\ \delta \end{Bmatrix}$$

$$= \bar{q} \left\{ \bar{A}_{0A} \begin{Bmatrix} \xi \\ \delta \end{Bmatrix} + \bar{A}_{1A} \begin{Bmatrix} \dot{\xi} \\ \dot{\delta} \end{Bmatrix} + \bar{A}_{2A} \begin{Bmatrix} \ddot{\xi} \\ \ddot{\delta} \end{Bmatrix} + \bar{A}_{3A} x_{A_1} + \ldots + \bar{A}_{(N_\ell+2)A} x_{A_{N_\ell}} \right\}$$

$$+ \bar{q} \left\{ \bar{A}_{0g} \left\{ \alpha_g \right\} + \bar{A}_{1g} \left\{ \dot{\alpha}_g \right\} + \bar{A}_{2g} \left\{ \ddot{\alpha}_g \right\} + \bar{A}_{3g} x_{g_1} + \ldots + \bar{A}_{(N_\ell+2)g} x_{g_{N_\ell}} \right\}$$

$$+ \begin{bmatrix} \Phi_\xi^T \beta_\Lambda & 0 \\ 0 & \beta_\delta \end{bmatrix} \begin{Bmatrix} u_{\Lambda s} \\ u_{\delta s} \end{Bmatrix} \tag{12.2.47}$$

The output equation is given by a simplified form of (12.2.44)

$$y = x_y + y_r \tag{12.2.48}$$

where x_y is given by (12.2.41) and (12.2.40) and y_r by (12.2.36). The output of the compensator u_c is related to the system input u (i.e. the physical output of the control computer) by (12.2.46), and the structural inputs which appear in (12.2.47) are given by (12.2.38) and (12.2.39). Finally, the performance is indicated by (12.2.35).

In order to prepare them for the form expected in subsequent reduction and control design operations, the equations of motion are now transformed into first order state space form. .The homogeneous aerodynamics are first incorporated into the left hand side of (12.2.47)

$$\bar{M} = \begin{bmatrix} M_{\xi\xi} & M_{\xi\delta} \\ M_{\xi\delta} & M_{\delta\delta} \end{bmatrix} - \bar{q} \bar{A}_{2A} \tag{12.2.49}$$

$$\bar{C} = \begin{bmatrix} C_{\xi\xi} & 0 \\ 0 & 0 \end{bmatrix} - \bar{q} \bar{A}_{1A} \tag{12.2.50}$$

$$\overline{K} = \begin{bmatrix} K_{\xi\xi} & 0 \\ 0 & 0 \end{bmatrix} - \overline{q}\overline{A}_{0A} \tag{12.2.51}$$

In order to derive a relatively simple model equation, it will further be assumed that there are no actuator, sensor or processor lags. The dynamics are therefore confined to the structural dynamics and aerodynamics, and the state vector can be simplified to

$$x = \begin{Bmatrix} \xi \\ \delta \\ \dot{\xi} \\ \dot{\delta} \\ x_A \\ x_g \end{Bmatrix} \tag{12.2.52}$$

Defining for convenience

$$\overline{\beta}_u = \begin{bmatrix} \Phi_\xi^T \beta_\Lambda & 0 \\ 0 & \beta_\delta \end{bmatrix} \tag{12.2.53}$$

$$\xi_A = \begin{Bmatrix} \xi \\ \delta \end{Bmatrix} \qquad x = \begin{Bmatrix} \xi_A \\ \dot{\xi}_A \\ x_A \\ x_g \end{Bmatrix} \tag{12.2.54}$$

Further assuming only a single aerodynamic lag per mode (N_ℓ=1), the model equations in the first order form are

$$\begin{Bmatrix} \dot{\xi}_A \\ \ddot{\xi}_A \\ \dot{x}_A \\ \dot{x}_g \end{Bmatrix} = \begin{bmatrix} 0 & I & 0 & 0 \\ -\overline{M}^{-1}\overline{K} & -\overline{M}^{-1}\overline{C} & \overline{q}\overline{M}^{-1}\overline{A}_{3A} & \overline{q}\overline{M}^{-1}\overline{A}_{3g} \\ 0 & I & -\overline{a}_1 I & 0 \\ 0 & 0 & 0 & -\overline{a}_1 I \end{bmatrix} \begin{Bmatrix} \xi_A \\ \dot{\xi}_A \\ x_A \\ x_g \end{Bmatrix}$$

$$+\begin{bmatrix} 0 \\ \overline{M}^{-1}\overline{\beta}_u \\ 0 \\ 0 \end{bmatrix} \begin{Bmatrix} u_\Lambda \\ u_\delta \end{Bmatrix} + \begin{bmatrix} 0 & 0 & 0 \\ \overline{q}\overline{M}^{-1}\overline{A}_{og} & \overline{q}\overline{M}^{-1}\overline{A}_{1g} & \overline{q}\overline{M}^{-1}\overline{A}_{2g} \\ 0 & 0 & 0 \\ 0 & I & 0 \end{bmatrix} \begin{Bmatrix} \alpha_g \\ \dot{\alpha}_g \\ \ddot{\alpha}_g \end{Bmatrix} \qquad (12.2.55)$$

and the accompanying output equations are

$$y = y_{sx} + y_r \qquad (12.2.56)$$

$$z = C_{zx}x + z_r \qquad (12.2.57)$$

Any such system equation can be written in the generalized condensed first order state space form

$$\dot{x} = Ax + B_u u + B_w w \qquad (12.2.58)$$

where in the case of (12.2.55) the third term of the right hand side includes the gust disturbance input, and would also include the reference command driving term if a servo loop with dynamics were included. The corresponding generalized forms of the output equations are

$$y = C_{yx}x + D_{yw}w \qquad (12.2.59)$$

$$z = C_{zx}x + D_{zw}w \qquad (12.2.60)$$

where the D terms may or may not be zero, depending on the degree to which the disturbances and reference commands directly influence the outputs and performance metrics.

Model reduction

The modeling procedure presented above will produce a relatively high order structural model which represents the closest analytical representation of physical truth. This model is referred to as the evaluation model, because it is used to evaluate the performance of compensators which are eventually designed. However, in general the order of this evaluation model is too high to immediately begin control design. For example, even a modest ten mode structural model (20 states) with 2 aerodynamic lags per mode (20 states), with two gust inputs and 2 lags per gust input (4 states), augmented by 2 actuators each with second order dynamics and a two pole Padé approximation to a computational delay (8 states) and 4 sensors with second order dynamics (8 states) will yield a model with 60 states. Actual models are often quite a bit larger.

The model reduction step is shown in the fourth row of Figure 12.3. The objective (in principle) of this step is to reduce this evaluation model to a tractable order for control design, such that the closed loop controller designed for the reduced model will deliver performance indistinguishable from that of a closed loop controller designed for the full evaluation model. However, such a criterion presupposes knowledge of the closed loop performance before the compensator is designed. Therefore, the objective (in practice) of model reduction is to reduce the model, such that the transfer functions important to the closed loop design are not impacted significantly in the bandwidth of interest. This statement still presupposes knowledge of the expected bandwidth and the transfer functions of interest, but these are usually known at the time of preliminary control design.

Thus, there are three identifiable reasons for model order reduction: to make the model size tractable during control design synthesis, to make the resulting compensator (whose order often scales with the size of the control design model) implementable in a real time processor, and to make a distinction between a higher order "evaluation model" and a reduced order "control design model" for the purposes of evaluation of the proposed compensator.

There are two approaches to model order minimization. One is to try to keep the order of the model to a minimum at each step of its derivation. Techniques such as the Minimum-State Method of fitting aerodynamics (12.2.39) have this as their objective. The second approach is to develop a large order model, and then, in a single operation, consistently and systematically reduce its order to the minimum possible size. Such a model reduction procedure consists of the following steps:

- Derive the higher order model with all dynamics which engineering judgment suggests are important.
- Rank the states of the model in order of decreasing importance by a formalized ranking process.

- Truncate the dynamics of the less important states.
- Retain the low frequency information of the truncated modes by a "static" correction.

The process for deriving the higher order model has already been presented. Several methods are available for ranking the relative importance of modes, including modal cost analysis [9] and internal balancing [10], which is described below. Options for the truncation and correction steps will also be presented.

The proper ranking of the states rests on an understanding of what makes a particular state important. A particular state or mode is important to a given transfer function if it has a large modal residue (the numerator in a partial fraction expansion of the transfer function). The residue is the product of the modal observability (by the sensor in question) times the controllability (by the actuator in question). For the aeroelastic system

$$\dot{x} = Ax + B_u u$$
$$y = C_y x \qquad (12.2.61)$$

a measure of the modal controllability is the controllability Grammian W_C, given by

$$0 = AW_c + W_c A^T + B_u \Sigma_{uu} B_u^T \qquad (12.2.62)$$

where Σ_{uu} is a quadratic norm of the magnitude of the expected value of the input u. One possible choice is the square of the maximum value of the actuator input. If the problem is stochastic, a second option is the covariance of u.. A measure of the observability is the observability Grammian W_O, given by

$$0 = A^T W_o + W_o A + C_y^T \Sigma_{yy} C_y \qquad (12.2.63)$$

where Σ_{yy} is a quadratic norm of the magnitude of the output y, with choices similar to those for u.

The problem is that some modes will have a large observability, but small controllability, and vice versa. Therefore, it is necessary to find a similarity transform

$$x_b = Tx \qquad (12.2.64)$$

which "balances" the system, such that the transformed observability and controllability

$$W_{cb} = TW_cT^T$$
$$W_{ob} = T^{-T}W_oT^{-1} \tag{12.2.65}$$

are balanced or equal for all modes, so that

$$W_{cb} = W_{ob} = \Sigma = diag\{\sigma_1, \sigma_2, .., \sigma_n\} \tag{12.2.66}$$

The $\sigma_i' s$ are the Hankel singular values. Large $\sigma_i' s$ correspond to highly controllable-observable modes. Low $\sigma_i' s$ correspond to almost unobservable-uncontrollable modes. A procedure for finding the transformation T is based on the fact that

$$\Sigma^2 = TW_cW_oT^{-1} \tag{12.2.67}$$

so that σ_i^2 is the ith eigenvalue of W_cW_o and T^{-1} is the matrix of eigenvectors of W_cW_o.

Conceptually, the procedure is to find the transformation T^{-1} from Equation (12.2.67), and order the columns of T^{-1} such that

$$\sigma_1 \geq \sigma_2 \geq ... \geq \sigma_n \tag{12.2.68}$$

that is to reorder the columns so that the corresponding Hankel singular values are in order from the most observable-controllable to the least. The system of Equations (12.2.61) is then transformed to

$$\dot{x}_b = A_bx_b + B_{u_b}u$$
$$y = C_{y_b}x_b \tag{12.2.69}$$

where the balanced system is defined by

$$A_b = TAT^{-1} \quad B_{u_b} = TB_u \quad C_{y_b} = C_yT^{-1} \tag{12.2.70}$$

which has the same input-output structure as (12.2.61) but a different internal parameterization.

In order to prepare for the model reduction, the vector of balanced states (which may now have no resemblance to the original states) is partitioned into those states to be retained x_R (those with higher $\sigma_i' s$) and those to be truncated or eliminated x_E

$$x_b = \begin{Bmatrix} x_R \\ x_E \end{Bmatrix} \tag{12.2.71}$$

In a corresponding manner, the system is partitioned

$$\begin{Bmatrix} \dot{x}_R \\ \dot{x}_E \\ y \end{Bmatrix} = \begin{bmatrix} A_R & A_{RE} & B_R \\ A_{ER} & A_E & B_E \\ C_R & C_E & 0 \end{bmatrix} \begin{Bmatrix} x_R \\ x_E \\ u \end{Bmatrix} \tag{12.2.72}$$

such that A_R is N_R x $N_{R,}$, and the remaining terms are partitioned appropriately. In practice, the transformation process described is intrinsically ill conditioned. However, other numerical schemes can be used to arrive at the balanced realization of (12.2.72) [11]. If the dynamics of the truncated modes are ruthlessly eliminated, the reduced order model is

$$\begin{aligned} \dot{x}_R &= A_R x_R + B_R u \\ y &= C_R x_R \end{aligned} \tag{12.2.73}$$

This simplification assumes that the retained and truncated dynamics are uncoupled and that the states of the truncated dynamics do not contribute in any manner to the output. While the first is a reasonably good assumption for lightly damped oscillatory systems, for which the normal modes are nearly the balanced modes, it may not be a good assumption for an aeroelastic system in which the normal structural modes are recoupled by the aerodynamics. The second assumption, that the truncated modes do not contribute to the output, is also not necessarily true. Often there is significant influence of the truncated dynamics on the static (i.e. very low frequency) value of the output and on the zeros of the transfer function in the bandwidth of interest.

A reduction of (12.2.72) which retains more of the information in the original model uses a Guyan reduction coupled with the retention of the static influence of the truncated modes. By assuming only that the dynamics of the truncated modes can be neglected, but that the static response is important, the term on the left hand side of the second of equations (12.2.72) is set to zero, yielding

$$x_E = -A_E^{-1} A_{ER} x_R - A_E^{-1} B_E u \tag{12.2.74}$$

Substituting into the first and third equations of (12.2.72) gives

$$\dot{x}_R = \left[A_R - A_{RE}A_E^{-1}A_{ER}\right]x_R + \left[B_R - A_{RE}A_E^{-1}B_E\right]u \tag{12.2.75}$$

$$y = \left[C_R - C_E A_E^{-1}A_{ER}\right]x_R - C_E A_E^{-1}B_E u \tag{12.2.76}$$

Note that the output Equation (12.2.76) now has a static feed through or "D" term present. This system will correctly represent the dynamics of the retained modes and the zero frequency static displacement response of the truncated modes. However, it will not roll off at high frequency and will not correctly represent the velocity response of the truncated modes. In order to capture these two features, the state vector can once again be augmented with pseudo states x_P. The number of such states can be the smaller of the number of inputs or outputs. Assuming the number of inputs is smaller, the system can be written as:

$$\begin{Bmatrix}\dot{x}_R \\ \dot{x}_P\end{Bmatrix} = \begin{bmatrix}A_R - A_{RE}A_E^{-1}A_{ER} & 0 \\ 0 & -A_P\end{bmatrix}\begin{Bmatrix}x_R \\ x_P\end{Bmatrix} + \begin{bmatrix}B_R - A_{RE}A_E^{-1}B_E \\ A_p\end{bmatrix}u \tag{12.2.77}$$

$$y = \left\lfloor C_R - C_E A_E^{-1}B_{ER} \quad -C_E A_E^{-1}B_E \right\rfloor\begin{Bmatrix}x_R \\ x_P\end{Bmatrix} \tag{12.2.78}$$

In this form, the direct feed through term of (12.2.76) has been eliminated. The transfer functions now all roll off, and the displacement *and* velocity output contributions of the higher frequency modes are present in the retained dynamics. The influence of the truncated dynamics is represented through the pseudo states, whose state dynamics can be chosen as a heavily damped set of poles which lie at frequencies above the cross over. The resonance and damping of the pseudo states can be adjusted to properly match the phase of key transfer functions at cross over. The inclusion of such velocity effects in the aeroelastic control problem can be important, because of the dependence of the unsteady aerodynamic forces on structural velocity [12].

The procedure described above gives a method for ranking the importance of the modes, truncating the less important modes, and retaining the low frequency information of the truncated modes. It does not address the absolute criteria of how many or which modes to truncate. This is a matter of engineering judgment.

A useful procedure is to calculate the Hankel singular values, and eliminate modes up to some fixed percent of the maximum singular value. Then examine the transfer functions of the truncated system (with corrections for truncation) for changes from the full order evaluation model. In general, a truncation is justified if:

1. It does not significantly change the magnitude of the disturbance-performance transfer function G_{zw} in the bandwidth.
2. It does not significantly change the magnitude or pole/zero structure of the disturbance-output or input-performance transfer functions G_{yw} and G_{zu} in the bandwidth .
3. It does not noticeably change the magnitude or pole-zero spacing of the input-output transfer function G_{yu} in the bandwidth.
4. It does not remove any poles from G_{yu} in the frequency range just outside the bandwidth which might yield greater than unity loop gain when cascaded with the compensator.
5. It does not significantly alter the plant's open loop stability parameters, such as divergence or flutter speed.

If a truncation is justified, another round of truncation and correction can be implemented, continuing as long as the criteria are not violated.

The model reduction outlined above can be applied to the entire evaluation model at the end of the modeling process, as suggested by Figure 12.3. However, experience has shown that it is desirable to apply the reduction process independently at three stages: first after the *in vacuo* structural model is derived, then again after the aerodynamics are added, then a third time after the other actuator-sensor-processor dynamics are added. All of the dynamics truncated at each step in deriving the control design model should be retained in an evaluation model, for eventual evaluation of the compensator.

The conclusion of the structural modeling process is the convergence on the control design model, which should then meet the requirements for the model enumerated at the beginning of the section: (1) that it contain enough states and information to accurately represent the important transfer functions in the bandwidth of interest; and (2) that its order be as small as possible, in order to not unnecessarily complicate the control design

Having derived the control design model, the process of compensator design can begin. Volumes have been written on the subject of control theory and design synthesis. The serious student of aeroelastic control should be familiar with both classical and optimal techniques. The objective of this chapter is not to repeat these texts on compensation, but to show the application of compensation to the aeroelastic problem. This will be done by first developing the control model equations of a typical section in Section 12.3, and then developing compensators for the system in Section 12.4.

12.3 Control modeling of the typical section

Typical section models, used in Chapters 2 and 3 to examine static and dynamic aeroelastic behavior, are now exploited to develop an understanding of aeroelastic control. In this section, the open loop dynamic modeling for control is performed. Included in the typical section model are the effects of anisotropic material, and the forces and moments produced by both structural actuators and conventional articulated control surfaces.

The modeling of a two degree-of-freedom typical section is instructive for three reasons. First, it exemplifies the modeling process described in the previous section. Second, it allows insight into the open loop dynamic features which must be understood prior to compensator design. It is found that the choice of actuators and sensors have a strong influence on the improvement in performance which can be expected by the addition of active control. Variation of structural and aerodynamic parameters also limits achievable performance. Third, it allows an exploration of the fundamental mechanisms governing the control of dynamic aeroelastic systems, and the development of a method for consistently comparing various control schemes in order to assess the advantages and limitations of each.

Typical section governing equations

Structural model: The typical section, shown in Figure 12.6, is given plunge h and pitch α degrees of freedom, as well as a leading and trailing edge flap. A force and moment per unit span act at the elastic axis, which represents the influence of structural actuation in bending and torsion and the aerodynamic forces and moments due to pitch, plunge, the motion of the flaps, and the unsteady gust. The structural restraints in bending and torsion appear at the elastic axis, which serves as a reference for all moments.

$$
\begin{aligned}
m\ddot{h} + mbx_\alpha\ddot{\alpha} + k_h h &= f_{ea_\Lambda} + f_{ea_A} \\
mbx_\alpha\ddot{h} + I_\alpha\ddot{\alpha} + k_\alpha\alpha &= m_{ea_\Lambda} + m_{ea_A}
\end{aligned}
\qquad (12.3.1)
$$

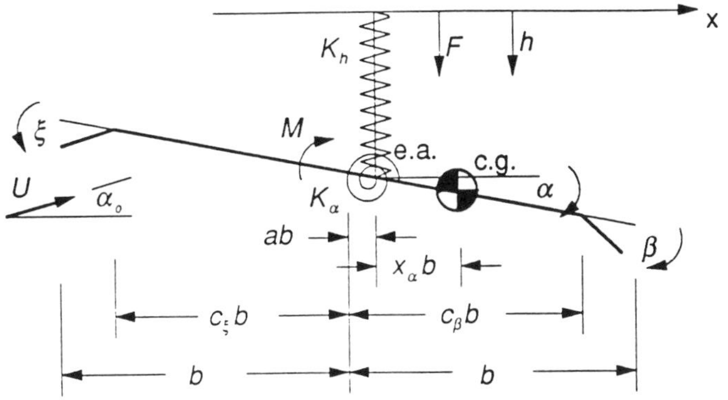

Figure 12.6: Typical section geometry.

The mass of the flaps is included in the inertial terms. The dynamics of the flaps will be modeled as commanded displacements, so that the dynamics associated with the flaps can be neglected. This is a valid assumption, provided that high authority servo loops are present which track commanded surface deflections at a bandwidth higher than the aeroelastic frequency range of interest, and that the flaps are mass balanced.

Using the semi-chord (b), the torsional frequency (ω_α), and the typical section mass (m), an appropriate set of non-dimensional parameters is obtained. These include the familiar radius of gyration $R_\alpha^2 = \dfrac{I_\alpha}{mb^2}$, mass ratio $\mu = \dfrac{m}{\pi\rho b^2}$, frequency ratio $\overline{\omega}_h = \dfrac{\omega_h}{\omega_\alpha}$, Laplace variable $p = \dfrac{s}{\omega_\alpha}$, and plunge displacement $\overline{h} = \dfrac{h}{b}$. The new variables are the nondimensional force and moment inputs: $\dfrac{f_{ea}b}{k_\alpha}$ and $\dfrac{m_{ea}}{k_\alpha}$. Note that this normalization of the forces and moments is consistent with the normalization of time by ω_α. The normalized equations are

$$\begin{bmatrix} \frac{1}{R_\alpha^2} & \frac{x_\alpha}{R_\alpha^2} \\ \frac{x_\alpha}{R_\alpha^2} & 1 \end{bmatrix} \begin{Bmatrix} \bar{h}p^2 \\ \alpha p^2 \end{Bmatrix} + \begin{bmatrix} \frac{\overline{\omega}_h^2}{R_\alpha^2} & 0 \\ 0 & 1 \end{bmatrix} \begin{Bmatrix} \bar{h} \\ \alpha \end{Bmatrix} = \begin{Bmatrix} \frac{f_{ea}b}{k_\alpha} \\ \frac{m_{ea}}{k_\alpha} \end{Bmatrix}_A + \begin{Bmatrix} \frac{f_{ea}b}{k_\alpha} \\ \frac{m_{ea}}{k_\alpha} \end{Bmatrix}_\Lambda \tag{12.3.2}$$

Aside from determining the inertial properties of the typical section, the remaining task in the structural modeling is to find k_h, k_α and a, the location of the elastic axis. The simplest general model which will capture these features for a wing built up of anisotropic materials is that of a laminated plate. The governing differential equation for an anisotropic plate-like lifting surface with induced strain actuation layers is [13, 14, and 15]

$$D_{11}\frac{\partial^4 w}{\partial x^4} + 2D_{12}\frac{\partial^4 w}{\partial x^2 \partial y^2} + 4D_{16}\frac{\partial^4 w}{\partial x^3 \partial y} + D_{22}\frac{\partial^4 w}{\partial y^4} + 4D_{26}\frac{\partial^4 w}{\partial x\, \partial y^3} +$$

$$4D_{66}\frac{\partial^4 w}{\partial x^2 \partial y^2} + m_{\Lambda_{11}}\frac{\partial^2 \Lambda_x}{\partial x^2} + m_{\Lambda_{12}}\frac{\partial^2 \Lambda_y}{\partial x^2} + m_{\Lambda_{12}}\frac{\partial^2 \Lambda_x}{\partial y^2} + m_{\Lambda_{22}}\frac{\partial^2 \Lambda_y}{\partial y^2} = \Delta p_A \tag{12.3.3}$$

where: $D_{ij} = \int Q_{ij} z^2 dz$
is the bending stiffness

$m_{\Lambda_{ij}} = \int Q_{ij} z dz$ for z where the actuator is present
is the induced strain actuation moment

Λ is the actuation strain

The induced strain actuation moment is due to a distributed strain actuator, such as a piezoelectric. The evaluation of these terms will be discussed in more detail below in the section on actuator-sensor modeling. The Δp_A term is due to the aerodynamic pressure, and will be modeled more explicitly in the aerodynamic model section.

The simplest possible approximate structural solution to the governing plate equations is a two mode Ritz model where the two modes are a simple parabolic bending mode and a linear torsional mode.

$$
\begin{aligned}
w(x,y,t) &= \sum_{i=1}^{2} \gamma_i(x,y) q_i(t) \\
\gamma_1(x,y) &= \frac{x^2}{\tilde{l}^2} \\
\gamma_2(x,y) &= \frac{xy}{\tilde{l}} \\
\text{where } \tilde{l} &= \frac{3}{4} l
\end{aligned}
\tag{12.3.4}
$$

The sign convention which makes the typical section and plate models consistent can be seen in Figure 12.7. When evaluated at the 3/4 span point (or $\tilde{l}$), these modes have unit displacement and unit twist at the midchord, and are equivalent to the midchord plunge h_c and pitch α_c of the typical section.

$$
\begin{aligned}
h_c(t) &= w(\tilde{l},0,t) = q_1(t) \\
\frac{\partial h_c}{\partial y} &= \alpha_c(t) = \frac{\partial w}{\partial y}(\tilde{l},0,t) = q_2(t)
\end{aligned}
\tag{12.3.5}
$$

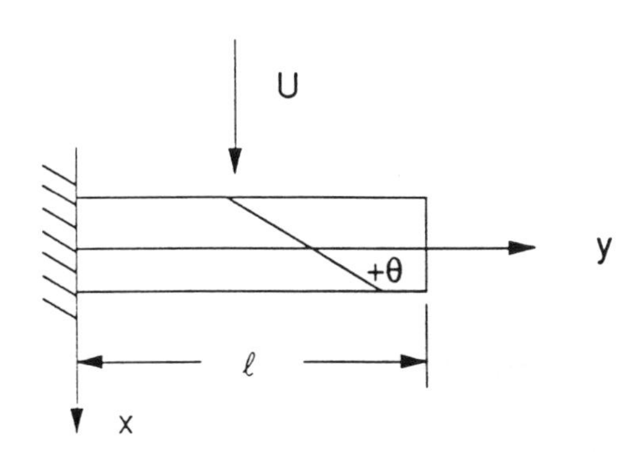

Figure 12.7: Geometry of the plate-like wing model.

These mode shapes are then integrated over the wing to obtain the stiffness matrix.

$$\begin{Bmatrix} F_c \\ M_c \end{Bmatrix} = \begin{bmatrix} \frac{8lb}{\bar{l}^4} D_{11} & \frac{8lb}{\bar{l}^3} D_{16} \\ \frac{8lb}{\bar{l}^3} D_{16} & \frac{8lb}{\bar{l}^2} D_{66} \end{bmatrix} \begin{Bmatrix} h_c \\ \alpha_c \end{Bmatrix} \tag{12.3.6}$$

Note that the terms on the left side have dimensions of force and moment, and that the equation is in general coupled. In order to incorporate this coupled stiffness matrix into the typical section equations, the expressions must be divided by the span l. An elastic axis location and the corresponding uncoupled stiffness matrix must be found. The transformation matrix which relates the displacements at the midchord and those at the elastic axis is

$$\begin{Bmatrix} q_1 \\ q_2 \end{Bmatrix} = \begin{Bmatrix} h_c \\ \alpha_c \end{Bmatrix} = \begin{bmatrix} 1 & -ab \\ 0 & 1 \end{bmatrix} \begin{Bmatrix} h \\ \alpha \end{Bmatrix} = \mathbf{T} \begin{Bmatrix} h \\ \alpha \end{Bmatrix} \tag{12.3.7}$$

The spring forces at the elastic axis are calculated through the appropriate transformation and the transformed equilibrium equation is [16]

$$\begin{Bmatrix} f_{ea} \\ m_{ea} \end{Bmatrix} = \left(\frac{1}{l} \right) \mathbf{T}^{\mathrm{T}} \mathbf{K} \mathbf{T} \begin{Bmatrix} h \\ \alpha \end{Bmatrix} \tag{12.3.8}$$

The distance of the elastic axis aft of the midchord ab and the uncoupled stiffnesses are found by setting the off-diagonal terms of $\mathbf{T}^{\mathrm{T}}\mathbf{K}\mathbf{T}$ to zero.

$$\begin{aligned} a &= \frac{K_{12}}{K_{11} b} = \frac{3}{4} \frac{l}{b} \frac{D_{16}}{D_{11}} \\ k_h &= \frac{K_{11}}{l} = \frac{8b}{\bar{l}^4} D_{11} \\ k_\alpha &= \frac{K_{22}}{l} \left(1 - \frac{K_{12}^2}{K_{11} K_{22}} \right) = \frac{8b}{\bar{l}^2} D_{66} \left(1 - \frac{D_{16}^2}{D_{11} D_{66}} \right) \end{aligned} \tag{12.3.9}$$

These are the sought after elastic constants which can be placed directly into the governing equations of motion for the typical section, Equation (12.3.1). Note that the elastic axis location depends on the elastic properties and the geometry of the wing.

Aerodynamic model: The aerodynamics are found by adapting the incompressible wing-aileron-tab lifting surface results obtained by Theodorsen and Garrick [13] to a leading edge flap-wing-trailing edge flap lifting surface via a coordinate transformation. Initially only the steady state aerodynamic terms are retained to simplify the problem. The addition of unsteady aerodynamics will be added below as a plant modeling parameter variation. The steady aerodynamic force and moment per span due to a leading edge flap, trailing edge flap, homogeneous angle of attack and gust input are

$$\begin{aligned} f_{ea_A} &= -\rho U^2 b C_{L_\beta}\beta - \rho U^2 b C_{L_\xi}\xi - \rho U^2 b C_{L_\alpha}(\alpha+\alpha_o) \\ m_{ea_A} &= \rho U^2 b^2 C_{M_\beta}\beta + \rho U^2 b^2 C_{M_\xi}\xi + \rho U^2 b^2 C_{M_\alpha}(\alpha+\alpha_o) \end{aligned} \tag{12.3.10}$$

Note that the definition of the moment coefficient in this equation differs from the standard definition by a factor of two. Dividing by the appropriate nondimensionalizing terms from Equation (12.3.2), the normalized form of the unsteady forces are

$$\begin{Bmatrix} \dfrac{f_{ea}b}{k_\alpha} \\ \dfrac{m_{ea}}{k_\alpha} \end{Bmatrix}_A = \begin{bmatrix} 0 & -\bar{q}C_{L_\alpha} \\ 0 & \bar{q}C_{M_\alpha} \end{bmatrix} \begin{Bmatrix} \bar{h} \\ \alpha \end{Bmatrix} + \begin{bmatrix} -\bar{q}C_{L_\beta} & -\bar{q}C_{L_\xi} \\ \bar{q}C_{M_\beta} & \bar{q}C_{M_\xi} \end{bmatrix} \begin{Bmatrix} u_\beta \\ u_\xi \end{Bmatrix} + \begin{bmatrix} -\bar{q}C_{L_\alpha} \\ \bar{q}C_{M_\alpha} \end{bmatrix} \alpha_o$$

(12.3.11)

where

U_α	=	$\dfrac{U}{\omega_\alpha b}$	Reduced Velocity
$\bar{q}$	=	$\dfrac{U_\alpha^2}{\pi\mu R_\alpha^2} = \dfrac{\rho U^2 b^2}{k_\alpha}$	Normalized Dynamic Pressure
u_β	=	β	Trailing Edge Flap Deflection
u_ξ	=	ξ	Leading Edge Flap Deflection

Actuator-sensor model: In order to complete the model, information on the actuator and sensor must be incorporated. In order to maintain generality, it is assumed that sensors are available whose output is exactly the states: plunge and plunge rate, pitch and pitch rate.

The aerodynamic actuators have already been modeled in the structural and aerodynamic modeling steps, but the structural actuators must now be included. Again for simplicity, it will be assumed that the structural actuator is a uniformly distributed induced strain actuator, such as a piezoelectric. The governing equation for such a plate-like structure with induced strain actuation was given in (12.3.3). Assuming a single simple bending mode solution, such as that of (12.3.4), the governing equation can be reduced to a model equation of the form

$$D_{11}\frac{\partial^4 w}{\partial x^4}+m_{\Lambda_{11}}\frac{\partial^2 \Lambda_x}{\partial x^2}=0 \tag{12.3.12}$$

Non-dimensionalizing this model plate equation will provide an additional scaling parameter governing strain actuation. The dimensional quantities will be non-dimensionalized in the following manner: the spanwise dimension x by the span l, the vertical displacement w by a reference displacement w_o, the substrate stiffnesses D_{ij} by a reference stiffness D_o, the piezoelectric moment terms $m_{\Lambda_{ij}}$ by a nominal reference m_{Λ_o}, and the piezoelectric strain Λ_x by a reference strain Λ_o. The resulting equation is

$$\frac{D_o w_o}{l^4}\overline{D}_{11}\frac{\partial^4 \overline{w}}{\partial \overline{x}^4}+\frac{m_{\Lambda_o}\Lambda_o}{l^2}\overline{m}_{\Lambda_{11}}\frac{\partial^2 \overline{\Lambda}_x}{\partial \overline{x}^2}=0 \tag{12.3.13}$$

where the bars designate the non-dimensional quantity. Dividing the dimensional part of the second term by the first gives the new non-dimensional critical parameter

$$C_r=\frac{m_{\Lambda_o}l^2\Lambda_o}{D_o w_o} \tag{12.3.14}$$

The introduction of a new physical element, the strain actuation, has introduced one new nondimensional number into the problem. This critical parameter expresses the relative authority of the strain actuation. As with other such parameters, it can be used in comparing various designs, normalizing data, and scaling results from models to full scale. Specification of this parameter, together with the other structural parameters (radius of gyration, frequency ratio, and mass ratio) and aerodynamic parameters (reduced velocity and Mach number) completely characterizes the aeroelastic scaling problem.

In order to make the non-dimensional group of (12.3.14) applicable to a specific problem, the reference length for w_o and the reference stiffness D_o must be chosen. There are three possible choices for w_o: the span l, the semichord b, and the thickness h. For beam-like problems, l might be the appropriate choice. For plate analysis of large deflections, h might be appropriate. But for aeroelastic problems in which the fundamental interest is in controlling the angle of attack of the wing, the choice of b is the natural one, since it equates the non-dimensional parameter C_r with the ability to induce a given twist in the wing, and is consistent with the normalization of the displacement by b in (12.3.2). The choice of D_o depends on the dominant stiffness of interest in a particular problem. As suggested by the second of Equation (12.3.9), D_{11} is appropriate for bending, while the reduced torsional stiffness given by the third of (12.3.9) is appropriate for twisting. Thus, for the typical section bending problem

$$C_r = \left. \frac{m_{\Lambda_o} l^2 \Lambda_o}{D_{11} b} \right|_{w_o = b,\ D_o = D_{11}} \tag{12.3.15}$$

Substituting this parameter into (12.3.13), assuming uniform stiffness and actuation along the spanwise direction, and integrating twice in that direction gives the differential equation for strain actuation induced bending of a platelike beam

$$\frac{\partial^2 \overline{w}}{\partial \overline{x}^2} + \frac{m_{\Lambda_o} l^2 \Lambda_o}{D_{11} b} = 0 \tag{12.3.16}$$

Assuming a uniform strain model [19] for a thin induced strain actuator of thickness t_a symmetrically placed on the top and bottom surface of the plate structure of thickness t_s, with the actuator causing bending, the critical parameter C_r can be evaluated as

$$\frac{m_{\Lambda_o} l^2 \Lambda_o}{D_{11} b} = \frac{2 E_a t_a b t_s l^2 \Lambda}{\left(E_a t_a b t_s^2 + \frac{E_s t_s^3 b}{6} \right) b} \tag{12.3.17}$$

which reduce s to

$$\frac{m_{\Lambda_o} l^2 \Lambda_o}{D_{11} b} = \frac{2 l^2 \Lambda}{b t_s} \frac{6}{(6 + \Psi)} \tag{12.3.18}$$

where

$$\Psi = \frac{E_s t_s}{E_a t_a}$$

Substituting into (12.3.16) and integrating along the span, the deflection at the 3/4 typical section gives

$$\bar{h} = \bar{w}\big|_{\bar{x}=3/4} = \frac{9}{16}\frac{l^2\Lambda}{bt_s}\frac{6}{(6+\Psi)} \tag{12.3.19}$$

The equivalent force per unit span f_{ea} (12.3.1) of such an actuator on the typical section can be found by equating the deflections created by the strain actuator and a force at the three quarter span. For the bending mode assumed in (12.3.4), the three-quarter span deflection due to a force at the three-quarter span is

$$\bar{h} = \frac{f_{ea}}{bk_h} \tag{12.3.20}$$

Equating (12.3.19) and (12.3.20), solving for the equivalent force f_{ea}, and then substituting into the expression for the normalized equivalent force u_h gives

$$u_h = \frac{f_{ea}b}{k_\alpha} = \frac{\Lambda b}{t_s}\frac{6}{6+\Psi}\frac{D_{11}}{D_{66}\left(1-\dfrac{D_{16}^2}{D_{11}D_{66}}\right)} \tag{12.3.21}$$

For an isotropic plate, D_{16} is zero and D_{11} and D_{66} are related, such that the normalized force reduces to

$$u_h = \Lambda\frac{b}{t_s}\frac{6}{6+\Psi}\frac{2(1+\nu)}{\left(1-\nu^2\right)} \tag{12.3.22}$$

Thus, the equivalent force depends only on the actuation strain, the structural thickness ratio and the relative stiffness of the actuation layer to the plate. The Poisson's ratio ν and the relative stiffness ratio Ψ are taken as 0.3 and 10.0 respectively in the numerical example below.

Calculating the equivalent moment u_α is not as straight forward as finding the equivalent force u_h because the problem is inherently two-dimensional. Creating an equivalent moment requires that either the strain actuator possesses a free shear strain term (not present in common strain actuation materials such as piezoceramics) or that the substructure exhibits bend-twist or extension-twist coupling. However, an effective shear strain term can be created in a piezoceramic by properly constraining the actuator in the longitudinal and transverse directions [15]. Imposing such constraints on a strain actuator produces an equivalent moment in the same manner (physically) as actuating an unconstrained piezoceramic bonded to a substructure with bend-twist coupling. Thus, the strain actuators are assumed to be fully constrained in the transverse direction, unconstrained in the longitudinal direction, and mounted at 45 degrees. Observing that the maximum in-plane shear strain given by Mohr's circle model yield is one half the maximum extension, this model suggests that an equivalent moment u_α can be created equal to one half the magnitude of the equivalent force u_h. It will be assumed in the analysis below that the strain actuation moment and force can be commanded independently.

The structural actuation vector is then

$$\begin{Bmatrix} \dfrac{f_{ea}b}{k_\alpha} \\ \dfrac{m_{ea}}{k_\alpha} \end{Bmatrix}_\Lambda = \begin{bmatrix} 1 & 0 \\ 0 & 1 \end{bmatrix} \begin{Bmatrix} u_h \\ u_\alpha \end{Bmatrix} \tag{12.3.23}$$

Control model: Assuming no lags in the actuator, sensor or processor, the control design model can be summarized as

$$\begin{bmatrix} \dfrac{1}{R_\alpha^2} & \dfrac{x_\alpha}{R_\alpha^2} \\ \dfrac{x_\alpha}{R_\alpha^2} & 1 \end{bmatrix} \begin{Bmatrix} \bar{h}p^2 \\ \alpha p^2 \end{Bmatrix} + \begin{bmatrix} \dfrac{\overline{\omega}_h^2}{R_\alpha^2} & \bar{q}C_{L_\alpha} \\ 0 & 1-\bar{q}C_{M_\alpha} \end{bmatrix} \begin{Bmatrix} \bar{h} \\ \alpha \end{Bmatrix} = \begin{bmatrix} 1 & 0 & -\bar{q}C_{L_\beta} & -\bar{q}C_{L_\xi} \\ 0 & 1 & \bar{q}C_{M_\beta} & \bar{q}C_{M_\xi} \end{bmatrix} \begin{Bmatrix} u_h \\ u_\alpha \\ u_\beta \\ u_\xi \end{Bmatrix} + \begin{bmatrix} -\bar{q}C_{L_\alpha} \\ \bar{q}C_{M_\alpha} \end{bmatrix} \alpha_o \tag{12.3.24}$$

where the homogeneous aerodynamic terms have been moved to the left hand side. Defining the state, control input and disturbance vectors,

$$x = \begin{Bmatrix} \bar{h} \\ \alpha \\ \dot{\bar{h}} \\ \dot{\alpha} \end{Bmatrix} \quad u = \begin{Bmatrix} u_h \\ u_\alpha \\ u_\beta \\ u_\xi \end{Bmatrix} \quad d = \alpha_o \tag{12.3.25}$$

the control problem is then transformed into the state space form of (12.2.61)

$$\dot{x} = \mathbf{A}x + \mathbf{B}_u u + \mathbf{B}_d d$$
$$y = \mathbf{C}_y x \tag{12.3.26}$$

where initially it will be assumed that all states are available as outputs, and therefore the C_y matrix is identity.

Open loop poles and zeros

Before attempting any control design on a system, it is valuable to examine its open loop characteristics from inputs to outputs, which can be represented in the frequency domain in terms of the poles and zeros of the frequency transfer functions. In the state space form, the input-output relation from the actuators to the measurements is

$$\mathbf{y} = \mathbf{C}_y (p\mathbf{I} - \mathbf{A})^{-1} \mathbf{B}_u \mathbf{u} \tag{12.3.27}$$

which represents a matrix of transfer functions. The open loop roots or poles, which are common to all the transfer functions, are found by setting the characteristic equation, the determinant of ($\boldsymbol{pI}$-$\boldsymbol{A}$), to zero.
For the typical section system (12.3.25) the characteristic equation is

$$\Delta(p) = \left[1 - \frac{x_\alpha^2}{R_\alpha^2}\right] p^4 + \left[\left(\overline{\omega}_h^2 + 1\right) - \overline{q}\left(C_{L_\alpha} x_\alpha + C_{M_\alpha}\right)\right] p^2 + \left[\overline{\omega}_h^2 \left(1 - \overline{q} C_{M_\alpha}\right)\right] \tag{12.3.28}$$

The roots of the system are dependent on the section geometry, structural properties and air speed, but are independent of the actuation method. The poles are the same as those of Equation 3.3.50, and are the indicators of flutter and divergence, as discussed in Section 3.3. These open loop poles are important indicators of the homogeneous stability of the system. However, in order to effect control, one must examine the way in which the actuators influence the system. This information is represented by the numerator of the transfer functions (12.3.27).

The transfer functions from the four control inputs (bending strain actuation u_h, torsion strain actuation u_α, trailing edge flap defection u_β, and leading edge flap deflection u_ξ) to the two output variables (plunge $\overline{h}$ and pitch α) are given by fully expanding (12.3.27) as

$$\begin{Bmatrix} \bar{h} \\ \alpha \end{Bmatrix} = \frac{R_\alpha^2}{\Delta(p)} \begin{bmatrix} \mathrm{n}_{hh}(p) & \mathrm{n}_{h\alpha}(p) & \mathrm{n}_{h\beta}(p) & \mathrm{n}_{h\xi}(p) \\ \mathrm{n}_{\alpha h}(p) & \mathrm{n}_{\alpha\alpha}(p) & \mathrm{n}_{\alpha\beta}(p) & \mathrm{n}_{\alpha\xi}(p) \end{bmatrix} \begin{Bmatrix} u_h \\ u_\alpha \\ u_\beta \\ u_\xi \end{Bmatrix}$$

$$\mathrm{n}_{hh}(p) = p^2 + 1 - \bar{q} C_{M_\alpha}$$

$$\mathrm{n}_{\alpha h}(p) = -\frac{x_\alpha p^2}{R_\alpha^2}$$

$$\mathrm{n}_{h\alpha}(p) = -\frac{x_\alpha p^2}{R_\alpha^2} - \bar{q} C_{L_\alpha}$$

$$\mathrm{n}_{\alpha\alpha}(p) = \frac{p^2}{R_\alpha^2} + \frac{\bar{\omega}_h^2}{R_\alpha^2}$$

$$\mathrm{n}_{h\beta}(p) = \bar{q} C_{L_\beta} \left(\bar{q} C_{M_\alpha} \left(1 - \frac{C_{L_\alpha} C_{M_\beta}}{C_{M_\alpha} C_{L_\beta}} \right) - 1 - p^2 \left(1 + \frac{C_{M_\beta} x_\alpha}{C_{L_\beta} R_\alpha^2} \right) \right)$$

$$\mathrm{n}_{\alpha\beta}(p) = \frac{1}{R_\alpha^2} \bar{q} C_{M_\beta} \left(p^2 \left(1 + \frac{C_{L_\beta} x_\alpha}{C_{M_\beta}} \right) + \bar{\omega}_h^2 \right)$$

$$\mathrm{n}_{h\xi}(p) = \bar{q} C_{L_\xi} \left(\bar{q} C_{M_\alpha} \left(1 - \frac{C_{L_\alpha} C_{M_\xi}}{C_{M_\alpha} C_{L_\xi}} \right) - 1 - p^2 \left(1 + \frac{C_{M_\xi} x_\alpha}{C_{L_\xi} R_\alpha^2} \right) \right)$$

$$\mathrm{n}_{\alpha\xi}(p) = \frac{1}{R_\alpha^2} \bar{q} C_{M_\xi} \left(p^2 \left(1 + \frac{C_{L_\xi} x_\alpha}{C_{M_\xi}} \right) + \bar{\omega}_h^2 \right)$$

(12.3.29)

where each element $\mathrm{n}_{ij}(p)$ of this two by four matrix relation represents response of the transfer function at one of the outputs (the first subscript) due to one of the inputs (the second subscript). The zeros of each individual single input single output (SISO) transfer function are found by setting each numerator $\mathrm{n}_{ij}(p)$ to zero. The zero locations are dependent on the section geometry and structural properties, as were the system poles. For the case of plunge measurement $\bar{h}$, the individual SISO transfer function zeros also depend on air speed U_α, through the normalized dynamic pressure $\bar{q}$. For the case of pitch measurement α, the zeros are not dependent on air speed U_α, at least in this model which incorporates only steady aerodynamics.

Equations (12.3.28) and (12.3.29) reveal that all of the pole and some of the zero locations are a function of air speed U_α, which indicates that a control law appropriate at one flight condition may not necessarily be appropriate at others. In particular, the rate of zero movement and the propensity for the pole-zero pattern along the imaginary axis to change (*i.e.* so called pole-zero flipping) is a particular indicator of the potential nonrobustness of the closed loop control scheme [20 and 21]. It will therefore be interesting to observe the relative and absolute movement of the system poles and the SISO transfer function zeros as a function of air speed. The pole-zero patterns are also strongly dependent on geometric and structural parameters of the wing, and implicitly on the modeling assumptions.

Reference typical section

As seen from Equations (12.3.28) and (12.3.29), the location of the system poles and transfer function zeros are highly dependent on the geometrical and structural properties of the typical section. Therefore, it is necessary to choose some reference parameters in order to proceed with illustrative analysis and control system design. The reference typical section is chosen to be representative of a simplified unswept rectangular wing of a high performance aircraft. The wing mass is assumed to be evenly distributed so that the center of mass lies at the midchord. In order to assure that flutter occurs before divergence, the elastic axis location is shifted ten percent forward of the midchord, which is representative of a 4.5 degree forward fiber sweep if constructed of common graphite epoxy materials in a unidirectional laminate. The flaps are both 10% of the chord. The resulting typical section has a frequency ratio $\bar{\omega}_h$ of one-fifth and a mass ratio μ of twenty, not atypical of built up wings. The model parameters are listed in Table 12.1.

Table 12.1 Parameters of the typical section model.

Parameter	Symbol	Value
Section Geometry	a	-0.2
	x_α	0.2
	R_α^2	0.25
	μ	20
	$\overline{\omega}_h$	0.2
Wing Parameters	$\frac{t}{2b}$	0.51%
	$\frac{l}{b}$	3.92
Steady Aerodynamic Coefficients	C_{L_α}	2π
	C_{M_α}	1.885
	C_{L_β}	2.487
	C_{M_β}	-0.334
	C_{L_ξ}	-0.087
	C_{M_ξ}	-0.146

Plant variation with airspeed

The variation with airspeed of the poles and the zeros of the structural plant will now be examined. The plot of the pole variation with airspeed is common in aeroelastic analysis, and is similar to a root locus in which the airspeed is the gain parameter. Figure 12.8 shows the root locus for this system as air speed is increased. Note that neither aerodynamic nor structural damping has been included in the model. The poles begin on the imaginary axis and eventually coalesce at the flutter point at $U_\alpha = 1.90$, followed by reversal and divergence. The frequencies of the poles can also be plotted against the airspeed explicitly, as was done in Figure 3.6. The equivalent graph for this typical section is given in Figure 12.9. The poles coalesce to flutter at $U_\alpha = 1.90$, then drop to divergence at $U_\alpha = 2.47$. The classical static divergence is obtained when both the imaginary and real part of the frequency reach zero, at $U_\alpha = 2.88$.

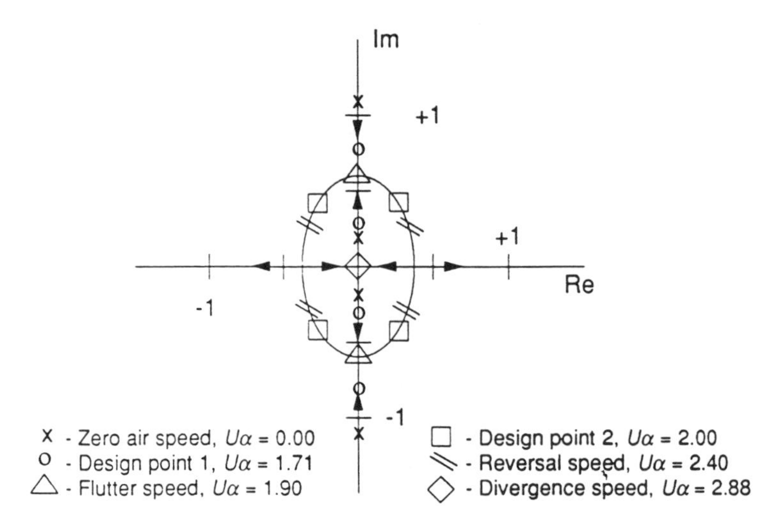

Figure 12.8: Location of the open loop poles as a function of normalized airspeed.

The new feature of Figure 12.9 is that the frequencies of the purely oscillatory zeros are also shown as a function of airspeed. For the case of plunge measurement or output $\bar{h}$, all four of the individual SISO transfer function zeros change with air speed, as indicated by Equation (12.3.29). The SISO transfer function zero associated with the bending strain actuator u_h is at the torsional natural frequency ω_α when the air speed is zero. This zero decreases quadratically to a frequency of zero at the static divergence speed ($U_\alpha = 2.88$). Likewise, the zero associated with the trailing edge flap actuator u_β decreases quadratically with air speed to a frequency of zero at the reversal speed ($U_\alpha = 2.40$). The air speed at which the oscillatory component of the individual SISO transfer function zeros goes to zero is especially significant to the aeroelastic control problem, since it is at this air speed that one of the zeros becomes non-minimum phase (*i.e.* moves into the right half of the Laplace-plane). The presence of a non-minimum phase zero indicates a fundamental limitation on the amount of control which can be applied to the system [22].

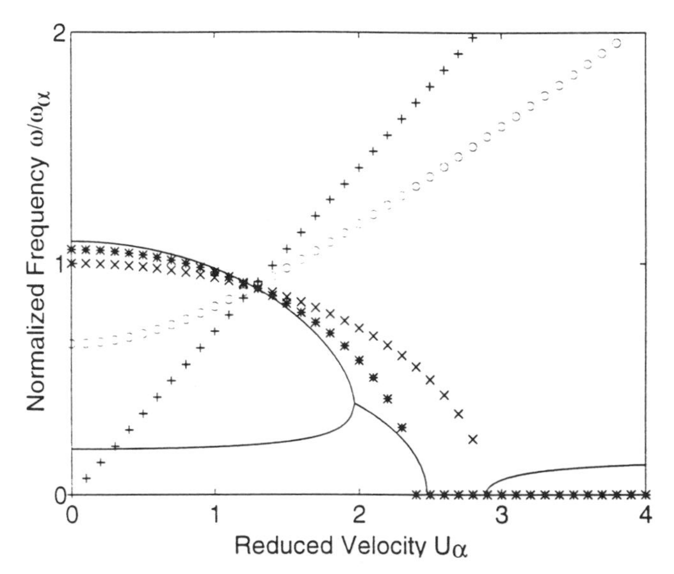

*Figure 12.9: Pole and SISO zero frequencies as a function of normalized airspeed for the reference typical section. Zeros are: [o = h/u_ξ, * = h/u_β, + = h/u_α, × = h/u_h].*

The zeros associated with the torsion strain actuator u_α and leading edge flap actuator u_ξ increase with air speed as shown in Figure 12.9. Note that all four of the zero frequencies intersect with the nominal torsion pole frequency at the same air speed. This indicates that a pole-zero cancellation has occurred due to a loss of observability of this nominal torsional mode. At this air speed, the aeroelastic system is such that the torsional mode can not be observed by measuring the displacement of the plunge variable $\bar{h}$. Physically, this indicates that the node line of this mode has moved to the elastic axis at this air speed.

The individual SISO transfer function zeros associated with pitch α measurement, which are independent of the air speed, are not shown in Figure 12.9. Examining Equation (12.3.29), it can be seen that the zeros associated with the bending strain actuator u_h are always at the origin, and that the zeros associated with the torsion strain actuator u_α are always on the imaginary axis at $\bar{\omega}_h$. In both cases this places the zeros below the lower (nominally plunge) pole of the system for all non zero airspeeds. The zeros associated with the leading edge flap u_x and trailing edge flap u_β actuators depend on $\bar{\omega}_h$ and the aerodynamic influence coefficients, and may be imaginary or real. In the event they are real, once again one would be a nonminimum phase zero indicating a limitation on achievable control.

Due to the significant movement of the system poles and zeros as a function of air speed, it is necessary to choose specific dynamic pressures to analyze the system and design feedback control schemes. Three air speed ranges could be chosen. The first is below the speed of the plunge output pole-zero cancellation; that is, up to $U_\alpha = 1.30$. In this speed range most or all of the plunge output zeros are between the poles, while the pitch output zeros are below the poles. The second possible speed range is above the plunge output pole-zero cancellation, but below flutter($U_\alpha = 1.90$). In this range the plant is still stable, but none of the zeros lie between the poles in frequency. Of course, the third possible speed range is above flutter, when the plant is unstable.

Aside from active flutter suppression problems, the most realistic range to examine is the high sub-flutter speed range. For the purposes of discussion, a specific design point is chosen in this range, designated as design point 1 ($U_\alpha = 1.71$) on Figure 12.8. Most of the discussion of compensation design in Section 12.4 will be for aeroelastic control at this design point. The aeroelastic plant which must be controlled at design point 1 is characterized by the pole-zero locations, which are show in Figure 12.10. Note that in no case for a SISO pair is there an imaginary zero between the imaginary poles, and for one pair there are real zeros, in a minimum and nonminimum phase pair.

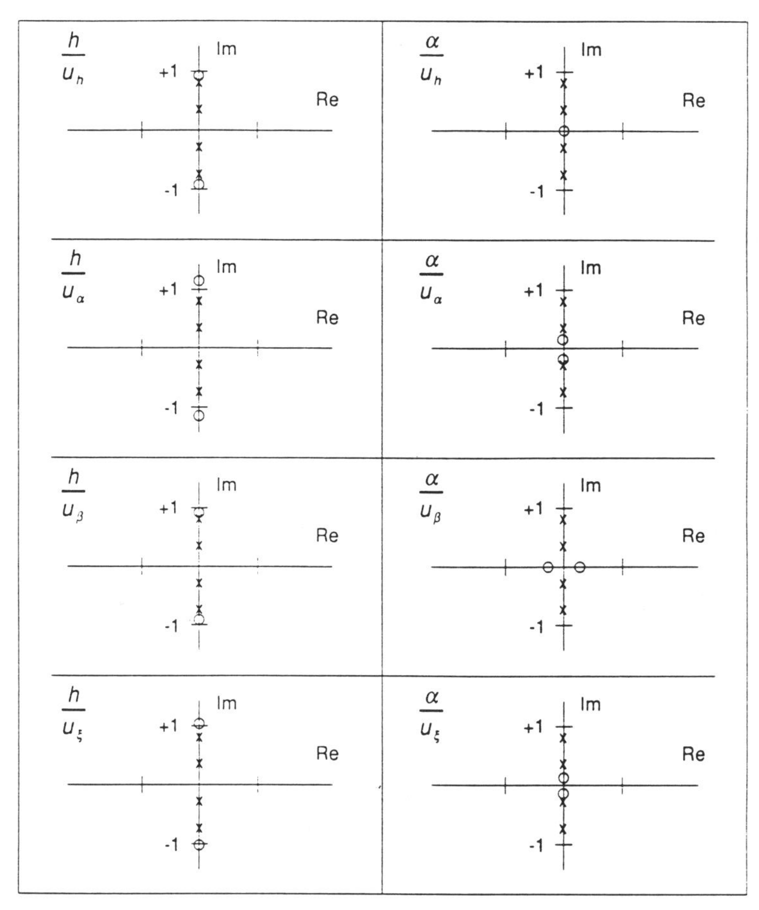

Figure 12.10: Pole and SISO zero locations for the reference typical section at design point 1.

In order to provide a reference for discussion of active stabilization, a second design point, about five percent above flutter (U_{α} = 2.00), is chosen and designated as design point 2 on Figure 12.8.

Plant variation with structural parameters

Since the main objective of the typical section model is to achieve better physical insight into the problem of aeroelastic control, it is important to check the robustness of the qualitative results to physical parameter variation and the assumptions of the modeling process.

In the aeroelastic problem, the ultimate influence of the structural parameters is to establish three speeds of interest: the divergence speed, the flutter speed, and the reversal speed of a trailing edge flap. Note that leading edge flaps and strain actuators in general do not "reverse." For the reference section of Table 12.1 and Figure 12.8, these occur in the order of flutter, reversal and finally divergence. If divergence occurs before flutter, the pole-zero pattern will be as shown in Figure 12.11. The parameters of this case are identical to the reference case, except that now $x_\alpha = -0.2$; i.e. the center of mass has moved in front of the elastic axis. In this case, divergence has occurred before flutter, as evidenced by the intersection of the lower pole with the zero imaginary frequency axis. Due to the parameters chosen, reversal has also occurred prior to divergence, as evidenced by the intersection of the trailing edge flap-plunge output zero with the zero imaginary frequency axis. The open loop behavior in this case, as evidenced by the pole-zero pattern along the imaginary axis, is quite different than the reference case. The resulting compensation would be also be quite different, although the process of compensator design and options open to the designer would be much the same.

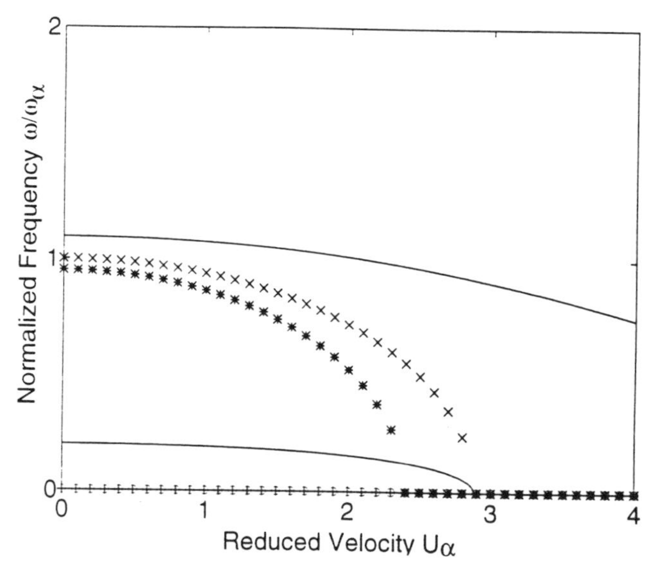

Figure 12.11: Pole and SISO zero frequencies as a function of normalized airspeed for a typical section with $x_\alpha = -0.2$. Zeros are: [o = h/u_ξ, = h/u_β, + = h/u_α, × = h/u_h].*

If divergence occurs after flutter, but reversal occurs before flutter, then the pole-zero pattern of Figure 12.12 is obtained. The parameters of this case are identical to the reference case, except that now a= -0.4; i.e. the elastic axis (and implicitly center of mass) have moved forward slightly. By comparison with Figure 12.9, this pole-zero pattern is quite similar to the reference case, except for the behavior of the trailing edge flap-plunge output zero. It starts above the torsion mode at low speed, moves below it, and then below the bending mode prior to reversal. After reversal, it forms a real pair which includes a nonminimum phase zero. Compensating this plant would be slightly different than that of the reference case, if the trailing edge flap were used in the speed range around flutter.

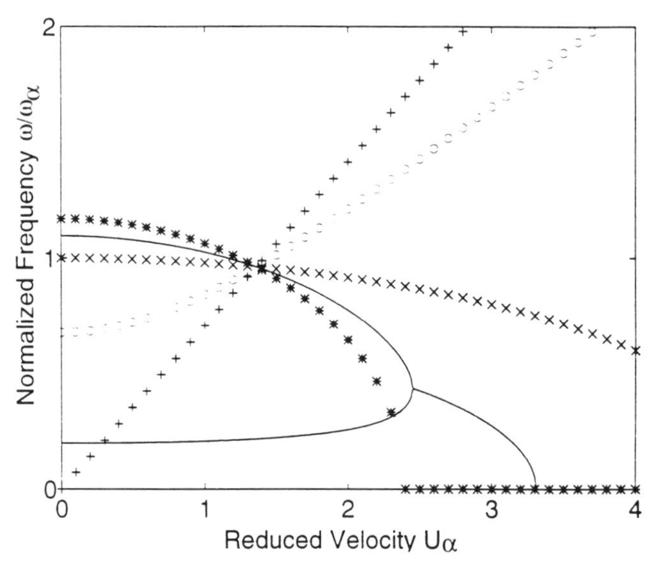

Figure 12.12: Pole and SISO zero frequencies as a function of normalized airspeed for a typical section with $a = -0.4$. Zeros are: [$o = h/u_\xi$, $* = h/u_\beta$, $+ = h/u_\alpha$, $\times = h/u_h$].

In summary, when divergence and reversal occur after flutter, the qualitative trends of the reference case hold. When divergence occurs after flutter, but reversal before flutter, the qualitative trends of the reference case still hold, except for the input subject to reversal near and above its reversal speed. If divergence occurs before flutter, a distinctly different plant and compensation approach are evident.

Plant variation with aerodynamic model

In order to further examine the generality of the qualitative results, the parameter robustness to aerodynamic modeling will be examined. In contrast to the reference section which includes only steady effects, full unsteady incompressible aerodynamics will be now be incorporated. The expressions for lift and moment (12.3.10) must be modified to include all of the non-circulatory terms, as well as any circulatory rate or acceleration terms. In addition, Theodorsen's function, $C(k)$, must be implemented with its complex frequency dependent nature. The resulting aerodynamic model is similar to (4.3.66), but with the additional unsteady forces and moments due to leading and trailing edge flaps [17].

The resulting aerodynamic model depends on an irrational function of the reduced frequency. In order to transfer this model into state space form, a rational approximation will be derived, using the approach outlined in Section 12.2. In order to capture the physics of the aerodynamic lags, a one pole approximation for Theodorsen's function is used

$$C(k) = 1 + \frac{-0.4544ik}{ik + 0.1902} \tag{12.3.30}$$

where k is the reduced frequency. This approximation is matched to the function at reduced frequencies of $k = 0$ and $k = 0.5$. The latter reduced frequency corresponds to the range of aeroelastic frequencies at design point 1. By representing Theodorsen's function (which appears in each of the aerodynamic operators) by a single pole approximation, new states must be added to the system to represent the aerodynamic lags. Recognizing that the second term of (12.3.30) is of the form of (12.2.24), the lag states can be appropriately defined and incorporated into the state space model. Unfortunately, the lags occur on the position and velocity of the flaps as well, so their dynamics must also be included in the model. A simple spring-mass-dashpot model is used with critical damping and a natural frequency 100 times greater than the torsional mode of the typical section. In this manner, the flap dynamics do not interfere with the main flutter phenomenon.

Solving the resulting complex matrix eigenvalue problem, the poles as a function of the airspeed are shown on Figure 12.13. Now a complex modal coalescence has occurred. Also shown are the zeros as a function of air speed for the imaginary zeros. For clarity, the leading edge flap zeros have been omitted. The movement of the open loop poles and SISO zeros with airspeed for the model with unsteady aerodynamics can be compared to the reference steady case, shown in Figure 12.9. Overall, the trends are quite similar and the differences subtle. When unsteady aerodynamics are included, two of the zeros from the pitch output transfer functions depend on airspeed. The most notable changes are in the trailing edge flap zeros. The trailing edge flap input-plunge output zero is initially above both poles and then crosses to lie between the poles before flutter. Additionally, the trailing edge flap input-pitch output zero, which was a real pair in the reference case, is now imaginary in the speed range of interest.

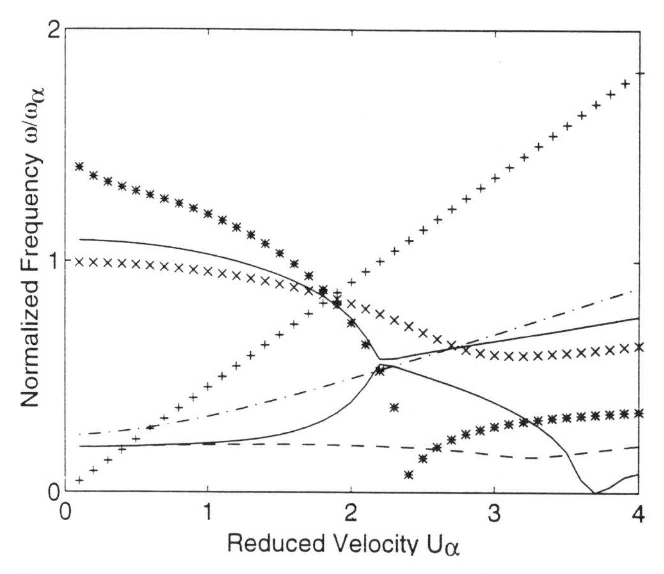

Figure 12.13: Pole and SISO zero frequencies as a function of normalized airspeed for the reference typical section with unsteady aerodynamics. Zeros are: [$*$ = h/u_β, $+$ = h/u_α, $\times$ = h/u_h, --- = α/u_α, -·- = α/u_β].

As in the homogeneous case, the essential features of the physics are captured in the steady model. Therefore, the following discussion of compensation, based on the reference section with steady aerodynamics, is illustrative of the opportunities and limitations of aeroelastic control.

12.4 Control design for the typical section

Compensation can now be designed for the aeroelastic plant represented by the reference typical section. Various static and dynamic compensation schemes will be developed for single and multiple inputs and outputs using both classical and optimal state space techniques. The closed loop plants resulting from these control laws will then be compared. As will be seen, the ultimate performance of the closed loop system depends not only on the compensator, but on the type and number of actuators and sensors. In order to choose the best combination of actuators and sensors for a particular lifting surface, it is essential to understand the fundamental mechanisms involved in aeroelastic control and the advantages and inherent limitations of different control schemes.

Before actually designing the compensation scheme for an aeroelastic system, several things must be considered: the control objective, the control topology, and the nature of the plant. The possible control objectives were outlined in Section 12.1. At design point 1, the reference typical section is stable. Possible objectives include regulating the plant for gust suppression, or incorporating it as part of a servo system for maneuver enhancement. Both objectives will be discussed below. At design point 2 of the reference typical section, the plant is unstable. This will allow demonstration of the third control objective: flutter suppression.

The control topology is determined by the choice of actuator and sensor combinations, as well as the feedback type: static or dynamic. The simplest case is single input single output (SISO) static feedback, which will be considered first. More elaborate cases of static compensation include single input multiple output (SIMO) and multiple input multiple output (MIMO). The limiting case occurs when all states are measured and fed back to all actuators. An optimal solution to this case is referred to as the Linear Quadratic Regulator (LQR). There is of course nothing which limits the designer to static compensation. Dynamic compensation can also be used with SISO systems, or with multiple inputs and outputs. An optimal solution of the MIMO dynamic compensator is the Linear Quadratic Gaussian (LQG) problem. All of these compensation schemes will be applied to the regulation problem. The servo design and compensation of the unstable plant will be examined only in the context of MIMO LQR.

The nature of the plant is established by the homogeneous dynamics, and the influence of the actuators and sensors. All of this information is stored in the SISO transfer functions, such as those in Figure 12.10. For this particular plant, the pole-zero diagrams reveal a marginally stable, multimode oscillatory plant, with a sufficient variety of actuator-sensor pairs that some combinations are certain to be effective in closed loop control.

Single-input static feedback

The combination of four actuator control inputs and two measurement outputs allows for eight SISO feedback control options to be considered. The transfer functions for the eight options are defined by Equation (12.3.29), and the location of the poles and zeros are illustrated graphically in Figure 12.10. At the air speed of design point 1, all of the poles lie on the imaginary axis indicating that the system is neutrally stable. The control objective is regulation. From both a classical and optimal design perspective, regulation translates into driving the oscillatory poles of the system first to the left in the Laplace plane, and then along a Butterworth pattern (at approximately a +/-135 degree angle from the origin) to infinity [23 and 24]. Thus, the initial control objective is to add damping to the two modes of the system, moving the poles to the left in the Laplace plane.

Compensating this plant consists of regulating two important interacting modes. All two mode interaction problems which involve lightly damped oscillatory structural modes can be thought of as belonging to one of four types, represented schematically in Figure 12.14. From SISO root locus rules, the angle of departure can be calculated for each of the four types with displacement feedback ($u=-fy$) and rate feedback ($u=-f\dot{y}$). The angle of departure is the direction in which the closed loop pole leaves the open loop pole when the gain is first increased from zero. Unfortunately, in only one case (the type with an imaginary zero which alternates between the two poles, designated as PZP) does static compensation by velocity feedback move the closed loop poles in the desired direction: to the left. Static displacement feedback on the other three types leads initially to no change in stability, and then to a static or dynamic instability. Rate feedback on the other types leads immediately to dynamic instability.

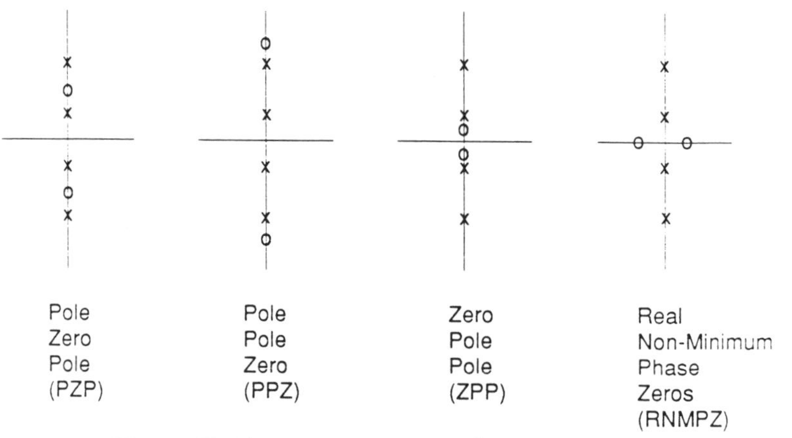

Figure 12.14: Representative pole-zero patterns.

Comparison of the eight pole-zero patterns of Figure 12.10 with the four types of Figure 12.14 reveals that none of the eight pole-zero patterns are of the PZP type. Therefore, static output feedback of neither displacement nor rate of either of the two states to any of the four inputs will succeed in producing a stable well regulated closed loop plant.

Since static SISO control does not succeed in regulation, the next option is to try multiple state measurements fed back to a single actuator, or SIMO control. This can be done either by making two independent measurements, or by placing a single translational sensor at some desirable point on the section so that a stabilizing combination of the plunge and pitch state variables is measured and fed back to the actuator. The output of a single translational sensor is fixed by its placement

$$y = \bar{h} + x_s \alpha \tag{12.4.1}$$

In the state space equations (12.3.26), this is equivalent to setting the output matrix to

$$\mathbf{C}_y = \begin{bmatrix} 1 & x_s & 0 & 0 \end{bmatrix} \tag{12.4.2}$$

If the sensor can be placed in a position x_s, which places a SISO transfer function zero between the open loop poles (i.e. makes the transfer function into one of the PZP type), the system can be stabilized using rate feedback of the output variable y. Equations (12.3.29) and (12.4.1) can be combined to give equations for the four SISO transfer functions between any of the four control inputs and the output measurement y. Setting the numerator of each transfer function to zero gives an expression for each SISO transfer function zero. These expressions can then be solved for the sensor position x_s, which yields the desired SISO transfer function zeros for each method of actuation. The sensor positions as a function of the desired zero locations for the four actuation schemes are

$$u_h \text{ Control: } x_s = \frac{R_\alpha^2}{x_\alpha}\left(1 - \frac{\overline{q}C_{M_\alpha} - 1}{p^2}\right)$$

$$u_\alpha \text{ Control: } x_s = \frac{x_\alpha p^2 + \overline{q}C_{L_\alpha}R_\alpha^2}{p^2 + \overline{\omega}_h^2}$$

$$u_\beta \text{ Control: } x_s = \frac{p^2\left(\dfrac{C_{M_\beta}x_\alpha}{C_{L_\beta}R_\alpha^2} + 1\right) + 1 - \overline{q}C_{M_\alpha}\left(1 - \dfrac{C_{M_\beta}C_{L_\alpha}}{C_{L_\beta}C_{M_\alpha}}\right)}{\dfrac{1}{R_\alpha^2}\left(p^2\left(\dfrac{C_{M_\beta}}{C_{L_\beta}} + x_\alpha\right) + \dfrac{C_{M_\beta}}{C_{L_\beta}}\overline{\omega}_h^2\right)}$$

$$u_\xi \text{ Control: } x_s = \frac{p^2\left(\dfrac{C_{M_\xi}x_\alpha}{C_{L_\xi}R_\alpha^2} + 1\right) + 1 - \overline{q}C_{M_\alpha}\left(1 - \dfrac{C_{M_\xi}C_{L_\alpha}}{C_{L_\xi}C_{M_\alpha}}\right)}{\dfrac{1}{R_\alpha^2}\left(p^2\left(\dfrac{C_{M_\xi}}{C_{L_\xi}} + x_\alpha\right) + \dfrac{C_{M_\xi}}{C_{L_\xi}}\overline{\omega}_h^2\right)}$$

(12.4.3)

where p is the desired frequency of the complex conjugate pair of zeroes. By choosing x_s such that the zero is between the poles (as in the PZP pattern of Figure 12.14), a stable closed loop system results, in principle, for rate feedback. For example, if a vertical rate sensor is placed at the leading edge of the section, and the output is fed back to bending strain control input u_h, the root locus of Figure 12.15 results. Using this input-output pair, stable compensation has been achieved, with good damping in the lower or bending mode and a small amount of damping in the upper or torsion mode.

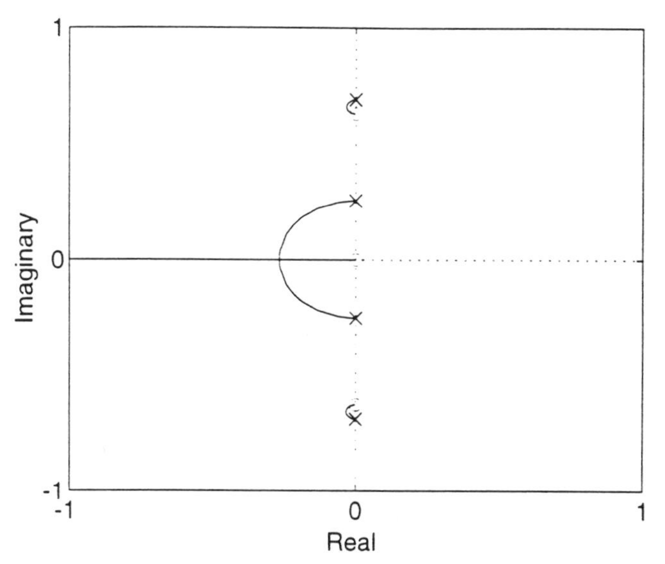

Figure 12.15: Root locus for a leading edge vertical rate sensor output and plunge input.

Unfortunately, in practice, sensor placement will not work for all configurations or choices of actuators. The sensor location needed for stable feedback is sometimes found to be physically off of the typical section. For example, using the reference typical section and design point 1 values, it is possible to find a stable sensor location for each method of actuation. However, all of the stabilizing sensor locations are found to be physically well off the section at design point 2.

Clever sensor placement underlies many practical vibration suppression designs. By appropriately choosing the location for an accelerometer, integrating to obtain rate, and feeding back to an actuator, successful gust alleviation has been implemented.

Full state static feedback

The problems associated with sensor placement can be avoided by measuring both the plunge and pitch variables independently, and feeding back some linear combination of these outputs which stabilizes the system. One method of implementing such a stable control scheme is the use of full state feedback schemes such as the Linear Quadratic Regulator (LQR). In full state feedback, the input to the system is

$$u = -\mathbf{F}x \tag{12.4.4}$$

where **F** is the sought after gain matrix, and x is the full state vector. Using full state feedback is particularly advantageous because control laws may be developed which utilize various combinations of both the displacement and rate variables, which can be scheduled to change with flight conditions, and are not limited to combinations corresponding physically to locations on the airfoil.

The optimal gains **F** for full state feedback can be found by solving the LQR problem, which entails minimizing the scalar cost functional J

$$\begin{aligned} J &= \int_0^{\infty} \left(x^T \mathbf{R}_{xx} x + \rho u^T \mathbf{R}_{uu} u \right) dt \\ \mathbf{R}_{xx} &= \mathbf{C}_z{}^T \mathbf{C}_z \end{aligned} \tag{12.4.5}$$

where $\mathbf{R}_{xx}$ is the penalty on the states, $\mathbf{C}_z$ is the performance metric defined in (12.2.4), and $\rho\mathbf{R}_{uu}$ is the penalty on the control inputs which combines a scale factor ρ and an input weighting matrix $\mathbf{R}_{uu}$. The solution for the gain matrix **F** is obtained by substituting the input **u** into the cost function, and minimizing the cost constrained by the system dynamics [20]. The minimization yields the equations for the gains

$$\begin{aligned} 0 &= \mathbf{A}^T \mathbf{S} + \mathbf{S}\mathbf{A} + \mathbf{R}_{xx} - (1/\rho)\mathbf{S}\mathbf{B}_u \mathbf{R}_{uu}^{-1} \mathbf{B}_u^T \mathbf{S} \\ F &= (1/\rho)\mathbf{R}_{uu}^{-1} \mathbf{B}_u^T \mathbf{S} \end{aligned} \tag{12.4.6}$$

The gains represent the optimal combination of the states to be fed back to each actuator. The methodology can be applied to single or multiple actuator input systems.

In the design point 1 example, the state cost penalty is an evenly weighted combination of the plunge and pitch displacement states normalized by their maximum values. The maximum values are determined from the deflection associated with the assumed modes of (12.3.4), and an allowable strain of one percent. Similarly, each control input is normalized by its maximum value. The peak deflection for the trailing edge flap is taken to be five degrees. The peak leading edge flap deflection is determined by equating its hinge moment with that of the trailing edge flap, and found to be about two and a half degrees. For the strain actuators, the maximum control input is computed using a peak actuation strain of $600\mu\varepsilon$. The resulting maximum control inputs in bending and twisting were calculated from (12.3.22) and (12.3.23). For systems with more than one actuator input, all of the normalized control inputs are weighted evenly.

Full state single input static feedback

The first application of full state feedback is to the single input case. An interesting result is found for LQR in the asymptotic limit of small gains, which corresponds conceptually to the angle of departure in the root locus analysis. It can be shown that in the low gain case for lightly damped oscillatory systems, the only non-zero gains are on state variables which represent rates. The gains on the displacements are zero [25]. Furthermore, since there is only a single actuator, the loop can be broken at the input, and the pole-zero structure of the loop transfer function examined. This loop transfer function treats the state outputs as a single virtual measurement, which combines the states in the ratio given by the gain $\mathbf{F}$. The resulting loop transfer function is given by $\mathbf{F}(p\mathbf{I}-\mathbf{A})^{-1}\mathbf{B}_u$. The pole-zero pattern of this loop will always resemble the alternating PZP pattern of Figure 12.14. Thus for a single input, the low gain LQR solution is equivalent to a sophisticated rate feedback sensor placement algorithm. Of course, as the control penalty scalar ρ decreases and gains increase, some displacement feedback will be added.

Figure 12.16 shows the movement of the LQR closed loop pole locations of the reference typical section at design point 1, for each of the four individual actuator inputs. The closed loop poles are plotted for control weights which decrease from a large value toward zero ($\rho=10^{+4}$ to $\rho=10^{-4}$). As the control weight decreases, the control gains tend to increase, and the poles move away from their open loop locations. At high gain, the poles of these root loci move towards the MIMO zeros of the system.

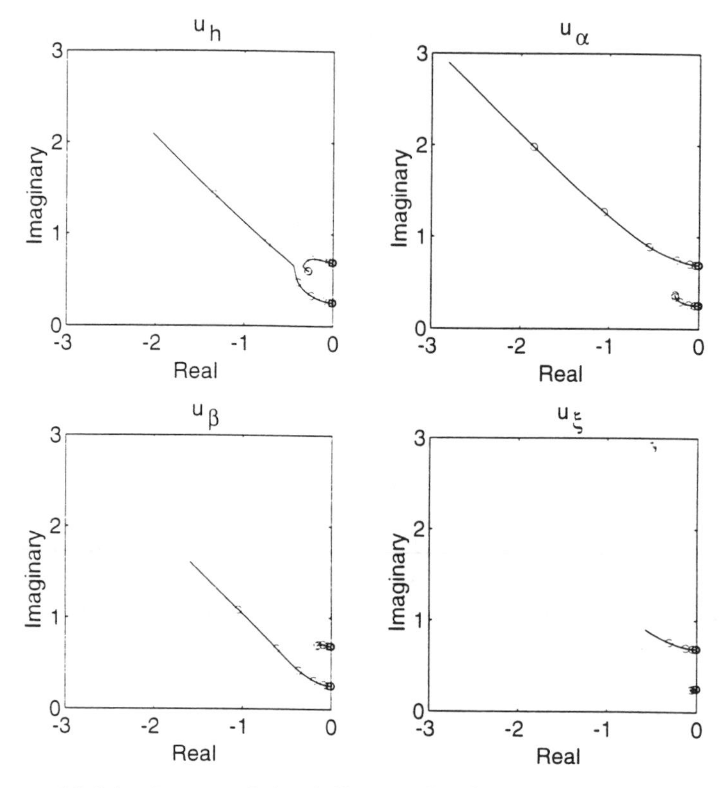

Figure 12.16: Locus of the full state feedback closed loop poles with increasing LQR gain for the four individual actuator inputs.

For purposes of determining their zeros, MIMO systems are distinguished as square or nonsquare. In this context, square implies that the number of states which contribute to the performance and the number of control inputs are equal. For a square system, the zeros are given by

$$Z(p) = \mathbf{C}_z(p\mathbf{I} - \mathbf{A})^{-1}\mathbf{B}_u \tag{12.4.7}$$

These zeros are either at infinity, or are finite transmission zeroes. In either case, the contribution to the state cost (the first term on the right hand side of (12.4.5)) diminishes as the closed loop poles approach the zeros. For those poles approaching infinity, the modal response decreases. For those poles approaching finite transmission zeros, the modal response is still finite, but the observability of the mode in the state cost diminishes.

For nonsquare systems, the high gain asymptotic limits for some of the closed loop poles are given by the finite stable zeros (or reflections of the nonminimum phase zeros) of the full Hamiltonian system

$$H(p)=\left[\mathbf{C}_z(-p\mathbf{I}-\mathbf{A})^{-1}\mathbf{B}_u\right]^T\left[\mathbf{C}_z(p\mathbf{I}-\mathbf{A})^{-1}\mathbf{B}_u\right] \tag{12.4.8}$$

when they exist. The remaining closed loop poles move along stable Butterworth patterns to the zeros at infinity. The finite zeros of H(p) are called compromise zeros for non-square systems [26]. While the contribution to the state cost of the closed loop modes which approach infinity diminishes, the state cost of the modes which approach the compromise zeros is finite even at high gain. Thus fundamental limitations on performance are present in systems with compromise zeros.

The systems whose loci are shown in Figure 12.16 all contain one actuator and a performance objective that penalizes both pitch and plunge. The asymptotic performance of the LQR compensator is limited for such nonsquare systems, even though all of the states are measurable. As shown in the figure, with each of the actuators acting separately, only one of the closed loop poles moves along a Butterworth pattern to a zero at infinity. The pole that moves to infinity is an indication of the relative effectiveness of the actuator on the two modes. The induced strain bending u_h and trailing edge flap u_β actuators are able to move the lower plunge pole along a stable Butterworth pattern. This indicates that plunge is relatively more controllable by these two actuators. The induced strain torsion u_α and leading edge flap u_ξ actuators are able to move the pitch pole along a stable Butterworth pattern, indicating pitch is primarily controllable by these two actuators. The rate at which the poles move along the loci is an indication of the absolute effectiveness of the actuator. The curves show that for this geometry, airspeed and assumptions on actuator authority, the induced strain bending and torsion actuators are more effective than the conventional control surfaces. It is also evident from the figure that the leading edge control surface is significantly less effective than the other actuators.

For systems with more performance variables than control inputs, the presence of compromise zeros indicates a limit in the achievable performance of the closed loop system. A choice of actuators with good authority over the states which contribute to the performance will cause all of the zeros to be at or near infinity, implying arbitrarily good closed loop performance.

Full state multiple input feedback

The limitation in performance associated with these compromise zeroes can be overcome by using more than one input. In this two mode model, two inputs provide sufficient degrees of freedom to place both poles independently. Figure 12.17 shows the closed loop pole locations for the cases of the actuators acting in pairs and all four controls acting together. In contrast with the single actuator case, this square system has no compromise zeros, and there are no limitations on closed loop performance.

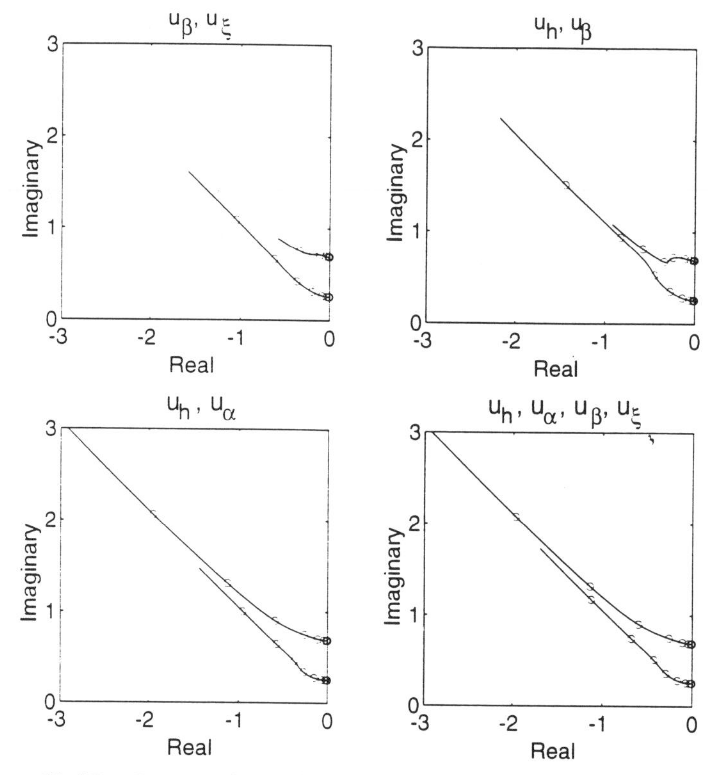

Figure 12.17: Locus of the full state feedback closed loop poles with increasing LQR gain for multiple actuator inputs.

Note that, although all pole pairs move to infinity, some actuator pairs are more effective than others. As before, the rate at which these poles move in the complex plane is directly related to the effectiveness of the multiple actuator system in question. The combination of the bending and torsion strain actuators provides better performance, as does the combination of trailing edge flap and torsion strain actuator (not shown). These are good combinations because they combine an actuator that effectively controls plunge (bending strain or trailing edge flap) with one that effectively controls pitch. In contrast, the bending strain and trailing edge flap actuator combination is less effective because both actuators tend to control only the plunge motion.

Finally, it is observed that the performance obtained using four actuators is only marginally better than that obtained using the best combination of two actuators. The two states present in the performance metric can both be controlled by two well chosen actuators. Little additional benefit is obtained when there are more actuators than the number of states in the performance metric.

Dynamic output feedback

While full state feedback combined with a sufficient number of actuator inputs provides arbitrarily good high gain closed loop performance, it is unrealistic to expect that all of the states of the system will be measured. This leads to the design of output feedback controllers in which only certain states or their fixed linear combinations are available for feedback. The static feedback of a subset of the states was found above to have severe performance limitations. The limitations of static output feedback can be overcome by using classical dynamic compensation, or dynamic compensators designed using optimal methods such as the Linear Quadratic Gaussian method.

Classical dynamic compensation seeks to add dynamics to the feedback path such that the Bode plot or root locus of the loop transfer function (plant times dynamic compensator) has desirable stability, performance and robustness characteristic. For low order damped plants such as servo mechanisms, the dynamic elements typically added to the compensator include lag, lead-lag and integrators. For lightly damped oscillatory structural plants, notch filters and complex pole-zero pairs are also used.

As an example of a classical design, a compensator will be developed for the ZPP type plant of Figure 12.14. This is representative of the cases of pitch output feedback to torsion strain control, and pitch to leading edge control, as shown in Figure 12.10. A possible compensation scheme would be to add a complex, lightly damped zero at a frequency between the two poles, and then add a real pole near the origin. This compensator transfer function then has two zeros, and one pole

$$K(p)=\frac{(p+j\overline{\omega}_z)(p-j\overline{\omega}_z)}{p} \tag{12.4.9}$$

and is improper, in that it does not roll off at high frequency. A pair of high frequency real poles can be added after crossover to provide a proper compensator

$$K(p)=\frac{(p+j\overline{\omega}_z)(p-j\overline{\omega}_z)}{p}\frac{1}{\left(p^2+2j\zeta\overline{\omega}_p+\overline{\omega}_p^2\right)} \tag{12.4.10}$$

An actual compensator design of this type is shown for the pitch output torsion strain input loop at design point 1 in Figure 12.18. The SISO root locus shows that the closed loop system is stable, and damping has been added to the poles. However, as in all single output systems in which the structure of the compensator is set and only the gain is varied, the extent the poles can be moved into the left half plane is limited.

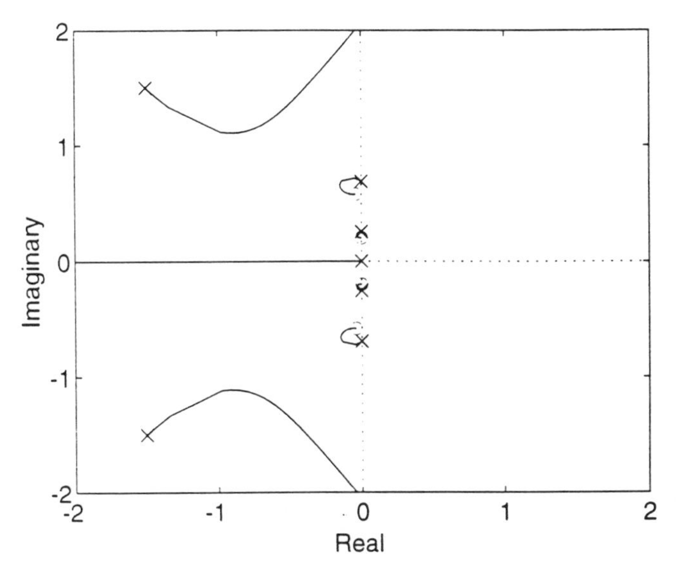

Figure 12.18: Root locus for classical compensation of the pitch output and torsion strain actuator input.

The extension of dynamic compensation to multiple loops is a difficult design task. An optimal MIMO feedback compensation may be obtained through a Linear Quadratic Gaussian method, which couples a Kalman filter with the already discussed optimal regulator. The architecture of LQG is shown in Figure 12.19. The plant is augmented by process and measurement noises

$$\begin{aligned} \dot{x} &= \mathbf{A}x + \mathbf{B}_u u + \mathbf{B}_w w \\ y &= \mathbf{C}_y x + v \end{aligned} \qquad (12.4.11)$$

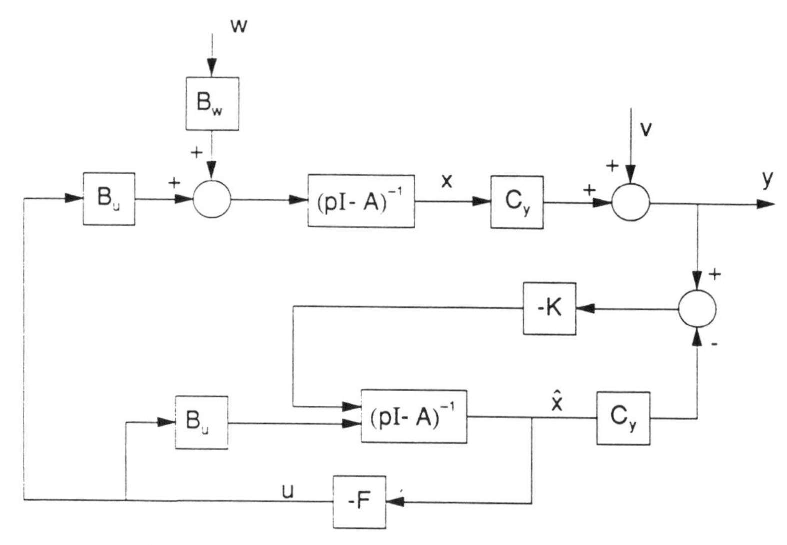

Figure 12.19: The system architecture of LQG.

The Kalman filter estimates the values of the states from a knowledge of the control inputs and a model of the plant. The estimation is kept convergent by comparing estimates of the measurements with the actual measurements, and feeding back the error through the Kalman filter gains **K**.

$$\dot{\hat{x}} = \mathbf{A}\hat{x} + \mathbf{B}_u u + \mathbf{K}\left[y - \mathbf{C}_y \hat{x}\right] \tag{12.4.12}$$

Using the state estimates, the controller may operate as though full state feedback has been achieved. It should be emphasized that despite the subtleties and assumptions made in its derivation, the Kalman filter, when cascaded with the LQR control gains, merely produces a dynamic compensator in the form of a transfer function from measurement to input. The form of this compensator is given as

$$K(p) = -\mathbf{F}(p\mathbf{I} - \{\mathbf{A} - \mathbf{B}_u\mathbf{F} - \mathbf{C}_z\mathbf{K}\})^{-1}\mathbf{K} \tag{12.4.13}$$

where **F** is the LQR gain derived above.

The design of the Kalman filter is the dual problem to the design of the full state feedback controller. The filter gains are chosen based on statistical knowledge of the disturbances on the plant, and noise in the measurements. The estimation error minimization leads to governing equations of the form

$$
\begin{aligned}
0 &= \mathbf{AP} + \mathbf{PA}^T + \mathbf{B}_w \Sigma_{ww} \mathbf{B}_w^T - \mathbf{PC}_y^T \Sigma_{vv}^{-1} \mathbf{C}_y \mathbf{P} \\
\mathbf{K} &= \mathbf{PC}_y^T \Sigma_{vv}^{-1}
\end{aligned}
\qquad (12.4.14)
$$

If the measurement noise covariance Σ_{vv} is set to be high relative to the process noise covariance Σ_{ww}, the measurements will be of less value and the state estimates will be more heavily based on the plant model. In contrast, if the process noise is high compared to the measurement noise, the measurements will be emphasized.

In order to compare with the previously designed classical compensator, an LQG compensator can be designed for the loop from pitch measurement to torsion strain control. The unsteadiness in the inflow angle, a 1 degree broadband white noise signal, constitutes the process noise. The measurement noise is calculated as a percentage of the maximum value for the given state, and is set to 1%.

The resulting compensator poles and zeros are shown in Figure 12.20 for a low LQR gain. The Kalman filter has placed a zero near the origin (to provide rate-like feedback from the displacement pitch measurement), a pair of complex poles below the first plant pole (to insure the proper angle of departure), a pair of nonminimum phase lightly damped zeros between the plant poles (to insure the proper angle of departure), and a pair of highly damped complex poles (to roll off the compensator). For comparison with the classical compensation, a classical (Evan's) root locus has been constructed using these low gain compensator poles and zeros. By comparison with the classical dynamic compensator of Figure 12.18, much the same performance has been achieved, by much the same means. Of course, the machinery of LQG allows much more complex MIMO cases to be handled just as directly. Furthermore, LQG adjusts the location of the compensator poles and zeros for each level of control authority demanded. The compensator poles and zeros of Figure 12.20 are only the ones corresponding to very low gain.

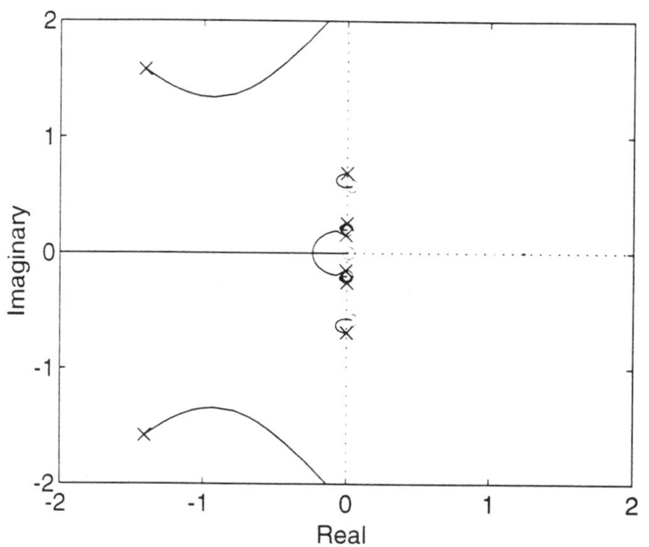

Figure 12.20: A root locus constructed using the plant and low gain LQG compensator poles and zeros.

A few trends in LQG are notable. The order of the compensation is always constrained to be exactly that of the plant model. This can lead to high order controllers, often higher than is necessary or can be implemented. LQG often returns compensators with features which are not robust to parameter variation, such as the nonminimum phase lightly damped zero pair of Figure 12.20. Unstable compensators are often derived which must be implemented very judiciously. Finally, a Kalman filter cannot invent information, it can only filter it. The more well-chosen measurements that are available and the more information they contain on the modes of importance to the performance, the more accurate the estimates of the important states will be and the better the overall system will perform.

Compensation for an aeroelastic servo

The preceding discussion of compensation has focused on regulating the plant. This achieves disturbance rejection for the purposes of gust load alleviation or ride comfort. Other possible control objectives include servo design and active stabilization, which conceptually builds upon regulation. A servo is a system designed to track a reference command input, usually in the form of a commanded displacement or velocity. An aeroelastic servo could be part of a flight control system, in which a maneuver is obtained or enhanced by commanding deflections to the wing.

One method of designing a servo is to introduce a minor variation into the Linear Quadratic Gaussian framework. In contrast to the standard LQG problem, the exogenous inputs are expanded to include reference command inputs. Additional integrator states are also introduced to ensure that there will be no steady state error in tracking the reference. Servos may be designed without these integrators, but they will exhibit a steady state error to a displacement input.

The servo architecture can be seen in Figure 12.21. In order to form the LQ servo problem, the reference inputs must be commands on actual states of the system. Assuming that there are reference commands on a portion of the system states, the state vector is separated into two parts: x_r, the states which correspond to reference inputs, and x_{nr}, those states which have no reference. In addition to the system states, integrator states are appended to the model. The integrator states integrate the error between the reference command and those states which correspond to the reference inputs.

$$\begin{Bmatrix} \dot{x}_r \\ \dot{x}_{nr} \\ \dot{x}_I \end{Bmatrix} = \left[\begin{array}{cc|c} \multicolumn{2}{c|}{\mathbf{A}} & \mathbf{0} \\ \hline -\mathbf{I} & \mathbf{0} & \mathbf{0} \end{array} \right] \begin{Bmatrix} x_r \\ x_{nr} \\ x_I \end{Bmatrix} + \begin{Bmatrix} \mathbf{B_u} \\ \hline \mathbf{0} \end{Bmatrix} u + \begin{Bmatrix} \mathbf{0} \\ \mathbf{0} \\ \mathbf{I} \end{Bmatrix} r \tag{12.4.15}$$

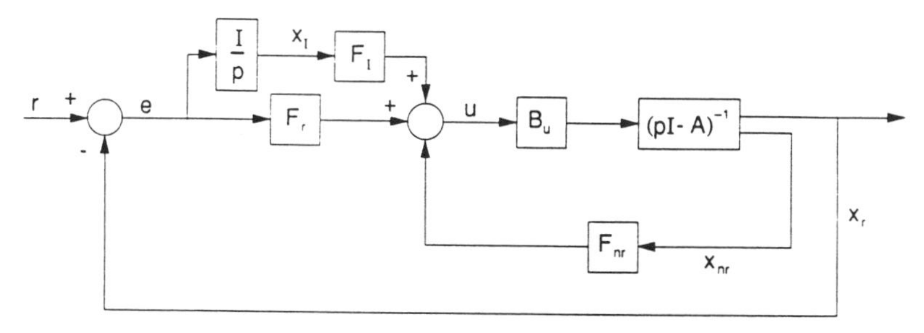

Figure 12.21: The LQR servo architecture.

To calculate the optimal gains, the augmented state and control influence matrices are used in the LQR problem. Note that the third term on the right hand side of (12.4.15), does not explicitly appear in the LQG design problem. The elements of the state weighting matrix $\mathbf{R_{xx}}$ on x_r and x_{nr} are chosen as in the standard LQG problem, and a nonzero weighting on the integrator states x_I is chosen to provide the desired settling time at the reference command value. Once the gains are calculated, the gain matrix may be separated into three sections corresponding to the three different portions of the state vector. The control input is augmented by the appropriate gain times the reference inputs.

$$u = -\begin{bmatrix} \mathbf{F_r} & \mathbf{F_{nr}} & -\mathbf{F_I} \end{bmatrix} \begin{Bmatrix} x_r \\ x_{nr} \\ x_I \end{Bmatrix} + \mathbf{F_r} r \qquad (12.4.16)$$

If this is rewritten as

$$u = -\mathbf{F} y$$
$$y = \mathbf{I} x + \begin{bmatrix} -\mathbf{I} \\ 0 \\ 0 \end{bmatrix} r \qquad (12.4.17)$$

then the relation with the general servo formulation of (12.2.36) and (12.2.44) can be seen.

As an example, a servo can be designed for the pitch output to torsion strain actuator input loop of the reference typical section at design point 1. The pitch output, related to the angle of attack, was selected since it is an important state in maneuver control. One integrator state is appended to the state vector and the LQR optimal gains are calculated. The response to a step input in pitch angle can be seen in Figure 12.22. With the specified weightings, the response tracks the reference within several periods. The performance of the LQ servo may be altered by changing the weightings of the states in the LQ formulation.

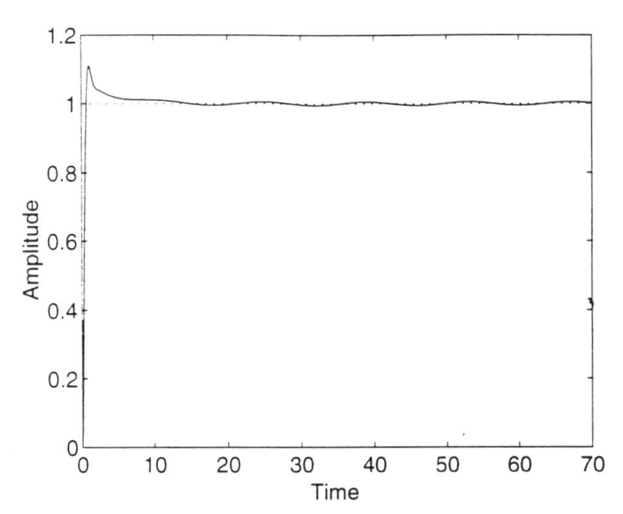

Figure 12.22: Response to a step input in pitch with an LQR servo, pitch output and torsion strain actuator input (time normalized by ω_α).

Compensation for an unstable system

A final possible application of aeroelastic control is control of dynamic instability, such as flutter. Especially in the case of mild instabilities, such control has the promise of extending the envelope of the aircraft, at modest cost and complexity.

In order to exemplify this objective, the reference typical section is examined at design point 2 with U_α=2.00, a speed 10% past the open loop flutter point. At this design point, there is a complex conjugate pair of poles in the left half plane and its mirror image in the right half plane, as can be seen in Figure 12.23. It is these complex poles in the right half plane which indicate dynamic instability.

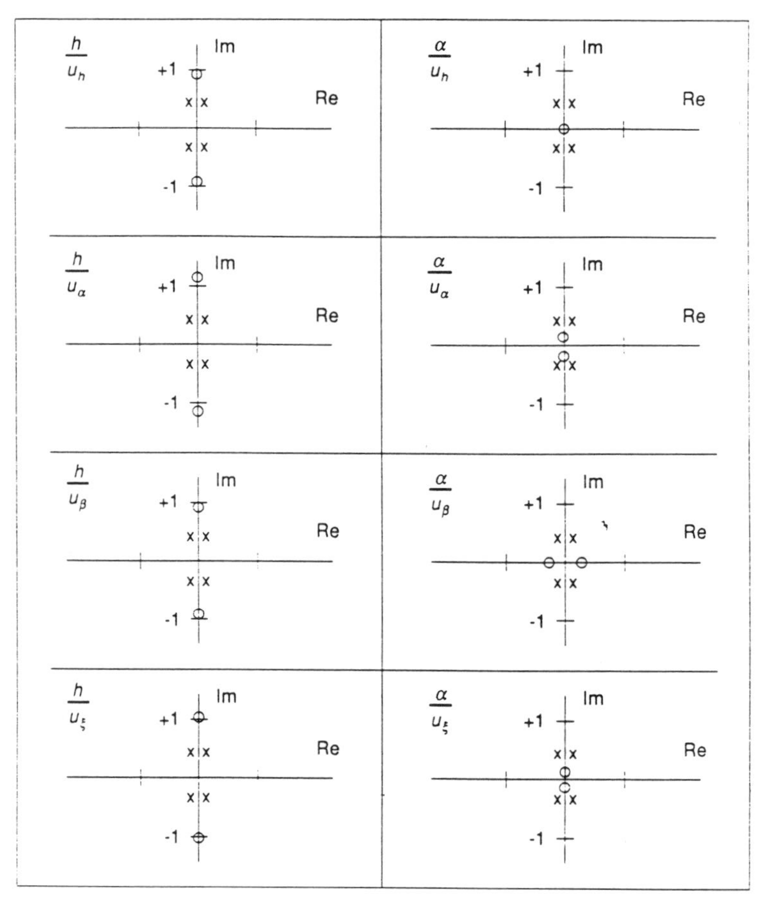

Figure 12.23: Pole and SISO zero location for the reference typical section at design point 2.

As for the stable plant, full state static feedback may be used to regulate and, in this case, stabilize the plant. The Linear Quadratic Regulator guarantees stability of the closed loop plant and provides the optimal feedback gains for the full state feedback problem.

Using the Linear Quadratic Regulator on the unstable plant, many of the same observations can be made as for the stable plant. The main difference in using LQR on an unstable plant is that there is a finite minimum amount of control that must be exerted to stabilize the system. Since LQR guarantees not only a stable closed loop system, but also minimum gain and phase margins, the first step it takes is to stabilize the plant by moving the right half plane poles to their mirror image in the left half plane.

The case of single actuators used to stabilize the post-flutter plant is shown in Figure 12.4.24, with the same choices for state and control weighting as were used for the stable design point 1. Once the plant has been stabilized, there are two identical pairs of complex poles in the left half plane. In all three cases, both pole loci start at this same location, which corresponds to the minimum control effort LQG will allow. With increasing gain, one closed loop pole moves toward infinity on a stable Butterworth pattern, while the second approaches a compromise zero. Since the poles are not distinguishable, the classification of actuators by which mode they control more effectively is no longer possible. As in the case of an initially stable plant, when only one actuator is used, there is still a fundamental limitation on performance.

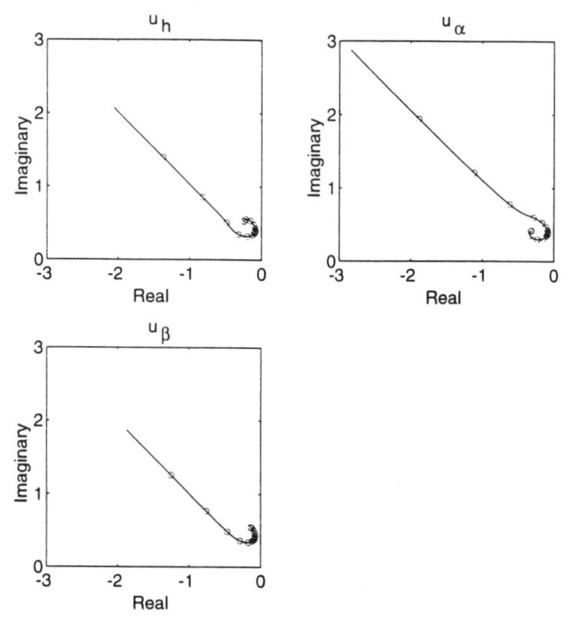

Figure 12.24: Locus of the full state feedback closed loop poles at design point 2 with increasing LQR gain for the three individual actuator inputs.

When multiple actuators are used, the performance limitation is removed, as can be seen in Figure 12.4.25. Two actuators are capable of effectively controlling both of the important modes involved in the flutter mechanism. It is also apparent that the best of the two actuator combinations performs nearly as well as the combination using all three actuators. This shows once again that diminishing benefit is obtained by having more actuators than the number of states which must be controlled.

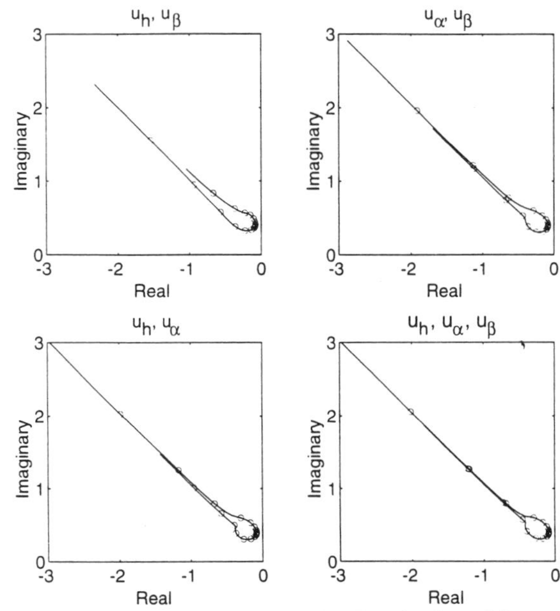

Figure 12.25: Locus of the full state feedback closed loop poles at design point 2 with increasing LQR gain for multiple actuator inputs.

Lessons learned from the typical section

As with traditional aeroelastic phenomena, many lessons can be learned from the closed loop behavior of a typical section. These lessons form the basis for understanding the fundamental mechanisms and limitations involved in aeroelastic control, and for comparing alternative control methods. They can be summarized in the observations which follow.

Both the poles and zeros of the individual SISO loop transfer functions shift rapidly with air speed. The general pattern of poles and zeros is established by the relative order of the divergence, flutter and reversal speeds. The zeros associated with trailing edge flap inputs are particularly sensitive to aerodynamic modeling assumptions.

The SISO pole-zero patterns are such that in some speed ranges simple output gain feedback of any single displacement or rate state variables will not stabilize the system. It is necessary to use some combination of the rate states in order to achieve a stable output feedback loop.

Stable static output feedback loops can be found using sensor placement techniques, or by the solution to the Linear Quadratic Regulator (LQR) problem. For the limiting case of low gain control, the LQR solution yields a solution similar to that of sensor placement, *i.e.* a stabilizing combination of the rate states. However, the LQR gain ratios can be varied with airspeed, and are not restricted to those which correspond to physically realizable sensor locations.

Even when all states are available for feedback, fundamental limitations can exist on closed loop performance. If fewer actuators are used than the number of states important in the performance metric, closed loop performance limitations are identified by the presence of finite MIMO compromise zeros. As gain is increased, some closed loop poles go to these zeros, rather than along stable Butterworth patterns. Such limitations can be avoided by utilizing at least as many control actuators as important performance states.

In the common case when all of the states are not available as outputs for feedback, dynamic compensation must be used. Such compensators can be derived by classical design, or by application of optimal techniques such as the Linear Quadratic Gaussian (LQG) compensation. For simple problems, the compensators derived by the two methods are quite similar. At low gain they place compensator poles and zeros along the imaginary axis to achieve a desirable alternating pole-zero pattern, and then add a pole or zero near the origin to achieve rate feedback. Classical techniques generally yield lower order, more robust compensators, but are difficult to extend to larger order problems. LQG returns a compensator which is of the same order as the plant model, and is often nonrobust to plant parameter uncertainty. However, LQG and other optimal techniques can be applied directly to higher order problems.

The fundamental control task for an aeroelastic system is regulation. The techniques of plant regulation can be applied directly to the task of flutter control, i.e. stabilizing an otherwise unstable plant. With minor modification, the methodology can incorporate aeroelastic control into a servo tracking system.

Other issues in compensator design

Two important problems in the design of compensation for aeroelastic systems have not been addressed: compensator order and compensator robustness. High compensator order comes about due to the high order of aeroelastic systems. Concern for robustness is due to low plant damping (in other than the first few modes) and significant variation in plant parameters due to change in configuration, flight Mach number and flight dynamic pressure.

Even after the model reduction described in Section 12.2, the order of the control design model will be large (typically 20 to 40 states). Three control design approaches are possible for this model, as depicted in Figure 12.26. The first and second reflect the fact that most optimal design methodologies return compensators of the order of the design model, which in practice are too large to run in real time. The first option is to further reduce the design model (by the technique described in Section 12.2) to the desired final order of compensator, then design for that model. The second, preferable option is to design the compensator based on the control design model, and then reduce the order of the compensator to the desired size, using techniques similar to those described above for model reduction [27].

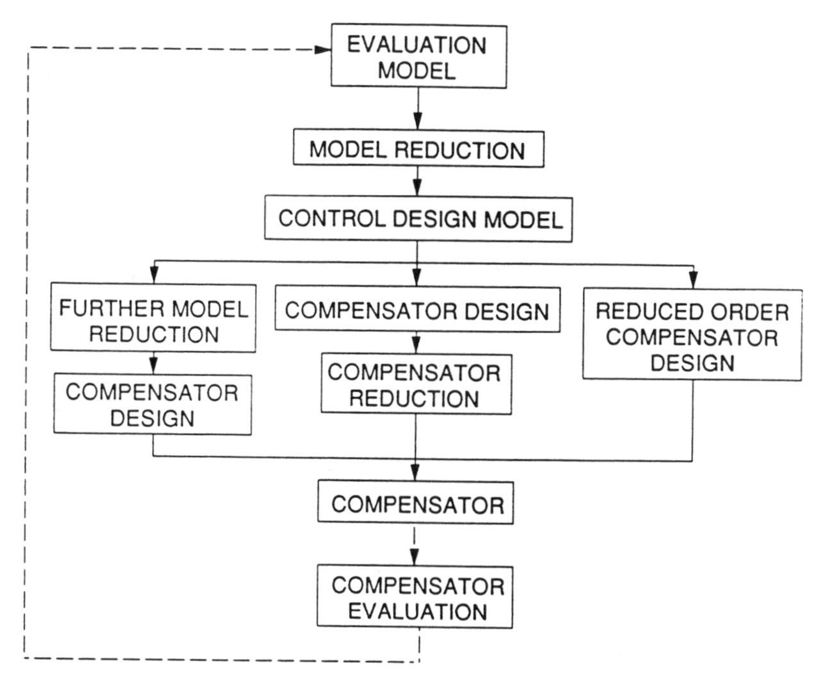

Figure 12.26: The process of deriving and evaluating a reduced order compensator.

A third option involves directly designing a reduced order compensator by techniques such as optimal projection [28], or by classical dynamic compensation, using successive loop closure [8]. In any case, the derived compensator should be exercised on the full order evaluation model, which is the best mathematical approximation of physical reality. This evaluation will expose potential problems due to over-truncation of the model or compensator.

A designer is wise to evaluate the compensator for several variations of the evaluation model, in order to assess compensator robustness. Robustness is the sensitivity of stability and performance to plant variation. Two types of aeroelastic plant variation mandate robustness. The first is uncertainty in the dynamics of a nominally time invariant plant due to simple modeling errors. Compensator robustification can be approached by using techniques which sensitize the compensator [29] or which bound the allowable uncertainty [30].

The second type of plant variation is due to the fact that the plant is not time invariant. Changes occur due to configuration change (stores, fuel load, wing geometry, etc.) and flight condition (Mach number and dynamic pressure). Approaches to dealing with these parameter variations include: the design of simple, very robust compensators; gain scheduling [31], in which the gains are scheduled to vary with flight parameters and configuration; and adaptive methods [32]. The wide and dramatic changes in the plant which occur due to changes in flight parameters ultimately place significant realistic limits on the performance of aeroelastic control in practical applications.

References for Chapter 12

[1] Noll, Thomas, "*Aeroservoelasticity*" to be published in Monograph on Flight Vehicle Materials, Structures, and Dynamics Technology-Assessment and Future Directions .

[2] Horikawa, H. and Dowell, E.H. "An Elementary Explanation of the Flutter Mechanism with Active Feedback Controls", *J. Aircraft,* (April 1979) pp. 225-232.

[3] Tiffany, S.H. and Adams, W.M. Jr., "Nonlinear Programming Extensions to Rational Function Approximation methods for Unsteady Aerodynamic Forces", NASA TP-2776 (July 1988).

[4] Karpel, M., "Design for Active Flutter Suppression and Gust Alleviation Using State-Space Aeroelastic Modeling", *J. of Aircraft*, Vol. 19, No. 3 (March 1982) pp. 221-227.

[5] Newsom, J. R., Abel, Irving, and Pototzky, Anthony, "Design of a Flutter Suppression System for an Experimental Drone Aircraft", *J. of Aircraft,* Vol. 2, No. 1 (May 1985).

[6] Johnson, E., Hwang, C., Joshi, D., Kesler, D. and Harvey, D., "Test Demonstration of Digital Control of Wing/Store Flutter", *AIAA J. of Guidance , Control, and Dynamics,* Vol. 6, No. 3 (May-June 1983) pp. 176-181.

[7] Bisplinghoff, R.L., and Ashley, H., *Principles of Aeroelasticity*, John Wiley and Sons, New York, 1962.

[8] Maciejowski, S.M., *Multivariable Feedback Design,* Addison-Wesley, Wokingham, Eng, C, 1989.

[9] Skelton, R.E. and Hughes, P.C., "Modal Cost Analysis for Linear Matrix-Second-Order Systems", *J. of Dynamic Systems, Measurement and Control,* Vol. 102 (September 1980) pp. 151-158.

[10] Moore, B.C., "Principal Component Analysis in Linear System: Controllability, Observability, and Model Reduction", *IEEE Transactions on Automatic Control,* AC-26 (1981) pp. 17-31.

[11] Chiang, R.Y. and Safonov, M.G., *Robust Control Toolbox,* The Mathworks, Inc., 1988.

[12] Karpel, M., "Reduced -Order Aeroelastic Models via Dynamic Residualization", *J. of Aircraft,* Vol. 27 (May 1990) pp. 449-455.

[13] Jones, Robert M., *Mechanics of Composite Materials,* Hemisphere Publishing Corporation, New York, 1975.

[14] Meirovitch, Leonard, *Elements of Vibration Analysis,* McGraw Hill Book Company, New York, 1986.

[15] Lazarus, K.B., and Crawley, E.F., "Induced Strain Actuation of Composite Plates", GTL Report No. 197, Massachusetts Institute of Technology, Cambridge, MA (1989).

[16] Strang, Gilbert, *Introduction to applied Mathematics,* Wellesley-Cambridge Press, Cambridge (1986).

[17] Theodorsen, T. and Garrick, I.E., "Nonstationary Flow About a Wing-Aileron-Tab Combination Inculding Aerodynamic Balance", NACA No. 736, 1942.

[18] Crawley, E.F. and de Luis, J., "Use of Piezoelectric Actuators as Elements of Intelligent Structures", *AIAA J.,* Vol. 25, No. 10 (October 1987) pp. 1373-1385.

[19] Barrett, R.M., "Intelligent Rotor Blade Actuation Through Directionally Attached Piezoelectric Crystals", *presented at the 46th Annual Forum of the American Helicopter Society,* (1990).

[20] Cannon, R.H. and Rosenthal, D.F., "Experiments in Control of Flexible Structures with Non-Collocated Sensors and Actuators", *J. of Guidance, Control, and Dynamics,* Vol. 7, No. 5 (September-October, 1984) pp. 546-553.

[21] Fleming, F.M. and Crawley, E.F., "The Zeros of Controlled Structures: Sensor/Actuator Attributes and Structural Modelling", AIAA Paper No. 91-0984, *Proceedings of the 32nd SDM Conference,* Baltimore, MD (April, 1991) pp. 1806-1816.

[22] Freudenberg and Looze, "Right Half Plane Poles and Zeros and Design Tradeoffs in Feedback Systems", *IEEE Transactions on Automatic Control,* Vol. AC-30, No. 6 (June, 1985) pp. 555-565.

[23] Miller, D.W., Jacques, R.N. and de Luis J., "Typical Section Problems for Structural Control Applications", AIAA Paper No. 90-1225, *Proceedings*

of the AIAA Dynamic Specialists Conference, Long Beach, CA (April 5-6, 1990) pp. 334-348.

[24] Kwakernaak, Huibert and Sivan, Raphael, *Linear Optimal Control Systems*, Wiley-Interscience, New York, 1972.

[25] Lazarus, Kenneth B. and Crawley, Edward F., "Multivariable High-Authority Control of Plate-Like Active Lifting Surfaces", SERC Report #14-92, Massachusetts Institute of Technology, (June 1992).

[26] Emami-Naeini, A. and Rock, S.M., "On Asymptotic Behavior of Non-Square Linear Optimal Regulators", *Proceedings of the 23rd Conference on Decision and Control*, Las Vegas, NV (December 1984) pp. 1762-1763.

[27] Yousuff, Ajmal andSkelton, Robert, "Controller Reduction by Component Cost Analysis", *IEEE Transactions on Automatic Control*, Vol. AC-29, No. 6 (June 1984) pp. 520-530.

[28] Hyland, David C. and Bernstein, Dennis, "The Optimal Projection Equations for Model Reduction and the Relationship Among the Methods of Wilson, Skelton, and Moore", *IEEE Transactions on Auto Control*, Vol. AC-30, No. 12 (December 1985) pp. 1201-1211.

[29] Yedavalli, R.K. and Skelton, R.E., "Controller Design for Parameter Sensitivity Reduction in Linear Regulators", *Optimal Control Applications and Methods*, Vol. 3, (1982) pp. 221-240.

[30] Peterson, I.R. and Hollot, C.V., "A Riccati Equation Approach to the Stabilization of Uncertain Systems", *Systems and Control Letters*, Vol. 8, (1987) pp. 351-357.

[31] Shamma, J. and Athans, M., "Guaranteed Properties for Nonlinear Gain Scheduled Control Systems" *Proceedings of the 27th IEEE Conference on Decision and Control*, (1988).

[32] Marendra, K.S. and Annaswarmy, A.M., *Stable Adaptive Systems*, Prentice Hall, New Jersey, 1989.

Appendix I

A primer for structural response to random pressure fluctuations*

Nomenclature

a	Plate length
b	Plate width
D	$Eh^3/12(1-\nu^2)$, Plate bending stiffness
E	Modulus of elasticity
H_n	Plate transfer function
h	Plate thickness
I_n; I	Plate impulse function; see equation (I.21)
K_n^2	$=\dfrac{m\omega_n^2 a^4}{D}$
M_m	Plate generalized mass
m	Plate mass/area
n	normal
p	Pressure on plate
Q_m	Generalized force on plate
q_n	Generalized plate coordinate
R	Correlation function
t	Time
w	Plate deflection
x, y, z	Cartesian coordinates
∇^2	Laplacian
Φ	Power spectral density
ρ_m	Plate density
σ	stress
τ	Dummy time
ζ_m	Modal damping
ω_m	Modal frequency

* This Appendix is based upon a report by E. H. Dowell and R. Vaicaitis of the same title Princeton University AMS Report No. 1220, April 1975.

Appendix I

Introduction

In this appendix we shall treat the response of a structure to a convecting-decaying random pressure field. The treatment follows along conventional lines after Powell [1] and others. That is, the pressure field is modelled as a random, stationary process whose correlation function (and/or power spectra) is determined from experimental measurements. Using this empirical description of the random pressure, the response of the structure is determined using standard methods from the theory of linear random processes [2, 3]. The major purpose of the appendix is to provide a complete and detailed account of this theory which is widely used in practice (in one or another of its many variants). A second purpose is to consider systematic simplifications to the complete theory. The theory presented here is most useful for obtaining analytical results such as scaling laws or even, with enough simplifying assumptions, explicit analytical formulae for structural response.

It should be emphasized that, if for a particular application the simplifying assumptions which lead to analytical results must be abandoned, numerical simulation of structural response time histories may be the method of choice [4, 5]. Once one is committed to any substantial amount of numerical work (e.g., computer work) then the standard power spectral approach loses much of its attraction.

Excitation-response relations for the structure

In the present section we derive the excitation-response relations for a flat plate. It will be clear, however, that such relations may be derived in a similar manner for any linear system.

The equation of motion for the small (linear) deformation of a uniform isotropic flat plate is

$$D\,\nabla^4 w + m\frac{\partial^2 w}{\partial t^2} = p \tag{I.1}$$

where w is the plate deflection, p the pressure loading and the other terms are defined in the Nomenclature. Associated with (I.1) are the natural modes and frequencies of the plate which satisfy

$$D\,\nabla^4 \psi_n - \omega_n^2 m \psi_n = 0 \tag{I.2}$$

where ω_n is the frequency and $\psi_n(x, y)$ the shape of the nth natural mode.

In standard texts it is shown that the ψ_n satisfy an orthogonality condition

$$\iint \psi_n \psi_m \, dx \, dy = 0 \quad \text{for} \quad m \neq n \tag{I.3}$$

If we expand the plate deflection in terms of the natural modes

$$w = \sum_n q_n(t)\psi_n(x, y) \tag{I.4}$$

then substituting (I.4) into (I.1), multiplying by ψ_m and integrating over the plate area we obtain

$$M_m[\ddot{q}_m + \omega_m^2 q_m] = Q_m \qquad m = 1, 2, \ldots \tag{I.5}$$

where we have used (I.2) and (I.3) to simplify the result. M_m and Q_m are defined as

$$M_m \equiv \iint m\psi_m^2 \, dx \, dy$$

$$Q_m \equiv \iint p\psi_m \, dx \, dy \tag{I.6}$$

$$\cdot \equiv d/dt$$

For structures other than a plate the final result would be unchanged, (I.5) and (I.6); however, the natural modes and frequencies would be obtained by the appropriate equation for the particular structure rather than (I.1) or (I.2). Hence, the subsequent development, which depends upon (I.5) only, is quite general.

Before proceeding further we must consider the question of (structural) damping. Restricting ourselves to structural damping only we shall include its effect in a gross way by modifying (I.5) to read

$$M_m[\ddot{q}_m + 2\zeta_m\omega_m\dot{q}_m + \omega_m^2 q_m] \equiv Q_m \tag{I.7}$$

where ζ_m is a (nondimensional) damping coefficient usually determined experimentally. This is by no means the most general form of damping possible. However, given the uncertainty in our knowledge of damping from a fundamental theoretical viewpoint (see [6]) it is generally sufficient to express our meager knowledge. If damping is inherent in the material properties (stress-strain law) of the structure, the theory of viscoelasticity may be useful for estimating the amount and nature of the damping.

However, often the damping is dominated by friction at joints, etc., which is virtually impossible to estimate in any rational way.

Now let us turn to the principal aim of this section, the stochastic relations between excitation (pressure loading), and response (plate deflection or stress). We shall obtain such results in terms of correlation functions and power spectra.

The correlation function of the plate deflection w is defined as

$$R_w(\tau; x, y) \equiv \lim_{T\to\infty} \frac{1}{2T} \int_{-T}^{T} w(x, y, t)w(x, y, t+\tau)\,\mathrm{d}t \tag{I.8}$$

Using (I.4) we obtain

$$R_w(\tau; x, y) = \sum_m \sum_n \psi_m(x, y)\psi_n(x, y)R_{q_mq_n}(\tau) \tag{I.9}$$

where

$$R_{q_mq_n}(\tau) \equiv \lim_{T\to\infty} \frac{1}{2T} \int_{-T}^{T} q_m(t)q_n(t+\tau)\,\mathrm{d}t \tag{I.10}$$

is defined to be the cross-correlation of the generalized coordinates, q_m. Defining power spectra

$$\Phi_w(\omega; x, y) \equiv \frac{1}{\pi} \int_{-\infty}^{\infty} R_w(\tau; x, y)\mathrm{e}^{-i\omega\tau}\,\mathrm{d}\tau \tag{I.11}$$

$$\Phi_{q_mq_n}(\omega) \equiv \frac{1}{\pi} \int_{-\infty}^{\infty} R_{q_mq_n}(\tau)\mathrm{e}^{-i\omega\tau}\,\mathrm{d}\tau \tag{I.12}$$

we may obtain from (I.9) via a Fourier Transform

$$\Phi_w(\omega; x, y) = \sum_m \sum_n \psi_m(x, y)\psi_n(x, y)\Phi_{q_mq_n}(\omega) \tag{I.13}$$

(I.9) and (I.13) relate the physical deflection, w, to the generalized coordinates, q_m.

Consider next similar relations between physical load p and generalized force Q_m. Define the cross-correlation

$$R_{Q_mQ_n}(\tau) \equiv \lim_{T\to\infty} \frac{1}{2T} \int_{-T}^{T} Q_m(t)Q_n(t+\tau)\,\mathrm{d}t \tag{I.14}$$

Using the definition of generalized force (I.6)

$$Q_m(t) \equiv \iint p(x, y, t)\psi_m(x, y)\,\mathrm{d}x\,\mathrm{d}y$$

$$Q_n(t+\tau) \equiv \iint p(x^*, y^*, t+\tau)\psi_n(x^*, y^*)\,\mathrm{d}x^*\,\mathrm{d}y^*$$

and substituting into (I.14) we obtain

$$R_{Q_mQ_n}(\tau)=\iiiint \psi_m(x,y)\psi_n(x^*,y^*)$$

$$\cdot R_p(\tau;x,y,x^*,y^*)\,\mathrm{d}x\,\mathrm{d}y\,\mathrm{d}x^*\,\mathrm{d}y^* \tag{I.15}$$

where we define the pressure correlation

$$R_p(\tau;x,y,x^*,y^*)\equiv\lim_{T\to\infty}\frac{1}{2T}\int_{-T}^{T}p(x,y,t)p(x^*,y^*,t+\tau)\,\mathrm{d}t \tag{I.16}$$

Note that a rather extensive knowledge of the spatial distribution of the pressure is required by (I.16).

Again defining power spectra

$$\Phi_{Q_mQ_n}(\omega)\equiv\frac{1}{\pi}\int_{-\infty}^{\infty}R_{Q_mQ_n}(\tau)\mathrm{e}^{-i\omega\tau}\,\mathrm{d}\tau \tag{I.17}$$

$$\Phi_p(\omega;x,y,x^*,y^*)\equiv\frac{1}{\pi}\int_{-\infty}^{\infty}R_p(\tau;x,y,x^*,y^*)\mathrm{e}^{-i\omega\tau}\,\mathrm{d}\tau \tag{I.18}$$

we may obtain from (I.15)

$$\Phi_{Q_mQ_n}(\omega)=\iiiint \psi_m(x,y)\psi_n(x^*,y^*)$$

$$\cdot\Phi_p(\omega;x,y,x^*,y^*)\,\mathrm{d}x\,\mathrm{d}y\,\mathrm{d}x^*\,\mathrm{d}y^* \tag{I.19}$$

Finally, we must relate the generalized coordinates to the generalized forces. From (I.7) we may formally solve (see [2], for example or recall Section 3.3)

$$q_n(t)=\int_{-\infty}^{\infty}I_n(t-t_1)Q_n(t_1)\,\mathrm{d}t_1 \tag{I.20}$$

where the 'impulse function' is defined as

$$I_n(t)\equiv\frac{1}{2\pi}\int_{-\infty}^{\infty}H_n(\omega)\mathrm{e}^{i\omega t}\,\mathrm{d}\omega \tag{I.21}$$

and the 'transfer function' is defined as

$$H_n(\omega)\equiv\frac{1}{M_n[\omega_n^2+2\zeta_n i\omega_n\omega-\omega^2]}$$

Also

$$H_n(\omega) = \int_{-\infty}^{\infty} I_n(t)e^{-i\omega t}\,dt$$

which is the other half of the transform pair, cf (I.21).

From (I.20) and (I.10)

$$R_{q_m q_n}(\tau) = \lim_{T\to\infty} \frac{1}{2T} \iiint_{-T}^{T} I_m(t-t_1)I_n(t+\tau-t_2)Q_m(t_1)Q_n(t_2)\,dt_1\,dt_2\,dt$$

Performing a change of integration variables and noting (I.14),

$$R_{q_m q_n}(\tau) = \iint_{-\infty}^{\infty} I_m(\xi)I_n(\eta)R_{Q_m Q_n}(\tau-\eta+\xi)\,d\xi\,d\eta \tag{I.22}$$

Taking a Fourier Transform of (I.22) and using the definitions of power spectra (I.12) and (I.17), we have

$$\Phi_{q_m q_n}(\omega) = H_m(\omega)H_n(-\omega)\Phi_{Q_m Q_n}(\omega) \tag{I.23}$$

Summarizing, the relations for correlation functions are (I.9), (I.15), and (I.22) and for power spectra (I.13), (I.19) and (I.23). For example, substituting (I.19) into (I.23) and the result into (I.13) we have

$$\Phi_w(\omega;\,x,y) = \sum_m \sum_n \psi_m(x,y)\psi_n(x,y)H_m(\omega)H_n(-\omega)$$

$$\cdot \iiiint \psi_m(x,y)\psi_n(x^*,y^*)$$

$$\cdot\, \Phi_p(\omega;\,x,y,x^*,y^*)\,dx\,dy\,dx^*\,dy^* \tag{I.24}$$

This is the *desired final result* relating the physical excitation to the physical response in stochastic terms.

Sharp resonance or low damping approximation

Often (I.24) is approximated further. Two approximations are particularly popular and useful. The first is the 'neglect of off-diagonal coupling'. This means omitting all terms in the double sum except those for which $m = n$. The second is the 'white noise' approximation which assumes that Φ_p is essentially constant relative to the rapidly varying transfer functions

$H_m(\omega)$. Making both of these approximations in (I.24) we may obtain the mean square response

$$\bar{w}^2(x, y) \equiv R_w(\tau = 0; x, y) = \int_0^\infty \Phi_w(\omega; x, y)\, d\omega$$

$$\approx \frac{\pi}{4} \sum_m \frac{\psi_m^2(x, y)}{M_m^2 \omega_m^3 \zeta_m} \iiint \psi_m(x, y)\psi_m(x^*, y^*)$$

$$\cdot \Phi_p(\omega_m; x, y, x^*, y^*)\, dx\, dy\, dx^*\, dy^* \qquad (I.25)$$

Of course, only one or the other of these approximations may be made, rather than both. However, both stem from the same basic physical idea: The damping is small and hence, H_m has a sharp maximum near $\omega = \omega_m$. That is

$$H_m(\omega_m)H_n(-\omega_m) \ll |H_m(\omega_m)|^2$$
$$H_m(\omega_n)H_n(-\omega_n) \ll |H_n(\omega_n)|^2$$

and the 'neglect of off-diagonal coupling' follows. Also

$$\int \Phi_p |H_m(\omega)|^2\, d\omega \approx \Phi_p(\omega_m) \int |H_m(\omega)|^2\, d\omega$$

and (I.25) follows by simple integration.

Note that if we take the spatial mean square of (I.24) then using orthogonality (for a uniform mass distribution) one may show that the off-diagonal terms do not contribute (see Powell [1]).

Finally note that if we desire stress rather than deflection, then it may be shown that analogous to (I.25) one obtains

$$\bar{\sigma}^2 = \frac{\pi}{4} \sum_m \frac{\sigma_m^2(x, y)}{M_m^2 \omega_m^3 \zeta_m} \iiiint \psi_m(x, y)\psi_m(x^*, y^*)$$

$$\cdot \Phi_p(\omega_m; x, y, x^*, y^*)\, dx\, dy\, dx^*\, dy^* \qquad (I.26)$$

where σ_m is stress due to $w = \psi_m$.

References for Appendix I

[1] Powell, A., Chapter 8 in book, *Random Vibration*, edited by S. H. Crandall, Technology Press, Cambridge, Mass., 1958.

[2] Laning, J. H. and Battin, R. H., *Random Processes in Automatic Control*, McGraw-Hill, New York, N.Y., 1956.

[3] Lin, Y. K., *Probabilistic Theory of Structural Dynamics*, McGraw-Hill, New York, N.Y., 1967.

[4] Dowell, E. H., *Aeroelasticity of Plates and Shells*, Noordhoff International Publishing, Leyden, The Netherlands, 1974.

[5] Vaicaitis, R., Dowell, E. H. and Ventres, C. S., 'Nonlinear Panel Response by a Monte Carlo Approach', *AIAA Journal*, Vol. 12, No. 5 (May 1974) pp. 685–691.

[6] Lazan, B. J., *Damping of Materials and Members in Structural Mechanics*, Pergamon Press, New York, N.Y., 1968.

Appendix II

Some example problems

Problems such as these have been used successfully as homework assignments. When used as a text, the instructor may wish to construct variations on these problems.

Chapter 2

Questions

Typical section with control surface

1. Compute q_{REVERSAL} for finite K_δ and show it is the same as computed in the text for $K_\delta \to \infty$.

2. Compute $q_{\text{DIVERGENCE}}$ explicitly in terms of K_α, K_δ, etc.

Beam-rod model

3. Compute $q_{\text{DIVERGENCE}}$ using one and two mode models with uniform beam-rod eigenfunctions.

Assume

$GJ = GJ_0[1 - y/l]$

How do these results compare to those for

$GJ = GJ_0 \sim \text{constant}$?

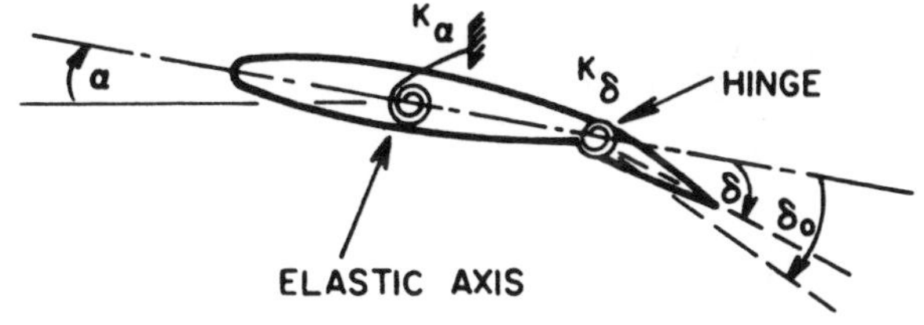

Answers 1. The two equations of static moment equilibrium are as follows:

$$eqS\left(\frac{\partial C_L}{\partial \alpha}\alpha + \frac{\partial C_L}{\partial \delta}\delta\right) + qSc\frac{\partial C_{MAC}}{\partial \delta}\delta - K_\alpha \alpha = 0 \quad \text{(about elastic axis)}$$

$$qSc\left(\frac{\partial C_H}{\partial \alpha}\alpha + \frac{\partial C_H}{\partial \delta}\delta\right) - K_\delta(\delta - \delta_0) = 0 \quad \text{(about hinge axis)}$$

These equations are given in matrix form as follows:

$$\begin{bmatrix} eqS\frac{\partial C_L}{\partial \alpha} - K_\alpha & eqS\frac{\partial C_L}{\partial \delta} + qSC\frac{\partial C_{MAC}}{\partial \delta} \\ qSC\frac{\partial C_H}{\partial \alpha} & qSc\frac{\partial C_H}{\partial \delta} - K_\alpha \end{bmatrix} \begin{bmatrix} \alpha \\ \delta \end{bmatrix} = \begin{bmatrix} 0 \\ -K_\delta \cdot \delta_0 \end{bmatrix}$$

Solving for α and δ, one obtains

$$\alpha = \frac{\begin{vmatrix} 0 & eqS\frac{\partial C_L}{\partial \delta} + qSc\frac{\partial C_{MAC}}{\partial \delta} \\ -K_\delta \cdot \delta_0 & qSc\frac{\partial C_H}{\partial \delta} - K_\delta \end{vmatrix}}{\Delta} = \frac{K_\delta \cdot \delta_0 qS\left(e\frac{\partial C_L}{\partial \delta} + c\frac{\partial C_{MAC}}{\partial \delta}\right)}{\Delta} \tag{II.1}$$

$$\delta = \frac{\begin{vmatrix} eqS\cdot\frac{\partial C_L}{\partial \alpha} - K_\alpha & \dot{0} \\ qSc\frac{\partial C_H}{\partial \alpha} & -K_\delta \cdot \delta_0 \end{vmatrix}}{\Delta} = \frac{-K_\delta \cdot \delta_0\left(eqS\frac{\partial C_L}{\partial \alpha} - K_\alpha\right)}{\Delta} \tag{II.2}$$

where

$$\Delta \equiv \begin{vmatrix} eqS\frac{\partial C_L}{\partial \alpha} - K_\alpha & eqS\frac{\partial C_L}{\partial \delta} + qSc\frac{\partial C_{MAC}}{\partial \delta} \\ qSc\frac{\partial C_H}{\partial \alpha} & qSc\frac{\partial C_H}{\partial \delta} - K_\delta \end{vmatrix} \tag{II.3}$$

If control surface reversal occurs when $q = q_R$, then

$$L = qS\left[\frac{\partial C_L}{\partial \alpha}\alpha + \frac{\partial C_L}{\partial \delta}\delta\right] = 0$$

for $q = q_R$, i.e.,

$$\left[\frac{\partial C_L}{\partial \alpha}\alpha + \frac{\partial C_L}{\partial \delta}\delta\right] = 0$$

$$\text{at } q = q_R \tag{II.4}$$

Substitution of (II.1) and (II.2) into (II.4) gives

$$0=\frac{\partial C_L}{\partial \alpha}\cdot\frac{K_\delta\cdot\delta_0}{\Delta}q_R S\left(e\frac{\partial C_L}{\partial \delta}+c\frac{\partial C_{MAC}}{\partial \delta}\right)-\frac{\partial C_L}{\partial \delta}\cdot\frac{K_\delta\,\delta_0}{\Delta}\left(eq_R S\frac{\partial C_L}{\partial \alpha}-K_\alpha\right)$$

$$=\frac{K_\delta\cdot\delta_0}{\Delta}\left(eq_R S\frac{\partial C_L}{\partial \alpha}\frac{\partial C_L}{\partial \delta}+q_R SC\frac{\partial C_L}{\partial \alpha}\frac{\partial C_{MAC}}{\partial \delta}-eq_R S\frac{\partial C_L}{\partial \alpha}\frac{\partial C_L}{\partial \delta}+K_\alpha\frac{\partial C_L}{\partial \delta}\right)$$

$$=\frac{K_\sigma\cdot\delta_0}{\Delta}\left(q_R Sc\frac{\partial C_L}{\partial \alpha}\frac{\partial C_{MAC}}{\partial \delta}+K_\alpha\frac{\partial C_L}{\partial \delta}\right)$$

Thus the reversal dynamic pressure q_R for finite K_δ is

$$q_R=\frac{\dfrac{K_\alpha}{Sc}\left(\dfrac{\partial C_L}{\partial \delta}\Big/\dfrac{\partial C_L}{\partial \alpha}\right)}{-\dfrac{\partial C_{MAC}}{\partial \delta}}$$

which is identical with q_R when $K_\delta\to\infty$!

2. The divergence dynamic pressure is determined by $\Delta=0$. That is,

$$\left(eqS\frac{\partial C_L}{\partial \alpha}-K_\alpha\right)\left(qSc\frac{\partial C_H}{\partial \delta}-K_\delta\right)-qSc\frac{\partial C_H}{\partial \alpha}\left(eqS\frac{\partial C_L}{\partial \delta}+qSc\frac{\partial C_{MAC}}{\partial \delta}\right)=0$$

$$\therefore\quad q^2S^2c^2\left(\bar{e}\cdot\frac{\partial C_L}{\partial \alpha}\frac{\partial C_H}{\partial \delta}-\bar{e}\frac{\partial C_H}{\partial \alpha}\frac{\partial C_L}{\partial \delta}-\frac{\partial C_{MAC}}{\partial \delta}\right)$$

$$-qSc\left(K_\alpha\cdot\frac{\partial C_H}{\partial \delta}+K_\delta\cdot\bar{e}\frac{\partial C_L}{\partial \alpha}\right)+K_\alpha K_\delta=0$$

$(\bar{e}\equiv e/c)$ (II.5)

If $A\neq 0$ (A is defined below),

$$q=\frac{1}{Sc}\cdot\frac{B\pm\sqrt{B^2-4AC}}{2A}$$

where

$$\begin{cases} A\equiv\bar{e}\left(\dfrac{\partial C_L}{\partial \alpha}\dfrac{\partial C_H}{\partial \delta}-\dfrac{\partial C_H}{\partial \alpha}\dfrac{\partial C_L}{\partial \delta}\right)-\dfrac{\partial C_{MAC}}{\partial \delta} \\ B\equiv K_\alpha\dfrac{\partial C_H}{\partial \delta}+K_\delta\bar{e}\dfrac{\partial C_L}{\partial \alpha} \\ C\equiv K_\alpha K_\alpha \end{cases}$$

Then divergence occurs when $AB > 0$ and $B^2 - 4AC \geq 0$ (for which e.g., (II.5) has two positive roots), and the divergence dynamic pressure q is

$$q_D = \min\left[\frac{1}{Sc}\cdot\frac{B+\sqrt{B^2-4AC}}{2A}, \frac{1}{Sc}\frac{B-\sqrt{B^2-4AC}}{2A}\right]$$

If $A = 0$, then divergence occurs when $B > 0$, and the divergence dynamic pressure q_D is

$$q_D = \frac{1}{Sc}\cdot\frac{C}{B}$$

To sum up, divergence occurs when

(a)

$$\begin{cases} \left\{\bar{e}\left(\frac{\partial C_L}{\partial\alpha}\frac{\partial C_H}{\alpha\delta} - \frac{\partial C_H}{\partial\alpha}\frac{\partial C_L}{\partial\delta}\right) - \frac{\partial C_{MAC}}{\partial\delta}\right\}\left(K_\alpha\frac{\partial C_H}{\partial\delta} + K_\delta\bar{e}\frac{\partial C_L}{\partial\alpha}\right) > 0 \\ \text{and} \\ \left(K_\alpha\frac{\partial C_H}{\partial\delta} - K_\delta\bar{e}\frac{\partial C_L}{\partial\alpha}\right)^2 + 4K_\alpha K_\delta\left(\bar{e}\frac{\partial C_H}{\partial\alpha}\frac{\partial C_L}{\partial\delta} + \frac{\partial C_{Mac}}{\partial\delta}\right) \geq 0 \end{cases}$$

and the divergence dynamic pressure q_D is

$$q_D = \frac{K_\alpha\frac{\partial C_H}{\partial\delta} + K_\delta\bar{e}\frac{\partial C_L}{\partial\alpha} - \sqrt{\left(K_\alpha\frac{\partial C_H}{\partial\delta} - K_\delta\bar{e}\frac{\partial C_L}{\partial d}\right)^2 + 4K_\alpha K_d\left(\bar{e}\frac{\partial C_H}{\partial\alpha}\frac{\partial C_L}{\partial\delta} + \frac{\partial C_{MAC}}{\partial\delta}\right)}}{2Sc\left\{\bar{e}\left(\frac{\partial C_L}{\partial\alpha}\frac{\partial C_H}{\partial\delta} - \frac{\partial C_H}{\partial\alpha}\frac{\partial C_L}{\partial\delta}\right) - \frac{\partial C_{MAC}}{\partial\delta}\right\}}$$

when

$$\bar{e}\left(\frac{\partial C_L}{\partial\alpha}\frac{\partial C_H}{\partial\delta} - \frac{\partial C_H}{\partial\alpha}\frac{\partial C_L}{\partial\sigma}\right) - \frac{\partial C_{MAC}}{\partial\delta} > 0$$

and

$$q_D = \frac{K_\alpha\frac{\partial C_H}{\partial\delta} + K_\delta\bar{e}\frac{\partial C_L}{\partial\alpha} + \sqrt{\left(K_\alpha\frac{\partial C_H}{\partial\delta} - K_\delta\bar{e}\frac{\partial C_L}{\partial\alpha}\right)^2 + 4K_\alpha K_\delta\left(\bar{e}\frac{\partial C_H}{\partial\alpha}\frac{\partial C_L}{\partial\delta} + \frac{\partial C_{MAC}}{\partial\delta}\right)}}{2Sc\left\{\bar{e}\left(\frac{\partial C_L}{\partial\alpha}\frac{\partial C_H}{\partial\delta} - \frac{\partial C_H}{\partial\alpha}\frac{\partial C_L}{\partial\delta}\right) - \frac{\partial C_{MAC}}{\partial\delta}\right\}}$$

when

$$\bar{e}\frac{\partial C_L}{\partial \alpha}\frac{\partial C_H}{\partial \delta}-\frac{\partial C_H}{\partial \alpha}\frac{\partial C_L}{\partial \delta}-\frac{\partial C_{MAC}}{\partial \delta}<0$$

or,

(b)

$$\begin{cases}\bar{e}\left(\dfrac{\partial C_L}{\partial \alpha}\dfrac{\partial C_H}{\partial \delta}-\dfrac{\partial C_H}{\partial \alpha}\dfrac{\partial C_L}{\partial \delta}\right)-\dfrac{\partial C_{MAC}}{\partial \delta}=0\\ \text{and}\\ K_\alpha\dfrac{\partial C_H}{\partial \delta}+K_\delta\bar{e}\dfrac{\partial C_L}{\partial \alpha}>0\end{cases}$$

and the divergence dynamic pressure q_D is

$$q_D=\frac{K_\alpha K_\delta}{Sc\left(K_\alpha\dfrac{\partial C_H}{\partial \delta}+K_\delta\bar{e}\dfrac{\partial C_L}{\partial \alpha}\right)}$$

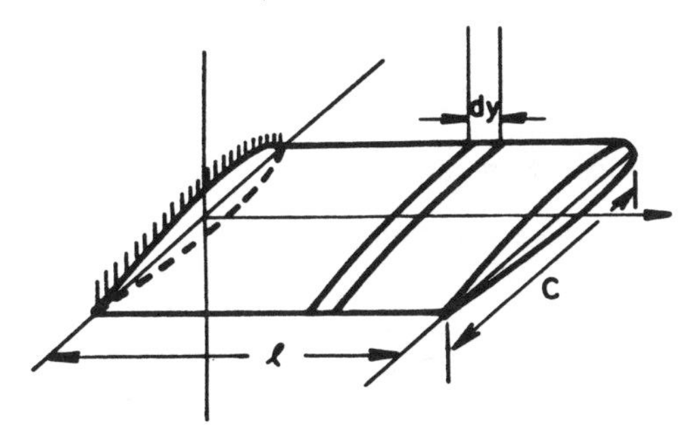

3. The equation of static torque equilibrium for a beam rod is

$$\frac{\mathrm{d}}{\mathrm{d}y}\left(GJ\frac{\mathrm{d}\alpha_e}{\mathrm{d}y}\right)+M_y=0 \tag{II.6}$$

where

$$\begin{aligned}M_y&=M_{AC}+L_e\\ &=qc^2C_{MAC_0}+eqc\frac{\partial C_L}{\partial \alpha}(\alpha_0+\alpha_e)\end{aligned} \tag{II.7}$$

If we put $\gamma=[1-y/l]$ and $y=l\bar{y}$, then, from (II.6) and (II.7), we have

$$\frac{\mathrm{d}}{\mathrm{d}\bar{y}}\left(\gamma\frac{\mathrm{d}\alpha_e}{\mathrm{d}\bar{y}}\right)+\frac{qcel^2\dfrac{\partial C_L}{\partial \alpha}}{GJ_0}\alpha_e=-\frac{qcl^2}{GJ_0}\left(c\cdot C_{MAC_0}+e\frac{\partial C_L}{\partial \alpha}\alpha_0\right) \tag{II.8}$$

(1) Eigenvalues and functions for constant wing properties. Putting

$$\lambda^2 \equiv \frac{qcel^2 \dfrac{\partial C_L}{\partial \alpha}}{GJ_0}$$

we have the characteristic equation as follows

$$\text{(II.7)} \rightarrow \frac{d^2\alpha_e}{d\bar{y}^2} + \lambda^2 \alpha_e = 0 \qquad (\gamma = 1 \text{ for constant wing properties})$$

Hence, $\alpha_e = A \sin \lambda \bar{y} + B \cos \lambda \bar{y}$.
As boundary conditions are

$$\begin{cases} \alpha_e = 0 \text{ at } \bar{y} = 0 \rightarrow B = 0 \\ \dfrac{d\alpha_e}{d\bar{y}} = 0 \text{ at } \bar{y} = 1 \rightarrow A\lambda \cos \lambda - B\lambda \sin \lambda \rightarrow \cos \lambda = 0 \end{cases}$$

(If $A\lambda = 0$ then $\alpha_e \equiv 0$, which is of no interest.)
So

$$\begin{cases} \text{Eigenvalues:} \quad \lambda_m = (2m-1)\dfrac{\pi}{2}, \; m = 1, 2, \ldots \\ \text{Eigenfunctions:} \; \alpha_m = \sin \lambda_m \bar{y} \end{cases}$$

We first find the divergence dynamic pressure for the wing with constant properties. Let

$$\alpha_e = \sum_m a_n \alpha_n, \qquad K \equiv -\frac{qcl^2}{(GJ)_0}\left(cC_{MAC_0} + e\frac{\partial C_L}{\partial \alpha}\alpha_0\right) = \sum_n A_n \alpha_n$$

Then

$$\sum_n a_n \left(\frac{d^2\alpha_n}{d\bar{y}^2} + \lambda^2 \alpha_n\right) = K$$

As

$$\frac{d^2\alpha_n}{d\bar{y}^2} = -\lambda_n^2 \alpha_n$$

so

$$\sum_n a_n(\lambda^2 - \lambda_n^2)\alpha_n = K$$

$$\sum_n a_n \int_0^1 (\lambda^2 - \lambda_n^2)\alpha_n \alpha_m \, d\bar{y} = \int_0^1 K\alpha_m \, d\bar{y} = \tfrac{1}{2}A_m$$

since

$$\int_0^1 \alpha_n \alpha_m \, d\bar{y} = \tfrac{1}{2}\delta_{mn}$$

$$= \tfrac{1}{2} m = n$$

$$= 0 m \neq n$$

Hence

$$\frac{a_m}{2}(\lambda^2 - \lambda_m^2) = \tfrac{1}{2}A_m$$

$$a_m = \frac{A_m}{\lambda^2 - \lambda_m^2}$$

Thus

$$\alpha_e = \sum_n \frac{A_n}{\lambda^2 - \lambda_n^2} \cdot \alpha_n$$

$\alpha_e \rightarrow \infty$ when

$$\lambda = \lambda_m = (2m-1)\frac{\pi}{2}$$

hence, the divergence dynamic pressure q_D, corresponds to the minimum value of λ_m, i.e., $\pi/2$. Thus

$$q_D = \frac{GJ_0}{cel^2 \dfrac{\partial C_L}{\partial \alpha}} \frac{\pi^2}{4}$$

for constant wing properties.

(2) $GJ = GJ_0(1 - y/l) = GJ_0(1 - \bar{y})$, variable wing properties. We assume for simplicity that only the torsional stiffness varies along span and that other characteristics remain the same.

Putting

$$\alpha_e = \sum_n b_n \cdot \alpha_n, \qquad K \equiv -\frac{qcl^2}{GJ_0}\left(cC_{MAC_0} + e\frac{\partial C_L}{\partial \alpha}\alpha_0\right) = \sum_n A_n \alpha_n$$

and

$$\lambda^2 \equiv \frac{qcel^2}{GJ_0}\frac{\partial C_L}{\partial \alpha}$$

we get from (II.8)

$$\sum_n b_n\left\{\frac{d}{d\bar{y}}\left(\gamma\frac{d\alpha_n}{d\bar{y}}\right)+\lambda^2\alpha_n\right\}=K$$

$$\therefore\quad \sum_n b_n\int_0^1\left\{\frac{d}{d\bar{y}}\left(\gamma\frac{d\alpha_n}{d\bar{y}}\right)+\lambda^2\alpha_n\right\}\alpha_m\,d\bar{y}=\int_0^1 K\alpha_m\,dy=\tfrac{1}{2}A_m$$

$$\therefore\quad [C_{mn}]\{b_n\}=\tfrac{1}{2}A_m \text{ (for finite } n) \tag{II.9}$$

where

$$C_{mn}=\int_0^1\left\{\frac{d}{d\bar{y}}\left(\gamma\frac{d\alpha_n}{d\bar{y}}\right)+\lambda^2\alpha_n\right\}\alpha_m\,d\bar{y}$$

$$=-\int_0^1\gamma\frac{d\alpha_n}{d\bar{y}}\frac{d\alpha_m}{d\bar{y}}\,d\bar{y}+\frac{\lambda^2}{2}\,\delta_{mn}$$

$\left(\gamma\frac{d\alpha_n}{d\bar{y}}\alpha_m=0\right.$ at $\bar{y}=0$ and 1 because of the boundary conditions for eigenfunctions.$\left.\right)$

(1) One mode model. The assumed mode is as follows:

$$\alpha_1=\sin\lambda_1\bar{y}=\sin\frac{\pi}{2}\bar{y}\rightarrow\frac{d\alpha_1}{d\bar{y}}=\frac{\pi}{2}\cos\frac{\pi}{2}\bar{y}$$

Equation (II.9) is

$$C_{11}b_1=\frac{A_1}{2} \tag{II.10}$$

where

$$C_{11}=-\int_0^1(1-\bar{y})\left(\frac{d\alpha_1}{d\bar{y}}\right)^2 d\bar{y}+\frac{\lambda^2}{2}$$

$$\therefore\quad C_{11}=\frac{\lambda^2}{2}-\frac{\pi^2+4}{16}$$

From (II.10),

$$b_1=\frac{A_1}{\lambda^2-\dfrac{\pi^2+4}{8}}$$

Then divergence occurs when

$$\lambda^2 = \frac{\pi^2 + 4}{8}$$

and

$$q_D = \frac{GJ_0}{cel^2 \frac{\partial C_L}{\partial \alpha}} \frac{\pi^2 + 4}{8} = (q_D)_{\substack{\text{const.}\\ \text{wing prop.}}} \times 0.703$$

(2) Two mode model. Assumed modes are

$$\begin{cases} \alpha_1 = \sin \lambda_1 \bar{y} = \sin \frac{\pi}{2} \bar{y} \Rightarrow \frac{d\alpha_1}{d\bar{y}} = \frac{\pi}{2} \cos \frac{\pi}{2} \bar{y} \\ \alpha_2 = \sin \lambda_2 \bar{y} = \sin \frac{3}{2} \pi \bar{y} \Rightarrow \frac{d\alpha_2}{d\bar{y}} = \frac{3}{2} \pi \cos \frac{3}{2} \pi \bar{y} \end{cases}$$

Equation (II.9) is as follows:

$$\begin{bmatrix} C_{11} & C_{12} \\ C_{21} & C_{22} \end{bmatrix} \begin{bmatrix} b_1 \\ b_2 \end{bmatrix} = \frac{1}{2} \begin{bmatrix} A_1 \\ A_2 \end{bmatrix} \tag{II.11}$$

where

$$C_{11} = -\int_0^1 (1-\bar{y}) \left(\frac{d\alpha_1}{d\bar{y}} \right)^2 d\bar{y} + \frac{\lambda^2}{2} = \frac{\lambda^2}{2} - \frac{\pi^2 + 4}{16}$$

$$C_{12} = -\int_0^1 (1-\bar{y}) \frac{d\alpha_2}{d\bar{y}} \frac{d\alpha_1}{d\bar{y}} d\bar{y} = -\frac{3}{4}$$

$$C_{21} = -\int_0^1 (1-\bar{y}) \frac{d\alpha_1}{d\bar{y}} \frac{d\alpha_2}{d\bar{y}} d\bar{y} = C_{12} = -\frac{3}{4}$$

$$C_{22} = -\int_0^1 (1-\bar{y}) \left(\frac{d\alpha_2}{d\bar{y}} \right)^2 d\bar{y} + \frac{\lambda^2}{2} = \frac{\lambda^2}{2} - \frac{9\pi^2 + 4}{16}$$

Then equation (II.11) is as follows:

$$\begin{bmatrix} \lambda^2 - \frac{\pi^2 + 4}{8} & -\frac{3}{2} \\ -\frac{3}{2} & \lambda^2 - \frac{9\pi^2 + 4}{8} \end{bmatrix} \begin{bmatrix} b_1 \\ b_2 \end{bmatrix} = \begin{bmatrix} A_1 \\ A_2 \end{bmatrix}$$

Thus divergence occurs when

$$\begin{vmatrix} \lambda^2 - \dfrac{\pi^2+4}{8} & -\dfrac{3}{2} \\ -\dfrac{3}{2} & \lambda^2 - \dfrac{9\pi^2+4}{8} \end{vmatrix} = 0$$

$$\therefore \quad \lambda^2 = \frac{5\pi^2+4}{8} \pm \frac{1}{2}\sqrt{\pi^4+9}$$

q_D is given by the smaller value of λ^2, i.e.,

$$q_D = \frac{GJ_0}{cel^2 \dfrac{\partial C_L}{\partial \alpha}} \times \left(\frac{5\pi^2+4}{8} - \frac{1}{2}\sqrt{\pi^4+9} \right)$$

$$= (q_D)_{\substack{\text{const.} \\ \text{wing prop.}}} \times 0.612$$

Question

Beam-rod model

4. For a constant GJ, etc. wing, use a two 'lumped element' model and compute the divergence dynamic pressure. Neglect rolling. Compare your result with the known analytical solution. How good is a one 'lumped element' solution?

Answer

4.

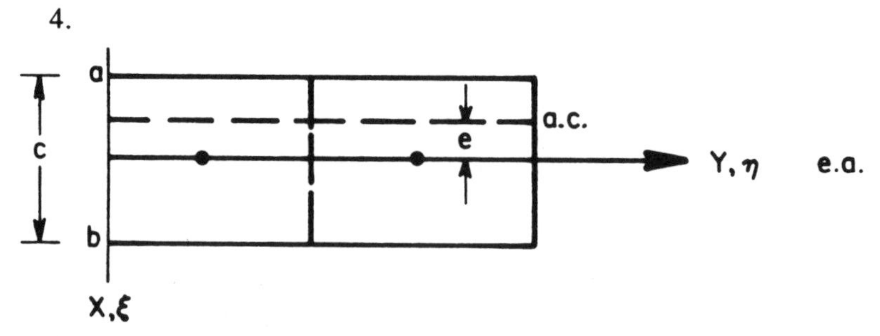

(a) Two lumped element model

$$\alpha(y) = \int_0^l C^{\alpha M}(y, \eta) M(\eta) \, d\eta \tag{II.12}$$

where

$C^{\alpha M}(y, \eta)$: twist about y axis at y due to unit moment at η

$$M(\eta) = \int_a^b p(\xi, \eta)\xi \, d\xi$$

Equation (II.12) in matrix form is

$$\{\alpha\} = [C^{\alpha M}]\{M\} \Delta\eta \tag{II.12}$$

where from structural analysis,

$$[C^{\alpha M}] = \begin{bmatrix} \dfrac{l/4}{GJ} & \dfrac{l/4}{GJ} \\ \dfrac{l/4}{GJ} & \dfrac{3l/4}{GJ} \end{bmatrix} \tag{II.13}$$

and $C^{\alpha M}(i, j)$ is the twist at i due to unit moment at j. Using an aerodynamic 'strip theory' approximation, the aerodynamic moment may be related to the twist,

$$\{M\} = qce\frac{\partial C_L}{\partial \alpha}\begin{bmatrix} 1 & 0 \\ 0 & 1 \end{bmatrix}\{\alpha\} = qce\frac{\partial C_L}{\partial \alpha}\{\alpha\} \tag{II.14}$$

From (II.12) and (II.14), one has

$$\{\alpha\} = [C^{\alpha M}]\{M\}\Delta\eta = qCe\frac{\partial C_L}{\partial \alpha}[C^{\alpha M}]\ \Delta\eta\{\alpha\}$$

or rewritten, using $\Delta\eta = l/2$,

$$\left[\begin{bmatrix} 1 & 0 \\ 0 & 1 \end{bmatrix} - \frac{l}{2}qce\frac{\partial C_L}{\partial \alpha}\frac{l}{4GJ}\begin{bmatrix} 1 & 1 \\ 1 & 3 \end{bmatrix}\right]\{\alpha\} = \begin{Bmatrix} 0 \\ 0 \end{Bmatrix} \tag{II.15}$$

Setting the determinant of coefficients to zero, gives

$$|\quad| = 0 \rightarrow 2Q^2 - 4Q + 1 = 0 \tag{II.16}$$

where

$$Q \equiv \frac{l^2}{8GJ}qce\frac{\partial C_L}{\partial \alpha}$$

Solving (II.16), one obtains

$$Q = \frac{2 \pm \sqrt{2}}{2}$$

The smaller Q gives the divergence q_D.

$$q_D = 4(2-\sqrt{2})\,\frac{GJ/l}{(lc)e\dfrac{\partial C_L}{\partial \alpha}}$$

$$\doteq 2.34\,\frac{GJ/l}{(lc)e\dfrac{\partial C_L}{\partial \alpha}}$$

(b) One lumped element model

$$\alpha = qce\frac{\partial C_L}{\partial \alpha}C^{\alpha M}\Delta\eta\alpha$$

where

$$\Delta\eta = l, \qquad C^{\alpha M} = \frac{l/2}{GJ}$$

$$\left(1 - qCe\frac{\partial C_L}{\partial \alpha}\frac{l/2}{GJ}\right)\alpha = 0$$

$$\therefore \quad q_D = 2\frac{GJ/l}{(lc)e\dfrac{\partial C_L}{\partial \alpha}}$$

Recall that the analytical solution is (cf. Section 2.2)

$$q_D = \left(\frac{\pi}{2}\right)^2\frac{GJ}{l}/(lc)e\frac{\partial C_L}{\partial \alpha}$$

$$= 2.46\cdots$$

A comparison of the several approximations is given below. In the two element model the error is about 5%.

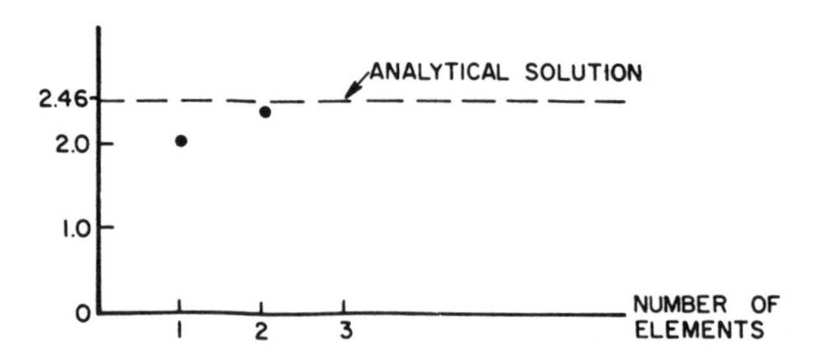

Question

5. Consider a thin cantilevered plate of length l and width b which represents the leading edge of a wing at supersonic speeds. See sketch.

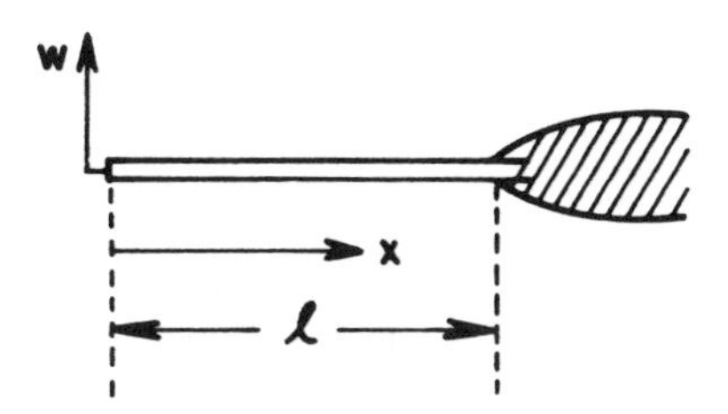

The aerodynamic pressure loading (per unit chord and per unit span) at high speeds is given by (Sections 3.4 and 4.2)

$$p = \frac{2\rho U^2}{(M^2-1)^{\frac{1}{2}}} \frac{\partial w}{\partial x}$$

Sign convention
p down
w up

where M is mach number and w is transverse deflection (not downwash!). Compute the divergence speed.

(1) Work out a formal mathematical solution, without numerical evaluation, using classical differential equation methods.

(2) How would you use Galerkin's method with an assumed mode of the form

$$w = a\{2(1-x/l)^2 - \tfrac{4}{3}(1-x/l)^3 + \tfrac{1}{3}(1-x/l)^4\}$$

to obtain a numerical answer? What boundary conditions on w does the assumed mode satisfy?

Answer

(1) Governing equilibrium equation is:

$$EI\frac{\partial^4 w}{\partial x^4} = -p = -\frac{2\rho U^2}{(M^2-1)^{\frac{1}{2}}} \frac{\partial w}{\partial x}$$

Define

$$K \equiv \left[\frac{2\rho U^2}{(M^2-1)^{\frac{1}{2}}}\right]\left[\frac{1}{EI}\right]$$

the equation above becomes

$$\frac{\partial^4 w}{\partial x^4} + K\frac{\partial w}{\partial x} = 0 \qquad \text{(II.17)}$$

The boundary conditions are:

$$w(l)=\frac{\partial w}{\partial x}(l)=\frac{\partial^2 w}{\partial x^2}(0)=\frac{\partial^3 w}{\partial x^3}(0)=0 \tag{II.18}$$

The characteristic equation of differential equation (II.17) is

$$\gamma^4+K\gamma=0 \tag{II.19}$$

The roots are $\gamma_1=0$ and γ_2, γ_3, γ_4 such that $\gamma^3=-K$. Now

$$(-K)^{\frac{1}{3}}=K^{\frac{1}{3}}e^{i\frac{1}{3}(\pi+2n\pi)}, \qquad n=0,1,2$$

and defining $K^1=K^{\frac{1}{3}}$ the roots γ_2, γ_3, γ_4 become

$$\gamma_2=K^1e^{i\pi/3}\;=K^1[\cos\pi/3+i\sin\pi/3]=K^1\left[\frac{1}{2}+i\frac{\sqrt{3}}{2}\right]$$

$$\gamma_3=K^1e^{i\pi}\;=K^1[\cos\pi+i\sin\pi]=K^1[-1]$$

$$\gamma_4=K^1e^{i5\pi/3}=K^1\left[\cos\frac{5\pi}{3}+i\sin\frac{5\pi}{3}\right]=K^1\left[\frac{1}{2}-i\frac{\sqrt{3}}{2}\right]$$

Therefore $w(x)$ has the form:

$$w(x)=b_1+b_2e^{-K^1x}+e^{K^1(x/2)}\left[b_3\cos\left(K^1\frac{\sqrt{3}}{2}x\right)+b_4\sin\left(K^1\frac{\sqrt{3}}{2}x\right)\right]$$

$$w'(x)=-b_2K^1e^{-K^1x}+\frac{K^1}{2}e^{K^1(x/2)}\left[(b_3+b_4\sqrt{3})\cos\left(K^1\frac{\sqrt{3}}{2}x\right)\right.$$
$$\left.+(b_4-b_3\sqrt{3})\sin\left(K^1\frac{\sqrt{3}}{2}x\right)\right]$$

$$w''(x)=b_2K^{1^2}e^{-K^1x}+\left(\frac{K^1}{2}\right)^2e^{K^1(x/2)}\left[(2\sqrt{3}b_4-2b_3)\cos\left(K^1\frac{\sqrt{3}}{2}x\right)\right.$$
$$\left.+(2\sqrt{3}b_3+2b_4)\sin\left(K^1\frac{\sqrt{3}}{2}x\right)\right]$$

$$w'''(x)=-b_2K^{1^3}e^{-K^1x}-K^{1^3}e^{K^1(x/2)}\left[b_3\cos\left(K^1\frac{\sqrt{3}}{2}x\right)+b_4\sin\left(K^1\frac{\sqrt{3}}{2}x\right)\right]$$

(II.20)

Using boundary conditions (II.18), we obtain from (II.20),

$$w(l)=0=b_1+b_2\mathrm{e}^{-K^1l}+\mathrm{e}^{K^1\frac{l}{2}}\left[(b_3\cos\left(K^1\frac{\sqrt{3}}{2}\,l\right)+b_4\sin\left(K^1\frac{\sqrt{3}}{2}\,l\right)\right]$$

$$\frac{\partial w}{\partial x}(l)=0=-b_2K^1\mathrm{e}^{-K^1l}+\frac{K^1}{2}\mathrm{e}^{K^1\frac{l}{2}}\left[(b_3+b_4\sqrt{3})\cos\left(K^1\frac{\sqrt{3}}{2}\,l\right)\right.$$

$$\left.+(b_4-b_3\sqrt{3})\sin\left(K^1\frac{\sqrt{3}}{2}\,l\right)\right]$$

$$\frac{\partial^2 w}{\partial x^2}(0)=0=b_2K^{1^2}+\left(\frac{K^1}{2}\right)^2(2\sqrt{3}b_4-2b_3)$$

$$\frac{\partial^3 w}{\partial x^3}(0)=0=-b_2K^{1^3}-b_3K^{1^3} \tag{II.21}$$

The condition for nontrivial solutions is that the determinant of coefficients of the system of linear, algebraic equations given by (II.21) be zero. This leads to

$$\mathrm{e}^{-\frac{3}{2}K''}=-2\cos\left(\frac{\sqrt{3}}{2}\,K''\right) \tag{II.22}$$

where

$$K''\equiv K^1l$$

In order to find the solution to equation (II.22), one would plot on the same graph as a function of K'' the right and left sides of this equation and note the points (if any) of intersection. The first intersection for $K''>0$ is the one of physical interest. Knowing this particular K'', call it K''_D, one may compute

$$U_D^2=\frac{K_D''^3(M^2-1)^{\frac{1}{2}}EI}{2\rho l^3}$$

to find the speed U at which divergence occurs.

(2) This is left as an excercise for the reader.

Questions

Sweptwing divergence

6. Derive the equations of equilibrium and associated boundary conditions, (2.6.1, 2.6.2, 2.6.11 and 2.6.12) from Hamilton's Principle. Note that Hamilton's Principle is the same as the Principle of Virtual Work for the present static case.

For a constant property sweptwing undergoing bending only, use classical solution techniques to compute the lowest eigenvalue corresponding to divergence. That is from (2.6.10), (2.6.11), (2.6.12), show that $\lambda_d = -6.33$.

- Now use Galerkins method to compute an approximate λ_D. For h, assume that

$$h = a_0 + a_1 \bar{y} + a_2 \bar{y}^2 + a_3 \bar{y}^3 + a_4 \bar{y}^4$$

From the boundary conditions (2.6.11), (2.6.12) show that

$$a_0 = b_0 = 0$$
$$a_3 = -4a_4$$
$$a_2 = 6a_4$$

and thus $h = a_4(\bar{y}^4 - 4\bar{y}^3 + 6\bar{y}^2)$. Using this representation for h, compute λ_D. How does this compare to the exact solution?

- Now consider both bending and torsion for a constant property wing. Assume

$$\alpha = b_0 + b_1 \bar{y}^2 + b_2 \bar{y}^3$$

Determine the possible form of α from the boundary conditions.

Determine λ_D for $GJ/EI = 1$, $\bar{e}/\bar{c} = 0.5$, $l/\bar{c} = 10$. Compare to the earlier result for bending only. Plot your results in terms of $\tilde{\lambda}_D$ vs Λ where $\tilde{\lambda}_D \equiv q(\partial \bar{C}_L/\partial\alpha)(\bar{c}l^3/EI)$.

Section 3.1

Question

Starting from

$$U = \frac{1}{2}\iiint [\sigma_{xx}\varepsilon_{xx} + \sigma_{xy}\varepsilon_{xy} + \sigma_{yx}\varepsilon_{yx} + \sigma_{yy}\varepsilon_{yy}]\, dx\, dy\, dz$$

and

$$\varepsilon_{xx} = -z\frac{\partial^2 w}{\partial x^2}$$

$$\varepsilon_{yy} = -z\frac{\partial^2 w}{\partial y^2}$$

$$\varepsilon_{xy} = -z\frac{\partial^2 w}{\partial x\, \partial y}$$

$$\sigma_{xx} = \frac{E}{(1-\nu^2)}[\varepsilon_{xx} + \nu\varepsilon_{xy}]$$

$$\sigma_{yy} = \frac{E}{(1-\nu^2)}[\varepsilon_{yy} + \nu\varepsilon_{xx}]$$

$$\sigma_{xy} = \frac{E}{(1+\nu)}\varepsilon_{xy} = \sigma_{yx}$$

$w = w(x, y)$ only

1. Compute $U = U(w)$
2. For $w = -h(y) - x\alpha(y)$

Compute $U = U(h, \alpha)$

3. Using a kinetic energy expression

$$T = \frac{1}{2}\iiint \rho\left(\frac{\partial w}{\partial t}\right)^2 \mathrm{d}x\, \mathrm{d}y\, \mathrm{d}z$$

Compute $T = T(h, \alpha)$

4. Assume $h(y) = q_h f(y)$

$$\alpha(y) = q_\alpha g(y)$$

where f, g are specified.

Determine equations of motion for q_h, q_α using Lagrange's Equations, where the virtual work done by aerodynamic pressure, p, is given by

$$\delta W = \iint p\delta w\, \mathrm{d}x\, \mathrm{d}y$$

5. Return to 1; now assume

$$w = \sum_m q_m\psi_m(x, y)$$

where

ψ_m is specified.
Determine equations of motion for q_m.

Answer

1. Potential energy U:

$$U=\frac{1}{2}\iiint(\sigma_{xx}\varepsilon_{xx}+\sigma_{xy}\varepsilon_{xy}+\sigma_{yx}\varepsilon_{yx}+\sigma_{yy}\varepsilon_{yy})\,dx\ dy\ dz$$

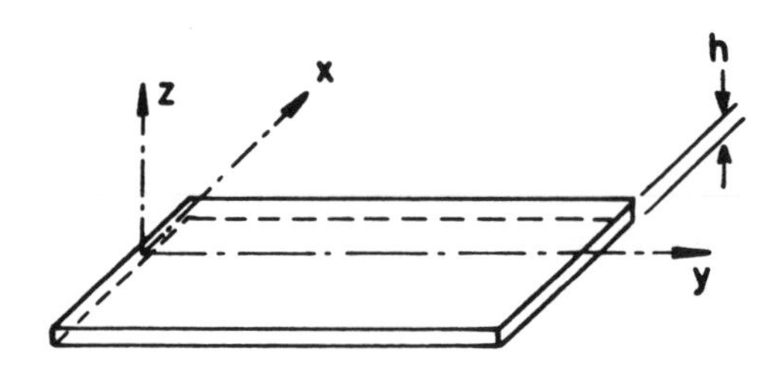

where

$$\begin{cases}\varepsilon_{xx}=-z\dfrac{\partial^2 w}{\partial x^2}\\ \varepsilon_{xy}=-z\dfrac{\partial^2 w}{\partial x\,\partial y}=\varepsilon_{yx}\\ \varepsilon_{yy}=-z\dfrac{\partial^2 w}{\partial y^2}\end{cases}$$

$$\begin{cases}\sigma_{xx}=\dfrac{E}{1-\nu^2}(\varepsilon_{xx}+\nu\varepsilon_{yy})\\ \sigma_{xy}=\dfrac{E}{1+\nu}\,\varepsilon_{xy}\\ \sigma_{yy}=\dfrac{E}{1-\nu^2}(\varepsilon_{yy}+\nu\varepsilon_{xx})\end{cases}$$

Thus

$$\sigma_{xx}\varepsilon_{xx} + \sigma_{xy}\varepsilon_{xy} + \sigma_{yx}\varepsilon_{yx} + \sigma_{yy}\varepsilon_{yy}$$

$$= \frac{E}{1-\nu^2}(\varepsilon_{xx} + \nu\varepsilon_{yy}) \cdot \varepsilon_{xx} + 2\frac{E}{1+\nu}\varepsilon_{xy}^2 + \frac{E}{1-\nu^2}(\varepsilon_{yy} + \nu\varepsilon_{xx})\varepsilon_{yy}$$

$$= \frac{E}{1-\nu^2}(\varepsilon_{xx}^2 + 2\nu\varepsilon_{xx}\varepsilon_{yy} + \varepsilon_{yy}^2) + \frac{2E(1-\nu)}{1-\nu^2}\varepsilon_{xy}^2$$

$$= \frac{E}{1-\nu^2}\left\{z^2\left(\frac{\partial^2 w}{\partial z^2}\right)^2 + 2\nu \cdot z^2\left(\frac{\partial^2 w}{\partial x^2}\right)\left(\frac{\partial^2 w}{\partial y^2}\right) + z^2\left(\frac{\partial^2 w}{\partial y^2}\right)^2 \right.$$

$$\left. + 2(1-\nu)z^2\left(\frac{\partial^2 w}{\partial x\,\partial y}\right)\right\}$$

$$= \frac{Ez^2}{1-\nu^2}\left\{\left(\frac{\partial^2 w}{\partial x^2}\right) + \left(\frac{\partial^2 w}{\partial y^2}\right)^2 + 2\nu\left(\frac{\partial^2 w}{\partial x^2}\right)\left(\frac{\partial^2 w}{\partial y^2}\right) + 2(1-\nu)\left(\frac{\partial^2 w}{\partial x\,\partial y}\right)^2\right\}$$

$$\therefore \quad U = \frac{1}{2}\iiint \frac{Ez^2}{1-\nu^2}\left[\left(\frac{\partial^2 w}{\partial x^2}\right)^2 + \left(\frac{\partial^2 w}{\partial y^2}\right)^2 + 2\nu\left(\frac{\partial^2 w}{\partial x^2}\right)\left(\frac{\partial^2 w}{\partial y^2}\right) + 2(1-\nu) \right.$$

$$\left. \times \left(\frac{\partial^2 w}{\partial x\,\partial y}\right)^2\right] \mathrm{d}x\,\mathrm{d}y\,\mathrm{d}z$$

$$= \frac{1}{2}\iint D\left[\left(\frac{\partial^2 w}{\partial x^2}\right)^2 + \left(\frac{\partial^2 w}{\partial y^2}\right)^2 + 2\nu\left(\frac{\partial^2 w}{\partial x^2}\right)\left(\frac{\partial^2 w}{\partial y^2}\right) + 2(1-\nu)\left(\frac{\partial^2 w}{\partial x\,\partial y}\right)^2\right] \mathrm{d}x\,\mathrm{d}y \tag{II.23}$$

where

$$D \equiv \frac{E}{1-\nu^2}\int z^2\,\mathrm{d}z$$

2. For $w = -h(y) - x\alpha(y)$

$$\begin{cases} w = -\dfrac{\partial^2 h}{\partial y^2} - x\dfrac{\partial^2 \alpha}{\partial y^2} \\[2ex] \dfrac{\partial^2 w}{\partial x\,\partial y} = -\dfrac{\partial \alpha}{\partial y} \\[2ex] \dfrac{\partial^2 w}{\partial x^2} = 0 \end{cases}$$

Hence, from (II.23), we have

$$U=\frac{1}{2}\iint D\left[\left(\frac{\partial^2 h}{\partial y^2}\right)+\left(\frac{\partial^2\alpha}{\partial y^2}x\right)^2+0^2+2\nu\cdot\left(\frac{\partial^2 h}{\partial y^2}+\frac{\partial^2\alpha}{\partial y^2}\cdot y\right)\cdot 0\right.$$

$$\left.+2(1-\nu)\left(\frac{\partial\alpha}{\partial y}\right)^2\right]\mathrm{d}x\,\mathrm{d}y$$

$$=\frac{1}{2}\iint D\left[\left(\frac{\partial^2 h}{\partial y^2}\right)^2+2\left(\frac{\partial^2 h}{\partial y^2}\right)\left(\frac{\partial^2\alpha}{\partial y^2}\right)\cdot x+\left(\frac{\partial^2\alpha}{\partial y^2}\right)^2x^2+2(1-\nu)\left(\frac{\partial\alpha}{\partial y}\right)^2\right]\mathrm{d}x\,\mathrm{d}y \tag{II.24}$$

Using the estimates,

$$\frac{\partial\alpha}{\partial y}\sim\frac{\alpha}{l},\quad x\sim c,\quad \frac{h}{l}\sim\alpha$$

we see the second and third terms can be neglected compared to the first and fourth for $c/l \ll 1$. Thus U becomes

$$U=\frac{1}{2}\int EI\left(\frac{\partial^2 h}{\partial y^2}\right)^2\mathrm{d}y+\frac{1}{2}\int GJ\left(\frac{\partial\alpha}{\partial y}\right)^2\mathrm{d}y$$

where

$$EI=\int D\,\mathrm{d}x,\qquad GJ=\int 2D(1-\nu)\,\mathrm{d}x$$

Note that if $\frac{h}{c}\sim\alpha$ is used as an estimate, then to deduce the final expression for U from (II.24), it is required that $\int Dx\mathrm{d}x=0$ which defines the "elastic axis".

3. For $w=-h(y,t)-\alpha(y,t)x$

$$T=\frac{1}{2}\iiint\rho\left(\frac{\partial w}{\partial t}\right)^2\mathrm{d}x\,\mathrm{d}y\,\mathrm{d}z$$

$$=\frac{1}{2}\int[m\dot{h}^2+2S_\alpha\dot{h}\dot{\alpha}+I_\alpha\dot{\alpha}^2]\,\mathrm{d}y$$

where: $m\equiv\int\rho\,\mathrm{d}x\,\mathrm{d}z$; $S_\alpha\equiv\int\rho x\,\mathrm{d}x\,\mathrm{d}z$; and $I_\alpha\equiv\int\rho x^2\,\mathrm{d}x\,\mathrm{d}z$.

Recall

$$\delta W=\int -L\,\delta h\,\mathrm{d}y+\int My\,\delta\alpha\,\mathrm{d}y$$

Using the above expressions for U, T and δW, one can derive the governing partial differential equations for h and α and the associated boundary conditions from Hamilton's principle.

4. $w(x,y,t)=q_h(t)\cdot f(y)+q_\alpha(t)g(y)x$

$$\therefore\quad \frac{\partial w}{\partial t}=\dot{q}_hf(y)+\dot{q}_\alpha g(y)x\qquad \dot{}\equiv\frac{\mathrm{d}}{\mathrm{d}t}$$

$$T = \frac{1}{2}\iiint \rho\left(\frac{\partial w}{\partial t}\right)^2 \mathrm{d}x\,\mathrm{d}y\,\mathrm{d}z$$

$$= \frac{1}{2}\iiint \rho(\dot{q}_h f(y) + \dot{q}_\alpha g(y)x)^2 \,\mathrm{d}x\,\mathrm{d}y\,\mathrm{d}z$$

$$= \frac{1}{2}\iiint \rho(\dot{q}_h^2\{f(y)\}^2 + 2\dot{q}_h\dot{q}_\alpha f(y)g(y)x + \dot{q}_\alpha^2\{g(y)\}^2x^2)\,\mathrm{d}x\,\mathrm{d}y\,\mathrm{d}z$$

$$= \frac{1}{2}\left[\dot{q}_h^2\iiint \rho\{f(y)\}^2\,\mathrm{d}x\,\mathrm{d}y\,\mathrm{d}z + 2\dot{q}_h\dot{q}_\alpha\iiint \rho f(y)g(y)x\,\mathrm{d}x\,\mathrm{d}y\,\mathrm{d}z\right.$$

$$\left. + \dot{q}_\alpha^2\iiint \{g(y)\}^2x^2\,\mathrm{d}x\,\mathrm{d}y\,\mathrm{d}z\right]$$

$$= \tfrac{1}{2}(\dot{q}_h^2 M_{hh} + 2\dot{q}_h\dot{q}_\alpha M_{h\alpha} + \dot{q}_\alpha^2 M_{\alpha\alpha}) \qquad \text{(II.25)}$$

where

$$M_{hh} \equiv \iiint \rho\{f(y)\}^2\,\mathrm{d}x\,\mathrm{d}y\,\mathrm{d}z$$

$$M_{h\alpha} \equiv \iiint \rho f(y)g(y)x\,\mathrm{d}x\,\mathrm{d}y\,\mathrm{d}z \qquad \text{(II.26)}$$

$$M_{\alpha\alpha} \equiv \iiint \rho\{g(y)\}^2x^2\,\mathrm{d}x\,\mathrm{d}y\,\mathrm{d}z$$

For $w = q_h(t)f(y) + q_\alpha(t)g(y)x$ the potential energy is given as follows:

$$' \equiv \frac{\mathrm{d}}{\mathrm{d}y}$$

$\left(\frac{\partial^2 h}{\partial y^2}\right) = q_h f''(y), \frac{\partial \alpha}{\partial y} = q_\alpha g'(y), \frac{\partial^2 \alpha}{\partial y^2} = q_\alpha g''(y)$ into (II.24)

$$U = \frac{1}{2}\iint D[\{q_h f''(y)\}^2 + 2q_h f''(y)q_\alpha g''(y)x + \{q_\alpha g''(y)\}^2\,x^2$$

$$+ 2(1-\nu)\{q_\alpha g'(y)\}^2]\,\mathrm{d}x\,\mathrm{d}y$$

$$= \frac{1}{2}\iint D\{q_h^2\{f''(y)\}^2 + 2q_h q_\alpha f''(y)g''(y)x + q_\alpha^2[\{g''(y)\}^2y^2$$

$$+ 2(1-\nu)\{g'(y)\}^2]\,\mathrm{d}x\,\mathrm{d}y$$

$$= \frac{1}{2}\left[q_h^2\iint D\{f''(y)\}^2\,\mathrm{d}x\,\mathrm{d}y + 2q_h q_\alpha\iint Df''(y)g''(y)x\,\mathrm{d}x\,\mathrm{d}y\right.$$

$$+ q_\alpha^2 \iint D\{g''(y)\}^2 x^2 + 2(1-\nu)\{g'(y)\}^2\Big] \,dx\,dy$$

$$= \tfrac{1}{2}[q_h^2 K_{hh} + 2q_h q_\alpha K_{h\alpha} + q_\alpha^2 K_{\alpha\alpha}] \tag{II.27}$$

where

$$\begin{cases} K_{hh} \equiv \iint D\{f''(y)\}^2 \,dx\,dy \\ K_{h\alpha} \equiv \iint Df''(y)g''(y)x \,dx\,dy \\ K_{\alpha\alpha} \equiv \iint D[\{g''(y)\}^2 x^2 + 2(1-\nu)\{g'(y)\}^2] \,dx\,dy \end{cases} \tag{II.28}$$

Virtual work

$$\delta W = \iint p\delta w \,dx\,dy$$

where

$$\delta w = \delta h + \delta\alpha x = f(y)\,\delta q_h + g(y)x\,\delta q_\alpha$$

$$\therefore \quad \delta W = \iint p(f(y)\,\delta q_h + g(y)x\,\delta q_\alpha)\,dx\,dy$$

$$= \delta q_h \iint pf(y)\,dx\,dy + \delta q_\alpha \iint pg(y)x\,dx\,dy$$

$$= Q_h \delta q_h + Q_\alpha \delta q_\alpha$$

where

$$\left.\begin{aligned} Q_h &\equiv \iint pf(y)\,dx\,dy \\ Q_\alpha &\equiv \iint pg(y)\cdot x\,dx\,dy \end{aligned}\right. \tag{II.29}$$

Lagrangian, $L \equiv T - U$, may be written

$$= \tfrac{1}{2}(\dot{q}_h^2 M_{hh} + 2\dot{q}_h\dot{q}_\alpha M_{h\alpha} + \dot{q}_\alpha^2 M_{\alpha\alpha})$$

$$- \tfrac{1}{2}(q_h^2 K_{hh} + 2q_h q_\alpha K_{h\alpha} + q_\alpha^2 K_{\alpha\alpha})$$

$$\frac{\partial L}{\partial \dot{q}_h} = \dot{q}_h M_{hh} + \dot{q}_\alpha M_{h\alpha}, \quad \frac{\partial L}{\partial q_h} = -q_h \cdot K_{hh} - q_\alpha K_{h\alpha}$$

$\therefore$

$$\frac{\partial L}{\partial \dot{q}_\alpha} = \dot{q}_h M_{h\alpha} + \dot{q}_\alpha M_{\alpha\alpha}, \quad \frac{\partial L}{\partial q_\alpha} = -q_h K_{h\alpha} - q_\alpha K_{\alpha\alpha}$$

Then Lagrange's equations of motion are

$$\frac{\mathrm{d}}{\mathrm{d}t}\left(\frac{\partial L}{\partial \dot{q}_h}\right)-\frac{\partial L}{\partial q_h}=Q_h \rightarrow M_{hh}\ddot{q}_h+M_{h\alpha}\ddot{q}_\alpha+K_{hh}q_h+K_{h\alpha}\cdot q_\alpha=Q_h$$

$$\frac{\mathrm{d}}{\mathrm{d}t}\left(\frac{\partial L}{\partial \dot{q}_\alpha}\right)-\frac{\partial L}{\partial q_\alpha}=Q_\alpha \rightarrow M_{h\alpha}\ddot{q}_h+M_{\alpha\alpha}\ddot{q}_\alpha+K_{h\alpha}q_h+K_{\alpha\alpha}q_\alpha=Q_\alpha \tag{II.30}$$

where M_{hh}, $M_{h\alpha}$, $M_{\alpha\alpha}$, K_{hh}, $K_{h\alpha}$, $K_{\alpha\alpha}$, Q_h and Q_α are given in (II.26), (II.28) and (II.29).

5. When

$$w(x, y, t)=\sum_m q_m(t)\psi_m(x, y)$$

$$\begin{cases}\dfrac{\partial^2 w}{\partial x^2}=\sum\limits_m q_m\dfrac{\partial^2\psi_m}{\partial x^2}\\[2ex] \dfrac{\partial^2 w}{\partial y^2}=\sum\limits_m q_m\dfrac{\partial^2\psi_m}{\partial y^2}\\[2ex] \dfrac{\partial^2 w}{\partial x\,\partial y}=\sum\limits_m q_m\dfrac{\partial^2\psi_m}{\partial x\,\partial y}\end{cases}$$

$$\therefore\begin{cases}\left(\dfrac{\partial^2 w^2}{\partial x^2}\right)=\sum\limits_m\sum\limits_n q_m q_n\dfrac{\partial^2\psi_m}{\partial x^2}\dfrac{\partial^2\psi_n}{\partial x^2}\\[2ex] \left(\dfrac{\partial^2 w^2}{\partial y^2}\right)=\sum\limits_m\sum\limits_n q_m q_n\dfrac{\partial^2\psi_m}{\partial y^2}\dfrac{\partial^2\psi_n}{\partial y^2}\\[2ex] \left(\dfrac{\partial^2 w^2}{\partial x^2}\right)\left(\dfrac{\partial^2 w}{\partial y^2}\right)=\sum\limits_m\sum\limits_n q_m q_n\dfrac{\partial^2\psi_m}{\partial x^2}\dfrac{\partial^2\psi_n}{\partial y^2}\\[2ex] \left(\dfrac{\partial^2 w^2}{\partial x\,\partial y}\right)=\sum\limits_m\sum\limits_n q_m q_n\dfrac{\partial^2\psi_m}{\partial x\,\partial y}\dfrac{\partial^2\psi_n}{\partial x\,\partial y}\end{cases}$$

Then from (II.23) the potential energy is

$$\begin{aligned}U=\frac{1}{2}\iint D\Bigg[&\sum_m\sum_n q_m q_n\frac{\partial^2\psi_m}{\partial x^2}\frac{\partial^2\psi_n}{\partial x^2}+\sum_m\sum_n q_m q_n\frac{\partial^2\psi_m}{\partial y^2}\frac{\partial^2\psi_n}{\partial y^2}\\ &+2\nu\sum_m\sum_n q_m q_n\frac{\partial^2\psi_m}{\partial x^2}\frac{\partial^2\psi_n}{\partial y^2}+2(1-\nu)\sum_m\sum_n q_m q_n\frac{\partial^2\psi_m}{\partial x\,\partial y}\frac{\partial^2\psi_n}{\partial x\,\partial y}\Bigg]\mathrm{d}x\,\mathrm{d}y\\ =\frac{1}{2}\sum_m\sum_n q_m q_n\iint D\Bigg[&\frac{\partial^2\psi_m}{\partial x^2}\frac{\partial^2\psi_n}{\partial x^2}+\frac{\partial^2\psi_m}{\partial y^2}\frac{\partial^2\psi_n}{\partial y^2}+2\nu\frac{\partial^2\psi_m}{\partial x^2}\frac{\partial^2\psi_n}{\partial y^2}\end{aligned}$$

$$+2(1-\nu)\frac{\partial^2\psi_m}{\partial x\,\partial y}\frac{\partial^2\psi_n}{\partial x\,\partial y}\bigg]\,dx\,dy$$

$$=\frac{1}{2}\sum_m\sum_n q_m q_n\cdot K_{mn} \tag{II.31}$$

where

$$K_{mn}\equiv\iint D\bigg[\frac{\partial^2\psi_m}{\partial x^2}\frac{\partial^2\psi_n}{\partial x^2}+\frac{\partial^2\psi_m}{\partial y^2}\frac{\partial^2\psi_n}{\partial y^2}+2\nu\frac{\partial^2\psi_m}{\partial x^2}\frac{\partial^2\psi_n}{\partial y^2}$$

$$+2(1-\nu)\frac{\partial^2\psi_m}{\partial x\,\partial y}\frac{\partial^2\psi_n}{\partial x\,\partial y}\bigg]\,dx\,dy$$

Note $K_{mn}\neq K_{nm}$!

Kinetic energy

$$\frac{\partial w}{\partial t}=\sum_m \dot q_m\psi_m(x,y)$$

$$\therefore\quad\left(\frac{\partial w}{\partial t}\right)^2=\sum_m\sum_n \dot q_m\dot q_n\psi_m\psi_n$$

$$T=\frac{1}{2}\iiint\rho\left(\sum_m\sum_n \dot q_m\dot q_n\psi_m\psi_n\right)dx\,dy\,dz$$

$$=\frac{1}{2}\sum_m\sum_n \dot q_m\dot q_n\iiint\rho\psi_m\psi_n\,dx\,dy\,dz$$

$$=\frac{1}{2}\sum_m\sum_n \dot q_m\dot q_n M_{mn} \tag{II.32}$$

where

$$M_{mn}\equiv\iiint\rho\psi_m\psi_n\,dx\,dy\,dz$$

Virtual work

$$\delta W=\iint pw\,dx\,dy$$

$$=\iint p\left(\sum_m\delta q_m\psi_m\right)dx\,dy$$

$$\delta W=\sum_m\delta q_m\iint p\psi_m\,dx\,dy$$

$$=\sum_m Q_m\delta q_m \tag{II.33}$$

where

$$Q_m \equiv \iint p\psi_m \, \mathrm{d}x \, \mathrm{d}y$$

Lagrangian:

$$L = T - U = \frac{1}{2}\sum_m \sum_n \dot{q}_m \dot{q}_n M_{mn} - \frac{1}{2}\sum_m \sum_n q_m q_n K_{mn}$$

$$\frac{\partial L}{\partial \dot{q}_j} = \frac{1}{2}\sum_n \dot{q}_n M_{jn} + \frac{1}{2}\sum_m \dot{q}_m M_{mj} = \frac{1}{2}\sum_m \dot{q}_m (M_{jm} + M_{mj})$$

$$= \sum_m \dot{q}_m M_{mj} \qquad (M_{mj} = M_{jm})$$

$$\frac{\partial L}{\partial q_j} = -\frac{1}{2}\left(\sum_n q_n K_{jn} + \sum_m q_m K_{mj}\right) = -\frac{1}{2}\sum_m (K_{mj} + K_{jm})$$

Lagrange's equations of motion

$$\frac{\mathrm{d}}{\mathrm{d}t}\left(\frac{\partial L}{\partial \dot{q}_j}\right) - \frac{\partial L}{\partial q_j} = \sum_m \ddot{q}_m M_{mj} + \frac{1}{2}\sum_m q_m (K_{mj} + K_{jm}) = Q_j \qquad (j = 1, 2, \ldots) \tag{II.34}$$

Note: $K_{mj} + K_{jm} = K_{jm} + K_{mj}$, i.e., coefficient symmetry is preserved in final equations.

Section 3.3

Question. Use the vertical translation of and angular rotation about the center of mass of the typical section as generalized coordinates.

a. Derive equations of motion.
b. Determine the flutter dynamic pressure and show that it is the same as discussed in text. Use steady or quasi-static aerodynamic theory.

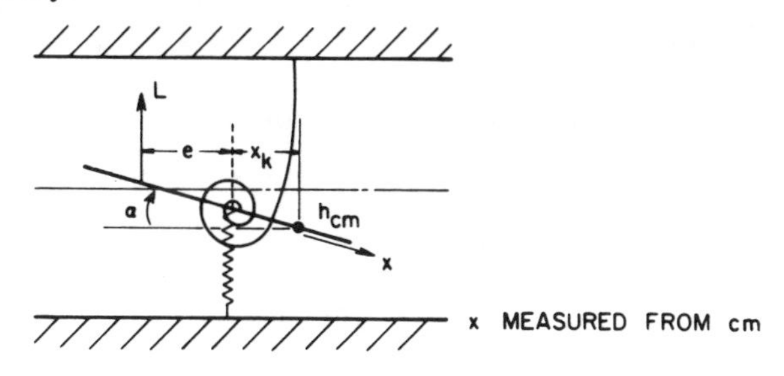

Answer

$$T=\frac{m}{2}\dot{h}_{cm}^2+\frac{I_{cm}}{2}\dot{\alpha}^2$$

$$U=\tfrac{1}{2}K_h(h_{cm}-\alpha x_k)^2+\tfrac{1}{2}K_\alpha\alpha^2$$

$$\delta W=\int p\delta w\,\mathrm{d}x,\qquad w=-h_{cm}-x\alpha,$$

vertical displacement of a point on airfoil

$$=\int p(-\delta h_{cm}-x\delta\alpha)\,\mathrm{d}x$$

$$=\delta h_{cm}\left(-\int p\,\mathrm{d}x\right)+\delta\alpha\left(-\int px\,\mathrm{d}x\right)$$

$$=\delta h_{cm}(-L)+\delta\alpha(M_y),\quad M_y \text{ is moment around c.m}$$

$$Q_{h_{cm}}=-L\equiv-\int p\,\mathrm{d}x$$

$$Q_\alpha=M_y\equiv-\int px\,\mathrm{d}x$$

$$T-U=\frac{m}{2}\dot{h}_{cm}^2+\frac{I_{cm}}{2}\dot{\alpha}^2-\frac{K_h}{2}(h_{cm}-\alpha x_k)^2-\frac{K_\alpha}{2}\alpha^2$$

From Lagrange's equations,

$$\begin{aligned}&-m\ddot{h}_{cm}-K_h(h_{cm}-\alpha x_K)-\int p\,\mathrm{d}x=0\\&-I_{cm}\ddot{\alpha}+K_hx_K(h_{cm}-\alpha x_K)-K_\alpha\alpha-\int px\,\mathrm{d}x=0\end{aligned}\tag{II.35}$$

Substituting

$$\int p\,\mathrm{d}x=qS\frac{\partial C_L}{\partial\alpha}\alpha,\qquad\int px\,\mathrm{d}x=-qS(e+x_k)\frac{\partial C_L}{\partial\alpha}\alpha,\qquad h=\bar{h}\mathrm{e}^{pt}$$

and $\alpha=\bar{\alpha}\mathrm{e}^{pt}$ into the above equations, we obtain

$$\begin{bmatrix}(mp^2+K_h) & -K_hx_k+qS\dfrac{\partial C_L}{\partial\alpha}\\ -K_hx_K & I_{cm}p^2+K_hx_k^2+K_\alpha-qS\left(\mathrm{e}+x_k\dfrac{\partial C_L}{\partial\alpha}\right)\end{bmatrix}\begin{Bmatrix}\bar{h}\mathrm{e}^{pt}\\ \bar{\alpha}\mathrm{e}^{pt}\end{Bmatrix}=\begin{Bmatrix}0\\0\end{Bmatrix}$$

The condition that the coefficient matrix is zero gives

$$Ap^4 + Bp^2 + C = 0 \tag{II.36}$$

where

$$A = mI_{cm} = mI_\alpha - S_\alpha^2 \qquad (I_{cm} = I_\alpha - mx_k^2, \quad S_\alpha = mx_k)$$

$$B = m\left[K_h x_k^2 + K_\alpha - qS(e + x_k)\frac{\partial C_L}{\partial \alpha}\right] + K_h I_{cm}$$

$$= m\left[K_\alpha - qSe\frac{\partial C_L}{\partial \alpha}\right] + K_h\alpha - S_\alpha qS\frac{\partial C_L}{\partial \alpha}$$

$$C = K_h\left[K_h x_k^2 + K_\alpha - qS(e + x_k)\frac{\partial C_L}{\partial \alpha}\right] + K_h x_k\left(-K_h x_k + qS\frac{\partial C_L}{\partial \alpha}\right)$$

$$= K_h\left[K_\alpha - qSe\frac{\partial C_L}{\partial \alpha}\right]$$

These A, B and C are the same as in equation (3.3.51), Section 3.3, in the text. Thus we have the same flutter boundary.

$$p^2 = \frac{-B + [B^2 - 4AC]^{\frac{1}{2}}}{2A}$$

(a) $B > 0$ ($A > 0$, $C > 0 \leftarrow$divergence free). If p^2 is complex (not real), then instability occurs.

$\therefore \quad B^2 - 4AC = 0$ gives flutter boundary, i.e.,

$$Dq_F^2 + Eq_F + F = 0$$

or

$$q_F = \frac{-E \pm [E^2 - 4DF]}{2D}$$

where

$$D \equiv \left\{(me + S_\alpha)S\frac{\partial C_L}{\partial \alpha}\right\}^2$$

$$E \equiv \left\{-2(me + S_\alpha)[mK_\alpha + K_h I_\alpha] + 4[mI_\alpha - S_\alpha^2]eKS\frac{\partial C_L}{\partial \alpha}\right\}$$

$$F \equiv [mK_\alpha + K_h I_\alpha]^2 - 4[mI_\alpha - S_\alpha^2]K_h K_\alpha$$

The smaller, real, and positive q_F is the flutter dynamic pressure.

(b) $B < 0$. Note that $B = 2\sqrt{AC}$ before $B = 0$ as q increases. Hence flutter always occurs for $B > 0$.

Question. Prove that

1.

$$\phi_{hF}(\tau) = \phi_{Fh}(-\tau)$$

and

2.

$$\Phi_{hF}(\omega) = H_{hF}(-\omega)\Phi_{FF}(\omega)$$

This is a useful exercise to confirm one's facility with the concepts of correlation function and power spectral density.

Answer

1. Prove that $\phi_{hF}(\tau) = \phi_{Fh}(-\tau)$. We start with the definition of the cross-correlation function:

$$\phi_{hF}(\tau) = \lim_{T\to\infty} \frac{1}{2T} \int_{-T}^{+T} h(t)F(t+\tau)\,\mathrm{d}t \tag{II.37*}$$

The response $h(t)$ is given by

$$h(t) = \int_0^t I_{hF}(t-\tau_1)F(\tau_1)\,\mathrm{d}\tau_1 \tag{II.38}$$

Here we have taken $h(t)$ in dimensional form and $I_{hF}(t)$ represents the response to an impulse. Substituting (II.38) into (II.37),

$$\phi_{hF}(\tau) = \lim_{T\to\infty} \frac{1}{2T} \int_{-T}^{+T} \int_{-\infty}^{+\infty} I_{hF}(t-\tau_1)F(\tau_1)F(t+\tau)\,\mathrm{d}\tau_1\,\mathrm{d}t$$

(One may change the limit $(0, t)$ in the inner integral to $(-\infty, +\infty)$ since the impulse will be zero for $(t-\tau_1)<0$.) Let $t' \equiv t-\tau_1 \Rightarrow \tau_1 = t-t'$ and interchange the order of integration. Then $\mathrm{d}\tau_1 = -\mathrm{d}t'$ and

$$\phi_{hF}(\tau) = -\int_{+\infty}^{-\infty} I_{hF}(t') \lim_{T\to\infty} \frac{1}{2T} \int_{-T}^{+T} F(t-t')F(t+\tau)\,\mathrm{d}t\,\mathrm{d}t'$$

$$= -\int_{+\infty}^{-\infty} I_{hF}(t')\phi_{FF}(\tau+t')\,\mathrm{d}t'$$

* A short proof goes as follows. Define $\eta \equiv t-\tau$. Then $\mathrm{d}\eta = \mathrm{d}t$ and $t = \eta - \tau$; using these and (II.37) the proof follows by inspection.

Thus

$$\phi_{hF}(\tau) = +\int_{-\infty}^{+\infty} I_{hF}(\lambda)\phi_{FF}(\tau+\lambda)\,\mathrm{d}\lambda \tag{II.39}$$

where $\lambda \equiv t' =$ dummy variable.

We follow the same procedure for $\phi_{Fh}(\tau)$.

$$\phi_{Fh}(\tau) = \lim_{T\to\infty}\frac{1}{2T}\int_{-T}^{+T} F(t)h(t+\tau)\,\mathrm{d}t$$

$$= \lim_{T\to\infty}\frac{1}{2T}\int_{-T}^{+T} F(t)\left\{\int_{-\infty}^{+\infty} I_{hF}(t+\tau-t_2)F(t_2)\,\mathrm{d}t_2\right\}\mathrm{d}t$$

let $t'' = t+\tau-\tau_2$. $\Rightarrow \mathrm{d}t'' = -\mathrm{d}\tau_2$, $\tau_2 = \tau + t - t''$

$$\phi_{Fh}(\tau) = -\int_{+\infty}^{-\infty} I_{hF}(t'')\left\{\lim_{T\to\infty}\frac{1}{2T}\int_{-T}^{+T} F(t-t''+\tau)F(t)\,\mathrm{d}t\right\}\mathrm{d}t''$$

$$= -\int_{+\infty}^{-\infty} I_{hF}(t'')\phi_{FF}(\tau-t'')\,\mathrm{d}t''$$

$$= \int_{-\infty}^{+\infty} I_{hF}(\lambda)\phi_{FF}(\tau-\lambda)\,\mathrm{d}\lambda$$

Let $\tau \to -\tau$:

$$\phi_{Fh}(-\tau) = +\int_{-\infty}^{+\infty} I_{hF}(\lambda)\phi_{FF}(-\tau-\lambda)\,\mathrm{d}\lambda$$

but $\phi_{FF}(\tau) = \phi_{FF}(-\tau)$ and hence

$$\phi_{Fh}(-\tau) = +\int_{-\infty}^{+\infty} I_{hF}(\lambda)\phi_{FF}(+\tau+\lambda)\,\mathrm{d}\lambda \tag{II.40}$$

Comparing (II.39) and (II.40) we see that

$$\phi_{hF}(\tau) = \phi_{Fh}(-\tau)$$

2. Prove that $\Phi_{hF}(\omega) = H_{hF}(-\omega)\Phi_{FF}$. By definition, the spectral density function is the Fourier transform of the correlation function.

We apply this to the cross correlation function defined by (II.39).

$$\Phi_{hF}(\omega) \equiv \frac{1}{\pi}\int_{-\infty}^{+\infty} \phi_{hF}(\tau)e^{-i\omega\tau}\, d\tau$$

$$\Phi_{hF}(\omega) = \frac{1}{\pi}\int_{-\infty}^{+\infty}\int_{-\infty}^{+\infty} I_{hF}(t)\phi_{FF}(\tau+t)e^{-i\omega\tau}\, dt\, d\tau$$

$$= \frac{1}{\pi}\int_{-\infty}^{+\infty}\int_{-\infty}^{+\infty} I_{hF}(t)e^{+i\omega t}\phi_{FF}(\tau+t)e^{-i\omega\tau - i\omega t} dt d\tau$$

$$= \int_{-\infty}^{+\infty} I_{hF}(t)\left\{\frac{1}{\pi}\int_{-\infty}^{+\infty} \phi_{FF}(\tau+t)e^{-i\omega(\tau+t)}\, d\tau\right\}e^{+i\omega t}\, dt$$

By definition

$$\frac{1}{\pi}\int_{-\infty}^{+\infty} \phi_{FF}(\tau')e^{-i\omega\tau'}\, d\tau' = \Phi_{FF}(\omega)$$

Let $\tau' \equiv \tau + t$, and substitute in RHS of equation for Φ_{hF}. Then

$$\Phi_{hF}(\omega) = \int_{-\infty}^{+\infty} I_{hF}(t)e^{+i\omega t}\Phi_{FF}(\omega)\, dt$$

Now, since

$$H_{hF}(\omega) = \int_{-\infty}^{+\infty} I_{hF}(t)e^{-i\omega t}\, dt \text{ it follows that}$$

$$\Phi_{hF}(\omega) = H_{hF}(-\omega)\Phi_{FF}(\omega)$$

Section 3.6

Typical section flutter analysis using piston theory aerodynamics

Pressure: $p = \rho a\left[\frac{\partial z_a}{\partial t} + \frac{\partial z_a}{\partial x}\right]$

Motion: $z_a = -h - \alpha(x - x_{EA})^*$

Upper surface: $p_U = \rho a[-\dot{h} - \dot{\alpha}(x - x_{EA}) - U\alpha]$

Lower surface: $p_L = -\rho a[-\dot{h} - \dot{\alpha}(x - x_{EA}) - U\alpha]$

---

* x is measured from airfoil leading edge; b is half-chord of airfoil.

Net pressure: $$p_L - p_U = \frac{4\rho U^2}{2M}\left[\frac{\dot{h}}{U} + \frac{\dot{\alpha}}{U}(x - x_{EA}) + \alpha\right]$$

Lift: $$L \equiv \int_0^{2b} (p_L - p_U)\,\mathrm{d}x$$

$$= \frac{4\rho U^2}{2M}\left\{\left[\frac{\dot{h}}{U} - \frac{\dot{\alpha} x_{EA}}{U} + \alpha\right]2b + \frac{\dot{\alpha}}{U}\frac{(2b)^2}{2}\right\}$$

Moment: $$M_y = -\int_0^{2b} (p_L - p_U)(x - x_{EA})\,\mathrm{d}x$$

$$= x_{EA}L - \frac{4\rho U^2}{2M}\left[\frac{\dot{h}}{U} - \dot{\alpha}\frac{x_{EA}}{U} + \alpha\right]\frac{(2b)^2}{2}$$

$$- \frac{4\rho U^2}{2M}\frac{\dot{\alpha}}{U}\frac{(2b)^3}{3} \tag{II.41}$$

Assume simple harmonic motion,

$$h = \bar{h}e^{i\omega t}$$

$$\alpha = \bar{\alpha}e^{i\omega t}$$

$$L = \bar{L}e^{i\omega t}$$

$$M_y = \bar{M}_y e^{i\omega t}$$

$$\bar{L} = \frac{4\rho U^2}{2M}\left\{\frac{i\omega}{U}2b\bar{h}\right.$$

$$\left. + \left[\frac{-i\omega x_{EA}}{U} + 1 + \frac{i\omega}{U}\frac{(2b)}{2}\right]2b\bar{\alpha}\right\}$$

$$\equiv 2\rho b^2\omega^2(2b)\left\{(L_1 + iL_2)\frac{\bar{h}}{b} + [L_3 + iL_4]\bar{\alpha}\right\}$$

Thus from equation (3.6.3) in Section 3.6,

$$L_1 + iL_2 = \frac{\dfrac{2\rho U^2}{M}\dfrac{i\omega\rho 2b}{U}}{2\rho b^2\omega^2(2b)\dfrac{1}{b}} = \frac{i}{M}\frac{U}{\omega b} \tag{II.42}$$

and

$$L_3 + iL_4 = \frac{\frac{2\rho U^2}{M}\left[\frac{-i\omega x_{EA}}{U} + 1 + \frac{i\omega(2b)}{U^2}\right]2b}{2\rho b^2\omega^2(2b)}$$

$$= \frac{1}{M}\left(\frac{U}{b\omega}\right)^2\left[\frac{-i\omega b}{U}\frac{x_{EA}}{b} + 1 + \frac{i\omega b}{U}\right] \qquad \text{(II.43)}$$

Questions

(1) Derive similar equations for

$M_1 + iM_2$ and $M_3 + iM_4$

(2) Fix $\frac{\omega_h}{\omega_\alpha} = 0.5$, $r_\alpha = 0.5$, $x_\alpha = 0.05$

$\frac{x_{ea}}{b} = 1.4$, $M = 2$

Choose several k, say $k = 0.1$, 0.2, 0.5, and solve for

$$\left(\frac{\omega}{\omega_\alpha}\right)^2 \quad \text{and} \quad \frac{m}{2\rho_\infty bS} \equiv \mu \qquad (S \equiv 2b)$$

from (3.6.4) using the method described on pp. 107 and 108. Plot k vs μ and ω/ω_α vs μ.

Finally plot $\frac{U}{b\omega_\alpha} \equiv \frac{\omega/\omega_\alpha}{k}$ vs μ. This is the flutter velocity as a function of mass ratio.

Answers

Recall equation (3.6.3) again from Section 3.6,

$$\bar{M}_y = -2\rho b^3\omega^2(2b)\left\{[M_1 + iM_2]\frac{\bar{h}}{b} + [M_3 + iM_4]\bar{\alpha}\right\}$$

Comparing the above and (II.41), one can identify

$$M_1 + iM_2 = \frac{iU}{Mb\omega}\left[1 - \frac{x_{ea}}{b}\right]$$

$$M_3 + iM_4 = \frac{1}{M}\left(\frac{U}{b\omega}\right)^2\left[1 - \frac{x_{ea}}{b}\right] + i\frac{1}{M}\frac{U}{b\omega}\left\{\left[1 - \frac{x_{ea}}{b}\right]^2 + \frac{1}{3}\right\} \qquad \text{(II.44)}$$

Recall the method described in Section 3.6 for determining the flutter boundary.

1. Evaluate real and imaginary parts of equation (3.6.4) and set each individually to zero.
2. Solve for $(\omega_\alpha/\omega)^2$ in terms of themass ratio, μ, from the imaginary part of the equation.
3. Substituting this result into the real part of the equation, obtain a quadratic in μ. Solve for possible values of μ for various k. To be physically meaningfull, μ must be positive and real.
4. Return to 2. to evaluate ω/ω_α
5. Finally determine $\dfrac{U}{b\omega_\alpha} = \dfrac{U}{b\omega}\dfrac{\omega}{\omega_\alpha} = \dfrac{1}{k}\dfrac{\omega}{\omega_\alpha}$.

In detail these steps are given below.

1. *Real part*

$$\mu^2\left\{\left[1-\left(\frac{\omega_\alpha}{\omega}\right)^2\left(\frac{\omega_h}{\omega_\alpha}\right)^2\right]r_\alpha^2\left[1-\left(\frac{\omega_\alpha}{\omega}\right)^2\right]-x_\alpha^2\right\}$$

$$+\mu\left\{\frac{-1}{k^2M}\left(1-\frac{x_{ea}}{b}\right)\left[1-\left(\frac{\omega_\alpha}{\omega}\right)^2\left(\frac{\omega_h}{\omega_\alpha}\right)^2\right]+\frac{x_\alpha}{k^2M}\right\}-\frac{1}{3k^2M^2}=0 \qquad \text{(II.45)}$$

Imaginary part

$$\mu r_\alpha^2\left[1-\left(\frac{\omega_\alpha}{\omega}\right)^2\right]-\frac{1}{k^2M}\left(1-\frac{x_{ea}}{b}\right)$$

$$+\left[\left(1-\frac{x_{ea}}{b}\right)^2+\frac{1}{3}\right]\left[1-\left(\frac{\omega_\alpha}{\omega}\right)^2\left(\frac{\omega_h}{\omega_\alpha}\right)^2\right]\mu$$

$$-\left[1-\frac{x_{ea}}{b}\right]\mu x_\alpha-\left[1-\frac{x_{ea}}{b}\right]\left[\mu x_\alpha-\frac{1}{k^2M}\right]=0 \qquad \text{(II.46)}$$

2. Solving for $(\omega_\alpha/\omega)^2$ from (II.46),

$$\left(\frac{\omega_\alpha}{\omega}\right)^2=\frac{r_\alpha^2+\left(1-\frac{x_{ea}}{b}\right)^2+\frac{1}{3}-2x_\alpha\left(1-\frac{x_{ea}}{b}\right)}{r_\alpha^2+\left(\frac{\omega_h}{\omega_\alpha}\right)^2\left[\left(1-\frac{x_{ea}}{b}\right)^2+\frac{1}{3}\right]} \qquad \text{(II.47)}$$

Note (II.47) is independent of μ and k; this is a consequence of using piston theory aerodynamics and would not be true, in general, for a more elaborate (and hopefully more accurate) aerodynamic theory.

Substituting the various numerical parameters previously specified into (II.47) gives

$$\left(\frac{\omega_\alpha}{\omega}\right)^2 = 2.099 \quad \text{or} \quad \frac{\omega}{\omega_\alpha} = 0.69 \tag{II.48}$$

3. Using (II.48) in (II.45) along with the other numerical parameters gives

$$-0.133\,\mu^2 + \frac{0.121}{k^2}\,\mu - \frac{1}{12k^2} = 0 \tag{II.49}$$

Solving for μ,

$$\mu_{1,2} = \frac{0.45}{k^2} \pm \frac{1}{k}\left[\frac{0.21}{k^2} - 0.63\right]^{\frac{1}{2}} \tag{II.50}$$

Note that there is a maximum value of k possible, $k_{max} = [0.21/0.63]^{\frac{1}{2}}$ Larger k give complex μ which are physically meaningless. Also note that $\mu \to 0.67, \infty$ as $k \to 0$.

4. ω/ω_α is evaluated in (II.48) and for these simple aerodynamics does not vary with μ or k.

5. From (II.48) and a knowledge of k, $U/b\omega_\alpha$ is known.

The above results are tabulated below.

k	μ_1	μ_2	$U/b\omega_\alpha$
0.0	0.67	∞	∞
0.1	0.69	89.6	6.9
0.2	0.72	22	3.45
0.3	0.75	9.28	2.3
0.4	0.81	4.3	1.73
0.5	0.937	2.66	1.38
0.57	1.39	1.39	1.21

From the above table (as well as equation (II.50)) one sees that for $\mu < 0.67$, no flutter is possible. This is similar to the flutter behavior of the typical section at incompressible speeds. At these low speeds mass ratios of this magnitude may occur in hydrofoil applications. Although no such applications exist at high supersonic speeds, it is of interest at least from a fundamental point of view that this somewhat surprising behavior at small μ occurs there as well.

Section 4.1

Questions.

1. Starting from Bernoulli's equation, show that

$$\frac{\hat{a}}{a_\infty} \sim M_\infty^2 \frac{\hat{u}}{U_\infty}$$

2. Previously we had shown that the boundary condition on a moving body is (within a linear approximation)

$$\left.\frac{\partial \hat{\phi}}{\partial z}\right|_{z=0} = \frac{\partial z_a}{\partial t} + U_\infty \frac{\partial z_a}{\partial x}$$

What is the corresponding boundary condition in terms of $\hat{p}$?

3. Derive approximate formulae for the perturbation pressure over a two-dimensional airfoil at supersonic speeds for very low and very high frequencies.

Answers.

1. Bernoulli's equation is

$$\frac{\partial \phi}{\partial t} + \frac{\nabla\phi \cdot \nabla\phi}{2} + \int_{p_\infty}^{p} \frac{dp_1}{\rho^1(p^1)} = \frac{U_\infty^2}{2}$$

Since

$$a^2 \equiv \frac{\partial p}{\partial \rho} \quad \text{and} \quad \frac{p}{\rho^\gamma} = \text{constant}$$

we may evaluate integral in the above to obtain

$$\frac{U_\infty^2}{2}\frac{\partial \phi}{\partial t} + \frac{u^2}{2} = \frac{a^2 - a_\infty^2}{\gamma - 1}, \qquad u \equiv |\nabla\phi|$$

Assume

$$a = a_\infty + \hat{a}$$
$$u = U_\infty + \hat{u}$$
$$\phi = U_\infty x + \hat{\phi}$$

where $\hat{a} \ll a_\infty$, etc. To first order

$$-M_\infty^2 \frac{\hat{u}}{U_\infty} - \frac{1}{a_\infty^2}\frac{\partial \hat{\phi}}{\partial t} = \frac{2}{\gamma - 1}\frac{\hat{a}}{a_\infty} + \text{terms } (\hat{a}^2, \text{etc})$$

This means that $M_\infty^2(\hat{u}/U_\infty)$ and $\hat{a}/a_\infty$ are quantities of the same order, at least for steady flow where $\partial\hat{\phi}/\partial t \equiv 0$.

2.

$$\left.\frac{\partial\hat{\phi}}{\partial z}\right|_{z=0} = Dz_a; \qquad D \equiv \frac{\partial}{\partial t} + U_\infty \frac{\partial}{\partial x} \tag{II.51}$$

By the linearized momentum equation

$$\frac{-\partial p}{\partial x} = \rho_\infty D\hat{u}$$

but

$$\hat{u} = \nabla_x \hat{\phi}$$

$$\therefore \quad \hat{p} = -\rho_\infty D\hat{\phi}$$

$$\therefore \quad -\frac{\partial\hat{p}}{\partial z} = \rho_\infty \frac{\partial}{\partial z}(D\hat{\phi})$$

$$= \rho_\infty \frac{\partial}{\partial z}(D(\hat{\phi})) = \rho_\infty D\left(\frac{\partial}{\partial z}\hat{\phi}\right)$$

From (II.51) and the above

$$\left.\frac{-\partial p}{\partial z}\right|_{z=0} = \rho_\infty D^2 z_a$$

$$\frac{\partial p}{\partial z} = -\rho_\infty D^2 z_a \quad \text{at} \quad z = 0 \tag{II.52}$$

3.

$$\nabla^2\phi - \frac{1}{a_\infty^2}\left[\frac{\partial}{\partial t} + U_\infty \frac{\partial}{\partial x}\right]^2 \phi = 0$$

where

$$\left.\frac{\partial\phi}{\partial z}\right|_{z=0} = \frac{\partial}{\partial t} z_a + U_\infty \frac{\partial}{\partial x} z_a \equiv w$$

off wing $\left.\frac{\partial\phi}{\partial z}\right|_{z=0} = 0$ thickness case — This does not matter here, because there are no disturbances ahead of wing in supersonic flow.

$\phi|_{z=0} = 0$ lifting case,

For a two dimensional solution, let $\phi(x, z, t) = \bar{\phi}(x, z)e^{i\omega t}$ and $w = \bar{w}e^{i\omega t}$. Thus

$$\frac{\partial^2 \bar{\phi}}{\partial x^2} + \frac{\partial^2 \bar{\phi}}{\partial z^2} - \frac{1}{a_\infty^2}\left[-\omega^2 \bar{\phi} + 2i\omega U_\infty \frac{\partial \bar{\phi}}{\partial x} + U_\infty^2 \frac{\partial^2 \bar{\phi}}{\partial x^2}\right] = 0$$

Recall $u, v, w = 0$ for $x \leq 0$ (leading edge) in supersonic flow. Taking Laplace transform (quiescent condition at $x = 0$)

$$\Phi \equiv \int_0^\infty \bar{\phi} e^{-px}\, dx$$

then

$$p^2\Phi + \frac{\partial^2 \Phi}{\partial z^2} - \frac{1}{a_\infty^2}[-\omega^2\Phi + 2i\omega p U_\infty \Phi + p^2 U_\infty^2 \Phi]$$

or

$$\frac{d^2\Phi}{dz^2} = \left[-p^2 - \frac{\omega^2}{a_\infty^2} + \frac{2i\omega p M}{a_\infty} + p^2 M^2\right]\Phi \equiv \mu^2 \Phi$$

Thus

$$\Phi = Be^{-\mu z}$$

Now

$$\left.\frac{d\Phi}{dz}\right|_{z=0} = W, \qquad W \equiv \int_0^\infty \bar{w} e^{-px}\, dx$$

Thus

$$\left.\frac{\partial \Phi}{\partial z}\right|_{z=0} = -\mu B, \qquad B = -\frac{W}{\mu}$$

Hence

$$\Phi = -\frac{W}{\mu} e^{-\mu z}$$

so

$$\bar{\phi}|_{z=0} = \int \mathscr{L}^{-1}\left\{-\frac{1}{\mu}\right\} \bar{w}(\xi, \omega)\, d\xi$$

For low frequencies, we can ignore ω^2 terms, so

$$\mu^2 \cong (M^2-1)\left\{p+\frac{iM\omega}{a_\infty(M^2-1)}\right\}^2$$

$$\frac{1}{\mu}=\frac{-1}{\sqrt{M^2-1}}\left(\frac{1}{p+\dfrac{iM\omega}{a_\infty(M^2-1)}}\right)$$

$$\mathscr{L}^{-1}\left(\frac{-1}{\mu}\right)=\frac{-1}{\sqrt{M^2-1}}\exp\left[-iM\omega x/a_\infty(M^2-1)\right]$$

and

$$\bar{\phi}\big|_{z=0}=\frac{-1}{\sqrt{M^2-1}}\int_0^x \exp\left[-iM\omega(x-\xi)/a_\infty(M^2-1)\right]\bar{w}(\xi,\omega)\,\mathrm{d}\xi$$

and if we select our coordinate system such that $w(0)=0$, for low frequencies the perturbation pressure, $\hat{p}$, is from Bernoulli's equation

$$\hat{p}=\frac{\rho_\infty e^{i\omega t}}{\sqrt{M^2-1}}\left[-\frac{i\omega\exp\left[i\omega(t-Mx/a_\infty(M^2-1))\right]}{(M^2-1)}\right.$$

$$\left.\times\int\exp\left[iM\omega\xi/a_\infty(M^2-1)\right]\bar{w}(\xi,\omega)\,\mathrm{d}\xi+U_\infty\bar{w}(x,\omega)\right]$$

$$\cong\frac{\rho_\infty e^{i\omega t}}{\sqrt{M^2-1}}U_\infty\bar{w}(x) \qquad \textit{for very low frequencies} \tag{II.53}$$

For high frequencies,

$$\frac{\mathrm{d}^2\Phi}{\mathrm{d}z^2}=\left[\frac{-\omega^2}{a_\infty^2}+\frac{2i\omega pM}{a_\infty}+(M^2-1)p^2\right]\Phi$$

$$\cong\left[\frac{i\omega}{a_\infty}+pM\right]^2\Phi$$

when we ignore the $(-p^2)$ term compared to those involving ω. Then,

$$\frac{-1}{\mu}\cong\frac{-1}{pM+\dfrac{i\omega}{a_\infty}}$$

and

$$\bar{\phi}\big|_{z=0}=\int\mathscr{L}^{-1}\left(\frac{-1}{\mu}\right)\bigg|_{x-\xi}\bar{w}(\xi,\omega)\,\mathrm{d}\xi$$

by the convolution theorem. Now

$$\mathscr{L}^{-1}\left[\frac{-1}{pM+\dfrac{i\omega}{a}}\right] = \frac{-1}{M}\exp\left(-i\omega x/a_\infty M\right)$$

so

$$\bar{\phi}|_{z=0} = -\frac{1}{M}\int_0^x \exp\left[-i\omega(x-\xi)/a_\infty M\right]\bar{w}(\xi, x)\,\mathrm{d}\xi$$

and from Bernoulli's equation

$$\begin{aligned}\therefore \quad \hat{p} = {} & \frac{\rho_\infty}{M} i\omega \exp\left[i\omega(x - x/a_\infty M)\right]\int_0^x \exp\left(i\omega\xi/a_\infty M\right)\bar{w}(\xi, \omega)\,\mathrm{d}\xi \\ & + \frac{\rho_\infty U_\infty}{M}\exp\left[i\omega(t - x/a_\infty M)\right]\exp\left(i\omega x/a_\infty M\right)\bar{w}(x, \omega) \\ & - \frac{\rho_\infty U_\infty}{M}\frac{i\omega}{a_\infty M}\exp\left[i\omega(t - x/a_\infty M)\right]\int_0^x \exp\left(i\omega\xi/a_\infty M\right)\bar{w}(\xi, \omega)\,\mathrm{d}\xi\end{aligned}$$

$$\hat{p} \cong \frac{\rho_\infty U_\infty}{M}\bar{w}(x, \omega)\mathrm{e}^{i\omega t} \quad \textit{for high frequencies.}$$

This is known as the (linearized, small perturbation) piston theory approximation. It is a useful and interesting exercise (also a rather straight forward one) to determine pressure distributions, lift and moment for translation and rotation of a flat plate using the piston theory.* The low frequency approximation considered earlier is also useful in this respect.

* Ashley, H. and Zartarian, G., 'Piston Theory—A New Aerodynamic Tool for the Aeroelastician', *J. Aero. Sciences*, *23* (December 1956) pp. 1109–1118.

Mechanics

SOLID MECHANICS AND ITS APPLICATIONS

Series Editor: G.M.L. Gladwell

Aims and Scope of the Series

The fundamental questions arising in mechanics are: *Why?, How?,* and *How much?* The aim of this series is to provide lucid accounts written by authoritative researchers giving vision and insight in answering these questions on the subject of mechanics as it relates to solids. The scope of the series covers the entire spectrum of solid mechanics. Thus it includes the foundation of mechanics; variational formulations; computational mechanics; statics, kinematics and dynamics of rigid and elastic bodies; vibrations of solids and structures; dynamical systems and chaos; the theories of elasticity, plasticity and viscoelasticity; composite materials; rods, beams, shells and membranes; structural control and stability; soils, rocks and geomechanics; fracture; tribology; experimental mechanics; biomechanics and machine design.

1. R.T. Haftka, Z. Gürdal and M.P. Kamat: *Elements of Structural Optimization.* 2nd rev.ed., 1990 ISBN 0-7923-0608-2
2. J.J. Kalker: *Three-Dimensional Elastic Bodies in Rolling Contact.* 1990 ISBN 0-7923-0712-7
3. P. Karasudhi: *Foundations of Solid Mechanics.* 1991 ISBN 0-7923-0772-0
4. *Not published*
5. *Not published.*
6. J.F. Doyle: *Static and Dynamic Analysis of Structures.* With an Emphasis on Mechanics and Computer Matrix Methods. 1991 ISBN 0-7923-1124-8; Pb 0-7923-1208-2
7. O.O. Ochoa and J.N. Reddy: *Finite Element Analysis of Composite Laminates.* ISBN 0-7923-1125-6
8. M.H. Aliabadi and D.P. Rooke: *Numerical Fracture Mechanics.* ISBN 0-7923-1175-2
9. J. Angeles and C.S. López-Cajún: *Optimization of Cam Mechanisms.* 1991 ISBN 0-7923-1355-0
10. D.E. Grierson, A. Franchi and P. Riva (eds.): *Progress in Structural Engineering.* 1991 ISBN 0-7923-1396-8
11. R.T. Haftka and Z. Gürdal: *Elements of Structural Optimization.* 3rd rev. and exp. ed. 1992 ISBN 0-7923-1504-9; Pb 0-7923-1505-7
12. J.R. Barber: *Elasticity.* 1992 ISBN 0-7923-1609-6; Pb 0-7923-1610-X
13. H.S. Tzou and G.L. Anderson (eds.): *Intelligent Structural Systems.* 1992 ISBN 0-7923-1920-6
14. E.E. Gdoutos: *Fracture Mechanics.* An Introduction. 1993 ISBN 0-7923-1932-X
15. J.P. Ward: *Solid Mechanics.* An Introduction. 1992 ISBN 0-7923-1949-4
16. M. Farshad: *Design and Analysis of Shell Structures.* 1992 ISBN 0-7923-1950-8
17. H.S. Tzou and T. Fukuda (eds.): *Precision Sensors, Actuators and Systems.* 1992 ISBN 0-7923-2015-8
18. J.R. Vinson: *The Behavior of Shells Composed of Isotropic and Composite Materials.* 1993 ISBN 0-7923-2113-8

Kluwer Academic Publishers – Dordrecht / Boston / London

Mechanics

SOLID MECHANICS AND ITS APPLICATIONS

Series Editor: G.M.L. Gladwell

19. H.S. Tzou: *Piezoelectric Shells.* Distributed Sensing and Control of Continua. 1993 ISBN 0-7923-2186-3
20. W. Schiehlen (ed.): *Advanced Multibody System Dynamics.* Simulation and Software Tools. 1993 ISBN 0-7923-2192-8
21. C.-W. Lee: *Vibration Analysis of Rotors.* 1993 ISBN 0-7923-2300-9
22. D.R. Smith: *An Introduction to Continuum Mechanics.* 1993 ISBN 0-7923-2454-4
23. G.M.L. Gladwell: *Inverse Problems in Scattering.* An Introduction. 1993 ISBN 0-7923-2478-1
24. G. Prathap: *The Finite Element Method in Structural Mechanics.* 1993 ISBN 0-7923-2492-7
25. J. Herskovits (ed.): *Advances in Structural Optimization.* 1995 ISBN 0-7923-2510-9
26. M.A. González-Palacios and J. Angeles: *Cam Synthesis.* 1993 ISBN 0-7923-2536-2
27. W.S. Hall: *The Boundary Element Method.* 1993 ISBN 0-7923-2580-X
28. J. Angeles, G. Hommel and P. Kovács (eds.): *Computational Kinematics.* 1993 ISBN 0-7923-2585-0
29. A. Curnier: *Computational Methods in Solid Mechanics.* 1994 ISBN 0-7923-2761-6
30. D.A. Hills and D. Nowell: *Mechanics of Fretting Fatigue.* 1994 ISBN 0-7923-2866-3
31. B. Tabarrok and F.P.J. Rimrott: *Variational Methods and Complementary Formulations in Dynamics.* 1994 ISBN 0-7923-2923-6
32. E.H. Dowell (ed.), E.F. Crawley, H.C. Curtiss Jr., D.A. Peters, R. H. Scanlan and F. Sisto: *A Modern Course in Aeroelasticity.* Third Revised and Enlarged Edition. 1995 ISBN 0-7923-2788-8; Pb: 0-7923-2789-6
33. A. Preumont: *Random Vibration and Spectral Analysis.* 1994 ISBN 0-7923-3036-6
34. J.N. Reddy (ed.): *Mechanics of Composite Materials.* Selected works of Nicholas J. Pagano. 1994 ISBN 0-7923-3041-2
35. A.P.S. Selvadurai (ed.): *Mechanics of Poroelastic Media.* 1995 (forthcoming) ISBN 0-7923-3329-2
36. Z. Mróz, D. Weichert, S. Dorosz (eds.): *Inelastic Behaviour of Structures under Variable Loads.* 1995 ISBN 0-7923-3397-7
37. R. Pyrz (ed.): *IUTAM Symposium on Microstructure-Property Interactions in Composite Materials.* Proceedings of the IUTAM Symposium held in Aalborg, Denmark. 1995 ISBN 0-7923-3427-2
38. M.I. Friswell and J.E. Mottershead: *Finite Element Model Updating in Structural Dynamics.* 1995 ISBN 0-7923-3431-0

Kluwer Academic Publishers – Dordrecht / Boston / London

Mechanics

From 1990, books on the subject of *mechanics* will be published under two series:

***FLUID* MECHANICS AND ITS APPLICATIONS**

Series Editor: R.J. Moreau

***SOLID* MECHANICS AND ITS APPLICATIONS**

Series Editor: G.M.L. Gladwell

Prior to 1990, the books listed below were published in the respective series indicated below.

MECHANICS: DYNAMICAL SYSTEMS

Editors: L. Meirovitch and G.Æ. Oravas

1. E.H. Dowell: *Aeroelasticity of Plates and Shells.* 1975 ISBN 90-286-0404-9
2. D.G.B. Edelen: *Lagrangian Mechanics of Nonconservative Nonholonomic Systems.* 1977 ISBN 90-286-0077-9
3. J.L. Junkins: *An Introduction to Optimal Estimation of Dynamical Systems.* 1978 ISBN 90-286-0067-1
4. E.H. Dowell (ed.), H.C. Curtiss Jr., R.H. Scanlan and F. Sisto: *A Modern Course in Aeroelasticity.* *Revised and enlarged edition see under Volume 11*
5. L. Meirovitch: *Computational Methods in Structural Dynamics.* 1980 ISBN 90-286-0580-0
6. B. Skalmierski and A. Tylikowski: *Stochastic Processes in Dynamics.* Revised and enlarged translation. 1982 ISBN 90-247-2686-7
7. P.C. Müller and W.O. Schiehlen: *Linear Vibrations.* A Theoretical Treatment of Multi-degree-of-freedom Vibrating Systems. 1985 ISBN 90-247-2983-1
8. Gh. Buzdugan, E. Mihăilescu and M. Radeş: *Vibration Measurement.* 1986 ISBN 90-247-3111-9
9. G.M.L. Gladwell: *Inverse Problems in Vibration.* 1987 ISBN 90-247-3408-8
10. G.I. Schuëller and M. Shinozuka: *Stochastic Methods in Structural Dynamics.* 1987 ISBN 90-247-3611-0
11. E.H. Dowell (ed.), H.C. Curtiss Jr., R.H. Scanlan and F. Sisto: *A Modern Course in Aeroelasticity.* Second revised and enlarged edition (of Volume 4). 1989 ISBN Hb 0-7923-0062-9; Pb 0-7923-0185-4
12. W. Szemplińska-Stupnicka: *The Behavior of Nonlinear Vibrating Systems.* Volume I: Fundamental Concepts and Methods: Applications to Single-Degree-of-Freedom Systems. 1990 ISBN 0-7923-0368-7
13. W. Szemplińska-Stupnicka: *The Behavior of Nonlinear Vibrating Systems.* Volume II: Advanced Concepts and Applications to Multi-Degree-of-Freedom Systems. 1990 ISBN 0-7923-0369-5
    Set ISBN (Vols. 12–13) 0-7923-0370-9

MECHANICS OF STRUCTURAL SYSTEMS

Editors: J.S. Przemieniecki and G.Æ. Oravas

1. L. Frýba: *Vibration of Solids and Structures under Moving Loads.* 1970 ISBN 90-01-32420-2
2. K. Marguerre and K. Wölfel: *Mechanics of Vibration.* 1979 ISBN 90-286-0086-8

Mechanics

3. E.B. Magrab: *Vibrations of Elastic Structural Members.* 1979 ISBN 90-286-0207-0
4. R.T. Haftka and M.P. Kamat: *Elements of Structural Optimization.* 1985
   Revised and enlarged edition see under Solid Mechanics and Its Applications, Volume 1
5. J.R. Vinson and R.L. Sierakowski: *The Behavior of Structures Composed of Composite Materials.* 1986 ISBN Hb 90-247-3125-9; Pb 90-247-3578-5
6. B.E. Gatewood: *Virtual Principles in Aircraft Structures.* Volume 1: Analysis. 1989
   ISBN 90-247-3754-0
7. B.E. Gatewood: *Virtual Principles in Aircraft Structures.* Volume 2: Design, Plates, Finite Elements. 1989 ISBN 90-247-3755-9
   Set (Gatewood 1 + 2) ISBN 90-247-3753-2

MECHANICS OF ELASTIC AND INELASTIC SOLIDS

Editors: S. Nemat-Nasser and G.Æ. Oravas

1. G.M.L. Gladwell: *Contact Problems in the Classical Theory of Elasticity.* 1980
   ISBN Hb 90-286-0440-5; Pb 90-286-0760-9
2. G. Wempner: *Mechanics of Solids with Applications to Thin Bodies.* 1981
   ISBN 90-286-0880-X
3. T. Mura: *Micromechanics of Defects in Solids.* 2nd revised edition, 1987
   ISBN 90-247-3343-X
4. R.G. Payton: *Elastic Wave Propagation in Transversely Isotropic Media.* 1983
   ISBN 90-247-2843-6
5. S. Nemat-Nasser, H. Abé and S. Hirakawa (eds.): *Hydraulic Fracturing and Geothermal Energy.* 1983 ISBN 90-247-2855-X
6. S. Nemat-Nasser, R.J. Asaro and G.A. Hegemier (eds.): *Theoretical Foundation for Large-scale Computations of Nonlinear Material Behavior.* 1984 ISBN 90-247-3092-9
7. N. Cristescu: *Rock Rheology.* 1988 ISBN 90-247-3660-9
8. G.I.N. Rozvany: *Structural Design via Optimality Criteria.* The Prager Approach to Structural Optimization. 1989 ISBN 90-247-3613-7

MECHANICS OF SURFACE STRUCTURES

Editors: W.A. Nash and G.Æ. Oravas

1. P. Seide: *Small Elastic Deformations of Thin Shells.* 1975 ISBN 90-286-0064-7
2. V. Panc: *Theories of Elastic Plates.* 1975 ISBN 90-286-0104-X
3. J.L. Nowinski: *Theory of Thermoelasticity with Applications.* 1978
   ISBN 90-286-0457-X
4. S. Łukasiewicz: *Local Loads in Plates and Shells.* 1979 ISBN 90-286-0047-7
5. C. Fiřt: *Statics, Formfinding and Dynamics of Air-supported Membrane Structures.* 1983 ISBN 90-247-2672-7
6. Y. Kai-yuan (ed.): *Progress in Applied Mechanics.* The Chien Wei-zang Anniversary Volume. 1987 ISBN 90-247-3249-2
7. R. Negruţiu: *Elastic Analysis of Slab Structures.* 1987 ISBN 90-247-3367-7
8. J.R. Vinson: *The Behavior of Thin Walled Structures.* Beams, Plates, and Shells. 1988
   ISBN Hb 90-247-3663-3; Pb 90-247-3664-1